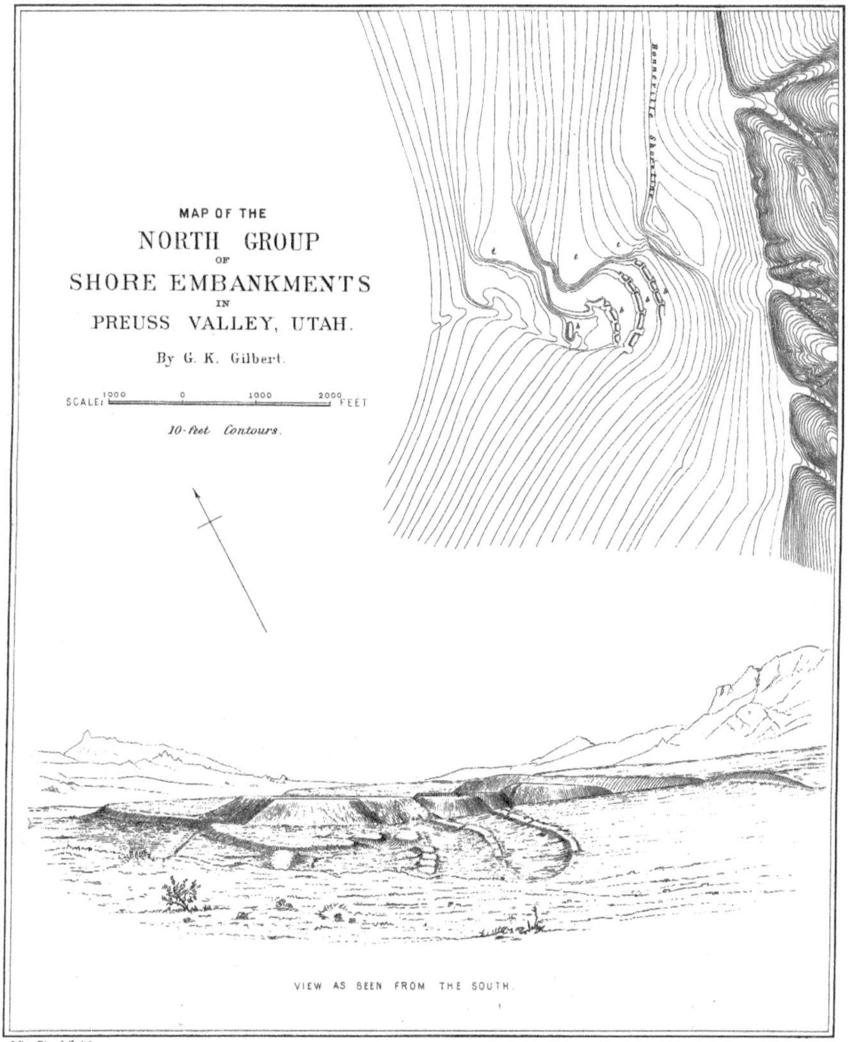
MAP OF THE
NORTH GROUP
OF
SHORE EMBANKMENTS
IN
PREUSS VALLEY, UTAH.
By G. K. Gilbert.
SCALE: 1000 0 1000 2000 FEET
10-feet Contours.
Bonneville Shoreline
VIEW AS SEEN FROM THE SOUTH.

Developments in Earth Surface Processes

Volume 20

Lake Bonneville: A Scientific Update

Developments in Earth Surface Processes, 20

Series Editor – J.F. Shroder Jr

For previous volumes refer
http://www.sciencedirect.com/science/bookseries/09282025

Developments in Earth Surface Processes

Volume 20

Lake Bonneville: A Scientific Update

Edited by

Charles G. Oviatt
Kansas State University,
Manhattan, KS, United States

John F. Shroder
University of Nebraska at Omaha,
Omaha, NE, United States

ELSEVIER
AMSTERDAM • BOSTON • HEIDELBERG • LONDON • NEW YORK • OXFORD
PARIS • SAN DIEGO • SAN FRANCISCO • SINGAPORE • SYDNEY • TOKYO

Front cover: Map of Lake Bonneville from Gilbert (1890, USGS Monograph 1, Plate III). Map shows the outline of Lake Bonneville at its highest level (the Bonneville shoreline) as interpreted by Gilbert, and the routes traveled through the basin by Gilbert and his assistants.

Frontispiece: Topographic map and landscape drawing of a set of transgressive-phase V-bars on the eastern side of Wah Wah Valley (called Pruess Valley in Gilbert's day), from Gilbert (1890, USGS Monograph 1, Plate XVI).

Elsevier
Radarweg 29, PO Box 211, 1000 AE Amsterdam, Netherlands
The Boulevard, Langford Lane, Kidlington, Oxford, OX5 1GB, United Kingdom
50 Hampshire Street, 5th Floor, Cambridge, MA 02139, United States

First edition 2016

ISBN: 978-0-444-63590-7
ISSN: 0928-2025

Publisher: Candice Janco
Acquisition Editor: Amy Shapiro
Editorial Project Manager: Tasha Frank
Production Project Manager: Vijayaraj Purushothaman
Cover Designer: Greg Harris

Typeset by SPi Global, India

Dedication

This book is dedicated to Donald R. Currey, who was a revered mentor and valued colleague to many people, and who studied Lake Bonneville with great intelligence and passion.

Photo of D.R. Currey in 1978 by Andrei Sarna-Wojcicki at Rye Patch Reservoir, Nevada

Contents

Contributors

Numbers in Parentheses indicate the pages on which the author's contributions begin.

K.D. Adams (145), Desert Research Institute, Reno, NV, United States

N.J. Anderson (1), Earth Science Education, Salt Lake City, UT, United States

G. Atwood (1), Earth Science Education, Salt Lake City, UT, United States

B.G. Bills (145), Jet Propulsion Laboratory, California Institute of Technology, Pasadena, CA, United States

B.B. Bowen (598), University of Utah, Salt Lake City, UT, United States

J.M. Broughton (292, 371), University of Utah, Salt Lake City, UT, United States

G.T. Carling (526), Brigham Young University, Provo, UT, United States

T.E. Cerling (165), University of Utah, Salt Lake City, UT, United States

M.A. Chan (570, 617), University of Utah, Salt Lake City, UT, United States

K. Cornwell (75), California State University, Sacramento, CA, United States

H.S. Godsey (617), University of Utah, Salt Lake City, UT, United States

J.S. Honke (221), U.S. Geological Survey, MS980, Denver Federal Center, Denver, CO, United States

S.U. Janecke (28), Utah State University, Logan, UT, United States

P.W. Jewell (88, 570, 598), University of Utah, Salt Lake City, UT, United States

D.S. Kaufman (28), School of Earth Sciences and Environmental Sustainability, Northern Arizona University, Flagstaff, AZ, United States

G. Komatsu (570), International Research School of Planetary Sciences, Università d'Annunzio, Pescara, Italy

B.J.C. Laabs (462), North Dakota State University, Fargo, ND, United States

M.S. Lachniet (551), University of Nevada Las Vegas, Las Vegas, NV, United States

N.A. Lifton (165), Purdue University, West Lafayette, IN, United States

T.C. Lowndes (75), 2301 NE Blakeley St., Apt 306, Seattle, WA, United States

K.D. Lupo (352), Southern Methodist University, Dallas, TX, United States

D.B. Madsen (504), Texas State University, San Marcos, TX, United States

A.L. Mayo (184, 526), Brigham Young University, Provo, UT, United States

J.H. McBride (526), Brigham Young University, Provo, UT, United States

J.P. McGeehin (221), U.S. Geological Survey, MS926A, 12201 Sunrise Valley Drive, Reston, VA, United States

D.M. Miller (60, 127), U.S. Geological Survey, Menlo Park, CA, United States

J.S. Munroe (462), Middlebury College, Middlebury, VT, United States

D.T. Nelson (598), University of North Carolina—Pembroke, Pembroke, NC, United States

S.T. Nelson (184, 526), Brigham Young University, Provo, UT, United States

R.Q. Oaks Jr. (28), Utah State University, Logan, UT, United States

J. O'Connor (105), U.S. Geological Survey, Portland, OR, United States

C.H. Okubo (570), U.S. Geological Survey, Tucson, AZ, United States

J. Ormö (570), Centro de Astrobiología (CSIC-INTA), Torrejón de Ardoz, Spain

C.G. Oviatt (75, 88, 221, 442), Kansas State University, Manhattan, KS, United States

T.J. Parker (570), Jet Propulsion Laboratory, California Institute of Technology, Pasadena, CA, United States

J.L. Pederson (28), Utah State University, Logan, UT, United States

V.A. Pedone (442), California State University, Northridge, CA, United States

G.A. Phelps (60), U.S. Geological Survey, Menlo Park, CA, United States

F.M. Phillips (165), New Mexico Tech, Socorro, NM, United States

L.W. Raming (598), University of Utah, Salt Lake City, UT, United States

M.C. Reheis (28), U.S. Geological Survey, Denver, CO, United States

K.A. Rey (184, 526), Brigham Young University, Provo, UT, United States

D. Rhode (420), Desert Research Institute, Reno, NV, United States

D.N. Schmitt (352), SWCA Environmental Consultants, Salt Lake City, UT, United States

J.F. Shroder (75), University of Nebraska at Omaha, Omaha, NE, United States

G.R. Smith (292), Museum of Zoology, University of Michigan, Ann Arbor, MI, United States

J.V. South (526), Brigham Young University, Provo, UT, United States

R.S. Thompson (221), U.S. Geological Survey, MS980, Denver Federal Center, Denver, CO, United States

D.G. Tingey (184, 526), Brigham Young University, Provo, UT, United States

T.J. Wambeam (1), Earth Science Education, Salt Lake City, UT, United States

A.L. Wolfe (371), University of Utah, Salt Lake City, UT, United States

Foreword

LETTING EARTH TELL ITS STORY

On the day after his birthday in 1871 a 28-year-old geologist arrived at Halleck Station, Nevada, to join the US Army's Geographical Survey West of the 100th Meridian. Commanded by Lt. George Wheeler, that survey did not begin well. Its initial major undertaking was a midsummer march into Death Valley, and that ordeal was followed by an exploration in the upstream direction along what turned out to be the least interesting portion of the Colorado River's Grand Canyon. Fortunately the subsequent field season was less dramatic, involving this young geologist, Grove Karl Gilbert, in survey work near Salt Lake City. Gilbert's research documented deposits and levels associated with what had been, at the height of Earth's last ice age, a great lake, now reduced to the much smaller Great Salt Lake and other remnants. Really a vast inland sea that was no more, this great paleolake had once covered about 50,000 km^2 of what is now northwestern Utah, extending into portions of northeastern Nevada and southeastern Nevada.

Gilbert's 1875 Wheeler Survey report named the ancient lake in honor of Captain B.L.E. Bonneville, whose western expeditions of the 1830s, including some involving the Great Salt Lake, had been made widely known in a book by the popular early American writer, Washington Irving. That same year Gilbert transferred to a survey headed by John Wesley Powell, who had become famous for his explorations of the Colorado River and the Grand Canyon, wisely accomplished in the downstream direction. Gilbert's scientific work blossomed with the Powell Survey, and with its successor organization, the United States Geological Survey (USGS). The latter was formed in 1879 through the unification of the former Wheeler and Powell Surveys, plus a third survey that was directed by Clarence King.

In the course of his long career with the USGS Gilbert became arguably the most respected of all American Geologists. In 1890, nearly two decades after his initial studies and while he was the USGS Chief Geologist, Gilbert's research on the great Utah paleolake was published as USGS Monograph Number One: *Lake Bonneville*. Gilbert's biographer, the well-known environmental historian, Stephen J. Pyne, has described this book as "one of the consummate works of nineteenth century geology."

Lake Bonneville: A Scientific Update reveals what has been learned in the one-and-one-quarter century since the publication of Gilbert's famous monograph. The original *Lake Bonneville* dealt with shorelines, lake deposits, glacial moraines, playas, the isostatic deformation induced by the paleolake's water load, and Gilbert's discovery that the immense lake had filled to a level defined by a well-developed shoreline before incising a deep spillway at its northern end, thereby lowering to a second prominent shoreline level. *Lake Bonneville: A Scientific Update* not only revisits all these topics, but its nearly two-dozen chapters also include treatments on vegetation, fishes, birds, mammals, and early human occupation. The update shows how technological and scientific advances since Gilbert's day are providing more precision and insight into the paleolake's sedimentology, geochronology, geochemistry, paleoclimate, and geophysics. It shows how the northern spillway incision, first recognized by Gilbert, is now understood to have generated a cataclysmic megaflood that coursed through the Snake and Columbia Rivers to the Pacific Ocean. But this updating of Gilbert's *Lake Bonneville* also reveals much more.

The intellectual position of geology and the Earth sciences has changed since the late 19th century. Owing in part to the explorations and discoveries in which Gilbert was one of many participants, 19th-century geology was at the peak of the intellectual hierarchy for the realms of the academy, political influence, and the public imagination. During the 20th century the frontiers of scientific exploration, discovery, and explanation moved in the direction of fundamental physics, operating at both subatomic and cosmic scales. While the late 20th century saw much scientific emphasis shifting to explorations in molecular biology, the new millennium appears to involve an awakening to critical issues that relate humankind to a changing Earth, and *Lake Bonneville: A Scientific Update* illustrates much of the kind of science that needs to be a part of this awakening.

Geology is commonly been viewed as dealing with very ancient Earth history. However, *Lake Bonneville: A Scientific Update*, like its 1890 predecessor, deals with those very recent portions of Earth history that are most relevant to human welfare. As Gilbert writes in the second paragraph of his *Lake Bonneville*:

> *When the work of the geologist is finished and his final comprehensive report is written, the longest and most important chapter will be upon the latest and shortest of the geologic periods.*

Gilbert ascribed the importance of the latest geological period to its rather complete recording of various processes of change through the products of those processes. The processes, in turn, are to be understood through a combination of present-day observations and, "…the study of freshly formed and perfectly preserved products, [such] that the relation of product to process is learned." Moreover, "It is through the study of the phenomena of the latest period that the connection between the present processes of change and the products of past changes is established." Gilbert used this argument to establish two purposes to his study of the long-gone phenomenon of Lake Bonneville. The first purpose is that of "…discovery of the local Pleistocene history." The second purpose is "...discovery of the processes by which the changes constituting this history were wrought."

Human beings are now witnessing a period of immense process changes for the world that they inhabit. These changes are so striking and pervasive that they have convinced many scientists that Earth's habitability for humankind is being progressively diminished. The products of these changing processes are even considered to be so intense that they justify the recognition of a new geological epoch, the Anthropocene, defined as the period in which human activities have come to have significant global impact on Earth's geological and ecological processes.

One approach to coping with the immense changes that are impacting Earth's human habitability is to employ the great advances in computational efficiency for generating mathematical solutions to simulate what could happen in the future. This program has great current appeal in the policy and political realms, where the perception of reality is more important for achieving immediate action than is the full understanding of that reality. The hard truth is that accurate prediction of anything in the future is a logical impossibility. Science can only be a guide to decisions relating to future action; it cannot be an infallible oracle. The only reality accessible to us is that evidenced by the products of past processes. Even the present is already past by the time that we access it. As Gilbert so wisely stated in his *Lake Bonneville*, we need to "discover the processes by which the changes" are wrought. The changes in water balances, climate, and related processes that influenced the evolution of Lake Bonneville are not the exact and specific changes that will happen in the future. However, they are real changes that reflect the operations of the real Earth, as opposed to being products of a simulation that requires highly fallible assumptions about that Earth in order to define the "system" that must be specified to make the simulation possible.

Rather than us telling, through our assumption-based model simulations, how the Earth will be, it is the more critical role of the Earth sciences to let Earth tell us. That is what Gilbert was reporting in 1890's *Lake Bonneville*, and that is what much more advanced methods for reading Earth's message has revealed since his time. *Lake Bonneville: A Scientific Update* provides a progress report on that message.

In *Lake Bonneville* Gilbert wrote, "The history of Bonneville is therefore the history of the ancient climate of Utah and is thereby closely linked to the material interests of the territory." He added that agricultural settlements had matured along Bonneville's ancient shorelines, and that "the problem of secular change" in climate was "of such vital importance to the agriculture of an arid domain" and that "the public domain presents no more important problem to the survey." These problems of aridity and climate change with regard to the arid American west greatly concerned survey director John Wesley Powell, who inserted his organization into the political domain by advocating that unrestricted irrigation and reservoir development be delayed until the appropriate science was accomplished to insure what we would now term "sustainability."

The US Congress was so infuriated by this intrusion of scientific considerations into profit-driven development that in 1890 it gutted Powell's proposed science programs, and soon afterward halved the entire budget of the USGS. With funding curtailed and having just completed *Lake Bonneville*, Gilbert took a leave of absence and initiated field work at what is now Meteor Crater, Arizona, while also making observations of the moon at Washington's Navel Observatory. This resulted in more than his seminal research publications in planetary geology. One US congressman stated, "So useless has the survey become that one of its most distinguished members has no better way to employ his time than to sit up all night gaping at the moon."

Gilbert would undoubtedly be both surprised and pleased to find, as described in the *Lake Bonneville: A Scientific Update* chapter by Chan and others, that his great paleolake has become a useful analog for advancing understanding of the newly discovered evidence for ancient lakes and oceans on other planetary surfaces in the solar system. It is surprises like this that form one of the many delights of science as a human enterprise, an enterprise that continues to usefully progress for the well-being of humankind. Unfortunately the same cannot be said for politics. As of this writing, one of the two major political party's remaining candidates for presidential nomination has as an accomplishment of his congressional tenure a 3-year, ideologically motivated campaign to totally eliminate the USGS.

I suggest that the Earth sciences, broadly construed, are emerging as intellectually the most important sciences for all humankind, and that this realization has profound implications for many issues of current societal concern, including the communication of scientific issues to the general public, science education in a democracy, and the future habitability of the planet. A critical component to realizing this vision is the kind of geological thinking exemplified in both *Lake Bonneville* books.

But there is an even more fundamental reason for the renewed importance of the Earth sciences: the sense of connection to nature that is intrinsic to our humanity. This can be sensed in a letter that Gilbert wrote in 1880 to then USGS director Clarence King, describing field studies of Lake Bonneville, "To the geologist, accustomed to speak familiarly of millions of years, it was the veriest yesterday when all these things were wrought; nor can anyone who stands on the quartzite shingles of one of the old beaches, and contemplates the rounded pebbles, gleaming with the self-same polish they received when the surf laid over them, fail to be impressed by the freshness of the record."

Similarly, one cannot fail to be impressed with what Earth has presented to us in the vast scope of Lake Bonneville's origin, rise, fluctuations, and demise—all a consequence of the kinds of process changes that are of direct concern for the future habitability of the planet. This is what *Lake Bonneville: A Scientific Update* presents to us. It is a story that merits our deep attention.

V. Baker
University of Arizona, Tucson, AZ, United States

Introduction

C.G. Oviatt* and J.F. Shroder Jr.†
*Kansas State University, Manhattan, KS, United States
†University of Nebraska at Omaha, Omaha, NE, United States

LAKE BONNEVILLE IN WESTERN NORTH AMERICA

Lake Bonneville was the largest late Pleistocene pluvial lake in the Great Basin of western North America. Its drainage basin encompassed much of what is now western Utah and parts of Wyoming, Idaho, and Nevada (Fig. 1). The shorelines and deposits of Lake Bonneville were probably recognized by the earliest inhabitants of the eastern Great Basin. Many people living there now or who have visited the basin see the shorelines on the mountain sides and speculate about the history of the ancient lake. Focused scientific research on Lake Bonneville began with G.K. Gilbert in the 1870s and 1880s, and his masterpiece on the lake was published in 1890.

Lake Bonneville grew and shrank during the last major episode of global glaciation, referred to as Marine Oxygen Isotope Stage (MIS) 2. The lake formed in the hydrographically closed basin occupied today by the Great Salt Lake and other shallow lakes in response to climate change that caused the inputs of water, as direct precipitation to the lake and as runoff from rain and snow in the mountains, to be greater than outputs by evaporation from the lake surface. Although it is tempting to think that Lake Bonneville grew in response to the input of glacial meltwater, the lake did not receive runoff from the large Laurentide and Cordilleran ice sheets that grew to cover much of northern North America during MIS 2, and the total volume of ice in mountain glaciers in valleys tributary to Lake Bonneville was less than 3.5% of the volume of the lake at its highest level (Chapter 17). Although fluxes have not been determined, groundwater inflow to the lake from the surrounding mountains must have been an important component of the water budget, and groundwater outflow from the basin would have been initiated as the lake reached elevations high enough that water could flow through permeable materials, such as tectonically shattered bedrock, alluvium, colluvium, and mass-movement deposits, on the basin rim. For most of its history Lake Bonneville was a closed-basin lake, so in that sense it was typical of the many other Pleistocene-age, closed-basin lakes in the Great Basin (Mifflin and

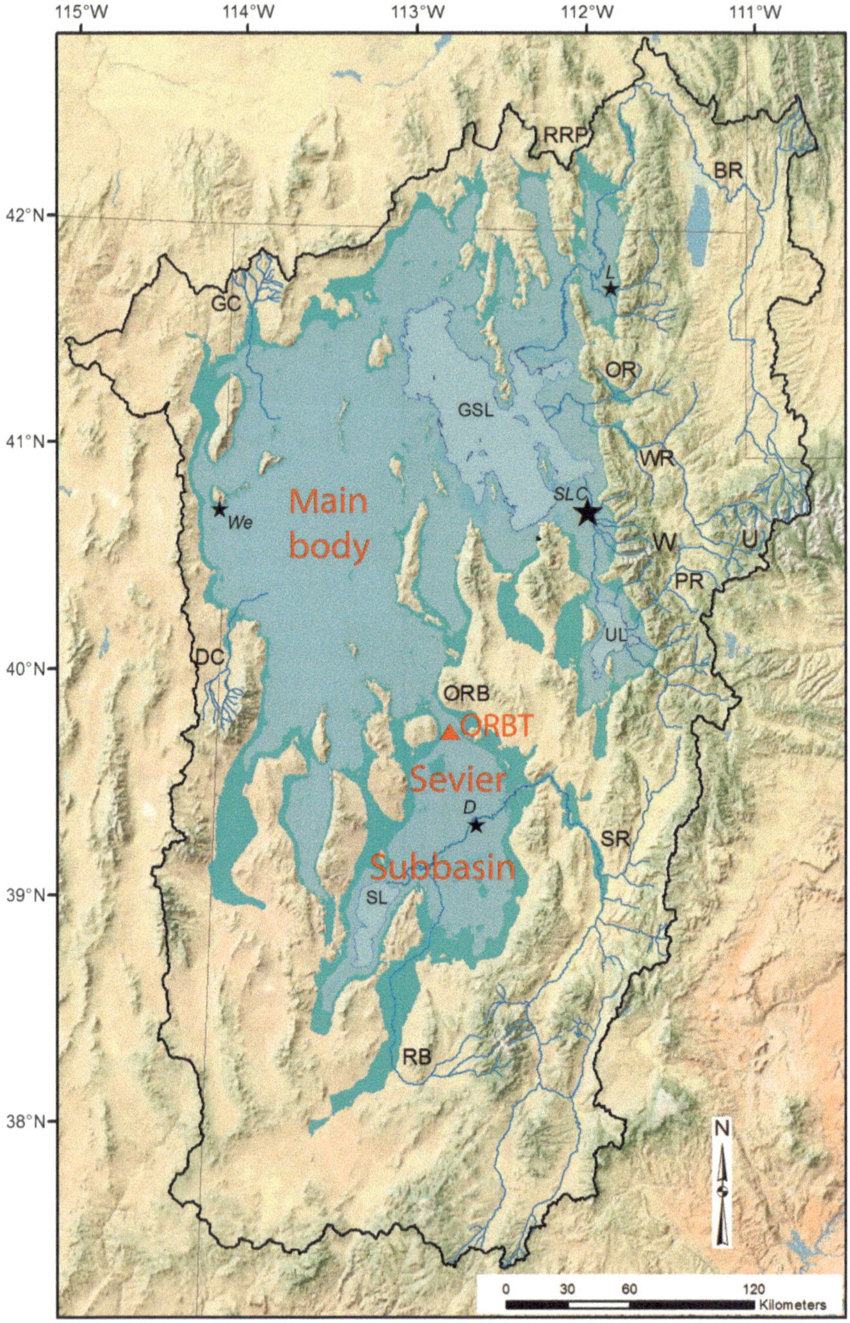

FIG. 1 Map of Lake Bonneville produced by Ken Adams. The *solid black line* represents the drainage divide of the Bonneville basin. *Three shades of blue* show lakes in the Bonneville basin; the *lightest shade* is for modern lakes, Great Salt Lake (*GSL*), Utah Lake (*UL*), and Sevier Lake (*SL*);

Wheat, 1979; Smith and Street-Perrott, 1983; Benson and Thompson, 1987; Adams and Redwine, 2016), but it was atypical in terms of its size. Lake Bonneville was much larger in surface area and deeper than the next-largest Pleistocene lake in the Great Basin, Lake Lahontan, in western Nevada (Morrison, 1991).

At its highest level, marked by the Bonneville shoreline, Lake Bonneville had inundated a number of smaller closed basins (Eardley et al., 1957). The two largest of the subbasins of Lake Bonneville are the main body in the north, which consists primarily of the Great Salt Lake subbasin and a number of smaller closed basins, and the Sevier body in the south (Fig. 1). The main body was fed by the major rivers Provo, Weber, and Bear, in addition to many smaller streams and rivers. The Sevier body was fed by the Sevier and Beaver Rivers. All of these major rivers headed in high, glaciated mountains along the eastern side of the basin. Apart from a few exceptional streams, such as Grouse Creek and Deep Creek, which today are ephemeral but which in Bonneville times produced recognizable deltas, little runoff was supplied to the lake from the west side of the basin.

During the time since G.K. Gilbert's tremendous work was published in 1890, many people have pursued research topics concerning Lake Bonneville and the Bonneville basin. Notable research has been undertaken and published by A.J. Eardley, D.R. Currey, and W.E. Scott. Not everything produced by these three people has withstood scientific tests, but they had a lasting impact on the evolving understanding of the lake, and many, many others have made important contributions that have built on and refined Gilbert's accomplishments.

LAKE BONNEVILLE CHRONOLOGY

In the following paragraphs we present an overview of Lake Bonneville chronology primarily based on the results of radiocarbon analyses from the Bonneville basin (Oviatt, 2015). Among previous summaries of the radiocarbon chronology are those by Broecker and Kaufman (1965), Scott et al. (1983), and Oviatt et al. (1992). The chronology presented by Oviatt (2015)

Lake Bonneville at the Provo shoreline is shown in the next *darker blue*, and the *darkest blue shade* shows the lake extent at the Bonneville shoreline. Two primary subbasins merged at lake levels higher than the Stansbury shoreline—the main body, which includes the Great Salt Lake basin and some smaller closed basins, and the Sevier body. The primary connection between the main body and the Sevier body was a strait at the modern topographic divide between the two bodies, the Old River Bed threshold (ORBT). Two additional straits southwest of the ORBT connected the two bodies at lake levels higher than the Provo shoreline. Major rivers tributary to Lake Bonneville were the Beaver River (*RB*), Sevier (*SR*), Provo (*PR*), Weber (*WR*), Ogden (*OR*), and Bear (*BR*). The highest and most extensive mountain ranges along the east side of the basin are the Uinta Mountains (U) and the Wasatch Mountains (W). Two tributaries on the west side of the basin are labeled: DC (Deep Creek) and GC (Grouse Creek). *SLC*, Salt Lake City, *D*, Delta, *We*, Wendover, *L*, Logan, and *RRP*, Red Rock Pass.

was based on an analysis of 368 radiocarbon ages from outcrops in the Bonneville basin obtained by many researchers since the 1950s. Many other ages from sediment cores were not included in the compilation. Even with hundreds of available ages some parts of the Bonneville chronology are not well dated, but despite uncertainties, certain features of the Bonneville radiocarbon chronology are well understood. New work on dating techniques in addition to radiocarbon (such as optically stimulated luminescence and surface exposure dating using cosmogenic radionuclides; Chapter 9) has potential to improve the accuracy and precision of the chronology.

Lake Bonneville began rising above elevations similar to modern Great Salt Lake about 30 cal ka BP (Fig. 2). This age estimate is dependent on a small number of radiocarbon ages with large uncertainties; it is a "nice round number" that is consistent with the available evidence, but future research will likely improve upon it (Chapter 11). Climate change probably initiated Lake Bonneville, but diversion of the upper Bear River into the Bonneville basin

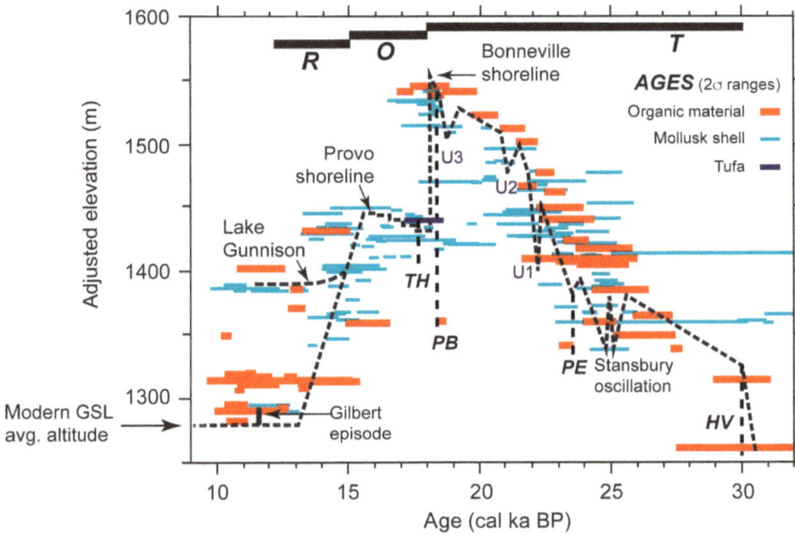

FIG. 2 Chronology of Lake Bonneville based on analysis of hundreds of radiocarbon ages. *T*, transgressive phase; *O*, overflowing phase; *R*, regressive phase. Lake Gunnison was in the Sevier basin. Only some post-Stansbury transgressive-phase oscillations, which must have been numerous in the closed basin, are shown (including three that have been documented based on stratigraphic evidence, U1, U2, and U3; Oviatt, 1997). It is likely that the lake oscillated in pre-Stansbury time, but stratigraphic evidence of them has not been found. Elevations are adjusted for differential isostatic rebound using the "Currey equation" (Oviatt, 2015). Basaltic volcanic ashes are useful stratigraphic markers in Lake Bonneville deposits: *HV*, Hansel Valley basaltic ash; *PE*, Pony Express basaltic ash; *PB*, Pahvant Butte basaltic ash; *TH*, Tabernacle Hill basaltic ash. Radiocarbon ages were calibrated using CALIB 7.0 and the IntCal13 curve (Stuiver and Reimer, 1993; Reimer et al., 2013). *GSL*, Great Salt Lake. *Modified from Oviatt, C.G., 2015. Chronology of Lake Bonneville, 30,000 to 10,000 yr B.P. Quat. Sci. Rev. 110, 166–171.*

prior to the beginning of the Bonneville lake cycle played a role (Chapter 2). Limited evidence from sediment cores from Great Salt Lake suggests that at least as far back as 40 cal ka BP the lake supported brine shrimp at times, and saline-tolerant ostracods (*Limnocythere staplini*) at other times (Chapter 11). Both brine shrimp and saline-tolerant ostracods require the lake to be shallow enough for the water to have the appropriate chemistry (salinity and ionic composition) to support them. As far as is known, these conditions prevailed during all of MIS 3 (29–60 cal ka BP; Oviatt et al., 2014).

The lake fluctuated and oscillated constantly in the closed basin during its transgressive phase (Gilbert, 1890; Jewell, 2016; Chapter 3). By about 25 cal ka BP it had risen about a third of the way to its highest elevation, when it experienced two or more oscillations referred to collectively as the Stansbury oscillation (Oviatt et al., 1990). The Stansbury shoreline, which actually consists of multiple shorelines within a narrow elevational range (less than 50 m), formed at this time. After the Stansbury oscillation the long-term trend of lake-level change was upward, punctuated by numerous fluctuations and oscillations, at an average rate of about 50 m/ka decreasing to about 13 m/ka at about 22 cal ka BP (~1500 m in elevation). The largest and most impressive depositional landforms in the basin, referred to by Gilbert (1890) as the "intermediate shorelines" because of their topographic position between the Bonneville and Provo shorelines, formed during the post-Stansbury transgressive phase. Some transgressive-phase oscillations have been dated (Oviatt, 1997; Nelson and Jewell, 2015), and evidence of others will probably be discovered and dated. Major transgressive-phase oscillations are important because they were caused by climate changes that might be hemispheric or global in scope.

As Currey (1990) and Wambeam (2001) noted, because of the approximately linear relationship between lake-surface elevation and surface area, lake elevation (lake level) is a proxy for both surface area and water volume. Lake elevation increased by over 300 m during the transgressive phase. By about 18 cal ka BP Lake Bonneville had risen high enough to reach the low point on its basin rim (Fig. 2). The Bonneville shoreline marks the highest elevation attained by the lake (Chapter 5) when it reached the threshold at Zenda just north of Red Rock Pass, Idaho, and overflowed into the Snake River drainage basin. Gilbert assembled the pieces of the geomorphic puzzle in the Red Rock Pass area and reasoned that lake overflow led to a catastrophic drawdown of the lake as the alluvium, landslide deposits, and nonresistant Neogene sediment at the outlet were washed away by what he called "the Bonneville flood" (Gilbert, 1890; Chapters 4 and 6). Lake level fell rapidly by 125 m or more (Currey, 1982; Miller et al., 2013), in probably less than a year (Jarrett and Malde, 1987; O'Connor, 1993; Chapter 6), from the Bonneville shoreline (determined by initial overflow at the Zenda threshold) to the Provo shoreline (controlled by overflow at the Red Rock Pass threshold). The large landslide on the west side of Red Rock Pass, which had been reactivated by the Bonneville flood (Chapter 4), continued to move, while the

Provo shoreline was forming and caused the Red Rock Pass threshold to rise by at least 15 m during the ensuing 3 ka (Godsey et al., 2011; Miller et al., 2013; Chapter 7). The Provo shoreline marks the overflowing (open-basin) phase of lake history (Fig. 2).

Climate change led to a shift in water budget, from inputs exceeding outputs during most of the transgressive phase and probably all of the overflowing phase, to evaporative outputs exceeding inputs during the regressive phase. As a result, the lake dropped from the Provo shoreline at about 15 cal ka BP to elevations similar to those of modern Great Salt Lake, about 200 m lower, by about 13 cal ka BP. This decline was rapid, with an average rate of about 100 m per thousand years. The lake occupied a closed basin during this 2-ka regressive phase and must have experienced multiple oscillations and fluctuations on its way down, but a stratigraphic record of those oscillations has not yet been found. Regressive-phase shoreline deposits are thin or do not exist in many places, compared to the thick and widespread transgressive-phase deposits.

The lake remained at low levels, probably similar to those of modern Great Salt Lake, for about 1.4 ka, until it rose by about 15 m during the Gilbert episode, which culminated at about 11.6 cal ka BP (Oviatt, 2014). The concept of a "Gilbert shoreline" has been around at least since Currey (1982) showed it on a map, yet evidence does not exist that the landforms at most of the sites mapped by Currey as the "Gilbert shoreline" are related to the Gilbert episode. Lacustrine landforms (mostly gravel barriers) at most of Currey's (1982) Gilbert-shoreline sites are too high in elevation to be related to the Gilbert episode, and some were formed during the transgressive phase of Lake Bonneville. The culmination of the Gilbert episode at 11.6 cal ka BP was at the tail end of the Northern-Hemisphere climate event known as the Younger Dryas (12.9–11.6 cal ka BP; Bromley et al., 2014). The physical link between the Younger Dryas and the Gilbert episode has not been firmly established.

Between about 13 and 11.6 cal ka BP, including the time of the Gilbert episode, the lake supported ostracods tolerant of high-alkalinity waters but not brine shrimp (Oviatt, 2014; Chapter 11). The Great Salt Lake has contained hypersaline water and has supported brine shrimp (no ostracods) since the end of the Gilbert episode. Holocene lake rises were less than 6 m above average historic levels, compared to 15 m for the Gilbert episode. See Chapter 1 for an overview of links between modern Great Salt Lake and Lake Bonneville.

THIS VOLUME

Our intention for this volume is to summarize the state of knowledge in the early 21st century concerning many topics related to Lake Bonneville and the Bonneville basin. In addition to the chronology and history of Lake Bonneville, many other scientific topics have been pursued in the basin and are reviewed in this volume. The history of scientific research in the Bonneville basin is an important subject in itself (Sack, 1989), partly because of Gilbert's extensive and intensive initial contributions, and partly because of the length of time—over a century—that the science has been evolving. Gilbert made major contributions to some of the topics discussed here, including isostatic deformation caused by the Lake Bonneville water load (Chapter 8), and mountain glaciation in the Bonneville basin (Chapter 17). Some important topics that Gilbert discussed in his monograph, such as Quaternary faulting and volcanism, are not reviewed in this volume.

Sedimentary history, geochemistry, paleoclimate, geophysics, and remote sensing in studies of Lake Bonneville are treated in this volume—see Chapters 10 and 11 on sediment cores, Chapter 16 on water geochemistry during the transgressive phase, Chapter 19 on geophysics and structural architecture of part of the Bonneville basin, Chapter 20 on paleoclimate of the eastern Great Basin derived from speleothem records, and Chapter 22 on the use of remote sensing and digital terrain tools. Gilbert addressed these topics only partially or not at all, mostly because of limitations imposed by the knowledge base and available technology in the late 19th century.

Tremendous knowledge has been gained since Gilbert's day about many aspects of the natural history of the Bonneville basin—see mammals in Chapter 13, fossil fishes in Chapter 12, prehistoric people in Chapter 18, vegetation in Chapter 15, pollen in Chapter 11, and birds in Chapter 14. Final chapters in the volume discuss how shorelines of Lake Bonneville are an excellent analog for water bodies on other planets (Chapter 21), and how many landforms and deposits of Lake Bonneville should be regarded as geosites (geoantiquities) in need of greater environmental protection (Chapter 23). Exploitation of resources embodied in the landforms left by the lake has a profound affect on future generations of people, both scientists and non-scientists—once deposits of Lake Bonneville have been dug up and hauled away, the bases for scientific investigations and aesthetic appreciation of the ancient lake are gone forever.

GILBERT'S CONTRIBUTIONS AND LEGACY

Gilbert was a superb scientist and his profound study of Lake Bonneville was among his many contributions to a multitude of geological topics (Pyne, 1980; Yochelson, 1980). Gilbert traveled throughout the basin and applied his observational skills to many fascinating geographic and geologic problems. Not only was Gilbert an excellent observer, but he interpreted his observations in ways that were accurate, innovative, and groundbreaking. His interpretations did not violate, but exceeded and built on the science of his day. This volume is not the first publication that recognizes Gilbert's important contributions to Lake Bonneville research (Machette, 1988), and probably will not be the last.

It is safe to say that few, if any, people who have worked on the lake since Gilbert have exceeded his exceptional abilities of observation and interpretation. He saw and understood more than any researcher of the 20th or 21st century. We now know that not all of Gilbert's scientific conclusions about the lake were correct, but most were. Scientific knowledge has advanced since 1890, but progress has not been monotonic—most scientific work has led to advancements, other work has led to dead ends or to interpretations that have slowed down or reversed forward progress. In most cases where new investigations are undertaken, it pays to take a look back at what Gilbert had to say about the topic. Gilbert himself was responsible for both correct and incorrect conclusions about lake history.

An example of an incorrect conclusion about lake history by Gilbert is his proposal that Lake Bonneville history consisted of two subcycles. He named the earlier of the two subcycles the "Yellow Clay Epoch" after the lithologic unit he had called the Yellow Clay, and the later subcycle the "White Marl Epoch" after the lithologic unit of that name (Fig. 3). Many researchers since Gilbert accepted the idea of a two-part Bonneville cycle, and different lake-level chronologies were built on that basic idea, culminating in a scenario by Morrison (1966), in which the two Bonneville subcycles were further divided into a total of eight high-amplitude transgressions and regressions (this elaborate lake-level reconstruction has not withstood scientific tests). Careful work by Scott et al. (1983) showed that the Bonneville lake cycle consisted of a single MIS 2 cycle, and laid the groundwork for an accurate radiocarbon chronology. In the 1980s it became apparent that the "Yellow Clay" and "White Marl" were deposited during a single lake cycle—they were different depositional facies of Lake Bonneville (Oviatt, 1987) (Fig. 3).

Gilbert belabored the question of the depositional environment of the "Yellow Clay" and arrived at a hypothesis that was consistent with the observations available to him at the time (Gilbert, 1890, pp. 200–209). As with his incorrect hypothesis about the origin of Meteor Crater in Arizona (he concluded it was a volcanic explosion crater), Gilbert's strength was not that he always got the right answer, although he usually did, but that he stringently followed the rules—to be accepted a hypothesis had to be consistent with the data at hand, regardless of how intuitively attractive a different conclusion might be.

An example of how one of Gilbert's original hypotheses was correct, yet later researchers diverged from his conclusions, is his interpretation of the geomorphic history of the Bonneville shoreline. Gilbert wrote (1890, p. 171):

> *Thirteen years ago I had the temerity to predict, first, that the position of the Bonneville shore-line would eventually be shown to have been determined by an overflow of the lake, and second, that the Provo shore-line would be found to have been similarly determined. The first of these predictions has been verified in its letter, but not in its spirit; the second has proved to have full warrant. My anticipation was based on the following consideration: A lake without overflow has its extent determined by the ratio of precipitation to evaporation within its basin; and since this ratio is inconstant, fluctuating from year to year and from decade to decade, it is highly improbable that the water level will remain constant long enough to permit its waves to carve a deep record. I failed to take account of the fact that the highest shore-mark of the series is conspicuous by*

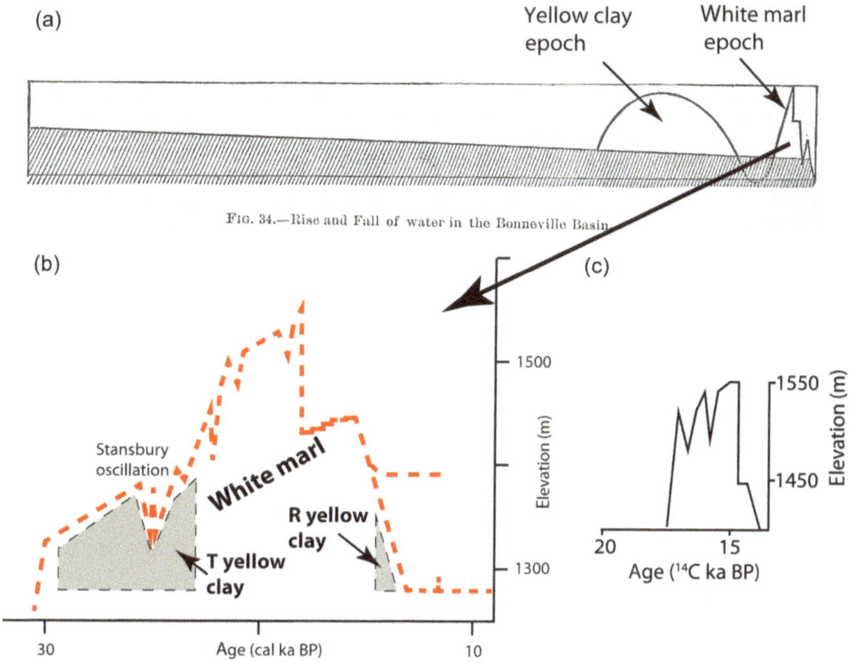

FIG. 3 Changing views on the chronology of Lake Bonneville. (A) Gilbert's summary diagram of the history of the Bonneville basin (1890, fig. 34, p. 262): "… the upper and lower horizontal lines represent the horizons of the Bonneville shore and the surface of Great Salt Lake. Horizontal distances represent time, counted from left to right. The curve represents the height of the oscillating water surface, and the shaded area indicates ignorance." He interpreted the stratigraphic unit he called the yellow clay as having been deposited during a separate high-lake cycle (the "Yellow Clay Epoch"). (B) A summary of the interpretations of Oviatt (1987, 2015) for the Bonneville lake cycle, which Gilbert called the "White Marl Epoch." Gilbert's yellow clay ("T yellow clay" or transgressive-phase yellow clay) was deposited early in the Bonneville lake cycle. At lower altitudes along the Old River Bed the white marl (the deep-water sediments of Lake Bonneville) is overlain by a regressive-phase "yellow clay" (R yellow clay), which Gilbert did not mention in his monograph. The yellow clay of Gilbert occurs only at the Old River Bed where he worked. Note that the lake-level curve in B (*dashed red line*) shows the lake transgressing to the Bonneville shoreline and dropping immediately in response to the Bonneville flood, similar to Gilbert's interpretation (as in A). In contrast, an example of how the chronology was interpreted for many years is shown in C, with the lake maintaining a constant elevation controlled by overflow into the Snake River drainage basin for hundreds of years as the Bonneville shoreline formed. *Panel (C): Modified from Oviatt, C.G., 1997. Lake Bonneville fluctuations and global climate change. Geology 25, 155–158.*

reason of the contrast there exhibited between land sculpture and littoral sculpture. We now know that the height of the Bonneville shore-line was determined in a certain sense by overflow, since a discharge limited the rise of the water; but the carving of the shore was essentially completed before the discharge; and as soon as that began, the water level fell. At the Provo horizon, on the contrary, a constant or nearly constant water-level was maintained by discharge for a very long time.

This is a clear statement of Gilbert's conclusions about the role of overflow in the formation of the Bonneville and Provo shorelines based on his extensive observations in the basin. Later researchers, however, formulated their own interpretations.

The idea that the lake overflowed for a prolonged period during the development of the Bonneville shoreline and prior to the Bonneville flood appeared sometime after 1890. The earliest publication of it we are aware of was by Pack (1939), although the idea might have been presented earlier than this. Prolonged overflow at the Bonneville shoreline became part of the lore of Lake Bonneville—everyone "knew" the lake had overflowed for hundreds of years, or longer, while a broad erosional platform was produced at the Bonneville shoreline (eg, Scott et al., 1983; Oviatt et al., 1992) (Fig. 3). In fact, however, no evidence exists for prolonged overflow at the Bonneville shoreline—there is no erosional platform at the Bonneville shoreline (Chapter 5). Gilbert's interpretation of the Provo shoreline, as the result of a long period of overflow at a nearly constant level on the regressive limb of the lake-cycle curve, has proved to be correct (Chapter 7).

Gilbert's legacy continues as his many hypotheses about Lake Bonneville are tested, most to be confirmed and accepted, others to be replaced with ideas that are consistent with the evolving science. We hope this book helps demonstrate that ideas about Lake Bonneville continue to improve as new information and techniques become available, and that the geology, geomorphology, hydrology, plants, animals, and people, of its basin should be deeply treasured. Lake Bonneville is unique because of its magnitude and the number of high-quality scientific investigations that have been devoted to its understanding for so long. This volume will help document where we are now and point toward important unanswered questions.

ACKNOWLEDGMENTS

We are grateful to Ken Adams of the Desert Research Institute for producing the map in Fig. 1, to K. Adams and D. Miller for reading a draft of this introduction and making helpful comments, and to the authors and reviewers of chapters in the volume for their positive responses to announcements, inquiries about the book, and reviews, and for submitting superb manuscripts. Future updates on Lake Bonneville will keep us informed, but we will probably always look back to G.K. Gilbert for insights from a true master.

REFERENCES

Adams, K.D., Redwine, J.L., 2016. Great Basin Paleoenvironmental Database. http:/www.dri.edu/gbped (accessed 1/3/2016).

Benson, L.V., Thompson, R.S., 1987. The physical record of lakes in the Great Basin. In: Ruddiman, W.F., Wright Jr., H.E. (Eds.), North America and Adjacent Oceans During the Last Deglaciation. Geological Society of America, Boulder, CO, pp. 241–260.

Broecker, W.S., Kaufman, A., 1965. Radiocarbon chronology of Lake Lahontan and Lake Bonneville II, Great Basin. Geol. Soc. Am. Bull 76, 537–566.

Bromley, G.R.M., Putnam, A.E., Rademaker, K.M., Lowell, T.C., Schaefer, J.M., Hall, B., Winckler, G., Birkel, S.D., Borns, H.W., 2014. Younger Dryas deglaciation of Scotland driven by warming summers. Proc. Natl. Acad. Sci. U.S.A 111 (17), 6215–6219.

Currey, D.R., 1982. Lake Bonneville: selected features of relevance to neotectonic analysis. U.S. Geological Survey Open-File Rep. 82-1070.

Currey, D.R., 1990. Quaternary paleolakes in the evolution of semidesert basins, with special emphasis on Lake Bonneville and the Great Basin, U.S.A. Palaeogeogr. Palaeoclimatol. Palaeoecol 76, 189–214.

Eardley, A.J., Gvosdetsky, V., Marsell, R.E., 1957. Hydrology of Lake Bonneville and sediments and soils of its basin. Geol. Soc. Am. Bull 68, 1141–1202.

Gilbert, G.K., 1890. Lake Bonneville. U.S. Geological Survey Monograph 1.

Godsey, H.S., Oviatt, C.G., Miller, D.M., Chan, M.A., 2011. Stratigraphy and chronology of offshore to nearshore deposits associated with the Provo shoreline, Pleistocene Lake Bonneville, Utah. Palaeogeogr. Palaeoclimatol. Palaeoecol 310, 442–450.

Jarrett, R.D., Malde, H.E., 1987. Paleodischarge of the late Pleistocene Bonneville Flood, Snake River, Idaho, computed from new evidence. Geol. Soc. Am. Bull 99, 127–134.

Jewell, P.W., 2016. Quantitative identification of erosional Lake Bonneville shorelines, Utah. Geomorphology 253, 135–145.

Machette, M.N. (Ed.), 1988. In the footsteps of G.K. Gilbert—Lake Bonneville and neotectonics of the eastern Basin and Range Province. Utah Geological and Mineral Survey Miscellaneous Publication 88-1.

Mifflin, M.D., Wheat, M.M., 1979. Pluvial lakes and estimated pluvial climates of Nevada. Nevada Bureau of Mines and Geology Bulletin 94.

Miller, D.M., Oviatt, C.G., McGeehin, J.P., 2013. Stratigraphy and chronology of Provo shoreline deposits and lake-level implications, late Pleistocene Lake Bonneville, eastern Great Basin, vol. 42. Boreas, USA, pp. 342–361.

Morrison, R.B., 1966. Predecessors of Great Salt Lake. In: Stokes, W.L. (Ed.), Utah Geological Society, Guidebook to the Geology of Utah, 20, Utah Geological Society, pp. 77–104.

Morrison, R.B., 1991. Quaternary stratigraphic, hydrologic, and climatic history of the Great Basin, with emphasis on Lakes Lahontan, Bonneville, and Tecopa. In: Morrison, R.B. (Ed.), Quaternary Nonglacial Geology: Conterminous U.S. The Geology of North America-K-2, Geological Society of America, Boulder, CO, pp. 283–320.

Nelson, D.T., Jewell, P.W., 2015. Transgressive stratigraphic record and possible oscillations of late Pleistocene Lake Bonneville, Northern Hogup Mountains, Utah, U.S.A. Palaeogeogr. Palaeoclimatol. Palaeoecol 432, 58–67.

O'Connor, J.E., 1993. Hydrology, hydraulics, and geomorphology of the Bonneville Flood. Geological Society of America Special Paper 274. 83 p.

Oviatt, C.G., 1987. Lake Bonneville stratigraphy at the Old River Bed. Am. J. Sci 287, 383–398.

Oviatt, C.G., 1997. Lake Bonneville fluctuations and global climate change. Geology 25, 155–158.

Oviatt, C.G., 2014. The Gilbert episode in the Great Salt Lake Basin, Utah. Utah Geological Survey Miscellaneous Publication 14-3.

Oviatt, C.G., 2015. Chronology of Lake Bonneville, 30,000 to 10,000 yr B.P. Quat. Sci. Rev 110, 166–171.

Oviatt, C.G., Currey, D.R., Miller, D.M., 1990. Age and paleoclimatic significance of the Stansbury shoreline of Lake Bonneville, northeastern Great Basin. Quat. Res 33, 291–305.

Oviatt, C.G., Currey, D.R., Sack, D., 1992. Radiocarbon chronology of Lake Bonneville, eastern Great Basin, USA. Palaeogeogr. Palaeoclimatol. Palaeoecol 99, 225–241.

Oviatt, C.G., Chan, M.A., Jewell, P.W., Bills, B.G., Madsen, D.B., Miller, D.M., 2014. Interpretations of evidence for large Pleistocene paleolakes in the Bonneville basin, western North America. Comment on "Bonneville Basin Shoreline Records of Large Lake During Marine Isotope Stage 3 and the Last Glacial Maximum," by Nishizawa et al. (2013). Palaeogeogr. Palaeoclimatol. Palaeoecol 401, 173–176.

Pack, F.J., 1939. Lake Bonneville: A popular treatise dealing with the history and physical aspects of Lake Bonneville. Bulletin of the University of Utah 30(4), Salt Lake City.

Pyne, S.J., 1980. Grove Karl Gilbert: A Great Engine of Research. University of Texas Press, Austin and London. 306 p.

Reimer, P.J., Bard, E., Bayliss, A., Beck, J.W., Blackwell, P.G., Ramsey, C.B., Grootes, P.M., Guilderson, T.P., Haflidason, H., Hajdas, I., Hatté, C., Heaton, T.J., Hoffmann, D.L., Hogg, A.G., Hughen, K.A., Kaiser, K.F., Kromer, B., Manning, S.W., Niu, M., Reimer, R.W., Richards, D.A., Scott, E.M., Southon, J.R., Staff, R.A., Turney, C.S.M., van der Plicht, J., 2013. IntCal13 and Marine13 radiocarbon age calibration curves 0–50,000 years cal BP. Radiocarbon 55, 1869–1887.

Sack, D., 1989. Reconstructing the chronology of Lake Bonneville: an historical overview. In: Tinkler, K.J. (Ed.), History of Geomorphology. Unwin Hyman, London, pp. 223–256.

Scott, W.E., McCoy, W.D., Shroba, R.R., Rubin, M., 1983. Reinterpretation of the exposed record of the last two cycles of Lake Bonneville, western United States. Quat. Res 20, 261–285.

Smith, G.I., Street-Perrott, F.A., 1983. Pluvial lakes of the Western United States. In: Porter, S.C. (Ed.), The Late Pleistocene. University of Minnesota Press, Minneapolis, MN.

Stuiver, M., Reimer, P.J., 1993. Extended 14C database and revised CALIB radiocarbon calibration program. Radiocarbon 35, 215–230. CALIB7.0: http:/calib.qub.ac.uk/calib/calib.html.

Wambeam, T.J., 2001. Modeling Lake Bonneville basin morphometry using digital elevation models: Salt Lake City. M.S. thesis, University of Utah.

Yochelson, E.L. (Ed.), 1980. The scientific ideas of G.K. Gilbert: an assessment on the occasion of the centennial of the United States Geological Survey (1879-1979). Geological Society of America Special Paper 183.

Chapter 1

The Present as a Key to the Past: Paleoshoreline Correlation Insights from Great Salt Lake

G. Atwood, T.J. Wambeam and N.J. Anderson
Earth Science Education, Salt Lake City, UT, United States

ABSTRACT

Reconstructions of the history of Pleistocene Lake Bonneville and its successor Great Salt Lake are largely based on correlations of paleoshorelines. Regional identification of evidence of a shoreline of a given age based on field evidence is complicated by at least three types of variability: elevation, materials, and age. Evidence of the 1986/87 high stillwater level of Great Salt Lake ranges from stillwater lake level to 3 m above stillwater lake level in response to patterns of wave energy. The variability can be inherited from initial conditions such as shorezone exposure to contrasting wave energies. Variability of paleoshorelines also can be due to postdepositional landform evolution such as tectonic deformation and sedimentary processes. Quantification of Great Salt Lake shoreline superelevation, the difference in elevation between the stillwater elevation of a lake and its shoreline evidence, can be used to reduce errors in correlating paleoshorelines of Lake Bonneville. Observations of constrictions and thresholds between bays of modern Great Salt Lake also provide insights about processes that result in variability along paleoshorelines of Lake Bonneville.

Keywords: Great Salt Lake, Lake Bonneville, Paleoshoreline, Shoreline correlation, Shoreline superelevation, Cutler Narrows, Coastal processes closed-basin lakes

1.1 INTRODUCTION

Determinations of paleolake elevations from ancient shoreline deposits, generally referred to as paleoshorelines, are used to calculate paleolake surface area and volume. In this sense, paleoshoreline refers to a landform created by shorezone processes, not the hypothetical interface of water and land. Determinations of paleoshorelines generally assume that the elevation of that landform represents the lake surface during periods of quiescence. Examination of modern Great Salt Lake shoreline deposits compared to recorded

Developments in Earth Surface Processes, Vol. 20. http://dx.doi.org/10.1016/B978-0-444-63590-7.00001-9

stillwater lake elevations brings that assumption into question because the evidence of the historic high level of Great Salt Lake varies as much as 3 m from the lake's stillwater high level.

Determination of historic elevations from preserved shoreline evidence is not straightforward. This chapter challenges the direct use of elevations on paleoshorelines to identify stillwater elevations of levels of Lake Bonneville. Understanding the processes that cause variability of shoreline elevation and materials can lead to more accurate determinations not only of paleolake stillwater level but also estimates of paleolake surface area and volume.

An accurate reconstructive climate history of paleolakes must be based on an accurate determination of paleolake stillwater surfaces (ie, the elevation of a lake's surface undisturbed by waves and referenced to a datum). Although general paleolake surface area and volume can be determined from a simple correlation among preserved shoreline deposits, this method is not sufficiently accurate to model lacustrine processes or reconstruct regional paleoclimate history. Variability of interpreted elevations of ancient shorelines can be caused by erroneous assumptions, misinterpretation of evidence of a paleoshoreline (that is, shoreline evidence), and changes due to tectonic deformation and sedimentary processes. Without an understanding of the relation between coastal processes and stillwater conditions, and changes due to tectonic deformation, the accuracy of paleolake water levels for research applications is questionable.

Shoreline evidence of modern and ancient lakes, even when clearly delineated, can be difficult to interpret (Fig. 1.1). A comparison of the 1986 stillwater lake level of Great Salt Lake and the elevation of storm debris preserved along the crests and margins of barrier beaches demonstrates that stillwater levels are below the elevation of deposits typically used to determine paleoshoreline elevations. Variability of coastal processes due to local conditions further complicates the accurate determination of stillwater lake levels. Reoccupation of stillwater levels adds complexity for assigning an age to a shoreline expression such as to a barrier beach. Accurate paleoshoreline correlation faces many of these same challenges. To reconcile equivocal but discontinuous paleoshoreline evidence, shorezone materials can be dated to clarify age relations.

Correlating discontinuous Lake Bonneville paleoshorelines is generally accomplished by detailed fieldwork incorporating topographic control, image analysis, and observations to assure that the evidence represents a specific lake level at a given time. However, field judgment is required including an awareness of the processes that account for shoreline modification and deformation.

1.2 GREAT SALT LAKE AND LAKE BONNEVILLE AS A SYSTEM

The modern Great Salt Lake and its older, larger, deeper Ice-Age predecessor Lake Bonneville are located on the eastern margin of the Basin and Range

FIG. 1.1 Photo of east shore of Antelope Island, Great Salt Lake, Utah. This photograph, taken in early Jun. 1986, documents the stillwater level of the official high level of Jun. 3, 1986. The shoreline of Great Salt Lake, defined as the interface of water and land, is clearly delineated. Complexities of shoreline correlation include challenges of age determination of shorezones occupied intermittently by Great Salt Lake. Over the past 190 years, the lake has occupied the stillwater level shown in this image for fewer than 30 days based on accounts by Gilbert (1890) for the 19th century and USGS monitoring of the 20th and 21st centuries. The duration and frequency of prior occupations are unknown. However, based on evidence of the 20th century, waves from an unknown but limited number of storm events modified the complex of barrier beaches and lagoons. Cross-cutting relationships indicate relative age and direction of lake currents. The elevation of shorezone evidence of high-lake level is higher than the lake's stillwater level. Photograph by Atwood. *Source: Atwood, G., 2006. Shoreline Superelevation: Evidence of Coastal Processes of Great Salt Lake, Utah. Utah Geological Survey, Misc Publication 06-9, 231 p., courtesy of UGS.*

physiographic province. Great Salt Lake has several bays that occupy the lowest areas of the Bonneville basin, is very shallow in that it averages only 9 m in its deepest places (Baskin and Allen, 2005; Baskin and Turner, 2006), and is hypersaline inasmuch as it reaches saturation (28% salinity) in Gunnison Bay (Loving et al., 2000). The map of Fig. 1.2 shows the extent of major levels of Lake Bonneville and Great Salt Lake based on mapping by Currey (1982) and Currey et al. (1984) and isostatic modeling (Wambeam, 2001). The map of Fig. 1.3 shows prominent locations associated with Lake Bonneville including locations referenced.

As expressions of a system, Lake Bonneville and Great Salt Lake share physical similarities such as geographic location, tectonic setting, and lake-level fluctuations due to climate and not to related tides. Lacustrine sediments and shorezone features of both lakes record lake-level changes. Both lakes have complex shapes with islands, thresholds, and bays. Contrasts between the lakes include physical characteristics such as size, extent, salinity, biota, and watersheds. Compared to Lake Bonneville, the bounding conditions

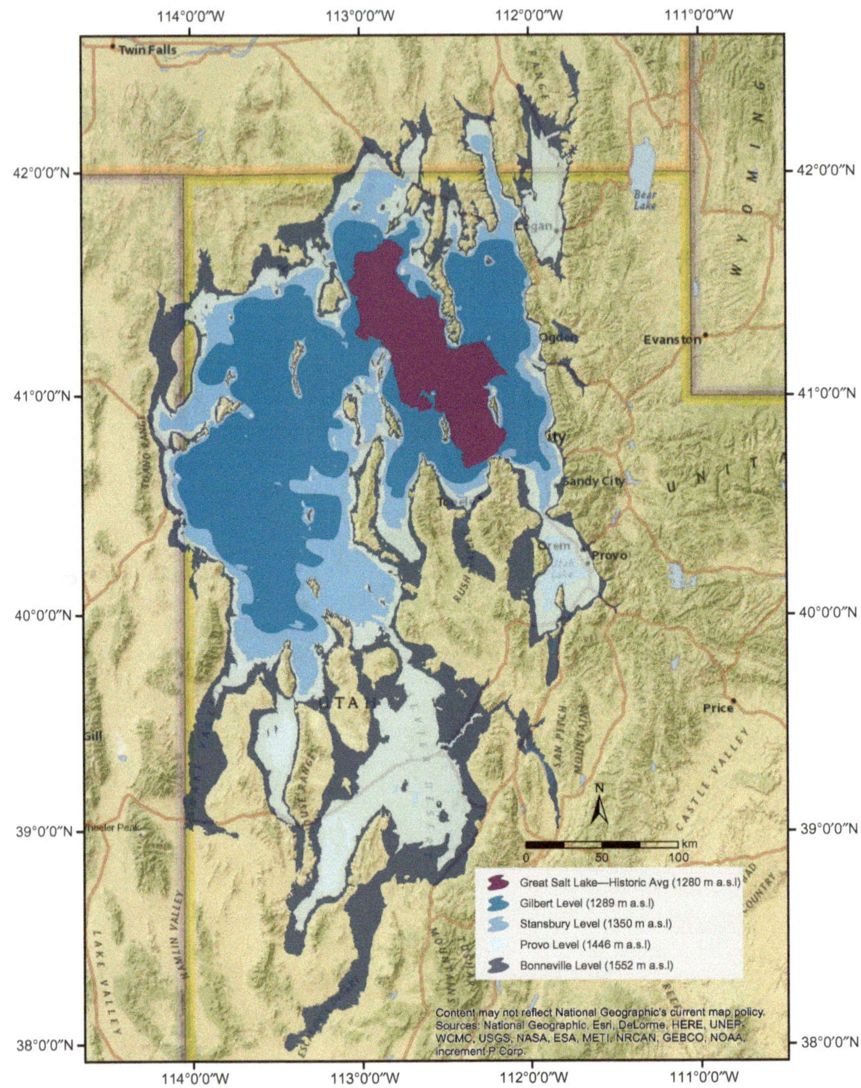

FIG. 1.2 Extent of major levels of Lake Bonneville and Great Salt Lake. The three major levels of Lake Bonneville (Marine Isotope Stage 2) after Wambeam (2001) are the Bonneville, Provo, and Stansbury levels in order of descending elevation. The two major levels of Great Salt Lake (Marine Isotope Stage 1) are the Gilbert level and the historic average level of Great Salt Lake. Depiction of the Gilbert level is based on Currey (1982). Historic average stillwater elevation of Great Salt Lake is based on Arnow and Stephens (1990). *GIS Data from Wambeam (2001).*

of Great Salt Lake are well defined including ranges of salinity, inflow of sediments and dissolved constituents, lake response to weather and climate, effects of contrasting storm systems, lake circulation patterns, inflow of groundwater, tectonic setting, and detailed hypsometry. The 170-year

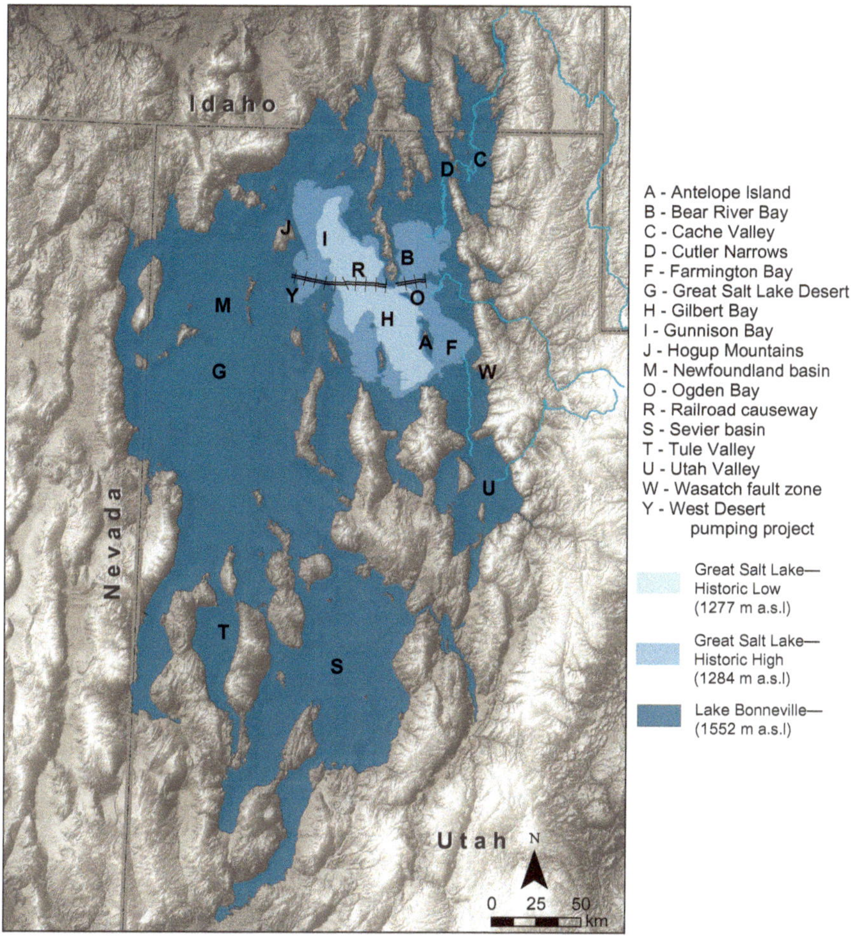

FIG. 1.3 Map of prominent features and locations referenced in the chapter. *Lake-level data from Wambeam (2001). Background data from USGS Digital Elevation Models.*

historic hydrograph of Great Salt Lake (Fig. 1.4) and the monitoring data collected by the State of Utah and the United States Geological Survey (USGS) records seasonal, annual, and decadal fluctuations in response to climate. Trends of lake level in Fig. 1.4 show the rapid rise and fall of two-decadal wet cycles.

This chapter suggests that research on Great Salt Lake shorelines can be used to reduce error in correlating the paleoshorelines of Lake Bonneville. Two lines of evidence are discussed:

(1) Conclusions from research along the 1986/87 shoreline of Great Salt Lake (Atwood, 2006) that documents variability of elevation and materials due to effects of uneven distribution of wave energy along shores of Great Salt Lake and

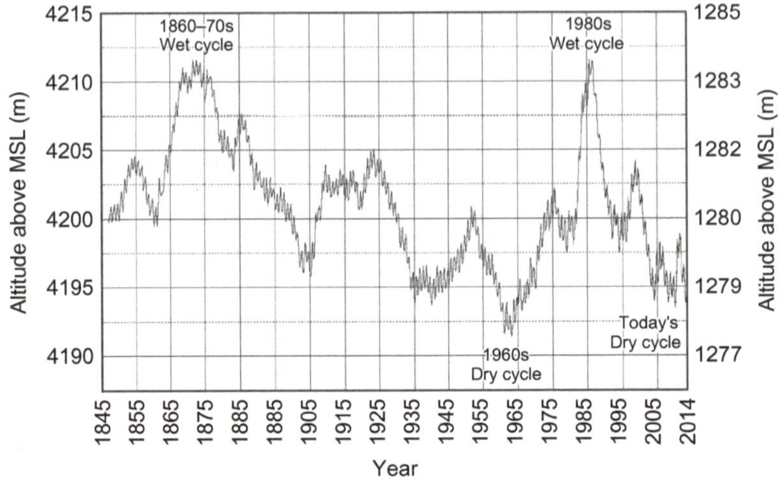

FIG. 1.4 Lake stillwater elevation trends of historic Great Salt Lake, 1845–2014. The Great Salt Lake hydrograph is based on USGS 170-year records. The high level of Great Salt Lake of the 1860s–70s was not monitored and was estimated as 1284 m (4212 ft) a.s.l. (Gilbert, 1890). The high level of the 1980s wet cycle included two peaks, in 1986 and in 1987 at approximately 1284 m (4212 ft) a.s.l. Trends of the 1960s and 2012–14 indicate two dry cycles lowering the lake to approximately 1277 m (4191 ft) a.s.l. Regional precipitation and lake evaporation are major factors affecting the water balance of Great Salt Lake, although interbasin diversions and consumptive water use complicate modeling. Seasonal and decadal trends of stillwater elevation in response to climate are comparatively well bounded for Great Salt Lake compared to the millennial trends of Lake Bonneville. *Source: Modified from Baskin, R.L., 2014. Occurrence and Spatial Distribution of Microbial Bioherms in Great Salt Lake, Utah. Salt Lake City, University of Utah, Ph.D. dissertation, data source USGS.*

(2) Observations of effects of bays, constrictions, and thresholds of historic Great Salt Lake on shorezone materials, lake levels, and lake chemistry.

1.3 CHALLENGES OF CORRELATING SHORELINES

Gilbert (1890) recognized the variability of elevation along Lake Bonneville shorelines. He surveyed paleoshorelines at major levels of Lake Bonneville and deduced that the Bonneville and Provo levels had threshold control whereas the Stansbury level did not. In the process, he recognized the challenges of correlating paleoshoreline evidence and wrote:

> *The discovery that the old water line is no longer of uniform height, and the fact that its variations of altitude afford a means of measuring the recent differential movements of the earth's crust within the basin, give occasion for great regret that the exact identification of the highest water stage is so difficult a matter.*
>
> Gilbert (1890, p. 125)

Since Gilbert's observations, researchers have refined the chronology of the lakes and experienced the challenges of correlating shorelines. In the 1950s and 1960s, researchers used field evidence of paleoshorelines supplemented with aerial photographs and evidence of offshore and deltaic deposits to define lake cycles, including the no-longer-recognized Alpine phase of the Bonneville cycle (Hunt et al., 1953; Feth et al., 1966). Their work highlights the challenges of correlating shoreline evidence with discontinuous packets of sediments.

From 1980 to 2000, Currey and students identified the extent and relative age of lake fluctuations from the late Pleistocene to the present using field identification of geomorphic expressions of erosional and depositional shorelines, supplemented with aerial photographs and 1:24,000-scale topographic maps, plus limited radiocarbon dating of shoreline evidence. Currey's correlations of paleoshorelines adjusted the recognized history of the Stansbury level to a transgressive phase of Lake Bonneville rather than a major regressive phase after the formation of the Provo shoreline (Currey et al., 1984). Currey and students continued to study the Stansbury level as fresh exposures of stratigraphic relationships provided opportunities to observe superposition among deposits. Currey's work to correlate paleoshorelines highlights challenges of correlating shoreline evidence that is largely buried by later lacustrine deposits. Currey correlated what appeared to be post-Bonneville, prehistoric shoreline evidence within the zone of 1291–1298 m a.s.l. and interpreted it as evidence of the Gilbert level (Currey, 1982; Currey et al., 1984). Oviatt (2014) questioned those correlations and their implications. Oviatt's work highlights the evolution of knowledge with increased stratigraphic evidence and age determinations for correlating low-level shorelines in zones reoccupied and reworked by subsequent lake fluctuations.

Work continues on specific sequences of shorelines. Godsey's (2012) research correlates multiple expressions of the Provo shorelines of Tooele County to explain effects of threshold control and tectonic adjustments. Nelson's (2012) work along the west shore of Great Salt Lake on multiple paleoshorelines intermediate between the Provo and Bonneville levels documented the magnitude of fluctuations during Lake Bonneville's transgression to the Bonneville level.

Shoreline correlation challenges researchers of Great Salt Lake as well as of Lake Bonneville. In the mid-1980s, as Great Salt Lake rose to its historic high levels (Fig. 1.4), the Utah Geological and Mineral Survey sponsored research to better understand the flooding hazards of the lake. Research included work by Murchison and Currey along the west shore of Great Salt Lake and on islands of Great Salt Lake (Murchison, 1989) to define the informally named Holocene high of 1286.6 m (4221 ft) a.s.l., work by Currey that correlated pre-Gilbert red beds (Currey et al., 1988), work by Mulvey on a series of shorelines along the east shore of Antelope Island (Murchison and Mulvey, 2000), and work by Atwood and Mabey on Antelope Island (2000) to estimate the frequency and magnitude of flooding events between historically high-lake levels and Murchison's Holocene high. That research concluded that Great Salt Lake had risen at least six times during the past 10,000 years to the level of the historic high level or higher. The 1980s work highlights the challenges of correlating historic and prehistoric shorelines of Great Salt Lake because of initial variability of shoreline evidence (Atwood and Mabey, 2000). Series of shoreline evidence were identified at 19 places around Antelope Island (Fig. 1.5). At none of these places were the same set of shorelines identified or their relative elevations consistent.

A century and a half of work to correlate paleoshorelines of Lake Bonneville has clarified the complexity of factors that result in differences in elevation, differences in materials, and inconsistencies of age for a given shoreline occupation. Challenges of paleoshoreline correlation stem from two types of factors: those contemporaneous with lake occupation such as coastal processes and those introduced after lake occupation such as changes in elevation due to fault displacement or changes in materials due to erosion. Initial and postoccupation processes include four classes of variations along shorelines: age, elevation, materials, and presence vs absence of evidence (Table 1.1).

Although determining the ages of shoreline materials has advanced shoreline correlation, age estimates of shoreline materials can be complicated beyond precision of analytical methods. Erosional shorelines such as abrasion platforms may have no datable material. Other sections of shore may have no datable material to analyze due to removal by postoccupation erosion. Depositional shores may have in situ datable materials, datable material that may be reworked, or introduced datable material. Inaccurate determination of age of paleoshorelines introduces potential error not only to interpretations within a lake basin, such as frequency of occupation, but also to correlations among lake basins.

Variability of elevations of shoreline evidence includes inherent differences due to erosional vs depositional processes (Gilbert, 1885). Variations of elevation along a given paleoshoreline introduce potential error in determining stillwater lake level and associated calculations of area, depth, volume, and derivative estimates, for example, of water balance and climate change.

Variations of materials along a paleoshoreline include differences in composition, shape, rounding, sorting, density, and clast size. Variations of materials introduce potential error if they are interpreted as indicative of wave energy when wave processes were not the primary cause of their variability. The presence or absence of shoreline evidence can be due to initial conditions dominated by erosion vs deposition, or by postoccupation effects of sedimentary processes.

All four classes of variability have been documented for shoreline evidence along levels of Lake Bonneville. All four types of variability were studied along the high-lake-level shoreline of 1986/87 on Antelope Island.

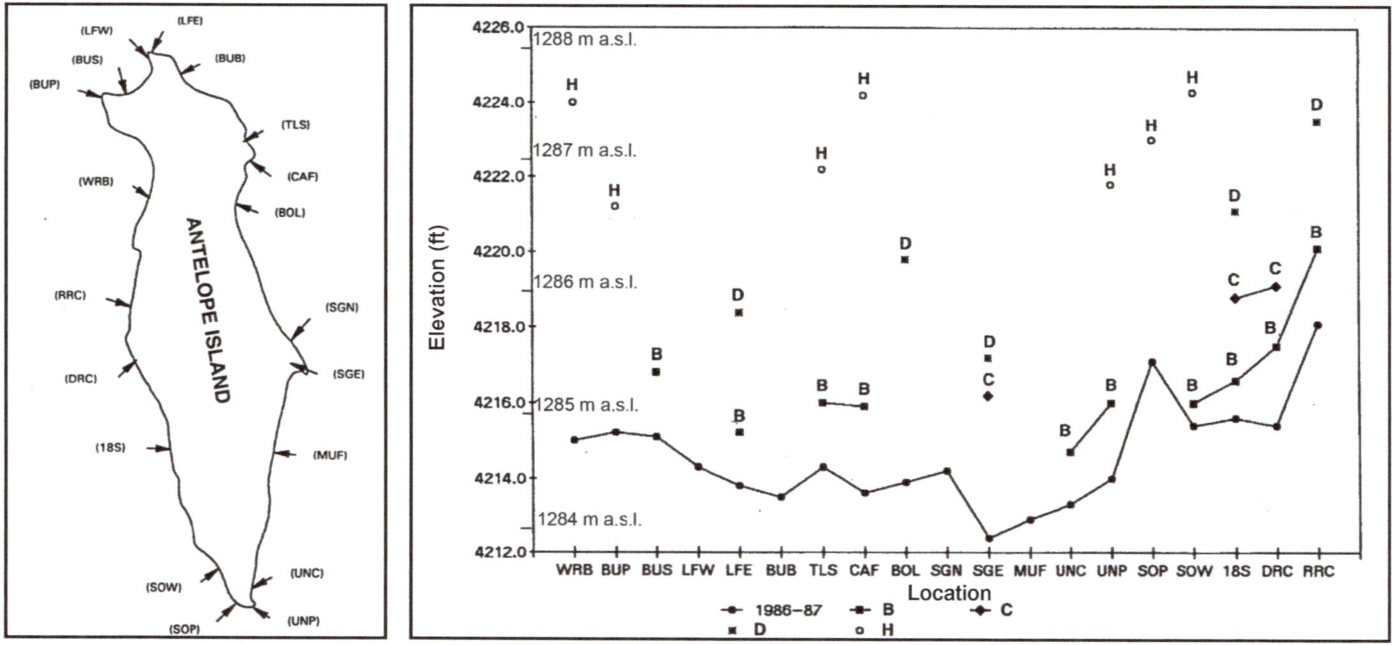

FIG. 1.5 Variability of elevation of evidence of Holocene shorelines of Great Salt Lake. The index map shows 19 locations around Antelope Island where evidence of the 1986/87 debris line of the historic high-lake level of Great Salt Lake was surveyed in 1987–88. At those places, evidence of higher elevation shorelines also were surveyed (B–D) up to the level of Murchison's (1989) Holocene high (H). The 1986/87 shoreline evidence varied almost 2 m from stillwater lake level. Evidence of older, higher lake levels was neither continuous nor at consistent relative height above the 1986/87 debris line. This data set, in spite of its uncertainties, provided information for policy makers about recurrence and frequency of high levels of Great Salt Lake. *Source: Atwood, G., Mabey, D.R., 2000. Shorelines of Antelope Island as evidence of fluctuations of the level of Great Salt Lake. In: King, J.K., Willis, G.C. (eds.), Geology of Antelope Island, Davis County, Utah. Utah Geological Survey, Misc Publication 00-1, pp. 85–97, courtesy of UGS.*

TABLE 1.1 The Factors and Magnitude of Change that Complicate Shoreline Correlation of Lake Bonneville (LB) and Great Salt Lake (GSL) Based on Initial Conditions and Variations Occurring Postoccupation

Agent of Change	Magnitude	Examples and References	Class[a]
Initial conditions			
Fluctuations of lake level			E, M
Daily due to wind	<1 m	GSL, USGS gage data; Lin and Wang (1978)	M
Seasonal patterns of precipitation and evaporation	0.6–2 m	GSL, USGS archives of lake level monitoring	E, M
Annual due to net water balance	0.3–2 m	GSL, Arnow and Stephens (1990)	E, M
Decadal and centennial due to variability of climate	7 m	GSL, Arnow and Stephens (1990), Atwood (2006)	E, M
Millennial due to changed climate	300 m	GSL and LB, Gilbert (1890)	E, M
Coastal processes (=shoreline superelevation)			E, M
Uneven distribution of wave energy	0–3 m	GSL, Atwood (2006), U.S. Army Corps of Engineers (1984)	E, M
Shoreline conditions (eg, slope and vegetation)	−0.1–2 m	GSL, Atwood (2006)	E, M
Wave environments (eg, erosional and swell)	0.1–2 m	GSL, Atwood (2006), Komar (1998)	E, M
Threshold control		LB, Sack (1990)	A, E, M
Erosion	up to 220 m	LB, Gilbert (1890)	A, E, M
Deposition	up to 20 m	LB, Oviatt (1997)	A, E, M
West Desert Pumping of Great Salt Lake	−2 m	GSL, Utah Division of Water Resources (1999)	A, E, M
Events			E
In-lake volcanic activity, faulting, or earth movement	Unknown		E

TABLE 1.1 The Factors and Magnitude of Change that Complicate Shoreline Correlation of Lake Bonneville (LB) and Great Salt Lake (GSL) Based on Initial Conditions and Variations Occurring Postoccupation—Cont'd

Agent of Change	Magnitude	Examples and References	Class
Postoccupation changes			
Tectonics			E
Isostatic displacements	74 m	LB, Crittenden (1963), Bills et al. (1994)	E
Fault displacements of Wasatch fault zone	30 m	Brigham City segment, Lund (2005)	E
Fault displacement of interior Basin and Range	1–12 m up to 5 m	LB, Machette (2005) GSL, Dinter and Pechmann (2014)	E
Deflection due to volcanism	6–9 m	LB, Oviatt (1991)	E
Sedimentation			P
Erosion	Eradication	GSL, Atwood and Mabey (2000)	P
Deposition	Burial	LB, Oviatt et al. (1990)	P
Human disturbance	Masking	LB, Chan and Godsey (2004)	P
Reoccupation		LB, Godsey (2012); GSL, Oviatt (2014)	A

[a]*Four classes of variations along shorelines:* A=*age;* E=*elevation;* M=*material;* P=*presence or absence of evidence.*

1.4 EVIDENCE FROM GREAT SALT LAKE

The first written records of Great Salt Lake date from explorers' accounts (Mabey, 1986). Monitoring of Great Salt Lake elevations began with Gilbert in 1875 and has continued with few interruptions to the present (Tibbetts et al., 2004). Historic Great Salt Lake has fluctuated about 7 m (20 ft) over the past 170 years. The historic average lake-level elevation of Great Salt Lake is 1280.2 m (4200 ft) a.s.l. whereas the historic low is 1277.5 m (4191.3 ft) a.s.l. and historic high is 1283.7 m (4211.6 ft) a.s.l. (Tibbetts et al., 2004). The lowest level of the lake prior to 2016 occurred in 1963. The highest levels occurred during a wet cycle of the 1860s–70s that predated lake monitoring, and again in 1986 and 1987 during the 1980s wet cycle. The gently sloping floor of the lake means that the 6 m change in historic elevation between the lowest and highest lake levels caused the surface area of the lake to more than double from 2460 to 5960 km^2, an increase of 3500 km^2 (Arnow and Stephens, 1990). Lake-level fluctuations dramatically change the lake's overall shape, surface area, and the shape of its five major bays; Gunnison, Bear River, Ogden, Gilbert, and Farmington bays.

Compared to Lake Bonneville, the boundary conditions of Great Salt Lake are better understood, including elevation, bathymetry, tectonics, chemistry, biology, weather, and shorezone dynamics. With time, as instrumentation becomes more advanced the boundary conditions are further refined. For example, during the 1986/87 high levels of Great Salt Lake, systematic monitoring of the lake included lake-level gaging at three places in the lake and monitoring of lake chemistry in the lake's major bays. However, monitoring of weather patterns during the 1986 and 1987 high-lake-level years lacked the in-lake coverage of weather monitoring put in place by 2002. Prior to satellite imagery, aerial photographs captured snapshots of the surface-area extent of Great Salt Lake. Subannual satellite imagery is now commonly available for Great Salt Lake and vicinity.

One shoreline of historic Great Salt Lake, that of the high level of 1986/87, appears unique for purposes of shoreline research. The high-lake level created shoreline evidence that could be identified with certainty, particularly along the shores of the lake's islands. Organic and terrigenous debris was virtually continuous from 1986 to 1996 and into the 21st century along long stretches (<5 km) of undisturbed regions of the lake's shore. No shoreline above this level has anthropogenic trash deposited with its terrigenous and organic debris. All shorelines lower than this level have anthropogenic trash in their debris. Its stillwater elevation also was known with certainty as it had been gaged by the USGS (Tibbetts et al., 2004).

Atwood (2006) mapped the shoreline's physical characteristics, including variability of the elevation of shoreline evidence. The 1986/87 shoreline of Great Salt Lake afforded an opportunity to research variability of elevation and materials of shoreline evidence virtually undisturbed by postdepositional factors of erosion, deposition, isostasy, or faulting. By documenting the variability of shoreline evidence associated with a monitored stillwater lake level, the research has provided methodology for reducing error associated with correlations of older shorelines of Great Salt Lake and Lake Bonneville.

1.5 INITIAL VARIABILITY OF SHORELINE EVIDENCE DUE TO COASTAL PROCESSES

Gilbert may have been the first to study the difference in the elevations of shoreline evidence of Great Salt Lake vs the stillwater elevation of the lake.

> *All about the lake shore there is a storm line marking the extreme advance of the water during gales in the summers of 1872, 1873, 1874… The height of this storm*

> *line above the contemporaneous stillwater surface varies with the locality being much greater on a shelving coast, over which the water is forced to a considerable distance by the winds, and especially small upon the islands. On the east side of Antelope Island it was found by measurement to be three feet above summer stage of the lake in 1877, or about one foot above the winter stage in 1872.*
>
> Gilbert (1890, p. 242)

Within 2 weeks prior to the historic 1986 high level of Great Salt Lake, Currey, Mabey, and Atwood reenacted Gilbert's shoreline survey on Antelope Island, which correlated the debris line of the 1870s high level with its stillwater level. From this survey, Gilbert deduced that the 1870s high level had reached 1283.7 m (4212 ft) a.s.l. The 1986 reenactment on Antelope Island concluded that the 1986 shoreline evidence was found at virtually the same elevation as the 1870s shoreline evidence and therefore the two high-lake levels were essentially equal (Mabey, 1986).

During 1993–97, Atwood (2006) mapped the entire 64-km extent of Antelope Island's 1986/87 shoreline classifying type, abundance, and size of the shorezone materials, and their associations with landforms and vegetation. In 1998, Atwood and Mabey surveyed 1228 elevations on the 1986/87 shoreline around Antelope Island (Fig. 1.6). The survey identified distinctive patterns of "shoreline superelevation," meaning the difference in elevation between shoreline evidence and stillwater elevation (Atwood, 2006). The patterns confirmed that the elevation of shoreline evidence did not equate with the stillwater elevation. One-third of surveyed shoreline evidence was at least 1 m above stillwater level. The difference in shoreline superelevation was not equally distributed geographically. Shoreline elevations ranged from 0.2 m below stillwater elevation in shallow bays with vegetation to 3.7 m above stillwater along rocky shores, averaging 0.9 m above stillwater elevation for the island as a whole (Atwood, 2006). The research quantified the variability of initial elevation differences due to coastal processes with confidence that tectonic and erosional processes had not affected evidence of initial variability of shoreline elevation. Had the evidence not been correlated to the 1986/87 stillwater level, the variability in elevations could have been misinterpreted as shorelines at different elevations formed at different times. The elevation variability could have caused overcounting of the number of high-level fluctuations and erroneous calculations of surface area, evaporative losses, and climate conditions.

Fig. 1.7 displays some of the results and patterns from Atwood (2006). Surveyed elevations were grouped as low, intermediate, and high shoreline superelevation. Patterns of superelevation were compared to patterns of diverse shorezone characteristics including aspect of the shoreline, shorezone slope, materials of the shorezone, and coastal setting. Patterns of superelevation were higher on the west side of the island; the windward, erosional coast. Lower superelevation along the east side of the island was associated with leeward exposures of an accreting coast. Higher superelevation was also

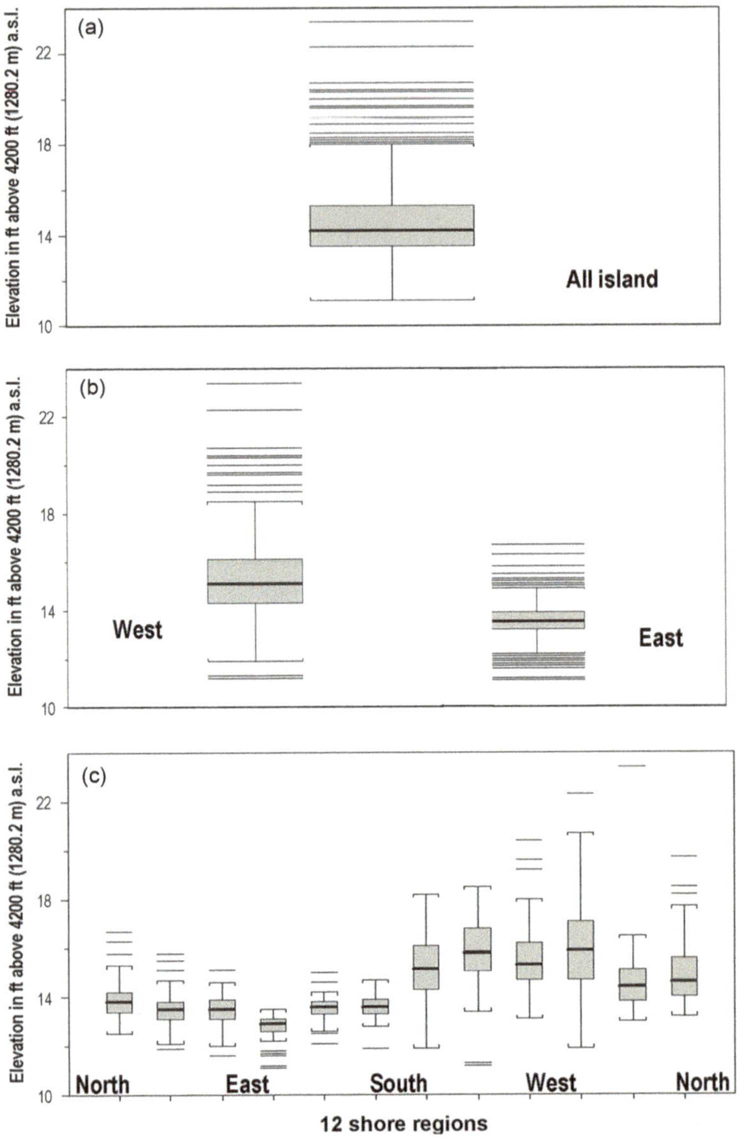

FIG. 1.6 Elevation variability of shoreline evidence of the 1986/87 shoreline on Antelope Island, Great Salt Lake. Elevations were surveyed at 1228 locations along continuous evidence of the 1986/87 shoreline. Box and whisker plots of the data show patterns of shoreline superelevation, elevation of debris above the stillwater lake level. Plot (a) shows data for the entire island, plot (b) for the west (windward) and east (leeward) sides of the island, and plot (c) for 12 regions of the shore of the island. The data indicate that shoreline evidence is rarely at the stillwater level of a closed basin lake. Patterns of shoreline superelevation define a surface that is not horizontal. One-third of the elevations of the debris line were more than 1 m above stillwater lake elevation. *Source: Atwood, G., 2006. Shoreline Superelevation: Evidence of Coastal Processes of Great Salt Lake, Utah. Utah Geological Survey, Misc Publication 06-9, 231 p., courtesy of UGS.*

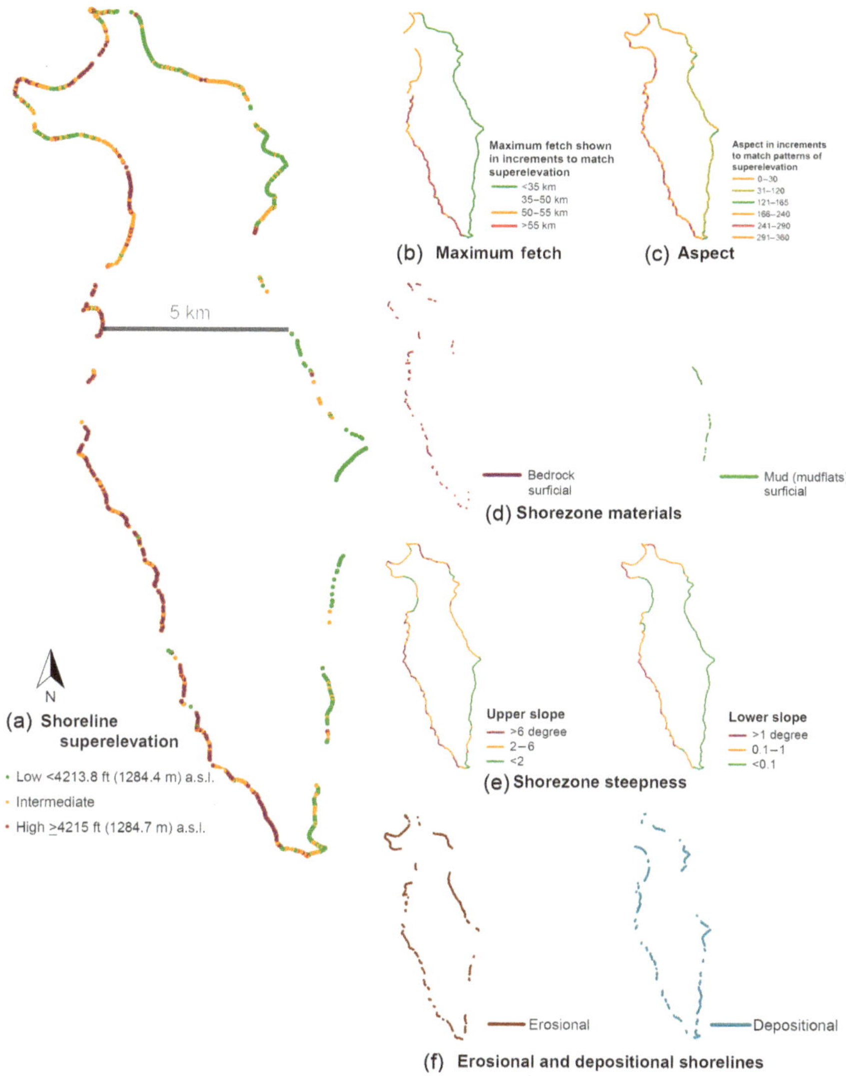

FIG. 1.7 Shoreline superelevation, wave energy, and geomorphic characteristics, Antelope Island. Map (a) displays 1228 surveyed locations along the 1986/87 shoreline evidence classified as low, intermediate, and high and interpreted to mean wave energy. Maps (b–f) display relationships between shoreline superelevation and geomorphic characteristics of the shorezone. *Source: Atwood, G., 2006. Shoreline Superelevation: Evidence of Coastal Processes of Great Salt Lake, Utah. Utah Geological Survey, Misc Publication 06-9, 231 p., courtesy of UGS.*

found to correlate with steep shorezone slopes, whereas lower superelevation was found to correlate with gentle shorezone and offshore slopes. Exposed bedrock was generally associated with higher superelevation and higher wave energy whereas depositional environments were associated with low superelevation and lower wave energy.

Variability of shorezone materials was mapped along the 1986/87 shoreline with inconsistent patterns. Larger material size and better sorting of shorezone debris did not consistently correlate with intermediate or high superelevation. The presence of cobbles and boulders along shorelines did not appear to be primarily due to depositional coastal processes, processes that act only intermittently along high-lake-level shorezones. The primary factor in size and rounding along shores was interpreted to be provenance, meaning the source of sediments, specifically, the relationships of cobbles and boulders to their bedrock source or to detritus in debris-flow deposits. "Boulder beaches" of cobbles were erosional remnants in the swash zone of the 1986/87 shorezone, winnowed and rounded by coastal processes but not evidence of coastal deposition (Fig. 1.8).

Greater wave energy causes higher wave runup, and less energetic wave environments result in lower wave runup (Komar, 1998). Shoreline superelevation is evidence of wave runup and therefore wave energy. Three factors contribute to high-energy wave environments: wind strength, storm duration, and fetch, that is to say the distance across open water to a shore (U.S. Army Corps of Engineers, 1984). If both wind strength and fetch contributed to the variability, then patterns of shoreline superelevation might provide insights not only to fetch but also storm wind strength and direction.

Gilbert (1890) examined that question for the variability in elevations of Lake Bonneville's paleoshorelines and admonished:

> *At an early stage of the investigation, the writer thought that the coasts facing in certain directions gave evidence of exceptional amounts of wave work, and imagined that he had discovered therein the record of prevalent westerly winds or westerly storms in ancient times. This belief was dissipated by further study; and he discovered, as students of modern shores long ago discovered, that there is a close sympathy between the magnitude of the shore features and the 'fetch' of the efficient waves. The greater the distance through which waves travel to reach a given coast, the greater the work accomplished by them. The highest cliffs, the broadest terraces, and the largest embankments are those wrought by the unobstructed waves of the main body; and opposite coasts appear to have been equally affected.*
>
> Gilbert (1890, p. 107)

Lakes can be fetch-dominated or fetch-limited. Coastal processes of fetch-dominated lakes differ from fetch-limited lakes because of their size (Komar, 1998). Fig. 1.9 presents hypothetical patterns of shoreline superelevation for fetch-dominated lakes vs fetch-limited lakes. For a fetch-dominated lake, the long distance across open water provides more than enough opportunity for wind energy to transfer into wave energy causing waves to become unstable and to cease gaining height in response to storm duration or wind strength. In fetch-dominated lakes, the pattern of superelevation is

FIG. 1.8 Shorezone materials due to factors other than wave energy. Material size and sorting along the 1980s-high shoreline are rarely a simple function of wave energy. (a) Shorezone materials decrease in size with distance from bedrock source. The boulders in the foreground have not moved appreciably (more than 10 cm) from their bedrock source. When Great Salt Lake rises to high levels it winnows away fines leaving boulders in place. Clast size is not a function of wave energy. (b) Shorezone materials from flash flood deposits. Similar to (a), some shorezone debris on the east side of Antelope Island has flash flood rather than coastal origins. Material in the deposit includes reworked shore materials from paleoshoreline of Lake Bonneville. Effects of winnowing by storm waves can result in boulder beaches of cobble debris in low-energy wave environments. (c) Beaches of Antelope Island cut into coarse fan deposits on the west side of Antelope Island. Clast size along the west side of Antelope Island appears to be both inherited from terrigenous processes as well as a function of coastal processes that act only intermittently and for short duration. The 1986/87 high level exposed these sediments to coastal processes for no more than 3 months and as few as 10 storms. Their previous exposure to coastal processes was over 100 years earlier. *Source: Atwood, G., 2006. Shoreline Superelevation: Evidence of Coastal Processes of Great Salt Lake, Utah. Utah Geological Survey, Misc Publication 06-9, 231 p., courtesy of UGS.*

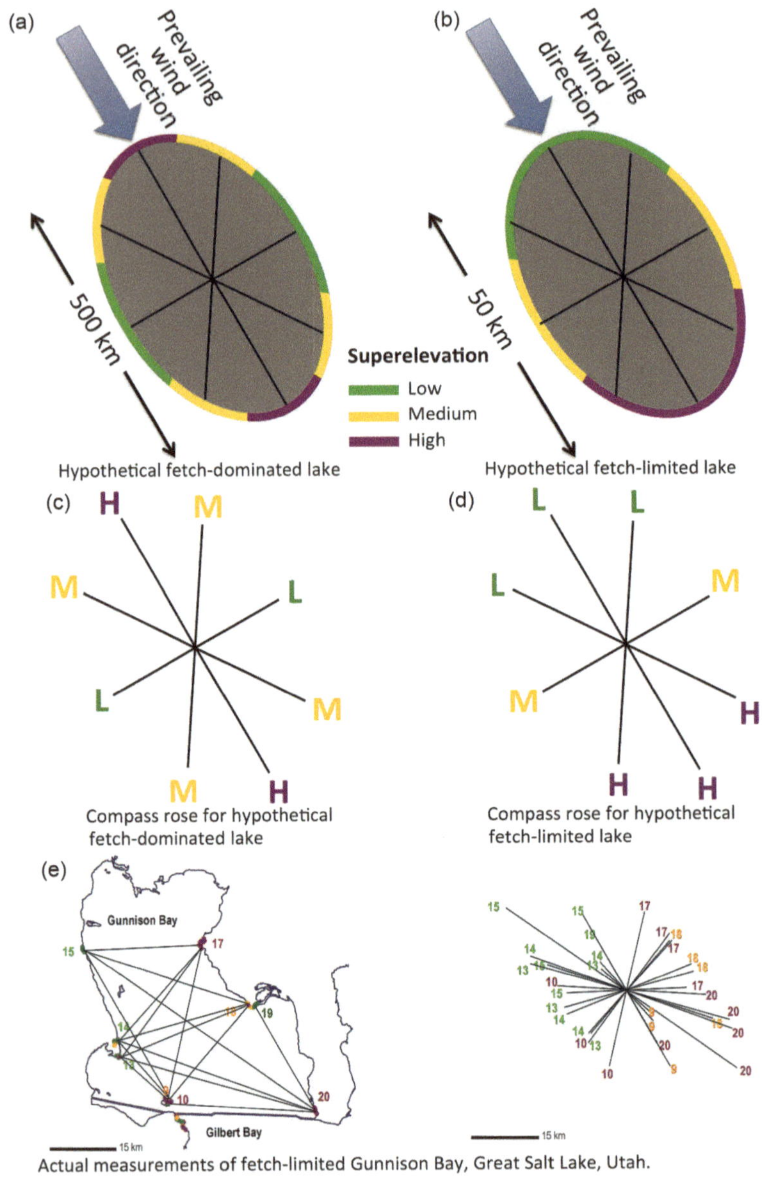

FIG. 1.9 Fetch and wave energy. Schematics indicate the hypothetical distribution of wave energy and shoreline superelevation along shores of a fetch-dominated lake with large surface area (a) and a fetch-limited lake (b). (c) and (d) show the patterns of wave energy and shoreline superelevation (H=high, M=intermediate, and L=low) expected for the hypothetical lakes (a) and (b) due to their contrast in fetch. In schematic (c) the pattern of high shoreline superelevation is symmetrical on opposite shores. In schematic (d) the pattern defines a surface with low superelevation upwind and high superelevation downwind. (e) Shows relative superelevation

symmetrical on opposing shores (Fig. 1.9c) because high wave energy caused overwhelmingly by long fetch masks the contributions of wind strength or storm duration. In a fetch-limited lake (Fig. 1.9b), the short distance across open water does not allow for the complete transfer of wind energy to wave energy, meaning the lake's waves do not become unstable. Therefore, wind strength and storm duration both contribute to the wave energy. The resulting pattern of shoreline superelevation for opposing shores should not be symmetrical for opposing shores but lower on the upwind shore and higher on the downwind shore as in Fig. 1.9d. The asymmetric pattern documents direction of the strongest storm winds.

In 2003, Atwood (2006) examined whether the pattern of superelevation on Antelope Island could be due to wind strength or simply to fetch. However, the direction of the strongest storm winds and the direction of the longest fetch on Antelope Island were both from the west and northwest. To test whether superelevation of shorelines of Great Salt Lake was fetch-dominated or fetch-limited, Atwood surveyed 608 points along the shores of Gunnison and Gilbert Bays. By 2003, much of the 1986/87 shoreline evidence of Gilbert Bay had been disturbed in contrast to that of Gunnison Bay. Fig. 1.9e shows a generalized map of surveyed locations of 1986/87 shoreline evidence on the perimeter of Gunnison Bay. Higher superelevation is shown in purple, intermediate in yellow, and low in green. Each surveyed locale has a line between it and all other locales across open water. The length of each line represents its fetch. In Fig. 1.9e, the lines from the map of Gunnison Bay are transposed so that the center of each line is the center of the diagram on the right maintaining direction and line length. The pattern of the diagram of Gunnison Bay (Fig. 1.9e) matches the hypothetical fetch-limited lake (Fig. 1.9d) and does not match the hypothetical fetch-dominated lake (Fig. 1.9c). For Great Salt Lake, because it is fetch-limited, its patterns of shoreline superelevation indicate evidence of relative storm wind strength and direction of the strongest storm winds.

Research concerning fetch on Great Salt Lake may help researchers correlate Lake Bonneville's paleoshorelines. For both lakes, wave energy affects variability of shoreline elevation. Because Lake Bonneville and Great Salt Lake have dramatically different surface area their fetch differs. Lake Bonneville at its largest had a surface area of approximately 50,500 km^2 (Wambeam, 2001), whereas Great Salt Lake's historic high level had a surface area of

at nine locations of 1986/87 shoreline evidence around the perimeter of Gunnison Bay. Numbers relate to original locations surveyed (Atwood, 2006). The patterns shown for Gunnison Bay match the fetch-limited lake schematic indicating that because Great Salt Lake is fetch-limited, the elevation variability of its shoreline evidence not only provides evidence of fetch but also strongest storm wind direction. *Source: Atwood, G., 2006. Shoreline Superelevation: Evidence of Coastal Processes of Great Salt Lake, Utah. Utah Geological Survey, Misc Publication 06-9, 231 p.*

approximately 5960 km^2 (Arnow and Stephens, 1990). As noted by Gilbert (1890), the main bodies of Lake Bonneville were fetch-dominated. Lake Bonneville's smaller bays such as in Tule Valley or Utah Valley may have been fetch-limited.

Effects of fetch for lakes can be calculated using models developed for shorezone protection (U.S. Army Corps of Engineers, 1984). These models have promised to reduce error in shoreline correlation due to initial variability along paleoshorelines. However, for large bays of Lake Bonneville, caution should be exercised when using patterns of contrasting wave energy to estimate wind strength and direction. Patterns of shoreline superelevation only for fetch-limited water bodies should be used as evidence of storm wind strength and direction.

1.6 BAYS, THRESHOLDS, AND CONSTRICTIONS

The second line of evidence that may be transferable from Great Salt Lake to Lake Bonneville addresses variability of shoreline evidence due to effects of shore complexity. Lake Bonneville and Great Salt Lake are both complex systems. The diversity of bays, constrictions, and thresholds of historic Great Salt Lake may provide as yet unrecognized evidence for understanding variability along historic shorelines of Great Salt Lake and paleoshorelines. Fig. 1.10 shows complexity of bays of Great Salt Lake. The Landsat images, taken 3 months apart during 2005, document seasonal migrations of shoreline positions. The satellite imagery documents complexities not evident in the monitoring of a single bay, such as the gage on Gilbert Bay.

In the 20th and 21st centuries, changes in the complexity of Great Salt Lake's bays have been largely due to humans and include factors of management. Documentation of their effects ranges from intense scrutiny, for example, constrictions between Gilbert and Gunnison Bays, to unpublished evidence of effects of flooding and desiccation of ponds that provide habitat for migratory birds.

Anthropogenic modifications to Great Salt Lake's bays, constrictions, and thresholds may hold insights to natural fluctuations of Lake Bonneville. Two examples from Great Salt Lake are explored to illustrate potential information transfer: effects of the constricted flow between Gunnison and Gilbert Bays due to the railroad causeway and effects of the West Desert Pumping Project that moved waters of Gunnison Bay from 1987 to 1989 into the West Desert (the Great Salt Lake Desert) to alleviate flooding associated with the 1980s wet cycle.

Great Salt Lake is a terminal lake with no natural outlet. The eastern part of Great Salt Lake is fed dominantly by fresh surface waters carried by rivers that head locally in the Wasatch Range and from the Uinta Mountains. The western part is not fed by any major rivers but receives its surface waters from the eastern part of Great Salt Lake. In 1959, a rock-fill railroad

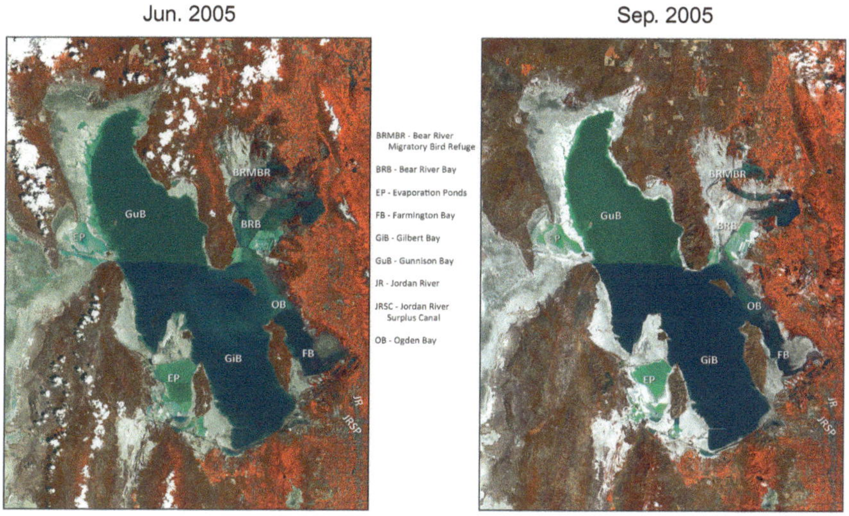

FIG. 1.10 Seasonal changes in bays of Great Salt Lake. Landsat images of Great Salt Lake for Jun. and Sep. 2005 display shoreline migration within bays and diked regions of Great Salt Lake. In Jun., runoff from snowmelt can flush the ponds of the Bear River Migratory Bird Refuge (BRMBR) in contrast to late summer when most of the refuge is dry. Mineral extraction ponds (EP) shrink in summer due to evaporation. The diversity of bays whether natural, artificial, large, or small provides evidence of the sensitivity of regions of the lake to effects of weather, salinity, and human management. *Source: Landsat imagery courtesy of NASA Goddard Space Flight Center and U.S. Geological Survey.*

causeway divided the west part of the lake into two bays, Gilbert and Gunnison. The causeway constricted flow to two culverts and caused Gunnison Bay to become more saline than Gilbert Bay and become the terminus of the Great Salt Lake drainage (Butts, 1980; Arnow and Stephens, 1990). Flow through the culverts became bidirectional with the denser, more saline water from Gunnison Bay flowing southward underneath the less dense, less saline water of Gilbert Bay that flowed northward (Fig. 1.11). A brine layer formed in the deepest areas of Gilbert Bay. As Gunnison Bay became more saline, the lake biota changed, including patterns of algae, brine shrimp, and microbial bioherms (Baskin, 2014). The constriction between the bays resulted in lake level differences. In 1983, the 1 m difference in stillwater elevation between Gilbert Bay and Gunnison Bay prompted a widening of the constriction between the two bays to reduce flooding along the shores of Gilbert Bay (Loving et al., 2000). Flows between Farmington and Gilbert Bays also are constricted with similar but less dramatic effects.

Effects of constrictions between bays of Great Salt Lake may provide analogues for effects of constrictions between bays of Lake Bonneville. Specifically, effects of historic constrictions between Gilbert and Gunnison Bays on salinity and stillwater lake elevation may provide insights to

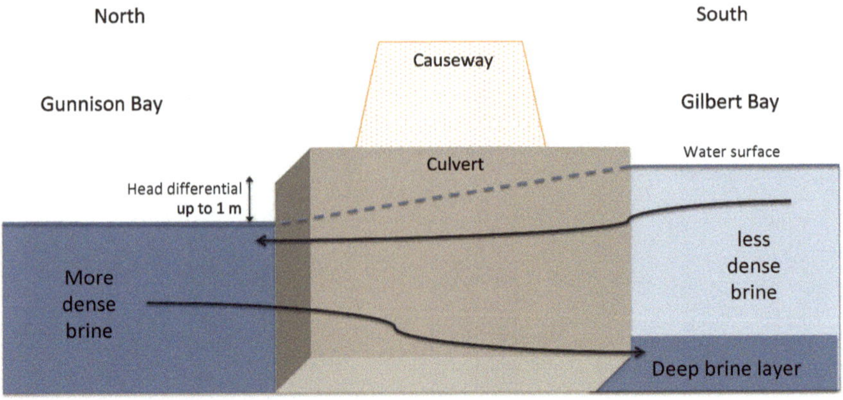

FIG. 1.11 Lake constrictions, Great Salt Lake. Constricted flow through culverts between Gilbert and Gunnison Bays alters lake chemistry of Great Salt Lake. Constricted flow through the rock-fill and culverts of the railroad causeway has caused the water of Gunnison Bay to become more saline and therefore denser than that of Gilbert Bay. This schematic exaggerates the head difference between the two water bodies. Density differences cause bidirectional flow. Less dense brine flows northward above the southward flowing denser brine from Gunnison Bay. The net effect is loss of salt from Gilbert Bay to Gunnison Bay. Gunnison Bay is the terminus of the Great Salt Lake drainage system. The southward flow of dense brine from Gunnison Bay into Gilbert Bay creates a deep, dense brine layer in lowest areas of Gilbert Bay. Not to scale. *Modified from Butts, D.S., 1980. Factors affecting the concentration of Great Salt Lake Brines. In: Gwynn, J.W. (ed.), Great Salt Lake, a Scientific, Historical, and Economic Overview. Utah Geological and Mineral Survey Bulletin, vol. 116, pp. 162–167.*

shoreline variability between the main body of Lake Bonneville and bays of Lake Bonneville such as in Cache Valley, Utah Valley, or the Sevier basin. Gilbert (1890, pp. 178–179) noted contrasts in paleoshoreline materials between the main body of Lake Bonneville and Cache Valley at the Provo level and attributed the abundance of calcium carbonate along shores of the main body to constricted flow through today's Cutler Narrows. Fig. 1.12A shows a cross-section and Fig. 1.12B shows a plan view of present-day topography at Cutler Narrows. According to shoreline elevation data from Currey (1982), the Bonneville level of Lake Bonneville is at approximately 1575 m a.s.l. and the Provo level at approximately 1462 m a.s.l. in the Cutler Narrows region. The elevation profile of Fig. 1.12A is an approximation that uses present topography to illustrate the magnitude of the change of cross-sectional flow from the Bonneville level to the Provo level. At the Bonneville level, four passages across the divide allowed for relatively open flow between the main body of Lake Bonneville and Cache Valley compared to the single, narrow, deep passage through present-day Cutler Narrows. A rough calculation indicates cross-sectional area was reduced from approximately 240,000 to 35,000 m^2. Further study of constrictions and flows between bays of Great Salt Lake may provide insights

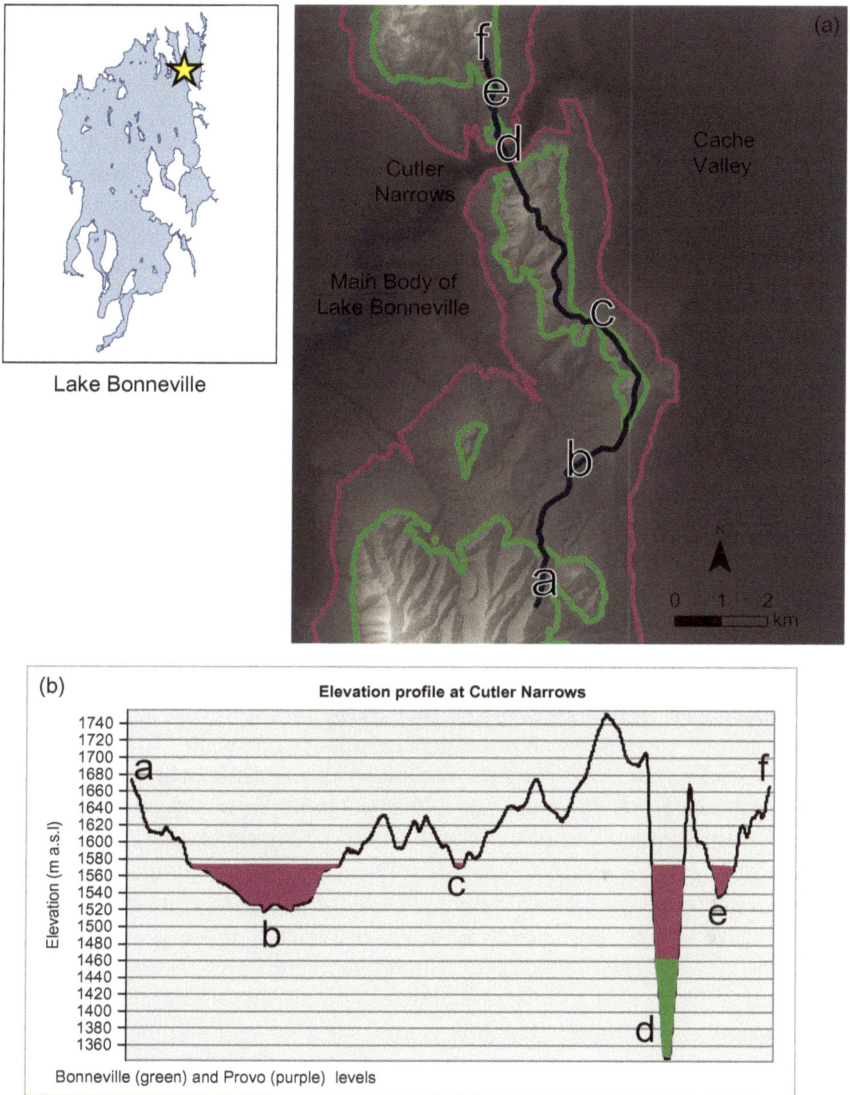

FIG. 1.12 Lake constrictions, Lake Bonneville. Two diagrams show modern Cutler Narrows and vicinity. Cross-section (a) shows a south to north elevation profile (a–f) indicated on map (b). Map (b) shows the watershed divide between Cache Valley to the east and the basin of present Great Salt Lake to the west, the former main body of Lake Bonneville. Elevation contours at 1575 and 1462 m a.s.l. indicate the approximate stillwater lake level of Lake Bonneville's Bonneville (purple) and Provo (green) levels. The schematic only approximates conditions as it uses unreconstructed modern topography and does not take into account effects of fault displacements and isostasy on stillwater levels of Lake Bonneville. It indicates that at the Bonneville level, four passes allowed lake circulation. The drop in stillwater level from the Bonneville level to the Provo level constricted flow to a single, deep, narrow passage. Hypothetical consequences include

(Continued)

to the variability of materials and elevations along paleoshorelines of isolated bays of Lake Bonneville.

Thresholds of Great Salt Lake also may have analogues for Lake Bonneville. As Great Salt Lake rose in the 1980s in response to above average precipitation and below average evaporation, the State of Utah undertook a massive project, the West Desert Pumping Project, to transfer water from Gunnison Bay westward across a natural threshold into the Great Salt Lake Desert and reduce flooding along the south shore of Gilbert Bay. With the water came salts from Gunnison Bay. The West Desert Pumping Project exported 2.73 million acre-feet of brines and 695 million tons of salt from Gunnison Bay into the Newfoundland basin—Great Salt Lake Desert (Loving et al., 2000). About 12% of the salts of Great Salt Lake were transferred to the Newfoundland basin. The Newfoundland basin became the temporary terminus of the Great Salt Lake drainage. The West Desert Pumping Project appears analogous to natural overflows, for example of transgressions to and regressions from the Gilbert level of prehistoric Great Salt Lake. Consequences of West Desert Pumping including variability of materials and shoreline evidence may provide insights to evidence of natural overflows across this and other thresholds.

1.7 CONCLUSIONS

This chapter has catalogued ways that variability presents challenges to paleoshoreline correlation. Variability introduces error in calculations of paleolake surface area and volume. Variability of shoreline evidence from historic Great Salt Lake illustrates ways that potential error can be recognized, quantified, and reduced. Exposures of the 1986/87 high shoreline of Great Salt Lake provide reliable evidence that shoreline superelevation varies systematically due to coastal processes. Patterns of shoreline superelevation reflect patterns of wave energy. Wave energy varies with fetch and storm wind strength for fetch-limited closed basin lakes. Shoreline evidence of the 1986/87 high-lake level of Great Salt Lake was higher and more variable on the windward, western side of Antelope Island that is characterized by rocky headlands and steep shorezone slope than along the accreting, sediment-dominated, gently sloping shorezones of the leeward, eastern side of the island.

Based on evidence from Great Salt Lake, it can be assumed that shoreline evidence of a closed basin lake rarely represents stillwater lake elevation. Researchers in the Bonneville basin can expect that interpretations of

FIG. 1.12—CONT'D bidirectional flow through Cutler Narrows with changed lake chemistry and resulting variability of materials along the Provo-level paleoshoreline. Specifically, the main body became a terminal water body with extensive calcium carbonate deposits along its shores while the water body of Cache Valley remained fresh. The constriction of Gilbert and Gunnison Bays of Great Salt Lake appear analogous to the constriction at Cutler Narrows of Lake Bonneville between Cache Valley and the main body of Lake Bonneville.

stillwater lake elevation of prehistoric Great Salt Lake and Lake Bonneville based on shoreline evidence will introduce error in calculations dependent on precise identification of stillwater lake level. Knowledge of the processes that contribute to variability of shoreline evidence can minimize error.

Shoreline superelevation, on the order of 1–3 m for historic Great Salt Lake, pales in magnitude to the variability of shoreline elevations due to tectonic displacement over the past 30,000 years. However, for research objectives where a precise horizontal datum is essential, such as to quantify displacements associated with specific faults, shoreline evidence cannot be assumed to define a horizontal plane. Initial variability of shoreline evidence may introduce significant error for studies along faults with small displacements.

Variability of materials along high-lake-level shorezones of historic Great Salt Lake indicates that more than one set of Earth processes affect shorezones that are only intermittently occupied by lakes. Materials along the 1986/87 shorezone do not consistently correlate with wave energy. Some sediments along the 1986/87 shorezone vary with distance from their bedrock origin and in response to both coastal and terrigenous processes. Based on evidence from shorelines of Great Salt Lake, the length of lake occupation of a given shoreline affects shorezone evolution. Great Salt Lake shorezones are nourished with upland materials, including from Bonneville paleoshorelines. Caution is advised when assuming clast size and rounding of sediments along paleoshorelines in the Bonneville basin result solely from wave environments.

Just as Lake Bonneville and Great Salt Lake together provide an exceptional record of climate change of western interior North America, the variability along their shorelines provides exceptional evidence of geomorphic processes and landscape evolution of the past 30,000 years. Great Salt Lake may provide analogues for Lake Bonneville with its bays, constrictions, thresholds, and shoreline migrations.

ACKNOWLEDGMENTS

This chapter has benefited from communication with Jonathan Allan, Marjorie Chan, Timothy A. Cross, David Handwerger, Paul Jewell, Don R. Mabey, Matthew A. Mabey, Leland J. Myers, C.G. (Jack) Oviatt, David Sheerer, and Thomas G. Tripp. We gratefully acknowledge an extensive editorial review by Robert L. Baskin.

REFERENCES

Arnow, T., Stephens, D., 1990. Hydrologic Characteristics of the Great Salt Lake, Utah: 1847-1986. U.S. Geological Survey Water, Supply Paper 2332.

Atwood, G., 2006. Shoreline Superelevation: Evidence of Coastal Processes of Great Salt Lake, Utah. Utah Geological Survey, Misc Publication 06-9. 231 p.

Atwood, G., Mabey, D.R., 2000. Shorelines of Antelope Island as evidence of fluctuations of the level of Great Salt Lake. In: King, J.K., Willis, G.C. (Eds.), Geology of Antelope Island, Davis County, Utah, pp. 85–97. Utah Geological Survey, Misc Publication 00-1.

Baskin, R.L., 2014. Occurrence and Spatial Distribution of Microbial Bioherms in Great Salt Lake, Utah. University of Utah, Salt Lake City. Ph.D. dissertation.

Baskin, R.L., Allen, D.A., 2005. Bathymetric Map of the South Part of Great Salt Lake. U.S. Geological Survey Scientific Investigation Map 2005-2894.

Baskin, R.L., Turner, J., 2006. Bathymetric Map of the North Part of Great Salt Lake. U.S. Geological Survey Scientific Investigation Map 2006-2954.

Bills, B.G., Currey, D.R., Marshall, G.A., 1994. Viscosity estimates for the crust and upper mantle from patterns of lacustrine shoreline deformation in the eastern Great Basin. J. Geophys. Res. 92, 11,493–11,508.

Butts, D.S., 1980. Factors affecting the concentration of Great Salt Lake Brines. In: Gwynn, J.W. (Ed.), Great Salt Lake, a Scientific, Historical, and Economic Overview, pp. 162–167. Utah Geological and Mineral Survey Bulletin 116.

Chan, M.A., Godsey, H., 2004. Geoantiquities: concepts and applications for education in the urban landscape. J. Geol. Educ. 52, 445–452.

Crittenden Jr., M.D., 1963. New Data on the Isostatic Deformation of Lake Bonneville (Utah). U.S. Geological Survey Professional Paper 454-E, p. E1–E31.

Currey, D.R., 1982. Lake Bonneville: Selected Features of Relevance to Neotectonic Analysis: U.S. Geological Survey Open-File Rep. 82-1070.

Currey, D.R., Atwood, G., Mabey, D.R., 1984. Major Levels of Great Salt Lake and Lake Bonneville. Utah Geological and Mineral Survey Map 73.

Currey, D.R., Berry, M.S., Green, S.A., Murchison, S.B., 1988. Very late Pleistocene red beds in the Bonneville basin, Utah and Nevada. Geological Society of America Abstracts with Programs 20 (6), 411.

Dinter, D.A., Pechmann, J.C., 2014. Paleoseismology of the Promontory Segment, East Great Salt Lake fault: Final Technical Rept. U.S. Geological Survey Award No. 02HQGR0105, 23 p, http://earthquake.usgs.gov/research/external/reports/02HQGR0105.pdf.

Feth, J.H., Barker, D.A., Moore, L.G., Brown, R.J., Veirs, C.E., 1966. Lake Bonneville: Geology and Hydrology of the Weber Delta District Including Ogden, Utah. U.S. Geological Survey Professional Paper 518, 76 p.

Gilbert, G.K., 1885. The Topographic Features of Lake Shores: U.S. Geological Survey 5th Annual Report. 69–123.

Gilbert, G.K., 1890. Lake Bonneville. U.S. Geological Survey Monograph, 1.

Godsey, H.S., 2012. Geomorphology, Stratigraphy and Chronology of Pleistocene Lake Bonneville Deposits, with Emphasis on the Provo Shoreline, Bonneville Basin, Utah. University of Utah, Salt Lake City. Ph.D. dissertation.

Hunt, C.B., Varnes, H.D., Thomas, H.E., 1953. Lake Bonneville: Geology of Northern Utah Valley, Utah. U.S. Geological Survey Professional Paper 257-A, 99 p.

Komar, P.D., 1998. Beach Processes and Sedimentation, second ed. Prentice Hall, Upper Saddle River, NJ. 544 p.

Lin, A., Wang, P., 1978. Wind tides of the Great Salt Lake. Utah Geol. 5 (1), 17–25.

Loving, B.L., Waddell, K.M., Miller, C.W., 2000. Water and Salt Balance of Great Salt Lake, Utah, and Simulation of Water and Salt Movement Through the Causeway, 1987-98: U.S. Geological Survey Water-Resources Investigations Report 00-4221.

Lund, W.R., 2005. Consensus Preferred Recurrence-Interval and Vertical Slip-Rate Estimates. Review of Utah Paleoseismic-Trenching Data by the Utah Quaternary Fault Parameters Working Group. Utah Geological Survey Bulletin 134.

Mabey, D.R., 1986. Notes on the historic high level of Great Salt Lake. Utah Geol. Miner. Surv. Surv. Notes 20 (2), 13–15.

Machette, M.N., 2005. Summary of the late Quaternary tectonics of the Basin and Range Province in Nevada, eastern California, and Utah. In: Lund, W.L. (Ed.), Proceedings Volume: Basin and Range Seismic Hazards Summit II. Utah Geological Survey, Misc. Publication 05-2.

Murchison, S.B., 1989. Fluctuation History of Great Salt Lake, Utah, During the Last 13,000 years. University of Utah, Salt Lake City. Ph.D. dissertation.

Murchison, S.B., Mulvey, W.E., 2000. Late Pleistocene and Holocene shoreline stratigraphy on Antelope Island. In: King, J.K., Willis, G.C. (Eds.), Geology of Antelope Island, Davis County, Utah, pp. 77–82. Utah Geological Survey, Misc Publication 00-1.

Nelson, D.T., 2012. Geomorphic and Stratigraphic Development of Lake Bonneville's Intermediate Paleoshorelines During the Late Pleistocene. University of Utah, Salt Lake City. Ph.D. dissertation.

Oviatt, C.G., 1991. Quaternary Geology of the Black Rock Desert, Millard County, Utah. Utah Geol. Miner. Surv. Spec. Stud. 73. 23 p.

Oviatt, C.G., 1997. Lake Bonneville and global climate change. Geology 25, 155–158.

Oviatt, C.G., 2014. The Gilbert Episode in the Great Salt Lake Basin, Utah. Utah Geological Survey Misc Publication 14-3, 20 p.

Oviatt, C.G., Currey, D.R., Miller, D.M., 1990. Age and paleoclimatic significance of the Stansbury shoreline of Lake Bonneville, Northeastern Great Basin. Quatern. Res. 33 (3), 291–305.

Sack, D., 1990. Quaternary Geologic Map of Tule Valley, West-Central Utah. Utah Geological and Mineral Survey Map 124, scale 1:100,000, 26 p.

Tibbetts, J.R., Enright, M., Wilberg, D.E., 2004. Water Resources Data, Utah, Water Year 2003: U.S. Geological Survey Water Report UT-03-1. CD and 495 p.

U.S. Army Corps of Engineers, 1984. Shore Protection Manual, fourth ed. USACE Coastal Engineering Research Center, Vicksburg. vol. 2.

Utah Division of Water Resources, 1999. The Great Salt Lake West Desert Pumping Project, its Design, Development and Operation. Utah Department of Natural Resources, Salt Lake City, Utah.

Wambeam, T.J., 2001. Modeling Lake Bonneville Basin Morphometry Using Digital Elevation Models. University of Utah, Salt Lake City. M.S. thesis.

Chapter 2

The Bear River's History and Diversion: Constraints, Unsolved Problems, and Implications for the Lake Bonneville Record

J.L. Pederson*, S.U. Janecke*, M.C. Reheis†, D.S. Kaufman‡ and R.Q. Oaks Jr.*
*Utah State University, Logan, UT, United States
†U.S. Geological Survey, Denver, CO, United States
‡School of Earth Sciences and Environmental Sustainability, Northern Arizona University, Flagstaff, AZ, United States

ABSTRACT

The shifting course of the Bear River has influenced the hydrologic balance of the Bonneville basin through time, including the magnitude of Lake Bonneville. This was first recognized by G.K. Gilbert and addressed in the early work of Robert Bright, who focused on the southeastern Idaho region of Gem Valley and Oneida Narrows. In this chapter, we summarize and evaluate existing knowledge from this region, present updated and new chronostratigraphic evidence for the Bear River's drainage history, and discuss implications for the Bonneville record as well as future research needs.

The Bear River in Plio-Pleistocene time joined the Snake River to the north by following the present-day Portneuf or Blackfoot drainages, with it likely joining the Portneuf River by middle Pleistocene time. An episode of volcanism in the Blackfoot-Gem Valley volcanic field, sparsely dated to ~100–50 ka, diverted the Bear River southward from where the Alexander shield volcano obstructed the river's path into Gem Valley. Previous chronostratigraphic and isotopic work on the Main Canyon Formation in southern Gem Valley indicates internal-basin sedimentation during the Quaternary, with a possible brief incursion of the Bear River ~140 ka. New evidence confirms that the Bear River's final diversion at 60–50 ka led to its integration into the Bonneville basin by spillover at a paleo-divide above present-day Oneida Narrows. This drove rapid incision before the rise of Lake Bonneville into the canyon and southern Gem Valley.

Developments in Earth Surface Processes, Vol. 20. http://dx.doi.org/10.1016/B978-0-444-63590-7.00002-0

Bear River diversion at 60–50 ka coincides with the end of the Cutler Dam lake cycle, at the onset of marine isotope stage 3. The Bear River subsequently contributed to the rise of Lake Bonneville, the highest pluvial lake known in the basin, culminating in the Bonneville flood. Key research questions include the prior path of the upper Bear River, dating and understanding the complex geologic relations within the Gem Valley-Blackfoot volcanic field, resolving evidence for possible earlier incursions of Bear River water into the Bonneville basin, and interpreting the sedimentology of the Main Canyon Formation.

Keywords: Bear River, Lake Bonneville, Drainage integration, Gem Valley, Paleohydrology

2.1 INTRODUCTION

Lake Bonneville was the largest pluvial lake in the Great Basin at the last glacial maximum, and its extent seems anomalously large (McCoy, 1987; Benson et al., 1990). Lake Bonneville had five times the volume of water of the next largest pluvial lake in western North America (Lake Lahontan) at its respective highest level (Karow and Hampel, 2010). Based on the lack of higher geomorphic markers and other lines of evidence (eg, Balch et al., 2005), Lake Bonneville was also larger than at least the three preceding pluvial lakes within its same basin. As Gilbert stated (1890, p. 94), "It marks the greatest expanse of the ancient lake…Above the Bonneville shoreline the whole aspect is that of the dry land." Perhaps most importantly, Lake Bonneville was the only one known to overflow its topographic threshold near Zenda, ~2.7 km north of Red Rock Pass (Gilbert, 1890), which enabled temporary external drainage and the catastrophic Bonneville flood. Indeed, we do not know how high Lake Bonneville would have reached if internal drainage had been retained. This contrasts with the Pleistocene record of alpine glaciation in the region, wherein moraines record the largest ice extent during the earlier Bull Lake glaciation (eg Munroe, 2005; Licciardi and Pierce, 2008), roughly at the time of the Little Valley lake cycle in the Bonneville basin. Regardless of how anomalous the size of Lake Bonneville was, researchers agree that the rise, fall, and extent of Lake Bonneville through time relates to climate change. Yet, studies sometime overlook that late Pleistocene drainage changes, especially the diversion of the Bear River that has the largest discharge of any river in the Great Basin, played a key role in the changing hydrologic balance of the system. Resolving the influence of drainage changes will foster a clearer understanding of each factor involved in the cyclic fluctuations of successive lake levels.

Gilbert (1890, pp. 218–219) recognized that the history of Lake Bonneville's shorelines is largely a story of the balance between water supply and evaporation, and also that the Bear River has been the most important and dynamic source of water. Key geographic relations occur where the Bear River swings north to the edge of the Great Basin divide, from the Uinta Mountains to Soda

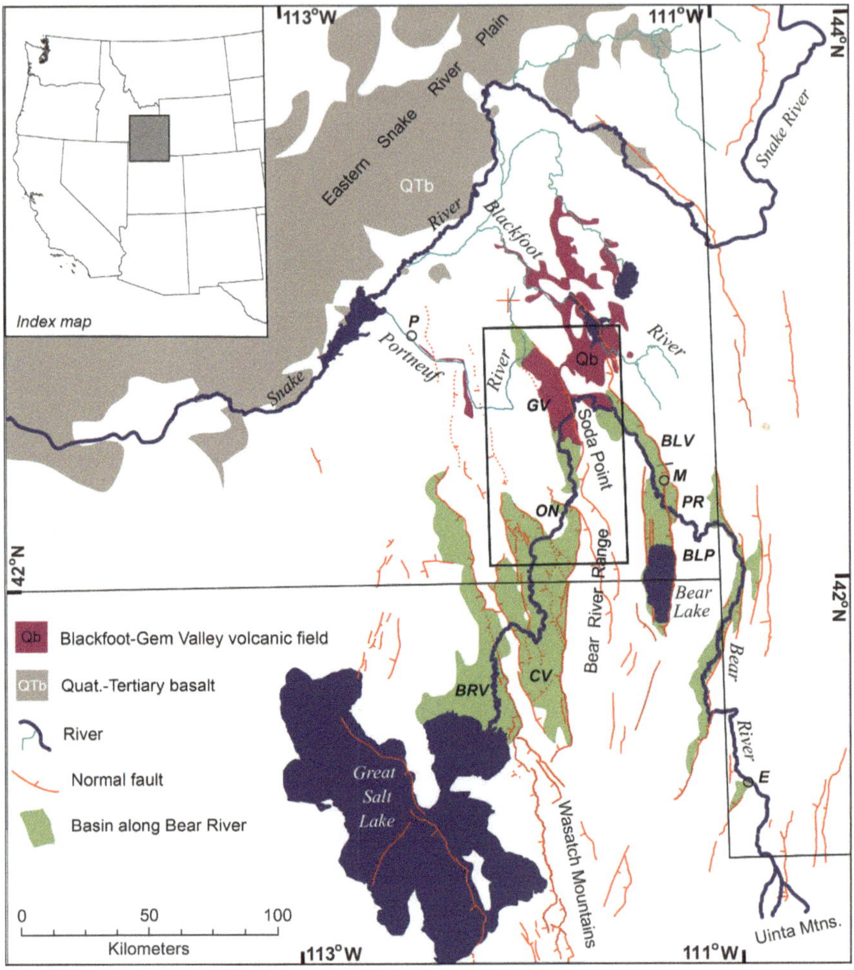

FIG. 2.1 Northeastern Bonneville basin and Bear River drainage region. *Rectangle outlines area* of Fig. 2.2. Key locations and features along the path of the Bear River are labeled: *E*, Evanston; *BLP*, Bear Lake Plateau; *PR*, Preuss Range; *M*, Montpelier; *BLV*, Bear Lake Valley; *GV*, Gem Valley; *P*, Pocatello; *ON*, Oneida Narrows; *CV*, Cache Valley; *BRV*, Bear River Valley.

Point, near Soda Springs, Idaho, between Bear Lake and Cache Valley, and then swings sharply south again (Fig. 2.1). Gilbert noted the very low divides between the Bear River and both the Portneuf and Blackfoot Rivers, both of which currently drain northwest to join the Snake River. He recognized that they and the Bear River may have been connected in earlier times, and perhaps were recently separated or diverted one way or another by the young basalt flows that built those drainage divides. Finally, in interpreting the overall hydrographic history of Lake Bonneville, Gilbert (1890, p. 263) explored the alternative hypothesis that rises and falls of the ancient lake might have been controlled by changes in the Bear River's drainage.

But Gilbert discussed how a former path of the Bear River to join the Snake River would have lowered lake level in the Bonneville basin, not whether recent diversion *into* the basin might have done the opposite.

Bright (1963) was the first to postulate that Lake Bonneville rose to its highest elevation of 1552 m partly in response to the first arrival of the Bear River into the Bonneville Basin. In fact, Bright grew up in the southeastern Idaho region in the middle of the geological records needed to test this hypothesis. His work, along with subsequent field and geochemical studies reviewed later, provide clear evidence that Pleistocene basalts erupted near the Bear River and eventually built a low divide across central Gem Valley in southeastern Idaho, thereby diverting the Bear River south from its connection to the Snake River basin into the Bonneville basin through the canyon of Oneida Narrows (Fig. 2.1; Bright, 1963; Bouchard et al., 1998; Link et al., 1999).

Our purpose in this chapter is to review the previous work in these key areas along the Bear River, present new results that help pinpoint the timing and mechanisms of this diversion, and discuss the implications for lake-level fluctuations in the Bonneville basin. We suggest future research that is needed to complete this understanding, including improved age control on basalt flows and sediments associated with the diversion of the Bear River, sedimentology of those same deposits, and investigation of possible earlier pathways of the upper Bear River.

2.2 SETTING AND EVOLUTION OF THE UPPER BEAR RIVER

The Bear River drains the northeast part of the Bonneville basin, which is itself at the northeastern corner of the Great Basin and formed as late Cenozoic extension created a set of closed basins inboard of the western North American plate margin. The Bonneville basin is fed largely by the Bear, Weber, Provo, and Sevier Rivers, which flow out of the Wasatch and western Uinta Mountains. The larger, modern-day Bear River drainage basin accounts for ~58% of the streamflow into the Bonneville basin (Tarboton, 2015). The 790-km-long Bear River exits the Bear Lake basin and turns west about half-way along its length southeast of Soda Springs, Idaho, there skirting the southern edge and eventually cutting through lava flows of the eastern (Blackfoot) portion of the Blackfoot-Gem Valley volcanic field (Fig. 2.1). Farther west, at a low divide across the center of Gem Valley built by basaltic cinder cones and lava flows, the river turns south and flows through the Black Canyon knickzone and southern Gem Valley (Fig. 2.2). Southern Gem Valley is also known in whole or in parts as Thatcher Basin, Gentile valley, and Mound valley (Bright's usage), but for clarity we will refer to Gem Valley as the entire structural basin and avoid these numerous local names. Farther south, the Bear River cuts through a low range between the main Portneuf and main Bear River Ranges, and enters the northern part of the Cache Valley graben through the ~300 m deep bedrock canyon of Oneida Narrows (Fig. 2.1).

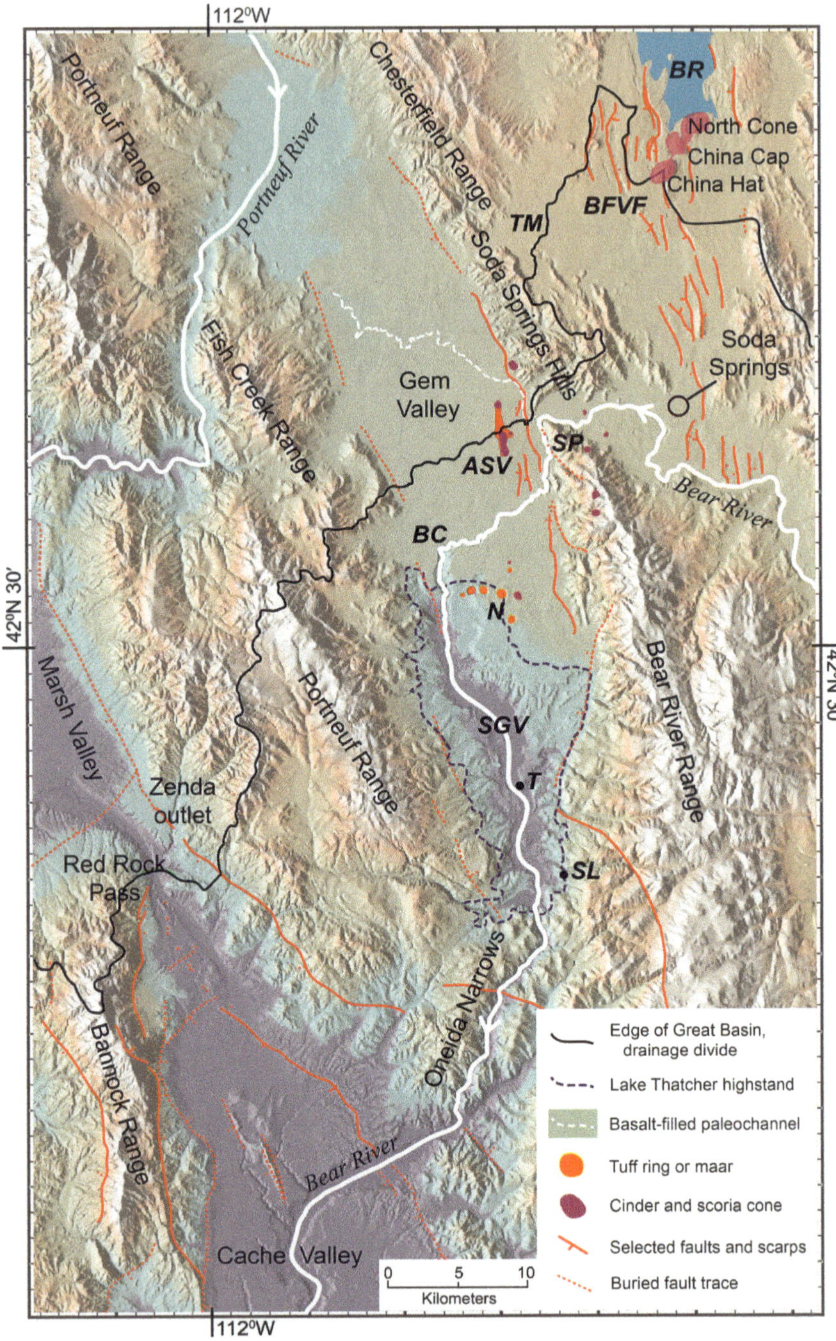

FIG. 2.2 Locations and features of the Gem Valley and Oneida Narrows area where the Bear River was diverted southward into the Bonneville basin. The highest shoreline of Lake Bonneville corresponds to the violet-to-blue transition of the background map at 1540 m, and the highest shoreline of Lake Thatcher is marked by the *dark blue dashed line* at 1660 m. *BR*, Blackfoot Reservoir; *BFVF*, Blackfoot volcanic field; *TM*, Tenmile shield volcano; *SP*, Soda Point; *ASV*, Alexander shield volcano; *BC*, Black Canyon; *N*, Niter cluster of volcanic vents; *SGV*, southern Gem Valley; *T*, Thatcher; *SL*, Smith locality of Fig. 2.4.

The geologic setting along this path of the Bear River is quite regular and systematic. From its headwaters in glaciated valleys on the north flank of the Uinta Mountains, the upper Bear River flows northwestward about 400 km (Fig. 2.1), stepping systematically westward through five east-tilted half grabens of the Basin and Range. The longest reach of the Bear River that cuts across the regional structural grain is northeast of Bear Lake, where the river crosses between the Bear Lake Plateau and the Preuss Range. After angling across Bear Lake Valley, the Bear River resumes a northward course as far as Soda Springs. The lower Bear River, downstream of Soda Point, changes to a southwest course, and from a left-stepping path to a right-stepping path through the extensional, mostly east-tilted basins of Gem Valley, Cache Valley, and Bear River Valley.

The integration of the upper Bear River is thought to postdate the Miocene-Pliocene Salt Lake Formation because that unit was deposited in basins of a significantly different landscape during early extension in the region (Janecke et al., 2003). The Salt Lake Formation is composed of tuffaceous and clastic lacustrine and fluvial sediments that have been distinctively tilted and folded by extensional structures (Evans and Oaks, 1996; Janecke and Evans, 1999; Janecke et al. 2003; Kruger et al., 2003; Long et al., 2006). Most of the Salt Lake Formation is 12–6 Ma, but uppermost conglomerates may be as young as ~3 Ma in some subbasins (Goessel et al., 1999; Oaks et al., 1999; Janecke et al., 2003; Keeley and Rodgers, 2015). The outward growth of the Basin and Range province documented by previous workers (Wernicke et al., 1987; Perkins et al., 1998) suggests that the undated exposures of the Salt Lake Formation along the NW-flowing, upper Bear River may be younger than those dated in basins to the west.

Early mappers (eg, Mansfield, 1927; Oriel and Platt, 1980; Taylor and Bright, 1987) interpreted the youngest sediments on the Bear Lake Plateau east of Bear Lake (Fig. 2.1) as Salt Lake Formation, and inferred that the Bear River was superimposed across structures there through a thick Miocene-Pliocene sedimentary fill. More recent geologic mapping has suggested that sedimentary rocks of the Salt Lake Formation are thin and fairly restricted areally on the Bear Lake Plateau (Coogan, 1992a,b, 1997a,b; Dover, 1995). This is pertinent because the oldest deposits mapped and interpreted as representing a through-flowing Bear River are preserved about 360 m above the river on the crest of the Bear Lake Plateau (Reheis, 2005; Fig. 2.1). Cross-bedded fluvial gravel and sand up to 5 m thick lie along a gentle north and east slope for at least 8 km along the drainage divide, toward the present Bear River. In one locality, nested channel fills show that this paleo-river incised as it migrated eastward (Reheis, 2005). These deposits are undated and are inset below the level of an Oligocene 28.8-Ma basaltic eruptive center (Coogan, 1997a). The gravels lack the distinctive pink and purple quartzite clasts derived from the Precambrian rocks of the Uinta Range, causing Reheis et al. (2009) to suggest that the earliest Bear River may have drained no farther south than about Evanston, Wyoming (Fig. 2.1), and only later captured its present headwaters, as previously suggested by Hansen (1985). An alternate possibility is that these high fluvial gravels predate the Bear River and are remnants of the Salt Lake Formation as mapped in the area by Coogan (1992a, appendix K), or perhaps they are both—earliest Bear River gravels in the uppermost Salt Lake Formation.

Evidence from terrace gravels and lacustrine deposits where the river enters Bear Lake Valley indicates that the upper Bear River was certainly integrated by late Pliocene to early Pleistocene time, after deposition of the youngest dated Salt Lake Formation (see Reheis et al., 2009, for details on the following relations). The highest terrace gravels mapped southeast of Montpelier, Idaho, are 80 m above the modern river, locally cap lacustrine deposits of a Plio-Pleistocene Bear Lake, and lie between two traces of the East Bear Lake Valley fault zone (Fig. 2.1). The second-highest river terrace in that area overlies several meters of fan-delta deposits, which in turn rest on deposits containing a tephra layer that is correlative with either the Bishop ash bed (765 ka; Zeeden et al., 2014) or upper Glass Mountain tephra (about 1 Ma; Sarna-Wojcicki, 2005).

Gilbert (1890) suggested that the low divides around Soda Springs and Gem Valley raise the possibility that the Blackfoot and Portneuf Rivers were once tributaries to the Bear River. He also briefly discussed the possibility, more pertinent here, that the Bear River once exited from Soda Springs through northwestern Gem Valley and then via the Portneuf River rather than southwest along its present course (Figs. 2.1 and 2.2). This Portneuf connection has been supported by geologic evidence (eg, Bright, 1963; Mabey, 1971; Janecke and Oaks, 2014). For example, Ludlum (1943), Bright (1963, 1967), and Scott et al. (1982) recognized that a pair of basalt flows follow a sinuous path westward and then southward down the Portneuf Valley from the Blackfoot-Gem Valley volcanic field. In addition, the Portneuf River appears underfit for its valley where it crosses through the Portneuf and Bannock ranges, supporting a paleo-Bear River addition to its discharge. More recently, Janecke and Oaks (2014) reported a separate basalt-filled meandering channel >4 km long, which trends west through central and west-central Gem Valley (thin dotted line in Fig. 2.2). Its east end projects to the modern Bear River at Soda Point, and the west end projects to the modern Portneuf River along the northwest margin of Gem Valley. Thus it may represent a basalt-filled channel of the paleo-lower Bear River. These lines of evidence strongly suggest that, at least during the middle to late Pleistocene, the paleo-Bear River flowed through northern Gem Valley and joined the Portneuf River to flow into the Snake River.

An older Plio-Pleistocene course directly north to the Snake River via the Blackfoot River cannot be ruled out, however. The modern divide between the Bear and Blackfoot drainages lies southeast of China Hat (Fig. 2.2) and is about 120 m above the Bear River near Soda Springs. Terrace remnants up to 40 m above the Bear River are recorded south of Soda Springs on the eastern, hanging-wall block of a normal fault (Reheis et al., 2009). If Quaternary volcanism resulted in 80 m or more of fill across the divide area, it is permissible that the river could have flowed north to the Blackfoot River during the early Pleistocene. In support of this, Mabey and Oriel (1970) indicated ~240–300 m of basalt in the area east of China Hat based on well and gravity data. However, fluvial gravels that might represent such a Bear River course have not been observed in a geothermal well drilled near Blackfoot Reservoir (eg, McCurry et al., 2011; Welhan et al., 2013), or along the lower valley of the Blackfoot River (Mansfield, 1929; Hladky et al., 1992).

Regardless of the exact pathway, the prior connection of the upper Bear River to the Snake River drainage is supported by consistent evidence from several species of organisms with aquatic biological affinity between the two basins (Taylor and Bright, 1987; Hershler and Sada, 2002; Smith et al., 2002; Mock et al 2006; Billman et al., 2010). However, the biological evidence does not accurately constrain the timing of aquatic connections or the location of the transfer. Miocene and Pliocene fossils of fish and molluscs provide general evidence of prior connections during these periods (Taylor and Bright, 1987; Smith et al., 2002), but former drainage patterns interpreted from fish do not always agree with those from molluscs. Distributions of several living species of fish in the Snake River and northern Bonneville-Bear River areas demonstrate interbasin transfers in the past, likely in both directions (eg, Minckley et al., 1986; Smith et al., 2002). In contrast to evidence from fish species, Taylor and Bright (1987) interpreted low endemism and a lack of obvious Snake River-related molluscs in the Bear Lake area to indicate that the lake was internally drained during the Pliocene and possibly not fed by an integrated Bear River until later.

Current research emphasizes the use of mitochondrial DNA (mtDNA) analysis to trace ancestry and vicariance among living species and to estimate time of divergence from geologic ages for related fossils, although estimated evolutionary rates vary among species (Smith et al., 2002). For example, the cutthroat trout, *Oncorhynchus clarki*, in the Bear River is estimated to have diverged from the upper Snake River population around 0.7 Ma based on mtDNA. However, these Bear River trout are more divergent from those in the main Bonneville basin, which suggests separate evolution during most of the Pleistocene. Similarly, *Lavinia* sp. (a cyprinid minnow) is reported from sediments thought to be about 6 Ma in both Cache Valley and north of Bear Lake, an age that supports a divergence time of ~6 Ma for the upper and lower Bear River as estimated independently from mtDNA evidence from living fish populations (Smith et al., 2002). The speckled dace (*Rhinichthys osculus*) is widespread in both the Bonneville and Snake River basins and occurs in two genetic groups corresponding to the southern Bonneville basin and the combined upper Snake River, northern Bonneville basin and the Bear River. The close similarity of populations in the northern group suggests fairly recent separation of the Bear and Snake River drainages (Billman et al., 2010). The same northern and southern division and similarity of Bear and Snake River populations exist in the Utah sucker (*Catostomus ardens*; Mock et al., 2006).

2.3 HISTORY OF FAULTING AND VOLCANISM IN GEM VALLEY

It is critical to understand the interplay of subsidence from normal faulting and volcanic constructional processes in Gem Valley because of its key role in diverting the Bear River to the Bonneville basin. The simple shape of Gem Valley itself contrasts with the geographic complexity of the five ranges along its margins (Figs. 2.1 and 2.2). Gem Valley basin is a structural graben ~2–3 km deep that lies on the southeast arm of the neotectonic parabola that is centered on the Yellowstone hot spot (Figs. 2.1 and 2.2; Pierce and Morgan, 1992; Piety et al., 1992). This graben is up to 70 km long, ~10 km wide, and is mostly filled by low-density sediments of the Miocene Starlight Formation and Miocene-Pliocene Salt Lake Formation (Mabey and Oriel, 1970). At most a few hundred meters of Quaternary sediment and basalt cap this thicker basin-fill sequence (Bright, 1963; Janecke and Oaks, 2014). The East Gem Valley fault zone has had enough activity to preserve fault scarps as high as 15 m across late Pleistocene basalt flows along at least 22 km of the central portion of its linear mountain front (Wong et al., 2012). A low slip rate of about 0.1 mm/year across the fault zone allowed two major breaks in the eastern mountain front to form at Tenmile Pass between the Chesterfield Range and Soda Springs Hills and along the Bear River between the Soda Springs Hills and Bear River Range. Lava flows entered Gem Valley from the Blackfoot field through both of these low points.

The bimodal, basalt-dominated, Quaternary Blackfoot-Gem Valley volcanic field is interpreted to have diverted the upper Bear River into its current southward route to the Bonneville basin in the middle to late Pleistocene (Bright, 1960, 1963, 1967; Bouchard et al., 1998). Tholeiitic porphyritic olivine basalt flows, cinder and spatter cones, and a small number of rhyolite domes erupted from a NNE-trending belt of ~100 discrete eruptive centers and fissures on the southeast margin of the eastern Snake River Plain (Mansfield, 1929; Oriel, 1968; Oriel and Platt, 1968, 1980; Armstrong, 1969; Perkins, 1979; Fiesinger et al., 1982; Pickett, 2004; Ford, 2005; Polun, 2011; Janecke and Oaks, 2014). Many of the basalt flows erupted from cones and fissures along N- to NNW-striking normal faults, fissures, and rift zones; the East Gem Valley fault zone, the Blackfoot rift zone, and the northern continuation of the East Bear Lake fault zone are among the most important conduits of the volcanic field (Oriel and Platt, 1968, 1980; Armstrong, 1969; Polun, 2011; Janecke and Oaks, 2014). Large-volume basaltic lavas flowed outward from these centers and produced a highly irregular map pattern of valley-filling basalt flows (Fig. 2.1; Lewis et al., 2012). Most of the volcanic rocks lie within a 250-km^2 rectangular area from central Gem Valley in the southwest to the Blackfoot Reservoir area in the northeast (Fig. 2.2), and far-travelled, valley-filling basalt flows, and satellite volcanic centers expand the outer edges of this volcanic field west to Pocatello, Idaho, and north to the margin of the eastern Snake River Plain (Mansfield, 1927; Ludlum, 1943, Bright, 1960, 1963, 1967; Oriel, 1965, 1968; Oriel and Platt, 1968, 1980; Armstrong, 1969; Fiesinger et al., 1982; Puchy, 1982; Coogan, 1992a; Pickett, 2004; Ford, 2005; Fig. 2.1).

The first K-Ar age determinations from the Blackfoot-Gem Valley volcanic field by Armstrong et al. (1975) yielded three whole-rock ages of ~100 ka on basalt flows in Gem Valley and three whole-rock and feldspar ages between 40 and 100 ka from the China Hat and China Cap rhyolite domes in the Blackfoot Reservoir area farther to the east (locations in Fig. 2.2). However, the analytical uncertainties were large and no tests were conducted to check for inherited Ar. In contrast, an older and reliable, whole-rock K-Ar age of 0.9 ± 0.2 Ma was determined by Armstrong et al. (1975) from a valley-filling basalt flow along the northern part of the Blackfoot River. This is the oldest dated basalt in the Blackfoot-Gem Valley volcanic field, with the early Pleistocene age supported by the reverse polarity (pre-0.781 Ma normal Bruhnes subchron) of other basalt flows in that northwest part of the Blackfoot-Gem Valley volcanic field (Mabey and Oriel, 1970).

Four subsequent $^{40}Ar/^{39}Ar$ and uranium-series dates from the volcanic field bring the early-to-late Pleistocene age of the volcanic field into better focus. A pair of stacked basalt flows known as the Portneuf (or Inkom) basalts were previously estimated by K-Ar to be ~140 ka (Armstrong et al., 1975) or ~600 ka (Scott et al., 1982), but a newer result of 430 ± 70 ka comes from a $^{40}Ar/^{39}Ar$ age on whole-rock from the lower lava (Rodgers et al., 2006). These two shoestring-like flows erupted from somewhere in the Blackfoot-Gem Valley volcanic field and flowed a minimum of 100 km west through canyons in the Portneuf and Bannock Ranges as well as across the Gem, Marsh and Portneuf valleys, reaching the south edge of Pocatello, Idaho (Figs. 2.1 and 2.2; Ludlum, 1943; Scott et al., 1982; Rodgers et al., 2006). These Portneuf basalt flows are the oldest yet dated in the Gem Valley part of the volcanic field.

The youngest basalt flows of the Gem Valley part of the volcanic field are late Pleistocene in age. This is based on a 90 ± 60 ka $^{40}Ar/^{39}Ar$ age on whole-rock from the Alexander scoria cone (Hughes and Pickett, in Pickett, 2004) at one of the subparallel fissures and rift zones at the center of the Alexander shield volcano. This cone lies 2.5 km north of the 1686-m divide that now separates the Columbia River drainage basin from the Bonneville basin (Fig. 2.2; Janecke and Oaks, 2014). Although there is a large uncertainty in the $^{40}Ar/^{39}Ar$ age determination arising from the low potassium content of all basalt flows in the field and this basalt's very young age, the 90 ± 60 ka age determination supports the late-middle to late Pleistocene ages inferred for many of the younger basalt flows nearby.

Volcanism to the east, in the Blackfoot part of the volcanic field, is better dated than in Gem Valley because the rhyolite domes there contain potassium-rich minerals (Fig. 2.2). Age control indicates that rhyolitic volcanism in the Blackfoot Reservoir area initiated between 2 and 1 Ma east of the Blackfoot Reservoir (Fig. 2.2; see summaries in Ford, 2005; McCurry and Welhan, 2012). At the youngest end, Pierce et al. (1982) redated the China Hat rhyolite dome to 61 ± 6 ka with the whole-rock K-Ar method. More recently the China Cap dome produced a 57 ± 8 ka $^{40}Ar/^{39}Ar$ crystallization age (Heumann, 1999), and uranium-series isochron dating of zircons provide an overlapping eruption age of 66 ± 7 ka from the same dome (Schmitt, 2011). Polun (2011) and Welhan et al. (2013) argued from all of these results that the genetically related rhyolite domes and maars in a NE alignment south of the Blackfoot Reservoir were erupted during a single episode. Altogether these geochronologic data suggest that basaltic and rhyolitic volcanism began together in the early Pleistocene and continued episodically into the late Pleistocene, ~100–50 ka.

2.3.1 Building of the Drainage Divide in Gem Valley

The Great Basin's drainage divide in Gem Valley has an irregular east-northeast trend across the Alexander shield volcano and a low point in the west at 1686 m (Fig. 2.2). The divide's surface, though covered with several meters of loess, preserves enough irregular topography to show that it is an original volcanic landscape that has not been modified by streamflow. The central part of Gem Valley has been paved over by basalts erupted from three volcanic centers. From north to south, these are the Tenmile (in the Blackfoot field), Alexander, and Niter clusters of volcanic vents (Fig. 2.2; Janecke and Oaks, 2014). The Alexander cluster of cinder cones and fissure vents, dated to 90 ± 60 ka (Pickett, 2004), in particular, may have built the current drainage divide across Gem Valley and thus eventually diverted the Bear River, as it appears to have erupted across the path of the river where it enters the east side of Gem Valley (Janecke and Oaks, 2014). In addition, a far-travelled flow from the Tenmile cluster of volcanic vents might have issued west through Tenmile Pass (Fig. 2.2), across northern Gem Valley, and played a role in confining or diverting the paleo-Bear River. Ongoing research in the volcanic rocks of Gem Valley may reveal the specific vent and basalt flow that diverted the Bear River southward.

The Niter volcano cluster is the southernmost cluster in the Blackfoot-Gem Valley volcanic field (Fig. 2.2). It is unusual in that it consists mostly of a half-dozen explosive phreatomagmatic maars and tuff rings, rather than cinder and scoria cones (Armstrong, 1969; Janecke and Oaks, 2014). The maars near the Bear River on the west side of this cluster give way eastward to tuff cones and finally to two cinder cones at higher elevations. The altitude of this transition between water-associated maars and normal cinder cones largely coincides with the ~1660-m highest shorelines of Lake Thatcher (Janecke and Oaks, 2014). Bright (1963) earlier documented pillow basalts nearby at this altitude. These phreatomagmatic maars and tuff rings suggest that most of the Niter cluster erupted when Lake Thatcher was at its highest levels. However, the westernmost two of these eruptive centers must have persisted somewhat longer because its low-relief ejecta ring lies in a lower, inset landscape position along the incised portion of southern Gem Valley, in association with basalt-capped terraces along the Bear River (Kackley and undifferentiated flows of Bright (1963) and Janecke and Oaks (2014)). This field relationship requires eruption of this westernmost maar after some of the incision of Oneida Narrows and southern Gem Valley, and thereby, the Bear River's integration south into the Bonneville basin.

2.4 CHRONOSTRATIGRAPHY OF SOUTHERN GEM VALLEY AND ONEIDA NARROWS

2.4.1 Prior Chronostratigraphy

The Pleistocene Main Canyon Formation constitutes the basin-fill record exposed in southern Gem Valley, and it provides critical information on the history of sedimentation and hydrology. The unit was first investigated in terms of stratigraphy and paleontology by Bright (1963, 1967). McCoy (1987), Hochberg (1996), and Bouchard et al. (1998) subsequently provided geochronologic and Sr-isotopic constraints as well as additional stratigraphic insights. Although detailed sedimentological work has yet to be done on the Main Canyon Formation, Bright (1963) interpreted it as Pleistocene sediments from local drainages with coarser alluvial fan sediments at the basin margins shed toward a fluctuating axial lake system with shorelines as high as 1660 m. A paleosol separates the lower and upper subunits of the ~150 m of exposed Pleistocene basin fill of the Main Canyon Formation (Hochberg, 1996). Bright (1963) and Hochberg (1996) interpreted the lower subunit as marshy and/or paludal sediments with numerous paleosols, and although the upper subunit is somewhat less organic-rich, it is still dominated by laminated to structureless silty sediment with numerous mollusc shells of lacustrine affinity.

Near its exposed base, the lower subunit contains the ~2 Ma Huckleberry Ridge tephra and the Lava Creek B tephra, the latter at 1535 m (~640 ka) (Izett, 1981; Hochberg, 1996; Bouchard et al., 1998; Lanphere et al., 2002). Hochberg (1996) discovered that the upper Main Canyon subunit includes beds geochemically correlated with an early Mt. St. Helens tephra, from a base at 1600 m and up to 28 m higher (Fig. 2.3). The tephra was first reported as ~110 ka (Bouchard et al., 1998), but it more broadly correlates with one of a series of early Mt. St. Helens tephra now thought to range from 120 to 70 ka (Kuehn and Negrini, 2010).

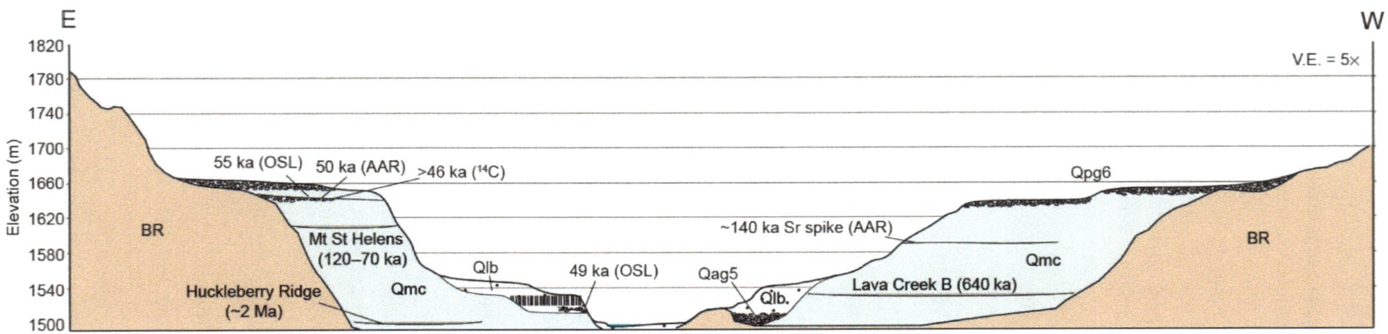

FIG. 2.3 Schematic cross section illustrating the chronostratigraphy of the Main Canyon Formation and other Quaternary deposits in southern Gem Valley, view looking south (downstream). *BR*, lower Paleozoic and Proterozoic sedimentary bedrock, *Qmc*, Quaternary Main Canyon Formation; *Qlb*, Quaternary sediment of Lake Bonneville overlying inset basalt flow (columnar hachures); *Qag5*, early-Bonneville-age channel gravel of the Bear River; *Qpg6*, late Pleistocene piedmont (alluvial fan) gravel capping the basin fill. The three corroborating ages by different methods (*AAR*, amino acid racemization) all come from the same marker near the top of the basin fill.

Initial work by Bright (1963, 1967) on the Main Canyon Formation included collection of a series of aquatic molluscs for radiocarbon dating, from near the exposed base to the top of the basin fill, ranging in age from 34 to 27 ka (uncalibrated radiocarbon years). In light of newer geochronology and given the potential for postdepositional isotopic exchange with shell carbonate, these are minimum ages, as recognized in McCoy's (1987) early work using aminostratigraphy. Regardless of knowledge of timing, Bright (1967) interpreted that, with the addition of water provided by the diverted Bear River into southern Thatcher Basin, spillover at the southern end through what is now Oneida Narrows fully integrated the Bear River into the Bonneville basin. Bright also recognized that subsequent incision of Oneida Narrows and southern Gem Valley by the Bear River was interrupted by the incursion of Lake Bonneville. This back-flooded the canyon and lower part of southern Gem Valley to the mouth of Black Canyon, north of the town of Thatcher (Fig. 2.2), and left fine-grained deltaic-lacustrine sediment lining the valley bottom.

Chronologic constraints indicate that sedimentation in the southern end of the Gem Valley graben either includes or proceeded after deposition of the 2-Ma Huckleberry Ridge ash and that sedimentation was episodic through deposition of the 640 ka Lava Creek B ash (Izett, 1981; Hochberg, 1996; Bouchard et al 1998). A maximum of ~150 m of dominantly fine, calcareous, organic-rich sediment accumulated at least through deposition of the 120–70 ka early Mount Saint Helens ash in the upper third of the Main Canyon Formation (Fig. 2.3). Average sediment accumulation rates in the Main Canyon Formation were very low, in part due to hiatuses represented by the multiple paleosols, and ranged from 0.07 to 0.20 mm/year for different parts of the sequence. These rates are a fraction of those calculated across similar timeframes from sediment cores drilled beneath the Great Salt Lake (0.38 mm/year; Davis, 1998) and Bear Lake (0.54 mm/year; Kaufman et al., 2009). Low sedimentation rates support the conclusion of Bouchard et al. (1998) that local streams provided the water and sediment for the bulk of the Main Canyon Formation. If the Bear River had been directly providing sediment for any extended period, its large discharge and sediment load would have resulted in a higher sediment accumulation rate as well as thicker, coarser, and redder-colored deposits.

2.4.2 History of Bear River Water Input to Southern Gem Valley

Bouchard et al. (1998) attempted to identify and date the arrival of Bear River water into the southern Gem Valley and its influence on deposition and the Main Canyon Formation through fingerprinting water sources by $^{87}Sr/^{86}Sr$ ratios. Two samples of Bear River water at Soda Springs yielded $^{87}Sr/^{86}Sr$ ratios of 0.70858 and 0.70922, varying seasonally, whereas the ratios from four local tributaries and a hot spring flowing into the southern Gem Valley were higher, ranging from 0.71037 to 0.71132. Because the Bear River has a discharge about twice that of the combined local sources, as estimated by Bouchard et al. (1998), they argued that its hydrologic contribution can be readily detected through relatively low $^{87}Sr/^{86}Sr$ ratios of mollusc shells found in the Main Canyon Formation. In the stratigraphically lowest and poorly exposed, early and middle Pleistocene part of this basin fill, they measured relatively high isotopic ratios (ave. = 0.71309) indicating no Bear River water entering the marshy setting of the basin. At the base of the upper Main Canyon Formation, stratigraphically between the ~640 ka Lava Creek B and the ~110 ka Mt. St. Helens tephra layers, Bouchard et al. (1998) reported a single mollusc sample with a lower $^{87}Sr/^{86}Sr$ ratio of 0.70987 and an amino acid racemization age of ~140 ka. They interpreted this lower Sr ratio as representing a first incursion of Bear River water into the basin, supported by Bright's (1963) suggestion that an interval of red sediment at that position in the basin fill is of Bear River provenance. Stratigraphically, this was followed by eight successively higher mollusc samples in the upper Main Canyon Formation, all with higher $^{87}Sr/^{86}Sr$ ratios indicative of internal drainage. At the top of the formation, they documented an isotopic shift in shell composition corresponding to a facies change, which they interpreted as the abrupt transgression of a larger lake caused by southward diversion of the Bear River. Bouchard et al. (1998) more specifically dated the final incursion of the Bear River based on shells in the same horizon as the isotopic shift by a minimum-accelerator mass spectrometry (AMS) radiocarbon date of 42.5 ± 1.5 cal ka (now recalibrated to 45.9 ± 2.8 cal ka utilizing; Reimer et al., 2013) and 50 ± 10 ka based upon amino acid dating (Table 2.1). Note that these were sampled at the same site as the ~27 ka radiocarbon date reported previously by Bright (1963).

It is worth building upon the hydrologic mass-balance calculations of Bouchard et al. (1998) to explore implications for paleohydrology. In addition to developing a mass-balance-and-mixing model to understand the sources and changes in their Sr isotopes, they pointed out that, even without a Bear River contribution, the local drainages and springs should have provided enough water to fill the southern Gem Valley basin over the timescale of centuries, and seemingly should have caused southward spillover (based on present-day discharges and evaporation rates). However, before construction of the volcanic divide separating northern and southern Gem Valley, presently at an altitude of 1686 m, any surface water may have joined the Bear River on a northern pathway out of Gem Valley. More importantly, groundwater would have found its own pathway downgradient through permeable basalts, either following the paleo-river's course northwest out of the basin, or flowing into the equally close Bonneville basin (Cache Valley, Fig. 2.2) to the south. Regardless, groundwater exit helped keep Gem Valley only marshy for most of the Pleistocene, as the preliminary sedimentologic interpretations of the Main Canyon Formation suggest. With the diversion of Bear River water, the more-restricted late Pleistocene southern basin would have filled rapidly—but how rapidly? Taking the historical record of Bear River

TABLE 2.1 Geochronology of the Upper Main Canyon Formation and Subsequent Integration of the Bear River

Method	Study	Sample	Stratigraphic Position	Dose Rate (Gy/ky)[a]	De (Gy)[b]	Age (ka)
Amino acid	Bouchard et al. (1998)	Molluscs lower in MCF	Earlier Bear River incursion (Sr)			~140
Tephra correlation	Hochberg (1996)	Early Mt. St. Helens tephra	Central-upper Main Canyon Fm.			~120–70
Amino acid	Bouchard et al. (1998)	Molluscs in lag gravel	Upper MCF lake transgression			50 ± 10
AMS radiocarbon	Bouchard et al. (1998)	Molluscs in lag gravel	Upper MCF lake transgression			45.9 ± 2.8[c]
OSL[d]	New	USU-846, Smith locality	Upper MCF lake transgression	3.35 ± 0.17	184.07 ± 9.29	55.0 ± 5.6[e]
OSL	New	USU-941, Sant Road cut	Inset Bear River, advanced incision	2.10 ± 0.11	102.43 ± 20.34	48.9 ± 6.9

[a] *Dose rates calculated from elemental chemistry using the conversion factors of Guérin et al. (2011) and cosmic-ray contribution following Prescott and Hutton (1994).*
[b] *OSL equivalent dose errors reported at 2σ.*
[c] *Radiocarbon age calibrated with IntCal03, errors reported at 2σ.*
[d] *OSL analyses use standard single-aliquot regenerative protocol (Murray and Wintle, 2000), calculated from between 17 and 43 accepted aliquots using the central-age model of Galbraith et al. (1999).*
[e] *OSL ages incorporate random and systematic errors reported at 1σ.*

discharge from the USGS gaging station at Pescadero (between Bear Lake and Soda Springs) as a proxy (1922–2014), we can confirm Bouchard's estimate of $\sim 5 \times 10^8$ m^3/year of water delivered by the Bear River to Gem Valley. As an order-of-magnitude estimate, one can generously represent the maximum volume of southern Gem Valley, partly filled with Main Canyon Formation, as a disk 5 km in radius and 100 m deep. Excluding evaporation rates, which are an order of magnitude slower than water input to this surface (Bouchard et al., 1998), we find that the basin could have filled with water and started spilling over in less than ~15 years.

In summary, the Main Canyon Formation's key role in recording evidence about the Bear River's diversion is well established, though further work is needed. The work of McCoy (1987), Hochberg (1996), and Bouchard et al. (1998) has made it clear that the short-lived lake basin in southern Gem Valley envisioned by Bright (1963, 1967) was instead a long-lived depositional basin that persisted as long as 2 m/year, generally matching the timing known for the volcanic field along its northern edge. Furthermore, the Bear River was a minor or secondary factor in the deposition of the Main Canyon Formation in southern Gem Valley, most of which has high Sr-isotopic signatures for mollusc shells indicative of internal drainage with only a singular occurrence of the red sediment diagnostic of the Bear River. All workers agree that the top of the formation records the Bear River's diversion southward into southern Gem Valley and the transgression of the highest Lake Thatcher. This southern incursion of the Bear River rapidly led to the hydrologic overtopping of the southern margin of Gem Valley, into the Bonneville basin.

2.4.3 New Age Control in the Upper Main Canyon Formation and Overlying Units

New geochronologic constraints on the timing of drainage diversion and incision of the Bear River are provided by mapping and optically stimulated luminescence (OSL) geochronology of Quaternary deposits above, within, and below Oneida Narrows. Only new data most pertinent to the Bear River's diversion are presented here. In southern Gem Valley, an exposure of the uppermost basin-fill strata has been sampled (Fig. 2.4). Observations there support previous interpretations of the sedimentology, with a mollusc-rich horizon of gravel lag marking the transgression of a standing-water lake over subaerial alluvial fan deposits derived from the eastern margin of the basin, which contain a well-developed buried soil (Fig. 2.4). This gravel lag, locally at ~1655 m, is interpreted to correlate with the horizon linked by Bouchard et al. (1998) to the shift to a Bear River-influenced Sr-isotopic signal. A new OSL age from a sample of the overlying, well-laminated lacustrine sandy silt, taken just 4–8 cm above the lag gravel, is 55.0 ± 5.6 ka (Fig. 2.4, Table 2.1). This date is consistent with the previous amino acid date of 50 ± 10 ka and the minimum ~46 ka AMS radiocarbon date reported by Bouchard et al. (1998). At the new

FIG. 2.4 Key relations and new OSL age in uppermost Main Canyon Formation found at the Smith locality, named for the local landowner (Fig. 2.2). Alluvial fan deposits and a buried soil profile marked by calcic-horizon development (as highlighted by gradational band) are slightly truncated and buried by a mollusc-rich, pebbly lag interpreted to mark lake transgression, as previously dated elsewhere in the basin. The new age determination is from directly overlying, laminated sandy silt interpreted as lake deposits preserved just before incision by the integrated Bear River.

study site, the dated sediment is, in turn, overlain by 4–5 m of thin-to-medium, tabular-bedded silt with dispersed molluscs and isolated, small lenses of locally derived gravel, here interpreted as littoral lacustrine deposits of the highest Lake Thatcher. Our rough hydrologic balance estimate and this date for the Bear River's incursion into the southern Thatcher Basin suggest that the rise of Lake Thatcher, spillover, and integration across the paleo-divide in Oneida Narrows would have occurred in just a few years at ~55 ka.

A new constraint on the timing of subsequent incision through southern Gem Valley is provided by an OSL date from sediment inset and near the bottom of the valley. The deposit is cross-bedded, pebbly sand of the Bear River that includes basaltic grains, and it underlies a basalt flow that stands above, but flanks, the modern, incised valley bottom between the town of Thatcher and the mouth of Black Canyon (Figs. 2.2 and 2.3). Both the deposit of the Bear River at 1520 m elevation and the overlying basalt flow are inset against the tuffaceous, early-middle Pleistocene, lower Main Canyon Formation. An OSL sample from the sandy Bear River sediment yields an age of

48.9 ± 6.9 ka (Table 2.1). Different beds of this road-cut exposure were likely sampled for geochronology by both Bright (1963) and McCoy (1987), with uncertain and mixed results. The basaltic lava flow that appears to disturb and overlie the newly dated fluvial deposit extends upstream a short distance and downstream along the eastern flank of the valley. The lava flow is, in turn, buried by a mantle of fine-grained Lake Bonneville deposits. This OSL age confirms that base-level lowering and much of the valley incision were accomplished just a few thousand years after river integration, well before the initial rise of Lake Bonneville, ~30 ka (Oviatt, 2015). Furthermore, this provides evidence for volcanism at ~50 ka, which in this case flowed into the actively eroding valley. As proposed by previous workers (Bright, 1963; Oriel and Platt, 1980; Bouchard et al., 1998), southern Gem Valley as well as the ~15-km-long Oneida Narrows canyon downstream was incised rapidly, mostly before it was back-flooded by the rising waters of Lake Bonneville (Fig. 2.5).

Previous estimates for the rate of this canyon incision have been hampered by the broad geochronologic constraints and by assumptions about the paleogeography and incision processes. Using a reconstruction of paleotopography based upon surficial mapping of the Main Canyon Formation basin-fill top and piedmont surfaces graded to it in southern Gem Valley, Pederson and King (2011) identified a paleo-drainage divide at ~1675 m (Fig. 2.5). This is less than half-way downstream through the canyon, at about the same position that Bright (1963) estimated. Thus, the spillover at ~55 ka took advantage of a low paleotopographic drainage divide and occupied north–south-directed local catchments (defined by the reconstruction of the mapped Qpg6 piedmont-upland surface) that had already defined and eroded the upper elevations of Oneida Narrows (Fig. 2.5). Only the narrower, inner gorge of Oneida Narrows needed to be incised by the Bear River after integration, representing a vertical maximum of ~225 m at the paleo-divide. Since new geochronology indicates most incision was accomplished before ~49 ka, the time-averaged vertical incision rate was 2–4 cm/year, depending upon whether one uses the central OSL ages or their reported errors to define the episode. Yet, it is likely that the vertical rate instead decayed over time and that incision was accomplished by an upstream-propagating and distributed knickzone across the 8-km distance through the resistant Proterozoic and Paleozoic strata of upper Oneida Narrows (Fig. 2.5). The evidence reviewed above for an earlier incursion of the Bear River into southern Gem Valley raises the possibility that Bear River water may have filled and spilled over the paleo-divide at that time as well, perhaps accomplishing prior erosion to the ~1675 m divide that was overcome at ~55 ka. So far, stratigraphic or geomorphic evidence for such an earlier integration and erosion episode have not been found. The present-day knickzone of Black Canyon in central Gem Valley represents the current position of the upstream-propagating wave of incision that was initiated by the final integration of the Bear River through Oneida Narrows.

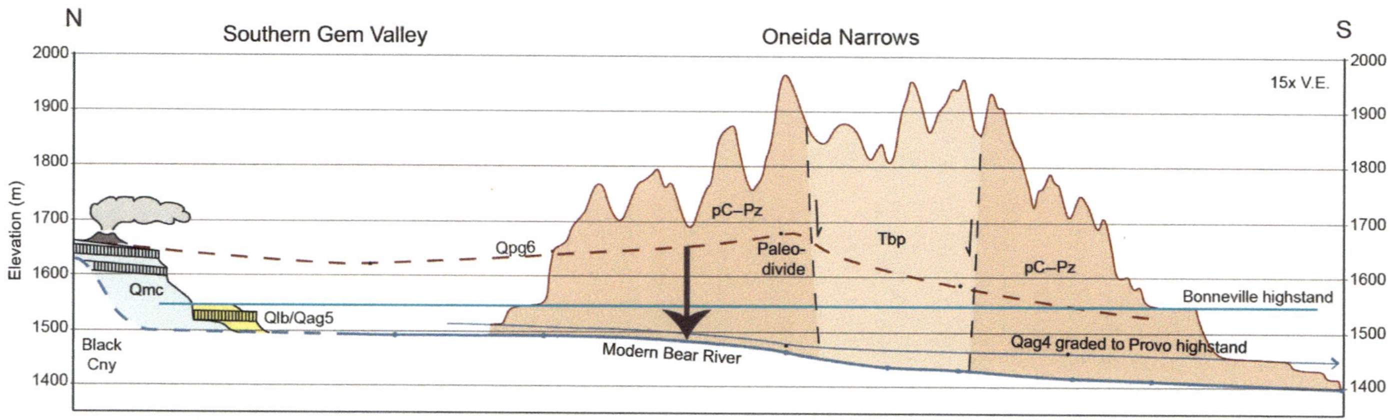

FIG. 2.5 Schematic longitudinal profile depicting geomorphic relations along the Bear River through southern Gem Valley and Oneida Narrows. Horizontal scale is shortened in Gem Valley to show relations. Qpg6 marks the reconstructed basin fill and paleo-drainage surface at the top of the Main Canyon Formation (Qmc), and Qlb/Qag5 are inset Bear River deposits, basalt flow (hachured columns), and Lake Bonneville deposits. Incision of Oneida Narrows through both Proterozoic/Paleozoic (pC–Pz) and Tertiary Salt Lake Formation (Tbp) bedrock below the reconstructed datum occurred rapidly; the propagating knickzone is currently located ~30 km upstream in Black Canyon.

2.5 PROXY EVIDENCE FROM THE BONNEVILLE BASIN

Although several paleoclimate and paleohydrologic reconstructions have been developed from the greater Bonneville basin, only two studies have interpretations regarding the history of the Bear River. First, as studied in the Main Canyon Formation, the input of the upper Bear River (above Oneida Narrows) may be detected by a shift in the Sr-isotope composition of carbonates formed in the terminal basin. Hart et al. (2004) interpreted the Sr-isotope values of lake water in the Bonneville basin as controlled by varying input from specifically the Bear River and Sevier River as well as from groundwater. They analyzed the $^{87}Sr/^{86}Sr$ value of molluscs from the primary shorelines of the Bonneville cycle and from the earlier Cutler Dam and Little Valley lake cycles and compared them to those of the modern Great Salt Lake. They found relatively low values in deposits of all lake cycles, generally tracking their magnitude, with Lake Bonneville deposits having the lowest value and the deposits of the Cutler Dam lake cycle and Currey's (1982) Gilbert shoreline having relatively higher Sr ratios. With their Sr-isotope mixing model, they tested for the effect of Bear River water, interpreting that the $^{87}Sr/^{86}Sr$ values measured in carbonates from all of the lake cycles are too low to have formed without input from the upper Bear River. This is partly consistent with Bouchard et al.'s (1998) Sr work reviewed above, in which one sample indicated the Bear River contributed water to southern Gem Valley during an earlier episode ~140 ka, potentially overflowing into the Bonneville basin at about the time of the Little Valley lake cycle. Yet, no incursion of Bear River water into Lake Thatcher was identified by Bouchard et al. (1998) at the time of the penultimate Cutler Dam lake cycle, which was dated by Kaufman et al. (2001) to 59 ± 5 ka, or during marine isotope stage (MIS) 4 (71–57 ka). The evidence described earlier indicates river integration just after this, in early MIS 3, yet the errors in numeric ages make it possible, strictly speaking, that Bear River integration and the Cutler Dam lake cycle overlapped in time. Lacustrine carbonates in the Bonneville basin older than the Little Valley lake cycle have not been analyzed for their Sr composition, and the $^{87}Sr/^{86}Sr$ values of molluscs in southern Gem Valley show that upper Bear River water was not a major contributor in the early and middle Pleistocene (Bouchard et al., 1998).

Davis (1998) and Davis and Moutoux (1998) analyzed the pollen content from drilled cores recovered around Great Salt Lake. They documented an abrupt increase in pollen concentration and sedimentation rate after 310 ka in the Indian Cove well, marking a change that they attributed to the initial diversion of the upper Bear River into the Bonneville basin. This does not match evidence from the Main Canyon Formation, which thus far indicates no influx of Bear River water or sediment into southern Gem Valley prior to about 140 ka. Age control in the Indian Cove core above the Lava Creek ash bed is by correlation of interglacial pollen assemblages with odd-numbered marine oxygen-isotope stages. Additional age control in the younger sequence in the Indian Cove core is needed to confidently assess the age of the transition to a higher sedimentation rate in the Bonneville basin, and more analysis is needed to determine if the increased sediment input can be ascribed to the Bear River vs a more local source.

2.6 DISCUSSION

2.6.1 Paleogeography and Diversion

Understanding the Pleistocene development of Gem Valley may be informed by taking a step back and considering the somewhat analogous relationships in the better-understood Bear Lake Basin ~75 km upstream. In both Gem Valley and Bear Lake Valley, the Bear River has either an antecedent or superimposed path west across the margin of an active structural basin, entering both north-trending valleys near their midpoints (Fig. 2.1). In Bear Lake Valley, the course of the river has alternated due to avulsion during the Quaternary between flowing: (1) south into the deeper subbasin of Bear Lake; (2) directly into the larger "open basin" area of maximum extent of the lake during glacial epochs (fig. 10 in Kaufman et al., 2009); and (3) along a route north of a smaller lake like the present one, and exiting the basin as it does today, when only man-made canals connect the river and lake (Robertson, 1978; Reheis, 2005; Kaufman et al., 2009). The path of the Bear River through Gem Valley during the early(?) to middle Pleistocene may have been broadly similar in map view to the modern Bear River in Bear Lake Valley. This analogy is consistent with other evidence that, for most of its history, the route of the Bear River was northwest across Gem Valley, toward the subsiding Snake River Plain, bypassing the depocenter on the south side of a growing divide created by flows from the Alexander shield volcano (Fig. 2.2).

Gem Valley contrasts with Bear Lake Valley in being less tectonically active and in having the coincidence of the Bear River entering the basin right where shield volcanoes and cinder cones were being constructed in the Pleistocene. No direct geologic evidence exists that an older "Lake Thatcher" ever occupied the northern part of Gem Valley, as Bright (1963, 1967) suggested, and instead the marshy sediments of the Main Canyon Formation are likely restricted to the southern basin. Due to the persistent presence of the Bear River, Pleistocene sedimentation rates in the southern Bear Lake Valley subbasin are significantly greater than in southern Gem Valley. The deposition of the lower and upper Main Canyon Formation was episodic, as indicated by paleosols, organic-rich intervals, and the significant unconformity that must lie between the basal Huckleberry Ridge Tuff (2 Ma) and the Lava Creek B ash (0.64 Ma) only 25 m above it (Hochberg, 1996). The slow accumulation rate is consistent with the conclusions of previous workers that only local catchments provided the water and sediment to the southern end of Gem Valley for the bulk of its history.

In Bear Lake Valley, groundwater connections exist between the river and Bear Lake even while surface water is disconnected. Likewise, during deposition of the Main Canyon Formation it is likely that a groundwater connection existed between the paleo-Bear River and the southern subbasin of Gem Valley. Some mixing and flow of paleo-groundwater was especially likely given the fractures and porous basalts along the river's path through the Alexander shield volcano. Thus, the water of the marshes in southern Gem Valley may have been mixed partly with water from the Bear River, even while the clastic sediment there came from the adjacent mountain ranges (Fig. 2.6). Likewise, a lack of a surface water connection of the Bear River into the Bonneville basin in earlier Pleistocene time does not preclude some upper Bear River water making it into the Great Basin through groundwater pathways.

Regarding the mechanism for diversion of the Bear River, although the geochronologic data are currently too sparse to be certain, the ~430-ka eruption and great volume of the two Portneuf basalt flows might reflect the onset of vigorous volcanism and more rapid building of a divide across Gem Valley. Such a scenario was previously proposed by Link et al (1999) to have taken place at 140 ka, based on Armstrong et al.'s (1975) erroneous age of the Portneuf basalt. The building of the mid-basin drainage divide in Gem Valley likely occurred in concert with the somewhat more rapid accumulation of marshy and lacustrine deposits of the upper Main Canyon Formation. The ancestral Bear River's path may have been influenced by lavas flowing downhill from the Tenmile Pass area or along structural depressions within the East Gem Valley fault zone (Janecke and Oaks, 2014). Yet the river's inferred northwestward path across Gem Valley toward the Snake River need not have been diverted toward the southern subbasin until near the end of deposition of the Main Canyon Formation.

2.6.2 Implications of Bear River Drainage Integration

Current chronostratigraphic constraints provide a 60–50-ka timeframe for full southward integration of the Bear River into the Bonneville basin. This interval falls just after, or perhaps overlaps the end of, the penultimate Cutler Dam (MIS 4) lake cycle and well before the most recent Bonneville (MIS 2) lake cycle. Based on lithofacies interpretation, ostracod zonation, and the age model of GLAD800 sediment cores, Balch et al. (2005, fig. 6) interpreted a freshening and rise in lake water of Lake Bonneville between ~40 and 20 ka, stronger than in previous pluvial cycles. Similarly, Nishizawa et al. (2013) presented evidence for unexpectedly high lakestands in the Bonneville basin during MIS 3, between 50 and 25 ka, though these interpretations are disputed (Oviatt et al., 2014). Globally as well as in the southwestern United States (eg, Wagner et al., 2010; Benson et al., 2011; Moseley et al., 2016), MIS 3 (57–24 ka) is recorded as a period of highly variable climate at the millennial timescale. It is possible that the strong climate fluctuations at the onset

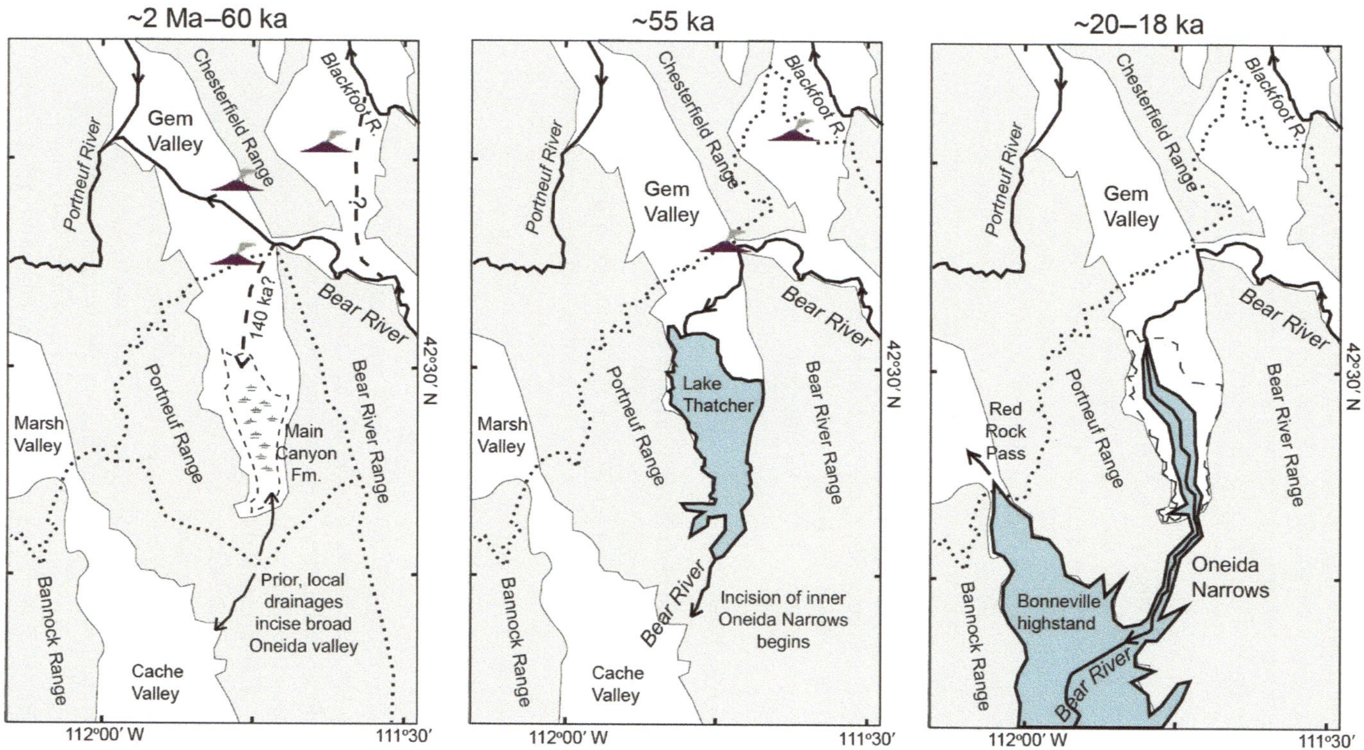

FIG. 2.6 Summary of Quaternary drainage evolution of Gem Valley and Oneida Narrows region in three phases (*dashed lines* represent changing drainage divide). In earlier Plio-Pleistocene time, the upper Bear River possibly could have joined the Blackfoot River, but evidence suggests it connected to the Portneuf River by the late Pleistocene, while internal-basin deposition dominated deposition of the Main Canyon Formation in southern Gem Valley. In the middle panel, eruption of the Alexander shield volcano diverted the river toward the south, raising Lake Thatcher at ~55 ka and causing spillover at the divide in Oneida Narrows, which integrated the river into the Bonneville basin. At right, subsequent incision of the inner gorge of Oneida Narrows and southern Gem Valley allowed Lake Bonneville to backflood far up the Bear River, while its hydrologic addition contributed to spilling of the lake at Red Rock Pass, culminating in the Bonneville flood.

of MIS 3 contributed a hydrologic push for the diversion and integration of the Bear River through southern Gem Valley. Yet, this discussion pertains to a singular, ~55 ka diversion of the Bear River, and not to the potential earlier episode of Bear River water incursion implied by Sr-isotope data from the more distal Bonneville basin, as well as in southern Gem Valley (Bouchard et al., 1998).

The southern Gem Valley record, and the addition of the Bear River's significant discharge to the Bonneville basin, have implications for the pace of Lake Bonneville's rise and its large extent. Although in some cases depicted as an incremental, stepwise progression that took ~10–15 ky (eg, Gilbert, 1890; Oviatt et al., 1992), the rise of Lake Bonneville was very rapid at times (Oviatt, 2015). The backflooding of Oneida Narrows and southern Gem Valley is striking evidence of this. The transgression fully defeated the largest river in the Great Basin and trapped deltaic sediment above the canyon. This is especially impressive considering the significant sediment load of the Bear River, as evidenced by the expansive size of the sandy delta formed later during the Bonneville cycle, both through Oneida Narrows and especially below that canyon in northeastern Cache Valley (eg, Lemons et al., 1996; Anderson and Link, 1998; Janecke and Oaks, 2011).

2.6.3 Primary Unsolved Problems

At the end of this review, it is clear that key gaps remain in our knowledge of the Bear River's history and diversion. For the river's earlier history, it remains unknown whether the Bear River once flowed north to join the present Blackfoot River drainage, and if so, when it was diverted westward toward its likely conjunction with the Portneuf River in Pleistocene time, before it ultimately connected with the Bonneville basin. More dating of key basalt flows and a search for gravel clasts and reddish sediment typical of the upper Bear River along the two proposed river courses could help illuminate this problem.

Of the linked components in the subsequent story of river diversion and integration, one of the least understood is the age and complex stratigraphy of the basaltic volcanism forming the drainage divide across central Gem Valley. These eruptions clearly span the time of Thatcher Basin filling and also after incision of southern Gem Valley due to drainage integration, but it is unknown which one of these eruptions, and of what age, finally did divert the Bear River southward from Soda Point.

A primary unresolved debate with respect to the Bear River's connection to Lake Bonneville is whether the upper river may have entered the Bonneville system prior to its final integration and cutting of the narrows at 60–50 ka. In Gem Valley, evidence in the upper Main Canyon Formation indicates a limited, earlier episode of Bear River water and sediment entering southern Gem Valley ~140 ka. If such an incursion involved the main channel of the

Bear River and lasted more than ~15 years into the restricted subbasin, it would have caused spillover, basin filling, drainage integration, and incision that would seem irreversible, precluding the subsequent and final spillover and integration at ~55 ka. Alternatively, the river could have been diverted from a northwest course into the southern Gem Valley for only a brief time during MIS 6. If so, it is possible that milder, earlier spillovers may have occurred and assisted in partial erosion of Oneida Narrows downstream of the most recent paleo-divide (Fig. 2.5). Indeed, the conclusion that incision of the resistant bedrock of Oneida Narrows was very rapid, stemming from the idea of a singular and later integration event, may itself be a subject for further inquiry. Still, no stratigraphic or sedimentologic evidence for an earlier lake transgression and spillover has been recognized yet in the Gem Valley record. In the greater Bonneville basin, Hart et al.'s (2004) analysis of Sr data suggests the Bear River had an earlier entrance into the basin at or before the Little Valley lake cycle of MIS 6. Yet, somehow the full integration of the river occurred at 60–50 ka according to the Gem Valley record. The resolution of these issues will require further work in southern Gem Valley and Oneida Narrows, especially on the sedimentology and chronostratigraphy of the Main Canyon Formation.

Finally, a related issue that deserves further exploration is whether the important data on Sr-isotope ratios are sensitive to other, unexplored changes of tributary and groundwater pathways. This may include changes related to the also-large Provo River and Weber River drainages, changing contributions from the hot and cold springs common along the Bear River's path, and even of groundwater pathways for some upper Bear River water into southern Gem Valley and perhaps the Bonneville basin earlier in Pleistocene time.

Although there is a long lineage of scientific thought and research recognizing the importance of the changing Bear River in the paleohydrology of the Lake Bonneville record, these and other key questions remain for future work to address. The timing of the final integration of the Bear River into the Great Basin is coming into focus at ~55 ka. Whether this was a singular entrance, or a more drawn-out arrival, the implications for the Bonneville paleoclimate record remain as profound as G.K. Gilbert and Robert Bright first recognized.

REFERENCES

Anderson, S.A., Link, P.K., 1998. Lake Bonneville sequence stratigraphy, Pleistocene Bear River delta, Cache Valley, Idaho. In: Pitman, J.K., Carroll, A.R. (Eds.), Modern and Ancient Lake Systems. Utah Geological Association Guidebook, vol. 26. Utah Geological Association, Salt Lake City, pp. 91–104.

Armstrong, F.C., 1969. Geologic map of the Soda Springs quadrangle, southeastern Idaho. U.S. Geological Survey, Miscellaneous Geologic Investigations Map I-557, scale 1:48,000.

Armstrong, R.L., Leeman, W.P., Malde, H.E., 1975. K-Ar dating Quaternary and Neogene volcanic rocks of the Snake River Plain, Idaho. Am. J. Sci. 275, 225–251.

Balch, D.P., Cohen, A.S., Schnurrenberger, D.W., Haskell, B.J., Valero Garces, B.L., Beck, J.W., Cheng, H., Edwards, R.L., 2005. Ecosystem and paleohydrological response to Quaternary climate change in the Bonneville basin Utah. Palaeogeogr. Palaeoclimatol. Palaeoecol. 221, 99–122.

Benson, L.V., Currey, D.R., Dorn, R.I., Lajoie, K.R., Oviatt, C.G., Robinson, S.W., Smith, G.I., Stine, S., 1990. Chronology of expansion and contraction of four great basin lake systems during the past 35,000 years. Palaeogeogr. Palaeoclimatol. Palaeoecol. 78, 241–286.

Benson, L.V., Lund, S.P., Smoot, J.P., Rhode, D.E., Spencer, R.J., Verosub, K.L., Louderback, L.A., Johnson, C.A., Rye, R.O., Negrini, R.M., 2011. The rise and fall of Lake Bonneville between 45 and 10.5 ka. Quat. Int. 235, 57–69.

Billman, E.J., Lee, J.B., Young, D.O., McKell, M.D., Evans, R.P., Shiozawa, D.K., 2010. Phylogenetic divergence in a desert fish: differentiation of speckled dace within the Bonneville, Lahontan, and upper Snake River basins. W. N. Am. Nat. 70 (1), 39–47.

Bouchard, D.P., Kaufman, D.S., Hochberg, A., Quade, J., 1998. Quaternary history of the Thatcher Basin, Idaho, reconstructed from the $^{87}Sr/^{86}Sr$ and amino acid composition of lacustrine fossils: implications for the diversion of the Bear River into the Bonneville basin. Palaeogeogr. Palaeoclimatol. Palaeoecol. 141, 95–114.

Bright, R.C., 1960. Geology of the Cleveland Area, Southeastern Idaho. M.S. thesis, University of Utah. 262 pp.

Bright, R.C., 1963. Pleistocene Lakes Thatcher and Bonneville, Southeastern Idaho. Ph.D. dissertation, University of Minnesota. 292 pp. (USU Special Collections 551.792 B768).

Bright, R.C., 1967. Late-Pleistocene stratigraphy in Thatcher Basin, southeastern Idaho. Tebiwa 10, 1–7.

Coogan, J.C., 1992a. Thrust Systems and Displacement Transfer in the Wyoming-Idaho-Utah Thrust Belt. Ph.D. dissertation, University of Wyoming. 240 pp., 17 plates.

Coogan, J.C., 1992b. In: Link, P.K., Kuntz, M.A., Platt, L.B. (Eds.), Structural Evolution of Piggyback Basins in the Wyoming-Idaho-Utah Thrust Belt, pp. 55–82. Geological Society of America, Memoir 179.

Coogan, J.C., 1997a. Geologic map of the Sheeppen Creek quadrangle, Rich County, Utah. Utah Geological Survey Miscellaneous Publication 97–2, scale 1:24,000.

Coogan, J.C., 1997b. Geologic map of the Bear Lake South quadrangle, Rich County, Utah. Utah Geological Survey, Miscellaneous Publication 97–1, scale 1:24,000.

Currey, D.R., 1982. Lake Bonneville: selected features of relevance to neotectonic analysis: U.S. Geological Survey Open-File Rep. 82–1070.

Davis, O.K., 1998. Palynological evidence for vegetation cycles in a 1.5 million year pollen record from the Great Salt Lake, Utah, USA. Palaeogeogr. Palaeoclimatol. Palaeoecol. 138, 175–185.

Davis, O.K., Moutoux, T.E., 1998. Tertiary and Quaternary vegetation history of the Great Salt Lake, Utah, USA. J. Paleolimnol. 19, 417–427.

Dover, J.H., 1995. Geologic map of the Logan $30' \times 60'$ quadrangle, Cache and Rich counties, Utah, and Lincoln and Uinta counties, Wyoming. U.S. Geological Survey, Miscellaneous Investigations Series Map I-2210, scale 1:100,000.

Evans, J.P., Oaks Jr., R.Q., 1996. Three-dimensional variations in extensional fault shape and basin form, The Cache Valley basin, eastern Basin and Range province, United States. Geol. Soc. Am. Bull. 108, 1580–1593.

Fiesinger, D.W., Perkins, W.D., Puchy, B.J., 1982. Mineralogy and petrology of Tertiary-Quaternary volcanic rocks in Caribou County, Idaho. Idaho Bur. Min. Geol. Bull. 26, 465–488.

Ford, M.T., 2005. The petrogenesis of Quaternary Rhyolite Domes in the Bimodal Blackfoot Volcanic Field, Southeastern Idaho. M.S. thesis, Idaho State University.

Galbraith, R.F., Roberts, R.G., Laslett, G.M., Yoshida, H., Olley, J.M., 1999. Optical dating of single and multiple grains of quartz from Jinmium Rock Shelter, northern Australia: Part I, Experimental design and statistical models. Archaeometry 41, 339–364.

Gilbert, G.K., 1890. Lake Bonneville. United States Geological Survey Monograph 1, 438 pp.

Goessel, K.M., Oaks Jr., R.Q., Perkins, M.E., Janecke, S.U., 1999. Tertiary stratigraphy and structural geology, Wellsville Mountains to Junction Hills, north-central Utah. In: Spangler, L.E., Allen, C.J. (Eds.), Geology of Northern Utah and Vicinity, pp. 45–69. Utah Geological Association Publication, vol. 27.

Guérin, G., Mercier, N., Adamiec, G., 2011. Dose-rate conversion factors Update. Ancient TL 29, 5–8.

Hansen, W.R., 1985. Drainage development of the Green River Basin in southwestern Wyoming and its bearing on fish biogeography, neotectonics, and paleoclimates. Mt. Geol. 22 (4), 192–204.

Hart, W.S., Quade, J., Madsen, D.B., Kaufman, D.S., Oviatt, C.G., 2004. The $^{87}Sr/^{86}Sr$ ratios of lacustrine carbonates and lake-level history of the Bonneville paleolake system. Geol. Soc. Am. Bull. 116, 1107–1119.

Hershler, R., Sada, D.W., 2002. Biogeography of Great Basin snails of the genus Pyrgulopsis. In: Hershler, R., Madsen, D.B., Currey, D.R. (Eds.), Great Basin Aquatic Systems History. Smithsonian Contributions to the Earth Sciences No. 33, Smithsonian Institution Press, Washington D.C., pp. 255–276.

Heumann, A., 1999. Timescales of Processes Within Silicic Magma Chambers. Ph.D. dissertation, Netherlands Research School of Sedimentary Geology. 197 pp. (NSF Publication No. 991001).

Hladky, F.R., Kellogg, K.S., Oriel, S.S., Link, P.K., Nielson, J.W., Amerman, R.E., 1992. Geologic map of the eastern part of the Fort Hall Indian Reservation, Bannock, Bingham, and Caribou counties, Idaho. U.S. Geological Survey IMAP 2006, scale 1:50,000.

Hochberg, A., 1996. Aminostratigraphy of Thatcher Basin, SE Idaho: Reassessment of Pleistocene Lakes. MS thesis, Utah State University. 107 pp.

Izett, G.A., 1981. Stratigraphic succession, isotopic ages, partial chemical analyses, and sources of certain silicic volcanic ash beds (4.0–0.1 m.y.) of the western United States: U.S. Geological Survey Open-File Rept. 81–763, 2 sheets.

Janecke, S.U., Evans, J.C., 1999. Folded and faulted salt lake formation above the miocene to pliocene new canyon and clifton detachment faults, malad and bannock ranges, Idaho. In: Hughes, S., Thackray, G. (Eds.), Field Trip Guide to the Deep Creek Half Graben and Environs. Guidebook to the Geology of Eastern Idaho, Idaho Museum of Natural History, Pocatello, ID, pp. 71–96.

Janecke, S.U., Oaks Jr., R.Q., 2011. New insights into the outlet conditions of Late Pleistocene Lake Bonneville, southeastern Idaho, USA. Geosphere 7 (6), 1369–1391.

Janecke, S.U., Oaks Jr., R.Q., 2014. Diversion(s) of the lower Bear River in Gem Valley, southeast Idaho, by a tectono-volcanic valve. In: Abstracts of the Workshop on Late Cenozoic to Recent Geologic and Biotic History of the Snake River, March 24–26, 2014, Pocatello, Idaho. http://geology.isu.edu/Papers/SRPProceedings.pdf.

Janecke, S.U., Carney, S.M., Perkins, M.E., Evans, J.C., Link, P.K., Oaks Jr., R.Q., Nash, B.P., 2003. Late Miocene-Pliocene detachment faulting and Pliocene-Pleistocene Basin-and-Range extension inferred from dismembered rift basins of the Salt Lake Formation, southeast Idaho. In: Raynolds, R.G., Flores, R.M. (Eds.), Cenozoic Systems of the Rocky Mountain Region.

Rocky Mountain Section of the Society of Economic Paleontologists and Mineralogists, Denver, CO, pp. 369–406. Special publication.

Karow, T., Hampel, A., 2010. Slip rate variations on faults in the Basin-and-Range Province caused by regression of Late Pleistocene Lake Bonneville and Lake Lahontan. Int. J. Earth Sci. 99 (8), 1941–1953.

Kaufman, D.S., Forman, S.L., Bright, J., 2001. Age of the cutler dam alloformation (late pleistocene), Bonneville basin, Utah. Quat. Res. 56 (3), 322–334.

Kaufman, D.S., Bright, J., Dean, W.E., Moser, K., Rosenbaum, J.G., Anderson, R.S., Colman, S.M., Heil, C.W., Jiménez-Moreno, G., Reheis, M.C., Simmons, K.R., 2009. A quarter-million years of paleoenvironmental change at Bear Lake, Utah and Idaho. In: Rosenbaum, J.G., Kaufman, D.S. (Eds.), Paleoenvironments of Bear Lake, Utah and Idaho, and Its Catchment, pp. 311–351. Geological Society of America Special Paper 450.

Keeley, J.A., Rodgers, D.W., 2015. Testing the Bannock detachment breakaway: negative results support moderate- to high-angle splay system and domino-style fault block rotation along the Valley fault, southern Portneuf Range, southeastern Idaho, USA. Rocky Mt. Geol. 50 (2), 119–151.

Kruger, J.M., Crane, T.J., Pope, A.D., Perkins, M.E., Link, P.K., 2003. Structural and stratigraphic development of Neogene basins in the Marsh Valley, Lava Hot Springs, and Wakley Peak areas, southeast Idaho: two phases of extension. Rocky Mountain Section. In: Raynolds, R.G., Flores, R.M. (Eds.), Cenozoic Systems of the Rocky Mountain Region. Rocky Mountain Section of the Society of Economic Paleontologists and Mineralogists, Denver, CO, pp. 407–457. Special publication, 499 pp.

Kuehn, S.C., Negrini, R.M., 2010. A 250 ky record of Cascade arc pyroclastic volcanism from late Pleistocene lacustrine sediments near Summer Lake, Oregon, USA. Geosphere 6 (4), 397–429.

Lanphere, M.A., Champion, D.E., Christiansen, R.L., Izett, G.A., Obradovich, J.D., 2002. Revised ages for tuffs of the Yellowstone Plateau volcanic field; assignment of the Huckleberry Ridge Tuff to a new geomagnetic polarity event. Geol. Soc. Am. Bull. 114, 559–568.

Lemons, D.R., Milligan, M.R., Chan, M.A., 1996. Paleoclimatic implications of late Pleistocene sediment yield rates for the Bonneville basin, northern Utah. Palaeogeogr. Palaeoclimatol. Palaeoecol. 123, 147–159.

Lewis, R.S., Link, P.K., Stanford, L.R., Long, S.P., 2012. Geologic Map of Idaho. Idaho Geological Survey Map 9, scale 1:750,000.

Licciardi, J.M., Pierce, K.L., 2008. Cosmogenic exposure-age chronologies of Pinedale and Bull Lake glaciations in greater Yellowstone and the Teton Range, USA. Quat. Sci. Rev. 27 (7), 814–831.

Link, P.K., Kaufmann, D.S., Thackray, G.D., 1999. Field guide to Pleistocene Lakes Thatcher and Bonneville and the Bonneville Flood, southeastern Idaho. In: Hughes, S.S., Thackray, G.D. (Eds.), Guidebook to the Geology of Eastern Idaho. Idaho Museum of Natural History, Pocatello, ID, pp. 251–266. 342 pp.

Long, S.P., Link, P.K., Janecke, S.U., Rodgers, D.W., Perkins, M.E., Fanning, C.M., 2006. Multiple phases of late Cenozoic extension and synextensional deposition of the Salt Lake Formation in an evolving supradetachment basin, Malad Range, southeast Idaho. Rocky Mt. Geol. 41, 1–27. http://dx.doi.org/10.2113/gsrocky.41.1.1.

Ludlum, J.C., 1943. Structure and stratigraphy of part of the Bannock Range, Idaho. Geol. Soc. Am. Bull. 54, 973–986.

Mabey, D.R., 1971. Geophysical data relating to a possible Pleistocene overflow of Lake Bonneville at Gem Valley, southeastern Idaho. U.S. Geological Survey Professional Paper 750-B, B122–B127.

Mabey, D.R., Oriel, S.S., 1970. Gravity and magnetic anomalies in the Soda Springs region, southeastern Idaho. U.S. Geological Survey Professional Paper 636-E, 15 pp., scale 1:125,000.

Mansfield, G.R., 1927. Geography, geology, and mineral resources of part of southeastern Idaho. U.S. Geological Survey Professional Paper 152, 453 pp.

Mansfield, G.R., 1929. Geography, geology, and mineral resources of the Portneuf quadrangle, Idaho. U.S. Geological Survey, Bulletin 803, 110 pp., scale 1:62,500.

McCoy, W.D., 1987. Quaternary aminostratigraphy of the Bonneville basin, western United States. Geol. Soc. Am. Bull. 98 (1), 99–112.

McCurry, M., Welhan, J., 2012. Do magmatic-related geothermal energy resources exist in southeast Idaho? GRC Trans. 36, 699–707.

McCurry, M., Welhan, J., Polun, S., Autenrieth, K., Rodgers, D.W., 2011. Geothermal potential of the Blackfoot Reservoir-Soda Springs volcanic field: a hidden geothermal resource and natural laboratory in SE Idaho. GRC Trans. 35, 917–924.

Minckley, W.L., Hendrickson, D.A., Bond, C.E., 1986. Geography of western North American freshwater fishes: description and relationships to intracontinental tectonism. The Zoogeography of North American Freshwater Fishes. Wiley, New York.

Mock, K.E., Evans, R.P., Crawford, M., Cardall, B.L., Janecke, S.U., Miller, M.P., 2006. Range-wide molecular structuring in the Utah sucker (*Catostomus ardens*). Mol. Ecol. 15, 2223–2238. http://dx.doi.org/10.1111/j.1365-294X.2006.02932.x.

Moseley, G.E., Edwards, R.L., Wendt, K.A., Cheng, H., Dublyansky, Y., Lu, Y., Boch, R., Spötl, C., 2016. Reconciliation of the Devils Hole climate record with orbital forcing. Science 351 (6269), 165–168.

Munroe, J.S., 2005. Glacial geology of the northern Uinta Mountains. In: Dehler, C.M., Pederson, J.L. (Eds.), Uinta Mountain Geology, vol. 33. Utah Geological Association Publication, Salt Lake City, pp. 215–234.

Murray, A.S., Wintle, A.G., 2000. Luminescence dating of quartz using an improved single aliquot regenerative-dose protocol. Radiat. Meas. 32, 57–73.

Nishizawa, S., Currey, D.R., Brunelle, A., Sack, D., 2013. Bonneville basin shoreline records of large lake intervals during Marine Isotope Stage 3 and the Last Glacial Maximum. Palaeogeogr. Palaeoclimatol. Palaeoecol. 386, 374–391.

Oaks Jr., R.Q., Smith, K.A., Janecke, S.U., Perkins, M.E., Nash, W.P., 1999. In: Spangler, L.E. (Ed.), Stratigraphy and tectonics of Tertiary strata of southern Cache Valley, north-central Utah, pp. 71–110. Utah Geological Association Publication vol. 27, 407 pp.

Oriel, S.S., 1965. Preliminary geologic map of the SW 1/4 of the Bancroft quadrangle, Bannock and Caribou Counties, Idaho. U.S. Geological Survey, Mineral Investigations Field Studies, Map MF-299, scale 1:24,000.

Oriel, S.S., 1968. Preliminary geologic map of Bancroft quadrangle, Caribou and Bannock Counties, Idaho: U.S. Geological Survey Open-File Report. OF-68-204, scale 1:48,000.

Oriel, S.S., Platt, L.B., 1968. Reconnaissance geologic map of the Preston quadrangle, southeastern Idaho: U.S. Geological Survey Open-File Report. OF-68-205, scale 1:62,500.

Oriel, S.S., Platt, L.B., 1980. Geologic map of the Preston 1 × 2 degree quadrangle, Idaho and Wyoming. U. S. Geological Survey Miscellaneous Investigations, Map I-1127, scale 1:250,000.

Oviatt, C.G., 2015. Chronology of Lake Bonneville, 30,000 to 10,000 yr B.P. Quat. Sci. Rev. 110, 166–171.

Oviatt, C.G., Currey, D.R., Sack, D., 1992. Radiocarbon chronology of Lake Bonneville, eastern Great Basin, USA. Palaeogeogr. Palaeoclimatol. Palaeoecol. 99, 225–241.

Oviatt, C.G., Chan, M.A., Jewell, P.W., Bills, B.G., Madsen, D.B., Miller, D.M., 2014. Interpretations of evidence for large Pleistocene paleolakes in the Bonneville basin, western North America: Comment on Bonneville Basin Shoreline Records of Large Lake During Marine Isotope Stage 3 and the Last Glacial Maximum, by Nishizawa et al. (2013). Palaeogeogr. Palaeoclimatol. Palaeoecol. 401, 173–176.

Pederson, J., King, J., 2011. New constraints on the integration of the Bear River and cutting of Oneida Narrows Canyon—implications for the Bonneville record. Geological Society of America, Abstracts with Programs 43 (4), 80.

Perkins, W.D., 1979. Petrology and Mineralogy of Quaternary Basalts, Gem Valley and Adjacent Bear River Range, Southeastern Idaho. M.S. thesis, Utah State University, Logan, Utah. 91 pp.

Perkins, M.E., Brown, F.H., Nash, W.P., Williams, S.K., McIntosh, W., 1998. Sequence, age, and source of silicic fallout tuffs in middle to late Miocene basins of the northern Basin and Range province. Geol. Soc. Am. Bull. 110 (3), 344–360.

Pickett, K.E., 2004. Physical Volcanology, Petrography, and Geochemistry of Basalts in the Bimodal Blackfoot Volcanic Field, Southeastern Idaho. M.S. thesis, Idaho State University, Pocatello, Idaho. 92 pp.

Pierce, K.L., Morgan, L.A., 1992. The track of the Yellowstone hot spot: volcanism, faulting, and uplift. In: Link, P.K., Kunz, M.A., Platt, L.B. (Eds.), Regional Geology of Eastern Idaho and Western Wyoming, pp. 1–53. Geological Society of America, Memoir 179, 312 pp.

Pierce, K.L., Fosberg, M.A., Scott, W.E., Lewis, G.C., Colmon, S.M., 1982. Loess deposits of southeastern Idaho: age and correlation of the upper two loess units. In: Bonnichsen, W., Breckenridge, R.M. (Eds.), Cenozoic Geology of Idaho, vol. 26. Idaho Bureau of Mines and Geology Bulletin, Moscow, ID, pp. 717–725.

Piety, L.A., Sullivan, J.T., Anders, M.H., 1992. Segmentation and earthquake potential of the Grand Valley fault, Idaho and Wyoming. In: Link, P.K., Kunz, M.A., Platt, L.B. (Eds.), Regional Geology of Eastern Idaho and Western Wyoming, vol. 179. Geological Society of America, Boulder, CO, pp. 155–182. 312 pp.

Polun, S.G., 2011. Kinematic Analysis of Late Pleistocene Faulting in the Blackfoot Lava Field, Caribou County, Idaho: Pocatello, Idaho, Idaho State University (M.S. thesis). 86 pp.

Prescott, J.R., Hutton, J.T., 1994. Cosmic ray contributions to dose rates for luminescence and ESR dating. Radiat. Meas. 23, 497–500.

Puchy, B.J., 1982. Mineralogy and Petrology of Lava Flows (Tertiary-Quaternary) in Southeastern Idaho and at Black Mountain, Rich County, Utah. M.S. thesis, Utah State University, Logan, Utah. 73 pp. Paper 3817, http://digitalcommons.usu.edu/etd/3817.

Reheis, M.C., 2005. Surficial geologic map of the upper Bear River and Bear Lake drainage basins, Idaho, Utah, and Wyoming. U.S. Geological Survey Scientific Investigations Map 2890, scales 1:150,000 and 1:50,000, http://pubs.usgs.gov/sim/2005/2890/.

Reheis, M.C., Laabs, B.J.C., Kaufman, D.S., 2009. Geology and geomorphology of Bear Lake Valley and upper Bear River, Utah and Idaho. In: Rosenbaum, J.G., Kaufman, D.S. (Eds.), Paleoenvironments of Bear Lake, Utah and Idaho, and Its Catchment. Geological Society of America Special Paper 450, 15–48. 351 pp.

Reimer, P.J., Bard, E., Bayliss, A., Beck, J.W., Blackwell, P.G., Ramsey, C.B., Buck, C.E., Cheng, H., Edwards, R.L., Friedrich, M., Grootes, P.M., 2013. IntCal13 and Marine13 radiocarbon age calibration curves 0–50,000 years cal BP. Radiocarbon 55 (4), 1869–1887.

Robertson III, G.C., 1978. Surficial geology, northern Bear Lake Valley, Idaho. M.S. thesis, Utah State University, Logan, Utah. 162 pp., map scale 1:48,000.

Rodgers, D.W., Long, S.P., McQuarrie, N., Burgel, W.D., Hersley, C., 2006. Geologic map of the Inkom Quadrangle, Bannock County, Idaho: Idaho Geological Survey Technical Report 06–2, scale 1:24,000.

Sarna-Wojcicki, A.M., 2005. Tephra layers of Blind Spring Valley and related Upper Pliocene and Pleistocene tephra layers, California, Nevada, and Utah: isotopic ages, correlation, and magnetostratigraphy. U.S. Geological Survey Prof. Paper 1701, U.S. Department of the Interior. 63 pp.

Schmitt, A.K., 2011. Uranium series accessory crystal dating of magmatic processes. Annu. Rev. Earth Planet. Sci. 39, 321–349.

Scott, W.E., Pierce, K.L., Bradbury, J.P., Forester, R.M., 1982. Revised Quaternary stratigraphy and chronology in the American Falls area, southeastern Idaho. In: Bonnichsen, B., Breckenridge, R.M. (Eds.), Cenozoic Geology of Idaho, vol. 26. Idaho Bureau of Mines and Geology. Bulletin, Moscow, ID, pp. 581–595.

Smith, G.R., Dowling, T.E., Gobalet, K.W., Lugaski, T., Shiozawa, D.K., Evans, R.P., 2002. Biogeography and timing of evolutionary events among Great Basin fishes. In: Hershler, R., Madsen, D.B., Currey, D.R. (Eds.), Great Basin Aquatic Systems History. Smithsonian Contributions to the Earth Sciences No. 33, 175–234.

Tarboton, D., 2015. The Great Salt Lake Water Budget: Presentation to the Great Salt Lake Council, 2011. http://www.gslcouncil.utah.gov/docs/2011/Mar/031611_budget.pdf.

Taylor, D.W., Bright, R.C., 1987. Drainage history of the Bonneville basin. In: Kopp, R.S., Cohenour, R.E. (Eds.), Cenozoic Geology of Western Utah, vol. 16. Utah Geological Association Publication, Salt Lake City, pp. 239–256, 684 pp.

Wagner, J.D.M., Cole, J.E., Beck, J.W., Patchett, P.J., Henderson, G.M., Barnett, H.R., 2010. Moisture variability in the southwestern United State linked to abrupt glacial climate change. Nat. Geosci. 3, 110–113.

Welhan, J.A., Garwood, D., Feeney, D., 2013. The Blackfoot volcanic field, southeast Idaho: a hidden high-T geothermal resource revealed through data mining of the National Geothermal Data Repository. GRC Trans. 37, 365–374.

Wernicke, B.P., England, P.C., Sonder, L.J., Christiansen, R.L., 1987. Tectonomagmatic evolution of Cenozoic extension in the North American Cordillera. Geol. Soc. Lond., Spec. Publ. 28 (1), 203–221.

Wong, I., Pezzopane, S., Dobrom, M., Oligo, S., Bott, J., Fabia, T., Seismic Hazards Group URS Corporation, 2012. Final Report Site-Specific Seismic Hazard Analyses for Soda Dam Project, Idaho: Report Prepared for PacifiCorp Energy, Portland, Oregon, 29 August 2012.

Zeeden, C., Rivera, T.A., Storey, M., 2014. An astronomical age for the Bishop Tuff and concordance with radioisotopic dates. Geophys. Res. Lett. 41, 3478–3484.

Chapter 3

The Pilot Valley Shoreline: An Early Record of Lake Bonneville Dynamics

D.M. Miller and G.A. Phelps
U.S. Geological Survey, Menlo Park, CA, United States

ABSTRACT
The Pilot Valley shoreline is named for distinctive gravel beaches on the eastern, northern, and western sides of Pilot Valley playa, Utah. The shoreline has been identified across the Bonneville basin where it is characterized by one to three beach crests between ~1305 and 1309 m elevation, all overlain by deep-water marl of Lake Bonneville. It thus represents the lowest and earliest recognized shoreline of Lake Bonneville. Features of the shoreline indicate that both high wave energy and high stream sediment discharge contributed to shoreline development. Basin hypsometry did not play a role in the development of the shoreline, which must have been caused by a combination of climatically driven hydrologic and storm factors, such as reduced precipitation that stabilized lake level and increase in storm-driven wave energy. The Pilot Valley shoreline is poorly dated at about 30 ka. If it is somewhat older, correlation with Greenland Interstadial 5.1 at 30.8–30.6 ka could explain the stabilization of lake level.

Keywords: Bonneville, Pilot Valley, Shoreline, Paleoclimate, Tectonics

3.1 INTRODUCTION

Gilbert (1890) identified three regionally continuous shorelines of Lake Bonneville (the Bonneville, Provo, and Stansbury) and argued that they represent lake-level control by both changing overflow thresholds and changing climate. About the latter, he wrote (p. 262):

> *"The oscillations of the lake were in all probability caused by oscillations of climate; and if we can satisfy ourselves as to the nature of the particular climate movements associated with the rise and the fall of the lake, we can immediately, by changing the notation of our curve, convert it into a record of geologic climate. For this purpose we revert once more to the fundamental conditions controlling the size of the closed lake. The size depends on the ratio between*

Developments in Earth Surface Processes, Vol. 20. http://dx.doi.org/10.1016/B978-0-444-63590-7.00003-2

the supply of water and the rate of evaporation. Rate of evaporation is purely a function of climate; but water supply depends quite as much on topographic configuration as on meteorologic conditions".

Although some interpretations about these shorelines have changed in the subsequent one and a half centuries of study, the three shorelines identified by Gilbert continue to be keys to interpreting Lake Bonneville. Despite their importance for interpreting hydrologic balance and paleoclimate, no other regional shorelines have been identified. The Lake Bonneville hydrograph (Fig. 3.1) illustrates the timing and elevations of the major shorelines identified by Gilbert (1890) following much careful study and the advent of radiometric dating

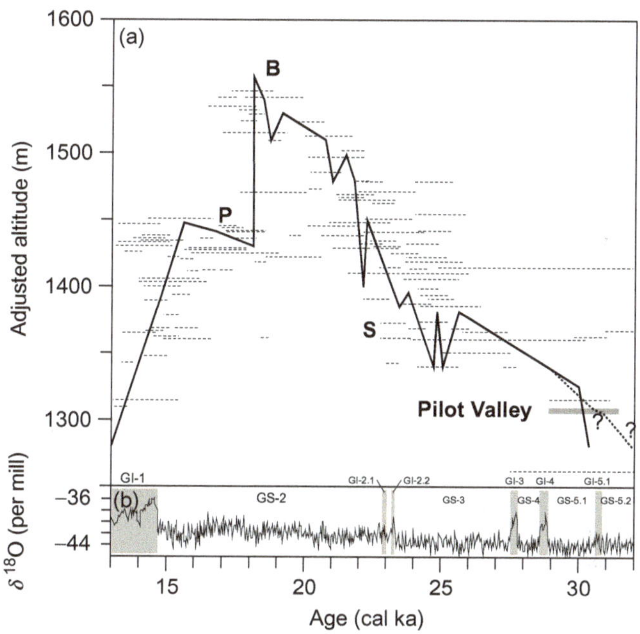

FIG. 3.1 (a) Hydrograph for Lake Bonneville showing calibrated radiocarbon ages (*thin dashed lines*, 2σ range shown) vs elevations adjusted for isostatic rebound. Some ages are limiting ages, not defining the lake level directly. Position and interpolated age for the Pilot Valley shoreline shown by *gray bar*. *Dotted line*, possible early lake hydrograph that is consistent with global climate. Major shorelines: *B*, Bonneville; *P*, Provo; *S*, Stansbury. (b) NGRIP oxygen isotopic record with ages adjusted from ice layer counting and matching abrupt transitions at beginnings of Greenland interstadials (GI). Interstadials (warm events) defined by record shown as well as by dust flux record, not shown. *GS*, Greenland stadial. *Panel (a): Modified from Oviatt, C.G., 2015. Chronology of Lake Bonneville, 30,000 to 10,000 yr B.P. Quat. Sci. Rev. 110, 166–171. Panel (b): Redrawn from Rasmussen, S.O., Bigler, M., Blockley, S.P., Blunier, T., Buchardt, S.L., Clausen, H.B., Cvijanovic, I., Dahl-Jensen, D., Hoek, W.Z., Lowe, J.J., Pedro, J.B., Popp, T., Seierstad, I.K., Steffensen, J.P., Svensson, A.M., Vallelonga, P., Vinther, B.M., Walker, M.J.C., Wheatley, J.J., Winstrup, M., 2014. A stratigraphic framework for abrupt climatic changes during the Last Glacial period based on three synchronized Greenland ice-core records: refining and extending the INTIMATE event stratigraphy. Quat. Sci. Rev. 106, 14–28.*

methods (Oviatt, 2015). A marl deposit that represents deep-water sedimentation when the lake was at high levels is a key stratigraphic marker for dividing early transgressive from late regressive beach deposits. Threshold control at an overflow outlet in southern Idaho was responsible for the formation of the Bonneville and Provo shorelines (Gilbert, 1890), although that threshold shifted somewhat to produce multiple beaches at the Provo shoreline (Gilbert, 1890; Currey, 1982; Miller et al., 2013). Prodigious beach deposits and landforms, along with tufa growth, mark the Stansbury shore zone, which was caused by two or more major oscillations in lake level early in its history (Fig. 3.1; Oviatt et al., 1990).

Regionally identified shorelines indicate significant hydrologic events that can be used to better understand geodynamic and lacustrine processes. For instance, the threshold-controlled Bonneville and Provo shorelines led to conclusions about the Bonneville flood (Gilbert, 1890; O'Connor, 1993). Together, these shorelines represent the time of maximum elevation of Lake Bonneville, which coincides with the maximum effective moisture in the watershed. The Stansbury shoreline has been interpreted as forming during a lake-level oscillation during a dry climatic episode (Oviatt et al., 1990; Oviatt, 2015). Wind patterns and fetch inferences can be derived from regional beach geomorphology (Schofield et al., 2004), and geodynamic properties that lead to isostatic rebound of shorelines can be quantified (Gilbert, 1890; Crittenden, 1963: Bills and May, 1987). In addition, study of shorelines can improve understandings of geomorphic processes such as rates of beach construction and erosion. Detailed study of the geomorphology and bed forms of beach deposits may be used to distinguish those formed during lake level change from those formed during static lake levels. Regional beaches also provide datums for evaluating tectonic deformation. This chapter focuses on a newly identified set of beach deposits at very low elevation that formed early in the Lake Bonneville cycle, examines the origin of the deposits, and makes inferences about lake dynamics that led to the beach deposits.

3.2 DESCRIPTION

Detailed mapping of Pilot Valley (located in Fig. 3.2) identified a mappable set of two to three, less commonly four, closely spaced gravelly shoreline deposits that were not mentioned by Gilbert (1890) or subsequent workers. These shoreline deposits are overlain by the Bonneville marl (Miller, 1990a,b, 1993; Miller et al., 1990) and occur at low elevations in several places around Pilot Valley playa (for which the shoreline deposits are named). Four locations where the deposits are prominent are shown in Fig. 3.2. The deposits underlie elongate benches or rounded ridge tops that are laterally nearly constant in elevation, which we interpret as beach crests. The crests lie at ~1305 to 1309 m (4280–4295 ft) elevation and are laterally extensive for as much as 3.2 km (Table 3.1). At locations in Pilot Valley where gravel

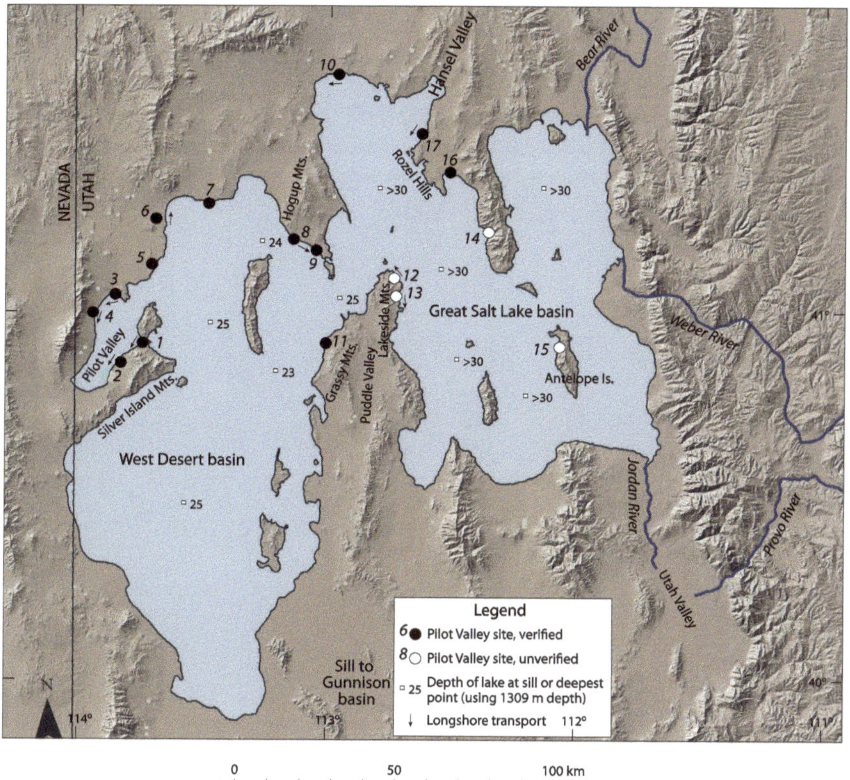

FIG. 3.2 Map of northern Utah area showing outline of Lake Bonneville at 1300 m lake level, slightly below the Pilot Valley shoreline, and the three major rivers feeding the lake. Localities at which the Pilot Valley shoreline deposits have been identified and inferred shown by *circles* with locality numbers keyed to Table 3.1. Longshore transport directions shown by *arrows*; minimum lake depth at highest Pilot Valley shoreline beach level shown for several basins and thresholds between basins.

deposits reach maximum volume, three or four beaches typically are evident, in close succession and in places overlapping (Fig. 3.3).

At Silver Island (Fig. 3.2, location 2) the elevations of the beach crests range from 1305 to 1307.6 m (~4280 ft to above 4290 ft on topographic maps). The elevations of beach crests are constant laterally. In these exposures, beach ridge morphology and shingled gravel deposits interbedded with pebbly sand confirm that the deposits represent beaches. The deposits are underlain by platy beds of cemented ooids. Adjacent Bonneville marl of the deep water lake can be traced onto the edges of the beaches. The maximum beach volume lies immediately south of a prominence, raising the possibility that gravel eroded from that source was transported south by long shore currents; however, proximity to a modern stream also provides the possibility that stream discharge accentuated beach development.

TABLE 3.1 Elevation, Location, and Slope Data for Selected Examples of the Pilot Valley Shoreline

Map No.	Site Name	Location		Beach Crest Elevation			Offshore Slope[c] (%)	Downslope Direction	Water Depth[d] (m)	Length (km)	Method	Confirmed
		UTM E[a]	*UTM N*[a]	*Lower*[b]	*Middle*	*Upper*						
1	Crater Island	265530	4544915	1310.4	1311.5	1312.2	5.0	300	12.0	3.2	Field	Yes
2	Silver Island	254924	4535018	1309.9	1312.0	1312.3	2.1	290	8.0	2.4	Field	Yes
3	Lemay Island	258085	4560933	1309.7			2.5	200	9.0	1	Field	Yes
4	Patterson Spring	250798	4555091	1305.0		1309.0	1.1	100	9	1.7	Field	No
5	Little Pigeon	269644	4569950	1308.0			6.0	190	10.5	1.1	Field	No
6	Pigeon north	271221	4584289	1309.6	1310.5	1311.9	1.8	90	4.1	3.6	Field	No
7	Terrace Mountain	287606	4588744	1308.0			1.6	170	4.9	1.4	Field	No
8	Hogup west	314043	4576649	1303.5	1305.2	1307.2	1.0	200	7.3	1.4	Field	No
9	Hopup east	320697	4572980	1303.0		1308.7	0.5	210	7.0	4.8	Field	Yes
10	Kelton	329815	4628490	1304.5	1307.1	1308.9	3.2	160	12.9	1.8	Field	Yes
11	Grassy	322891	4542920	1305.0		1308.0	4.9	340	13.4	1.5	Field	No
12	Lakeside Pt.	345147	4562568	1306.5		1311.8	3.0	10	20.7	0.8	Field	Yes
13	Lakeside N	345622	4557246	1309.7		1312.8	6.5	80	30.2	1.7	Field	No
14	Promontory Mts.	375329	4576615	1307.5		1309.6	3.5	270	16.9	0.7	Air photo	No
15	Antelope Island	396320	4539396	1308.3			3.7	280	26.3	1.7	Air photo	No
16	Rozel E	363822	4596228	1304.0	1305.0	1306.0	1.2	190	2.1	3.6	Field	No
17	Rozel W	355880	4608461	1306.2		1308.2	4.1	330	9.8	1.5	Field	Yes

[a] *Universal Transverse Mercator, zone 12, NAD 83.*
[b] *Lower beach crest or crest of single beach.*
[c] *Offshore slope measured for 100 m below lower shoreline.*
[d] *Difference in elevation between upper beach crest and offshore position with slope less than 1%.*

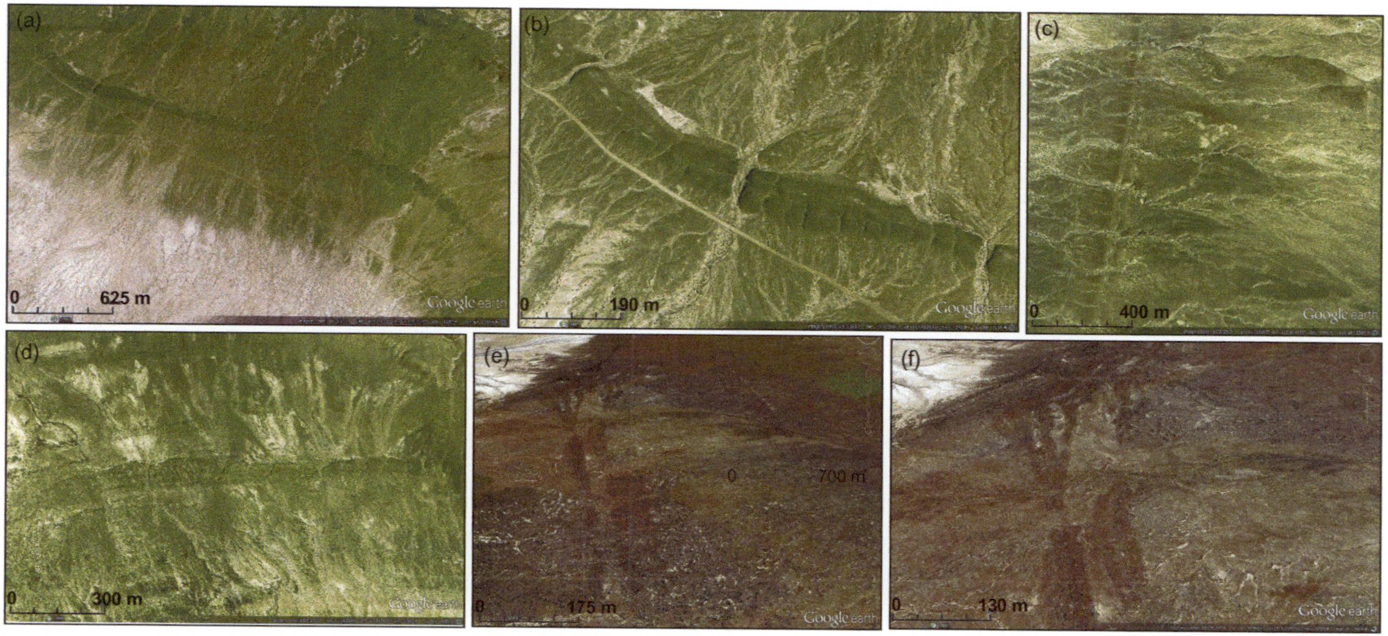

FIG. 3.3 Images of representative Pilot Valley shoreline deposits from Google Earth. Note the multiple beaches in close proximity and lack of nearby similarly developed beaches. (a) Silver Island (site 2, Fig. 3.2), viewed to NE; (b) Silver Island detail; (c) Crater Island (site 1, Fig. 3.2), viewed to north; (d) Lemay Island (site 3, Fig. 3.2), viewed to north; (e) Hogup west (site 8, Fig. 3.2), viewed to west, showing two main beach sets in the east branching to four in the west where the modern stream crosses; and (f) Hogup west detail.

A similar set of beach deposits was identified at Little Pigeon (Miller and Glick, 1987), Pigeon, Terrace, and Hogup Mountains, as well as near Kelton (Fig. 3.2, sites 5–10), and these were mapped carefully at the Terrace and Kelton sites (Miller and McCarthy, 2002; Felger et al., 2016). At all of these locations, deposits have characteristics of beaches such as shingled pebbles and elongate morphology with horizontal crests, and the altitude range is similar to those in Pilot Valley, with which they are correlated. The beach sets at Hogup and Kelton were tentatively identified as unusually high-elevation locations of a different (Gilbert) shoreline from aerial photograph study by Currey (1982), but overlap of the beach deposits by marl disproves this assignment (Oviatt and Miller, 1997; Felger et al., 2016).

Shoreline gravel deposits are best displayed at the Hogup Mountains (Fig. 3.3). In the west occur three main beach landforms and a fourth minor beach that form a gently cuspate set of ridges, and in the east are beach gravels forming a nearly straight bench at the same elevation as beaches to the west; this bench was quarried for railroad construction materials. At the western site (Fig. 3.4, beaches 1 and 2) two cuspate lower beach ridges form a ramp-and-flat geomorphology with a gentle swale locally intervening between the two crests. These beach ridges join eastward to form a gravel accumulation with two crests. A northern beach 3 that encroaches upon the two

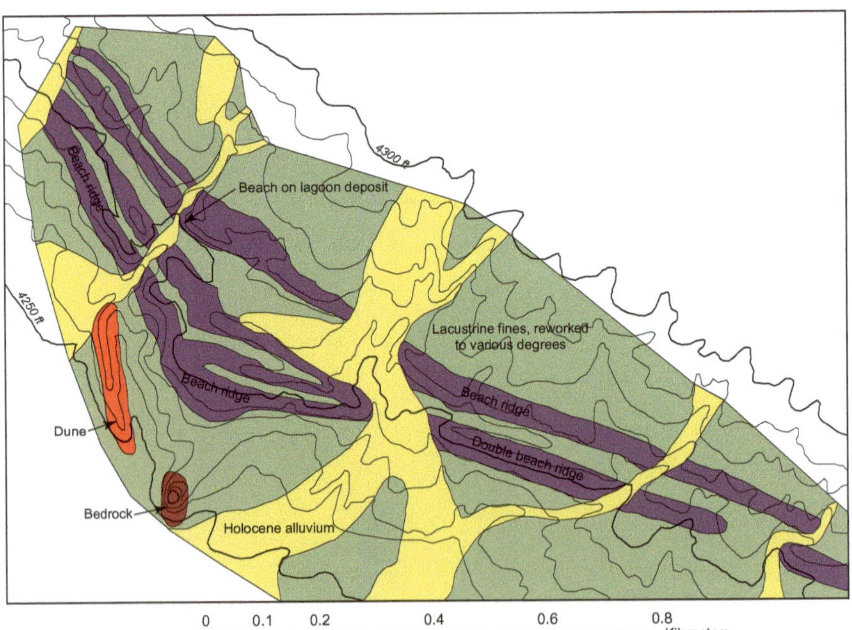

FIG. 3.4 Western Hogup Mountains exposures of the Pilot Valley shoreline deposits (compare to aerial view in Fig. 3.3E). Sketch map of the beaches shows topography in 5-ft contours from the Meadow Spring 7.5′ quadrangle map; the area is within Section 21, T.7N, R.11W.

southern beaches is partly overlapped by alluvial fan deposits. All beach landforms at this site are underlain by moderately sorted pebble gravel in a sand and silt matrix that exhibit foresets and backsets, and are greater than 3 m thick; the bases of the two southern beaches were not observed. Between beach 3 and beach 2 lies a poorly sorted, vaguely bedded yellowish silt and fine sand deposit in a position that is consistent with its interpretation as a lagoon deposit behind beach 2 (Fig. 3.4). Beach 3 gravels lie on this silt and sand deposit. Beaches have approximately constant elevation laterally; the crests for beaches 1 to 3 are approximately 1303.5, 1305.2, and 1307.2 m, respectively. The deposits are laterally extensive to the eastern quarry, where well-sorted gravel with sand matrix underlies marl and overlies alluvial fan deposits. The marl unit consists of basal 1.5 m of pebbly and sandy silt overlain by >2 m of laminated calcareous clay and silt (Stop 4 of Oviatt and Miller, 1997). The underlying fan deposits contain a strongly developed soil and overlie deposits of lake cycles predating Lake Bonneville.

Field investigations of shoreline gravel deposits at the Kelton, Rozel, Lemay, and Grassy sites (Fig. 3.2, sites 10, 17, 16, 3, and 11) confirm that they are directly overlain by Bonneville marl. Sites in Fig. 3.2 labeled as "unverified" either represent field locations we investigated that lack the marl to confirm this stratigraphic relation, or sites identified from a combination of aerial photography and 5-m DEMs (Table 3.1); these sites are probable, but unconfirmed, Pilot Valley-type shoreline deposits.

3.3 LAKE MARGIN CHARACTERISTICS

The regional correlation of beach crests at ~1305 to 1309 m elevation with the Pilot Valley shoreline, if correct, establishes that an early, low-elevation Lake Bonneville created a set of unusually thick and coarse shoreline deposits across much of its northern reach. Such a regional expression can be evaluated to learn what processes contributed to the shoreline development. We constructed metrics of offshore lake bottom slope, water depth, and wave direction from 5-m DEMs for such an evaluation (Table 3.1).

Elevation analysis reveals that the three beaches climb systematically with overlaps where they are closely spaced, indicating that higher beaches overlie lower. Beach sets are within ~4 m elevation, commonly 1305–1309 m, at most sites (Table 3.1). With available elevation control from topographic maps, it is not possible to discern regional changes that might be attributed to tectonic deformation. Increased offshore-slope and deeper water both provide a capacity for greater wave energy, but these parameters do not correlate with beach elevations. Beach landform length, a rough guide for shoreline deposit volume, also does not correlate with water depth or offshore-slope, and therefore the length is more likely controlled by long shore currents and local sediment supply. A weak correspondence between beach-deposit volume and shallow water depth in the offshore (measured where offshore slope decreases to 1%, Table 3.1) exists in that the five longest beaches tend to be associated with water depths less than 12 m, and four of five beaches are associated with depths of 8 m or less. The association between shallow offshore water depth and large-volume beach deposits may point to barrier beach origins for these sites. Sites 2, 13, and 17 lie adjacent to wave-cut notches, indicating the importance of local sediment supply and longshore transport for these sites. Sites 2 and 8 show local bulking near modern stream systems, suggesting local input of stream sediments accentuated the beach development. Sites 5, 11, and 15 are barrier beaches pinned by local topography to form bay mouth bars.

In summary, one to three beach gravel ridges ascend as indicated by overlapping relations and by higher gravels that lie on lagoon deposits associated with lower beaches. This group of shoreline deposits lie on either alluvial fan or beach rock (probable beach rock)-cemented ooid deposits and are capped in many locations by deepwater marl. No soils are evident within the beach gravels where they underlie the marl, indicating that lake basin filling during the Lake Bonneville cycle occurred soon after beach formation.

3.4 EARLY LAKE HYDROGRAPH

As a lake fills, it may reach internal thresholds that temporarily stabilize lake level, such as spilling into a formerly enclosed basin to form a new arm of the lake. To test for possible basin hypsometry causes for the Pilot Valley shoreline, we analyzed USGS 10-m and 90-m DEMs for Lake Bonneville's northern basin at elevations between modern Great Salt Lake (1280 m) and 1380 m, which is below the ~1385 m sill to the southern (Gunnison [Sevier]) basin. We iteratively calculated lake area and volume at 1 m elevation increments, adding subsidiary basins such as Utah and Puddle valleys as their thresholds were reached and exceeded. Previous hypsometry has been limited to evaluating Lake Bonneville at the major shoreline levels. This earlier estimation of hypsometry suggested that lake level, area, and volume varied linearly above levels of ~1295 m (Currey et al., 1984).

Our results (Fig. 3.5) show that lake area rapidly expands with small lake-level increases between 1280 and 1300 m, with a large area increase at the ~1285 threshold between the Great Salt Lake basin and the West Desert basin (Fig. 3.2), also known as the Great Salt Lake Desert. Lake area expansion per unit elevation gain gradually decreases to ~1310 m, and then behaves nearly linearly. Two notable expansions in area occur at 1364 m (Puddle Valley) and 1371 m (Utah Valley). Puddle Valley was catastrophically flooded (Oviatt and Miller, 1997), whereas Utah Valley was probably always full, fed by the Provo River, so its lake area and volume were effectively part of Lake Bonneville throughout lake rise to the level of Utah Valley. Lake volume grows slowly at lowest elevations and behaves quasi-linearly with lake level at elevations above ~1300 m. These relations are summarized on the area

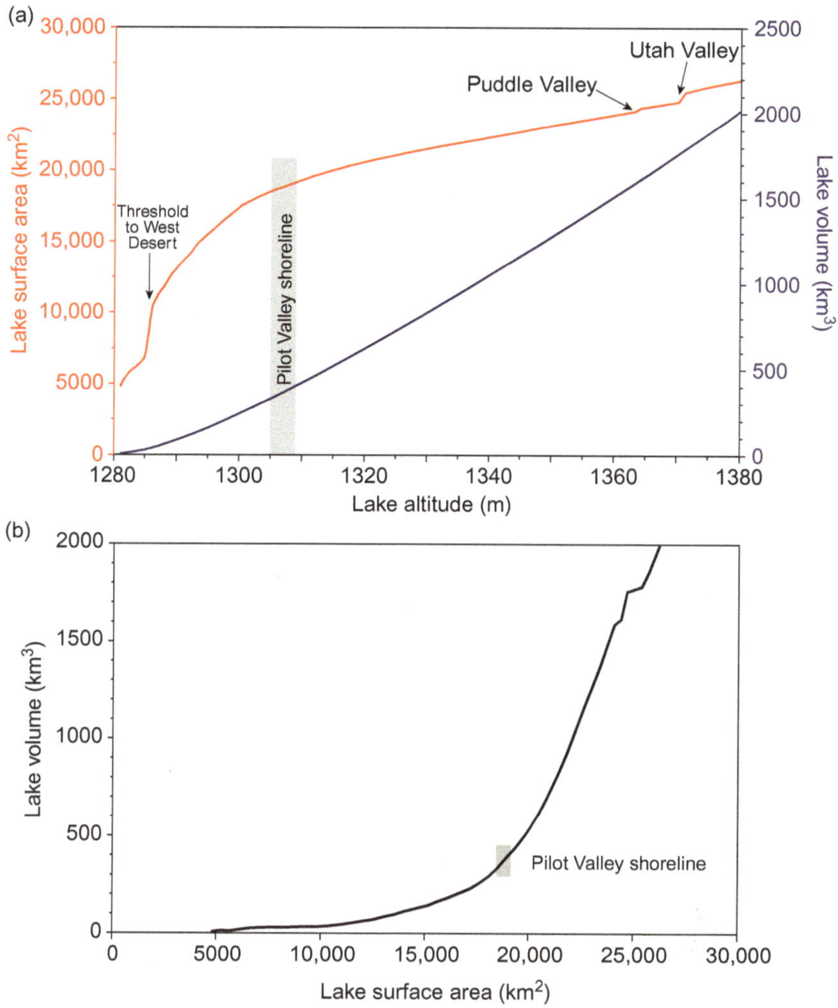

FIG. 3.5 Plots of lake level, surface area, and volume from analysis of 90-m DEMs. Analysis is only for lake levels below about 1380 m, avoiding most interbasin thresholds. (a) Lake area and lake volume vs elevation of lake surface; (b) lake volume vs lake area. Elevation range of Pilot Valley shoreline indicated.

vs volume plot (Fig. 3.5B), which shows volumes expanding slowly with lake-level change early in the lake rise history, and the two measures varying quasi-linearly near or slightly above the Pilot Valley shoreline elevation. No distinct changes in hypsometry are evident at or near the Pilot Valley shoreline elevation range. At the elevation range of the Pilot Valley shoreline, about 18 km^3 of water must be stored to raise the lake level 1 m.

Our analysis of basin hypsometry ignores sediment accumulated in basin bottoms during and after Lake Bonneville time, and therefore minimizes lake volume. This sedimentation change in basin geometry probably has greatest effects for broad basin bottoms at ~1284 m in the Great Salt Lake Desert (Fig. 3.2) and lower elevations for the basin around Great Salt Lake. Its effect is probably minimal for lake hypsometry near the Pilot Valley shoreline lake levels.

3.5 DISCUSSION

The Pilot Valley shoreline deposits are interpreted as beach deposits formed as a series of climbing beaches in a lake with ascending water level. Maximum beach development (measured by largest volume of shoreline deposits) in several places coincides with wave-cut cliffs and erosion of local prominences (Pigeon Mtn., Silver Island, Kelton; sites 5, 2, and 10) or input from streams (Silver Island, Hogup Mts.; sites 2 and 8). Maximum beach development of the Pilot Valley shoreline is most evident in northern reaches of water bodies, such as those associated with Pilot Valley playa. Erosional shorelines are difficult to identify in many places as uniquely belonging to the Pilot Valley shoreline, but in a few cases wave-cut notches pass laterally to beaches; an excellent example is the Rozel West locality (site 17).

Evidence for an early Lake Bonneville age for the Pilot Valley shoreline comes from the stratigraphic association with the overlying Bonneville marl. The early lake at this elevation formed about 30–33 cal ka BP (Oviatt, 2015) but this part of the age–depth curve is poorly constrained (two radiocarbon ages; Fig. 3.1). At the time of Pilot Valley shoreline development, water depth over much of the basin was greater than 25 m by an unknown amount. Uncertainty stems from the unknown thickness of the ~30 cal ka BP of subsequent sediment infilling. Therefore, the estimate of lake depth at the time of shoreline development using modern topography is an underestimate.

Beaches associated with the Pilot Valley shoreline formed between ~1305 and 1309 m elevation. Although no directly dated materials have been found associated with the Pilot Valley shoreline deposits, they lie ~10 m below the beaches and lagoon deposits draped by the Hansel Valley ash (Miller et al., 2008), indicating that they are older than this ~28 ka eruptive event.

Beaches apparently were not formed as a result of stabilization of lake level caused by basin hypsometric effects. Our analysis of the early lake hydrograph reveals no hypsometric irregularities at or near the Pilot Valley shoreline that would cause a slowing of lake-level rise, increasing the amount of time for shorelines to develop. Although river switching such as the Bear River diverting to the Snake River drainage could slow or reverse lake-level rise (Gilbert, 1890), no evidence exists for this. Gilbert noted that the Provo and Weber Rivers could have joined at a location west of the Uinta Mountains. It is possible that the basin in Utah Valley began receiving water as a result of a diversion of the Provo River away from the Weber, which would add a significant volume of water before spillover to the main lake, and increase evaporative potential through an increase in surface area. Such a speculative change in basin hydrography would rely on Utah Lake not overflowing and contributing to Lake Bonneville until the Provo River was captured, and could be tested by studies in Utah Valley.

Groundwater transfer into adjacent basins is not likely at this low elevation because most adjacent basins are at higher base levels. Metrics for shoreline volume (beach length) show little correspondence to offshore slopes and facing direction, and water depths. Nor does shoreline volume correspond with wave energy metrics such as water depth, facing direction, and steepness of offshore slopes, although there is a weak correspondence of the largest beach deposits with shallow water depth, possibly indicating a significant component of barrier beach processes. Therefore, the most likely cause for the larger beach deposits is local stream sediment input and erosion of local promontories, suggesting that sediment supply was key to shoreline deposit development. If stream sediment delivery was indeed important, the relatively small watersheds that fed the shoreline deposits may have been extremely active. In a few cases, cuspate geometry of beaches suggests sediment delivery from two directions.

Lack of isostatic rebound deformation of the Pilot Valley shoreline is consistent with a small lake load at the time of shoreline development. As a result, careful measurement of the Pilot Valley shorelines may reveal subsequent tectonic deformation that is greater than the 1–2 m variation in beach crests owing to wave dynamics (Atwood, 1994; Adams and Wesnousky, 1998). The Hansel Valley (Fig. 3.2) fault and related faults that pass southward across Great Salt Lake are the principal post-Bonneville faults (McCalpin et al., 1992; Dinter and Pechman, 2014). Measurements of beach elevation from DEM data are insufficient to resolve possible tectonic deformation. The highest Pilot Valley beaches (approximately 1308 m) east of this fault system, at Rozel and Antelope (Fig. 3.2, sites 17, 15), are lower than many, but not all, beaches farther west. Beaches at Hogup and Kelton (Fig. 3.2, sites 8, 10) are approximately the same elevation as those at Rozel and Antelope, indicating that no conclusions can be drawn without improved elevation measurements.

We interpret the origin of the Pilot Valley shoreline as a result of climatic slowdown in lake-level rise. No internal threshold of the basin, no hydrograph inflection, and no hydrologic diversions that change river supply are known at this time. Also, although beaches can form in very short time periods, similar groups of beach deposits that can be correlated regionally are not known, which suggests that factors for Pilot Valley beach development were unusual. Therefore, climatic effects such as reduced precipitation and enhanced evaporation (from reduced winter ice cover and/or warmer summers) are the likely causes for slower lake-level rise and enhanced shoreline development. Climate effects could be magnified somewhat by Utah Valley ceasing to overflow to Lake Bonneville if high evaporative output and (or) low river input occurred, further limiting lake-level rise. Studying the Jordan Narrows overflow point, which is currently close to 1375 m and above the Pilot Valley lake level, as well as early Lake Bonneville history in Utah Valley is needed to evaluate connections with Lake Bonneville.

Regional, hemispheric, and global climate variability during the time of Lake Bonneville are becoming increasingly clear. Wolff et al. (2010) and Rasmussen et al. (2014), for instance, demonstrated that global climate shifts correlate with abrupt shifts at the onset of Dansgaard-Oeschler events (Greenland interstadials GI; Fig. 3.1B) by using the fairly precise chronology for Greenland ice-core climate proxies. For the time period near 30 ka, suggested as the time for initial rise of Lake Bonneville (Oviatt, 2015 and references therein), the major climate events include a stadial (GS-5) from 32.04 to 28.90 ka with a short interstadial nested within it at 30.84–30.60 ka. Given the sparse age control on this early part of Lake Bonneville, one possible scenario for early Lake Bonneville is lake-level rise during GS-5, with a slower rise during the interstadial (GI-5.1) that could have produced the Pilot Valley shoreline features. This scenario is shown as a queried dotted line in Fig. 3.1A. Improving the chronology of early Lake Bonneville would help to establish the relation of lake level to global climate variability.

Evidence that the Pilot Valley shorelines are enhanced at sites with locally high sediment input, both from streams and by erosion of coastal promontories, indicates that more intense storms and enhanced wave energy accompanied shoreline development. Enhanced storms generally are not compatible with reduced precipitation causing still stands of lake level, although reduced winter precipitation and enhanced summer monsoon-like storms could result in this combination of slower lake-level rise and enhanced stream and wave energy. The highest of the Pilot Valley beaches is the smallest and the least consistently developed, indicating that it formed during a brief lake level stabilization before lake levels returned to their generally rapid rise (Fig. 3.1).

3.6 SUMMARY AND CONCLUSIONS

A shoreline accumulation of beach gravel and sand, whose crests consistently lie in the 1305–1309 m elevation range, is named the Pilot Valley shoreline. The gravel deposits of this shoreline are overlain by the deepwater marl of Lake Bonneville and thus formed early in the lake cycle. In many places, the Pilot Valley beach deposits form a rising set of two to three crests and have accentuated development near modern erosional prominences and streams, indicating that their formation may have been driven by relatively higher sediment supply. Two of the documented Pilot Valley shoreline sites were previously misidentified as isostatically deflected Gilbert shorelines by Currey (1982).

The elevation range of the Pilot Valley shoreline does not match any internal basin threshold or other hypsometric factor that might have stabilized the lake level, and therefore the hydrologic budget and weather, driven by both local and regional climate, are responsible for the formation of the shoreline. Enhanced evaporation and decreased river inputs, perhaps accompanying more intense summer storm activity, over a period of time as brief as several decades, could account for quasi-stable lake levels. We suggest that the slowing of lake-level rise responsible for the Pilot Valley shoreline may correlate with climate event GI-5.1, between 30.8 and 30.6 ka.

ACKNOWLEDGMENTS

We thank Don Currey and Jack Oviatt for field visits to some of the Pilot Valley shoreline sites, and Tracey Felger for her insights into the Kelton site. We appreciate the reviews of Andy Cyr and Ken Adams, which led to greater clarity of presentation and content, and comments from the editors.

REFERENCES

Adams, K.A., Wesnousky, S.G., 1998. Shoreline processes and the age of the Lake Lahonton highstand in the Jessup embayment, Nevada. Geol. Soc. Am. Bull. 110, 1318–1332.

Atwood, G., 1994. Geomorphology applied to flooding problems of closed-basin lakes—specifically Great Salt Lake. Geomorphology 10, 197–219.

Bills, B.G., May, G.M., 1987. Lake Bonneville: constraints on lithospheric thickness and upper mantle viscosity from isostatic warping of Bonneville, Provo, and Gilbert stage shorelines. J. Geophys. Res. 92, 11493–11508.

Crittenden Jr., M.D., 1963. New Data on the Isostatic Deformation of Lake Bonneville: U.S. Geological Survey Professional Paper 454-E. U.S. Geological Survey, pp. E1–E31.

Currey, D.R., 1982. Lake Bonneville: Selected Features of Relevance to Neotectonic Analysis: U.S. Geological Survey open-file report 82-1070. 30 pp., map 1:500,000.

Currey, D.R., Atwood, G., Mabey, D.R., 1984. Major Levels of Great Salt Lake and Lake Bonneville: Utah Geological and Mineral Survey Map 73.

Dinter, D.A., Pechman, J.C., 2014. Paleoseismology of the Promontory Segment, East Great Salt Lake Fault: Final Technical Report to U.S. Geological Survey. 23 pp.

Felger, T.J., Miller, D.M., Langenheim, V.E., Fleck, R.J., 2016. Geologic and Geophysical Maps of the Kelton Pass SE and Monument Peak SW Quadrangles, Box Elder County, Utah: Utah Geological Survey Miscellaneous Publication 16-1DM. 34 pp., scale 1:24,000.

Gilbert, G.K., 1890. Lake Bonneville. U.S. Geological Survey Monograph, vol. 1. 438 pp.

McCalpin, J., Robison, R.M., Garr, J.D., 1992. Neotectonics of the Hansel Valley-Pocatello Valley corridor, northern Utah and southern Idaho. In: Gori, P.L., Hays, W.W. (Eds.), Assessment of Regional Earthquake Hazards and Risk Along the Wasatch Front. U.S. Geological Survey, Utah. Professional Paper 1500, pp. G1–G18.

Miller, D.M., 1990a. Geologic Map of the Crater Island SW Quadrangle, Box Elder County, Utah: Utah Geological and Mineral Survey Map 129. 8 pp., scale 1:24,000.

Miller, D.M., 1990b. Geologic Map of the Lucin4 SW Quadrangle, Box Elder County, Utah: Utah Geological and Mineral Survey Map 130. 13 pp., 2 sheets, scale 1:24,000.

Miller, D.M., 1993. Geologic Map of the Crater Island NW Quadrangle, Box Elder County, Utah: Utah Geological and Mineral Survey Map 145. 13 pp., 2 sheets, scale 1:24,000.

Miller, D.M., Glick, L.L., 1987. Geologic Map of the Jackson Quadrangle, Box Elder County, Utah: Utah Geological and Mineral Survey Map 95. 7 pp., scale 1:24,000.

Miller, D.M., McCarthy, P.T., 2002. Geologic Map of the Terrace Mountain West Quadrangle, Box Elder County, Utah: Utah Geological Survey Miscellaneous Publications 02-3. 13 pp., scale 1:24,000.

Miller, D.M., Jordan, T.E., Allmendinger, R.W., 1990. Geologic Map of the Crater Island Quadrangle, Box Elder County, Utah: Utah Geological and Mineral Survey Map 128. scale 1:24,000, 16 pp.

Miller, D.M., Oviatt, C.G., McGeehin, J.P., 2013. Stratigraphy and chronology of Provo shoreline deposits and lake-level implications. Late Pleistocene Lake Bonneville, eastern Great Basin, USA: Boreas 42, 342–361.

Miller, D.M., Oviatt, C.G., Nash, B.P., 2008. Late Pleistocene Hansel Valley basaltic ash, northern Lake Bonneville, Utah, USA. Quat. Int. 178, 238–245.

O'Connor, J., 1993. Hydrology, Hydraulics, and Geomorphology of the Bonneville Flood. Geological Society of America Special Paper 274. 83 pp.

Oviatt, C.G., 2015. Chronology of Lake Bonneville, 30,000 to 10,000 yr B.P. Quat. Sci. Rev. 110, 166–171.

Oviatt, C.G., Miller, D.M., 1997. New explorations along the northern shores of Lake Bonneville. In: Link, P.K., Kowallis, B.J. (Eds.), Mesozoic to Recent Geology of Utah: Brigham Young University Geology Studies, vol. 42. Brigham Young University, Provo, Utah. Part II, pp. 345–371.

Oviatt, C.G., Currey, D.R., Miller, D.M., 1990. Age and paleoclimatic significance of the Stansbury Shoreline of Lake Bonneville, northeastern Great Basin. Quat. Res. 33, 291–305.

Rasmussen, S.O., Bigler, M., Blockley, S.P., Blunier, T., Buchardt, S.L., Clausen, H.B., Cvijanovic, I., Dahl-Jensen, D., Hoek, W.Z., Lowe, J.J., Pedro, J.B., Popp, T., Seierstad, I.K., Steffensen, J.P., Svensson, A.M., Vallelonga, P., Vinther, B.M., Walker, M.J.C., Wheatley, J.J., Winstrup, M., 2014. A stratigraphic framework for abrupt climatic changes during the Last Glacial period based on three synchronized Greenland ice-core records: refining and extending the INTIMATE event stratigraphy. Quat. Sci. Rev. vol. 106, 14–28.

Schofield, I., Jewell, P., Chan, M., Currey, D., Gregory, M., 2004. Shoreline development, long-shore transport and surface wave dynamics, Pleistocene Lake Bonneville, Utah. Earth Surf. Process. Landforms 29, 1675–1690.

Wolff, E.W., Chappellaz, J., Blunier, T., Rasmussen, S.O., Svensson, A., 2010. Millenial-scale variability during the last glacial: the ice core record. Quat. Sci. Rev. 29, 2828–2838.

Chapter 4

Landslides, Alluvial Fans, and Dam Failure at Red Rock Pass: The Outlet of Lake Bonneville

J.F. Shroder*, K. Cornwell†, C.G. Oviatt‡ and T.C. Lowndes§
*University of Nebraska at Omaha, Omaha, NE, United States
†California State University, Sacramento, CA, United States
‡Kansas State University, Manhattan, KS, United States
§2301 NE Blakeley St., Apt 306, Seattle, WA, United States

ABSTRACT
The outlet area of Lake Bonneville has had over a century and a quarter of science and speculation about its configuration and character and many disparate interpretations have appeared in the literature. In this chapter, we offer our interpretations of the outlet, which are based on field research begun in the 1980s and review of the pertinent literature. The Marsh Creek alluvial fan near Red Rock Pass, which formed the natural dam for Lake Bonneville, was built westward from the Portneuf Range to where it abutted a large landslide complex on the eastern slope of the Bannock Range. At the low point on the crest of the alluvial fan the Zenda threshold of Lake Bonneville determined the elevation of the Bonneville shoreline and was washed out by the Bonneville flood. The landslide includes deformed Neogene lacustrine sediments (the Salt Lake Formation) and blocks of Paleozoic bedrock. Landforms characteristic of large-scale mass movements on the Red Rock Pass landslide complex include hummocky topography and multiple scarps at the heads of smaller slumps. The older part of the landslide mass was emplaced long before Lake Bonneville in response to increasing tectonic relief during the Neogene and Quaternary. It sits higher on the eastern slope of the Bannock Range adjacent to a younger part of the landslide mass, which moved in response to catastrophic dewatering and erosional removal of lateral support by the Bonneville flood. Valleys with bowl-shaped heads and U-shaped cross-valley profiles on the north-sloping surface of the Marsh Creek alluvial fan, were formed by groundwater flow from Lake Bonneville late in its transgressive phase through the fan and other surficial deposits. Headward erosion of the valleys by spring sapping led to the Bonneville flood, a dam failure involving processes typical of failures of man-made and natural dams throughout the world.

Keywords: Bonneville, Red rock pass, Bonneville flood, Dam failure

4.1 INTRODUCTION

In this chapter, we describe observations and interpretations of the landforms, geology, and hydrology at Red Rock Pass (Fig. 4.1). The most important published work on Red Rock Pass is that of Gilbert, primarily in his monograph on Lake Bonneville (Gilbert, 1878; 1880; 1890; Chapter 6). Gilbert recognized that Red Rock Pass was the outlet for Lake Bonneville, that the overflow threshold was on the crest of the Marsh Creek alluvial fan, that the Bonneville flood caused the lake to drop rapidly from the Bonneville shoreline to the Provo shoreline, and that the position of the overflow threshold, i.e., the divide between the Bonneville basin and the Snake River drainage basin, shifted to the south during the flood. He also recognized that the postflood overflow threshold at Red Rock Pass was similar to the elevation of the Provo shoreline (although accurate topographic data were not available to Gilbert), and that overflow across that threshold for a prolonged period allowed the lake to produce the Provo shoreline, which is better developed than any other shoreline in the basin. He saw that Holocene alluvium had filled in the Marsh Creek valley and buried the Red Rock Pass threshold. An advantage that Gilbert had over many later workers who have attempted to understand Red Rock Pass is that he came to the outlet after studying the shorelines within the lake basin. He had a profound understanding of geomorphology—that is, he understood and was deeply curious about processes and history of landforms at Earth's surface. These perspectives allowed him to view Red Rock Pass as an integral part of Lake Bonneville.

During the 20th Century, some publications reiterated Gilbert's conclusions and others diverged from that work. Pack (1939) and Ives (1948) added a few observations to those of Gilbert but did little to advance knowledge of the outlet. Malde (1968) utilized an idea, first proposed by Pack (1939), and later promoted by Bright (1963), and others (e.g., INQUA, 1965), that Lake Bonneville rose above the Bonneville shoreline and fell at some later time to the elevation of the Bonneville shoreline. Malde (1968) thought the Paleozoic outcrops at Red Rock Pass (which we interpret as landslide blocks) represented remnants of a bedrock floor at the level of the Bonneville shoreline. Jarrett and Malde (1987) estimated the duration of the flood to be less than a year, in part based on a revised interpretation of the geomorphology at Red Rock Pass (that during the flood the lake level dropped approximately 108 m rather than Malde's, 1968, estimate of about 30 m). O'Connor (1993) and Chapter 6 refined the maximum flood-discharge estimate (about $1 \times 10^6\ m^3\ s^{-1}$) and conducted a thorough study of the hydraulics of the flood downstream along the Snake River. He also observed the bowl-shaped valley heads of ephemeral streams on the north-facing slope of the Marsh Creek alluvial fan and interpreted them as groundwater-sapping features. Janecke and Oaks (2011a,b) published

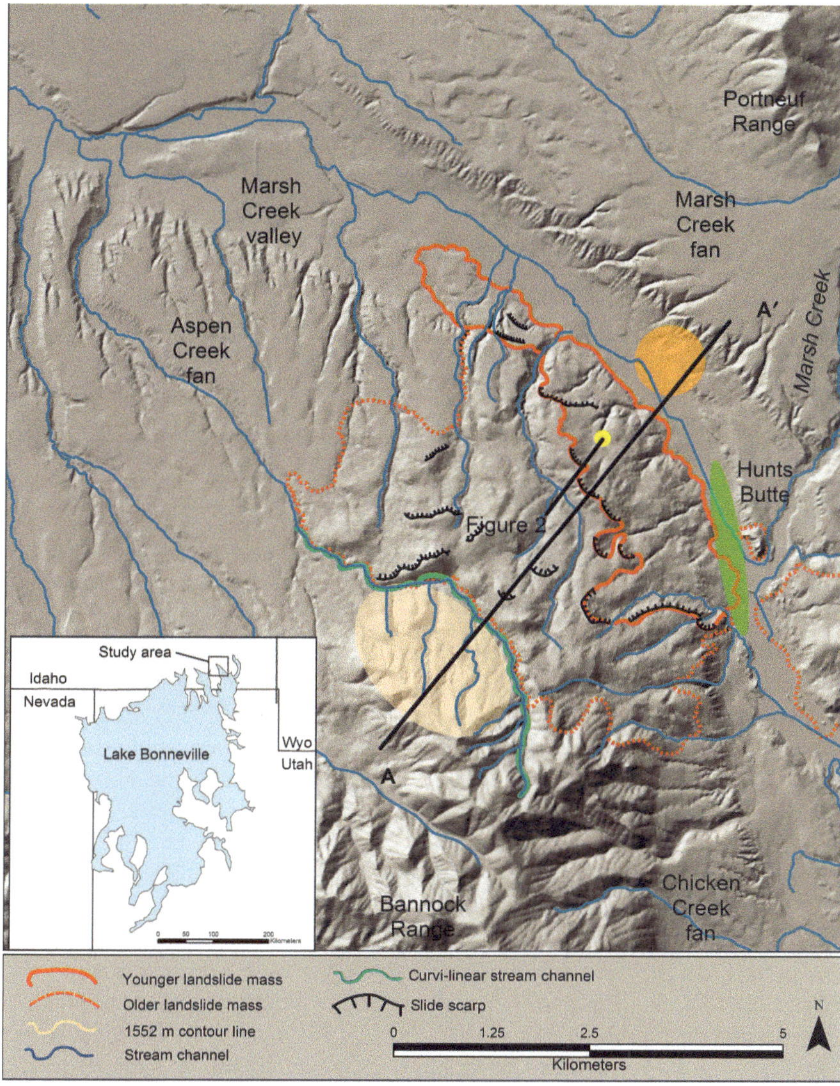

FIG. 4.1 Red Rock Pass in the northeastern corner of the Bonneville basin showing the younger and older landslide masses and the three dominant alluvial fans (Chicken Creek, Aspen Creek, and Marsh Creek fans) in the outlet area. The Zenda threshold was somewhere within the area encompassed by the *orange circle* before it was eroded away by the Bonneville flood; the Red Rock Pass threshold was somewhere within the area encompassed by the *green oval*, buried beneath Holocene alluvium in the valley bottom; the *yellow oval* marks the slide scarp at the head of the older landslide mass. The 1552-m contour line is at the elevation of the Zenda threshold. Profile A-A′ is shown in Fig. 4.3.

interpretations of the Red Rock Pass area that differ from those presented in this chapter. Some of our work on Red Rock Pass has been informally published in abstracts and theses (Sewell, 1980; Sewell and Shroder, 1981; Shroder and Sewell, 1981; Shroder and Lowndes, 1989a,b; Cornwell and Shroder, 1990), and here we summarize observations and interpretations derived from that work combined with more recent field observations and review of the literature.

The Red Rock Pass area had a complicated geologic history prior to Lake Bonneville. Tectonically deformed Proterozoic and Paleozoic rocks underlie the Bannock and Portneuf Ranges (DeVecchio et al., 2002; Long and Link, 2007). In late Neogene time, a graben beneath Red Rock Pass was partly filled with tuffaceous, fluvial, and lacustrine, fine-grained to coarse-grained clastic and carbonate rocks and sediments of the Salt Lake Formation (DeVecchio et al., 2002). Sometime in the late Neogene or early Quaternary, after the deposition of the Salt Lake Formation, massive landslides moved down the east face of the Bannock Range. The landslide deposits contain blocks of Paleozoic rocks and Salt Lake Formation. During Quaternary time, the Marsh Creek alluvial fan, the Aspen Creek fan, and the Chicken Creek fan (Fig. 4.1), in addition to smaller accumulations of alluvium, covered the preexisting topography and deposits.

Modern Red Rock Pass is close to the topographic divide between Marsh Creek, which now flows to the north to the Snake River drainage basin, and the northern part of Cache Valley (to the south). Currey (1982) distinguished Pleistocene Red Rock Pass, which was the overflow threshold for Lake Bonneville during Provo-shoreline time, from the Zenda threshold, which was the drainage divide on the crest of the Marsh Creek alluvial fan at the elevation of the Bonneville shoreline. For discussions in this chapter, Currey's (1982) elevations for the Zenda and Red Rock Pass thresholds, 1552 m and 1444 m, respectively, are used, although we recognize that these elevations may be revised with future work. The Bonneville flood eroded the toes of the Marsh Creek alluvial fan and the Red Rock Pass landslide and removed the Zenda threshold. It is not possible to determine the exact position of the Zenda threshold, so on the maps in Figs. 4.1 and 4.5 we symbolize it as a small circle with indefinite boundaries. Previous efforts to reconstruct the elevation and location of the Zenda threshold using trend-surface analysis of numerous points on the surface of the Marsh Creek alluvial fan and the Red Rock Pass landslide (Lowndes, 1989), produced a range of possible elevations that included Currey's (1982) elevation of 1552 m, and a location for the threshold about 3 km southeast of the point suggested by Currey (1982).

The Marsh Creek alluvial fan is over 60 m (200 ft) thick where it overlies deformed tuffaceous beds of the Salt Lake Formation near the hamlet of Zenda (42.38° N, 112.05° W) (DeVecchio et al., 2002, cross section C-C'). The alluvial-fan deposits in the Aspen Creek fan in the vicinity of (42.39° N, 112.12° W) (DeVecchio et al., 2002, cross section B-B') are approximately

90 m (300 ft) thick. These points are on the distal parts of the fans, and the fans probably thicken toward their apexes.

4.2 MASS MOVEMENTS

Extensive mass movement deposits, covering more than 18.6 km^2, flank the eastern slopes of the Bannock Range near the Zenda threshold in the Red Rock Pass area. They have characteristic features of mass-movement deposits and landforms (Dikau et al., 1997; Shroder and Bishop, 1998), including hummocky topography, displaced blocks of Lower Paleozoic carbonates and quartzites (Swan Peak Quartzite, Garden City Formation, and Saint Charles Formation), rotated and deformed blocks of Tertiary tuffaceous conglomerates, sandstones, siltstones, and mudstones of the Salt Lake Formation (Fig. 4.2), altered drainage, numerous small slide scarps (less than 100 m tall and 1 km long), and a large slip surface at the head of the slide mass.

We infer two ages of landslide deposits marked by a distinct topographic break (Fig. 4.1). The younger mass-movement deposits occupy lower elevations on and adjacent to the western margin of the Marsh Creek valley and appear to be to have been derived from the larger, older landslide that was reactivated during and after the Bonneville flood. We interpret large blocks of Paleozoic bedrock, including Hunts Butte and the Red Rock itself, as remnants of the landslide mass that were too large to be moved by the Bonneville flood. These large blocks are partially or completely brecciated, their internal bedding dips in inconsistent directions from block to block, and they are similar to other large blocks of bedrock that are part of the large landslide masses nearby.

A topographic profile from the top of the Bannock Range in the west to the Marsh Creek alluvial fan in the east shows the topographic characteristics and inferred slide geometry relative to the Zenda threshold elevation and present topography (Fig. 4.3). Upslope from the hummocky topography is a substantial slide scarp at the head of the older landslide mass (Fig. 4.1), which previously had been mapped as Quaternary alluvial fan deposits (Fig. 4.4; Long and Link, 2007). Surficial deposits on the landslide mass consist of fine-grained silts and clays, which we interpret to be primarily loess and colluvium, and coarse-grained sands, gravels, and boulders.

The boundary between the slide scarp and the slide mass produces an unusual curvilinear drainage pattern (Fig. 4.1). Although most drainage channels on the eastern slope of the Bannock Range drain eastward toward Marsh Creek valley, this channel flows to the north. The initial downslope movement of the older slide mass created the slide scarp and surface flow was diverted northward by the slide mass into the Aspen Creek alluvial-fan deposits, and ultimately into Marsh Creek far to the north.

The exact mechanism of failure that led to deposition of the older slide mass is unknown, although increasing relief of the Bannock Range created by

FIG. 4.2 Photograph of vertically dipping beds of the Salt Lake Formation in the younger part of the Red Rock landslide. See Fig. 4.1 for location.

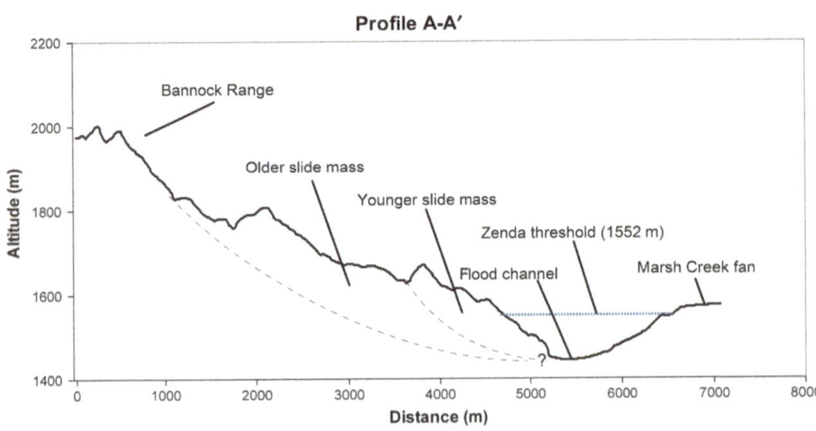

FIG. 4.3 Topographic profile through the Red Rock landslide mass from the Bannock Range in the west to Marsh Creek fan in the east. The map in Fig. 4.1 shows the location of the A-A′ profile line.

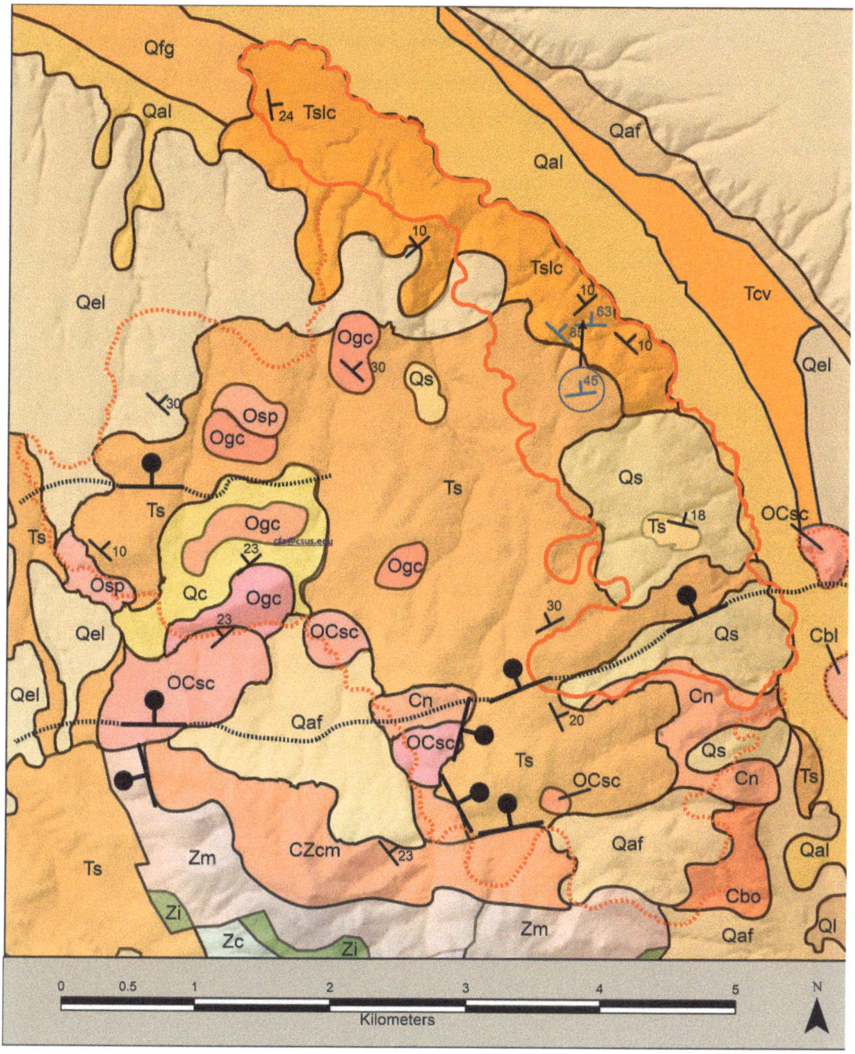

FIG. 4.4 Geologic map of the study area, with the younger and older parts of the Red Rock Pass mass movement superimposed (in solid and *dotted red lines*—see Fig. 4.1). Three strike-and-dip symbols in *blue* are for outcrops of tuffaceous Salt Lake formation, not reported by Long and Link (2007). *Modified from Long, S.P., Link, P.K., 2007. Geologic map compilation of the Malad City 30° by 60° quadrangle, Idaho. Idaho Geological Survey, Technical report 07-1; 1 sheet; 1:24,000 scale.*

tectonic deformation in response to late Cenozoic extension (Hintze and Kowallis, 2009) would have been involved. Removal of lateral support and dewatering of the landslide materials probably led to failure of the younger slide mass. If the slide scarp at the head of the older slide is characteristic of the underlying slip surface then the older slide mass averages about 68 m in

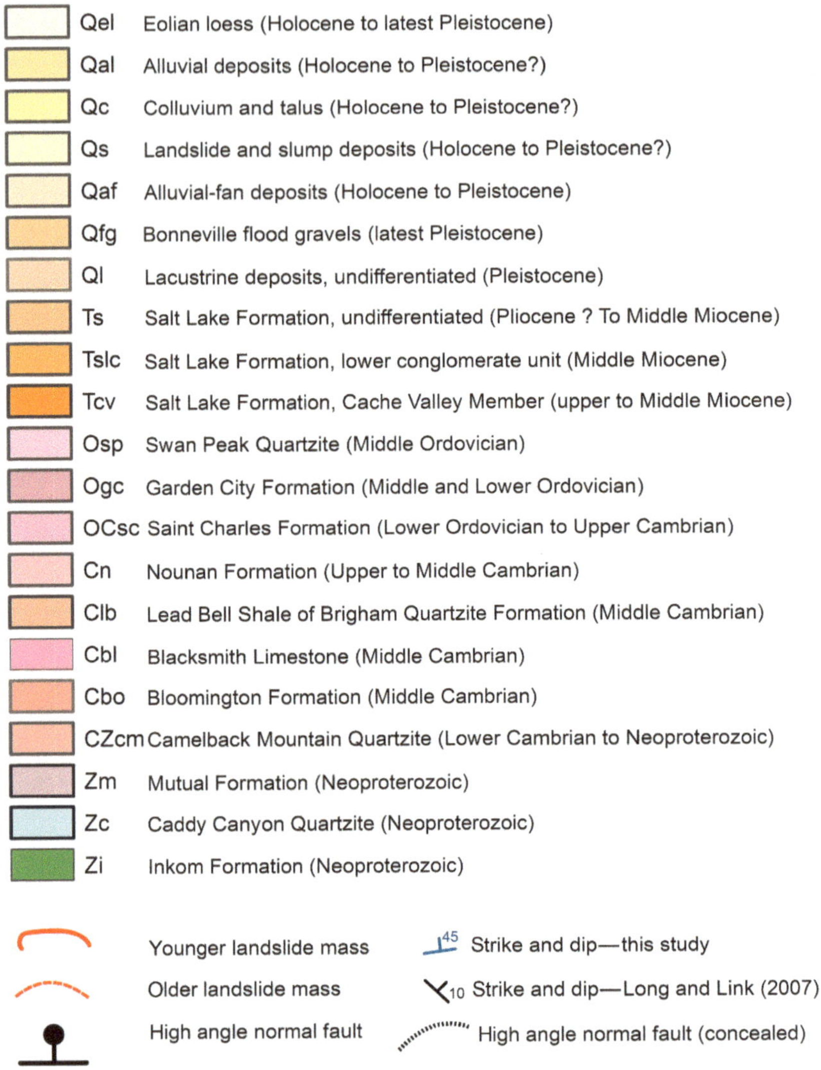

FIG. 4.4—CONT'D

thickness and probably moved in a rotational manner. Whether this mass initially moved in a series of failures is not clear. Multiplying the mean slide thickness (68 m) by the GIS-calculated area of the entire slide mass (2×10^7 m^2) produces a total slide volume of about 1×10^9 m^3. The depth of the failure surface on the younger landslide mass is unknown, although it must be lower than the Red Rock Pass threshold. If the ratio of the entire slide area to its thickness is applied to the younger slide mass, this yields a mean thickness of the young slide mass of about 21 m and a volume of about 1×10^8 m^3.

The positive water balance in Lake Bonneville caused it to continue overflowing across the Red Rock Pass threshold after the flood, for a period possibly as long as 3000 year (Godsey et al., 2011; Miller et al., 2013). The Provo shoreline formed during this period. Movement of the younger part of the Red Rock Pass landslide into the path of the overflowing lake caused the lake threshold to rise. As a result, the highest landforms associated with the Provo shoreline within the lake basin are 15 m higher than the lowest Provo-shoreline deposits (Miller et al., 2013; Chapter 7).

4.3 GROUNDWATER

As transgressing Lake Bonneville approached the 1552 m Zenda threshold about 18 cal ka BP, its hydraulic head created a north-sloping hydraulic gradient through coarse and fine-grained sediments of the Marsh Creek alluvial fan. The hydraulic conductivity of the landslide masses would have been heterogeneous, but groundwater that had significant hydraulic head would not have been stopped by the landslide materials.

Valleys of streams that originate on the north slope of the Marsh Creek alluvial fan have bowl-shaped heads (Fig. 4.5). Valleys are U-shaped, 100–200 m wide in cross-valley profiles, and roughly 10 m deep. Modern streams in these valleys are ephemeral, indicating that the water table is lower than the valley bottoms. The bowl-shaped heads of these valleys were noted by O'Connor (1993) and interpreted by him as possibly caused by groundwater sapping during Bonneville time. We agree with this interpretation. Note in Fig. 4.5 that the bowl-shaped valley heads are on the east side of the U-shaped valleys, suggesting that the groundwater flow entered those valleys from the east. We interpret this asymmetry as the result of sapping from groundwater that originated in the Portneuf Range and flowed downslope through alluvial sediments. The water table in the fan was elevated because of the presence of Lake Bonneville, and the sloping water table of groundwater from the Portneuf Range was higher than that. Marsh Creek is now entrenched over 50 m into its alluvial fan and the underlying Salt Lake Formation, but during Bonneville times the base level would have been higher, entrenchment would have been much less, and the water table in the fan would have been much higher than today.

In the space now occupied by the trench produced by Bonneville flood erosion (ie, Marsh Creek valley), directly down gradient from the Zenda threshold on the north-sloping surface of the Marsh Creek alluvial fan, a southward-migrating channel head would have been produced by groundwater sapping during Bonneville time. The valley would have had a symmetrical bowl-shaped valley head that would have migrated southward toward the Zenda threshold as Lake Bonneville rose. The geomorphic effects of

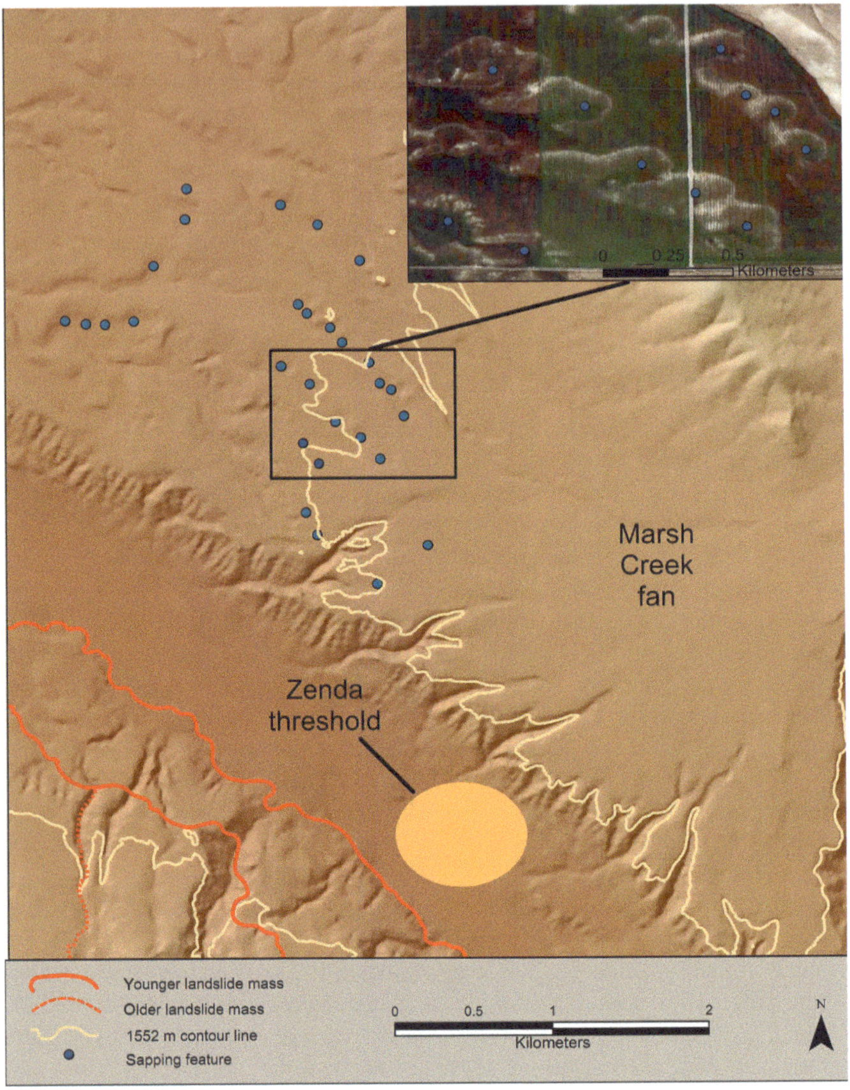

FIG. 4.5 Bowl-shaped valley heads (marked by *blue dots*) on the north flank of the Marsh Creek alluvial fan. The *light brown* contour line is at 1550 m, which is close to the estimated elevation of the Zenda threshold at 1552 m. In agreement with O'Connor (1993), we interpret the bowl-shaped valley heads and associated U-shaped valleys as having been produced by spring sapping caused by groundwater flow through the Marsh Creek alluvial fan from Lake Bonneville (south of the map area).

groundwater sapping are well known (Kochel and Piper, 1986; Higgins, 1984; Nash, 1996; Lamb et al., 2008; Gulick, 2001; Rosenberry, n.d.). The migration of the valley head and the continual rise of the lake ultimately led to failure of the Marsh Creek fan-dam and the catastrophic Bonneville flood

(O'Connor, 1993, p. 7). Modern earthen dams (such as Teton dam in Idaho, Hope Mills dam in North Carolina, Lawn Lake dam in Rocky Mountain National Park, Taum Sauk reservoir in Missouri) have failed by processes similar to those proposed here for the failure of the Marsh Creek fan-dam (Costa and Schuster, 1987; Shivers et al., 2015).

4.4 SUMMARY

G.K. Gilbert recognized the importance of Red Rock Pass and documented the fundamental landforms at the pass and their relationships with the major shorelines and geologic history of Lake Bonneville. Gilbert's map of the Red Rock Pass area (Chapter 6; Figs. 4.1 and 4.2) showed the large landslide complex just west of the pass, but in print he did not interpret the feature as a landslide. Most subsequent workers have either not recognized the landslide or down-played its importance in Lake Bonneville history.

Lake Bonneville rose to its highest elevation, marked by the low point on the crest of the Marsh Creek alluvial fan (the Zenda threshold), and overflowed into the Snake River drainage basin at 18 cal ka BP. Groundwater sapping and channeling of the surficial deposits at the Zenda threshold caused the failure of the Marsh Creek alluvial-fan dam for Lake Bonneville. The surficial materials at the threshold, which consisted of fan alluvium and landslide-deformed blocks of Paleozoic bedrock units and sediments of the Tertiary Salt Lake Formation, were rapidly eroded by the overflowing lake, resulting in the catastrophic Bonneville flood. Rapid erosion and removal of lateral support for the large Red Rock Pass landslide caused the landslide to be rejuvenated and to move downslope into the space created by flood erosion. After the main flood discharge had waned and the Provo shoreline had begun forming within the lake basin, continued movement of the landslide caused the Red Rock Pass threshold to rise, causing basin-wide transgression during the formation of the Provo shoreline.

ACKNOWLEDGMENTS

Thanks to colleagues and former students for valuable discussions and assistance, and to Jim O'Connor for helpful review comments on an earlier draft of this chapter.

REFERENCES

Bright, R.C., 1963. Pleistocene Lakes Thatcher and Bonneville, Southeastern Idaho (Ph.D. thesis). University of Minnesota.

Cornwell, K., Shroder Jr., J.F., 1990. Computer modeling of the Lake Bonneville breakout flood through the Red Rock Pass threshold. Geol. Soc. Am. Abs. Prog. 22, A23.

Costa, J.E., Schuster, R.L., 1987. The formation and failure of natural dams: U.S. Geological Survey Open-File report 87-392. U.S. Geological Survey, Reston, VA. 39 pp.

Currey, D.R., 1982. Lake Bonneville: selected features of relevance to neotectonic analysis: U.S. Geological Survey open-file rep. 82-1070.

DeVecchio, D.E., Oriel, S.S., Link, P.K., 2002. Geologic map of the Downey East Quadrangle and regions of the Swan Lake, Oxford, and Cottonwood Peak Quadrangles, Bannock County, Idaho: Technical report T-03-2. Idaho Geological Survey, Moscow, ID (Scale 1:24,000).

Dikau, R., Brunsden, D., Schrott, L., Ibsen, M.-L. (Eds.), 1997. Introduction. Landslide Recognition: Identification, Movements and Causes, John Wiley & Sons, Chichester, pp. 1–11.

Gilbert, G.K., 1878. The ancient outlet of the Great Salt Lake. Am. J. Sci. 15 (3rd Series), 256–259.

Gilbert, G.K., 1880. The outlet of Lake Bonneville. Am. J. Sci. 19, 341–349.

Gilbert, G.K., 1890. Lake Bonneville. U.S. Geological Survey Monograph 1. 438 pp.

Godsey, H.S., Oviatt, C.G., Miller, D.M., Chan, M.A., 2011. Stratigraphy and chronology of offshore to nearshore deposits associated with the Provo shoreline, Pleistocene Lake Bonneville, Utah. Palaeogeogr. Palaeoclimatol. Palaeoecol. 310, 442–450.

Gulick, V.C., 2001. Origin of the valley networks on Mars: a hydrological perspective. Geomorphology 37, 241–268.

Higgins, C.G., 1984. Piping and sapping: development of landforms by groundwater outflow. In: LaFleur, R.G. (Ed.), Groundwater as a Geologic Agent. Boston, MA, Allen and Unwin, pp. 18–58.

Hintze, L.F., Kowallis, B.J., 2009. Geologic History of Utah. (Brigham Young University Geology Studies Special Publication 9). Brigham Young University, Provo, UT. 225 p.

INQUA, 1965. Part H: Lake Bonneville. In: Guidebook for Field Conference E, Northern and Middle Rocky Mountains. International Union for Quaternary Research, VIIth Congress. Nebraska Academy of Sciences, Lincoln, NE, pp. 104–107.

Ives, R.L., 1948. The outlet of Lake Bonneville. Sci. Mon. 67, 415–426.

Janecke, S.U., Oaks Jr., R.Q., 2011a. Reinterpreted history of latest Pleistocene Lake Bonneville: Geologic setting of threshold failure, Bonneville flood, deltas of the Bear River, and outlines for two Provo shorelines, southeastern Idaho, USA. In: Lee, J., Evans, J.P. (Eds.), Geologic Field Trips to the Basin and Range, Rocky Mountains, Snake River Plain, and the Terranes of the U.S. Cordillera. The Geological Society of America. Field Guide, vol. 21, pp. 195–222.

Janecke, S.U., Oaks Jr., R.Q., 2011b. New insights into the outlet conditions of late Pleistocene Lake Bonneville, southeastern Idaho, USA. Geosphere 7 (6), 1369–1391. http://dx.doi.org/10.1130/GES00587.

Jarrett, R.D., Malde, H.E., 1987. Paleodischarge of the late Pleistocene Bonneville Flood, Snake River, Idaho, computed from new evidence. Geol. Soc. Am. Bull. 99, 127–134.

Kochel, R.C., Piper, J.F., 1986. Morphology of large valleys on Hawaii: evidence for groundwater sapping and comparisons with Martian valleys. J. Geophys. Res. 91, 175–192.

Lamb, M.P., Dietrich, W.E., Aciego, S.M., Depaolo, D.J., Manga, M., 2008. Formation of Box Canyon, Idaho, by megaflood: implications for seepage erosion on Earth and Mars. Science 320 (5879), 1067–1070.

Long, S.P., Link, P.K., 2007. Geologic map compilation of the Malad City 30° by 60° quadrangle, Idaho: Technical Report 07-11. Idaho Geological Survey (sheet; 1:24,000 scale).

Lowndes, T.C., 1989. Development of a Threshold Model for Red Rock Pass, Idaho, by Trend Surface Analysis of Remnant Alluvial Fan Altitudes. University of Nebraska at Omaha (unpublished senior thesis in Geology, 60 pp).

Malde, H.E., 1968. The Catastrophic Late Pleistocene Bonneville Flood in the Snake River Plain, Idaho. (U.S. Geological Survey Professional Paper 596; 52 pp).

Miller, D.M., Oviatt, C.G., McGeehin, J.P., 2013. Stratigraphy and chronology of Provo shoreline deposits and lake-level implications, late Pleistocene Lake Bonneville, eastern Great Basin, USA. Boreas 42, 342–361.

Nash, D.J., 1996. Groundwater sapping and valley development in the Hackness Hills, North Yorkshire, England. Earth Surf. Process. Landforms 21, 781–795.

O'Connor, J., 1993. Hydrology, Hydraulics, and Geomorphology of the Bonneville Flood. (Geological Society of America Special Paper 274; 83 pp).

Pack, F.J., 1939. Lake Bonneville: a popular treatise dealing with the history and physical aspects of Lake Bonneville. Bull. Univ. Utah 30 (4) (Salt Lake City; 112 pp).

Rosenberry, D., n.d., Ground-water sapping and the generation of natural ampitheaters. http://www.colorado.edu/geography/geomorph/rosenberry (accessed on 31 December 2015).

Sewell, R., 1980. Newly discovered landslip at RedRock Pass, Idaho: implications for threshold-controlled shorelines of Pleistocene Lake Bonneville. Great Plains—Rocky Mountain Division of the Association of American Geographers. Geogr. J. 9, 100.

Sewell Jr., R.E., Shroder Jr., J.F., 1981. Red Rock landslip: massive failure initiated by late Pleistocene Bonneville Flood. Geol. Soc. Am. Abs. Prog. 13, 189.

Shivers, M.J., Smith, S.J., Grout, T.S., Lewis, J.M., 2015. Dam-breach analysis and flood-inundation mapping for selected dams in Oklahoma City, Oklahoma, and near Atoka, Oklahoma: Scientific Investigations report 2015-5052. http://pubs.usgs.gov/sir/2015/5052/pdf/sir2015-5052.pdf (accessed 20.12.15).

Shroder Jr., J.F., Bishop, M.P., 1998. Mass movement in the Himalaya: new insights and research directions. Geomorphology 26, 13–35 (special volume on "Mass Movement in the Himalaya"; ed. J.F. Shroder, Jr).

Shroder Jr., J.F., Lowndes, T., 1989a. Threshold modeling, slope failure and the Bonneville Flood. Geol. Soc. Am. Abs. Prog. 21, A60.

Shroder Jr., J.F., Lowndes, T., 1989b. Slope Failures and the Bonneville Flood in Red Rock Pass, Idaho. 190 (Association of American Geographers, Baltimore, MD, Program Abstract).

Shroder Jr., J.F., Sewell Jr., R.E., 1981. Diamicton Differentiation; La Sal Mountains and High Plateaus, Utah, and Red Rock Pass, Idaho. 553 (Geological Society of America Abstracts with Programs).

Chapter 5

The Bonneville Shoreline: Reconsidering Gilbert's Interpretation

C.G. Oviatt* and P.W. Jewell†

*Kansas State University, Manhattan, KS, United States

†University of Utah, Salt Lake City, UT, United States

ABSTRACT

G.K. Gilbert thought that the Bonneville shoreline marked the highest elevation Lake Bonneville reached, limited by the elevation of the overflow point on the basin divide, and that the lake did not spill across the divide for a prolonged period prior to the Bonneville flood. Many researchers who have followed Gilbert have stated that the lake overflowed for a prolonged period (hundreds of years, or longer) after reaching its highest elevation and prior to the Bonneville flood, but observations of the shoreline geomorphology at a number of localities in the basin support Gilbert's interpretation. In some places the Bonneville shoreline appears to consist of a broad erosional platform, and this appearance may have led to the interpretation that the lake overflowed for a prolonged period prior to the flood. However, the geomorphic features that appear to be erosional platforms, are instead depositional platforms that consist of sediment deposited offshore beneath steep erosional cliffs cut in bedrock or in pre-Bonneville surficial deposits. Depositional platforms were deposited at higher and higher elevations as Lake Bonneville transgressed, and the waves gradually cut higher and deeper into the rocks and deposits that formed the pre-Bonneville piedmont landscape. The Bonneville flood occurred as soon as the transgressing lake reached the low point on the alluvial-fan dam just north of Red Rock Pass, Idaho. Because field observations do not support the view that the lake overflowed at the Bonneville shoreline for a prolonged period prior to the Bonneville flood, and those observations do support Gilbert's original hypothesis, a return to Gilbert's interpretation of lake history at the Bonneville shoreline is appropriate.

Keywords: Lake Bonneville, G.K. Gilbert, Shoreline geomorphology

5.1 INTRODUCTION

The highest shoreline of Lake Bonneville (Fig. 5.1a), named by Gilbert (1890) the Bonneville shoreline, is locally well preserved on the landscape. The processes

Developments in Earth Surface Processes, Vol. 20. http://dx.doi.org/10.1016/B978-0-444-63590-7.00005-6

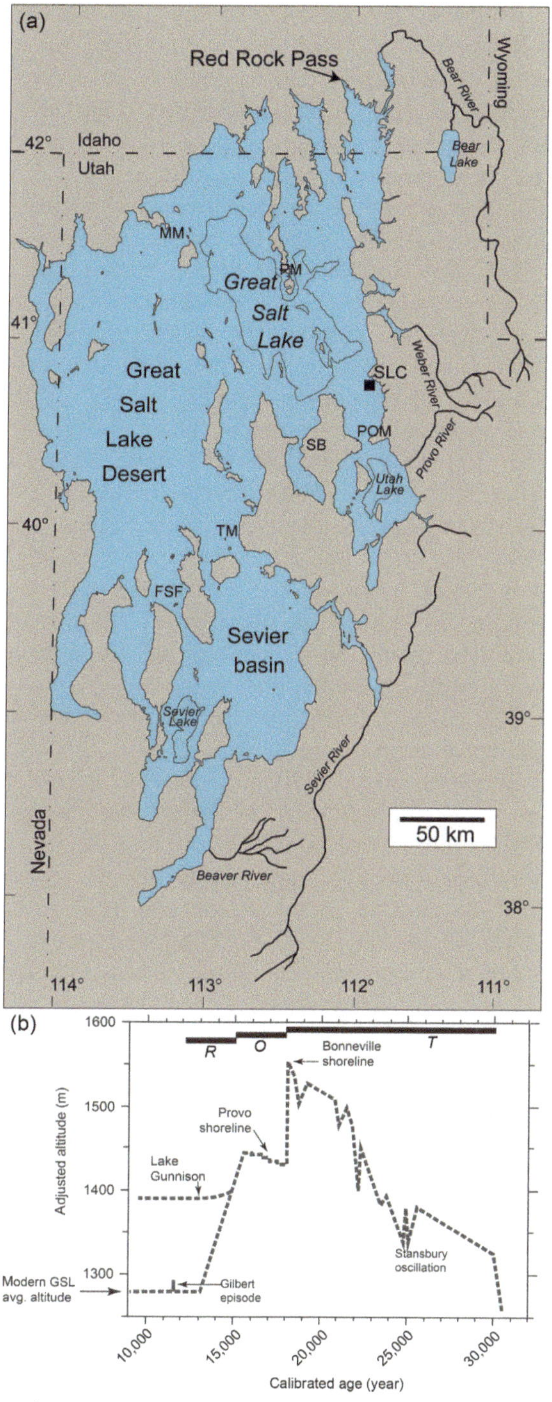

FIG. 5.1 See legend on next page.

that formed the Bonneville shoreline have been of great interest since Gilbert's time, and many suggestions have been proposed for its origin and history. The purpose of this chapter is to point out that the available geomorphic evidence supports Gilbert's hypothesis about its formation. Specifically, he thought that the Bonneville shoreline marked the level where the transgressing lake reached its highest elevation, where the lake stopped rising because it had reached the low point on the basin rim and began to overflow to the Snake River drainage basin, then dropped almost immediately during the catastrophic Bonneville flood to a more stable threshold elevation. Most researchers since Gilbert have replaced his interpretation with the idea that lake level stabilized for a prolonged period while the Bonneville shoreline formed. This chapter includes a description of the geomorphology of the Bonneville shoreline and the evidence for rising lake level (not stabilization) as the Bonneville shoreline formed, and a suggestion that it is now time to return to the interpretation of Gilbert (1890).

The following brief account of Lake Bonneville history is summarized from various sources (Gilbert, 1890; Oviatt, 2015) (Fig. 5.1b). About 30 cal ka BP Lake Bonneville began to rise above elevations similar to the average level of historic Great Salt Lake. The long-term trend of the transgressing lake in the closed basin was to rise, although it oscillated and fluctuated repeatedly (Fig. 5.1b), until about 18 cal ka BP when it reached the lowest elevation on the basin rim, the crest of the Marsh Creek alluvial fan just north of Red Rock Pass, Idaho, and began to overflow into the Snake River drainage. The ensuing Bonneville flood caused lake level to rapidly drop ~125 m (Miller et al., 2013) as the natural dam was washed out during the catastrophic Bonneville flood (Malde, 1968; O'Connor, 1993; O'Connor, 2016; Shroder et al., 2016). Following the flood, continued overflow at Red Rock Pass kept lake level relatively stable for possibly as long as 3 cal ka as the Provo shoreline formed, although lake level probably rose about 15 m during this period because of a rise in the overflow threshold elevation (Miller et al., 2013; Shroder et al., 2016). The lake returned to closed-basin hydrology after about 15 cal ka BP as climate dried (Godsey et al., 2011), and by about 13 cal ka BP it had dropped to elevations similar to those of historic Great Salt Lake.

FIG. 5.1 (a) Map of Lake Bonneville at its highest elevation, the Bonneville shoreline. *FSF*, Fish Springs Flat; *MM*, Matlin Mountains; *PM*, Promontory Mountains; *POM*, Point of the Mountain (Steep Mountain bench); *SB*, Stockton Bar; *SLC*, Salt Lake City; *TM*, Table Mountain. (b) Generalized chronology of the Bonneville lake cycle. Elevations are adjusted for differential isostatic rebound in the basin (Oviatt, 2015). *T*, transgressive phase; *O*, overflowing phase; *R*, regressive phase. Lake Gunnison is in the Sevier basin. *Panel a: After Currey, D.R., 1982. Lake Bonneville: Selected Features of Relevance to Neotectonic Analysis. U.S. Geological Survey Open File Report 82–1070. Panel b: Modified from Oviatt, C.G., Currey, D.R., Sack, D., 1992. Radiocarbon chronology of Lake Bonneville, eastern Great Basin, USA. Palaeogeogr. Palaeoclimatol. Palaeoecol. 99, 225–241; Oviatt, C.G., 2015. Chronology of Lake Bonneville, 30,000 to 10,000 yr B.P. Quat. Sci. Rev. 110, 166–171.*

The isostatic response of the basin to the Lake Bonneville water load is not addressed in this chapter (see Gilbert, 1890; Bills et al., 2002; Adams and Bills, 2016). A factor that should be considered, but so far has not been precisely enumerated, is the isostatic subsidence of the basin that must have occurred contemporaneously with the formation of shorelines during the late transgressive phase. Isostatic subsidence would have been greatest near the center of the basin where the water load was greatest, and least around the margin of the basin where the water load was minimal. Near the center of the basin the effect would have caused a component of transgression independent of the ongoing basin-wide transgression caused by climate change. D.R. Currey (Burr and Currey, 1988) anticipated this effect in his interpretation of the geomorphology of the Stockton Bar. The geomorphic and sedimentologic effects of isostatically induced transgression have not yet been distinguished from the effects of climatically induced transgression. The purpose of this chapter is to point out that geomorphic evidence of lake-level stabilization at the Bonneville shoreline has not been observed, and that a long-term trend of transgression, whatever the cause, and an immediate lake-level drop when the lake reached the overflow point on its basin rim, is an adequate explanation of the observed geomorphology.

5.2 PREVIOUS WORK

Gilbert (1890) hypothesized that the Bonneville flood occurred as soon as Lake Bonneville attained an elevation high enough to spill across the Marsh Creek alluvial-fan dam: "Uncemented alluvium is easily and rapidly torn up and removed, and as soon as a current began to flow across the divide, it must have commenced the excavation of a channel…" (Gilbert, 1890, p. 175). He also wrote "… the highest shore-mark of the series [the Bonneville shoreline] is conspicuous by reason of the contrast there exhibited between land sculpture and littoral sculpture. We now know that the height of the Bonneville shore-line was determined in a certain sense by overflow, since a discharge limited the rise of the water; but the carving of the shore was essentially completed before the discharge; and as soon as that began, the water level fell" (Gilbert, 1890, p. 171). Gilbert's fig. 34 (Gilbert, 1890, p. 262; Fig. 5.2a) shows the lake rising to its highest level, determined by the elevation of the crest of the Marsh Creek alluvial fan, and dropping immediately when the Bonneville flood commenced.

The idea that Lake Bonneville overflowed for a prolonged period prior to the Bonneville flood has been in the literature for a long time; the earliest publication we have found is that of Pack (1939). "Here ["one thousand feet above the present level"—at the Bonneville shoreline] it [the transgressing lake] came to a halt and remained stationary for a very long time, long enough, indeed, to construct a terrace much larger than it had done at any of the lower levels" (Pack, 1939, p. 31). He also wrote: "It is well to correct the popular notion that at this stage the lake had an outlet to the sea and that

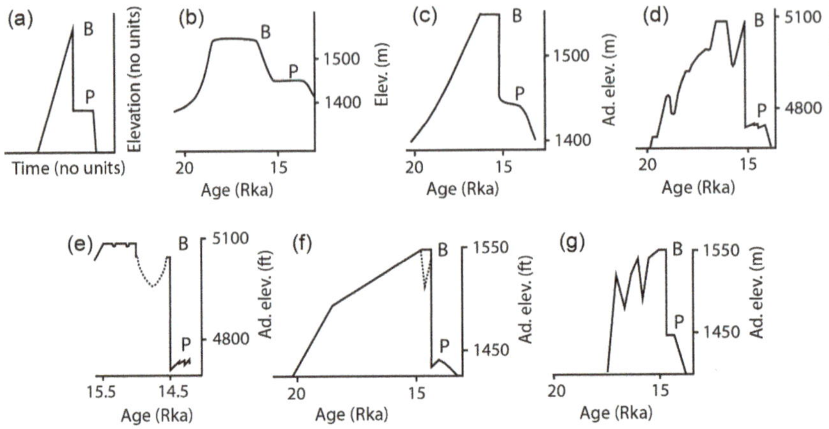

FIG. 5.2 Interpretations of Lake Bonneville chronology during the time of development of the Bonneville shoreline, Bonneville flood, and the initiation of the Provo shoreline. In all drawings, *B*, Bonneville shoreline; *P*, Provo shoreline; *Rka*, thousands of radiocarbon BP; time flows from left to right in all drawings; ad. elev., elevation (altitude) adjusted for differential isostatic rebound in the basin. (a) Gilbert (1890); (b) Bright (1963) (note vertical scale is unadjusted elevation); (c) Scott et al. (1983); (d) Currey and Oviatt (1985); (e) Currey and Burr (1988); (f) Oviatt et al. (1992); and (g) Oviatt (1997).

its surface remained stationary because of overflow. The truth is that the lake continued at this level, as it has done at previous levels, purely as the result of balance between intake and evaporation" (Pack, 1939, p. 33).

Williams (1952) wrote: "The well-developed shore features throughout the Bonneville Basin at the highest level of the lake seem to require that the highest lake was maintained at constant level by outflow."

Bright (1963, p. 131) wrote "the still stand [of Lake Bonneville at its highest level] lasted until the sill was reduced in length, and then the lake lowered rapidly to the Provo level" (Fig. 5.2b). In this view, the length of the sill (the overflow threshold on the Marsh Creek alluvial fan) was shortened by headward erosion by the overflowing river across the fan.

Malde (1968) and INQUA (1965) thought that Lake Bonneville had risen to an elevation higher than the Bonneville shoreline before the lake dropped catastrophically to the Bonneville shoreline in the Bonneville flood, and then stabilized to form the shoreline. "When the lake level fell [during the flood] to bedrock at the Bonneville shoreline, the resistant rocks impeded erosion, and discharge dwindled" (Malde, 1968, p. 9). He did not elaborate on how long he thought the lake lingered at the elevation of the Bonneville shoreline.

Scott et al. (1983) thought the lake overflowed for about 1000 year before catastrophically dropping during the Bonneville flood (Fig. 5.2c). They thought southward-directed headward erosion by the overflowing river on the northward-sloping surface of Marsh Creek alluvial fan had intersected the lake and caused catastrophic discharge (the Bonneville flood) to wash out the dam (Scott et al., 1983, pp. 276–277). In their interpretation "the 10- to 20-m-wide wave-cut benches that were locally cut in bedrock at the Bonneville shoreline require a stand of some duration at the highest level" (Scott et al., 1983, p. 277).

Currey's ideas evolved through time as information became available to him, and his interpretations became fairly elaborate (Fig. 5.2d and e) (Currey, 1990; Currey and Oviatt, 1985; Oviatt et al., 1992; Godsey et al., 2005). In general he thought the lake reached its highest elevation and began to overflow the basin divide on the Marsh Creek alluvial fan at about 15 ^{14}C ka BP, and that the Bonneville flood occurred at about 14.5 ^{14}C ka BP. Currey partly based his conclusion of a prolonged period of overflow (in this case, 0.5 ^{14}C ka) on his geomorphic interpretation of the Stockton Bar, a large depositional feature (Fig. 5.1a; Gilbert, 1890). Currey's hypothesis was that the three highest barriers and spits at the Stockton Bar formed while the lake was overflowing and the lake basin was isostatically subsiding in response to the water load (Burr and Currey, 1988, 1992; Currey and Burr, 1988).

Other authors who have thought that a prolonged period of overflow and a stable lake level were necessary for the formation of the Bonneville shoreline include Ives (1948), Oviatt et al. (1992) (Fig. 5.2f), Oviatt (1997) (Fig. 5.2g), and Janecke and Oaks (2011). None of these authors presented compelling geomorphic, stratigraphic, or geochronologic evidence for their conclusions.

In this chapter, we describe erosional slopes and depositional platforms at the Bonneville shoreline. Well-developed barriers and spits at the Bonneville shoreline, including features Gilbert (1890) referred to as "embankments," might also yield geomorphic information relevant to the overflow history of the lake at the Bonneville shoreline, but they are not discussed in this chapter.

5.3 METHODOLOGICAL APPROACH AND SHORELINE GEOMORPHOLOGY

Fig. 5.3 shows basic geometric relationships and definitions of geomorphic terms used in this chapter. Note that some geomorphic terms, such as "platform" and "slope," require modifying adjectives, such as depositional or erosional, if they are used as interpretative terms. This chapter is not intended to be a thorough treatment of the geomorphology of erosional shorelines in lakes. The discussion below is given as a partial explanation of our thinking about what evidence to look for if one were to conclude there had been a prolonged period of overflow and stable lake level while the Bonneville shoreline was forming. Further work is clearly required to fully document and test the hypotheses presented here.

In this chapter, we make the distinction between "erosional slope," "erosional platform," and "depositional platform." We use the term "erosional slope" for a slope formed by erosion of bedrock or some other surficial material, such as alluvium, which has a gradient greater than 1 degree (a tangent

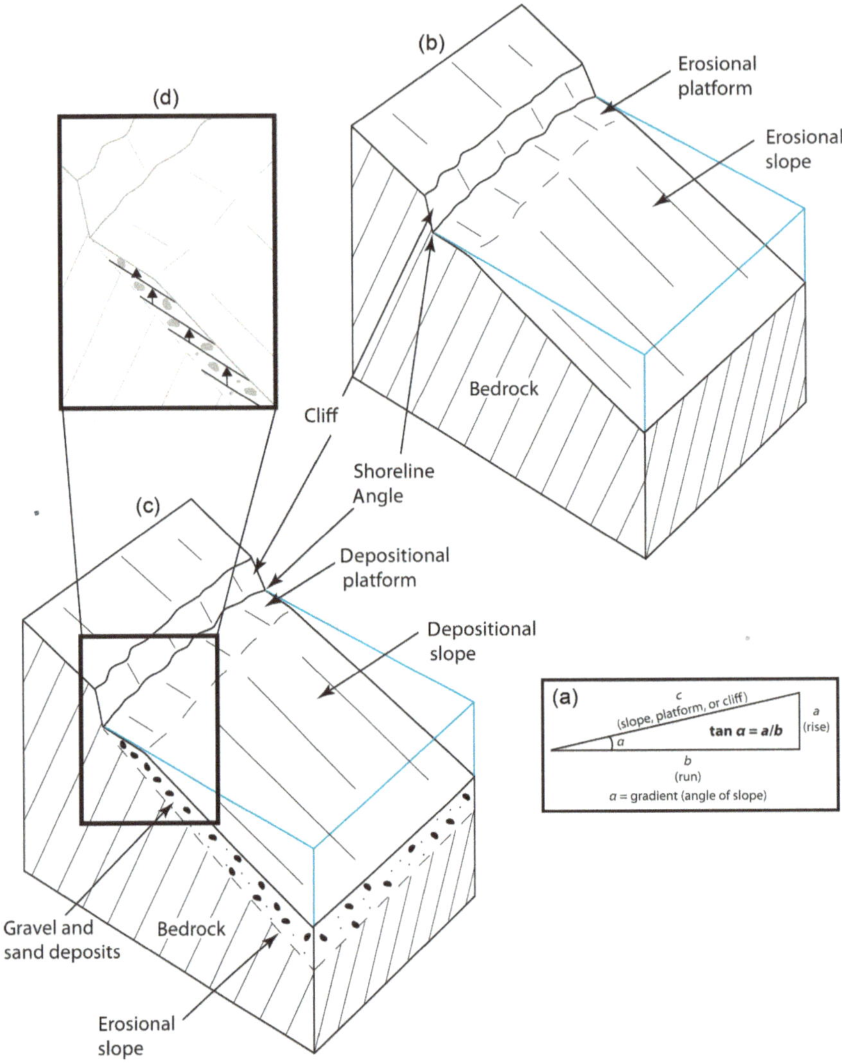

FIG. 5.3 Geomorphic and geometric relationships of lacustrine landforms discussed in this chapter (a). Schematic block diagrams show an erosional platform and erosional slope (b), a depositional platform and slope (c), and a view of the part of block diagram c showing hypothetical earlier positions of the depositional platform and successively higher platforms produced as the lake transgressed (d). The uppermost depositional platform, the one we see today, is the last platform to be produced and preserved—if the overflow threshold had been higher and the lake had transgressed to that threshold, the uppermost platform would have been higher.

of >0.017); in some cases the gradient of erosional slopes is greater than 10 degrees (Table 5.1). The term "erosional platform" is used for a surface cut by waves on bedrock or some other surficial material with a very low gradient, less than 1 degree (a tangent of <0.017). One degree is used in this chapter as

TABLE 5.1 Examples of Erosional Slopes in the Bonneville Basin

Location	Surficial Material	Gradient (Tangent)	Gradient (°)	USGS 7.5 min Quadrangle	Contour Interval (ft)	Longitude °W	Latitude °N
Promontory Mountains	Quartzite bedrock	0.0833	4.8	Pokes Point 1968	20	112.477	41.264
So. of Tooele, UT	Quartzite bedrock	0.0666	3.8	Stockton 1980	40	112.316	40.508
So. of Tooele, UT	Alluvium	0.0555	3.2	Tooele 1955	20	112.330	40.490
Simpson Mountains, NW piedmont	Alluvium	0.0555	3.2	Coyote Springs 1955	40	112.804	39.979
North slope of Table Mountain	Quartzite bedrock	0.2222	13	Table Mountain 1955	40	112.889	39.942
North slope of Table Mountain[a]	Quartzite bedrock	0.0489	2.8	Table Mountain	–	112.896	39.948
Southern Fish Springs Flat[a]	Volcanic bedrock	0.0419	2.4	Sand Pass NE	–	113.261	39.660

[a]*Examples of platforms at the Provo shoreline, surveyed in the field with hand-level and rod. Note that these platforms, which have a gradient of greater than 1 degree, formed during a period of overflow across the Red Rock Pass threshold, but that the overflow threshold elevation rose by 15 m while the lake was overflowing (Miller et al., 2013). Lake level rose throughout the basin by that amount, although the details of lake-level history, including the effects of isostatic deformation, during overflow have not yet been deciphered.*

an arbitrary boundary between erosional platforms and erosional slopes because most marine erosional platforms have gradients of less than 1 degree (Bradley, 1958; Bradley and Griggs, 1976; Pethick, 1984). We assume that: (1) most erosion by waves, particularly erosion of resistant bedrock, such quartzite, takes place in the zone where waves break; and (2) large waves move water at depths of many meters with enough velocity to move sediments, but not with enough velocity to erode bedrock.

An "erosional platform" is equivalent to an abrasion platform or a wave-cut platform in a marine setting. Erosional slopes and erosional platforms intersect the much steeper cliff (sea cliff) with an abrupt increase in gradient, referred to as the shoreline angle (Davis, 1932; Adams and Wesnousky, 1998) (Fig. 5.3b and c). In most cases, the shoreline angle for the shoreline of a Pleistocene lake, which marks the former still-water plane (Gilbert, 1890, Fig. 5.3), is buried by postlake colluvium. A "depositional platform" looks similar to an "erosional platform" but is made of sediments—it is a constructional landform rather than an destructional landform.

In marine environments shoreline erosional processes include abrasion, mechanical wave erosion, weathering, and solution (Pethick, 1984). The growth of ice crystals in cracks, and abrasion by clasts frozen into shore ice that is pushed and heaved by waves and tides (Stone et al., 1996), are also likely processes in areas where temperatures are cold enough. These potential processes will produce erosional platforms or erosional slopes in lakes, and are affected by such factors as salinity, temperature, stability of water level, and wave energy. An erosional platform can form when lake level remains stable for a sufficiently long period, depending on such factors as wave strength and bedrock resistance (Gilbert, 1885, 1890; Pethick, 1984). Following from the hypothesis presented in this chapter, an erosional slope is more likely to form where erosive energy is high but lake level is not constant, typical conditions in closed-basin lake basins. During the transgressive phase of Lake Bonneville the long-term trend of lake level was to rise, punctuated by numerous lake-level oscillations and fluctuations (Fig. 5.1b). Periods of stabilized lake level, if they occurred, would have been short-lived.

Lake level in a closed basin is determined by water balance, the ratio of water inputs to outputs plus or minus storage changes. Inputs can be from direct precipitation, surface water, or groundwater, and outputs are by evaporation or groundwater. All these are determined by weather and climate, which are constantly changing, so that water balance and lake level (a measure of storage change) is constantly changing. An overflowing or open-basin lake maintains a relatively constant level, with small lake-level changes occurring in response to changes in the rate of discharge across the overflow threshold. Based on the thinking presented above, well-developed erosional platforms can only form where lake level is stable long enough for waves to cut into the bedrock or other surficial material at a constant elevation so that the erosional cliff intersects with a low-gradient erosional platform, the width of which is dependent on such factors as time, wave strength, bathymetry, and erosional resistance. It is unlikely that well-developed erosional platforms will form in closed-basin lake settings because the amplitude of lake-level changes can be on the order of 10s of meters or more, and lake level is constantly fluctuating (for example, see the historic record of water-level changes in Great Salt Lake; Arnow and Stephens, 1990). Incoming waves with their tremendous erosive power can erode large volumes of rock, forming cliffs that retreat to higher elevations as the lake rises during a long-term transgression. Erosional slopes below the cliffs may or may not be covered by sediments generated in the surf zone (Fig. 5.3c).

We have observed narrow erosional platforms in the Bonneville basin that formed during the transgressive phase on soft (nonresistant) materials, such as mudstones or unconsolidated alluvium. An example of this is exposed in the wall of a gully below the Bonneville shoreline along the mountain front near the University of Utah hospital in Salt Lake City (40.773°N, 111.832°W). At this site, a platform cut approximately 5 m below the Bonneville shoreline on the Triassic Woodside Shale (Van Horn, 1969) is overlain by coarse gravel and boulders. Presumably, the platform formed on the soft shale late in the transgressive phase as the lake was approaching its highest elevation and maintained a relatively constant elevation for a short time. Harder rocks, such as limestones, in the same area do not show evidence of erosional platform formation.

It might be reasonable to expect well-developed erosional platforms to be present at the Provo shoreline because that shoreline formed during a long period of overflow at Red Rock Pass, possibly as long as 3 ka (Gilbert, 1890; Godsey et al., 2011; Miller et al., 2013; Oviatt, 2015). However, the overflow threshold rose by at least 15 m while the Provo shoreline was forming (Miller et al., 2013), so even the Provo shoreline formed while lake level was not stable. Topographic profiles of well-developed platforms at the Provo shoreline, one at Table Mountain cut on quartzite bedrock, and one in Fish Springs Flat cut on volcanic bedrock, have gradients of 2.8 degree and 2.4 degree, respectively (Table 5.1). These platforms do not fit the definition of erosional platform used in this chapter (that erosional platforms have a gradient less than 1 degree). Because of the isostatic response of the basin (rebounding) to the nearly instantaneous unloading of 125 m of water during the Bonneville flood and the rise of the overflow threshold by 15 m during this time (Miller et al., 2013), the response of lake level was complicated and not constant. The effects of these processes on Provo shoreline geomorphology have not yet been determined. We have not observed any erosional platforms at the Bonneville shoreline.

Depositional platforms form where waves have eroded the bedrock or other surficial materials, and the resulting erosional slopes are overlain by sediment derived from wave erosion at a higher elevation. The general pattern of unconsolidated sediment overlying an erosion surface is similar to that of a fluvial strath terrace where alluvium overlies an erosion surface cut on bedrock. As lake level rises, sediment is swept offshore to aggrade where wave energy is less than in the breaker zone. From a distance, depositional platforms look like erosional platforms, but on close examination it can be seen that depositional platforms are composed of sediments, and that they are not erosion surfaces. They do not represent prolonged lake-level stabilization—rather, they represent the uppermost level of the aggrading sediment in a rising lake. As the lake transgressed to higher elevations, the depositional platform rose with it (Fig. 5.3d). As with erosional slopes, depositional platforms are marked by shoreline angles (Fig. 5.3c). In many cases it is not easy to distinguish between depositional platforms and erosional platforms, but careful field investigations of exposures in gullies and outcrops of the materials that make up a platform can reveal their characteristics and origin.

Net erosion at the shore of a closed-basin lake is very effective in some places, in part because the wave zone crosses the landscape multiple times as lake level fluctuates during a transgression. Once rock is removed it is not replaced. Wave erosion, at places where wave energy was very high in Lake Bonneville, removed a tremendous amount of bedrock and alluvial-fan gravel. For example, Lifton et al. (2015) estimated that "several dozen meters," representing about 1.2×10^7 $m^3$of quartzite (S. Marrero, written communication via Lifton et al., 2015), were removed by the transgressing lake on one of the spurs of the Promontory Mountains (at the Pokes Point locality in Table 5.1).

5.4 THE BONNEVILLE SHORELINE

Examples of short segments of the Bonneville shoreline are shown in Fig. 5.4. In each of these cases a visual interpretation of the image leads the observer to the conclusion that an erosional platform, cut into bedrock, characterizes the shoreline segment. However, on the ground it can be seen that the platforms are made up of sediments—the cliffs are erosional but the platforms are depositional. At each site a tremendous volume of bedrock was eroded by waves as the lake gradually rose. The erosional slope thus produced, was buried by coarse sediment as the surf zone moved upward and water depth increased.

Table Mountain (Fig. 5.4a) is on the southern margin of the main body of the lake, and large waves approached the shore from the north or northwest. As the lake transgressed, sediments produced by wave erosion on the north slope of the mountain were moved laterally by long shore currents into large depositional landforms, primarily on the east side of the mountain (Gilbert, 1890, plate XXV) (Fig. 5.4a), or were swept offshore by the undertow to depths where wave agitation was not sufficient to transport the debris. Large boulders, some over a meter in diameter, coarse-grained gravel in thin accumulations, and scattered bedrock outcrops, are present on the surface of the erosional slope. As lake level rose the water depth at any one point on the erosional slope became deeper and deeper, and coarse sediment

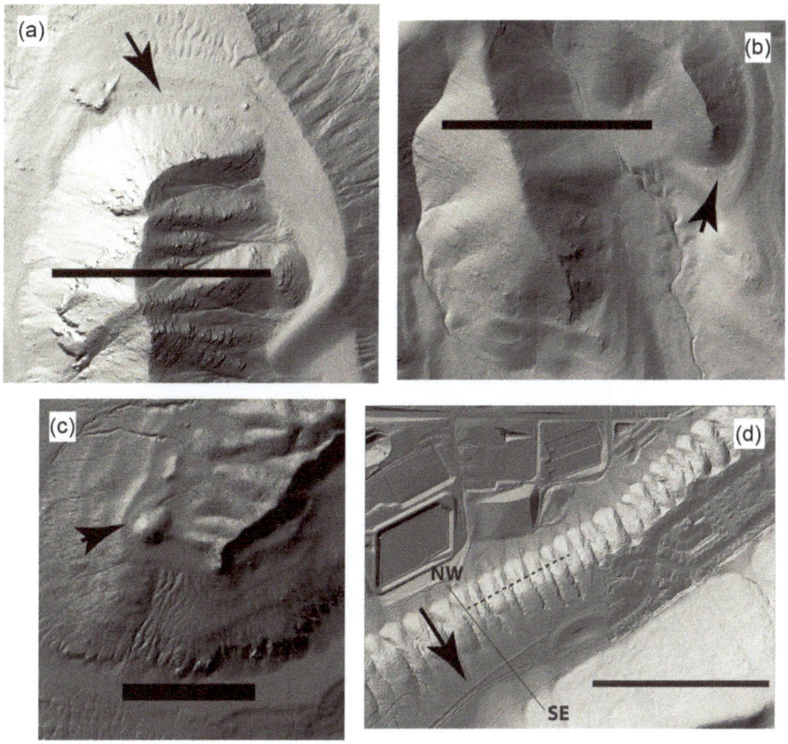

FIG. 5.4 LiDAR images showing examples of the Bonneville shoreline. In each of these examples the *arrow* marks a platform that could be interpreted as an erosional platform, but that is actually either an erosional slope (as in a) or a depositional platform (as in b, c, and d). The *black bar* in each photograph is 0.5 km. Refer to Fig. 5.1a for general locations. North is to the top in each image. (a) Table Mountain (39.943°N, 112.889°W); (b) Promontory Mountains (41.455°N, 112.456°W); (c) Matlin Mountains (41.585°N, 113.319°W); (d) Steep Mountain bench, directly east of Point of the Mountain (40.472°N, 111.895°W). The *dashed line* in d marks the approximate contact between the underlying tectonically shattered quartzite bedrock and the overlying sand and fine gravel deposited in Lake Bonneville. Refer to Fig. 5.5 for the profile along line NW–SE, and a photograph of the outer edge of the platform.

at that point was preserved. The average gradient of the erosional slope on the north flank of Table Mountain is 13 degree (Table 5.1).

At the sites on the Matlin Mountains and the Promontory Mountains (Fig. 5.4b and c), platforms that appear on initial examination to be erosional platforms are in fact, depositional platforms. These sites are typical of Bonneville shoreline localities throughout the basin.

At Steep Mountain, just to the east of Point of the Mountain at the south end of Salt Lake Valley (Fig. 5.1a and d), a depositional platform is well displayed. This platform has been commonly regarded as a good example of a well-developed erosional platform at the Bonneville shoreline (Shelton, 1966,

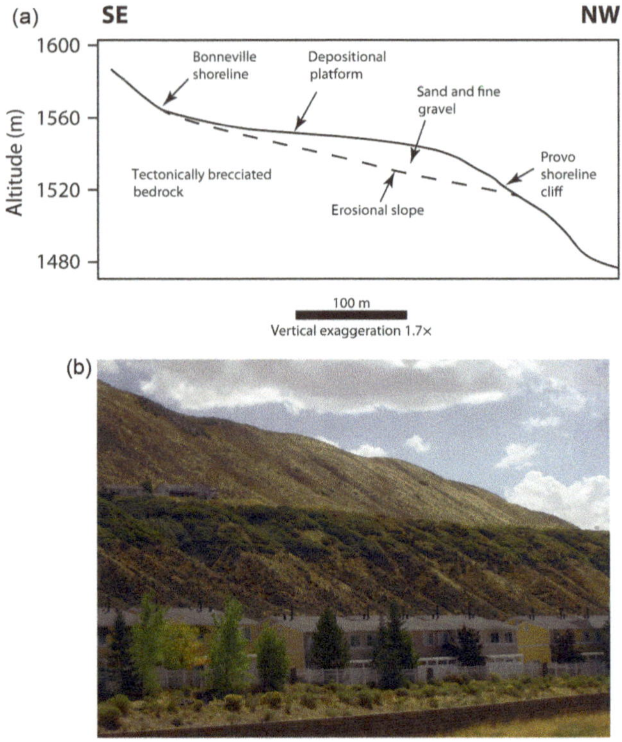

FIG. 5.5 (a) Topographic profile and geologic cross section across Steep Mountain bench, directly east of Point of the Mountain (refer to Figs. 5.1a and 5.4d for profile location). The shoreline angle is at the Bonneville shoreline. (b) Photograph of the Steep Mountain bench from the northeast. The contact at the base of the sand, where it overlies tectonically shattered bedrock, is marked by the lower limit of green shrubby vegetation on the dissected cliff of the Provo shoreline. The upper houses are on the depositional platform below the Bonneville shoreline, and the lower houses are on alluvium and colluvium overlying the erosional surface at the Provo shoreline. The deposit of sand and fine gravel at the outer margin of the depositional platform is about 20 m thick.

fig. 133), however the platform is not erosional in origin, it is depositional. A topographic profile and geologic cross section across the platform at this site (Fig. 5.5a) shows the cliff face, the shoreline angle at the Bonneville shoreline, the depositional platform composed of sand and fine gravel, and the erosional slope (in the subsurface) cut on tectonically shattered quartzite (Pennsylvanian Bingham Mine Formation of the Oquirrh Group; Biek, 2005). The contact between the Bonneville sand and gravel and the underlying easily eroded bedrock can be seen on the face of the cliff above the Provo shoreline from places nearby and from the valley floor many kilometers away because it is marked by the lower limit of shrubby vegetation, which thrives in the sandy sediment of the depositional platform but not in the underlying bedrock below the erosional slope (Fig. 5.5b). The sand and gravel are about 20 m thick near the outer edge of the depositional platform and thin to a feather edge at the shoreline angle at the Bonneville shoreline.

Lake Bonneville shorelines formed during the transgressive phase, including the shorelines Gilbert (1890) referred to as the intermediate shorelines, were formed quickly while lake level was unsteady in the closed basin. The Bonneville shoreline has more in common geomorphically and genetically with the intermediate shorelines than with the Provo shoreline.

5.5 BASIN-WIDE CONSISTENCY

The hypothesis that the Bonneville shoreline marks the level of Lake Bonneville when it reached its highest elevation in the basin and began to overflow across the lowest point on the basin rim, and that the lake did not overflow for a prolonged period prior to the flood, is consistent with other information from the basin. We have traversed much of the basin on numerous field excursions during the last four decades and we have not seen stratigraphic or geomorphic evidence for prolonged overflow at the Bonneville shoreline.

Geochronometric information does not support the idea of prolonged overflow. Two-sigma ranges of radiocarbon ages that limit the age of the late transgressive phase when the lake was within 7 m of its highest elevation overlap with two-sigma age ranges of radiocarbon ages from the Provo shoreline, leaving little or no time for prolonged overflow, although radiocarbon ages in this time interval are not precise enough to provide an unequivalent answer to the question of whether or not overflow was prolonged (Oviatt, 2015).

The geomorphology and geology of alluvial fans, Neogene sedimentary deposits (eg, the Salt Lake Formation), and massive landslides at the overflow threshold in the Red Rock Pass area are consistent with Gilbert's interpretation. This includes round-headed valleys on the north slope of the Marsh Creek alluvial fan that are probably groundwater-sapping features (O'Connor, 1993; Shroder et al., 2016). Hydraulic heads on groundwater in the unconsolidated deposits of the fan and landslide would have been much greater on the southern, lake side of the fan than on the northern side of the fan, and would have caused groundwater flow to increase through the permeable sediments as lake level rose. When the transgressing lake finally reached the topographically low point on the crest of the alluvial fan, overflow would have caused rapid erosion of the fan sediment, and lake level would have dropped catastrophically as the Bonneville flood ensued.

5.6 SUMMARY

The Bonneville shoreline marks the upper limit of the transgressing Lake Bonneville. From a distance, broad platforms at the Bonneville shoreline appear to be abrasion platforms, but close examination shows them to be depositional platforms. They formed as the rising lake eroded bedrock or other surficial materials, such as alluvial fans, then swept debris basin ward where the coarsest fragments accumulated and aggraded at the outer limit of the surf zone and smaller particles were moved farther offshore and down slope.

The view that appeared after Gilbert's (1890) observations and interpretations has been that the Bonneville shoreline formed during a period of stable lake level caused by prolonged overflow. However, Gilbert's interpretation that the Bonneville flood occurred as soon as the lake began to spill into the Snake River drainage basin is consistent with available evidence. Further work should focus on this question. For the time being, we suggest it would be appropriate to return to Gilbert's interpretation. Future work will help determine the relative importance of transgression caused by isostatic subsidence in response to the Lake Bonneville water load in the formation of erosional slopes and depositional platforms at the Bonneville shoreline.

ACKNOWLEDGMENTS

Field work to obtain information discussed in this chapter was partially supported by the Utah Geological Survey and by NSF grant EAR1053129. We are grateful to K. Adams, T. Fisher, and D. Wilkins for their reviews that led to the improvements in the manuscript, and to numerous colleagues who, over many years, have shared field experiences and insights with us.

REFERENCES

Adams, K.D., Bills, B.G., 2016. Chapter 8. Isostatic rebound and palinspastic restoration of the Bonneville and Provo shorelines in the Bonneville basin, UT, NV, and ID. In: Oviatt, C.G., Shroder Jr., J.F. (Eds.), Lake Bonneville: A Scientific Update, vol. 20. Elsevier. pp. 145–164.

Adams, K.D., Wesnousky, S.G., 1998. Shoreline processes and the age of the Lake Lahontan high stand in the Jessup embayment, Nevada. Geol. Soc. Am. Bull. 110, 1318–1332.

Arnow, T., Stephens, D., 1990. Hydrologic Characteristics of the Great Salt Lake, Utah: 1847–1986. (U.S. Geological Survey Water-Supply Paper 2332, Reston, VA).

Biek, R.F., 2005. Geologic Map of the Jordan Narrows Quadrangle, Salt Lake and Utah Counties, Utah. (Utah Geological Survey Map 208, Salt Lake City, UT).

Bills, B.G., Wambeam, T.J., Currey, D.R., 2002. Geodynamics of Lake Bonneville. In: Gwynn, J.W. (Ed.), Great Salt Lake: An Overview of Change. Department of Natural Resources, Utah Geological Survey, Salt Lake City, pp. 7–32 (special publication).

Bradley, W.C., 1958. Submarine abrasion and wave-cut platforms. Geol. Soc. Am. Bull. 69, 967–974.

Bradley, W.C., Griggs, G.B., 1976. Form, genesis, and deformation of central California wave-cut platforms. Geol. Soc. Am. Bull. 87, 433–449.

Bright, R.C., 1963. Pleistocene Lakes Thatcher and Bonneville, Southeastern Idaho (Ph.D. thesis). University of Minnesota.

Burr, T.N., Currey, D.R., 1988. The Stockton Bar. In: Machette, M.N. (Ed.), In the Footsteps of G.K. Gilbert—Lake Bonneville and Neotectonics of the Eastern Basin and Range

Province, pp. 66–73 (Utah Geological and Mineral Survey Miscellaneous Publication 88-1, Salt Lake City).

Burr, T.N., Currey, D.R., 1992. Hydrographic Modeling at the Stockton Bar, pp. 207–219 (Utah Geological Survey Miscellaneous Publication 92-3, Salt Lake City, UT).

Currey, D.R., 1990. Quaternary paleolakes in the evolution of semidesert basins, with special emphasis on Lake Bonneville and the Great Basin, USA. Palaeogeogr. Palaeoclimatol. Palaeoecol. 76, 189–214.

Currey, D.R., Burr, T.N., 1988. Linear model of threshold-controlled shorelines of Lake Bonneville. In: Machette, M.N. (Ed.), In the Footsteps of G.K. Gilbert—Lake Bonneville and Neotectonics of the Eastern Basin and Range Province, pp. 104–110 (Utah Geological and Mineral Survey Miscellaneous Publication 88-1, Salt Lake City).

Currey, D.R., Oviatt, C.G., 1985. Durations, average rates, and probable causes of Lake Bonneville expansions, still stands, and contractions during the last deep lake cycle, 32,000 to 10,000 years ago. In: Kay, P.A., Diaz, H.F. (Eds.), Problems of and Prospects for Predicting Great Salt Lake Levels. University of Utah Center for Public Affairs and Administration, Salt Lake City, pp. 9–24.

Davis, W.M., 1932. Glacial epochs of the Santa Monica Mountains, California. Proc. Natl. Acad. Sci. U. S. A. 18, 659–665.

Gilbert, G.K., 1885. The topographic features of lake shores. U.S. Geol. Surv. Annu. Rep. 5, 75–123.

Gilbert, G.K., 1890. Lake Bonneville. (U.S. Geological Survey Monograph 1; 438 pp, Reston, VA).

Godsey, H.S., Currey, D.R., Chan, M.A., 2005. New evidence for an extended occupation of the Provo shoreline and implications for regional climate change, Pleistocene Lake Bonneville, Utah, USA. Quat. Res. 63, 212–223.

Godsey, H.S., Oviatt, C.G., Miller, D.M., Chan, M.A., 2011. Stratigraphy and chronology of offshore to nearshore deposits associated with the Provo shoreline, Pleistocene Lake Bonneville, Utah. Palaeogeogr. Palaeoclimatol. Paleoecol. 310, 442–450.

INQUA, 1965. Part H: Lake Bonneville. In: Guidebook for Field Conference E: Northern and Middle Rocky Mountains. International Union for Quaternary Research, VIIth Congress, pp. 104–107.

Ives, R.L., 1948. The outlet of Lake Bonneville. Sci. Month. 67, 415–426.

Janecke, S.U., Oaks Jr., R.Q., 2011. New insights into the outlet conditions of late Pleistocene Lake Bonneville, southeastern Idaho, USA. Geosphere 7 (6), 1369–1391.

Lifton, N., Caffee, M., Finkel, R., Marrero, S., Nishiizumi, N., Phillips, F.M., Goehring, B., Gosse, J., Stone, J., Schaefer, J., Theriault, B., Jull, A.J.T., Fifield, K., 2015. In situ cosmogenic nuclide production rate calibration for the CRONUS-earth project from Lake Bonneville, Utah, shoreline features. Quat. Geochronol. 26, 56–69.

Malde, H.E., 1968. The Catastrophic Late Pleistocene Bonneville Flood in the Snake River Plain, Idaho. (U.S. Geological Survey Professional Paper 596; 52 pp, Reston, VA).

Miller, D.M., Oviatt, C.G., McGeehin, J.P., 2013. Stratigraphy and chronology of Provo shoreline deposits and lake-level implications, late Pleistocene Lake Bonneville, eastern Great Basin, USA. Boreas 42, 342–361.

O'Connor, J.E., 1993. Hydrology, Hydraulics, and Geomorphology of the Bonneville Flood. (Geological Society of America Special Paper 274, Boulder, CO).

O'Connor, J., 2016. Chapter 6. The Bonneville flood —A veritable débâcle. In: Oviatt, C.G., Shroder Jr., J.F. (Eds.), Lake Bonneville: A Scientific Update, vol. 20. Elsevier. pp. 105–126.

Oviatt, C.G., 1997. Lake Bonneville fluctuations and global climate change. Geology 25, 155–158.

Oviatt, C.G., 2015. Chronology of Lake Bonneville, 30,000 to 10,000 yr B.P. Quat. Sci. Rev. 110, 166–171.

Oviatt, C.G., Currey, D.R., Sack, D., 1992. Radiocarbon chronology of Lake Bonneville, eastern Great Basin, USA. Palaeogeogr. Palaeoclimatol. Palaeoecol. 99, 225–241.

Pack, F.J., 1939. Lake Bonneville: a popular treatise dealing with the history and physical aspects of Lake Bonneville. Bulletin of the University of Utah, vol. 30(4) University of Utah Press, Salt Lake City.

Pethick, J., 1984. An Introduction to Coastal Geomorphology. Arnold/Chapman & Hall, New York. 260 pp.

Scott, W.E., McCoy, W.D., Shroba, R.R., Rubin, M., 1983. Reinterpretation of the exposed record of the last two cycles of Lake Bonneville, western United States. Quat. Res. 20, 261–285.

Shelton, J.S., 1966. Geology Illustrated. WH Freeman, San Francisco/London.

Shroder, J.F., Cornwell, K., Oviatt, C.G., Lowndes, T.C., 2016. Chapter 4. Landslides, alluvial Fans, and Dam Failure at Red Rock Pass: The Outlet of Lake Bonneville. In: Oviatt, C.G., Shroder Jr., J.F. (Eds.), Lake Bonneville: A Scientific Update, vol. 20. Elsevier. pp. 75–87.

Stone, J., Lambeck, K., Fifield, L.K., Evans, J.M., Cresswell, R.G., 1996. A lateglacial age for the Main Rock Platform, western Scotland. Geology 24 (8), 707–710.

Van Horn, R., 1969. Preliminary Geologic Map of the Southern Half of the Fort Douglas Quadrangle. (U.S. Geological Survey Open-File 69-308, Reston, VA).

Williams, J.S., 1952. Red Rock Pass, outlet of Lake Bonneville. Geol. Soc. Am. Bull. 63 (12), 1357–1376.

Chapter 6

The Bonneville Flood— A Veritable Débâcle

J. O'Connor
U.S. Geological Survey, Portland, OR, United States

ABSTRACT

The Bonneville Flood was one of the largest floods on Earth. First discovered by G.K. Gilbert in the 1870s during his inspection of the outlet at Red Rock Pass, it was rediscovered in the 1950s by Harold Malde and coworkers, leading to mapping and assessment of spectacular flood features along Marsh Creek, Portneuf River, and Snake River for over 1100 km between the outlet and Lewiston, Idaho. The cataclysmic flood—from the rapid ~115 m drop of Lake Bonneville from the Bonneville level to the Provo level—was nearly 200 m deep in places and flowed at a maximum rate of about 1 million m^3s^{-1}; about 100 times greater than any historical Snake River flood. Along its route the Bonneville Flood carved canyons and cataract complexes and built massive boulder bars. These flood features have been a rich source for understanding megaflood processes. Yet it still offers much more with new and developing techniques for hydrodynamic modeling and landscape analysis.

Keywords: Bonneville Flood, Snake River, Snake River Plain, Megaflood, Idaho

6.1 INTRODUCTION

In far southeastern Idaho on the morning of Wednesday, Aug. 16, 1876, G.K. Gilbert rides north from Oxford: "I can see the Red Rock and beyond it. It surely looks as though there has been a current through but it also looks as though the beach line at points nearby was higher than the pass." Riding closer..."The red rock is only one of a number that are here *exceptionally* bared, in testimony to the stream that has washed them in degrading this pass."

Thus in field notes written 140 years ago, Gilbert gives the first account of the Bonneville Flood.[a] He affirms the flood in his 1890 monograph, *Lake Bonneville*, suggesting the flow was as great as "the flood volume of the Missouri" (Gilbert, 1890, p. 177).

[a] Following Malde's (1968) usage, I capitalize "Flood."

But then the Bonneville Flood was lost for several decades. Not until Harold A. Powers and Harold E. Malde of the US Geological Survey (USGS) began mapping on the Snake River Plain in the mid-1950s were the spectacular flood features recognized downstream (Malde, 1960). The flood effects on the Snake River Plain were then wonderfully described and analyzed by Malde in his classic 1968 USGS Professional Paper, *The Catastrophic Late Pleistocene Bonneville Flood in the Snake River Plain, Idaho.*

The pace picked up after Malde's report, with many workers including myself assessing the flood and its deposits. Yet the flood remains intriguing for many reasons: its lost aspect is curious; it is one of the largest known freshwater floods on earth; and its huge depositional and erosional features are fantastic and cry out for more work.

6.2 "FOR A TIME THE OUTPOURING WAS A VERITABLE Débâcle"

On his 1876 ride to water-washed Red Rock Pass, Gilbert ponders the possibilities in two plausible scenarios: The first involves two different outlets, one higher and resistant to form the higher Bonneville shoreline but abandoned when tectonic tilting diverted flow to the Red Rock Pass outlet. This enabled overflow and incision of Red Rock Pass to the Provo shoreline level, ~115 m lower (from his measurement of the Bonneville shoreline 340 ft above the Red Rock Pass threshold [1890, p. 175] and an estimated 30 ft of post-flood valley fill at Red Rock Pass [p. 177]). The second hypothesis: "there was only one outlet, vis. that at Red Rock Pass," the strongly formed Bonneville shoreline a consequence of little overflow "while wave action will beat at its maximum hence a lingering under conditions favorable to beach carving…This critical period passed, the surplus of water becoming greater cuts the outway rapidly…"

Thus while still entertaining multiple hypotheses, Gilbert within a few hours of first seeing Red Rock Pass describes in the second scenario our modern understanding of conditions for the Bonneville Flood. After more field work in 1877 to discount the first hypothesis, he settles on the second, describing it on page 175 of his *Lake Bonneville* monograph:

> *The relations of these various features will be better understood by reference to the map in [Figs. 6.1 and 6.2]. The Bonneville shore-line is traceable continuously about Cache Valley to the vicinity of the pass. On the east side its most northerly vestige is upon a butte a mile south of Hunt's Ranch. …Its height above the marsh is between Marsh and Five-Acre creeks is 340 feet. The nearest point at which the Provo shore-line was observed is about eight miles farther south, in the vicinity of the town of Oxford.*
>
> *Marsh Creek issues from its canyon in the mountains about two and one-half miles east of the old channel. The intervening space is occupied by a sloping alluvial plain terminating in a bluff. It is evident that this is an alluvial fan or*

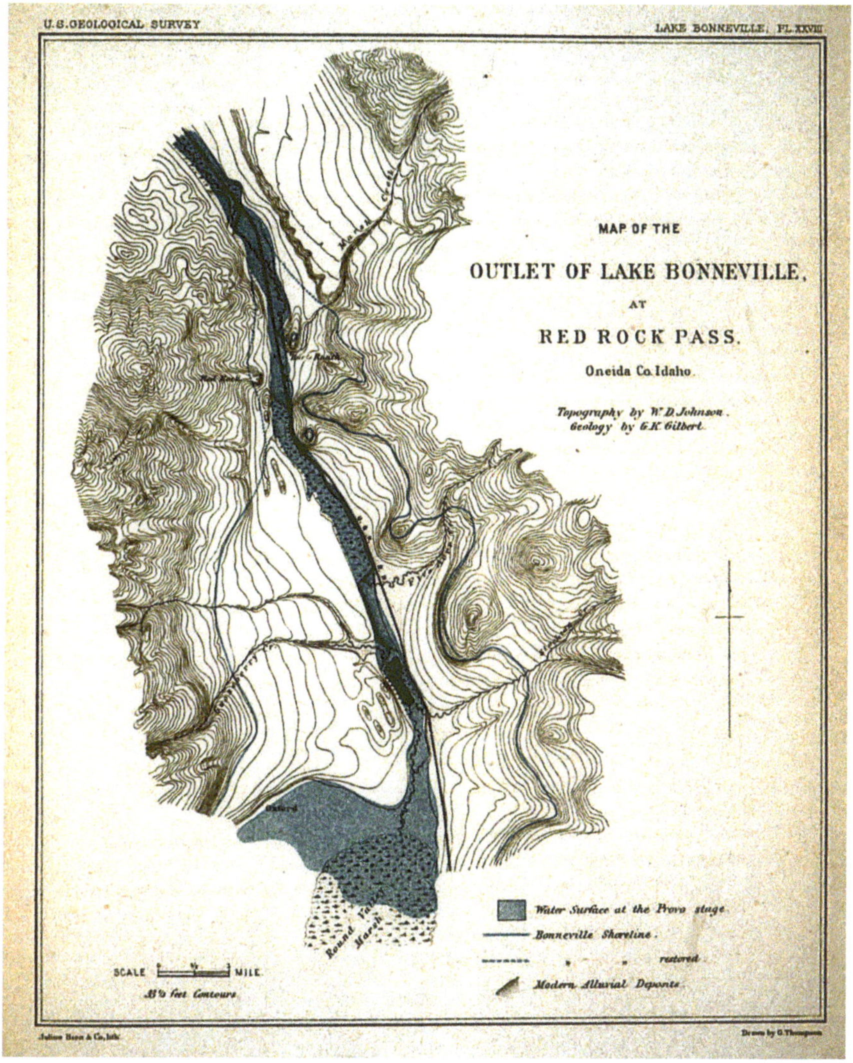

FIG. 6.1 Topographic map of Red Rock Pass (in the area of Red Rock and Hunts Ranch) by G.K. Gilbert and W.D. Johnson (Gilbert, 1890). The incised Marsh Creek fan slopes west in the north part of the map area. The stream labeled Five Acre Creek is now known as Swan Lake Creek, and most of the area in the map is now in Bannock County, Idaho.

alluvial cone constructed by the creek before excavation of the Bonneville outlet. It was afterward partially eroded by the outflowing river, and also by Marsh Creek, which has excavated a passage several hundred feet in depth.

Where this old alluvial plain approaches nearest to the Bonneville channel, its edge is fifty feet higher than the nearest terrace of the Bonneville shore, and

FIG. 6.2 Lithograph from Gilbert's (1890) Lake Bonneville monograph illustrating the view south from the surface of Marsh Creek fan (Fig. 6.1). Red Rock Pass is the area of bedrock outcroppings, Lake Bonneville exited through the gap toward the viewer. The hummocky terrain of the Red Rock landslide from the southwest encroaches into the flood-trenched channel of Marsh Creek.

> *a restoration of its profile indicates that it coalesced with slopes from the opposite mountain range at about the level of the Bonneville shore. A careful study of the ground has satisfied the writer that the base or outer margin of the alluvial cone was part of the ancient water parting, and was the point at which the out flow was initiated.*
>
> *The fact that the Bonneville water discharged at first over a barrier of alluvium instead of solid rock had much to do with the subsequent history of the lake. Uncemented alluvium is easily and rapidly torn up and removed, and as soon as a current began to flow across the divide, it must have commenced the excavation of a channel. As the channel increased, the volume of the escaping water became greater, and this increase of volume reacted on the power of erosion. In a short time a mighty river was formed, and the lowering of the lake surface resulted. For a time the outpouring was a veritable débâcle, and it could not have assumed the phase of an ordinary river commensurate with the inflow of the lake until the alluvial barrier was completely demolished and the resistance of the limestone reef was called into play. When the corrasion of the channel had proceeded so far as to give the river a bed of limestone, the process of excavation was changed from the mere transportation of loose detritus to the corrasion of solid rock, and the rate of excavation was greatly diminished. We have here the explanation of the rapidity of the final recession of the lake from the Bonneville level to the Provo.*

This is as about as clear and logical statement of the sequence of events at the lake outlet as could be. It is curious how controversial this scenario became, a controversy I do not really understand.

Outlet debates began with its first discovery. Gilbert had anticipated two outlets (though not necessarily at the same place) while exploring the basin with the Wheeler Survey in 1872. He argued that the well-formed Bonneville and Provo shorelines corresponded to overflow elevations (Gilbert, 1875, p. 91; Gilbert, 1890, pp. 17, 173). But Peale (1878) suggested a shoreline 300–600 ft above the Bonneville level and an outlet at then-called Portneuf Pass, dividing the Portneuf River basin from the Great Basin drainages 30 km north-northeast of Red Rock Pass. Gilbert dissented (1890, pp. 95, 173), noting that several back-and-forth parries had already consumed "a more voluminous literature than perhaps it deserved."

Yet more literature was expended in the 1960s, reviving the idea of a late Pleistocene lake higher than the Bonneville shoreline. This prompted speculation that the Bonneville Flood was the result of the lake dropping from this higher level (which must have been less than 35 m higher, the level of unaffected Kelton Pass 100 km west of Red Rock Pass). This interpretation seems to owe to R.C. Bright and M. Rubin, who described "wave-washed debris and scour features … 30 feet above the Bonneville shoreline at Red Rock Pass." Bright also recognized "six minor shorelines in northern Cache Valley between the Bonneville and Provo levels," to him indicating "sporadic downcutting to the Provo level" instead of rapid drawdown (in Richmond et al., 1965, p. 106).

Harold Malde (Fig. 6.3) adopted this view in his *Bonneville Flood* (1968, p. 11):

FIG. 6.3 Harold (Hal) E. Malde. Undated USGS photograph probably from the late 1950s or early 1960s when he worked on the Bonneville Flood.

Whatever the age assigned to the Bonneville Flood, geologists now agree that it was undoubtedly older than the formation of the Bonneville shoreline—a level later controlled by the height of resistant bedrock in Red Rock Pass…The true height of overflow is not determinable, but it could have been 100 feet above the Bonneville shoreline. Overflow may have been caused by a sudden failure of an alluvial barrier, as Gilbert (1890, pp. 175–178) supposed, but the actual deposit that formerly blocked the pass has never been identified. Probably it was entirely removed by the catastrophic outflow, and Lake Bonneville was thus drastically lowered—presumably as far as the bedrock lip at the Bonneville shoreline. Erosion thereafter would have been at a much slower rate and not of catastrophic magnitude. It follows that downcutting to the Provo shoreline, which was accomplished by the wearing away of hard rocks in Red Rock Pass…was in no way responsible for the Bonneville Flood. This conclusion differs from a long-standing idea of Gilbert that erosion of Red Rock Pass to the Provo shoreline caused the flood.

This revised scenario likely derived partly from Bright's higher-than-Bonneville shorelines (which seems not to have been confirmed) but also from early radiocarbon dating indicating the Bonneville Flood was about 30 ka, apparently older then the Bonneville shoreline then dated to about 18 ka and the Provo to 13.5 ka (Malde, 1968, pp. 10–11).

By 1987, however, Malde had reverted to Gilbert's original scenario (Jarrett and Malde, 1987):

Lake Bonneville fluctuated greatly in altitude, last reaching its highest level, the Bonneville shoreline, about 15,000 years ago (Scott et al., 1983). At that time, the lake overtopped its rim at Red Rock Pass and catastrophically discharged a vast volume of water down the Snake River…an event known as the Bonneville Flood (Malde, 1968). Red Rock Pass consequently was eroded to the level of the Provo shore-line, and the flood ended.

I am not sure what spurred Malde's change in thinking; perhaps better dating of the Lake Bonneville deposits (he critically considered new information in conducting the USGS review of the Scott et al. report), and possibly lack of confirmation of higher shorelines. Touring the terrain with him as a young graduate student in 1987, I could not muster the courage to ask.

But Gilbert's (1890) scenario is now basically accepted. Recent studies—such as Janecke and Oaks (2011) and Oviatt (2015), summarized by Shroder et al. (2016)—add details and chronology to conditions at Red Rock Pass, but G.K. Gilbert's original interpretation still mostly stands.

6.3 MELON GRAVEL

In 1955, amused by a whimsical billboard [Fig. 6.4] that advertised one patch of boulders as 'petrified watermelons,' we (Malde and Powers, 1962, p. 1216) applied to them the descriptive geologic name, Melon Gravel.

FIG. 6.4 "Petrified watermelons" advertised on a Bonneville Flood bar along Highway 30, southeast of King Hill, Idaho (on the back of a Stinker Gas Station billboard). This sign inspired the "Melon Gravel" formation name assigned by Malde and Powers (1962). It still stood in 1987, prompting a gift to my mother-in-law. *Photograph by Karen A. Demsey.*

The most spectacular Bonneville Flood features are downstream of Red Rock Pass (Figs. 6.5 and 6.6). The Melon Gravel, scabland, and dry cataracts of the Snake River Plain attest to the cataclysmic consequences of the flood. Malde (1968, p. 6), also described that before beginning his fieldwork in the Snake River Plain in 1955, H.A. Powers had informed him of the boulder deposits near Hagerman and "their catastrophic origin." Thus the flood was reborn, first in a two-page brief (Malde, 1960), and then comprehensively in 1968 with *The Catastrophic Late Pleistocene Bonneville Flood.*

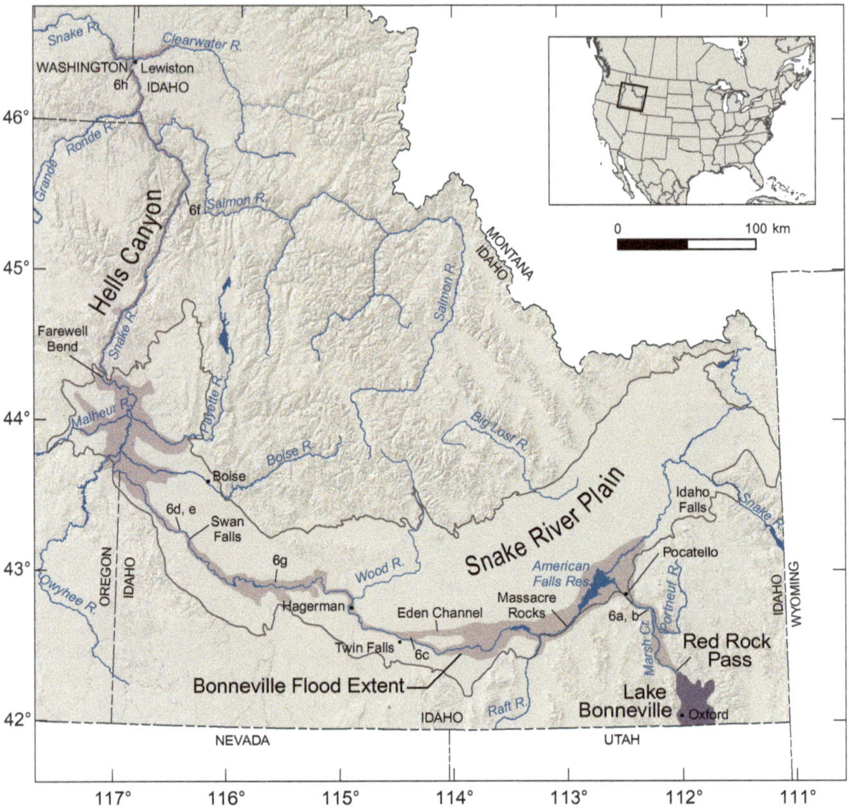

FIG. 6.5 Map of the Bonneville Flood route and approximate area of inundation between Red Rock Pass and Lewiston. Also shown are approximate locations of photographs of Fig. 6.6. *Redrafted from O'Connor, J.E., 1993. Hydrology, hydraulics, and geomorphology of the Bonneville flood. Geological Society of America Special Paper 274, 83 p.*

But between Gilbert's description of the débâcle in 1890, and Malde's resurrection in the 1960s, the Bonneville Flood was lost. This despite several studies of the Snake River Plain in the early 20th century (Russell, 1902; Stearns, 1936; Stearns et al., 1938) and even though the Russell and Stearns et al. reports show photographs of large flood-transported boulders deposited in Hagerman Valley.

Particularly curious is that I.C. Russell did not connect the flood features in the Snake River Plain to Lake Bonneville. Israel Cook Russell served as chief assistant for Gilbert's Great Basin project during 1880–1883. He assisted Gilbert in the Lake Bonneville basin in 1880, then investigated the Pleistocene lake basins farther west in 1881 and 1882, culminating in his report on Lake Lahontan (Pyne, 1980, pp. 112, 115–117; Russell, 1885; Gilbert, 1890, p. 18). Twenty years later, Russell spent the summer of 1901 investigating the *Geology and Water Resources of the Snake River Plains of*

FIG. 6.6 Photographs of Bonneville Flood features: (a) Aerial view east and upstream toward Inkom in the Portneuf River valley, floored by flood-eroded basalt flow. (b) Ground view of fluted and polished Portneuf Basalt near the scabland in a. Shovel handle 0.5 m. (c) Aerial view north from east of Twin Falls, showing scabland and cataracts carved by flow from the Eden Channel reentering the Snake River Canyon. (d) View downstream of immense boulder bar, mostly filling 1.5 km wide canyon downstream of the Swan Falls constriction. (e) Close-up view of bedload on the bar of (d). Many clasts have long diameters exceeding 10 m. (f) View north (downstream) in Hells Canyon, showing immense point bar (with large Ponderosa Pines along river's edge) and flood-swept valley slope to nearly 200 m above the Snake River. (g) Sand and fine gravel deposited from suspended load in an eddy bar east of Hagerman, Idaho. Pick is 0.7 m long. (h) Horizontally bedded Missoula Flood deposits on top of foreset bedded Bonneville Flood gravel in the Tammany Creek valley near Lewiston, Idaho. Here about 20 Missoula flood beds were left by that number of floods backing up the Snake River (Waitt, 1985). The foresets in the Bonneville Flood gravel dip up Tammany Creek valley, indicating up-valley flow as the bar formed. *Photographs by Jim E. O'Connor.*

FIG. 6.6—CONT'D

Idaho. This was the first extensive geologic reconnaissance of the Snake River Plain, summarized in a 202-page report. Russell (1902, p. 16) does describe a "washing of debris from the uplands and its deposition in alluvial cover and widely extended sheets of sand, gravel, and silt in the valleys." He also referred to the "young topography" (p. 124) west of American Falls in the Massacre Rocks area and attributed the deep alcoves near Twin Falls "to the action of great springs which come out at their heads" (p. 128). Russell photographed the immense rounded boulders left by the flood near Hagerman (Plate 25B). He recognized them as "water-worn" but concluded them to be "blocks of lava rounded by the river; not transported by the river or by the glacier" (as reported by Malde, 1968, p. 2). From his work and close relationship with Gilbert, Russell must have known about the Bonneville outflow and river. But why he did not connect the overflow with the spectacular features downstream remains a mystery.

Harold Stearns also missed the flood early in his career. His first extensive reports on the Snake River Plain in the 1930s also attributed the side-canyon alcoves to spring sapping (1936, pp. 444–447; Stearns et al., 1938, pp. 144–151). He even ascribed the immense and boulder-studded Bonanza Bar at the mouth of Lake Channel—a 7-km-long blind channel heading in a basalt-rimmed amphitheater—as the "detritus left in the wake of the receding spring" (Stearns et al., 1938, p. 149). The 1938 report also shows the immense rounded boulders in Hagerman Valley (plate 12A) but notes that they are "at the toes of [lava] dams" (p. 89), presumably resulting from breaching of lava-dams. After Malde's (1960) report, Stearns (1962) retracted the 1938 lava-dam interpretation, recognizing boulder deposits near King Hill, Idaho, and as far downstream as Hells Canyon as being from the Bonneville Flood.

Interpretations change with time as evidence accumulates. But it seems the flood should have been anticipated from Gilbert's work on the outlet, particularly since I.C. Russell was on the scene both in the Bonneville Lake basin and on the Snake River Plain. And it is relevant because in an ironic twist, the lost flood entered into the contentious Channeled Scabland debate of the 1920s and 1930s. In 1938, Richard Foster Flint argued against Bretz' (1923, 1925, 1927, 1928) Missoula Flood origin for the channeled scabland of eastern Washington, proposing instead that the scablands were carved by "leisurely streams with normal discharge, aggrading and then degrading." Flint's (1938) evidence included the scabland of the Snake River Plain:

> *At Shoshone Falls, just east of the city of Twin Falls, Idaho, on the Snake River plains, the Snake River has incised a canyon more than 500 feet deep...A broad shelf on the south wall about 100 feet below the rim consists of scabland with small anastomosing channels, isolated buttes, and shallow rock basins. The thin, vertically jointed flows yielded to the hydraulic force of Snake River as similar flows on the Columbia Plateau yielded to the hydraulic force of proglacial streams, yet I am not aware that unusual floods have been held to have affected the upper Snake River.*

Malde and Powers revived Gilbert's "Bonneville river" in the 1950s and energized it into the "Bonneville Flood." Soon other geologists recognized flood features along its course (Figs. 6.5 and 6.6). These included Trimble and Carr (1961a,b) and Trimble (1976) near Pocatello, Scott (1982) in the eastern Snake River Plain, Stearns (1962) in the western Snake River Plain and downstream through Hells Canyon, Vallier (1998) and Webster et al. (1982) in Hells Canyon, and Kehew (1977) and Webster et al. (1982) downstream near Lewiston, Idaho.

Yet Malde's (1968) *The Catastrophic Late Pleistocene Bonneville Flood* was the signature contribution. His work on the Snake River Plain has largely held up despite misinterpreting conditions at the outlet. Malde mapped flood features for about 300 km of the Snake River Plain, defined the water-surface profile, estimated discharge and duration, and quantified volumes of eroded and deposited materials. He showed the deep alcoves and great cataracts formed by water spilling out and then back into the Snake River Canyon. And he linked erosion and deposition to local hydraulic and geologic conditions. His connections between flood landforms and processes inspired my Ph.D thesis to investigate the flood more quantitatively (O'Connor, 1990, 1993). I benefited immensely from Malde's encouragement from the beginning, including a long field tour in Sep. 1987.

Malde's mapping and that of others show the extent and effects of the Bonneville Flood for 1100 km (Figs. 6.5–6.7). From Red Rock Pass it flowed north down Marsh Creek and into the Portneuf River valley (Fig. 6.6a and b) and debouched onto the Snake River Plain at Pocatello. There it left a boulder-studded fan before regathering to carve a new canyon at the downstream end of Pleistocene American Falls Lake, where American Falls Reservoir now lies (Trimble and Carr, 1961a,b; Trimble, 1976; Scott, 1982, Scott et al., 1982). From here the flood followed the vast and incised volcanic Snake River Plain west for 500 km. Within the plain, the flood descended in drops through constrictions and narrow canyon segments. A particularly steep section with abundant scabland and large boulder bars is the 100 km from east of Twin Falls to Hagerman. In this reach the flood split, most flow broadly crossing the uplands in the Eden channel north of the Snake River

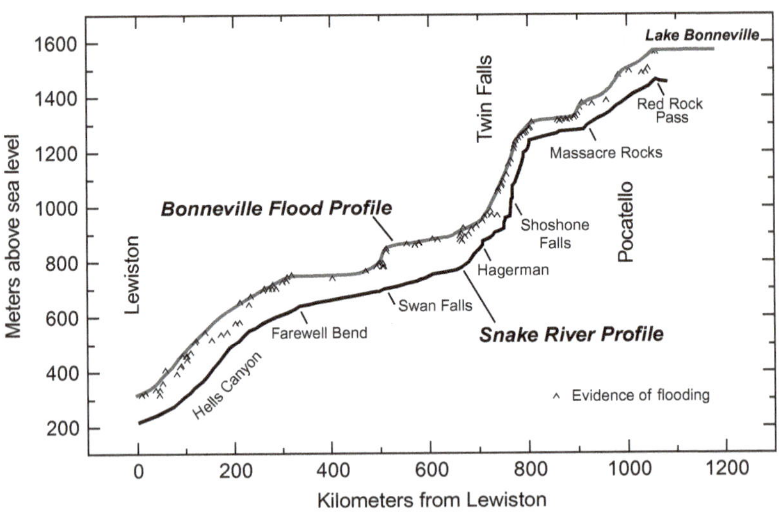

FIG. 6.7 Profile of March Creek, Portneuf River, and Snake River from Red Rock Pass to Lewiston Idaho, showing elevations of highest Bonneville Flood features and approximate profile of the maximum stage of the Bonneville Flood. *Redrafted and modified from O'Connor, J.E., 1993. Hydrology, hydraulics, and geomorphology of the Bonneville flood. Geological Society of America Special Paper 274, 83 p.*

Canyon and reentering in several spectacular cataracts and blind canyons carved into the north rim east of Twin Falls (Fig. 6.6c).

At the west end of the Snake River Plain the flood turned north through the constriction at Farewell Bend. The flood then descended steeply and deeply through confined valleys, particularly narrow Hells Canyon. Here the flood deposited gravel more than 175 m above present river level (Fig. 6.6f; O'Connor, 1993). Bonneville flood deposits extend north to Lewiston and a few kilometers beyond (Webster et al., 1982) but become more and more buried by deposits of Missoula floods that backed up the Snake River as far as the Salmon River confluence (Fig. 6.6h). The flood surely followed the Snake River west into the Columbia River and then to the Pacific Ocean but its deposits were mostly eroded or covered by the Missoula floods.

At the time of Malde's (1968) report, the age of the flood enjoyed "some latitude" (p. 10). Now that the outlet scenario is more or less settled, the timing of the Bonneville Flood is set by the chronology of Lake Bonneville. This chronology summarized by Oviatt (2015) shows Lake Bonneville reached peak level at or after 18.4 ka, (calendar years) but dropped to the Provo level before 17.1 ka. Oviatt (2015) concluded "a reasonable flood age near the middle of the overlapping range is 18 ka with an uncertainty of several hundred calibrated years."

The radiocarbon dating of Lake Bonneville is consistent with some cosmogenic dates from flood-transported boulders (Amidon and Clark, 2015) and flood-eroded basalt (Lamb et al., 2008, 2014; Amidon and Farley, 2011). Cosmogenic ^{3}He and ^{21}Ne dating of five clasts of Portneuf basalt resting on top of the eroded basalt flow along the Portneuf River 50 km below of Red Rock Pass (near the view of Fig. 6.6a) gave a combined result of 17.5 ± 0.9 ka (Amidon and Clark, 2015). Several analyses of flood-transported boulders give similar ages in the cataracts and alcoves of Blue Lakes and Devils Corral (Fig. 6.6c) near Twin Falls (Amidon and Clark, 2015), as do some cosmogenic ^{3}He analyses of flood-transported boulders and eroded basalt surfaces in box canyons and alcoves near Hagerman (Lamb et al., 2008, 2014; Amidon and Clark, 2015).

But the cosmogenic dating suggests a more complex history of erosion associated with some of these alcoves and blind canyons. Flooding at about 46 ka, probably from Wood River, may have partly incised gorges and blind canyons into the Snake River canyon rim near Hagerman (Lamb et al., 2008, 2014). And older ^{3}He and ^{21}Ne cosmogenic ages for boulders on some surfaces in the Devils Corral and Blue Lakes cataract complexes suggest a slightly earlier episode of erosion—21 ± 2 ka—that Amidon and Clark (2015) correlated with a glacial flood in the Big Lost River north of the Snake River Plain. They further proposed that Big Lost River flooding contributed to scabland erosion by infiltrating into the northern edge Snake River Plain, and reemerging downstream to incise the alcoves and canyons. This scenario harkens back to the spring-sapping hypotheses in these terrains by Russell (1902) and Stearns (1936), although analyses by Lamb et al (2006) and Lamb and Dietrich (2009) suggest this mechanism is implausible.

6.4 "THE FLOOD VOLUME OF THE MISSOURI"

From the work of Malde and later researchers we know the timing, extent and effects of the flood. But how big was it? For this we return to G.K. Gilbert. He revisits Red Rock Pass in 1877, one year after his first look, this time astride his horse Kit. Having found no other outlets, he focuses on hypothesis number two—one incised outlet—and closely examines outflow conditions. On Sep. 22 he records a cryptic paleohydraulic calculation for the Bonneville flood: "here the channel between soft banks is $9/2 \times 234$ [Kit footsteps] = 1053 ft and between hard and soft is $9/2 \times 134 = 603$ ft. A comparison of these with that of Jordan gap and Bear R. and Niagara will give an idea of the magnitude of the outflow." He carries it farther in the 1890 *Lake Bonneville* (p. 177):

> *Let us assume that at the time the Bonneville river traversed the lava-bed [within the Portneuf valley, 40 km below Red Rock Pass]...Let us assume that the slope of the water surface in flowing over the lava was 2½ feet to the mile, or one foot in 2,000. If now we assume in addition that the discharge equaled that of the Niagara River, we have all the data necessary for computing the mean depth; and we obtain for that depth 9 feet. To one who stands upon the lava bed and notes the scale of the carvings which ornament its surface, this determination appears far too small. Twenty feet would better accord with the phenomena, and twenty feet would discharge the flood volume of the Missouri.*

Not a refined paleohydraulic calculation, but a start. The largest flood on the Missouri River flood in Gilbert's time was about 20,000 $m^3\ s^{-1}$ (The peak flow in 1844 at USGS streamflow measurement station 06934500, Missouri River at Hermann, MO).

After Malde and Powers recognized downstream flood features in the 1950s and 1960s, discharge estimates poured in. Stearns (1962) proposed a peak flow of 0.3 million $m^3\ s^{-1}$ (10 million $ft^3\ s^{-1}$) at the head of Hells Canyon, 15 times greater than Gilbert's "flood volume of the Missouri" estimate. Stearns' (1962) value was probably by the Manning equation (Malde, 1968, p. 12). Malde (1968, pp. 24–25) also using the Manning equation in canyon reaches made several estimates ranging from 0.3 to 6 million $m^3\ s^{-1}$. His preferred estimate in 1968 (p. 12) was a critical-depth (contracted opening) calculation for the tight constriction near Swan Falls. This computation by Clifford T. Jenkins gave a range of 0.35–0.5 million $m^3\ s^{-1}$, leading Malde (p. 12) to conclude the peak flow was about 0.4 million $m^3\ s^{-1}$ (15 million $ft^3\ s^{-1}$). Later finding higher flood deposits just upstream of the Swan Falls constriction, Jarrett and Malde (1987) revised the estimate upward to a range of 0.7–1.0 million $m^3\ s^{-1}$ and a preferred value of 0.935 million $m^3\ s^{-1}$.

This peak flow estimate by Jarrett and Malde at Swan Falls is consistent with those I made soon after. From step-backwater calculations with the HEC-2 code developed by the Hydrologic Engineering Center (1985; but now superceded by HEC-RAS (http://www.hec.usace.army.mil/)), I calculated peak discharge for nine separate reaches between Red Rock Pass and Lewiston (O'Connor, 1993, pp. 14–16). These ranged from 1 million $m^3\ s^{-1}$ near the outlet at Red Rock Pass to about 0.6 million 1100 km downstream near Lewiston. I estimated a maximum discharge of 0.82–0.85 million $m^3\ s^{-1}$ at Swan Falls, about 10% less than Jarrett and Malde (1987). These results show the Bonneville Flood to be nearly 100 times larger than any historical flood on the Snake River, and more than 50 times Gilbert's original estimate by pacing the channel width on horseback.

Probable attainment of critical flow at Red Rock Pass allows calculation of the maximum discharge of the Bonneville Flood. For the present geometry of the outlet channel at the Marsh Creek fan—the valley bottom is ~1457 m above sea level—and the highest Lake Bonneville level at Red Rock Pass of 1552 m (Currey, 1990), a maximum of about 1.1 million $m^3\ s^{-1}$ could funnel through the outlet as it incised and widened to its present geometry. This value slightly exceeds but is consistent with a minimum estimate of 0.85–1.0 million $m^3\ s^{-1}$ calculated from highwater marks immediately downstream (O'Connor, 1993). The discharge through Red Rock Pass at critical flow would be less if the lake level had dropped significantly while excavating the outlet, but that is unlikely with a lake so voluminous (Walder and O'Connor, 1997). More likely is that the peak discharge was slightly larger than 1.1 million $m^3\ s^{-1}$ because post-flood valley fill has reduced the present outlet depth—as much as 50% greater if the total lake-level drop was 120–125 m (and implying 25–30 m of valley fill) as suggested by Miller et al. (2013) and Miller (2016).

These discharge calculations spurred estimates of flood duration. They ranged from Gilbert's (1890, p. 177) "less than twenty-five years" to Jarrett and Malde's (1987) "about 8 weeks." From rudimentary dam-break modeling fit to downstream estimates of peak flow, I estimated flow exceeded 0.5 million $m^3\ s^{-1}$ for about 300 days at Red Rock Pass (O'Connor, 1993, pp. 36–40).

The time is ripe for new hydrodynamic analysis. The discharge calculations and dam-break modeling of nearly three decades ago were appropriate for the time and improved upon earlier Manning-equation estimates. But technology and tools are much better now. Step-backwater calculations in the 1980s relied on cross sections hand-measured from paper maps. They were premised on the hydraulic assumption of steady, gradually varied flow—computationally feasible and applicable for parts of the flood route but not for energetic reaches. Now it is possible to model the entire lake and flow route dynamically as a single system, using digital elevation models and two-dimensional (maybe even three dimensional) flow models. Recent models, such as GeoClaw (eg, Berger et al., 2011), can better and efficiently account for hydraulic shocks in the transitions through critical flow, which were many for the Bonneville Flood in the steeper reaches (O'Connor, 1993). These types of models, already applied to the Missoula Floods (Denlinger and O'Connell, 2010), would readily work for the hydraulically simpler Bonneville Flood. New models would give better discharge estimates, improve knowledge of the evolving lake drainage and flood, and offer more realistic flow dynamics, which in turn could better explain erosional and depositional processes and features.

6.5 "LAKE BURSTS"

In Oct. 1879 Gilbert returns a third time to Red Rock Pass, this time venturing further down Marsh Creek and Portneuf River valleys nearly to Pocatello (Hunt, 1982). Evaluating erosional landforms, he again considers the magnitude of the flood. On Oct. 16th, he studies a scarp 120 m (400 ft) above the Portneuf River bottom. He concedes that "it does not need the hypothesis of the Bonneville river to account for it; but it accords with that theory and sustains it in a general way." The next day he climbs atop the Portneuf basalt flow on the valley bottom and sees large "flutings and potholes" (Fig. 6.6a and b).

But even before G.K. Gilbert, the ultimate anti-catastrophist Lyell (1830, p. 192) understood that floods formed landscapes:

> *It is evident, therefore, that when we are speculating on the excavating force which running water may have exerted in a particular valley, the most important question is not the volume of the existing stream, nor the present levels of the river channel, nor the size of the gravel, but the probability of a succession of floods, at some period since the time when some of the land in question may have been first elevated…*

Moreover, "lake bursts" could be the source of the "excavating force" (pp. 192–193):

> *The power which running water may exert, in the lapse of ages, in widening and deepening a valley, does not so much depend on the volume and velocity of the stream usually flowing in it, as on the number and magnitude of the obstructions which have, at different periods, opposed its free passage.*

The Bonneville "lake burst" was one of the largest on Earth in peak discharge and total flood volume (Fig. 6.8). Its peak discharge was matched or exceeded by just a few Quaternary floods—all resulting from natural dam failures; several of the Missoula floods (O'Connor and Baker, 1992; Benito and O'Connor, 2003), by floods from breached ice dams in Siberia (Herget, 2005), by rapid emptying of Alaskan calderas Aniakchak and Okmok (Waythomas et al., 1996; Wolfe and Begét, 2002), and breaching of a debris

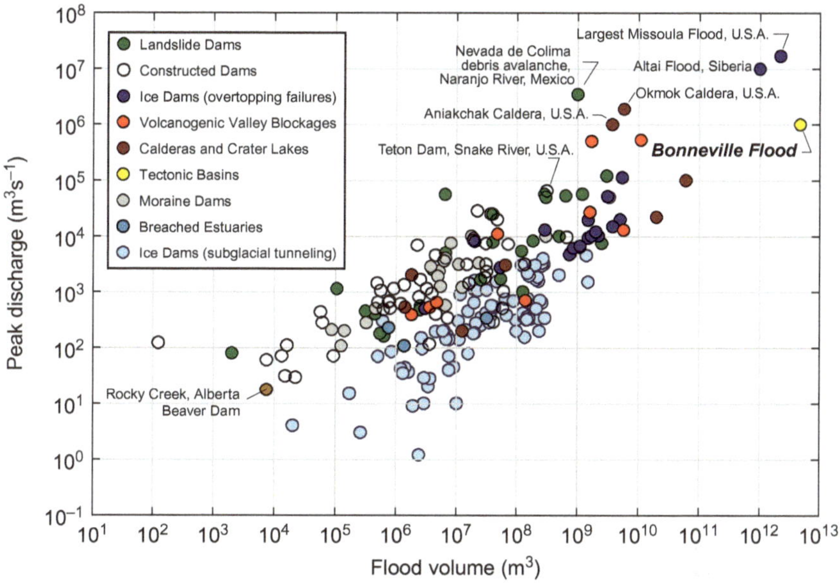

FIG. 6.8 Outburst floods for which flood volume and peak discharge are known. *Data from Walder and Costa (1996) and O'Connor, J.E., Beebee, R.A., 2009. Floods from natural rock-material dams. In: Burr, D., Carling, P., Baker, V. (Eds.), Megaflooding on Earth and Mars. Cambridge University Press, New York, pp. 128–171. Modified from O'Connor, J.E., Clague, J.J., Walder, J.S., Manville, V., Beebee, R.A., 2013. Outburst floods. In: Shroder, J. (Ed. in Chief), Wohl, E.E., (Ed.), Treatise on Geomorphology (Fluvial Geomorphology), vol. 9. Academic Press, San Diego, California, pp. 475–510.*

avalanche formed on the Naranjo River, Mexico, during the eruption of Nevada de Colima (Capra and Macías, 2002). The Bonneville Flood volume exceeds all known freshwater floods except some from large lakes like glacial Lake Agassiz in basins dammed by continental ice sheets (O'Connor et al., 2013).

Despite the size and volume of the Bonneville Flood, features in Marsh and Portneuf valleys, just downstream from the outlet, are subtle, particularly viewed from the ground–perhaps explaining Gilbert's hesitant notetaking and the lack of the word "flood" in the final monograph. But down on the Snake River Plain the "intricately dissected lava surface of branching channels, dry falls, and rock basins" stands out as carved by exceptional flow (Malde, 1968, p. 2). This "scabland," like that of the Missoula-Flood scabland of eastern Washington, attests to the erosional capability of outburst floods. Also prominent are the immense boulder bars covered with large water-worn boulders, owing to the tremendous flow competence (Fig. 6.6d and e).

Malde mapped the scabland and Melon Gravel in detail. He used this mapping to determine the water-surface profile and to quantify volumes of erosion and deposition for parts of the flood route on the Snake River

Plain. I followed up by attempting to calculate peak discharge in several places and using flow modeling to estimate erosional and depositional thresholds (O'Connor, 1993, pp. 44–66). In these analyses both Malde and I benefited from relatively simple conditions of the Bonneville Flood compared with many large outburst floods. The flood was mostly confined to a single pathway. And the flow was steadier than for most outburst floods because of the large volume of water released relative to the peak discharge. The Bonneville Flood had a peak discharge of about 1 million $m^3\ s^{-1}$ and a total flow volume of about 5000 km^3; the largest Missoula Flood had a peak discharge more than twenty times greater but a volume only about half the Bonneville Flood.

The 1980s–1990s work on erosional and deposition features and processes did identify associated hydraulic and energy conditions. But it too would benefit from reanalysis. Analysis of digital topography, updated geologic mapping, and better hydrodynamic modeling could address key questions and issues:

- What are the thresholds and processes of bedrock-channel erosion for different rock types and different hydrodynamic environments?
- What situations lead to canyon incision instead of widening? And at what rates?
- How do the alcoves and cataracts form and develop?
- What are the main controls on sediment transport and deposition? What is the fate of flood-transported sediment? How does deposit stratigraphy relate to flow dynamics?

Such analyses would apply to the ever-expanding domain of mega-flood erosion and deposition. Because of its simple hydrodynamic and geomorphic setting, the Bonneville Flood remains a nearly ideal situation for studying these phenomena.

Besides these general process questions, intriguing topics of special interest remain that are specific to the Bonneville Flood. These include the incised canyon near Massacre Rocks, downstream of Pleistocene American Falls Lake that was probably entirely formed during the flood (Scott et al., 1982). And the narrow and steep canyon reach east of Twin Falls—it holds Twin Falls, Shoshone Falls, Pillar Falls, and Cauldron Lynn—may have largely formed during the flood. How much did the Bonneville Flood contribute to the present morphology and depth of Hells Canyon? Mapping and application of newer geochronologic and erosion-rate techniques might help identify locations and effects of canyon- and alcove-carving flows on the Snake River Plain from other sources, as Lamb et al. (2008, 2014) suggest. Could elevated groundwater flow have actually helped carve some of the cataract and alcove complexes near Twin Falls before the Bonneville Flood as Amidon and Clark (2015) conclude? Also unknown is the full relationship between Missoula flooding and the Bonneville Flood. Exposures near Lewiston show at least

20 Missoula Floods left layers of sand and silt above coarse Bonneville Flood deposits (Waitt, 1985; Fig. 6.6h). But were there Missoula Floods prior to the ~18 ka Bonneville Flood? So far, Missoula flood deposits have not been found stratigraphically below Bonneville-Flood deposits, but I think searching has been scant.

These questions have broader implications. The earth-science community is gaining awareness of the potential for outburst floods to create spectacular planetary landscapes (Burr et al., 2009). Such floods derive from many sources and in many environments, with long-lasting geomorphic consequences (O'Connor et al., 2013). In some environments, as in the Himalaya, floods from natural-dam failures may be the main force incising valleys (Cenderelli and Wohl, 2003; Montgomery et al., 2004). The Snake River Plain studies of Lamb et al. (2006, 2008, 2014) and Amidon and Clark (2015) were motivated in part to understand Martian landscapes.

When Gilbert ventured to Red Rock Pass three times in 1876–1879, he was simply looking for the outlet of the vast inland sea that had cut and planed the prominent shorelines far above Great Salt Lake. He did not anticipate the rich lost-and-found history his "veritable débâcle" generated as it transitioned into the Bonneville Flood. The flood in turn has contributed significantly to understanding of fluvial landscapes. But most is important is that the story continues: Gilbert's work, Malde's work, my work, and that of many others is just a beginning for what should be even better understanding of the effects of large floods on landscapes. Not just on the Snake River Plain, but for our planet and others as well.

ACKNOWLEDGMENTS

Gilbert's field notes for the *Lake Bonneville* study were transcribed and edited by Hunt (1982), a huge service to science. My work on the Bonneville Flood in the 1980s was encouraged and guided by Vic Baker and Hal Malde. I was assisted in the field by Kirk Vincent, Sara Rathburn, Karen Demsey, and Lisa Ely. John Costa, William Scott, Richard Waitt, Kurt Othberg, Bob Jarrett, Michael Lamb, Tracy Vallier, Jack Oviatt, and William Bull provided advice, information, and reviews. This contribution was urged by Jack Oviatt and was mostly written during a residency at The Playa at Summer Lake, Oregon. Any use of trade, firm, or product names is for descriptive purposes only and does not imply endorsement by the US Government.

REFERENCES

Amidon, W.H., Clark, A.C., 2015. Interaction of outburst floods with basaltic aquifers on the Snake River Plain: implications for Martian canyons. Geol. Soc. Am. Bull. 127, 688–701. http://dx.doi.org/10.1130/B31141.1.

Amidon, W.H., Farley, K.A., 2011. Cosmogenic 3He dating of apatite, zircon and pyroxene from Bonneville flood erosional surfaces. Quatern. Geochronol. 6, 10–21. http://dx.doi.org/10.1016/j.quageo.2010.03.005.

Benito, G., O'Connor, J.E., 2003. Number and size of last-glacial Missoula floods in the Columbia River valley between the Pasco Basin, Washington, and Portland, Oregon. Geol. Soc. Am. Bull. 115, 624–638.

Berger, J.J., George, D.L., LeVeque, R.J., Mandli, K.T., 2011. The GeoClaw software for depth-averaged flows with adaptive refinement. Adv. Water Resour. 34, 1195–1206. http://dx.doi.org/10.1016/j.advwatres.2011.02.016.

Bretz, J.H., 1923. Glacial drainage on the Columbia plateau. Geol. Soc. Am. Bull. 34, 573–608.

Bretz, J.H., 1925. The Spokane Flood beyond the Channeled Scabland. J. Geol. 33, 97–115.

Bretz, J.H., 1927. Channeled Scabland and the Spokane Flood. J. Wash. Acad. Sci. 17, 200–211.

Bretz, J.H., 1928. Bars of the channeled scabland [Washington]. Geol. Soc. Am. Bull. 39, 643–701.

Burr, D.M., Carling, P.A., Baker, V.R. (Eds.), 2009. Megaflooding on Earth and Mars. Cambridge University Press, Cambridge, United Kingdom.

Capra, L., Macías, J.L., 2002. The cohesive Naranjo debris-flow deposit (10 km^3): a dam breakout flow derived from the Pleistocene debris-avalanche deposit of Nevado de Colima Volcano (México). J. Volcanol. Geotherm. Res. 117, 213–235.

Cenderelli, D.A., Wohl, E.E., 2003. Flow hydraulics and geomorphic effects of glacial-lake outburst floods in the Mount Everest region, Nepal. Earth Surf. Process. Landf. 28, 385–407.

Currey, D.R., 1990. Quaternary palaeolakes in the evolution of semidesert basins, with special emphasis on Lake Bonneville and the Great Basin, U.S.A. Palaeogeography, Palaeoclimatology, and Palaeoecology 76, 189–214.

Denlinger, R.P., O'Connell, D.R.H., 2010. Simulations of cataclysmic outburst floods from Pleistocene Glacial Lake Missoula. Geol. Soc. Am. Bull. 122, 678–689.

Flint, R.F., 1938. Origin of the Cheney-Palouse scabland tract. Geol. Soc. Am. Bull. 49, 461–524.

Gilbert, G.K., 1875. Explorations and Surveys West of the One Hundredth Meridian (Wheeler). vol. 3. Geology.

Gilbert, G.K., 1890. Lake Bonneville. U.S. Geological Survey Monograph 1. 438 p.

Herget, J., 2005. Reconstruction of Pleistocene Ice-Dammed Lake Outburst Floods in the Altai Mountains, Siberia. Geological Society of America Special Paper 386, 118 p.

Hunt, C.B (Ed.), 1982. Pleistocene Lake Bonneville, ancestral Great Salt Lake, as described in the notebooks in G.K. Gilbert, 1875–1880. 29, Brigham Young University Geology Studies, 225 p.

Hydrologic Engineering Center, 1985. HEC-2 Water Surface Profiles Users Manual. U.S. Army Corps of Engineers Hydrologic Engineering Center, Davis, California. 37 p.

Janecke, S.U., Oaks, R.Q., 2011. New insights into the outlet conditions of late Pleistocene Lake Bonneville, southeastern Idaho, USA. Geosphere 7, 1369–1391. http://dx.doi.org/10.1130/GES00587.1.

Jarrett, R.D., Malde, H.E., 1987. Paleodischarge of the late Pleistocene Bonneville, Flood, Snake River, Idaho, computed from new evidence. Geol. Soc. Am. Bull. 99, 127–134.

Kehew, A.E., 1977. Environmental Geology of Lewiston, Idaho, and Vicinity. University of Idaho, Moscow, Idaho. unpublished Ph.D dissertation.

Lamb, M.P., Dietrich, W.E., 2009. The persistence of waterfalls in fractured rock. Geol. Soc. Am. Bull. 121, 1123–1134. http://dx.doi.org/10.1130/B26842.1.

Lamb, M.P., Howard, A.D., Johnson, J., Whipple, K., Dietrich, W.E., Perron, J.T., 2006. Can springs cut canyons into rock? J. Geophys. Res. Planets 111, E07002. http://dx.doi.org/10.1029/2005JE002663.

Lamb, M.P., Dietrich, W.E., Aciego, S.M., DePaolo, D.J., Manga, M., 2008. Formation of Box Canyon, Idaho, by megaflood: implications for seepage erosion on Earth and Mars. Science 320 (5879), 1067–1070. http://dx.doi.org/10.1126/science.1156630.

Lamb, M.P., Mackey, B.H., Farley, K.A., 2014. Amphitheater-headed canyons formed by megaflooding at Malad Gorge, Idaho. Proc. Natl. Acad. Sci. U. S. A. 111, 57–62. http://dx.doi.org/10.1073/pnas.1312251111.

Lyell, C., 1830. Principles of Geology, vol. 1. John Murray, London. 511 p.

Malde, H.E., 1960. Evidence in the Snake River Plain, Idaho, of a catastrophic flood from Pleistocene Lake Bonneville. U.S. Geological Survey Professional Paper, In: Short papers in the geological sciences. 400-B, pp. B295–B297.

Malde, H.E., 1968. The Catastrophic Late Pleistocene Bonneville Flood in the Snake River Plain, Idaho. U.S. Geological Survey Professional Paper 596, 52 pp.

Malde, H.E., Powers, H.A., 1962. Upper Cenozoic stratigraphy of western Snake River Plain, Idaho. Geol. Soc. Am. Bull. 73, 1197–1220.

Miller, D.M., 2016. Chapter 7. The Provo Shoreline of Lake Bonneville. In: Oviatt, C.G., Shroder Jr., J.F. (Eds.), Lake Bonneville: A Scientific Update, vol. 20. Elsevier. pp. 127–144.

Miller, D.M., Oviatt, C.G., McGeehin, J.P., 2013. Stratigraphy and chronology of Provo shoreline deposits and lake-level implications, late Pleistocene Lake Bonneville, eastern Great Basin, USA. Boreas 42, 342–361. http://dx.doi.org/10.1111/j.1502-3885.2012.00297.x.

Montgomery, D.R., Hallet, B., Liu, Y., Finnegan, N., Anders, A., Gillespie, A., Greenberg, H.M., 2004. Evidence for Holocene megafloods down the Tsangpo River gorge, southeastern Tibet. Quatern. Res. 62, 201–207.

O'Connor, J.E., 1990. Hydrology, Hydraulics, and Sediment Transport of Pleistocene Lake Bonneville Flooding on the Snake River, Idaho. University of Arizona, Tucson, Arizona. unpublished Ph.D dissertation, 192 p.

O'Connor, J.E., Clague, J.J., Walder, J.S., Manville, V., Beebee, R.A., 2013. Outburst Floods. In: Shroder, J, Wohl, E.E. (Eds.), Treatise on Geomorphology 9 (Fluvial Geomorphology). Academic Press, San Diego, California, pp. 475–510.

O'Connor, J.E., 1993. Hydrology, Hydraulics, and Geomorphology of the Bonneville Flood. Geological Society of America Special Paper 274, 83 pp.

O'Connor, J.E., Baker, V.R., 1992. Magnitudes and implications of peak discharges from glacial Lake Missoula. Geol. Soc. Am. Bull. 104, 267–279.

Oviatt, C.G., 2015. Chronology of Lake Bonneville, 30,000 to 10,000 yr B.P. Quatern. Sci. Rev. 110, 166–171.

Peale, A.C., 1878. The ancient outlet of Great Salt Lake. Am. J. Sci. 15, 439–444. 3rd Series.

Pyne, S.J., 1980. Grove Karl Gilbert: A Great Engine of Research. University of Texas Press, Austin, Texas. 306 p.

Richmond, G.M., Fryxell, R., Montagne, J., Trimble, D.E., 1965. Guidebook for Field Conference E, Northern and Middle Rocky Mountains: International Association of Quaternary Research (INQUA) 7th Cong., Denver and Boulder, Colorado. 129 p.

Russell, I.C., 1885. Geological history of Lake Lahontan—a quaternary lake of northwestern Nevada. U.S. Geological Survey Monograph, 11. 288 p.

Russell, I.C., 1902. Geology and water resources of the Snake River Plains of Idaho. U.S. Geological Survey Bulletin, vol. 199. 192 p.

Scott, W.E., 1982. Surficial Geologic Map of the Eastern Snake River Plain and Adjacent Areas, 111 Degrees to 115 Degrees W., Idaho and Wyoming. U.S. Geological Survey Miscellaneous Field Investigations Map I-1372, scale 1:250,000.

Scott, W.E., Pierce, K.L., Bradbury, J.P., Forester, R.M., 1982. Revised quaternary stratigraphy and chronology in the American Falls area, southeastern Idaho. In: Bonnichsen, B., Breckenridge, R.M. (Eds.), Cenozoic Geology of Idaho. Idaho Bureau of Mines and Geology Bulletin, vol. 26. 581–595.

Scott, W.E., McCoy, W.D., Shroba, R.R., Rubin, R., 1983. Reinterpretation of the exposed record of the last two cycles of Lake Bonneville. Western United States. Quat Res. 20, 261–285.

Shroder, J.F., Cornwell, K., Oviatt, C.G., Lowndes, T.C., 2016. Chapter 4. Landslides, Alluvial Fans, and Dam Failure at Red Rock Pass: The Outlet of Lake Bonneville. In: Oviatt, C.G., Shroder Jr., J.F. (Eds.), Lake Bonneville: A Scientific Update, vol. 20. Elsevier. pp. 75–87.

Stearns, H.T., 1936. Origin of the large springs and their alcoves along the Snake River in southern Idaho. J. Geol. 44, 429–450.

Stearns, H.T., 1962. Evidence of Lake Bonneville flood along Snake River below King Hill, Idaho. Geol. Soc. Am. Bull. 73, 385–387.

Stearns, H.T., Crandall, L., Steward, W.G., 1938. Geology and Ground-Water Resources of the Snake River Plain in Southeastern Idaho. U.S. Geological Survey Water-Supply Paper 774, 268 p.

Trimble, D.E., 1976. Geology of the Michaud and Pocatello Quadrangles, Bannock and Power Counties, Idaho. U.S. Geological Survey Bulletin, vol. 1400. 88 p.

Trimble, D.E., Carr, W.J., 1961a. The Michaud delta and Bonneville River near Pocatello, Idaho. U.S. Geological Survey Professional Paper, Short papers in the geological sciences, 424-B, pp. B164–B166.

Trimble, D.E., Carr, W.J., 1961b. Late quaternary history of the Snake River in the American Falls region, Idaho. Geol. Soc. Am. Bull. 72, 1739–1748.

Vallier, T., 1998. Islands and Rapids—A Geologic Story of Hells Canyon. Confluence Press, Lewiston, Idaho. 151 p.

Waitt, R.B., 1985. Case for periodic, colossal jökulhlaups from Pleistocene glacial Lake Missoula. Geol. Soc. Am. Bull. 96, 1271–1286.

Walder, J.S., Costa, J.E., 1996. Outburst floods from glacier-dammed lakes: The effect of mode of lake drainages on flood magnitude. Earth Surface Processes and Landforms 21, 701–723.

Walder, J.S., O'Connor, J.E., 1997. Methods for predicting peak discharge of floods caused by failure of natural and constructed earthen dams. Water Resour. Res. 33, 2337–2348.

Waythomas, C.F., Walder, J.S., Mcgimsey, R.G., Neal, C.A., 1996. A catastrophic flood caused by drainage of a caldera lake at Aniakchak Volcano, Alaska, and implications for volcanic hazards assessment. Geol. Soc. Am. Bull. 108, 861–871.

Webster, G.D., Kyhns, M.J.P., Waggoner, G.L., 1982. Late Cenozoic gravels in Hells Canyon and the Lewiston basin, Washington and Idaho. In: Bonnichsen, B., Breckenridge, R.M. (Eds.), Cenozoic Geology of Idaho, Idaho Bureau of Mines and Geology Bulletin, vol. 26, pp. 669–683.

Wolfe, B.A., Begét, J.E., 2002. Destruction of an Aleut village by a catastrophic flood release from Okmok Caldera, Umnak Island, Alaska. Geol. Soc. Am. Abstr. Programs 34 (6), 126.

Chapter 7

The Provo Shoreline of Lake Bonneville

D.M. Miller
U.S. Geological Survey, Menlo Park, CA, United States

ABSTRACT

G.K. Gilbert studied the Bonneville basin 150 years ago and his findings have largely stood the test of time: The Provo shoreline, the most prominent geomorphic feature of Lake Bonneville, reflects threshold-stabilized overflow of the lake after the Bonneville flood and before a drier climate caused the lake to shrink. Subsequent refinements in chronology allow the Provo lake to be identified as about 18.2–14.8 cal ka BP, and stratigraphic studies show that the lake was gradually growing deeper during that time. Because the lake deepened through time as isostatic rebound occurred, individual landforms in general reflect processes operating for a small part of the ~3400 year of Provo time. Opportunities remain to improve our knowledge of the Provo lake; topics include (1) refinement of lake levels using delta and beach stratigraphy; (2) improved understanding of lake water chemistry and its role in determining deep-water sediment and cave deposits, which have disparate interpretations; (3) identifying processes at the threshold that caused the lake level to rise; and (4) identifying climate variability signals during Provo time.

Keywords: Lake Bonneville, Provo shoreline, Paleoclimate

7.1 INTRODUCTION

The Provo shoreline was described by Gilbert (1890) who noted that its massive geomorphic features make it the most prominent shoreline in the Bonneville basin. The shoreline lies about equidistant between the Stansbury and Bonneville shorelines and has been determined to postdate the Bonneville shoreline (Gilbert, 1890). In terms of the modern hydrograph (Fig. 7.1), the Provo shoreline represents a period of overflowing lake bracketed by periods of closed-basin lake conditions (Oviatt, 2015). Gilbert (1890) demonstrated that the Provo shoreline formed while the lake was overflowing at a threshold near Red Rock Pass, Idaho (Fig. 7.2).

Developments in Earth Surface Processes, Vol. 20. http://dx.doi.org/10.1016/B978-0-444-63590-7.00007-X
2016 Published by Elsevier B.V.

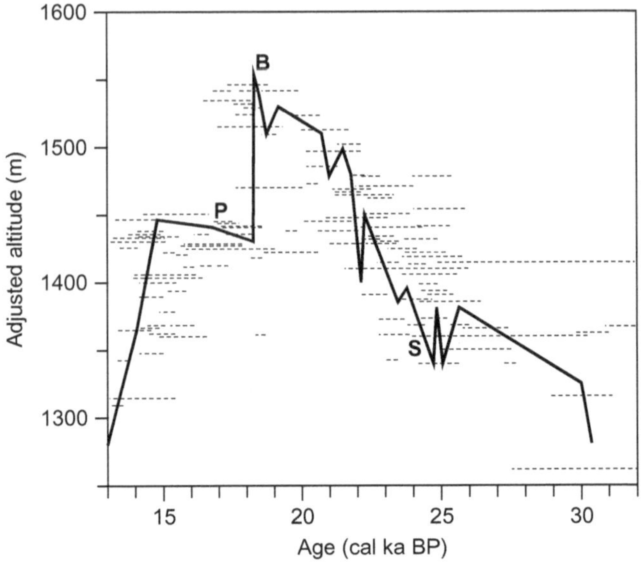

FIG. 7.1 Hydrograph for Lake Bonneville showing calibrated radiocarbon ages (*thin dashed lines*, 2σ range shown) vs elevations adjusted for isostatic rebound. Some ages are limiting ages, and do not define the lake level directly. Major shorelines: *B*, Bonneville; *P*, Provo; *S*, Stansbury. *Modified from Oviatt, C.G., 2015. Chronology of Lake Bonneville, 30,000 to 10,000 yr B.P. Quatern. Sci. Rev. 110, 166–171.*

Improved understanding of the Provo shoreline is important for understanding rates of geomorphic change, dynamics of enormous lakes, and details of lake response to climatic variability. Understanding the behavior of the threshold during long periods of overflow may yield insights on alluvial/fluvial systems. Understanding the duration of the Provo shoreline is fundamental to constraining isostatic rebound models and knowledge of regional lithosphere dynamics, as well as the timing and locus of rebound that is expressed in regional shorelines.

Although Gilbert made unparalleled advances in understanding the lake systems, several questions remain unresolved, including the controls on lake level at the threshold, details of the duration of the Provo lake, and the extent to which overflow variability reflects climatic variability (as used herein, the period during which Lake Bonneville overflowed after the Bonneville flood is "Provo time," and the lake during this time is the "Provo lake"). Carbonate isotopes on underwater cave carbonate and sediment cores yield lake-level interpretations that lack synchrony with each other and with the shoreline record. Finding measures of climatic variability during Provo time is a priority effort. These issues are addressed in this chapter.

7.2 G.K. GILBERT ON THE PROVO SHORELINE

Gilbert's (1890) extensive study of Lake Bonneville allowed him to arrive at several conclusions summarized here. He named the Provo shoreline for a

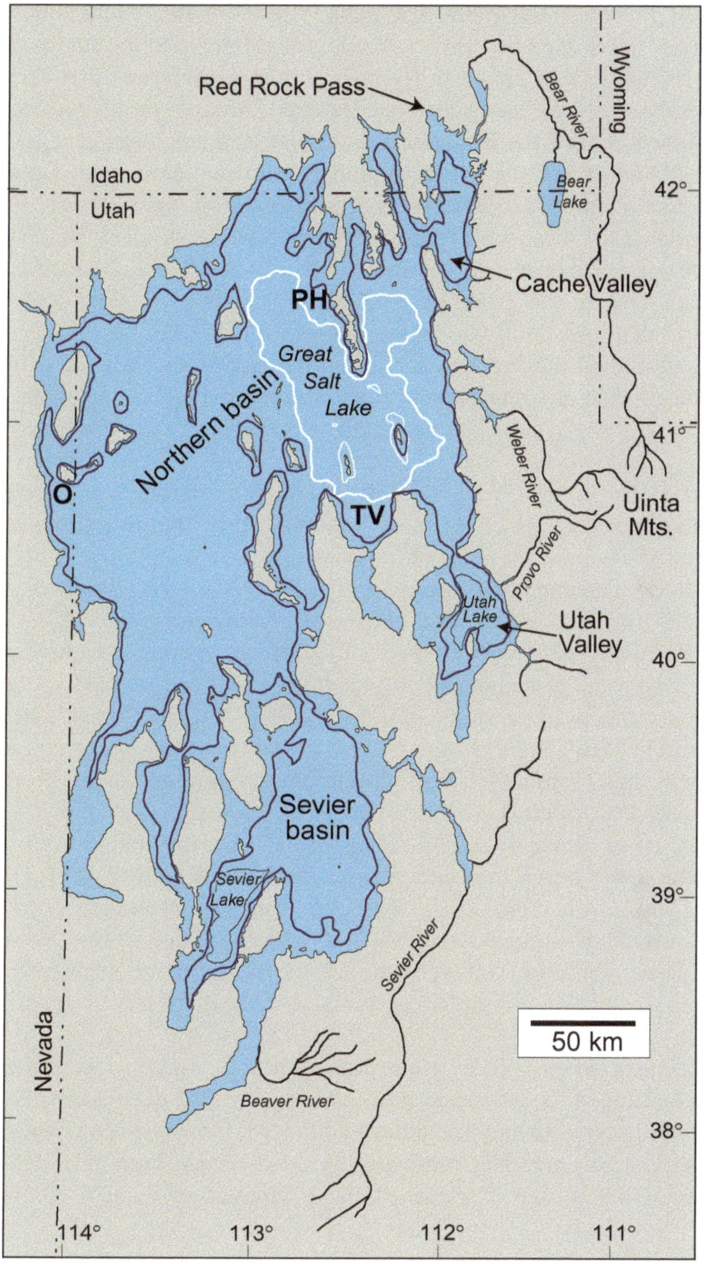

FIG. 7.2 Map of northern Utah area and Lake Bonneville with locations of detailed studies mentioned in the text. Bonneville shoreline outlines the extent of the lake, and inset *thick dark line* indicates the Provo shoreline. The *white line* indicates modern Great Salt Lake. *PH*, Promontory Hollow; *O*, Ola; *TV*, Tooele Valley.

prominent shoreline and delta that flanked the Wasatch Mountains in Utah Valley near the village of Provo and noted that the shoreline stands out from the other shorelines in the basin for being broader, more massive, and having more deeply cut cliffs. He determined that the Provo shoreline occurred about 375 ft (114 m) below the Bonneville shoreline, and postulated that more time was needed to create the Provo shoreline than the Bonneville because the smaller lake at the Provo level would have had less wave energy. He noted the abundance of tufa, which formed a few feet to as many as 30 ft (9 m) below beach crests both as a coating and as beach cement. He estimated that the surface area of the Provo lake was 13,000 mi^2 (33,700 km^2), of which a little more than one-tenth was in the Sevier arm (Fig. 7.2).

Gilbert studied the Provo shoreline in the Red Rock Pass area (Fig. 7.2), where he tracked the shoreline to near the town of Oxford, about 8 miles (13 km) south of the pass. Its elevation was similar to the floor of the pass, allowing him to deduce that the Bonneville flood reduced lake level from the Bonneville to the Provo shoreline, which therefore was formed in a threshold-controlled lake. He also reasoned that the overflow threshold migrated from somewhat north of the pass to a point south of Swan Lake during the flood. He noted that tufa is lacking in the Cache Valley, and reasoned that Bear River input into Cache Valley may have been greater than the overflow at Red Pass. This would have resulted in less-evaporated water in Cache Valley than in the remainder of the lake, an effect accentuated by restricted circulation caused by a single narrow passageway between Cache Valley and the main body.

Gilbert also wrote about the multiple beaches present at the Provo level, which generally present as two beaches but in a few places as many as 65 beaches were counted. He described the upper beach as the better developed, and termed the lower beach the "underscore," noting that it lay 5–20 ft (1.5–6 m) lower. Double wave-cut cliffs were observed in erosional settings that matched the double beaches found in most depositional settings. He assumed that the lower shoreline was younger and of shorter duration than the upper, but noted that no stratigraphic relations had been observed to prove this inference.

Although a few of Gilbert's interpretations with regard to the Provo shoreline have changed as chronometric methods, study of deep-water sediments, and knowledge of global climate have advanced, his conclusions have largely stood the test of time. The following sections review current understanding and present a few new hypotheses in an attempt to explain contradictory studies.

7.3 SHOREZONE GEOMORPHOLOGY AND DEPOSITS

The Provo shoreline of Lake Bonneville outlines an area of about 38,000 km^2, indicating a lake with a maximum depth of about 220 m and estimated

TABLE 7.1 Comparison of the Provo Lake with Modern Freshwater Lakes

Name	Area (km^2)	Volume (km^3)	Maximum Depth (m)	Continent
Provo	38,000	5300	220	North America
Superior	82,100	11,600	450	North America
Baikal	31,700	23,600	1640	Asia
Tanganyika	32,900	18,900	1470	Africa
Michigan	58,000	4920	280	North America
Huron	59,600	3540	230	North America
Victoria	68,800	2750	85	Africa
Great Bear	31,150	2240	445	North America
Malawi	29,600	7725	705	Africa
Vostok	15,700	5400	1000	Antarctica

volume of 5300 km^3 (Currey et al., 1984; updated by Adams and Bills (2016). The Provo lake ranks among the largest of modern freshwater lakes, similar to some of the Great Lakes, but was significantly shallower than most areally large lakes (Table 7.1). Many of the large lakes in Table 7.1 are also several million years old, with the Great Lakes being the most recently formed. The Great Lakes are unlike the Provo lake in that they have extensive watersheds with low relief and are in a warmer, nonglacial environment. Nonetheless, some aspects of the lakes in Table 7.1 may be useful as analogs for the geomorphology and deposits of the Provo lake.

The prominent geomorphic features associated with the Provo shoreline are illustrated in Fig. 7.3. The following summary is largely derived from Currey et al. (1983), Currey and Burr (1988), Currey (1990), and Oviatt and Miller (1997). The Provo lake formed in a broad basin traversed by many mountain ranges, from which small to very large streams flowed. As a result the source of shorezone sediment was highly variable, ranging from locally eroded materials to sediment delivered by rivers. Wave energy, in general, was tremendous, but varied greatly because of the complex basin geography. Wave-cut cliffs are common, as are broad, gently lakeward-sloping terraces of both erosional and constructional origin. Spits, recurved beaches, and tombolos are common constructional features, although bay-mouth barrier beaches are perhaps the most impressive and widespread landforms. Small to massive-sized deltas formed, with the largest developing along the western

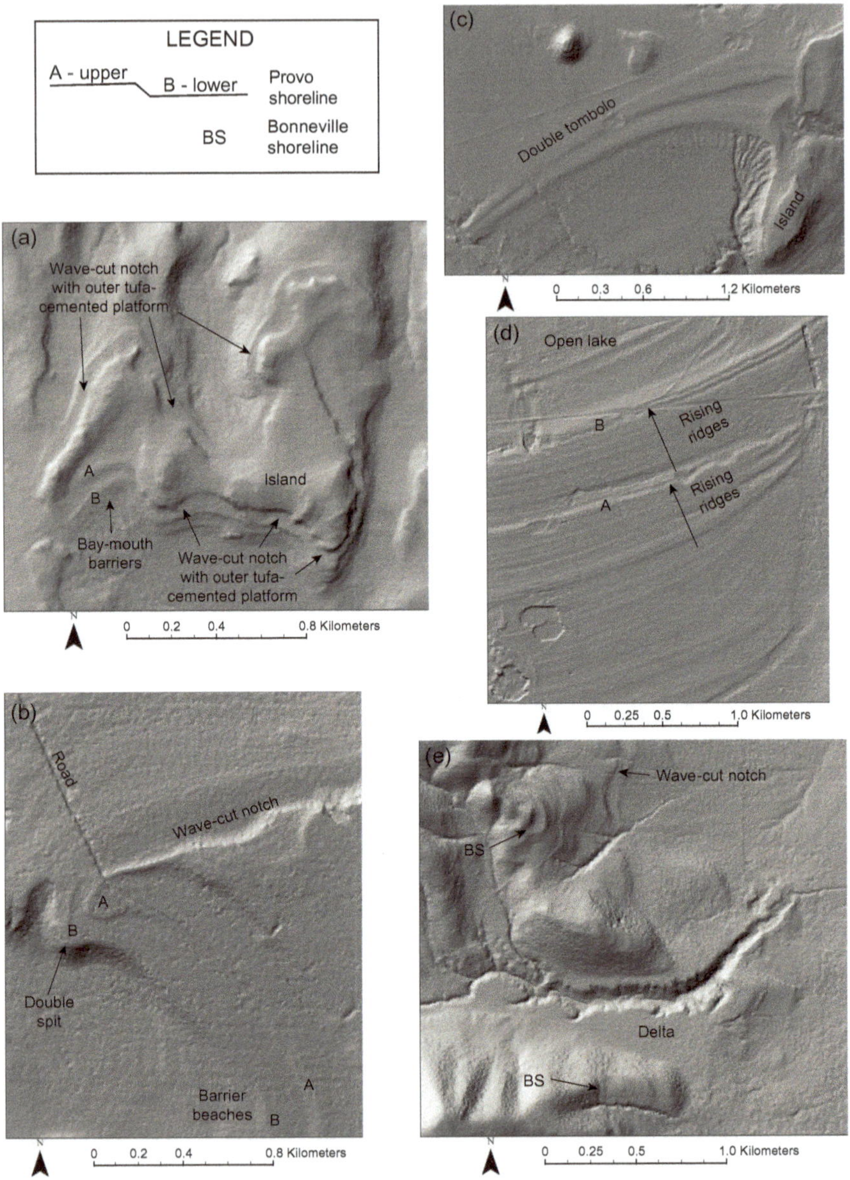

FIG. 7.3 Hillshade images for several Provo shoreline features. (a) Island formed by the Wildcat Hills, Curlew Valley. (b) Double spit and beaches, northern Hogup Mountains. (c) Double tombolo, Dove Creek Hills. (d) Multiple beach ridges, Tooele Valley. (e) Delta whose upper surface grades to the Provo level, western Malad Valley at Broad Canyon.

front of the Wasatch Mountains where there was ample sediment supply and active erosion.

Features in nearshore and offshore deposits serve to identify the Provo as well. Extensive tufa in beach deposits form beachrock and farther offshore, tufa-draped bedrock and sediment effectively armor underlying materials from erosion. Yet farther offshore, a broad zone of white-weathered marl is commonly visible from great distances, indicating thick accumulations. In offshore sediment, the beginning of Provo time is generally identifiable as marl overlying a thin, coarse-grained bed representing the Bonneville flood.

As noted by Gilbert (1890), the Provo shoreline in many places consists of two distinct features. Double beaches and double wave-cut notches are vertically separated by approximately 3 m (Miller et al., 2013). The upper feature is the younger of the two as demonstrated by stratigraphic and geomorphic relations, and in some locations even older underlying aggradational beach gravel occurs. In total, these stratigraphic data point to aggradation and lake-level rise during Provo time. Complicating interpretation of this lake-level rise is the differential rebound of the basin caused by removal of load during the Bonneville flood. Rebound varied from 0 to ~59 m (Currey, 1982), generally increasing from lake margin to basin center. Shorelines forming during this rebound would tend to decline in elevation with time so aggradational shoreline sequences represent lake-level rise exceeding the local rebound and/or postdating the rebound. Total lake-level rise may have been as great as 15 m, and probably was caused by aggradation of the threshold near Red Rock Pass by landslide deposits.

Exaggerated shoreline development occurred in a few places, and helps to clarify Provo lake dynamics. Gilbert (1890) counted as many as 65 beach ridges at the Provo shoreline. Study of two locations with numerous beach ridges, Tooele Valley and Promontory Hollow (Fig. 7.2), by Burr and Currey (1988) and Godsey et al. (2005) led to a nomenclature of alternating master high ridges and low swales (P1–P9), where ridges are assigned odd numbers and swales are assigned an even number. This system devolved the ridges into a more generalizable sequence of five ridges, and of these ridges, two (P1 and P3) were thought to represent the upper beach, and another set of two (P5 and P7) to represent the lower beach, of the typical two-beach Provo exposures described by Gilbert (Fig. 7.3d). The ridges form two sets of landforms that each aggrade and rise in elevation 3–6 m toward the open lake. This geometry can be explained as resulting from progressive lake-level rise with higher ridges stepping toward the lake until the whole beach complex is overtopped and new beaches form landward of the former lagoon, and at a higher elevation.

The P9 beach lies offshore of, and ~10 m below the two main Provo beaches, and has neither been identified in more than three locations (Godsey et al., 2005), nor tied to threshold control. Miller et al. (2013) suggested that the P9 represents a regressive closed-basin lake level following abandonment of the Provo shoreline. Janecke and Oaks (2012), however, identified a possible stream-meander feature roughly 10 m below the threshold south of the Red Rock Pass area and suggested that the lake at the P9 beach may have overflowed through a later-modified threshold at this elevation. Lack of an overflow route at this P9 level makes an alternative interpretation of the meanders as deep-lake scour during the Bonneville flood more attractive.

Another example of exaggerated Provo development is at Ola (Fig. 7.2), where a huge tombolo complex formed. A railroad cut exposes a post-Bonneville flood succession of gravel about 18 m thick that exhibits three internal packages of beach gravel, each topped by a flat tufa-cemented surface, and all of which are overlain by the two Provo beaches that are widespread in the basin (Miller et al., 2013). This stratigraphy was interpreted as representing ~15 m of lake-level rise during Provo time.

Despite the complex geography of the Bonneville basin and its many subbasins, internal control on Provo lake level was not a factor. Internal thresholds are only known well below and above the Provo shoreline (Sack, 2002). The Sevier basin (Fig. 7.2) was a wide, shallow arm of the Provo lake that may have had restricted circulation with the main lake. Water in Cache Valley also had restricted circulation with the main lake via a narrow notch (Gilbert, 1890) and water in Utah Valley may have had restricted circulation with the main water body as well.

7.4 NEARSHORE AND OFFSHORE SEDIMENTS

Offshore from beaches, sediments are finer grained due to increase in water depth and attendant decrease in wave energy. Although offshore sediment is rarely traceable to beaches, its stratigraphic context commonly can be defined by identification of the Bonneville flood bed. The flood bed is an unusually coarse bed contained within white marl, and commonly is marked by abundant ostracod shells (Oviatt et al., 1994). Study of offshore sediment in outcrop provides evidence for continuous occupation of the Provo shoreline with no "mid-Provo" drawdown (Godsey et al., 2011).

Offshore sediments such as the white marl offer continuous detailed archives to explore changes in sediment associated with lake hydrology and climate (Oviatt et al., 1994). Although ambiguities for causes of changes in sediment properties are inherent (Oviatt et al., 1994; Reheis et al., 2014), with careful correlation to the outcrop record and consideration of external forcings (eg, river diversions and threshold failures), the record contained in offshore sediment adds much detail to the history of the Provo lake. Studies of sediment and other deep-lake records are considered in more detail in the next section.

Delta records offer precise measures of water levels where topsets meet foresets (Reheis et al., 2014) and carry a wealth of information about the interactions of major rivers and the lake. Anderson and Link (1998), Milligan and Lemons (1998), and Lemons and Chan (1999) have studied deltas associated with major rivers along the east side of the basin.

7.5 CHRONOLOGY AND DYNAMICS

7.5.1 History of Chronologic Studies

The age of the Provo lake was weakly constrained as "glacial age" by Gilbert (1890) prior to radiometric dating techniques. After about 30 years of various researchers collecting radiometric ages for fossils and other materials, Oviatt et al. (1992) summarized the chronology of Lake Bonneville. They considered the Bonneville flood to be <15.3 ^{14}C ka BP and the duration of Provo lake to be about 15.0–14.0 ^{14}C ka BP. This result and more recent studies that benefited from additional dated material are given in Table 7.2. Godsey et al. (2005, 2011) dated nearshore and offshore sediment that could be tied to the Provo shoreline, and showed that the Provo lake existed for several thousand years.

Benson et al. (2011) and McGee et al. (2012) dated different materials, a sediment core and cave carbonate, respectively, and had to interpret what part of the dated sections corresponded to the Provo lake because direct ties with geomorphic features were lacking. Their interpretations conflict with the outcrop record and with each other, but they pointed to the possibility that outcrop dating had flaws or that events not identified from outcrop had affected the lake during Provo time. The core data of Benson et al. (2011; Fig. 7.4c) provide nonunique lake proxies that would be extremely difficult to interpret without the outcrop record. An abrupt increase in total inorganic carbon (TIC) at ~18.5 ka was interpreted as the beginning of overflow at the Bonneville shoreline but could also represent the Bonneville flood (Reheis et al., 2014). Benson et al. (2011) interpreted a decrease in TIC at

TABLE 7.2 Comparison of Ages for Provo Time (All in cal ka BP)

Reference	Methods	Materials	Provo Start	Provo End
Oviatt et al. (1992)	^{14}C various	Outcrops	17.7	17.0
Godsey et al. (2005)	^{14}C on mollusks	Outcrops	17.7	13.8
Godsey et al. (2011)	^{14}C on mollusks	Outcrops		15.0
Benson et al. (2011)	^{14}C and paleomagnetic secular variation	Sediment core	17.0	15.2
McGee et al. (2012)	^{14}C and U-series	Cave carbonate	18.2	16.4
Miller et al. (2013)	^{14}C on mollusks	Outcrops	18.1	
Oviatt (2015)	^{14}C various	Outcrops	18.0	15.0

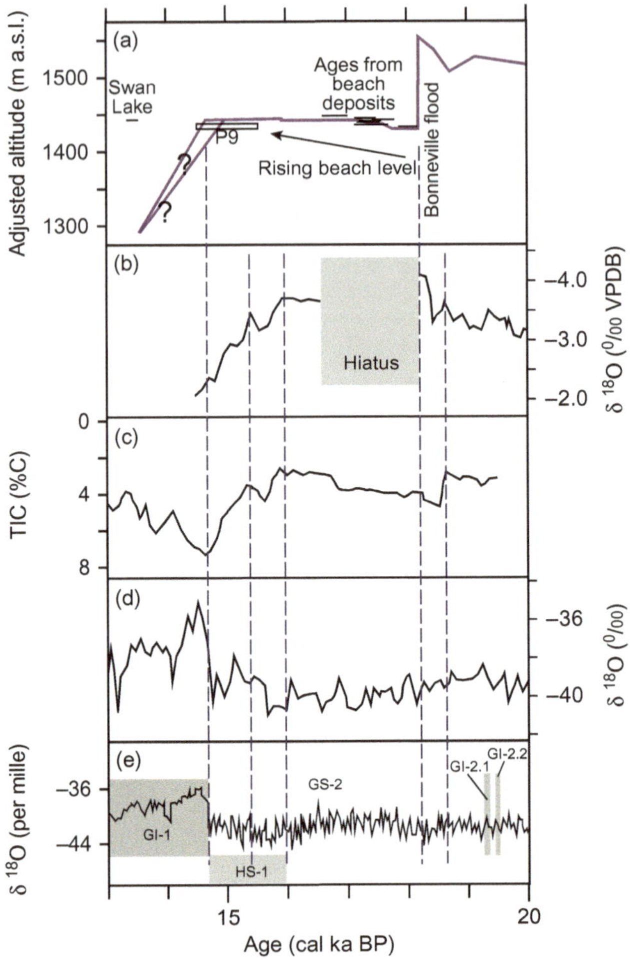

FIG. 7.4 Plots of lake levels and climate proxies. (a) Hydrograph for Lake Bonneville from 20 to 13 ka. *Black horizontal lines* are AMS radiocarbon ages for the Provo beach deposits shown with 2σ errors. Box labeled "P9" encloses several gastropod ages that were all collected from same post-Provo beach outcrop, but prepared and analyzed in different labs (Godsey et al., 2011). (b) Plot of $\delta^{18}O$ from cave carbonate. (c) Plot of TIC measured from sediment core. (d) GISP2 Greenland ice core isotope record (Stuiver et al., 1995). (e) NGRIP oxygen isotopic record with ages adjusted from ice layer counting and matching abrupt transitions at beginnings of Greenland interstadials (GI). Interstadials (warm events) defined by this record as well as by dust flux record (not shown). *GS*, Greenland stadial. *Panel (a): Modified from Miller, D.M., Oviatt, C.G., McGeehin, J.P., 2013. Stratigraphy and chronology of Provo shoreline deposits and lake-level implications, Late Pleistocene Lake Bonneville, eastern Great Basin, USA. Boreas 42, 342–361. doi:10.1111/j.1502-3885.2012.00297.x. ISSN 0300-9483. Panel (b): After McGee, D., Quade, J., Edwards, R. L., Broecker, W.S., Cheng, H., Reiners, P.W., and Evenson. P., 2012. Lacustrine cave carbonates: novel archives of paleohydrologic change in the Bonneville Basin (Utah, USA). Earth Planet. Sci. Lett., 351–352, 182–194. Panel (c): After Benson, L.V., Lund, S.P., Smoot, J.P., Rhode, D.E., Spencer, R.J., Verosub, K.L., Louderback, L.A., Johnson, C.A., Rye, R.O., Negrini, R.M., 2011. The rise and fall of Lake Bonneville between 45 and 10.5 ka. Quatern. Int., 235, 57–69. Panel (e): After Rasmussen, S.O., Bigler, M., Blockley, S.P. Blunier, T., Buchardt, S.L., Clausen, H.B., Cvijanovic, I., Dahl-Jensen, D., Hoek, W.Z., Lowe, J.J., Pedro, J.B., Popp, T., Seierstad, I.K., Steffensen, J.P., Svensson, A.M., Vallelonga, P., Vinther, B.M., Walker, M.J.C., Wheatley, J.J., Winstrup, M., 2014, A stratigraphic framework for abrupt climatic changes during the Last Glacial period based on three synchronized Greenland ice-core records: refining and extending the INTIMATE event stratigraphy, Quatern. Sci. Rev., 106, 14–28.*

~17.0 ka as the Bonneville flood, and an increase at ~15.2 ka to be the end of the Provo overflow. McGee et al. (2012) interpreted a decrease in $\delta^{18}O$ at ~18.9 ka as the beginning of overflow at the Bonneville shoreline (Fig. 7.4b). A hiatus in calcite deposition from 18.2 to 16.4 ka was interpreted as Provo overflow, which reduced calcite saturation.

Miller et al. (2013) combined stratigraphic and geomorphic study of Provo beach deposits to suggest that the Bonneville flood was older than had been assumed, and the Provo lake began as early as 18.3 cal ka BP. In a recent review paper, Oviatt (2015) suggested that the Provo lake existed from ~18 to ~15 cal ka BP. A presentation of the age data currently available is given in Fig. 7.4a. This representation shows a Bonneville flood age of 18.2–18.3 cal ka BP, which is derived from the cluster of best ages for the Bonneville shoreline and the several ages in the 17.5–18.1 cal ka BP range that represent early beach accumulation at Ola. The P7 beach is dated only by offshore sediment. The P9 beach, which apparently postdates the Provo shoreline, is dated by gastropods within the beach at about 14.8 cal ka BP and serves as a minimum-limiting age for the Provo. By this interpretation, the Provo lake existed from about ~18.2 to 14.8 cal ka BP.

7.5.2 Integrated Model

Although published conclusions for timing of the Provo lake are disparate, a solution that integrates and honors all studies may be possible. The several divergent chronologies (Table 7.2) employ different materials, use several different dating methods, and investigate outcrop and deep-lake archives. Possible limitations to each of these studies exist, and they range from uncertainties associated with dating lacustrine mollusks and tufa (Miller et al., 2013; Oviatt, 2015), to uncertainties in tying offshore sediments to specific lake levels (Godsey et al., 2011; Reheis et al., 2014), to ambiguities in applying proxies for lake behavior based on presumably well-dated sediment-core and cave-carbonate archives.

Recent dating of beach sediment has suggested that the Provo lake began as early as 18.2–18.5 cal ka BP. Oviatt (2015) suggested that this earlier age can be understood if one accepts that the Bonneville shoreline was cut by a rising lake, not a threshold-controlled lake. The Bonneville shoreline is primarily visible as a distinct break between topography above that was created by fluvial processes and topography below that was primarily created by lacustrine processes (Gilbert, 1890). By this view, the lake did not stabilize at the Bonneville shoreline; rather, when the lake overtopped an ancestral divide near Red Rock Pass (Fig. 7.2), it immediately failed catastrophically and rapidly entrenched as much as 125 m (Miller et al., 2013), stabilizing on a bedrock threshold to begin the Provo lake. Ages for the Bonneville shoreline would therefore coincide with the Bonneville flood (Fig. 7.4). This finding fits with the cave record of McGee et al. (2012) if their

18.2 ± 0.3 cal ka BP hiatus (which represents lake mixing and/or reduced saturation of waters) represents the beginning of overflow and the flood. This date is constrained by U–Th methods as well as radiocarbon, helping to reduce uncertainties caused by radiocarbon dating of mollusks. This finding may also fit with the ~18.5 cal ka BP sharp increase in sediment TIC measured by Benson et al. (2011), which could represent the Bonneville flood (Reheis et al., 2014). By this reasoning the Provo shoreline was initiated between 18.5 and 18.2 cal ka BP.

The end of the Provo lake is constrained by nearshore- and offshore-dated sediments (Godsey et al., 2005, 2011) and by post-overflow sediment in the threshold area (Miller et al., 2015) as older than 13.5 cal ka BP and probably older than a ~14.5 cal ka BP cluster of ages in sediment offshore from the outermost beach of the Provo shoreline. It appears that the "P9" beach, ~10 m below the Provo shoreline, formed at about 14.8 cal ka BP (Fig. 7.4), further limiting the end of the Provo lake (Godsey et al., 2011). This timing of ~14.8 cal ka BP for the termination of the Provo aligns well with the McGee et al. (2012) conclusion that cave carbonate deposition ended at 14.7 cal ka BP, although it differs from their interpretation of that event. This timing is somewhat younger than the ending at 15.3 cal ka BP interpreted by Benson et al. (2011), but that core record can also be interpreted as indicating an overflowing lake until ~15.1–14.8 cal ka BP.

The interpretations of lake sediment (Benson et al., 2011) and cave carbonate deposit (McGee et al., 2012) chemistry differ in several aspects from each other and from the timing proposed earlier. These discrepancies may be partly reconciled by considering Gilbert's (1890) recognition that Cache Valley, the locus of both Bear River freshwater input and discharge through Red Rock Pass, had restricted circulation with the remainder of the Provo water body. If the Bear River discharge was similar to Provo lake discharge at Red Rock Pass, little exchange between Cache Valley and the main body of the lake would occur. Chemical and isotopic measures of lake water in the main body would in this case reflect a highly evaporative condition similar to a closed-basin lake. This may be reflected in the 16.2–14.7 ka "closed basin" record interpreted by McGee et al. (2012), as well as the ~15.9 ka reduction in overflow interpreted by Benson et al. (2011). In addition, stepped increases in lake level caused by rises in the threshold may have temporarily caused strictly closed lake behavior during lake level rise to the new threshold. Such events would be indistinguishable from the drop below the threshold proposed by Benson et al. (2011) but not observed in the outcrop record (Godsey et al., 2011).

7.5.3 Isostatic Rebound

The entire Bonneville basin region was depressed by the load of Lake Bonneville, and rebounded as lake load decreased (Gilbert, 1890). An event such as the Bonneville flood, with its nearly instantaneous release of ~125 m of water load, must have caused isostatic rebound that was distinct from the rebound following drawdown from the Provo shoreline. Miller et al. (2013) analyzed this rebound by evaluating correspondence between (a) the vertical separation between the two Provo shorelines and (b) the amount of total rebound marked by the bowed Bonneville shoreline. Correspondence is extremely low, suggesting that there is no differential rebound observed in the two shorelines. This was interpreted as indicating that most of the rebound from the Bonneville flood was accomplished during the early part of Provo time, before the two shorelines occurring so widely around the basin were formed.

Shorelines formed during rebound shortly after the Bonneville flood might be poorly preserved because the shorelines would be slowly uplifted and continually reworked. Toward the margins of the basin, earlier shorelines might be preserved because isostatic rebound was lower magnitude, but none have thus far been identified. It is possible that beaches identified as P9 in some places are actually early Provo beaches.

7.6 PALEOCLIMATE INTERPRETATIONS

7.6.1 Evolution of Thought

Gilbert (1890) argued that records of closed-basin lakes serve as paleoclimate proxies by establishing that historical Great Salt Lake levels changed as a function of river discharge to the lake. He reasoned that the rise and fall of Lake Bonneville similarly reflected increased discharge followed by reduced discharge. Much later, Spencer et al. (1984) examined sediment cores from Great Salt Lake to interpret lake history and paleoclimate, but they interpreted the Provo lake as part of the transgressive history predating the Bonneville shoreline. Oviatt et al. (1992, 1994) laid out the radiocarbon chronology and marl chemistry for Lake Bonneville, and later, a paleoclimate interpretation (Oviatt, 1997) by comparing chronology and lake levels to global climate records. The latter study argued that major oscillations in lake level during the rise to the Bonneville shoreline coincided with warm, dry periods, some of which are global in extent such as Dansgaard–Oeschger interstadial events, and that some rapid lake-level rises coincided with global ice-rafting (Heinrich) stadial events. Godsey et al. (2005) argued that the Provo shoreline existed for much longer than previously appreciated, which would require that it coincided with the last Heinrich stadial (Fig. 7.4).

Sediment core and cave carbonate records (Benson et al., 2011; McGee et al., 2012) are important because they can show nuances of closed-basin and overflowing dynamics of the lake during Provo time, when the shoreline record is mainly reflecting threshold elevation changes that are likely unrelated to climate events (Miller et al., 2013). The interpretations must be tempered by the knowledge that the lake sediment mineralogy and isotope chemistry change as a function of location within the basin (Oviatt et al.,

1994) and therefore lake mixing can cause changes that might be interpreted as climatic changes. McGee et al. (2012) argued that long residence time of lake waters (centuries) and small changes in $\delta^{18}O$ in rainfall over time together can be used to interpret $\delta^{18}O$ as largely driven by the balance between river discharge and evaporation in the basin. However, changes in mixing of Sevier basin (Oviatt et al., 1994) and Cache Valley (Gilbert, 1890) waters with the main body, as well as temporary closed-basin behavior driven by threshold elevation increases, also may drive the lake chemistry during Provo time. Dynamics between the main body and Cache Valley may be key to understanding these records and thereby Provo-lake discharge. The main rivers feeding Lake Bonneville, apart from the Sevier River, all head in the western Uinta Mountains (Fig. 7.2). When the sum of Provo and Weber river discharge exceeded evaporation in the main body, flow from the main body into Cache Valley occurred and "overflowing" conditions held in the main body. Reduced discharge of these rivers would cause flow to instead come *from* Cache Valley to the main body, creating a "closed basin" signature for the main body. Because the Bear River is the greatest of the three rivers, it is possible for threshold discharge of the lake to be maintained while the main body no longer discharges to Cache Valley.

The most pronounced changes in cave carbonate during Provo time are an extended hiatus from 18.2 to 16.4 ka, followed by signatures of enhanced evaporation to 14.7 ka. It is suggested here that the hiatus represents lake mixing caused by the Bonneville flood (McGee et al., 2012), which disrupted carbonate deposition during early Provo time. Later calcite deposition may represent varying ratios of threshold discharge and flow to Cache Valley. Benson et al. (2011) described deep-water marl deposited during Provo time and interpreted an increase in TIC at ~18.5 ka as the beginning of overflow, which is about the time of the Bonneville flood (Oviatt, 2015). Changes in TIC and $\delta^{18}O$ during Provo time, especially at ~16.8, ~15.8, and ~15.2 ka (Fig. 7.4), may indicate changes in water mixing and/or threshold elevation increases. For instance, the spike in TIC at ~15.8 ka may represent a threshold rise. Reduced TIC at ~16.8–15.9 ka may reflect greater river discharge and lake overflow, after which effective closed-basin conditions held until the end of Provo time.

The interpretive history of Provo time outlined above compares favorably with northern hemisphere climate records. Two of the Greenland ice records are shown in Fig. 7.4: GISP2 (Stuiver et al., 1995) and a tuned Greenland stadial/interstadial timing (Rasmussen et al., 2014; Steffensen et al., 2014). Greater overflow after ~16.8 ka, if the interpretation of the Benson et al. (2011) data is correct, may coincide with the onset of Heinrich 1. This is in keeping with the interpretations of McGee et al. (2012) that earlier Heinrich events correlate with rapid increase in lake volume, but conflicts with the Benson et al. (2011) interpretation of Heinrich 2 as a dry period. The ~14.8 ka termination of the Provo lake and rapid desiccation are closely synchronous with the beginning of the Bølling-Allerød dry period.

7.6.2 Comparison with Great Basin Lake Records

The Provo shoreline formed during a long time interval that coincides with the highstands of all central and northern Great Basin lakes. Reheis et al. (2014) summarized many studies of lakes in the Great Basin, concluding that all highstands occurred between 18 and 15 cal ka BP, and that many were in the latter half of that time interval. The timing for many highstands therefore broadly coincides with the last Heinrich stadial and is consistent with the sediment core record of Benson et al. (2011) (as interpreted herein) which showed that the 16.8–15.9 ka Provo lake may have had greatest overflow.

7.6.3 Overflow

Overflow conditions during Provo time have not been investigated directly but studies of deep-water sediment and cave deposits (described previously) provide hints of variable discharge during Provo time. Godsey et al. (2005) suggested that dated mollusks from offshore sediment required a mid-Provo lake-level fall, and cessation of discharge near Red Rock Pass, but later (Godsey et al., 2011) showed that sedimentological data do not support this interpretation. Benson et al. (2011) interpreted TIC in cored sediment to indicate a halt in overflow at ~15.9 ka. However, as described earlier, variable mixing with Cache Valley and threshold elevation increase might provide signals that can be misinterpreted as enhanced evaporation similar to a closed-basin lake.

7.7 QUESTIONS AND FUTURE STUDIES

Behavior of the overflow threshold and lake-level controls during Provo time needs investigation. Incision of the threshold by lake overflow and opposing channel infill by tributary channels and landsliding apparently caused fine adjustments to threshold levels during the long Provo period of threshold control. Understanding the interplay of geomorphic factors that raise and lower the threshold may yield more general insights applicable to closed-basin lakes. In addition, resolving the origin of meander (?) features set well below the Provo level (Janecke and Oaks, 2012) south of Red Rock Pass is needed, whether caused by overflow at the threshold or alternatively as deep-water scours formed during the Bonneville flood or by some mechanism that postdates Lake Bonneville entirely. Possible effects of earthquakes on threshold elevations (Janecke and Oaks, 2012) is another topic needing study. Although the P9 beach clearly postdates the main Provo beaches at Promontory, it is not known whether it represents threshold control at a reduced elevation of the Red Rock Pass area or stillstand of a post-Provo closed-basin lake.

Offshore sediments and carbonate cave deposits yield well-dated records of chemical and sediment properties that are interpreted differently than the outcrop record. Reconciling these records, whether by mechanisms involving Cache Valley dynamics as proposed herein, or by other means, is an important topic. The resolved records will provide better estimates of Provo chronology as well as fine-scale variations that may be interpretable in a regional-climate context. Studying sediment chemistry in Cache Valley (recorded in delta and offshore deposits) may be particularly fruitful, especially if compared to nearby marl deposits of the main body.

Two, and in a few places as many as 5–65, sets of beaches formed in the basin, and together comprise the Provo shoreline landform (Godsey et al., 2005; Miller et al., 2013). How widespread is evidence for more than two beaches, and what are the geometries of each package of sediment? Ground-penetrating-radar studies of select delta and beach deposits may answer these questions and help define lake levels for each using distinctive sediment transitions (Reheis et al., 2014).

Improving paleoclimate interpretations for Provo time depends partly on improved chronology for outcrop records and deep-water archives but also requires better information on the evolution of lake waters in several subbasins. Opportunities to improve dating of the Bonneville flood and the end of Provo lake overflow are needed, whether in deposits internal to Lake Bonneville or deposits associated with the flood's downstream record. Can cosmogenic exposure ages on beach deposits using water shielding of cosmogenic signal improve relative chronology for the beaches?

7.8 SUMMARY AND CONCLUSIONS

The Provo lake built the most prominent shoreline features of Lake Bonneville and deposited distinctive offshore sediments. The geomorphology and sediments have been studied and yielded diverse interpretations. A model for the Provo lake that integrates many features of these studies and emphasizes potential controls by restricted circulation among subbasins is proposed to stimulate studies that will reconcile disparate interpretations.

The lake lasted about 3400 years, from ~18.2 ka (the Bonneville flood) to ~14.8 ka (the beginning of the Bølling-Allerød dry period). During this time lake-level rose due to threshold control and aggradational beach deposits formed in response. The early Provo lake was strongly affected by differential isostatic rebound, nearly 60 m in the center of the basin, caused by removal of the lake load during the Bonneville flood.

Conflicting interpretations of deep-water sediment and cave-carbonate records, along with the outcrop record, must be resolved before the climatic implications for Provo time will be evident. More study of lake levels are needed using sediment transitions in deltas and beach deposits, improved chronology, and thorough investigation of varying deep-water records in subbasins.

ACKNOWLEDGMENTS

I thank many scholars of Lake Bonneville who have generously shared insights and corrected my mistakes: Jack Oviatt and Don Currey; Ken Adams, Genevieve Atwood, Holly Godsey, Susanne Janecke, David Madsen, Jim O'Connor, and Tammy Rittenour. Reviews by Holly Godsey and Marith Reheis, as well as editorial comments by Jack Oviatt, are appreciated.

REFERENCES

Adams, K.D., Bills, B.G., 2016. Chapter 8. Isostatic rebound and palinspastic restoration of the Bonneville and Provo shorelines in the Bonneville basin, UT, NV, and ID. In: Oviatt, C.G., Shroder Jr., J.F. (Eds.), Lake Bonneville: A Scientific Update, vol. 20. Elsevier, pp. 145–164.

Anderson, S.L., Link, P.K., 1998. Lake Bonneville sequence stratigraphy, Pleistocene Bear River delta, Cache Valley, Idaho. In: Pitman, J.K., Carroll, A.R. (Eds.), Modern and Ancient Lake Systems: New Problems and Perspectives, pp. 91–104. Utah Geological Association Guidebook 26.

Benson, L.V., Lund, S.P., Smoot, J.P., Rhode, D.E., Spencer, R.J., Verosub, K.L., Louderback, L.A., Johnson, C.A., Rye, R.O., Negrini, R.M., 2011. The rise and fall of Lake Bonneville between 45 and 10.5 ka. Quatern. Int. 235, 57–69.

Burr, T.N., Currey, D.R., 1988. The Stockton bar. In: Machette, M.N. (Ed.), In the Footsteps of G. K. Gilbert—Lake Bonneville and Neotectonics of the Eastern Basin and Range Province, pp. 66–73. Geological Society of America Annual Meeting, Field Trip Guidebook, Utah Geological and Mineral Survey Miscellaneous Publication, 88-1.

Currey, D.R., 1982. Lake Bonneville: selected features of relevance to neotectonic analysis: U.S. Geological Survey Open-File Report 82-1070. 30 p, map 1:500,000.

Currey, D.R., 1990. Quaternary paleolakes in the evolution of semidesert basins, with special emphasis on Lake Bonneville and the Great Basin, U.S.A. Palaeogeogr. Palaeoclimatol. Palaeoecol. 76, 189–214.

Currey, D.R., Burr, T.N., 1988. Linear model of threshold-controlled shorelines of Lake Bonneville. In: Machette, M.N. (Ed.), In the Footsteps of G. K. Gilbert—Lake Bonneville and Neotectonics of the Eastern Basin and Range Province, pp. 104–110. Geological Society of America Annual Meeting, Field Trip Guidebook, Utah Geological and Mineral Survey Miscellaneous Publication 88-1.

Currey, D.R., Oviatt, C.G., Plyler, G.B., 1983. Lake Bonneville stratigraphy, geomorphology, and isostatic deformation in west-central Utah. In: Gurgel, K.D. (Ed.), Geologic Excursions in Neotectonics and Engineering Geology in Utah, pp. 63–82. Guidebook-Part IV, Utah Geological and Mineralogical Survey Special Studies 62.

Currey, D.R., Atwood, G., Mabey, D.R., 1984. Major Levels of Great Salt Lake and Lake Bonneville. Utah Geological and Mineral Survey Map 73.

Gilbert, G.K., 1890. Lake Bonneville: U.S. Geological Survey Monograph 1. 438 p.

Godsey, H.S., Currey, D.R., Chan, M.A., 2005. New evidence for an extended occupation of the Provo shoreline and implications for regional climate change, Lake Bonneville, Utah, USA. Quatern. Res. 63, 212–223.

Godsey, H.S., Oviatt, C.G., Miller, D.M., Chan, M.A., 2011. Stratigraphy and chronology of offshore to nearshore deposits associated with the Provo shoreline Pleistocene Lake Bonneville, Utah, USA. Palaeogeogr. Palaeoclimatol. Palaeoecol. 310, 442–450.

Janecke, S.U., Oaks Jr., R.Q., 2012. New insights into the outlet conditions of Late Pleistocene Lake Bonneville, southeastern Idaho, USA. Geosphere 7, 1369–1391.

Lemons, D.R., Chan, M.A., 1999. Facies architecture and sequence stratigraphy of fine-grained lacustrine deltas along the eastern margin of Late Pleistocene Lake Bonneville, northern Utah and southern Idaho. Am. Assoc. Pet. Geol. Bull. 83, 635–665.

McGee, D., Quade, J., Edwards, R.L., Broecker, W.S., Cheng, H., Reiners, P.W., Evenson, P., 2012. Lacustrine cave carbonates: novel archives of paleohydrologic change in the Bonneville Basin (Utah, USA). Earth Planet. Sci. Lett. 351–352, 182–194.

Miller, D.M., Oviatt, C.G., McGeehin, J.P., 2013. Stratigraphy and chronology of Provo shoreline deposits and lake-level implications, Late Pleistocene Lake Bonneville, eastern Great Basin, USA. Boreas 0300-9483. 42, 342–361. http://dx.doi.org/10.1111/j.1502-3885.2012.00297.x.

Miller, D.M., Wahl, D.B., McGeehin, J.P., Rosario, J., Oviatt, C.G., Anderson, L., Presnetsova, L., 2015. Limiting age for the Provo shoreline of Lake Bonneville. Quatern. Int. 387, 99–105. http://dx.doi.org/10.1016/j.quaint.2015.01.001.

Milligan, M.R., Lemons, D.R., 1998. A sequence stratigraphic overview of sandy and gravelly lacustrine deltas deposited along the eastern margin of late Pleistocene Lake Bonneville, northern Utah. In: Pitman, J.K., Carroll, A. (Eds.), Modern and Ancient Lake Systems: New Problems and Perspectives, pp. 105–129. Utah Geological Association Guidebook 26.

Oviatt, C.G., 1997. Lake Bonneville fluctuations and global climate change. Geology 25, 155–158.

Oviatt, C.G., 2015. Chronology of Lake Bonneville, 30,000 to 10,000 yr B.P. Quatern. Sci. Rev. 110, 166–171.

Oviatt, C.G., Miller, D.M., 1997. New explorations along the northern shores of Lake Bonneville. In: Link, P.K., Kowallis, B.J. (Eds.), Mesozoic to Recent Geology of Utah, Brigham Young University Geology Studies. 42, pp. 345–371. Part II.

Oviatt, C.G., Currey, D.R., Sack, D., 1992. Radiocarbon chronology of Lake Bonneville, eastern Great Basin, USA. Palaeogeogr. Palaeoclimatol. Palaeoecol. 99, 225–241.

Oviatt, C.G., Habiger, G.D., Hay, J.E., 1994. Variation in the composition of Lake Bonneville marl: a potential key to lake-level fluctuations and paleoclimate. J. Paleolimnol. 11, 19–30.

Rasmussen, S.O., Bigler, M., Blockley, S.P., Blunier, T., Buchardt, S.L., Clausen, H.B., Cvijanovic, I., Dahl-Jensen, D., Hoek, W.Z., Lowe, J.J., Pedro, J.B., Popp, T., Seierstad, I.K., Steffensen, J.P., Svensson, A.M., Vallelonga, P., Vinther, B.M., Walker, M.J.C., Wheatley, J.J., Winstrup, M., 2014. A stratigraphic framework for abrupt climatic changes during the Last Glacial period based on three synchronized Greenland ice-core records: refining and extending the INTIMATE event stratigraphy. Quatern. Sci. Rev. 106, 14–28.

Reheis, M.C., Adams, K.D., Oviatt, C.G., Bacon, S.N., 2014. Pluvial lakes in the Great basin of the western United States—a view from the outcrop. Quatern. Sci. Rev. 97, 33–57.

Sack, D., 2002. Fluvial linkages in Lake Bonneville subbasin integration. In: Hershler, R., Madsen, D.B., Currey, D.R. (Eds.), Great Basin Aquatic Systems History, pp. 129–144. Smithsonian Contributions to Earth Sciences No. 38.

Spencer, R.J., Baedecker, M.J., Eugster, H.P., Forester, R.M., Goldhaber, M.B., Jones, B.F., Kelts, K., McKenzie, J., Madsen, D.B., Rettig, S.L., Rubin, M., Bowser, C.J., 1984. Great Salt Lake and precursors, Utah: The last 30,000 years. Contrib. Mineral. Petrol. 86, 321–334.

Steffensen, J.P., Svensson, A.M., Vallelonga, P., Vinther, B.M., Walker, M.J.C., Wheatley, J.J., Winstrup, M., 2014. A stratigraphic framework for abrupt climatic changes during the Last Glacial period based on three synchronized Greenland ice-core records: refining and extending the INTIMATE event stratigraphy. Quatern. Sci. Rev. 106, 14–28.

Stuiver, M., Grootes, P.M., Braziunas, T.F., 1995. The GISP2 $\delta^{18}O$ climate record of the past 16,5000 years and the role of the sun, ocean, and volcanoes. Quatern. Res. 44, 341–354.

Chapter 8

Isostatic Rebound and Palinspastic Restoration of the Bonneville and Provo Shorelines in the Bonneville Basin, UT, NV, and ID

K.D. Adams* and B.G. Bills†

*Desert Research Institute, Reno, NV, United States

†Jet Propulsion Laboratory, California Institute of Technology, Pasadena, CA, United States

ABSTRACT

This study uses modern digital elevation models (DEMs) along with existing Bonneville and Provo shoreline elevation measurements to palinspastically restore the landscape to its vertical position at the times these two shorelines were forming. The new data products produced in the study include a raster portraying the magnitude of rebound associated with the Bonneville shoreline as well as DEMs with the rebound associated with the Bonneville and Provo shorelines removed and derivative contours that more accurately trace the paths of these two shorelines on the ground. These products are useful in not only making more accurate maps, lake-surface area estimates, and lake-volume estimates at discrete times, they can also be used to correct the elevations of dating samples and more confidently identify the locations of lacustrine deposits and landforms related to the Bonneville and Provo lake stands for further study.

Keywords: Lake Bonneville, Shorelines, Isostatic rebound, Palinspastic restoration, Provo, Mapping

8.1 INTRODUCTION

Lake Bonneville was the largest pluvial lake in the Great Basin of the western US during the late Pleistocene. Over the last ~30,000 years the water surface elevation of Lake Bonneville has fluctuated through an elevation range of almost 350 m, mainly in response to changes in precipitation and evaporation (eg, Gilbert, 1890; Currey, 1990; Oviatt et al., 1992; Oviatt, 1997, 2015). As

Developments in Earth Surface Processes, Vol. 20. http://dx.doi.org/10.1016/B978-0-444-63590-7.00008-1

the water surface elevation changed, the erosion and deposition associated with wave-driven sediment transport formed a series of prominent shorelines, which are still readily visible on the landscape of the basin. When these shorelines formed they were essentially level, except for following the curvature of the Earth, corresponding closely to the prevailing still-water elevation of the water surface.

By the time the rising lake level reached the low spot (~1552 m) on the rim of its basin and began to spill into the Snake River drainage, the water load impounded by Lake Bonneville was massive; sufficiently so that it depressed the crust of the Earth by many tens of meters in the center of the basin. Shortly after overflow began, the alluvial dam near Red Rock Pass catastrophically failed at about 18 cal ka, releasing the upper 108–125 m of the lake in the massive Bonneville flood that may have lasted up to 1 year (Gilbert, 1890; Malde, 1968; O'Connor, 1993; Miller et al., 2013). The deeper Earth also responded to the sudden release of this water load, isostatically rebounding to a degree that was proportionate to the removed load (Bills and May, 1987; Bills et al., 1994, 2002).

Lake Bonneville continued to overflow at the Provo level for as long as 2500 years (Godsey et al., 2011; Miller et al., 2013, 2015; Oviatt, 2015), which allowed ample time for a suite of prominent shoreline features to develop at about 1444 m. The amount of time that the lake spent at the Provo level is also thought to have been sufficient for most of the rebound associated with the Bonneville flood to have occurred while the lake continued to spill through Red Rock Pass (Passey, 1981; Miller et al., 2013). A second period of isostatic rebound ensued when the lake receded from Red Rock Pass at some point between 16 and 15 cal ka due to climate change (Godsey et al., 2005, 2011; Miller et al., 2013, 2015; Oviatt, 2015), when the volume of water evaporating from the lake surface exceeded inflow (Gilbert, 1890).

The amount by which a particular point on a given shoreline is "de-leveled," or removed from its initial formation elevation by isostatic rebound, depends upon the time and place of its formation. More specifically, it depends upon the amount of isostatic subsidence, or downward deflection that had already occurred at the time and place of formation, and how much rebound has occurred since formation.

Currey (1982) collected 181 elevation measurements on the Bonneville shoreline and 112 on the Provo shoreline. Both of these data sets document broad domal patterns of rebound. The highest elevated Bonneville shorelines near the center of the basin are 74 m higher than the same age shorelines near the periphery of the basin, where the water load was significantly less. Isostatic rebound of the Provo shoreline approaches 59 m (Currey, 1982) with the highest elevated shorelines also near the center of the basin where the water load was at a maximum.

The deflection of the Bonneville and Provo shorelines from the horizontal presents challenges when trying to create accurate maps of the areal extents of the lake at these two levels. One cannot simply trace a given contour line or contour a certain elevation in a digital elevation model (DEM) because the actual traces of these shorelines cross many tens of meters of elevation, reflecting the broad domal uplift of these once horizontal features.

In this chapter, we attempt to briefly summarize what is currently known about the spatial and temporal patterns of rebound, recorded on the two most prominent shorelines in the Bonneville basin. As the body of literature on this topic is large, we mainly point the reader to earlier reviews. We also present a new methodology to palinspastically restore the landscape to its elevations when the Bonneville and Provo shorelines were forming by subtracting the signals of isostatic rebound from the modern landscape, which allows accurate reconstructions of the areal extent and volume of the lake at these two different levels. We also discuss the implications of these new maps and digital products for documenting the history and processes of isostatic rebound in the Bonneville basin.

8.2 PREVIOUS WORK

Advances in analyzing and modeling the isostatic rebound of Lake Bonneville over the last 125 years were greatly dependent on improvements of topographic data and surveying techniques that allowed more precise and basin-wide characterization of the rebound patterns associated with the Bonneville and Provo shorelines. This trend will likely continue with the recent production of 5 m DEMs that cover the entire state of Utah and the increasing availability of Light Detection and Ranging data for specific regions within the Bonneville basin.

Gilbert (1890) performed the earliest work on isostatic rebound in the Bonneville basin. He mapped elevation variations on the Bonneville and Provo shorelines using spirit levels and barometers to determine the heights of these shorelines above the elevation of the Great Salt Lake. Gilbert (1890) concluded that the most reasonable explanation for the deformation pattern expressed by the shorelines is that the strength of the Earth, in response to long-term loads, depends strongly on water depth. A shallow, cold, strong layer (later called the "lithosphere") supports short wavelength loads, and a warmer, deeper, weaker layer (later called the "asthenosphere") viscously flows in response to long-term stresses (Watts, 2001). The lake load and associated pattern of rebound on the Bonneville shoreline was used to derive the first published estimate of continental lithospheric thickness, which was about 50 km (Gilbert, 1890).

The next major effort to document and model the rebound pattern was by Crittenden (1963a), in which he compiled a database of 90 locations and elevations of points on the Bonneville shoreline based on aerial photograph and topographic map interpretation (1:24,000–1:250,000 scale). Crittenden (1963b) then applied a more rigorous model to the rebound process than was available to Gilbert. By this time, ages of the shorelines were reasonably well constrained (eg, Broecker and Orr, 1958). A major result of this modeling effort was that the relaxation timescale of the asthenosphere in the eastern Great Basin, was at least 10 times shorter (implying lower viscosity) than had been inferred from Fennoscandia glacial isostatic rebound studies (Niskanen, 1948).

Passey (1981) measured the elevations of 24 additional points on the Bonneville and Provo shorelines based on hand leveling from benchmarks and spot elevations labeled on 1:24,000 and 1:62,500-scale topographic maps and compared the rebound patterns on these different shorelines. Based on the larger magnitude of rebound associated with the Bonneville shoreline, Passey (1981) concluded that at least 90% of the rebound associated with removal of the water in the Bonneville flood had occurred prior to the Provo shoreline being abandoned.

A significant advance in rebound data compilation was made by Currey (1982) who documented the locations and elevations of 181 Bonneville shoreline points and 112 Provo shoreline points. His approach was to examine aerial photographs and moderate resolution topographic maps (1:24,000 and 1:62,500-scale) to identify and estimate the elevation of candidate shoreline points and then visit some of them in the field to verify the identification and perform hand level or spirit level surveys. New modeling efforts (Bills and May, 1987; Bills et al., 1994) used the improved data set of Currey (1982) and a modeling approach that allowed viscosity to vary with depth (McConnel, 1965; Gilbert and Backus, 1966). The primary conclusion of that research was that there is indeed an effectively elastic lithosphere, extending to 30 km depth, and that the sub-lithospheric viscosity is of order 10^{18} Pa s, or 100 times less than global average estimates for upper mantle viscosity (Peltier, 1974). Bills et al. (2002) presented an atlas of computed deflection maps for the Bonneville basin at 2000 year increments for the past 26 ka, which showed the progression of isostatic depression and rebound as the lake filled and desiccated over this period.

In addition to compiling a radiocarbon chronology for the Bonneville lake cycle, Oviatt et al. (1992) also used an empirical formula ("Currey equation" from Currey and Oviatt, 1985) to linearly adjust the elevations of radiocarbon samples to account for the effects of differential rebound in different parts of the basin. These adjustments allow the direct comparison of elevations and radiocarbon ages from samples deposited at locations with different rebound histories in order to develop a more reliable chronology (Oviatt et al., 1992). Jewell (2016) modified the "Currey equation" to more accurately adjust the elevations of intermediate shorelines, based on the difference between measured Bonneville and Provo shorelines. This approach assists in basin-wide correlations of the intermediate shorelines by minimizing the effects of spatially variable rebound (Jewell, 2016).

Lake Bonneville rebound demonstrates that the eastern Great Basin is anomalous with respect to lithospheric thickness and upper mantle viscosity, compared to the global average, but similar to values derived from the Lake Lahontan basin in the western Great Basin (Bills et al., 2007). A thin lithosphere and low upper mantle viscosity were already indicated for the Great Basin because of its extensional nature (eg, Stewart, 1978; Dickinson, 2002), recent volcanism (eg, Condie and Barsky, 1972), and high heat flow values (eg, Lachenbruch, 1978).

Historical maps of the various lakestands in the Bonneville basin range from small cartoon-like representations to maps plotted on small-scale topographic bases (eg, Snyder et al., 1964; Currey, 1982; Currey et al., 1985). Geographic information systems (GIS) software was likely used to produce more recent maps (eg, Miller et al., 2013, 2015; Oviatt, 2015) but as far as we know, only Jewell (2010) has attempted to correct his map for isostatic rebound, based on the analysis of Bills et al. (2002).

8.3 METHODS

In normal usage, a palinspastic reconstruction attempts to map geological units in the form or position in which they were originally deposited, removing effects of subsequent faulting and folding (eg, Dunbar and Cook, 2003). In the present case, we attempt to reconstruct the topography of the Bonneville basin at the times of formation of the Bonneville and Provo shorelines by removing the rebound signals associated with the removal of these loads.

To palinspastically restore the landscape of the Bonneville basin to the elevations which pertained when Lake Bonneville was present and to make accurate maps of the Bonneville and Provo shorelines we utilized a series of moderate resolution (30 m) DEMs and existing shoreline elevation data sets (Currey, 1982) using tools within ArcGIS v. 10.2. The basic procedure was to subtract the domal patterns of rebound associated with the Bonneville and Provo shorelines from the modern topography of the basin, which essentially palinspastically restores each part of the landscape to its vertical position when the respective shoreline was forming. All of the geoprocessing for this project was performed with files that have been projected to the UTM NAD83 Z12 coordinate system. An overview of the procedure is presented below, but details are available as (Supplemental Material S-1 (http://dx.doi.org/10.1016/B978-0-444-63590-7.00008-1).

The first step in this process is to assemble a single DEM that covers the entire Bonneville basin and extends well past (50+km) the limits of the former lake. In our case, we chose to mosaic thirty six 30 m DEMs (1 × 1 rasters) into a single 30 m DEM that covered 6 degrees of latitude and longitude. This level of detail represents a balance between higher resolution 10 m DEMs, which are difficult to work with because of the increased computer processing time, and lower resolution 90 m DEMs, which do not provide the level of detail that the 30 m DEMs provide. The 30 m DEM of the basin represents modern topography and serves as the base layer from which the rebound is subtracted.

We use the current elevations of 181 survey points on the Bonneville shoreline and 112 points on the Provo shoreline, as mapped by Currey

(1982), and linearly interpolate those spatial patterns. In Currey's data collection, he chose the elevationally highest Provo shoreline at each location in order to be consistent. For each shoreline, the domal pattern of rebound is then subtracted from the 30 m DEM of the modern topography to create two new 30 m DEMs that represent the elevation of the landscape at Bonneville and Provo times, respectively. Even though they are probably more precise, we did not include the Provo elevations surveyed by Miller et al. (2013) in this study because we wanted to use a single, basin-wide data set for consistency.

The amount of rebound on the Bonneville shoreline is assumed to be the difference between present-day elevation and 1552 m, which is the elevation of the reconstructed threshold (Currey, 1982). This is also the elevation of the nearest Bonneville shoreline to the outlet and is the lowest measured Bonneville shoreline in the basin (Currey, 1982). Based on these observations and the modeling results of Bills et al. (1994, 2002), which suggest that essentially no rebound has occurred at the outlet, we take the 1552 m elevation to be the original elevation of the Bonneville shoreline prior to isostatic deformation. We recognize, however, that in many places the rebound signal extended out past the extent of the Bonneville shoreline, particularly along the west side of the basin (eg, Bills et al., 1994, 2002). Similarly, the reconstructed Red Rock Pass threshold responsible for the Provo shoreline was initially at 1444 m (Currey, 1982) and modeling of the Provo deformation signal (Bills et al., 1994, 2002) also shows no rebound near the outlet. Therefore, we use 1444 m as the original elevation of the Provo shoreline prior to deformation.

The surface volume tool was used to calculate the surface areas and volumes of the lake at the Bonneville level and at the Provo level using the "rebound removed" 1552 m and 1444 m contours, respectively, with the appropriate "rebound removed" DEMs. This procedure basically calculates the volume between a landscape surface and a horizontal plane of your choosing. For comparison, the surface areas and volumes of the lake were also calculated using the modern 1552 m and 1444 m contours along with a modern topography DEM. To account for the volumes of water contained within the Great Salt Lake and Utah Lake, which are not well represented in the DEMs, we used the estimates of Baskin (2005, 2006) and Jackson and Stevens (1981), respectively, which were added to all of the volume estimates.

8.4 RESULTS

The geoprocessing performed in this study resulted in a series of DEMs, contours, and polygons that represent the elevation of the land surface at three times during the history of Lake Bonneville. The first period of time is when the lake had just reached its natural rim at about 1552 m while forming the Bonneville shoreline. The second period of time is after the Bonneville flood dropped the surface elevation of the lake by about 108–125 m and the Provo shoreline was forming as the lake continued to spill across Red Rock Pass at an elevation of about 1444 m (Currey, 1982; Miller et al., 2013). The third period of time is the modern era when the combined rebound signal of the Bonneville and Provo lake loads is present.

Using these products, a new map of the extents of the Bonneville and Provo shorelines is presented in Fig. 8.1. This map is presented at a scale of about 1:2,000,000, which makes it difficult to visualize how the lake outlines are different than maps made by simply contouring the 1552 m and 1444 m elevations on a 30 m DEM of the modern topography. Fig. 8.1 also shows the locations of Figs. 8.2 through 8.7 that focus on a few parts of the lake shore in greater detail to illustrate the differences in the paths of the contours derived from modern topography (rebound included) versus contours derived from the "rebound removed" topography. In all cases examined, the "rebound removed" 1552 m contour overlies or is close to overlying shoreline features that represent the Bonneville shoreline. In contrast, the modern 1552 m contour almost always lies some distance down slope from Bonneville shoreline features, sometimes by as much as several kilometers (Figs. 8.2–8.7). Exceptions include several of the far northern and southern extremities of Lake Bonneville where the "rebound removed" 1552 m and the modern 1552 m contours are nearly coincident (Fig. 8.5). Similarly, the "rebound removed" 1444 m contour accurately tracks the path of the Provo shoreline, whereas the modern 1444 m contour does not (Figs. 8.2–8.7).

The lateral distance on the ground between these two contours is dependent on the magnitude of rebound at a location as well as the slope on which the shoreline is developed. In many places, the lateral distance between the contours is 1–2 km (Figs. 8.2–8.7) but in the Rush Valley, south of the Stockton Bar, the difference on the ground between the two contours ranges up to 4–5 km.

New surface area and volume estimates for the Bonneville and Provo stillstands were derived from the "rebound removed" DEMs and are listed in Table 8.1, along with estimates derived from DEMs of modern topography.

8.5 DISCUSSION AND CONCLUSIONS

Map making has become progressively easier and more accurate as high resolution data sources have become increasingly available and mapping tools such as ArcGIS have become increasingly sophisticated. This is true when both mapping modern features such as the courses of rivers or ancient features such as Pleistocene shorelines surrounding now arid basins. Prior to GIS software, ancient shorelines were most commonly mapped from aerial photographs or by tracing the appropriate contour from topographic maps (eg, Snyder et al., 1964; Mifflin and Wheat, 1979; Currey, 1982; Currey et al., 1985). Now, it is a relatively simple procedure to contour a DEM at a given elevation to represent a former lake shoreline. In the case of larger lakes, however, which have undergone substantial isostatic rebound or other

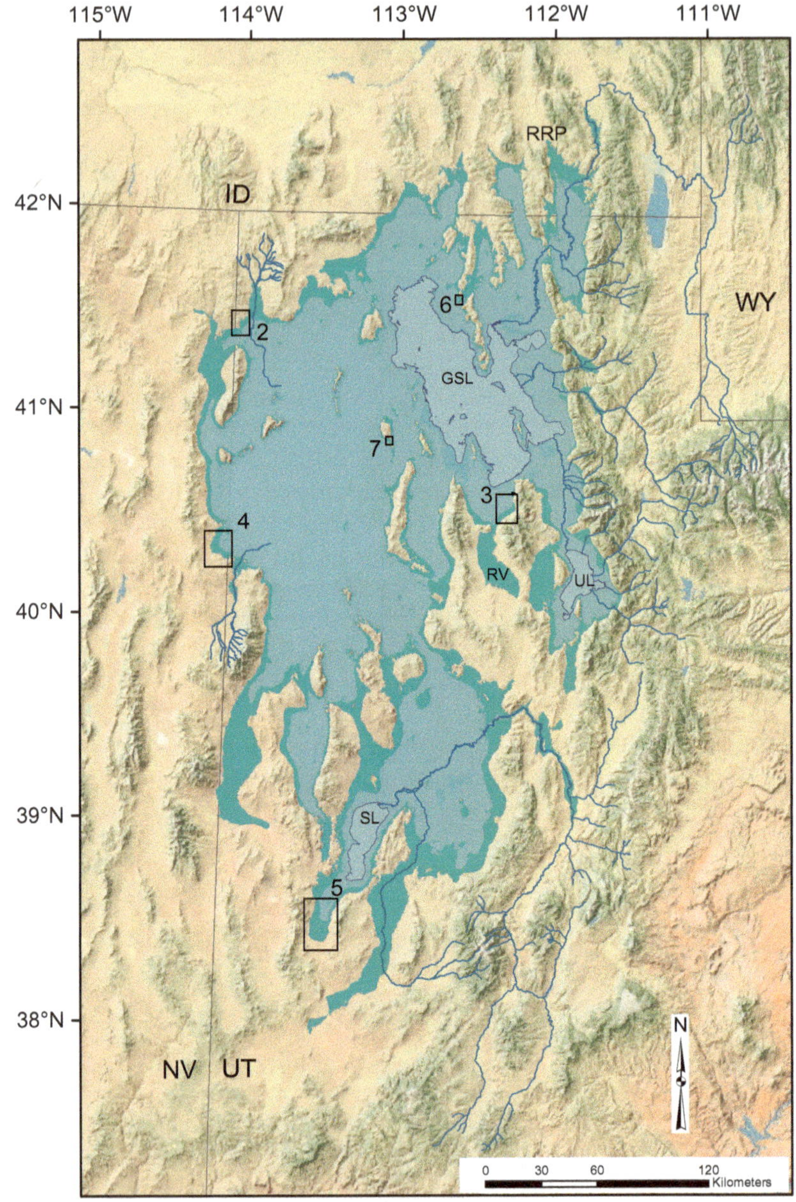

FIG. 8.1 Map of Lake Bonneville when the Bonneville shoreline was forming (*dark blue*) and when the Provo shoreline was forming (*light blue*). Also shown are the major rivers flowing into the basin and the historical extents of the Great Salt Lake (GSL), Utah Lake (UL), and Sevier Lake (SL). The numbered boxes indicate the locations of Figs. 8.2 through 8.7. RRP, Red Rock Pass; RV, Rush Valley.

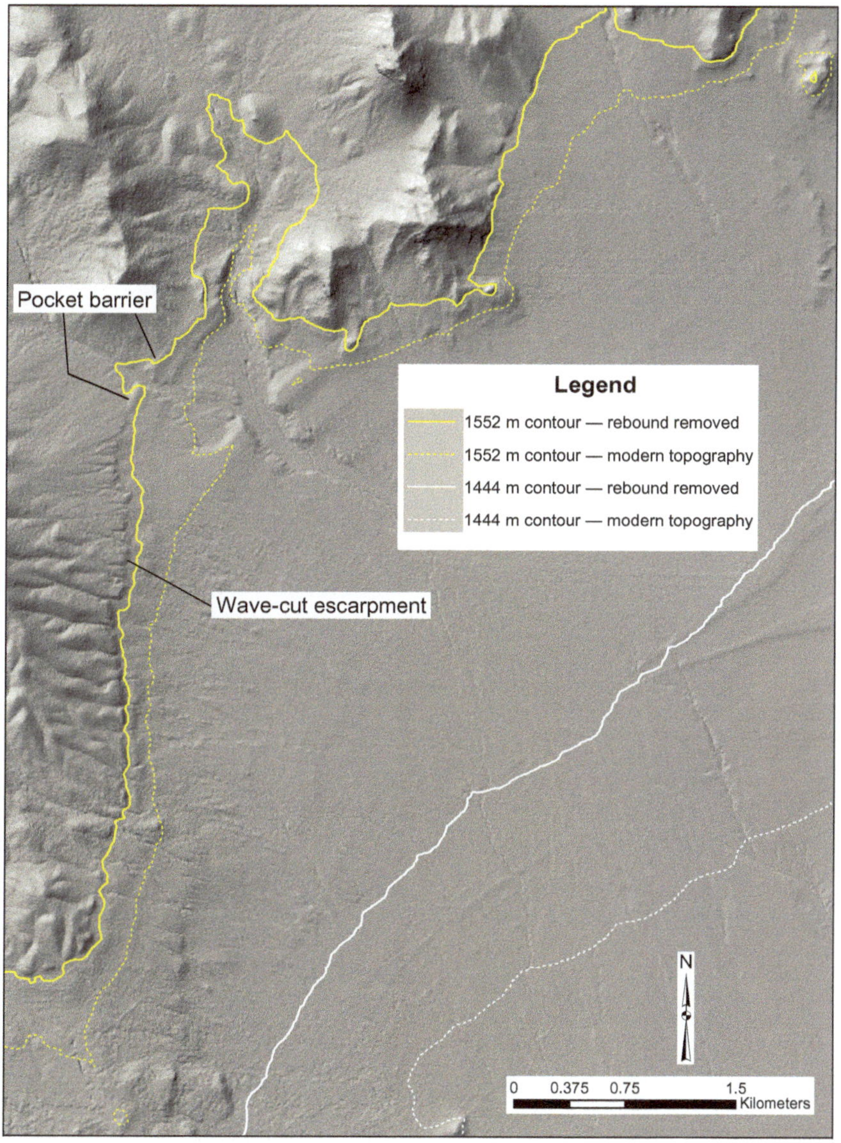

FIG. 8.2 Details of the paths of the "rebound removed" 1552 m and 1444 m contours compared to the paths of the modern topography 1552 m and 1444 m contours. Note that the "rebound removed" 1552 m contour accurately represents a small, dissected pocket barrier whereas the contour is slightly deflected away from the base of the wave-eroded escarpment directly to the south. The base for the image is a hillshaded 5 m DEM. Location is shown in Fig. 8.1.

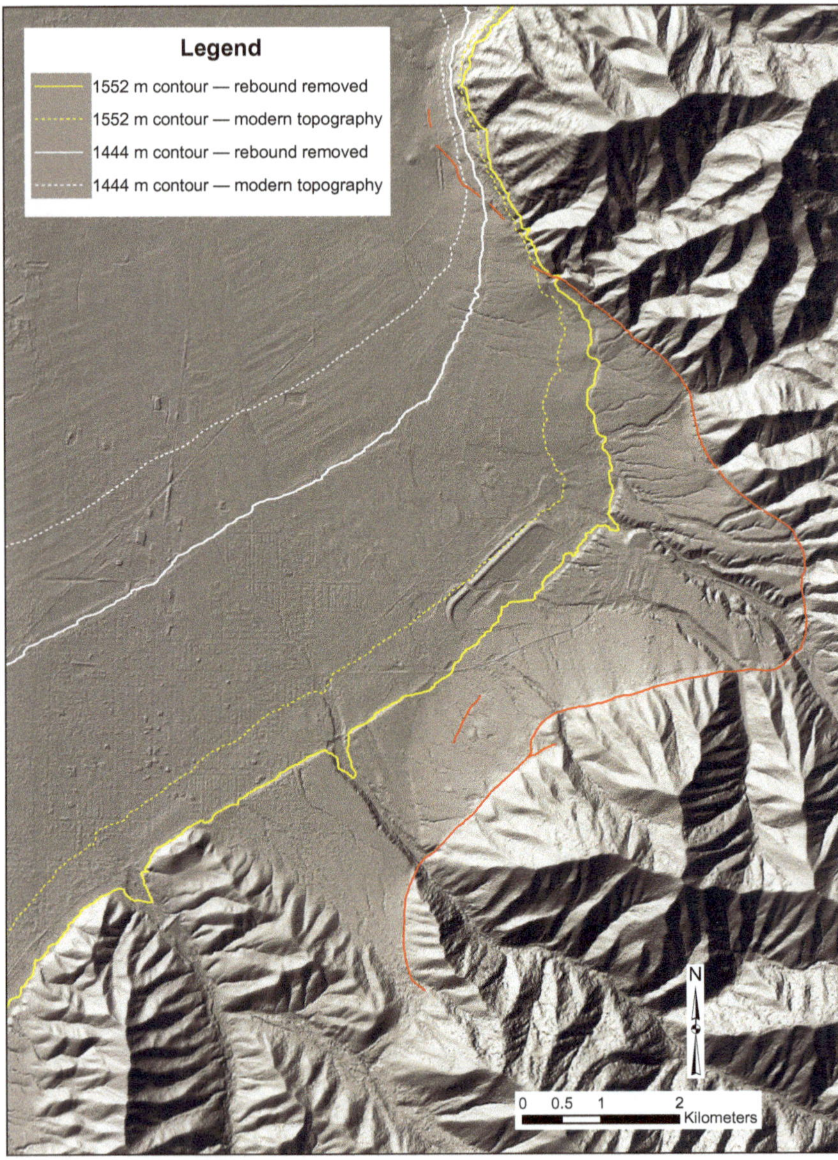

FIG. 8.3 Map of the western side of the Oquirrh Mountains showing the paths of the "rebound removed" shoreline contours compared to the modern topography contours. Also shown is the Oqurrih fault zone (*red*), which has been active during the Holocene (Black et al., 2004). The base for the image is a hillshaded 5 m DEM. Location is shown in Fig. 8.1.

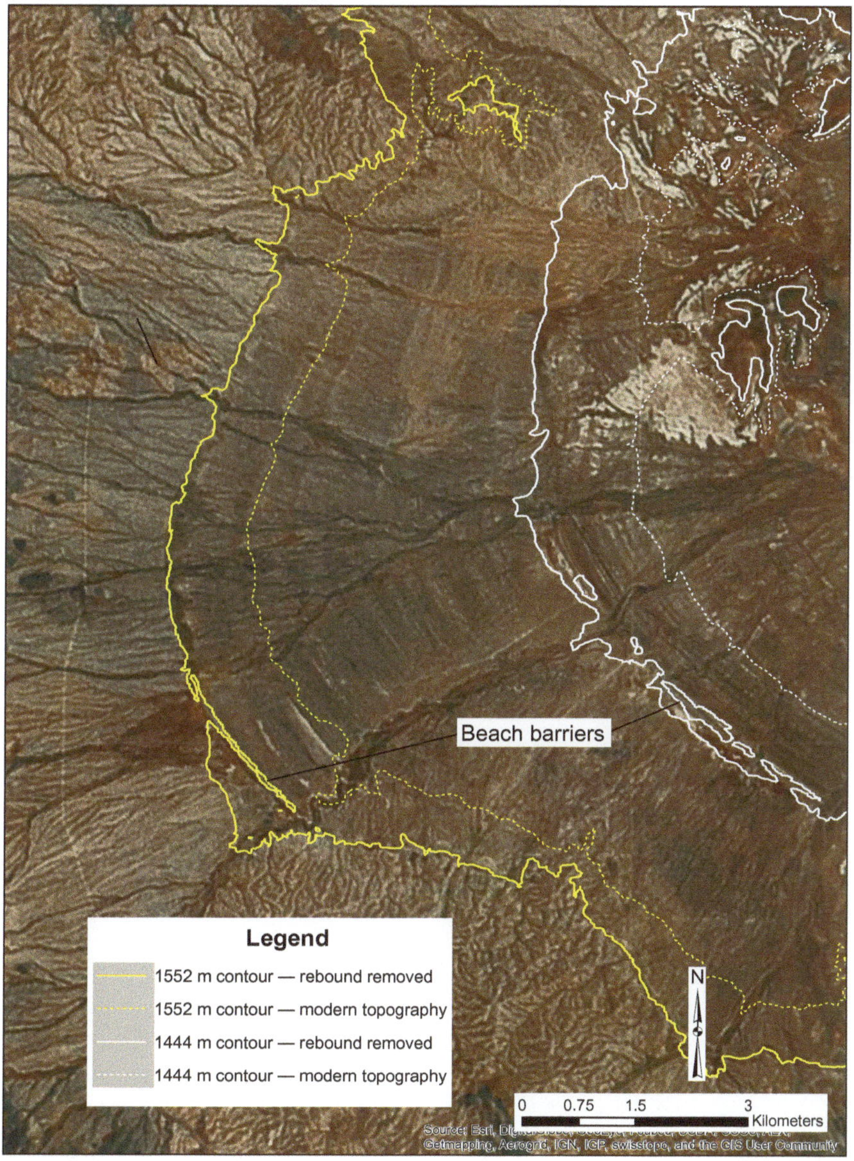

FIG. 8.4 Map of a small embayment in the western part of the basin (see Fig. 8.1 for location) that shows the details of the "rebound removed" 1552 m and 1444 m contours that clearly define the crests of beach barriers that represent the Bonneville and Provo shorelines, respectively. Base is ESRI imagery.

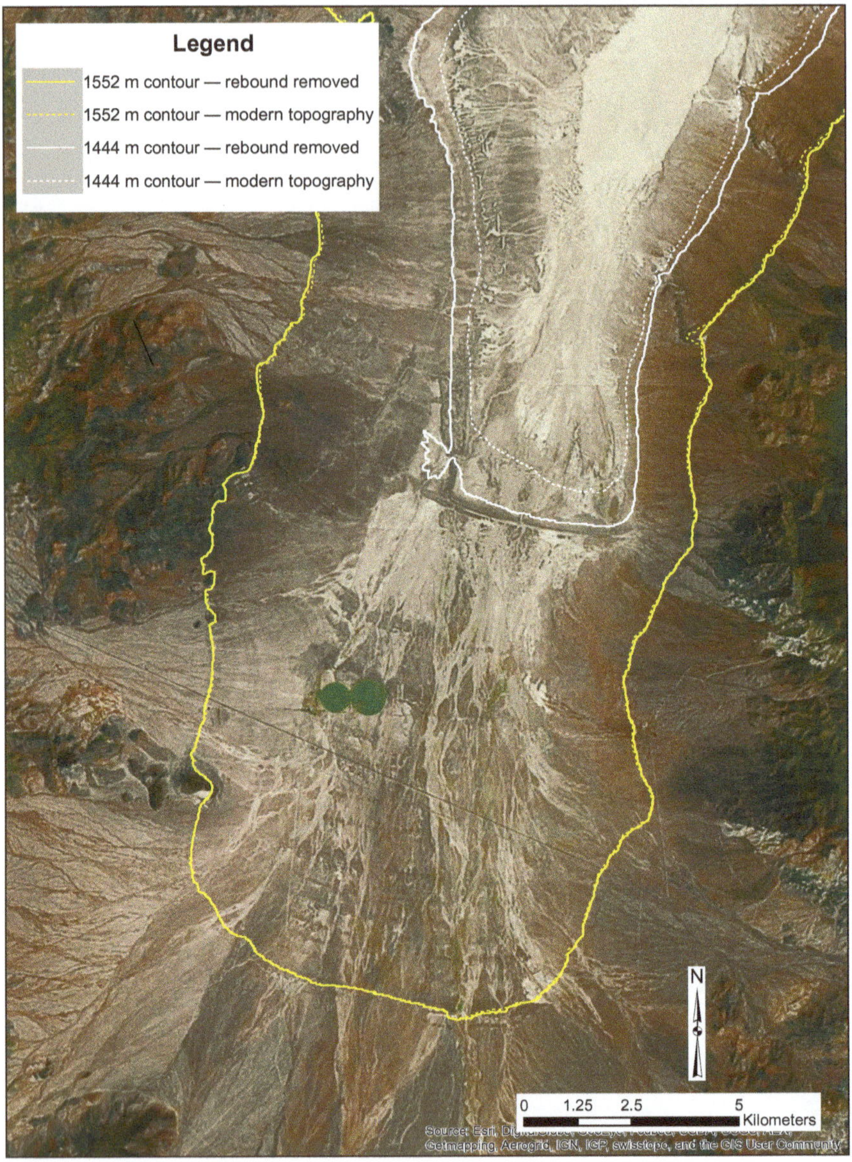

FIG. 8.5 Map of an embayment in the southwestern part of the basin (see Fig. 8.1 for location) that shows that the modern topography 1552 m contour is nearly coincident with the "rebound removed" 1552 m contour. In this embayment, there is still a separation between the 1444 m contours. Base is ESRI imagery.

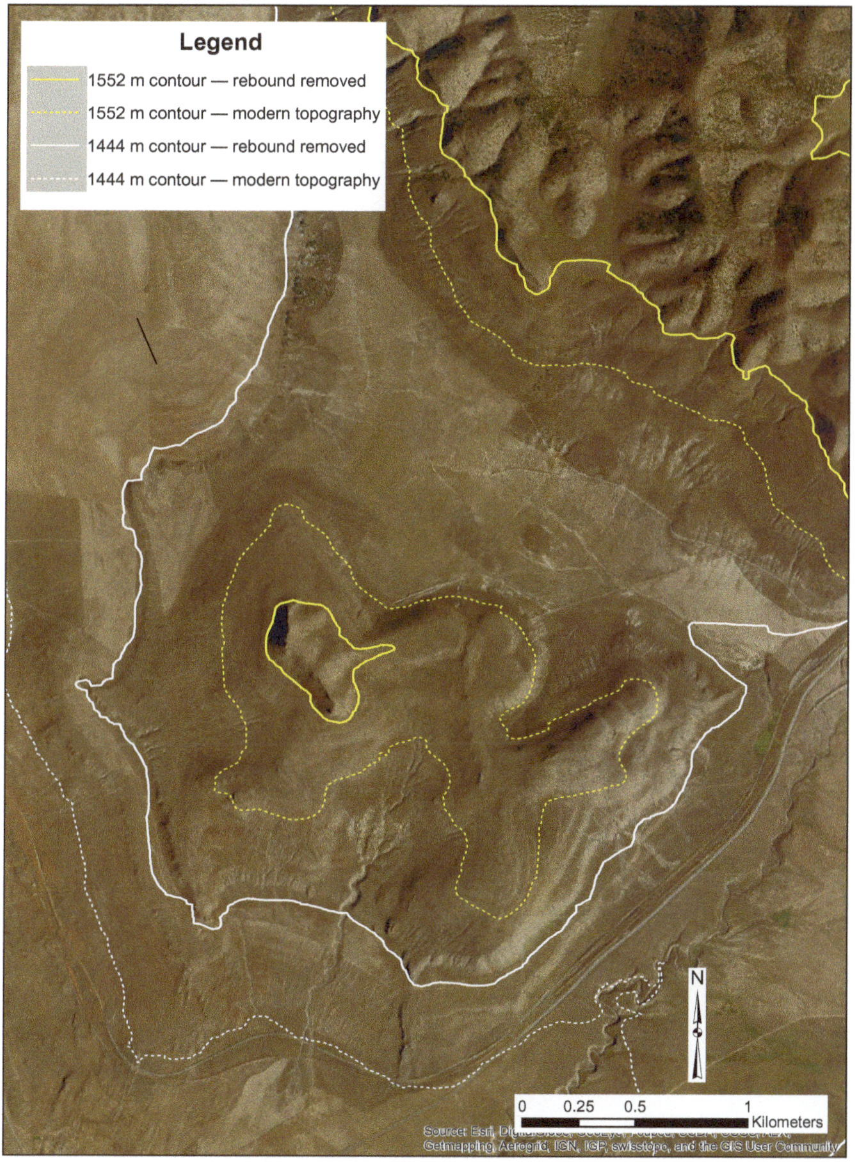

FIG. 8.6 Map of a small island in the northern part of the basin (see Fig. 8.1 for location) that shows the differences in location of the "rebound removed" contours compared to the modern topography contours. In this case, the island was actually much smaller at the Bonneville highstand than is represented by the modern topography contour. Base is ESRI imagery.

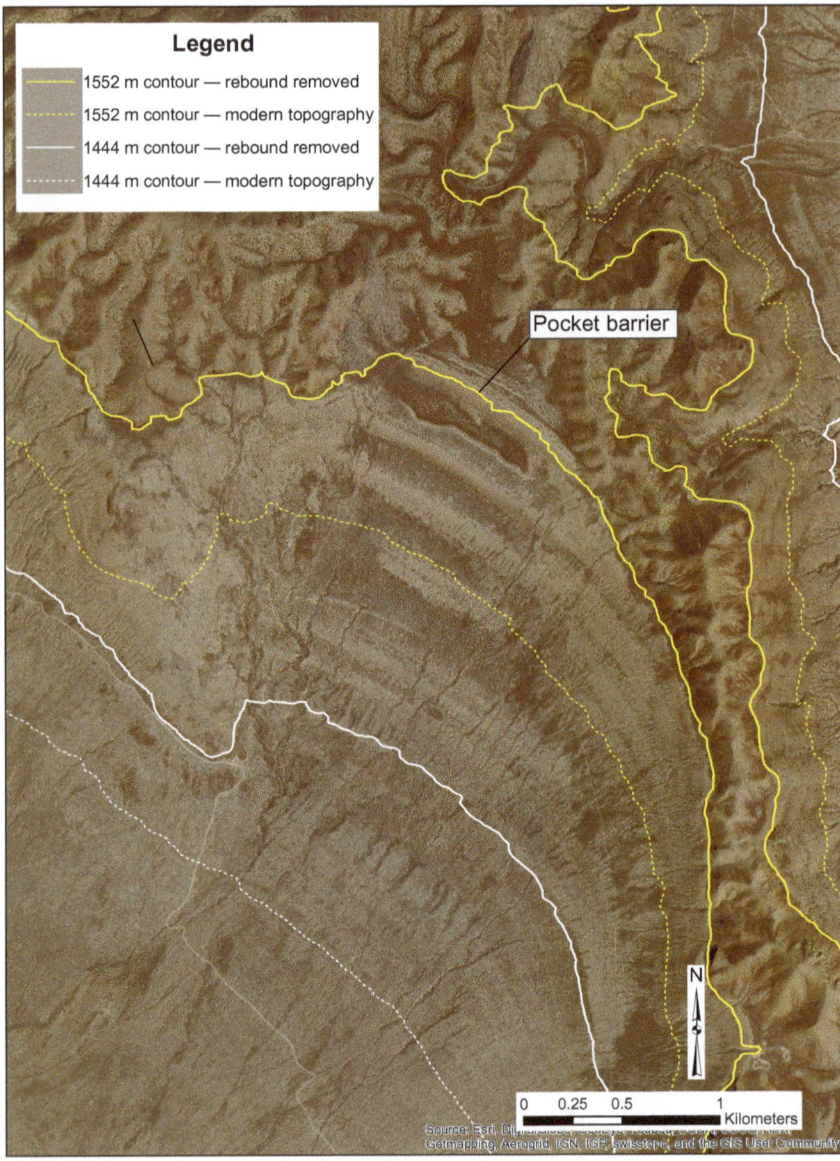

FIG. 8.7 Map of a site near the center of maximum loading (see Fig. 8.1 for location) that shows the "rebound removed" 1552 m contour exactly overlaying the crest of a Bonneville highstand pocket barrier. Presently, this pocket barrier is at an elevation of about 1625 m. Note that the Provo shoreline here is not particularly well developed. Base is ESRI imagery.

TABLE 8.1 Comparison Between Surface Areas and Volumes of Lake Bonneville Using a DEM with the Rebound Signal Removed and a Modern Topography DEM

Lake Stage	Area (km^2)	Volume (km^3)[a]
Rebound signal removed		
Bonneville	52,110	10,420
Provo	38,150	5290
Modern topography		
Bonneville	48,860	8620
Provo	35,510	4050

[a]The modern volumes of Great Salt Lake (~20 km^3) (Baskin, 2005, 2006) and Utah Lake (~1 km^3) (Jackson and Stevens, 1981) were added to the total volumes measured from the DEMs because the DEMs do not account for the volumes of these lakes.

neotectonic deformation, the shorelines are no longer horizontal and contouring a DEM of the modern topography will result in sometimes substantial errors in the mapped path of the shoreline. This paper presents an approach to restore a deformed shoreline to its pre-rebounded geometry.

In this chapter, we have presented a methodology for removing the isostatic rebound signal from modern DEMs to create new DEMs that can be contoured and otherwise spatially analyzed. At the scale shown in Fig. 8.1 (~1:2,000,000), this map does not look very different than previously published maps of Lake Bonneville (eg, Bills et al., 1994, 2002; Currey, 1982; Currey et al., 1985; Gilbert, 1890; Jewell, 2010; Miller et al., 2013; Oviatt, 2015). In detail, however, significant differences are exhibited. The new 1552 m Bonneville and 1444 m Provo shoreline shapefiles can be directly compared to the locations of these same contours that were generated from the modern topography DEM. Except for areas in the far extremities of the basin, the modern 1552 m and 1444 m lie downslope and within the "rebound removed" 1552 m and 1444 m contours, respectively. The observation that the "rebound removed" and modern contours are nearly coincident in the distal arms of the basin (Fig. 8.5), far from the maximum load imposed by Lake Bonneville, supports the modeling results of Bills et al. (1994, 2002) that isostatic deformation was minimal in these regions. Further, these observations support the selection of the 1552 and 1444 m elevations as approximate initial elevations for the Bonneville and Provo shorelines, respectively.

The "goodness of fit" of the new polygon shapefiles can also be evaluated by overlaying them on top of high resolution imagery or hillshaded versions of a modern DEM to compare the locations of the mapped shorelines to where the traces of the shorelines actually occur on the surface of the Earth (Figs. 8.2–8.7). In all of these maps, the Bonneville and Provo shorelines are well represented by the 1552 m and 1444 m contours, respectively, that were generated from the "rebound removed" DEMs.

When examining the paths of the "rebound removed" 1552 m and 1444 m contours in detail with respect to visible shoreline features, it is noted that the contours do not always perfectly align with the crests of beach ridges or along the bases of wave-cut escarpments. At least two reasons exist for this, the first of which is the natural variability of shoreline heights. In large lakes, with complicated, crenulated shorelines with different fetch lengths and offshore slopes, the amount of wave energy reaching a particular shore is variable. This variable wave energy directly influences the height at which shorelines of the same age form above a given water plane, which can approach several meters on larger lakes (eg, Atwood, 1994, 2006; Adams and Wesnousky, 1998).

In the case of erosional shorelines, a second reason that the paths of the "rebound removed" 1552 m and 1444 m contours do not always exactly follow the bases of wave-cut escarpments is that commonly a colluvial wedge was deposited after the shoreline angle was abandoned. This is particularly common where escarpments are cut into unconsolidated sediments, such as on the left side of Fig. 8.2. Evidence for sediment deposition at the base of wave-cut escarpments also includes the common occurrence of small rills cut into the truncated alluvial fan surfaces.

In addition to altering the mapped shape of the lake at Bonneville and Provo times, the differences between the "rebound removed" and modern topography DEMs also affect area and volume calculations. The newly calculated area and volume values are about 52,110 km^2 and 10,400 km^3 for the Bonneville shoreline and about 38,150 km^2 and 5,270 km^3 for the Provo shoreline (Table 8.1). These values are significantly larger than previous estimates and will improve boundary conditions for isostatic rebound modeling (eg, Bills et al., 1994) and paleohydrologic modeling (eg, Adams, 2003; Matsubara and Howard, 2009; Adams et al., 2015). Using these new values, the volume of water lost during the Bonneville flood is also substantially larger (~5135 km^3) than that estimated by O'Connor (1993)) (~4750 km^3), which could affect estimates of the duration or discharge for that event.

These new maps will also be useful for locating relevant deposits and landforms from particular lake stages in the Bonneville basin. For example, topset–foreset transitions in Gilbert delta sequences represent sedimentary evidence for water level at the time of their deposition (Adams, 2007; Reheis et al., 2014). As delta sequences commonly contain abundant organic matter for radiocarbon dating, it will be useful to have accurate maps of the locations of these features at Bonneville and Provo times so that the timing and duration of these episodes can be better constrained, a task recommended by Miller et al. (2013).

The new maps will also be helpful for locating Bonneville and Provo shorelines in the field. In addition to accurately tracing the shorelines around the perimeter of the basin, the new maps also portray many islands that were not mapped previously. According to these maps, there were 1118 islands (>100 m^2) at Bonneville time and 1368 islands at Provo time. Many of these islands are found in the main body of the lake, near where maximum loading occurred. If new data sets of shoreline elevation measurements were to be collected for future isostatic rebound modeling, then these island localities become important for locating and documenting the elevations of widely separated shoreline features and accurately reconstructing the geometry of the rebound signal at discrete epochs in time.

The 30 m Bonneville "magnitude of rebound" raster can be used to correct the elevations of radiocarbon and other dating samples, similar to using the rebound correction formula known as the "Currey equation" (Currey and Oviatt, 1985; Oviatt et al., 1992). To correct the elevation of a sample for the effects of rebound, one need only subtract the magnitude of rebound for the sample location from the current (modern) elevation of the sample. This approach should probably be used with caution when correcting the elevations of samples found near the basin floor because these samples would have been isostatically depressed when lake levels were high, but would have rebounded to about their original elevation when lake levels were low, as in modern times (Oviatt et al., 1992).

The techniques developed in this study to remove rebound signals from modern DEMs in order to make more accurate maps, better locate outcrops and landforms of interest, and correct elevations of dating samples can also be applied to other basins that have undergone isostatic rebound, so long as adequate topographic coverage and identifiable shoreline features occur. By removing rebound signals from the modern landscape, paleo-horizontal features such as shorelines can be used to better evaluate the effects of isostatic rebound and other neotectonic deformation, such as active faulting or regional tilting.

ACKNOWLEDGMENTS

Partial funding for this research was provided by NSF EAR Grant 1349414 and by the Desert Research Institute. A potion of this work was conducted at the Jet Propulsion Laboratory, California Institute of Technology, under contract with NASA. We thank the reviewers and editors for their constructive comments.

SUPPLEMENTAL DATA

Detailed methods on removing isostatic rebound signals from DEMs are available as Supplemental Material S-1 (http://dx.doi.org/10.1016/B978-0-

444-63590-7.00008-1). Digital copies of the DEMs and derivative shapefiles of the 1552 and 1444 m contours are available on request from the lead author.

REFERENCES

Adams, K.D., 2003. Age and paleoclimatic significance of late Holocene lakes in the Carson Sink, NV, USA. Quatern. Res. 60, 294–306.

Adams, K.D., 2007. Late Holocene sedimentary environments and lake-level fluctuations at Walker Lake, Nevada, USA. Geol. Soc. Am. Bull. 119, 126–139.

Adams, K.D., Wesnousky, S.G., 1998. Shoreline processes and the age of the Lake Lahontan highstand in the Jessup embayment, Nevada. Geol. Soc. Am. Bull. 110, 1318–1332.

Adams, K.D., Negrini, R.M., Cook, E.R., Rajagopal, S., 2015. Annually resolved late Holocene paleohydrology of the southern Sierra Nevada and Tulare Lake, California. Water Resour. Res. 51, 9708–9724. http://dx.doi.org/10.1002/2015WR017850.

Atwood, G., 1994. Geomorphology applied to flooding problems of closed-basin lakes—Specifically Great-Salt-Lake, Utah. Geomorphology 10, 197–219.

Atwood, G., 2006. Shoreline Superelevation: Evidence of Coastal Processes of Great Salt Lake, Utah. Utah Geological Survey Miscellaneous Publication 06-9, Salt Lake City, UT. 231 pp.

Baskin, R.L., 2005. Calculation of Area and Volume for the South Part of Great Salt Lake, Utah: U.S. Geological Survey open-file report 2005-1327. 6 pp.

Baskin, R.L., 2006. Calculation of Area and Volume for the North Part of Great Salt Lake, Utah: U.S. Geological Survey open-file report 2006-1359. 6 pp.

Bills, B.G., May, G.M., 1987. Lake Bonneville: constraints on lithospheric thickness and upper mantle viscosity from isostatic warping of Bonneville, Provo, and Gilbert stage shorelines. J. Geophys. Res. 92, 11,493–11,508.

Bills, B.G., Currey, D.R., Marshall, G.A., 1994. Viscosity estimates for the crust and upper mantle from patterns of lacustrine shoreline deformation in the eastern Great Basin. J. Geophys. Res. 99, 22,059–22,086.

Bills, G.G., Wambeam, T.J., Currey, D.R., 2002. Geodynamics of Lake Bonneville. In: Gwynn, J.W. (Ed.), Great Salt Lake: A Scientific, Historical & Economic Overview. Utah Geological and Mineral Survey, Salt Lake City, UT, pp. 7–32.

Bills, B.G., Adams, K.D., Wesnousky, S.G., 2007. Viscosity structure of the crust and upper mantle in western Nevada from isostatic rebound patterns of Lake Lahontan shorelines. J. Geophys. Res. 112. http://dx.doi.org/10.1029/2005JB003941.

Black, B.D., and DuRoss, C.B., and McDonald, G.N., and Hecker, S., compilers, 2004, Fault number 2398, Oquirrh fault zone. In: Quaternary Fault and Fold Database of the United States: U.S. Geological Survey. Website http://earthquakes.usgs.gov/hazards/qfaults, accessed 09/01/2015 11:19 AM.

Broecker, W.S., Orr, P.C., 1958. Radiocarbon chronology of Lake Lahontan and Lake Bonneville. Geol. Soc. Am. Bull. 69, 1009–1032.

Condie, K.C., Barsky, C.K., 1972. Origin of Quaternary basalts from Black Rock Desert region, Utah. Geol. Soc. Am. Bull. 83, 333–345.

Crittenden, M.D., 1963a. New Data of the Isostatic Deformation of Lake Bonneville. U.S. Geological Survey Professional Paper 454, pp. E1–E31.

Crittenden, M.D., 1963b. Effective viscosity of the Earth derived from isostatic loading of Pleistocene Lake Bonneville. J. Geophys. Res. 68, 5517–5530.

Currey, D.R., 1982. Lake Bonneville: Selected Features of Relevance to Neotectonic Analysis: U.S. Geological Survey open-file report 82-1070. 31 pp.

Currey, D.R., 1990. Quaternary palaeolakes in the evolution of semidesert basins, with special emphasis on Lake Bonneville and the Great Basin, U.S.A. Palaeogeogr. Palaeoclimatol. Palaeoecol. 76, 189–214.

Currey, D.R., Oviatt, C.G., 1985. Durations, average rates, and probable causes of Lake Bonneville expansions, stillstands, and contractions during the last-deep-lake cycle, 32,000 to 10,000 years ago. Geogr. J. Korea 10, 1085–1099.

Currey, D.R., Atwood, G., Mabey, D.R., 1985. Major Levels of Great Salt Lake and Lake Bonneville. Utah Geological Survey, Salt Lake City, UT. Map 73, 1:750,000 scale.

Dickinson, W.R., 2002. The Basin and Range Province as a composite extensional domain. Int. Geol. Rev. 22, 1–38.

Dunbar, J.A., Cook, R.W., 2003. Palinspastic reconstruction of structure maps: an automated finite element approach with heterogeneous strain. J. Struct. Geol. 25, 1021–1036.

Gilbert, G.K., 1890. Lake Bonneville. U.S. Geological Survey Monograph, vol. 1. 438 pp.

Gilbert, F., Backus, G.E., 1966. Propagator matrices in elastic wave and vibration problems. Geophysics 31, 326–333.

Godsey, H.S., Currey, D.R., Chan, M.A., 2005. New evidence for an extended occupation of the Provo shoreline and implications for regional climate change, Pleistocene Lake Bonneville, Utah, USA. Quatern. Res. 63, 212–223.

Godsey, H.S., Oviatt, C.G., Miller, D.M., Chan, M.A., 2011. Stratigraphy and chronology of offshore to nearshore deposits associated with the Provo shoreline, Pleistocene Lake Bonneville, Utah, USA. Palaeogeogr. Palaeoclimatol. Palaeoecol. 310, 442–450.

Jackson, R.H., Stevens, D.J., 1981. Physical and cultural environment of Utah Lake and adjacent areas. Great Basin Nat. Mem. 5, 3–23.

Jewell, P.W., 2010. River incision, circulation, and wind regime of Pleistocene Lake Bonneville, USA. Palaeogeogr. Palaeoclimatol. Palaeoecol. 293, 41–50.

Jewell, P.W., 2016. Quantitative identification of erosional Lake Bonneville shorelines, Utah. Geomorphology 253, 135–145.

Lachenbruch, A.H., 1978. Heat flow in Basin and Range Province and thermal effects of tectonic extension. Pure Appl. Geophys. 117, 34–50.

Malde, H.E., 1968. The Catastrophic Late Pleistocene Bonneville Flood in the Snake River Plain, Idaho. U.S. Geological Survey Professional Paper 596, 52 pp.

Matsubara, Y., Howard, A.D., 2009. A spatially explicit model of runoff, evaporation, and lake extent: application to modern and late Pleistocene lakes in the Great Basin region, western United States. Water Resour. Res. 45, WO6435. http://dx.doi.org/10.1029/2007WR005953.

McConnel, R.K., 1965. Isostatic adjustment in a layered Earth. J. Geophys. Res. 70, 5171–5180.

Mifflin, M.D., Wheat, M.M., 1979. Pluvial lakes and estimated pluvial climates of Nevada. Nevada Bureau of Mines and Geology Bulletin, vol. 94. Mackay School of Mines, University of Nevada, Reno, NV. 57 pp.

Miller, D.M., Oviatt, C.G., McGeehin, J.P., 2013. Stratigraphy and chronology of Provo shoreline deposits and lake-level implications, Late Pleistocene Lake Bonneville, eastern Great Basin, USA. Boreas 42, 342–361.

Miller, D.M., Wahl, D.B., McGeehin, J.P., Rosario, J., Oviatt, C.G., Anderson, L., Presnetsova, L., 2015. Limiting age for the Provo shoreline of Lake Bonneville. Quat. Int. 387, 99–105.

Niskanen, E., 1948. On the viscosity of the Earth's interior. Ann. Acad. Sci. Fenn. A III 15. 22 pp.

O'Connor, J.E., 1993. Hydrology, hydraulics, and geomorphology of the Bonneville flood. Geol. Soc. Am. Spec. Pap. 247. 83 pp.

Oviatt, C.G., 1997. Lake Bonneville fluctuations and global climate change. Geology 25, 155–158.

Oviatt, C.G., 2015. Chronology of Lake Bonneville, 30,000 to 10,000 yr B.P. Quat. Sci. Rev. 110, 166–171.

Oviatt, C.G., Currey, D.R., Sack, D., 1992. Radiocarbon chronology of Lake Bonneville, eastern Great Basin, USA. Palaeogeogr. Palaeoclimatol. Palaeoecol. 99, 225–241.

Passey, Q.R., 1981. Upper mantle viscosity derived from the difference in rebound of the Provo and Bonneville shorelines, Lake Bonneville Basin. J. Geophys. Res. 86, 11,701–11,708.

Peltier, W.R., 1974. Impulse response of a Maxwell Earth. Rev. Geophys. 12, 649–669.

Reheis, M.C., Adams, K.D., Oviatt, C.G., Bacon, S.N., 2014. Pluvial lakes in the Great Basin of the western United States: a view from the outcrop. Quat. Sci. Rev. 97, 33–57.

Snyder, C.T., Hardman, G., Zdenek, F.F., 1964. Pleistocene Lakes in the Great Basin. U.S. Geological Survey. Miscellaneous Geologic Investigations Map I-416.

Stewart, J.H., 1978. Basin and range structure in western North America. A review. In: Smith, R.B., Eaton, G.P. (Eds.), Cenozoic Tectonics and Regional Geophysics of the Western Cordillera. Geological Society of America Memoir, vol. 152, pp. 1–32.

Watts, A.B., 2001. Isostasy and Flexure of the Lithosphere. Cambridge University Press, New York, NY. 458 pp.

Chapter 9

Using Lake Bonneville Features to Calibrate In Situ Cosmogenic Nuclide Production Rates

N.A. Lifton*, F.M. Phillips† and T.E. Cerling‡
*Purdue University, West Lafayette, IN, United States
†New Mexico Tech, Socorro, NM, United States
‡University of Utah, Salt Lake City, UT, United States

ABSTRACT

Well-dated bedrock surfaces, lava flows, and deposits of very large boulders in northwestern Utah and southern Idaho associated with the highstand and subsequent catastrophic draining of Pleistocene Lake Bonneville, Utah, during the Bonneville flood have long been recognized as excellent locations for terrestrial in situ cosmogenic nuclide (TCN) production-rate calibration. The well-constrained independent chronology of the late Pleistocene lake cycle, in concert with negligible erosion of continually exposed surfaces and a wide variety of rock types providing appropriate target elements for TCN production, provides a firm basis for these calibrations. All commonly measured TCNs have been calibrated using landforms associated with Lake Bonneville, from some of the earliest studies to some of the most recent. This enables direct comparison of production rates for different nuclides from sites with similar exposure histories, imposing critical constraints on theoretical and empirical models of spatial and temporal variation in TCN production globally. Lake Bonneville thus continues to influence the field of TCN analysis in significant ways.

Keywords: Terrestrial cosmogenic nuclides, Bonneville flood, Lake shorelines, Production rate calibration

9.1 INTRODUCTION

Applications involving nuclides produced directly in mineral lattices via the interaction of secondary atmospheric cosmic rays with terrestrial materials (terrestrial in situ cosmogenic nuclides, or TCNs) have grown tremendously over the last 25 years (Granger et al., 2013). The production rate of a given nuclide is one of the key parameters underlying these applications—as such,

Developments in Earth Surface Processes, Vol. 20. http://dx.doi.org/10.1016/B978-0-444-63590-7.00009-3

the potential precision and accuracy of such applications is limited by how well the nuclide production rates are known. TCN production rates can be estimated theoretically, but such estimates can contain significant systematic uncertainties due to uncertainties in the underlying nuclear physics models and parameters (eg, Reedy, 2011). As such, TCN production rates are more commonly calibrated by measuring TCN concentrations in surficial rocks with well-constrained exposure histories.

Much effort has been directed toward identifying and sampling calibration sites globally for this purpose that meet three stringent criteria for a "well-constrained exposure history." These criteria include: (1) a well-defined and accurate independent age constraint for initial exposure of the sampled surface; (2) no significant preexisting nuclide inventory arising from exposure of the surface prior to this independent age; and (3) stable geometry of the surface and its surroundings following the initial exposure (this includes no significant erosion of, deposition on, or intermittent shielding of, the surface). In addition, a fourth requirement is the availability of appropriate mineral targets in the rock that yield the nuclide(s) of interest when subjected to the secondary cosmic-ray flux. Well-dated bedrock surfaces, lava flows, and deposits of very large boulders in northwestern Utah and southern Idaho associated with Pleistocene Lake Bonneville have long been recognized as meeting all of these criteria, providing a robust opportunity for TCN production-rate calibration (eg, Cerling, 1990; Oviatt et al., 1992; Lifton et al., 2001, 2015). These surfaces have been used to estimate late Quaternary production rates for essentially the entire suite of commonly measured CNs, most recently as part of the Cosmic-Ray-prOduced NUclide Systematics on Earth Project (CRONUS-Earth Project) (Lifton et al., 2015). Nuclides calibrated using Lake Bonneville features include ^{3}He (Cerling, 1990; Cerling and Craig, 1994; Goehring et al., 2010; Amidon and Farley, 2011), ^{21}Ne (Poreda and Cerling, 1992), ^{10}Be, ^{26}Al (Gosse and Klein, 1996; Lifton et al., 2015), ^{36}Cl (Zreda et al., 1991; Phillips et al., 1996; Stone et al., 1996; Lifton et al., 2015), and ^{14}C (Handwerger et al., 1999; Lifton et al., 2001, 2015; Miller et al., 2006); all of the commonly employed cosmogenic nuclides. This chapter describes how the features of Lake Bonneville satisfy the criteria that make it a robust area for calibration studies, and briefly summarizes recent site production rates.

9.2 ADVANTAGES OF LAKE BONNEVILLE FOR CALIBRATION

Lake Bonneville (Gilbert, 1890), a large, closed-basin lake during the Late Pleistocene in the region that is now occupied by the Great Salt Lake (Fig. 9.1), presents two key advantages in a TCN calibration site. First, the basin is characterized by a series of widespread wave-cut bedrock benches and depositional shoreline features that formed during lake-level fluctuations. These extensive and well-preserved shorelines and associated features cut across a wide variety of rock types—including those bearing quartz,

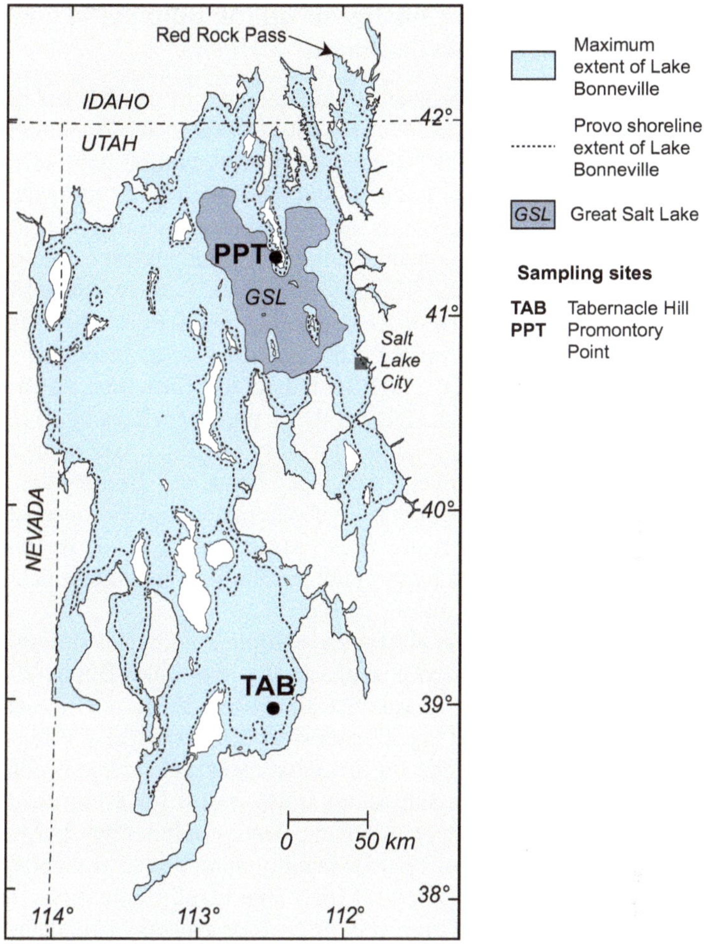

FIG. 9.1 Location of Lake Bonneville in relation to the current Great Salt Lake and the sampling locations (*TAB*, Tabernacle Hill, *PPT*, Promontory Point). Also shown is the location of Red Rock Pass, Idaho. *Modified after Miller, D.M., Oviatt, C.G., Mcgeehin, J.P., 2013. Stratigraphy and chronology of Provo shoreline deposits and lake-level implications, Late Pleistocene Lake Bonneville, eastern Great Basin, USA. Boreas 42, 342–361.*

carbonates, and mafic minerals such as olivine, pyroxene, and Ca-plagioclase. These shorelines, in turn, are understood in the context of a robust lake-level history, tightly constrained at present by over 360 radiocarbon ages (Oviatt, 2015). This enables straightforward comparison of production rates for a variety of nuclides, from a wide variety of mineral targets in spatially related samples, all with the same high quality independent age control. This rare combination of factors makes Lake Bonneville a notable TCN production-rate calibration locality.

9.2.1 Geochronological Constraints on the Bonneville and Provo Shorelines, and the Bonneville Flood

Although the Lake Bonneville basin is characterized by several major shoreline features, the two most extensive shorelines in the Lake Bonneville basin are of particular interest for TCN production-rate calibrations, as they are closely linked in time. The Lake Bonneville highstand (known as the Bonneville level) is well preserved as the Bonneville shoreline. This shoreline is unusual in that it was abandoned essentially instantaneously due to a catastrophic failure of the lake threshold at Red Rock Pass in southern Idaho, resulting in the Bonneville flood. The resulting ~100 m drop in lake level (modeled altitudes dropped from ca. 1552 to 1444 m, adjusted for isostatic rebound after the flood (Oviatt, 2015)), led to the formation of the Provo shoreline. The tight time constraints on these two shorelines form the basis of subsequent TCN production-rate calibrations. The Provo shoreline tends to be broader and more extensively developed than the Bonneville, leading Gilbert (1890) to suggest a longer period of occupation, a view since supported by recent work (eg, Godsey et al., 2011; Miller et al., 2013; Reheis et al., 2014), but the postflood Provo exposure chronology is more complicated than that of Bonneville (eg, Miller et al., 2013).

The Bonneville basin has experienced multiple lake cycles during the last ca. 780 ka, but available evidence suggests that the Lake Bonneville cycle during Marine Isotope Stage 2 (ca. 29–14 cal ka; 25–12 ^{14}C ka (Lisiecki and Raymo, 2005)) was the most extensive (Oviatt et al., 1999; Oviatt, 2015). The lake-level chronology for this most recent lake cycle is considered to be one of the most reliable and robust in the world for a Pleistocene lake (Reheis et al., 2014; Oviatt, 2015), having been studied since the late 19th century (Gilbert, 1890). Oviatt (2015) recently synthesized the current state of knowledge on the chronology of the lake-level fluctuations between 30 and 10 cal ka (ca. 26–9 ^{14}C ka) (Fig. 9.2). Key aspects of this chronology germane to TCN production-rate calibrations are summarized below, focusing on the timing of the Bonneville and Provo shoreline occupations. Oviatt (2015) converted radiocarbon ages (^{14}C ka) to calendar ages using CALIB 7.0 (Reimer et al., 2013); we quote those calibrated ages here (cal ka).

The first transgression events in the recent paleolake history, between approximately 30–25 cal ka (30–22 ^{14}C ka), have little chronological information compared to the later phases (Oviatt, 2015) (Fig. 9.2). Considerable controversy exists regarding lake levels during Marine Isotope Stage 3 (Nishizawa et al., 2013, 2014; Oviatt et al., 2014). After a period of relatively shallow occupation, the lake levels rose relatively quickly from a low level, probably at or below the modern lake levels, to a mid-range lake level around 29 cal ka (ca. 26.5 ^{14}C ka). The lake level oscillated over a vertical distance of about 45 m between ca. 25 and 22 cal ka (ca. 22–18 ^{14}C ka). The Stansbury shoreline was formed during these fluctuations, denoted the Stansbury

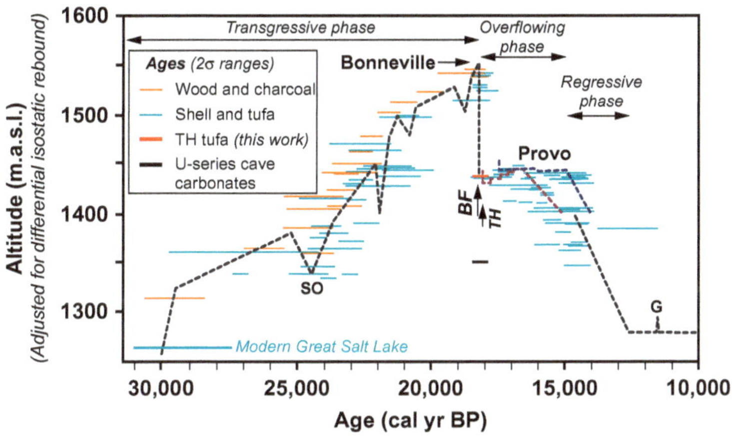

FIG. 9.2 Time–altitude diagram illustrating water level of Lake Bonneville through time, including age constraints. *Dashed gray line* shows the interpretation of lake-level history in the late Pleistocene (Reheis et al., 2014). *Dark blue* and *red dashed lines* following the Bonneville flood show recent alternative interpretations of Provo shoreline history from Miller et al. (2013). *TH*, Tabernacle Hill; *BF*, Bonneville flood; *SO*, Stansbury oscillation; *G*, Gilbert episode. Modern level of Great Salt Lake is indicated as well. *Modified after Oviatt, C.G., 2015. Chronology of Lake Bonneville, 30,000 to 10,000 yr B.P. Quatern. Sci. Rev. 110, 166–171; Reheis, M.C., Adams, K.D., Oviatt, C.G., Bacon, S.N., 2014. Pluvial lakes in the Great Basin of the western United States—a view from the outcrop. Quatern. Sci. Rev. 97, 33–57.*

oscillation (SO in Fig. 9.2). The Stansbury shoreline has not previously been used for TCN calibration.

The final transgression leading to the Bonneville stage began around 22 cal ka (18 ^{14}C ka) (Oviatt, 2015). Based on lake cores and shoreline stratigraphy, several smaller scale (30–50 m) fluctuations in lake level occurred during the transgression (Oviatt, 1997). At the end of this transgression, the lake reached its highest elevation at 1552 m (adjusted for isostatic rebound (Oviatt, 2015)). The Bonneville shoreline formed at this elevation as the lake level reached the low point on the basin divide at Zenda, Idaho, just north of Red Rock Pass (Oviatt, 2015). No evidence is recognized within the Lake Bonneville basin for any prior occupation at or near the Bonneville shoreline level (eg, Gilbert, 1890; Nishizawa et al., 2013; Reheis et al., 2014; Oviatt, 2015). At this time, the lake was large, having an area of more than 51,000 km^2 (Currey et al., 1984) and was ca. 350 m deep (Bills and May, 1987).

The period of overflow at the alluvial fan threshold at Red Rock Pass near Zenda, Idaho was likely brief (Shroder et al., 2016; O'Connor, 2016). Soon after lake level reached the alluvial fan sill at Red Rock Pass (Figs. 9.1 and 9.2), the fan failed catastrophically and the Bonneville flood inundated the Snake River plain to the north of the lake, eroding basaltic bedrock in its path and depositing large boulders downstream ("Melon Gravels") near Pocatello,

Idaho (Malde and Powers, 1962; Malde, 1968; O'Connor, 2016). During this flood event, the lake level rapidly dropped approximately between 108 and 125 m, releasing ca. 4750 km^3 of water (Currey, 1982; O'Connor, 1993; Miller et al., 2013). The lake level then stabilized at the elevation of Red Rock Pass and the Provo shoreline formed (Gilbert, 1890; Godsey et al., 2011; Janecke and Oaks, 2011; Miller et al., 2013; Shroder et al., 2016). During this development, the lake overflowed to the Snake River drainage.

Several radiocarbon ages constrain the time of occupation of the Bonneville shoreline, but the most reliable of these are from charcoal collected from a pre-Bonneville soil about 7 m below the highstand shoreline (15.25 ± 0.16 ^{14}C ka, ETH-3518; 15.32 ± 0.14 ^{14}C ka, AA-100116) and a wood sample 15 m below the Bonneville shoreline (15.10 ± 0.14 ^{14}C ka, W-5261) (Oviatt et al., 1992; Oviatt, 2015). These ages, averaging 15.22 ± 0.11 ^{14}C ka (1σ), calibrate to 18.5 ± 0.3 cal ka—we thus take that as a maximum age for the start of the Bonneville highstand given their stratigraphic context. No known shoreline evidence exists in the Bonneville basin for any earlier occupation of that level or higher (Gilbert, 1890; Oviatt et al., 1992; Reheis et al., 2014; Oviatt, 2015).

Miller et al. (2013) questioned the validity of dates on gastropods from Lake Bonneville based on new radiocarbon ages from the Provo shore zone that provide significantly older ages (18–17 cal ka; ca. 15–14 ^{14}C ka) than corresponding shell radiocarbon ages from offshore sands (eg, Godsey et al., 2005, 2011), or total inorganic carbon and $\delta^{18}O$ in sediment cores (Benson et al., 2011) (ca. 17–15 cal ka; ca. 14–12.5 ^{14}C ka). Another new dataset (McGee et al., 2012) provided U/Th ages of carbonate deposits in caves of various altitudes in the Bonneville basin. The results indicate that carbonate deposition ceased at 18.12 ± 0.15 ka in a cave in the Fish Springs Range that is 40 m below the Bonneville shoreline (McGee et al., 2012). By inference, this provides a minimum limiting age on the Bonneville flood. These new dates and questions about some of the original radiocarbon dates of Oviatt et al. (1992) and Oviatt and Nash (1989) led to an extensive reanalysis of the age of the Bonneville flood (Lifton et al., 2015). This reanalysis included a focus on the Tabernacle Hill basalt flow—a locale of particular significance to the Bonneville flood chronology.

The Tabernacle Hill flow is a small ($\sim$17 km^2), circular basalt feature that erupted into Lake Bonneville when the lake level was at or near the Provo shoreline level, as indicated by the presence of glassy basalt pillows, tufa, a shoreline bench, and beach deposits around the flow margins (Fig. 9.3). It is clear from pillow basalts around the margin of the flow and other geologic evidence that the basalt erupted into the lake approximately at the Provo shoreline (Gilbert, 1890; Oviatt and Nash, 1989). The top of the basalt flow shows no evidence of eruption into water, indicating that the eruption occurred after the Bonneville flood event when the water was at the level of the Provo shoreline.

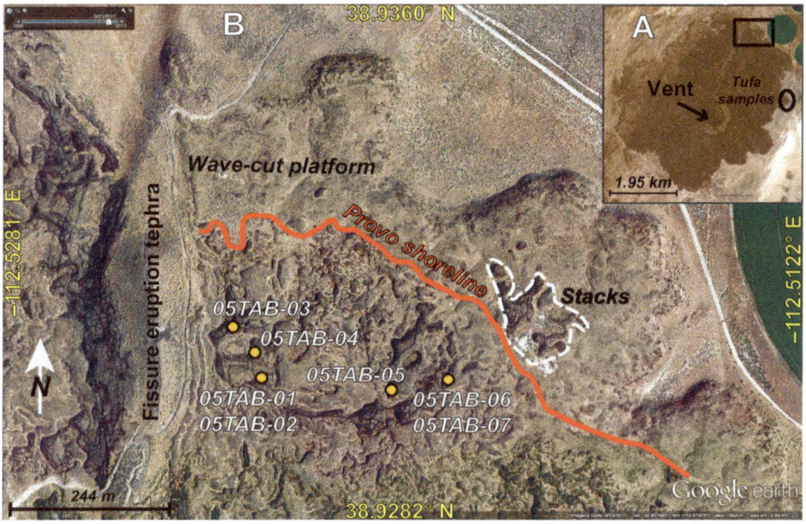

FIG. 9.3 (a) Google Earth image of Tabernacle Hill showing the whole flow and the area in which the samples were collected (*black rectangle*). The *black oval* indicates the tufa sampling location. (b) Close-up of the sampling area showing individual sample locations and other features, such as the Provo shoreline (altitude ca. 1440 m), wave-cut platform, and the tephra deposit.

Tufa encrustations on the margins of the flow provide a radiocarbon constraint on the youngest possible age of the basalt. A single radiocarbon date from this tufa of 17.3 ± 0.2 cal ka before 2010 (14.3 ± 0.09 ^{14}C ka; Beta-23803) (Oviatt and Nash, 1989) served as the most direct constraint on the minimum age of the Bonneville flood since it was first published, until the tufa was resampled for radiocarbon analysis as described by Lifton et al. (2015). Dense samples were collected from overhanging ceilings of shallow rock shelters on the flow margin, to avoid problems with recrystallization (often an issue with tufas), and carefully pretreated to remove atmospheric/organic contamination prior to analysis by accelerator mass spectrometry (Lifton et al., 2015). McGee et al. (2012) estimated radiocarbon reservoir effects in Lake Bonneville at <200 years based on 2σ agreement between U–Th and ^{14}C ages of lacustrine carbonates lining a rock shelter which bracket the Bonneville flood age. Indeed, this agreement is good enough that McGee et al. (2012) use calibrated ^{14}C ages with no reservoir correction to date carbonate samples where high detrital Th would yield unreliable U–Th ages. However, Lake Bonneville was enormous and a reservoir age estimate from a single location might not apply to the entire basin. The sudden drop in lake level after the flood would have likely temporarily increased groundwater discharge into the lake due to a steeper hydraulic gradient, and some of that additional groundwater could have been depleted in ^{14}C relative to the atmosphere at the time. However, it can also be argued that CO_2 exchange could have rapidly brought the shallow dissolved CO_2 in the lake into equilibrium with the atmosphere. Since direct evidence of a significant reservoir age in the vicinity of Tabernacle Hill when it erupted is currently lacking, we consider the calibrated tufa ages (with no reservoir correction) as the current best estimate of the formation age of the encrustations (see also Benson et al., 2011; Miller et al., 2013).

The mean radiocarbon age is 14.9 ± 0.2 ^{14}C ka, which yields a calibrated age range of 18.2 ± 0.3 cal ka (assuming no reservoir correction). This is significantly older than the previous age assignment of Oviatt and Nash (1989), although an unrecognized reservoir effect would render this a maximum age. Because all evidence points to eruption into a Provo-level lake, the flow has to be older than the tufa deposits and younger than the Bonneville flood. The limited time period of tufa deposition on the flow margin early in the Provo history is consistent with the carbonate record of McGee et al. (2012), who attributed the observed hiatus in carbonate deposition between 18.1 ± 0.3 and 16.4 ± 0.2 ka (U–Th) to freshening of the lake from overflow at the Provo level and/or from postflood mixing.

The new tufa ages provide the oldest postflood constraints; however, the new postflood ages overlap with the youngest preflood ages from recent studies (McGee et al., 2012; Miller et al., 2013; Reheis et al., 2014). Using only published pre-Bonneville ages from material other than shells, it seems clear that the new ages, including two different types of dating techniques, are consistent with an older age of the Bonneville flood than previously accepted (eg, Oviatt et al., 1992; Lifton et al., 2001). Based on the overlap of the 2σ ranges of the new ages and the pre-Bonneville soil charcoal age, when considered with the U-series data of McGee et al. (2012) (which are not directly associated with shoreline features), we place the age of the Bonneville flood (and resulting abandonment of the Bonneville shoreline) at 18.3 ± 0.3 cal ka. Similarly, an age of 18.2 ± 0.3 cal ka is adopted for the Tabernacle Hill basalt flow.

Current models of Provo-level history tend to favor approximately one to three thousand years of outflow with multiple minor shorelines reflecting changing outlet conditions at or near Red Rock Pass (Godsey et al., 2011; Janecke and Oaks, 2011; Miller et al., 2013, 2015). However, the exposure history of the Tabernacle Hill basalt is unaffected by these minor fluctuations in lake level during Provo time. The eruption occurred very early in the Provo shoreline history and an eruption during the transgressive phase would have shown evidence of eventual cover by water from the Bonneville shoreline occupation.

9.3 SAMPLING LOCATIONS

Two sites in particular within the Bonneville basin that are tightly associated with the pre- and postflood chronology have been exploited as prime calibration localities. The first of these is the Tabernacle Hill basalt flow (Cerling, 1990; Cerling and Craig, 1994; Stone et al., 1996), whereas the second is a prominent wave eroded, gently sloping bedrock bench associated with the Bonneville shoreline on the west side of Promontory Point (Gosse and Klein, 1996; Lifton et al., 2001, 2015) (Figs. 9.1 and 9.2). Additional sites that have been used for production-rate calibration studies include basalt bedrock surfaces scoured by the Bonneville flood in southern Idaho, and associated large boulder bars (Cerling, 1990; Cerling and Craig, 1994; Amidon and Farley, 2011), as well as Provo shorelines cut into limestone bedrock (Handwerger et al., 1999).

9.3.1 Tabernacle Hill

Samples have been collected from the north side of the volcanic flow in the vicinity of a fissure-eruption tephra (Fig. 9.3) (Cerling, 1990; Goehring et al., 2010). In addition, samples have been collected from the tops of tumuli to eliminate topographic shielding corrections and to reduce the possibility of cover by soil or ash (Fig. 9.4). Previous work by Stone et al. (1996) showed little evidence of tephra cover in the vicinity of these samples. The original surface texture of pahoehoe ropes is well preserved in most samples (Marrero, 2009)—useful in distinguishing samples that had undergone very little erosion.

FIG. 9.4 Photo looking south from the sampled northern portion of the Tabernacle Hill basalt flow toward the central cinder cone. Note prominent tumuli (pressure ridges). The grass is growing on thin Holocene eolian silty sediment that has accumulated on low-lying portions of the flow surface.

9.3.1.1 Site-Specific Considerations

At Tabernacle Hill, two factors that have the potential to affect measured concentrations in the flow are surface erosion and ash cover. Basalts erode through two distinct mechanisms: spalling of entire horizontal flow sheets at the centimeter-plus scale and gradual erosion. Pahoehoe flows can consist of a series of horizontal layers; spalling of the surface layer reveals a new layer beneath that is essentially indistinguishable from the original surface (Cerling and Craig, 1994). These secondary surfaces can have pahoehoe ropes and other indications of an uneroded surface. Rubble around the pahoehoe outcrops might indicate that the surface is not pristine. In the case of gradual erosion, the surface textures are worn down gradually through time. Pahoehoe ropes, glassy surface textures, or other indicators of an uneroded surface may be degraded or completely removed. In the case of a fresh basalt flow, the gradual erosion must first strip off the glassy, friable, outer layer until the denser, inner material is reached.

Erosion rates on basalts are difficult to quantify (Cerling, 1990). Cerling and Craig (1994) estimated that ropes cannot remain distinct with more than 1 cm of total erosion (a rate of 0.55 mm/kyr at Tabernacle Hill). On the other hand, Dunbar (1999) and Dunbar and Phillips (2004) estimated that erosion on very young basalt flows could be as high as 5 mm/kyr during the process of removing the glassy rind, but would slow significantly when the crystalline basalt surface was reached, after a few thousand years. The average rate over the ~18 ka exposure of the Tabernacle Hill flow is probably closer to the lower bounding erosion rate discussed above than the upper (Lifton et al., 2015).

Ash cover is possible in volcanic environments as a source of decreased cosmogenic production. This is especially worth consideration in this case because tuff cones such as those at Tabernacle Hill produce this ash as part of the eruption sequence (Oviatt and Nash, 1989). Samples were taken at or near the tops of tumuli where wind action is likely to erode loose material to avoid problems with accumulation of soil, ash, or other eolian material. We consider ash cover unlikely to have affected any of the samples considered here.

9.3.2 Promontory Point

The wave-eroded bedrock benches at Promontory Point are among the largest in the Bonneville basin, measuring hundreds of meters in width, with approximately 50–70 m of overburden removed (based on topographic map estimates below—see Section 9.3.2.1) (Figs. 9.5 and 9.6). The size and depth of the lake during the Bonneville highstand allowed large waves to be generated along the fetch of the lake. These waves had sufficient energy to erode bedrock; they were able to cut deeply into the surrounding bedrock during the final transgression to the highstand despite the relatively brief occupation

FIG. 9.5 (a) Photo of the sampled Bonneville-level wave-cut bench at Promontory Point, Utah. Bench is approximately 300 m wide. Toyota 4Runner indicated for scale. (b) View looking north from bench in (a), at Bonneville and Provo-level shoreline complex.

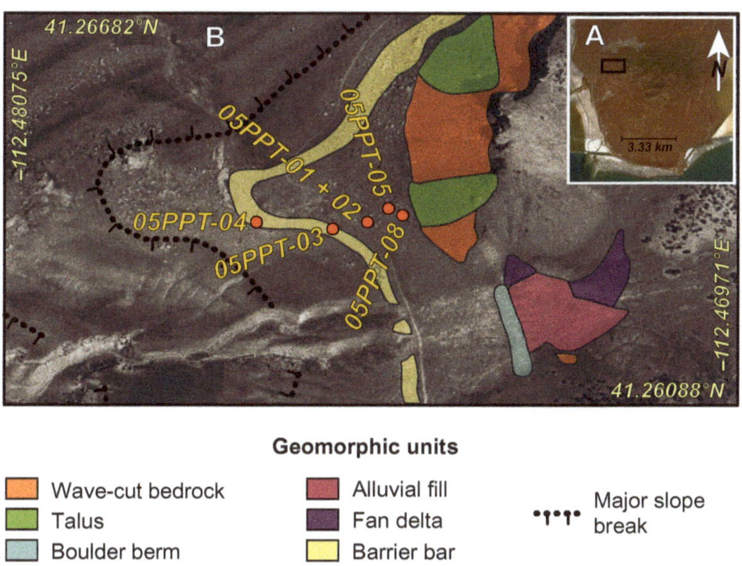

FIG. 9.6 (a) Google Earth image showing location of panel (b) (*black rectangle*) at the southern tip of Promontory Point. (b) Geomorphic map of the Promontory Point sampling site, showing sample locations on transect from outer edge of wave-cut bench to base of cliff.

of the shoreline. Promontory Point is at the southern end of a narrow peninsula (actually an island during the Bonneville highstand—Fig. 9.1) extending from the northern lake margin into what was the deepest part of the lake, exposing bedrock on this peninsula to intense wave action. The quartzite bedrock at the Bonneville level evinces the effects of the energy transported by these waves (Fig. 9.5).

These large-scale erosional features limit the possibility of significant cosmogenic nuclide inheritance. The surfaces are well preserved, exhibiting primary wave rounding/polishing. Quartzites and quartzite conglomerates of the late Precambrian Mutual Formation and the early Cambrian Tintic Quartzite Formation crop out at the site. Numerous joints within the quartzite likely facilitated the erosion (Lifton et al., 2001).

9.3.2.1 Site-Specific Considerations

Two factors occur at Promontory Point that could potentially affect measured concentrations: surface erosion since exposure, and a remnant signal derived from prior exposure to cosmic rays (inheritance). We argue that the former is insignificant at this location due to the well-preserved primary wave-rounded and polished surfaces on both the bedrock outcrops and boulders that were sampled. The potential for nuclide inheritance, however, deserves more consideration.

The samples from the Bonneville shoreline wave-eroded bench at Promontory Point conceivably could have experienced nuclide production at depth by muons under either rock or water, or both, prior to shoreline abandonment. Although accurate reconstruction of the pre-Bonneville shoreline topography is impossible, one can use existing topography to estimate plausible preshoreline configurations (and hence, potential overburden thicknesses) (Fig. 9.7).

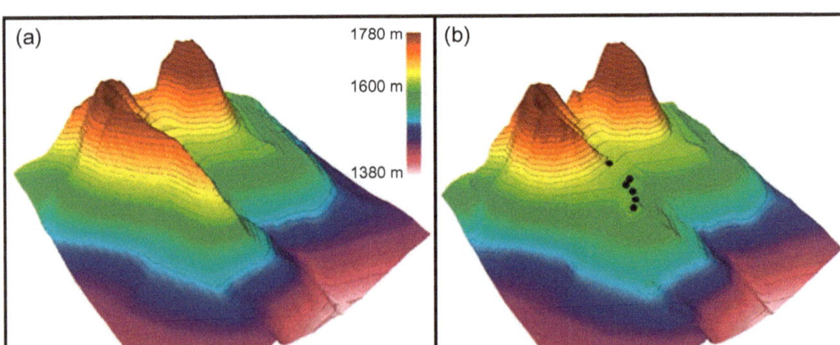

FIG. 9.7 (a) Speculative reconstruction of Promontory Point (based on current topographic trends) prior to erosion of the wave-cut bench, and (b) the current topography. Only the sampled ridge (on the left in each figure) was reconstructed. Approximate sample locations are shown as *black dots* in the image on the right. Samples were collected in a ca. 200-m transect stretching from near the base of the cliff to near the far end of the bench (Lifton et al., 2015).

Plausible reconstructed overburden thicknesses over the samples from this analysis ranged from ca. 50 to 70 m, which would reduce muogenic production by approximately a factor of 10^{-5} from the surface value. Lifton et al. (2015) collected samples along a ca. 200-m transect approximately perpendicular to the wave-cut notch to test for variable subsurface production; they did not find evidence of any significant effects in this regard.

Lifton et al. (2001) surveyed the maximum elevations of estuarine deposits at the site relative to sample locations similar to those of this study and concluded that the samples were submerged under ca. 17 m of water at the highest level of Lake Bonneville. Field evidence noted as part this study suggests >10 m of water was present over the bench. Spallogenic neutrons are rapidly attenuated with depth in water—assuming an effective attenuation length of 140 g/cm^2, 10–17 m of water would lead to a reduction in nuclide production by spallation of approximately 4–5 orders of magnitude over subaerial values. Combined with the brief duration of the occupation at the Bonneville highstand (eg, Shroder et al., 2016; O'Connor, 2016), and low surficial production rates from muons at the site (ca. 0.2 at g^{-1} y^{-1} for ^{10}Be), we argue that subaqueous production would have been negligible.

9.3.3 Flood-Scoured Basalts and Melon Gravels in Idaho

Lake Bonneville overflowed at Red Rock Pass in southern Idaho, causing a lowering of the lake by ca. 100 m; peak discharge at Red Rock Pass was estimated to be ca. 10^6 m^3/s, with flood deposits occurring downstream for >1000 km over an elevation gradient from ca. 1500 to 200 m above sea level (O'Connor, 1993, 2016). Basalt boulder bars occur beginning in Marsh Valley, an underfit river valley that is sourced in the Red Rock Pass overflow region. Today, Marsh Creek is a very minor tributary the Portneuf River, which joins the Snake River near Pocatello Idaho. Boulder bars occur along the length of this flow path from Inkom, Idaho, to Lewiston, Idaho, with basalt boulders being, by far, the predominant rock type in the Snake River Plain portion of the floodpath. However, sporadic rhyolites in the region provide the opportunity to determine production rates from acidic volcanic rocks as well as basaltic volcanic rocks (Amidon and Farley, 2011).

The boulder deposits are commonly referred to as the Melon Gravels, named for a ca. 50 m thick and 1 km long boulder bar near Melon Valley, Idaho (Malde and Powers, 1962; O'Connor, 2016). Boulders vary in size in the different reaches of the flood path, related to the basalt outcrops in the immediate vicinity; in the Marsh Creek—Portneuf River region, the largest fluvial boulders are between 1 and 2 m in diameter for the intermediate axis, whereas downstream near Hagerman, Idaho, boulders up to 4 m occur (O'Connor, 1993). Many boulders in these boulder bar deposits are smooth, with well-rounded boulders still having water-fluted surface features, indicating minimal erosion.

9.4 A SUMMARY OF NUCLIDE CALIBRATIONS AT LAKE BONNEVILLE

Geomorphic features associated with Lake Bonneville shorelines have thus proven to be extremely valuable targets for calibrations of TCN production rate. Indeed, some of the earliest calibrations were established here, and they continue to be generated from these sites. Cerling (1990) was the first study to make use of olivine phenocrysts from Tabernacle Hill and flood-deposited boulders in deriving a production rate for ^{3}He. Poreda and Cerling (1992) revisited the Tabernacle Hill samples for a ^{21}Ne production-rate estimate in both olivine and plagioclase phenocrysts, as did Cerling and Craig (1994) for ^{3}He. Zreda et al. (1991) and Phillips et al. (1996) also used Tabernacle Hill samples in their calibration of whole-rock production rates for ^{36}Cl, whereas Stone et al. (1996) evaluated ^{36}Cl production from Ca in plagioclase mineral separates from Tabernacle Hill basalt samples. Gosse and Klein (1996) estimated the ^{10}Be production rates from quartzites on the Provo shoreline at Promontory Point. Handwerger et al. (1999) analyzed limestone samples from wave-eroded Provo shoreline benches for in situ ^{14}C production rates, whereas Lifton et al. (2001) and Miller et al. (2006) generated comparable in situ ^{14}C production rates in quartz from the Bonneville shoreline cut benches at Promontory Point. More recently, Amidon and Farley (2011) analyzed ^{3}He production from accessory minerals such as pyroxene, zircon, and apatite in Bonneville flood-scoured surfaces cut into rhyolite along the Snake River in Idaho. These ^{3}He production rates have been recalculated for site-specific values and our preferred chronology in Table 9.1, along with mineral-specific ^{21}Ne values of Poreda and Cerling (1992), demonstrating the compositional dependence of each. ^{3}He production-rate estimates, derived as above from pyroxenes in Bonneville flood-transported boulders in Inkom, Idaho (Amidon and Clark, 2015), are also presented in Table 9.1.

As part of the CRONUS-Earth effort, Lifton et al. (2015) published updated production-rate estimates for all major CNs (except ^{21}Ne) from Promontory Point for nuclides produced in quartz, and Tabernacle Hill for nuclides produced in mafic minerals. Samples from bedrock outcrops and large boulders exhibiting primary wave-polish were analyzed for ^{10}Be, ^{26}Al, and ^{14}C from Promontory Point (Fig. 9.6). Tabernacle Hill was resampled for updated production rates for ^{3}He and ^{36}Cl (Goehring et al., 2010; Lifton et al., 2015) (Fig. 9.3). Descriptions for each sample, including composition and other important sample information, and sample analytical results for each nuclide have been included in Lifton et al. (2015). Current best estimates of site production rates for these nuclides derived for CRONUS-Earth are presented in Table 9.2 (Goehring et al., 2010; Lifton et al., 2015), recalculated using the newest chronology.

TABLE 9.1 Accessory Mineral ^{3}He (Amidon and Farley, 2011; Amidon and Clark, 2015) and ^{21}Ne Production Rates (Poreda and Cerling, 1992) Associated with Lake Bonneville

Mineral	Mean Latitude[a] (°N)	Mean Longitude[a] (°E)	Mean Altitude[a] (m)	Age-Corrected Site Production Rate[b] (at g^{-1} a^{-1})
^{3}He—zircon *Snake River, Idaho*	42.6043	245.5451	980	224±25
^{3}He—apatite *Snake River, Idaho*	42.6041	245.5444	981	278±20
^{3}He—pyroxene *Snake River, Idaho*	42.6030	245.5487	982	265±16
^{3}He—pyroxene *Inkom, Idaho*	42.7945	247.7678	1424	345±4
^{21}Ne—olivine (Fo_{81}) *Tabernacle Hill, Utah*	38.9299	247.4801	1455	128±8
^{21}Ne—plagioclase (Ab_{36}) *Tabernacle Hill, Utah*	38.9299	247.4801	1455	48±11

[a]*Mean of sample locations.*
[b]*Corrected to present best estimate of Bonneville flood (^{3}He) and Tabernacle Hill (^{21}Ne) ages. Thickness and shielding corrections from original publications.*

TABLE 9.2 Recent Lake Bonneville-Derived Production Rates (Lifton et al., 2015)

Nuclide	Mean Latitude[a] (°N)	Mean Longitude[a] (°E)	Mean Altitude[a] (m)	Site Production Rate (at g^{-1} a^{-1})
^{3}He[b]	38.9302	247.4790	1459	353.4±2.6
^{10}Be[c]	41.2637	247.5245	1603	15.1±0.4
^{14}C[c]	41.2637	247.5245	1603	50.9±2.4
^{26}Al[c]	41.2637	247.5245	1603	100.4±9.0
^{36}Cl[d]	38.9302	247.4790	1459	130.0±5.9

[a]*Mean of sample locations.*
[b]*Olivine (Fo_{81}).*
[c]*Quartz.*
[d]*Plagioclase (Ab_{36}) mineral separates.*

It is beyond the scope of this chapter to scale the site production rates to reference production rates at sea level and high latitude. Aside from delving into the complexities of which reference production rate is associated with which scaling model (eg, Balco et al., 2008; Lifton et al., 2014), CRONUS-Earth reference production rates were derived as a global fit, yielding a single global value for each production pathway. Interested readers are referred to Borchers et al. (2016), Phillips et al. (2016), and Lifton (2016) for details. We follow Borchers et al. (2016) in preferring nuclide-specific scaling as described for the LSD model of Lifton et al. (2014).

9.5 CONCLUSIONS

Landforms associated with Lake Bonneville have thus proven to be key sites for TCN production-rate calibration. The well-constrained independent chronology of the late Pleistocene lake cycle in general, and in particular that associated with the timing of the Bonneville flood, provides a firm basis for these calibrations. In addition, the wide range of rock types comprising these landforms enables production rates of multiple TCNs from different target elements to be compared directly at the same location, imposing critical constraints on theoretical and empirical models of spatial and temporal variation in TCN production globally. From the early days of TCN research to today's increasingly sophisticated applications, Lake Bonneville continues to influence the field of TCN analysis in significant ways.

ACKNOWLEDGMENTS

The authors gratefully acknowledge support from NSF grants EAR-0345150 (Lifton) and EAR-0345949 (Phillips). We also thank Mr. John Young for allowing access to the Promontory Point site. Mark Kurz, Darryl Granger, Bob Finkel, Marc Caffee, Kuni Nishiizumi, and Terry Swanson helped to collect samples and provided valuable field insights. Adam Hudson processed the Tabernacle Hill tufas for radiocarbon dating, and Dylan Rood contributed significantly to sample analyses. Finally, we thank Will Amidon for thoughtful and thorough comments that greatly improved this chapter.

REFERENCES

Amidon, W., Clark, A.C., 2015. Interaction of outburst floods with basaltic aquifers on the Snake River Plain: implications for Martian canyons. Geol. Soc. Am. Bull. 127, 688–701.

Amidon, W.H., Farley, K.A., 2011. Cosmogenic ^{3}He production rates in apatite, zircon and pyroxene inferred from Bonneville flood erosional surfaces. Quatern. Geochronol. 6, 10–21.

Balco, G., Stone, J.O., Lifton, N.A., Dunai, T.J., 2008. A complete and easily accessible means of calculating surface exposure ages or erosion rates from Be-10 and Al-26 measurements. Quatern. Geochronol. 3, 174–195.

Benson, L.V., Lund, S.P., Smoot, J.P., Rhode, D.E., Spencer, R.J., Verosub, K.L., Louderback, L.A., Johnson, C.A., Rye, R.O., Negrini, R.M., 2011. The rise and fall of Lake Bonneville between 45 and 10.5 ka. Quatern. Int. 235, 57–69.

Bills, B., May, G., 1987. Lake Bonneville: constraints on lithospheric thickness and upper mantle viscosity from isostatic warping of Bonneville, Provo, and Gilbert stage shorelines. J. Geophys. Res. 0148-022792, 11493–11508.

Borchers, B., Marrero, S., Balco, G., Caffee, M., Goehring, B., Lifton, N., Nishiizumi, K., Phillips, F., Schaefer, J., Stone, J., 2016. Geological calibration of spallation production rates in the CRONUS-Earth project. Quatern. Geochronol. 31, 188–198.

Cerling, T., 1990. Dating geomorphologic surfaces using cosmogenic ^{3}He. Quatern. Res. 33, 148–156.

Cerling, T., Craig, H., 1994. Cosmogenic ^{3}He production rates from 39 N to 46 N latitude, western USA and France. Geochim. Cosmochim. Acta 58, 249–255.

Currey, D.R., 1982. Lake Bonneville: selected features of relevance to neotectonic analysis, US Geological Survey Open-File Report 82-1070, 30 pp.

Currey, D., Atwood, G., Mabey, D., 1984. Major Levels of Great Salt Lake and Lake Bonneville.

Dunbar, N.W., 1999. Cosmogenic ^{36}Cl-determined age of the Carrizozo lava flows, south-central New Mexico. N. M. Geol. 21, 25–29.

Dunbar, N.W., Phillips, F.M., 2004. Cosmogenic ^{36}Cl ages of lava flows in the Zuni–Bandera volcanic field, northcentral New Mexico, U.S.A.. New Mex. Bur. Geol. Mineral Resour. Bull. 160, 309–318.

Gilbert, G.K., 1890. Lake Bonneville. United States Geological Survey Monograph 1. United States Geological Survey, Washington, D.C., 438 pp.

Godsey, H., Currey, D., Chan, M., 2005. New evidence for an extended occupation of the Provo shoreline and implications for regional climate change, Pleistocene Lake Bonneville, Utah, USA. Quatern. Res. 63, 212–223.

Godsey, H.S., Oviatt, C.G., Miller, D.M., Chan, M.A., 2011. Stratigraphy and chronology of offshore to nearshore deposits associated with the Provo shoreline, Pleistocene Lake Bonneville, Utah. Palaeogeogr. Palaeoclimatol. Palaeoecol. 310, 442–450.

Goehring, B.M., Kurz, M.D., Balco, G., Schaefer, J.M., Licciardi, J., Lifton, N., 2010. A reevaluation of in situ cosmogenic ^{3}He production rates. Quatern. Geochronol. 5, 410–418.

Gosse, J., Klein, J., 1996. Production rate of in situ cosmogenic ^{10}Be in quartz at high altitude and mid latitude. Radiocarbon 38, 154–155.

Granger, D.E., Lifton, N.A., Willenbring, J.K., 2013. A cosmic trip: 25 years of cosmogenic nuclides in geology. Geol. Soc. Am. Bull. 125, 1379–1402.

Handwerger, D., Cerling, T., Bruhn, R., 1999. Cosmogenic ^{14}C in carbonate rocks. Geomorphology 27, 13–24.

Janecke, S.U., Oaks, R.Q., 2011. New insights into the outlet conditions of late Pleistocene Lake Bonneville, southeastern Idaho, USA. Geosphere 7, 1369–1391.

Lifton, N., 2016. Implications of two Holocene time-dependent geomagnetic models for cosmogenic nuclide production rate scaling. Earth Planet. Sci. Lett. 433, 257–268.

Lifton, N., Jull, A., Quade, J., 2001. A new extraction technique and production rate estimate for in situ cosmogenic ^{14}C in quartz. Geochim. Cosmochim. Acta 65, 1953–1969.

Lifton, N., Sato, T., Dunai, T.J., 2014. Scaling in situ cosmogenic nuclide production rates using analytical approximations to atmospheric cosmic-ray fluxes. Earth Planet. Sci. Lett. 386, 149–160.

Lifton, N., Caffee, M., Finkel, R., Marrero, S., Nishiizumi, K., Phillips, F.M., Goehring, B., Gosse, J., Stone, J., Schaefer, J., Theriault, B., Jull, A.J.T., Fifield, K., 2015. In situ cosmogenic nuclide production rate calibration for the CRONUS-Earth project from Lake Bonneville, Utah, shoreline features. Quatern. Geochronol. 26, 56–69.

Lisiecki, L.E., Raymo, M.E., 2005. A Pliocene-Pleistocene stack of 57 globally distributed benthic δ^{18}O records. Paleoceanography 20, 17.

Malde, H.E., 1968. The Catastrophic Late Pleistocene Bonneville Flood. US Geological Survey Professional Paper 596, pp. 1–69.

Malde, H.E., Powers, H.A., 1962. Upper Cenozoic stratigraphy of western Snake River Plain, Idaho. Geol. Soc. Am. Bull. 73, 1197–1220.

Marrero, S., 2009. Chlorine-36 Production Rate Calibration Using Shorelines from Pleistocene Lake Bonneville, UT. Earth and Environmental Dept., New Mexico Tech, Socorro, New Mexico.

McGee, D., Quade, J., Edwards, R.L., Broecker, W.S., Cheng, H., Reiners, P.W., Evenson, N., 2012. Lacustrine cave carbonates: novel archives of paleohydrologic change in the Bonneville Basin (Utah, USA). Earth Planet. Sci. Lett. 351–352, 182–194.

Miller, G., Briner, J., Lifton, N., Finkel, R.C., 2006. Limited ice-sheet erosion and complex exposure histories derived from in situ cosmogenic ^{10}Be, ^{26}Al, and ^{14}C on Baffin Island, Arctic Canada. Quatern. Geochronol. 1, 74–85.

Miller, D.M., Oviatt, C.G., Mcgeehin, J.P., 2013. Stratigraphy and chronology of Provo shoreline deposits and lake-level implications, Late Pleistocene Lake Bonneville, eastern Great Basin, USA. Boreas 42, 342–361.

Miller, D.M., Wahl, D.B., McGeehin, J.P., Rosario, J., Oviatt, C.G., Anderson, L., Presnetsova, L., 2015. Limiting age for the Provo shoreline of Lake Bonneville. Quatern. Int. 387, 99–105. http://dx.doi.org/10.1016/j.quaint.2015.01.001.

Nishizawa, S., Currey, D.R., Brunelle, A., Sack, D., 2013. Bonneville basin shoreline records of large lake intervals during Marine Isotope Stage 3 and the Last Glacial Maximum. Palaeogeogr. Palaeoclimatol. Palaeoecol. 386, 374–391.

Nishizawa, S., Sack, D., Brunelle, A., 2014. Reply to the comment on "Bonneville basin shoreline records of large lake intervals during Marine Isotope Stage 3 and the Last Glacial Maximum" by S. Nishizawa, D. R. Currey, A. Brunelle, D. Sack. Palaeogeogr. Palaeoclimatol. Palaeoecol. 401, 177–180.

O'Connor, J., 1993. Hydrology, Hydraulics, and Geomorphology of the Bonneville Flood. Geological Society of America Special Paper 274, 1-84.

O'Connor, J., 2016. Chapter 6. The Bonneville flood—A veritable débâcle. In: Oviatt, C.G., Shroder Jr., J.F. (Eds.), Lake Bonneville: A Scientific Update, vol. 20. Elsevier. pp. 105–126.

Oviatt, C., 1997. Lake Bonneville fluctuations and global climate change. Geology 25, 155–158.

Oviatt, C.G., 2015. Chronology of Lake Bonneville, 30,000 to 10,000 yr B.P. Quatern. Sci. Rev. 110, 166–171.

Oviatt, C., Nash, W., 1989. Late Pleistocene basaltic ash and volcanic eruptions in the Bonneville basin, Utah. Bull. Geol. Soc. Am. 101, 292–303.

Oviatt, C., Currey, D., Sack, D., 1992. Radiocarbon chronology of Lake Bonneville, eastern Great Basin, USA. Palaeogeogr. Palaeoclimatol. Palaeoecol. 99, 225–241.

Oviatt, C., Thompson, R., Kaufman, D., Bright, J., Forester, R., 1999. Reinterpretation of the Burmester core, Bonneville Basin, Utah. Quatern. Res. 52, 180–184.

Oviatt, C.G., Chan, M.A., Jewell, P.W., Bills, B.G., Madsen, D.B., Miller, D.M., 2014. Interpretations of evidence for large Pleistocene paleolakes in the Bonneville basin, western North America: comment on: Bonneville basin shoreline records of large lake intervals during marine isotope stage 3 and the last glacial maximum, by Nishizawa et al. (2013). Palaeogeogr. Palaeoclimatol. Palaeoecol. 401, 173–176.

Phillips, F., Zreda, M., Flinsch, M., Elmore, D., Sharma, P., 1996. A reevaluation of cosmogenic ^{36}Cl production rates in terrestrial rocks. Geophys. Res. Lett. 23, 949–952.

Phillips, F.M., Argento, D.C., Balco, G., Caffee, M.W., Clem, J., Dunai, T.J., Finkel, R., Goehring, B., Gosse, J.C., Hudson, A.M., Jull, A.J.T., Kelly, M.A., Kurz, M., Lal, D., Lifton, N., Marrero, S.M., Nishiizumi, K., Reedy, R.C., Schaefer, J., Stone, J.O.H.,

Swanson, T., Zreda, M.G., 2016. The CRONUS-Earth Project: a synthesis. Quatern. Geochronol. 31, 119–154.

Poreda, R., Cerling, T., 1992. Cosmogenic neon in recent lavas from the western United States. Geophys. Res. Lett. 19, 1863–1866.

Reedy, R.C., 2011. Cosmogenic-nuclide production rates: reaction cross section update. Nucl. Instrum. Methods Phys. Res. B 294, 470–474.

Reheis, M.C., Adams, K.D., Oviatt, C.G., Bacon, S.N., 2014. Pluvial lakes in the Great Basin of the western United States—a view from the outcrop. Quatern. Sci. Rev. 97, 33–57.

Reimer, P.J., Bard, E., Bayliss, A., Beck, J.W., Blackwell, P.G., Ramsey, C.B., Buck, C.E., Cheng, H., Edwards, R.L., Friedrich, M., Grootes, P.M., Guilderson, T.P., Haflidason, H., Hajdas, I., Hatté, C., Heaton, T.J., Hoffmann, D.L., Hogg, A.G., Hughen, K.A., Kaiser, K.F., Kromer, B., Manning, S.W., Niu, M., Reimer, R.W., Richards, D.A., Scott, E.M., Southon, J.R., Staff, R.A., Turney, C.S.M., van der Plicht, J., 2013. IntCal13 and Marine13 radiocarbon age calibration curves 0–50,000 years cal BP. Radiocarbon 55, 1869–1887.

Shroder Jr., J.F., Cornwell, K., Oviatt, C.G., Lowndes, T., 2016. Chapter 4. Landslides, alluvial fans, and dam failure at Red Rock Pass: The outlet of Lake Bonneville. In: Oviatt, C.G., Shroder Jr., J.F. (Eds.), Lake Bonneville: A Scientific Update, vol. 20. Elsevier. pp. 75–87.

Stone, J.O., Allan, G.L., Fifield, L.K., Cresswell, R.G., 1996. Cosmogenic chlorine-36 from calcium spallation. Geochim. Cosmochim. Acta 60, 679–692.

Zreda, M.G., Phillips, F.M., Elmore, D., Kubik, P.W., Sharma, P., Dorn, R.I., 1991. Cosmogenic chlorine-36 production rates in terrestrial rocks. Earth Planet. Sci. Lett. 105, 94–109.

Chapter 10

Late Pleistocene to Early Holocene Sedimentary History of the Lake Bonneville Pilot Valley Embayment, Utah-Nevada, USA

K.A. Rey, A.L. Mayo, D.G. Tingey and S.T. Nelson
Brigham Young University, Provo, UT, United States

ABSTRACT
Multiple shallow-sediment cores were obtained from the Pilot Valley playa, the location of a Lake Bonneville sub-basin. Analyses of stratigraphy, ostracods, mineralogy, chemistry, total inorganic carbon, total organic carbon, and stable isotopes were performed to place these sediments into proper context with respect to the Lake Bonneville cycle. Results showed that the Pilot Valley playa contains a full sequence of Lake Bonneville deposits, in addition to sediments deposited before and after the Lake Bonneville cycle. Within the Bonneville marl is a sequence of algal-laminated mud. Five calibrated ^{14}C ages, ranging from 25.5 to 17.0 cal ka BP, determined from algae filaments in the laminated marl, in addition to its stratigraphic position near the base of the sequence, place it the early transgressive phase of Lake Bonneville. Oolitic sand occurring below the Bonneville marl indicates the existence of a shallow ($<\sim$5 m), saline lake in the Pilot Valley embayment prior to the transgression of Lake Bonneville ($\sim$32 k ^{14}C years). Analysis of sediments deposited during the late regressive phase of Lake Bonneville indicates the lake may have fallen to levels below the elevation of Pilot Valley prior to transgression of the Gilbert episode.

Keywords: Lake Bonneville, Algal-laminated mud, Pilot Valley playa, Pilot Valley stratigraphy

10.1 INTRODUCTION

The Bonneville basin, located in the northeastern Great Basin of the western United States, has been episodically occupied by pluvial lakes during the last

Developments in Earth Surface Processes, Vol. 20. http://dx.doi.org/10.1016/B978-0-444-63590-7.00010-X

800 ka (Eardley et al., 1973; Oviatt et al., 1999). The most recent large lake, Lake Bonneville, has the Great Salt Lake as a present-day remnant (Gilbert, 1890; Currey, 1990). The story of Lake Bonneville has been largely interpreted from ancient shorelines and attendant outcrops (Gilbert, 1890; Morrison, 1965; Oviatt et al., 1990, 1994a; Oviatt, 1991, 1997; Godsey et al., 2005, 2011). Lake Bonneville shorelines, occurring as wave-cut terraces along most mountain ranges throughout the basin, are relatively easy to identify and represent low, middle, and high lake stands. Shorelines allow lake volume calculations and preserve evidence of the transgressive/regressive character of the lake.

Traditionally, four major Lake Bonneville shorelines have been identified (Fig 10.1). From oldest to youngest they are the Stansbury, Bonneville, Provo, and Gilbert, with approximate elevations of 1360,[a] 1555, 1444, and 1300 m above sea level (m asl), respectively (Gilbert, 1890; Currey, 1980, 1982; Oviatt, 1987, 2015a; Oviatt et al., 1990, 1992). Stansbury formed during the initial transgressive phase. The Bonneville shoreline represents the high stand, and the greatest water depth, attained by Lake Bonneville. Following the catastrophic outflowing of water known as the Bonneville flood (O'Connor, 1993, 2016), an event that caused the water level to drop by over 100 m, the Provo shoreline formed during a period of relatively stable lake level when the water level was modulated by spilling at the natural threshold into the Snake River drainage in the northeast corner of the lake. The Gilbert episode, traditionally viewed as the latest regressive phase of Lake Bonneville, has recently been interpreted to be separate and distinct from Lake Bonneville. It occurred as a separate transgression after Lake Bonneville had dropped to levels similar to those of the current Great Salt Lake and remained at low levels between about 13 and 11.6 cal ka BP (Oviatt, 2014, 2015a).

Much of Lake Bonneville research has focused on shoreline outcrops. Because the lake occupied broad and relatively flat basins, lake-bed sediments have great potential to shed light on the lake history. Published work based on lake-bed cores is not extensive. Eardley et al. (1973) studied the Burmester core, obtained near the south arm of the Great Salt Lake, in which they identified multiple lake cycles over the past 800 ka. Oviatt et al. (1999) re-examined the Burmester core and reported only four major lake cycles during the same time period. Spencer et al. (1984) collected many cores from Great Salt Lake and analyzed them for stratigraphic and geochemical features and attempted to correlate them with outcrop-based studies. Thompson et al. (1990, 2016) and Pedone and Oviatt (2016) analyzed chemistry, mineralogy, crustaceans, and undertook AMS ^{14}C dating of organic-rich mud from Lake

[a]The Stansbury stage of Lake Bonneville consists of several visible shorelines in the Bonneville basin. The elevation used in the paper follows an average value from the conclusions of Oviatt et al. (1990).

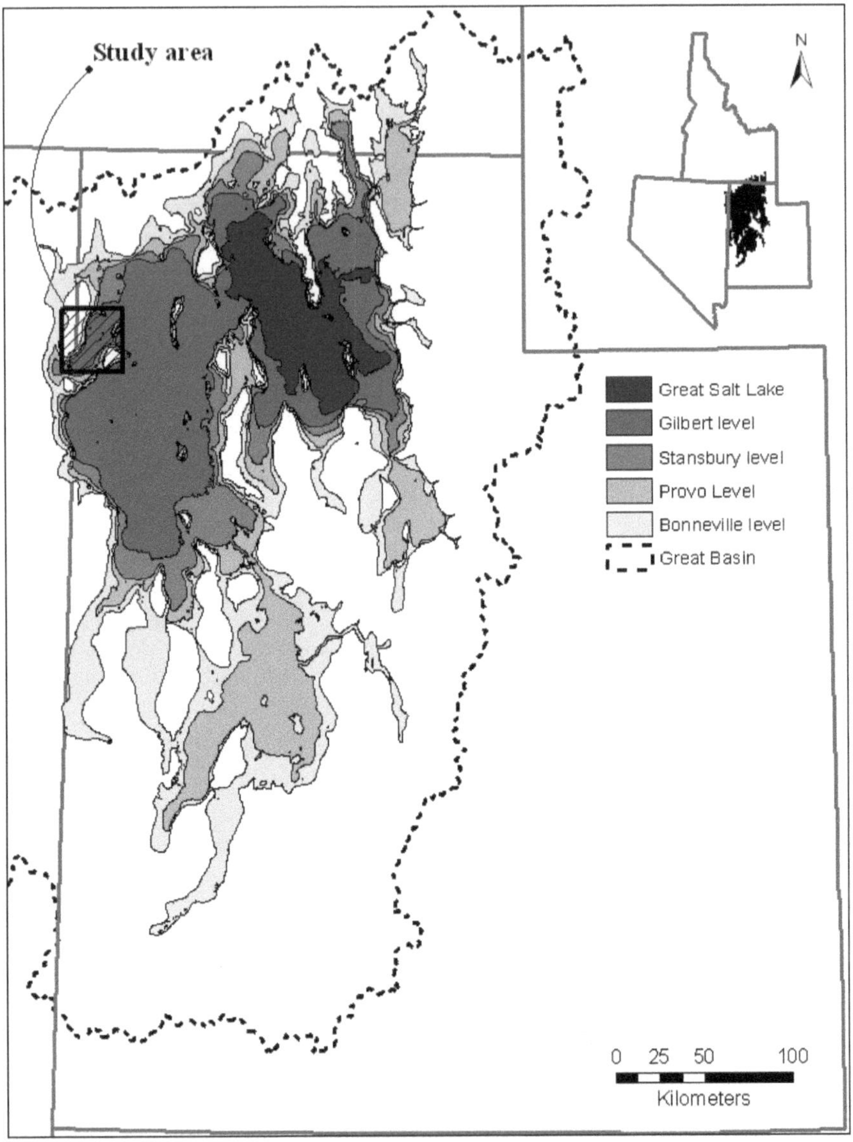

FIG. 10.1 Map showing major Lake Bonneville levels and study area. *Modified from Currey, D.R., 1982. Lake Bonneville: selected features of relevance to neotectonic analysis. U.S. Geological Survey Open File Report 82-1070.*

Bonneville sediments. Benson et al. (2011) collected and analyzed two cores from Blue Lake in western Utah.

Lake-bed sediment studies have identified a sequence of dark brown to black colored organic-rich marl and mud (referred to as algal-laminated mud in this chapter) near the base of the Bonneville sequence. Coring by Benson et al. (2011) and Spencer et al. (1984), and outcrop studies by Oviatt et al. (1990) suggests that the sequence of finely algal-laminated mud formed during the early transgression of Lake Bonneville, perhaps within the time frame known as the Stansbury oscillation. The Stansbury oscillation was a period during the early transgressive phase of Lake Bonneville when lake level fluctuated by up to 45 m (Oviatt et al, 1990). Alternatively, the algal-laminated mud might have been deposited prior to the Stansbury oscillation during the transgressive phase of the lake.

The purpose of this investigation is to characterize and interpret the sequence of Lake Bonneville sediments deposited in the restricted Pilot Valley arm of the lake, described herein as the Pilot Valley embayment, from 22 shallow-sediment cores. Our shallow cores from the Pilot Valley playa exhibit a similar sequence of algal-laminated mud and massive to mottled marl as found in previous investigations of lake-bed Bonneville sediment. Objectives of this investigation are to:

1. characterize the shallow stratigraphy in the Pilot Valley embayment from Lake Bonneville time to present;
2. evaluate the chemistry and mineralogy of the sediment in cores;
3. determine ^{14}C ages for multiple stratigraphic layers in the algal-laminated mud; and
4. correlate Pilot Valley embayment core stratigraphy with published biostratigraphy data.

Knowing the spatial distribution and age and character of the Lake Bonneville sediments from Pilot Valley allows for a greater understanding of factors related to lake depth, climate, and water chemistry.

10.2 SITE DESCRIPTION

Pilot Valley (Fig. 10.2) is a hydrologically closed basin filled today by a playa and alluvial fans. It is mostly located in Utah but straddles the Utah and Nevada state line in the northeastern Great Basin (Figs. 10.1 and 10.2). The basin, approximately 33 km long in a north–south direction and 8–15 km wide, is bounded to the west and east by the Pilot Range and Silver Island Mountains, respectively. The Pilot Range reaches an elevation of over 3200 m asl and is composed of quartzite, schist, and phyllite (Miller, 1990a). The Silver Island Mountains are considerably lower in elevation with the highest point approximately 2300 m asl and are composed of Paleozoic limestone, dolomite, and shale with lesser amounts of Mesozoic granitoid rocks (Miller, 1990b).

Both the Pilot Range and the Silver Island Mountains are flanked by extensive alluvial fans (Miller, 1990a, 1993). The Bonneville and Provo shorelines can be seen as wave-cut terraces throughout the valley and are especially visible along the flanks of the Silver Island Mountains. The Stansbury

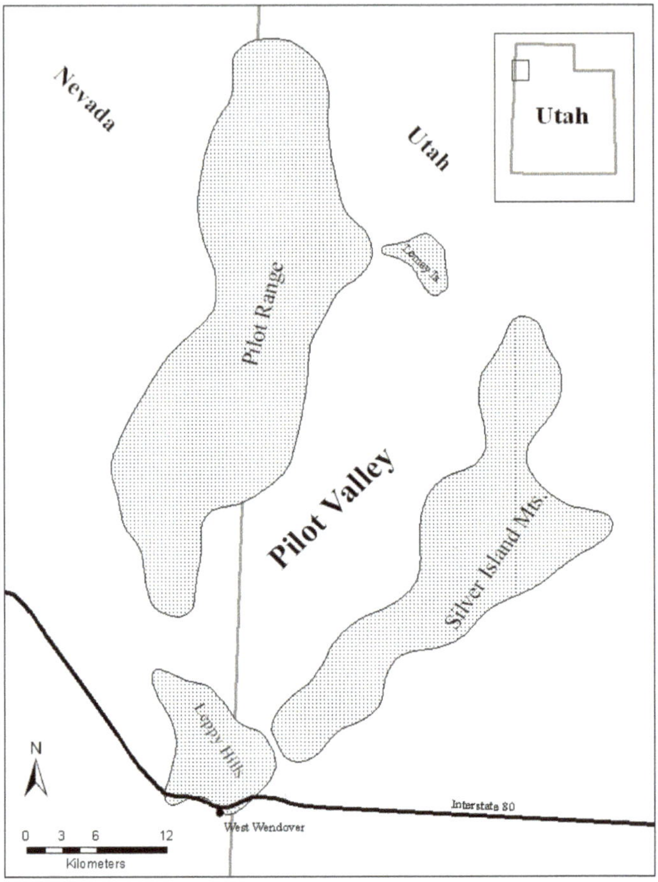

FIG. 10.2 Pilot Valley and surrounding area.

and Gilbert shorelines (as mapped by Currey, 1982) are locally discernable. The near-surface sedimentary and topographic transition between the Pilot Range alluvial fans and playa sediments is abrupt at the base of the fans, whereas the transition between the Silver Island Mountains alluvial fans and the playa sediments extends horizontally for hundreds of meters. Locally these fans interfinger with lacustrine deposits. Further down-gradient the basin is filled with thousands of meters of lacustrine and clastic sediment (Carling et al., 2012).

10.3 METHODS

Twenty-two cores were collected from the margins and interior of the Pilot Valley playa using a hand-driven percussion corer fitted with 5 cm diameter

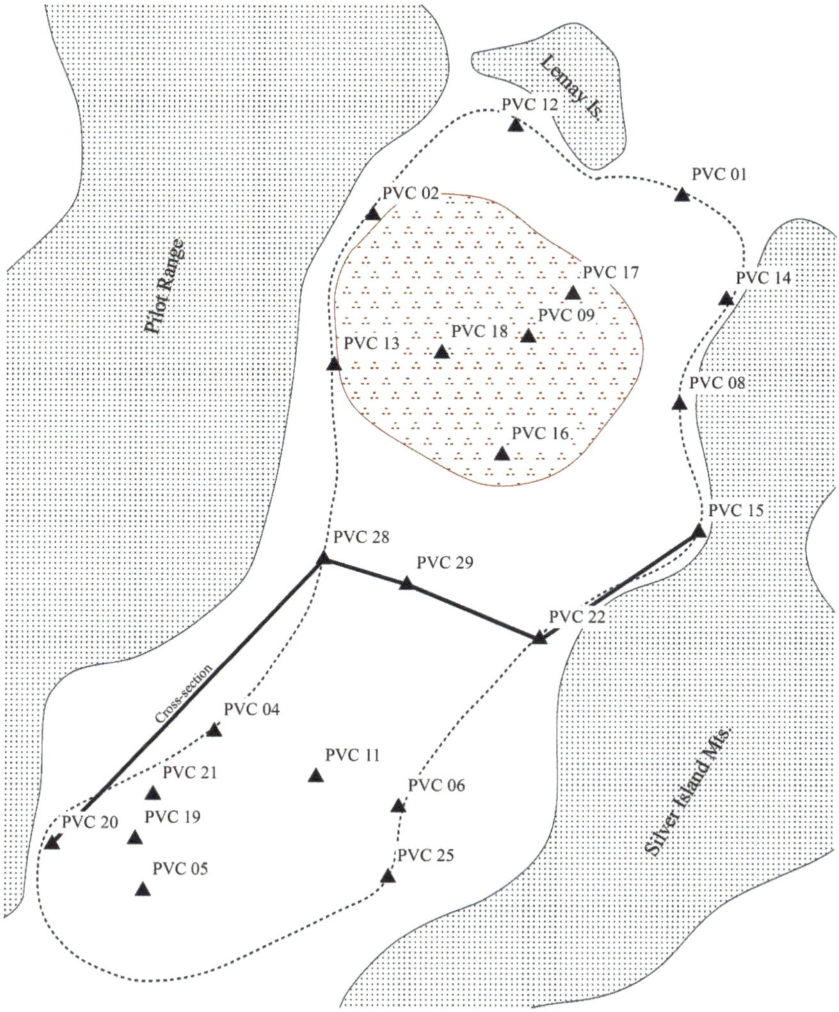

FIG. 10.3 Core locations in Pilot Valley. Cross-section is represented by *bold black line*.

thin-wall plastic liners (Fig. 10.3; Table 10.1). Cores were collected in sections up to 60 cm in length and core retrieval averaged about 85%.

All cores were prepared at the Brigham Young University (BYU) Hydrogeochemistry Laboratory using the procedure of Schnurrenberger et al. (2003). Cores were split in half lengthwise with one half of each core wrapped in cellophane, vacuum sealed, and archived. The other half was scraped clean, photographed and examined for stratigraphic features. Cumulative depth for each core was calculated by taking into account disturbed bedding, gaps, etc., and adjusting as necessary. All reported depths, unless otherwise noted,

TABLE 10.1 UTM Coordinates of Core Locations in Pilot Valley Used in the Cross-Section in Fig. 10.3

	UTM[a]		
Core ID	***Zone***	***Easting***	***Northing***
PVC 15	12T	264,930	4,545,077
PVC 20	12T	241,242	4,533,605
PVC 22	12T	258,842	4,541,197
PVC 28	12T	251,063	4,544,106
PVC 29	12T	254,080	4,543,172
Pilot Valley 09-6	12T	260,154	4,554,396

Datum is WGS 84.
[a]*WGS 84 datum.*

TABLE 10.2 Table of Stratigraphic Units from Pilot Valley Playa with Distinguishing Features as well as other Minor Features

Unit	Distinguishing Feature	Other Features	Depth (cm)
I	Tan silt to stand	Weak laminations, halite in top of some cores, ash bed in two cores	0–99
II	Massive, yellowish gray marl	Weakly laminated; silt/sand	99–138
III	Massive, olive-gray marl	Several white laminations, sand bed at 167 cm	138–265
IV	Dark brown laminated marl	Olive-gray bedded marl in middle of unit	265–313
V	White to gray oolitic sand	Little to no clay	313–324

Depth intervals are for core PVC 15.

were taken from core PVC 15 (Table 10.2) and are listed in centimeters (cm) below ground surface (bgs). Core PVC 15 was chosen for detailed analysis of stratigraphy, ostracods, mineralogy, chemistry, and stable isotopes because it contained the most complete section of pre- to post-Lake Bonneville sediments.

Laboratory analyses were performed on 60 samples from PVC 15 with an average interval of 5 cm. Each sample was analyzed for mineralogy, ostracods, major and trace elements, total inorganic carbon (TIC), total organic carbon (TOC), and stable isotopes. Samples for ^{14}C analysis were collected from core PVC 22 as it displayed excellent preservation of the algal-laminated mud.

Mineralogy was determined from powder X-ray diffraction (XRD) patterns using RockJock11 (Eberl, 2003) with an average error of ±2%. Quartz, orthoclase, and oligoclase are lumped as silicates and are considered to be clastic detritus from the surrounding mountains. For the purpose of this investigation dolomite is considered to be clastic detritus, although its origin is uncertain. Major and trace elements were analyzed by X-ray fluorescence (XRF). Major element results are listed as oxides (eg, CaO for calcium) and are reported in weight percent (%). Trace elements are reported in parts per million (ppm). Error for major elements is 1% and error for trace elements is <5%.

TIC and TOC were determined following the method of Heiri et al. (2001). Samples for stable isotope analysis were prepared following the sealed vessel method of Swart et al. (1991), and the samples were analyzed on a Finnigan Delta^{plus} isotope ratio mass spectrometer utilizing a CTC Analytics CombiPAL gas bench. Both $\delta^{18}O$ and $\delta^{13}C$ were measured on $CO_{2(g)}$ evolved from carbonate minerals. Isotope ratios are reported with respect to Vienna Standard Mean Ocean Water (V_{SMOW}) for oxygen and Pee Dee Belemnite (PDB) for carbon.

Ostracod samples were prepared and analyzed by Charles G. "Jack" Oviatt at Kansas State University. The first and last appearances of ostracod species were identified, but valve counts were not performed.

Sediment ages are based on ^{14}C analysis from the algal-laminated mud. Four ^{14}C samples were taken from the algal-laminated mud in core PVC 22 and one additional auger hole a few miles north of PVC 22. The depth of the sample is equivalent to Unit IV in PVC 15. In addition, a sample of ooids was obtained from an auger hole immediately adjacent to core PVC 22 (McGuire, 2014) and was analyzed for ^{14}C as a way to constrain the earliest timing of the filling of the Pilot Valley embayment prior to Lake Bonneville. All ^{14}C samples were analyzed at the Center for Applied Isotope Studies at the University of Georgia (CAIS-UG), whereas $\delta^{13}C$ was analyzed at BYU. Radiocarbon ages were calibrated with CALIB 7.1 using the InCal13 data set (Reimer et al., 2013).

10.4 RESULTS

10.4.1 Stratigraphy and Sedimentology

In the Pilot Valley embayment cores, five stratigraphic layers were identified based on mineralogy, lithology, color, texture, organic content and other

FIG. 10.4 Schematic stratigraphic column illustrating generalized marl section of sediments from Pilot Valley playa. Total height of column is approximately 3 m, but sections are not to scale.

factors (Fig. 10.4). From oldest to youngest the layers are herein designated as Units V–I. Unit V is a pre-Bonneville deposit, Units IV, III, and II are Bonneville deposits, and Unit I is a post-Bonneville deposit. The Bonneville sequence in the Pilot Valley cores exhibit the full sedimentary sequence typical of Lake Bonneville deep water marl (Oviatt et al., 1994a; Oviatt, 2011, oral communication).

Unit V consists of oolitic sand and occurs in all cores with sufficient depth, typically about 3 m below the top of Unit I. In core PVC 15, Unit V, 313–324 cm, is predominately white to light gray colored oolitic sand with little to no clay except in its upper part (Fig. 10.5). The ooids are typically rod shaped but round ooids are not uncommon. The oolitic sand in other cores has some variability in lithology. For example, cores PVC 28 and PVC 29 (Fig. 10.3) have alternating clay and oolitic sand with small of amounts of clastic sand and silt. The thickness of the oolitic sand is not well defined but a total thickness of about ~1.5 m occurs in PVC 22 located on the eastern playa margin. A total thickness of ~1 m occurs in cores in the southern part of the playa. The ooids have ages ranging from 32.3 to 28.1 ^{14}C ka BP (McGuire, 2014).

FIG. 10.5 Unit V: PVC 15 oolitic sand.

FIG. 10.6 Unit IV: PVC 15 algal laminated marl.

Unit IV is dark brown colored, finely laminated organic-rich marl (referred to in this chapter as algal-laminated mud). In some cores, several thin beds of olive-gray colored marl occur in the middle of the unit. Unit IV is present in most cores though not all contain a complete section. In core PVC 15, Unit IV, 265–313 cm, consists of olive-gray to dark brown colored laminated marl containing abundant algal filaments (Figs. 10.6 and 10.7). This same algal-laminated mud has been noted at other low elevation sites

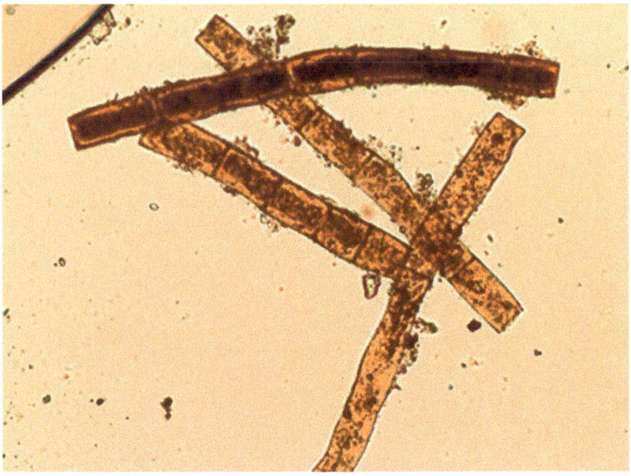

FIG. 10.7 Photo micrograph of algae filaments from algal laminated marl of Unit IV.

around the basin at, or near, the base of the Bonneville section (Oviatt, oral communication). The algal filaments have been identified in other parts of the Bonneville basin as probably belonging to the genus *Cladophora* (Oviatt et al., 1990). Laminations ranging from a few millimeters to <1 mm thick are separated by several centimeters of weakly laminated to bedded olive-gray colored marl. The contact between Units V and IV is locally gradational.

The depth to the top of Unit IV varies by location. In the north and west-central playa, the top is about 2 m bgs and about 3 m bgs in the east-central and southern playa. The unit thickness varies from about 0.5 to 1 m. The organic content is oxidized on the west side of the playa resulting in light gray to white coloration.

Unit III is massive to mottled olive-gray colored marl occurring throughout the playa. It has an average thickness of ~1.5 m, but is at least 2 m thick in core PVC 20 from the southern playa. In cores from the eastern, central, southern, and southwestern playa, Unit III contains a thin sand layer. The sand layer varies in thickness from several millimeters (mm) to several centimeters. In cores PVC 07 and PVC 29 it is only a few mm thick, whereas in cores PVC 22 and PVC 20 it is several cm thick showing a general trend of being thicker near the base of the surrounding mountain ranges. White-colored laminations are common below the sand layer, as well as silt and a few pebbles. In core PVC 15, Unit III, 138–265 cm, is characterized by massive light olive-gray to yellow-gray colored marl (Fig. 10.8). Weak to very weak laminations and/or bedding may occur and mottling is common. White-colored laminations, which contain abundant aragonite crystals, are present at intervals 237, 240, and 249 cm. Small amounts of disseminated sand and/or silt and as wells as small (<5 cm) pebbles locally occurs. At interval 167 cm a distinct sand bed has sharp contacts with the overlying and underlying marl (Fig. 10.9).

FIG. 10.8 Unit III: PVC 15 olive-gray massive marl with white aragonitic laminations.

FIG. 10.9 Sand bed in PVC 20 in Unit III. Individual marks on meter stick are equivalent to one mm. Where shown in other figures, it is labeled as "sand bed."

Unit II is massive to mottled yellow-gray colored marl. Disseminated sand and/or silt are common, especially in the upper part. Thickness is typically 0.3–0.5 m, though it reaches nearly 1 m in the playa center in core PVC 29. In cores near the playa margin, Unit II occurs 0.5–1.5 m bgs and is exposed at the surface in the central parts of the playa. In core PVC 15, Unit II, 99–138 cm, consists of yellowish gray, massive marl (Fig. 10.10). Bedding and/or laminations, when present, are weak to moderate. Disseminated silt or sand is common as is mottling. This unit has increased silt and sand contents near the top. The contact with Unit III, at depth 138 cm and as indicated by a change from olive-gray (Unit III) to yellow-gray (Unit II), is slightly gradational to sharp.

Unit I consists of tan silty to sandy sediments occurring in cores on the western and eastern playa margins as well as cores in the southern playa. The thickness ranges from about 0.3 to 1.5 m. The unit is thickest in the south and thicker on the east than the west. Unit I is not present in cores from the central parts of the playa where Unit II is exposed at the surface. In the northeastern part of the playa, Unit I consists of a post-Lake Bonneville salt pan that overlies and is intercalated with Unit II. The salt in cores PVC 09, 16,

FIG. 10.10 Unit II: PVC 15 yellow-gray massive marl. Yellow-gray marl on the left is Unit II whereas olive-gray marl on the right is Unit III and marks the Bonneville Flood in this core.

FIG. 10.11 Unit I: PVC 15 laminated silt and sand bed. Laminated silty clay is on the *right*. Ooids appear in sand bed at 42 cm which is 94 cm cumulative depth.

17, and 18 consists of abundant euhedral halite crystals up 5 mm in size that occur to a depth of ~1 m bgs. In core PVC 15, Unit I, 0–99 cm, consists of yellowish gray to moderate olive brown, silty to sandy sediments and tends to grade upward from Unit II (Fig. 10.11). The lowest part of this unit is laminated and the remainder is bedded to massive. A thin ash bed is at 24 cm depth, and may be the Mazama ash which occurs in other locations in northwestern Utah (Mehringer et al., 1971; Grey and Bennett, 1972; Murchison, 1989; Louderback and Rhode, 2009). At 94 cm a sand bed appears to uncomfortably overlie laminated sediments. Rod-shaped ooids occur in the sand bed, though only in minor amounts. Above the sand bed massive silty clay dominates. Based on the lithology in five cores, including PVC 15, an east–west cross-section of the shallow playa sediments, extending from the Silver Island Mountains to the Pilot Range, has been constructed (Fig. 10.12).

10.4.2 Ostracods

The ostracod faunal succession is similar to that recognized in other Bonneville cores and outcrops (Fig. 10.13; Spencer et al., 1984; Forester, 1987; Thompson

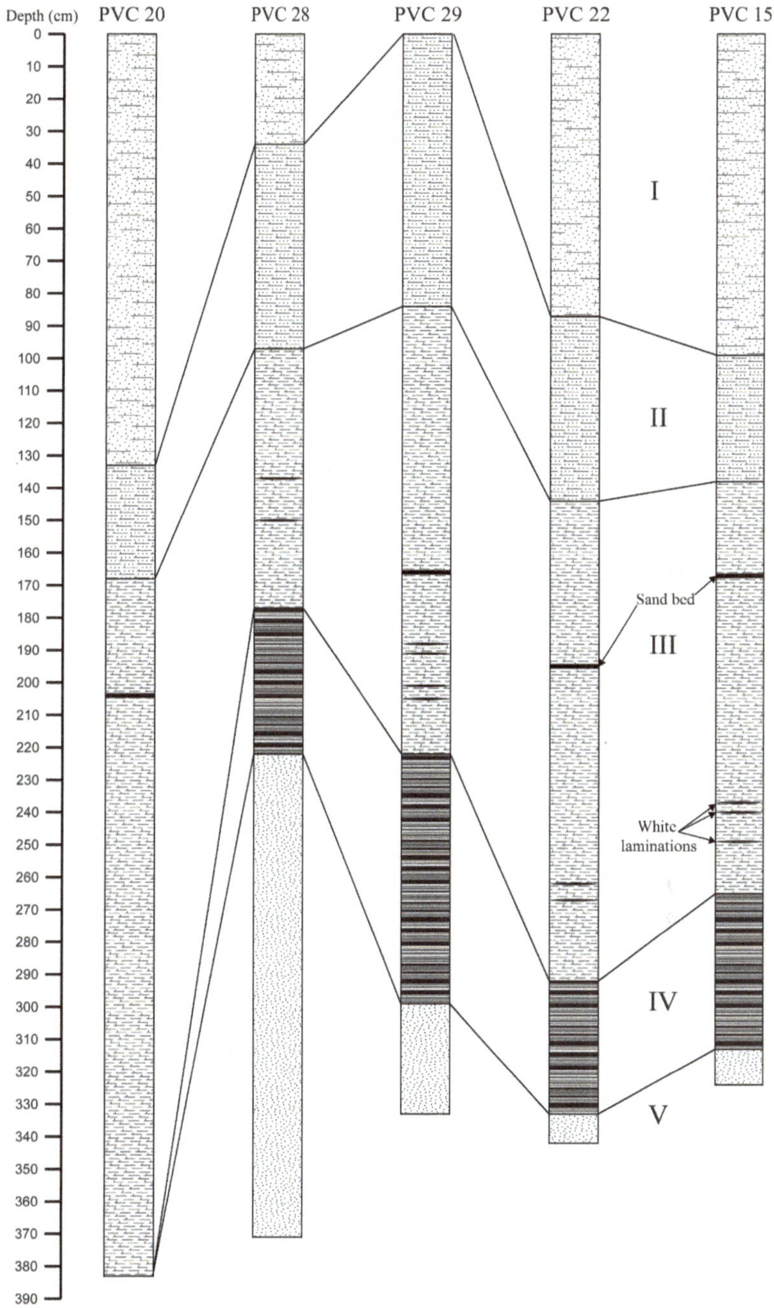

FIG. 10.12 Stratigraphic cross-section across Pilot Valley from west to east (*left* to *right*) along the line shown in Fig. 10.3. Depth is in centimeters. Cores PVC 15, 22, and 28 each contain the full sequence of Units I–V. Core PVC 29 does not contain Unit I while core PVC 20 does not contain Units IV and V. Sand bed is shown as *bold, dark line* in Unit III of cores. However, it is missing in core PVC 28.

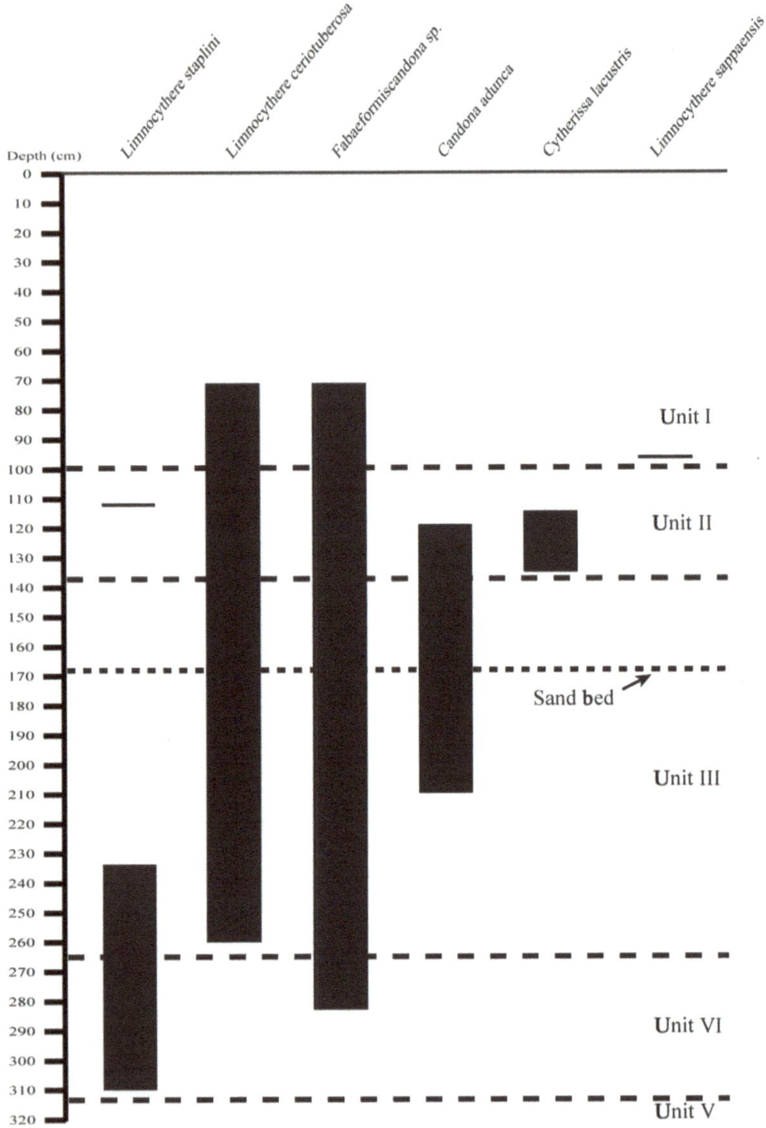

FIG. 10.13 Bar graph of ostracods, showing first and last appearances vs depth in core PVC 15. Valve counts were not made, and the graph simply shows the presence or absence of ostracod species. Unit boundaries are *short dashed lines*. Sand bed is *labeled*.

et al., 1990; Oviatt, 1991; Oviatt, 2015b). Unit V is devoid of ostracods. Unit IV is dominated by *Limnocythere staplini* with a few *Fabaeformiscandona* sp. in the upper part of the unit. The lower part of Unit III contains *L. staplini* and *L. ceriotuberosa* and the upper part also contains *C. adunca*.

Unit II contains *L. sappaensis*, *Cytherissa lacustris*, *L. ceriotuberosa*, *Candona adunca*, *F.* sp. and *L. staplini*. Unit I contains *F.* sp., *L. ceriotuberosa*, and *L. sappaensis* at 97 cm depth. From 92 to 72 cm ostracods valves appear reworked and no ostracods occur above 72 cm.

10.4.3 Mineralogy

Mineralogy includes nonclays and clays. Nonclays are primarily quartz, orthoclase, oligoclase, calcite, aragonite, and dolomite with minor amounts of halite and pyrite. Clays are smectite and muscovite. Quartz, orthoclase, and oligoclase generally vary directly with each other and so are plotted together in Fig. 10.14.

The silicate content in Unit V is very low, ranging from 0 to a few weight %. A large silicate spike occurs at the base of Unit IV to almost 70%, followed by a decrease to an average value of ~35%. Silicate concentrations in Unit III have an average value of about 50%. At the base of Unit II, silicate concentration abruptly decreases to about 15% and then rises gradually to about 40% at the top of the unit. Silicate concentrations are relative constant in Unit I and vary between 30% and 40%.

Carbonate mineral concentrations of calcite, aragonite, and dolomite average 19%, 14%, and 7%, respectively, in PVC 15. The carbonate minerals are both primary and secondary (Mueller and Foerstner, 1972). Calcite and aragonite are principally primary carbonate minerals that precipitated out of the water column. However, the surrounding mountains do contain limestone, so some of the calcite is likely detrital in origin. Dolomite may have originated from the surrounding mountains as clastic debris, but its origin is uncertain.

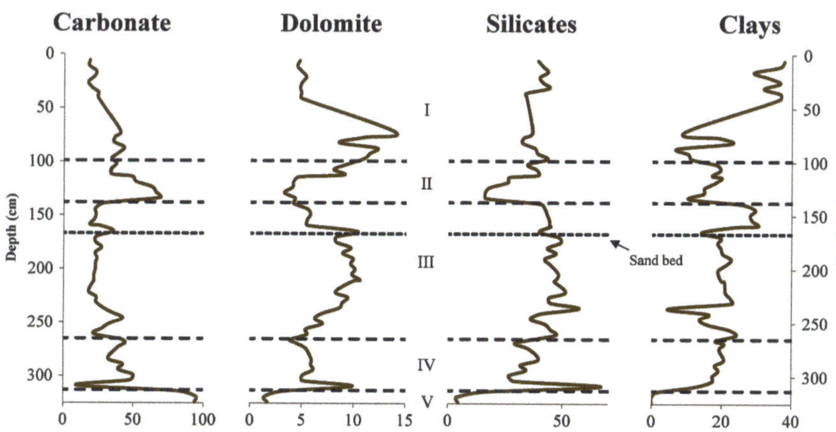

FIG. 10.14 PVC 15 core depth vs mineralogy. Carbonate is combined calcite and aragonite. "Silicates" is combined quartz, orthoclase, and oligoclase. *Vertical axis* is depth in cm, *horizontal axes* are wt%. Unit boundaries are *short dashed lines*. Sand bed is *labeled*.

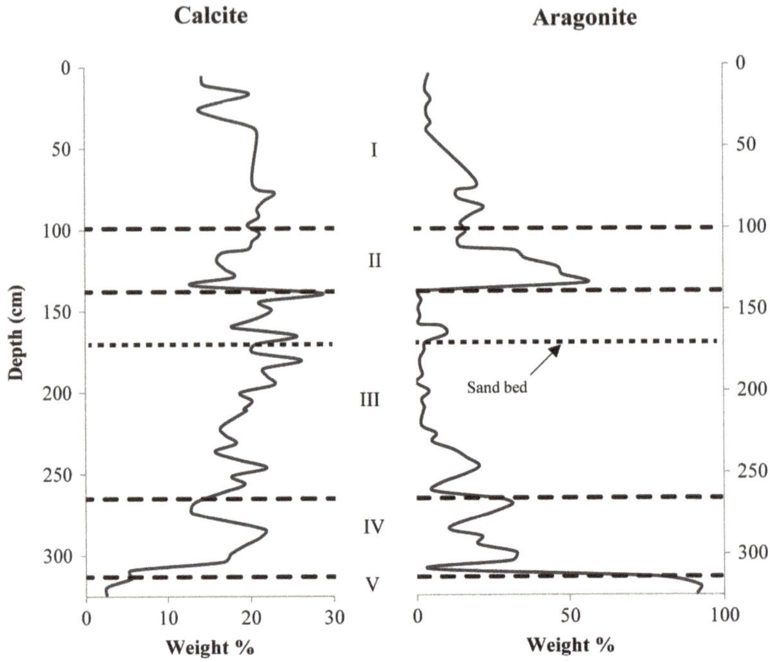

FIG. 10.15 PVC 15 depth vs calcite and aragonite. Unit boundaries are *short dashed lines*. Sand bed is *labeled*.

Calcite abundance varies from 3% to ~5% in Unit V. In Unit IV it increases from 5% at the base, to ~20% in the middle of the unit, then decreases to ~13% near the top. In Unit III calcite concentration is relatively constant at ~20% with a few excursions. In Unit II calcite concentration abruptly decreases at the base from ~30% to 13%. It gradually increases to ~21% at the top of the unit. In Unit I, calcite is relatively constant at ~21% from 97 to 36 cm, after which it decreases to ~16% (Fig. 10.15).

Greater than 90% of Unit V is aragonite (Fig. 10.15). In Unit IV, aragonite is about 4.5% where the silicate concentration is high, then varies between 15% and 30% over the rest of the unit. In Unit III, aragonite forms only ~1% of the sediment, except for two excursions to a maximum of 10%. At the base of Unit II, aragonite abruptly increases to >55% and then gradually decreases to <15% just below Unit I. The aragonite content of the lower half of Unit I ranges from ~15% to 22%, after which it decreases to ~4%.

Dolomite has lower overall abundance than either calcite or aragonite, ranging from ~2% to 14%. The lowest abundance of dolomite is in Unit V where it averages 2%. Dolomite abruptly increases to 10% at the base of Unit IV, then decreases abruptly to ~5% for the remainder of that unit. The dolomite concentration just above the sand bed (167 cm depth) is 10%. Dolomite at the base of Unit II is ~4% but steadily increases to 11% near the contact with Unit I. The abundance in Unit I from 97 to 72 cm averages 11% and above 72 cm the concentration decreases to ~5%.

Other nonclay minerals include halite and pyrite. Except for the salt pan in the northeast part of the playa, only small amounts of halite occur in four samples and are likely the result of these samples not having been rinsed thoroughly during initial preparation. Pyrite occurred in Units IV and III with an average abundance of ~1%.

The primary clay mineral is smectite with lesser amounts of muscovite. Variations in clay mineral abundance typically mirror those of silicates, though the values differ. It should be noted that RockJock11 does not do a good job of analyzing for smectite or other clays as their maximum two-theta peaks typically occur below the 20 degree two-theta lower limit of RockJock11 (Eberl, 2003). However, these clay minerals do have peaks above 20 degree two-theta which allows RockJock11 to distinguish clay minerals from nonclay minerals above its two-theta lower limit.

10.4.4 Bulk Chemistry—Major and Trace Elements

Silicon averages 42.7% for the entire core (Fig. 10.16). It has a spike of about 60% at the base of Unit IV coincident with the spike in silicates. Silicon

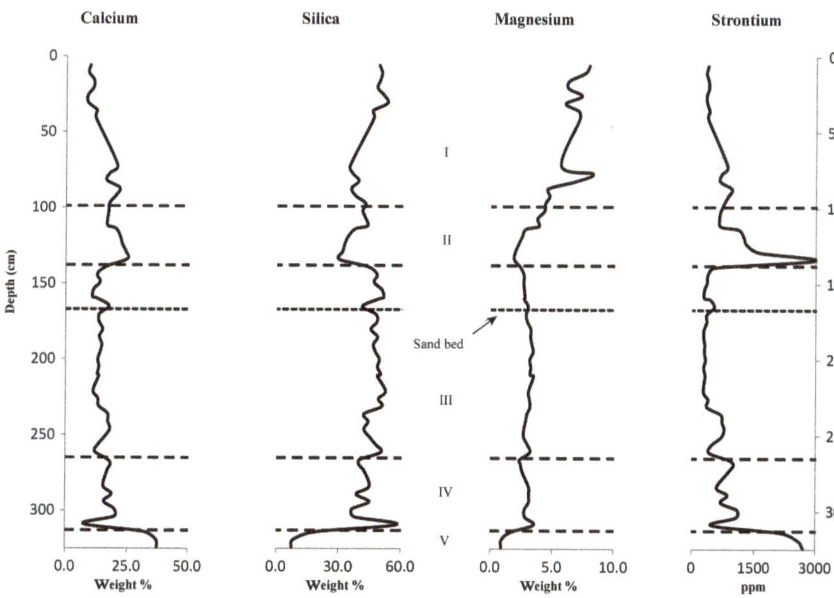

FIG. 10.16 Plots of XRF results for calcium, silica, magnesium, and strontium in PVC 15. Plots are depth vs weight percent (%) for Ca, Si, and Mg. Plot for Sr is depth vs ppm. Unit boundaries are *short dashed lines*. Sand bed is *labeled*.

abruptly decreases at the boundary between Units II and III to about 30%, then gradually increases to its average value of ~43%.

Calcium averages 16.3% for the entire core (Fig. 10.16). It has a spike of about 37% at the base of Unit V coincident with a spike in carbonates. Calcium directly decreases to about 8% at the base of Unit IV followed by an increase to about 20%. Calcium gradationally decreases to its average value of ~16% through the rest of Unit IV. It decreases again in the middle of Unit III to ~13% followed by an increase to ~25% at the base of Unit II. Calcium continues to decrease through the remainder of Units II and I.

Magnesium averages around 4% though the entire core. The lowest concentration is in Unit V where it is <2%. Magnesium rises to and remains near 3% at the top of Unit II until it reaches 8% near the top of Unit I (Fig. 10.16).

Strontium concentrations covary positively with aragonite concentrations (Fig. 10.15). Strontium exhibits a large range of values varying from a low of 295 ppm to a high of nearly 3000 ppm with an average of ~730 ppm. (Fig. 10.16).

10.4.5 Total Inorganic Carbon and Total Organic Carbon

TIC is one measure of carbonate content. TIC has a fair amount of variability (Fig. 10.17) and this is reflected in the changes in the carbonate and dolomite curves (Fig. 10.14). At the base of the core TIC approaches 50% in Unit V before steeply decreasing to ~6.5% at 309 cm. TIC remains between 9% and 17% between 304 and 170 cm in Units IV and III. Above 170 cm TIC concentration starts to vary more with values ranging between 7.7% and 25% in Units II and I. Average TIC is around 15.6% for the entire core. However, neglecting the three spikes in Unit V, TIC averages 14.1%.

TOC exhibits less variability than TIC (Fig. 10.17). The lowest TOC value is ~4% in Unit V. In Unit IV, TOC ranges from 5.8% to 11.5%, with an average of 8.6%. In Unit III, TIC values have a relatively narrow range between 5% and 6%. In Units II and I, TIC varies from 3.8% to 11.3%. Most variability appears to occur between 149 and 41 cm. TOC varies little from 36 cm to the top of the core.

10.4.6 Stable Isotopes—$\delta^{18}O$ and $\delta^{13}C$

The $\delta^{18}O$ values range from a high of 27.1‰ in Unit V to a low of 15.3‰ near the top of Unit I, with values typically varying 0‰ to 2‰ between adjacent samples (Fig. 10.18). An abrupt decrease in $\delta^{18}O$ occurs at the base of Unit IV from 27‰ to 19‰. The average $\delta^{18}O$ values in Unit IV, 22‰, are slightly higher than those in Unit III, 20‰. Above the sand layer in Unit III the $\delta^{18}O$ declines to about 19‰. An abrupt increase in $\delta^{18}O$ occurs at the base of Unit II from 19‰ to 25‰. The upper part of Unit II has lower $\delta^{18}O$ values than the lower part. An abrupt increase occurs at the base of Unit I from

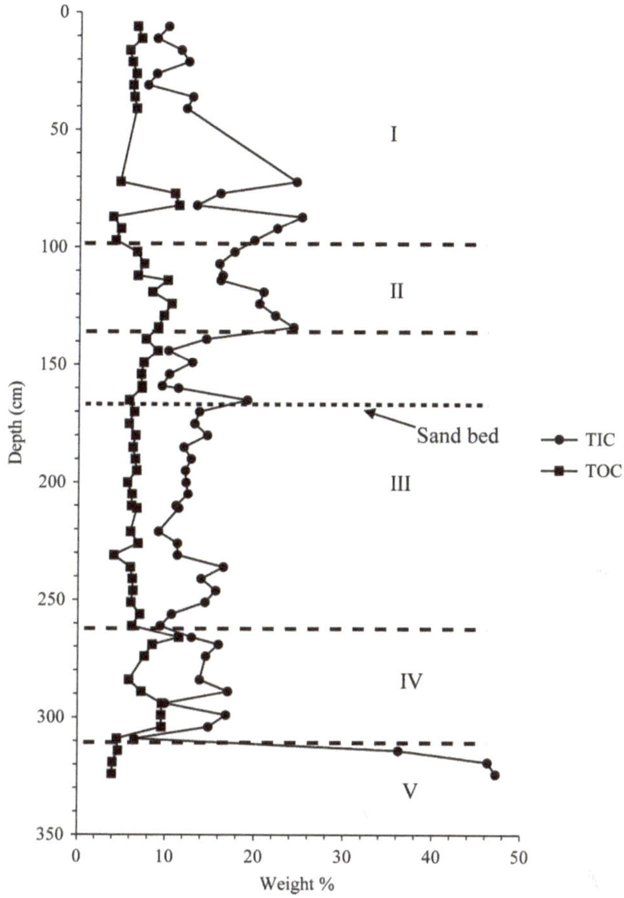

FIG. 10.17 PVC 15 TIC and TOC as a function of depth. *Black dots* are TIC and *black squares* are TOC. TIC and TOC are given in weight %, depth is in centimeters. *Dashed lines* are approximate boundaries between stratigraphic units. TIC exhibits a fair amount of variability while TOC exhibits less overall variability. Unit boundaries are *short dashed lines*. Sand bed is *labeled*.

~21‰ to 24‰. The $\delta^{18}O$ in Unit I, remains at this value up to 80 cm, where it gradually decreases to values less than 20‰. The $\delta^{13}C$ values range from a high of 3.2‰ to a low of −3.4‰. The $\delta^{13}C$ and $\delta^{18}O$ values positively covary, with an R^2 of 0.66 (Fig. 10.19).

10.4.7 Dating

Biostratigraphy is a useful tool for correlating deposits of similar age. No formal time series has been made for ostracods in the Bonneville basin, but they are useful in understanding changes in water chemistry that can be correlated to changes in lake level through time (Oviatt, 2015b). The ostracod

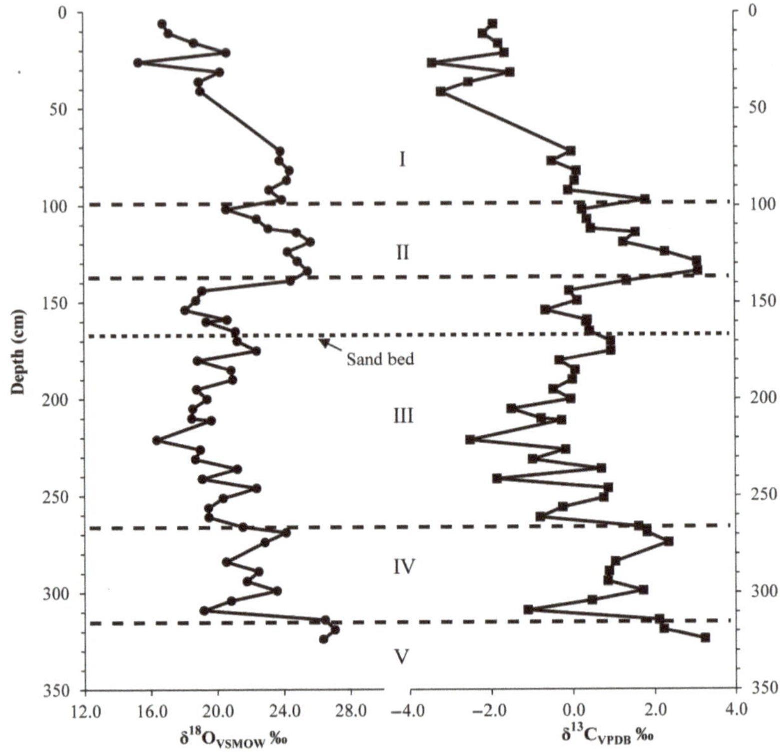

FIG. 10.18 Depth vs $\delta^{18}O$ and $\delta^{13}C$ in PVC 15 . Unit boundaries are *short dashed lines*. Sand bed is *labeled*.

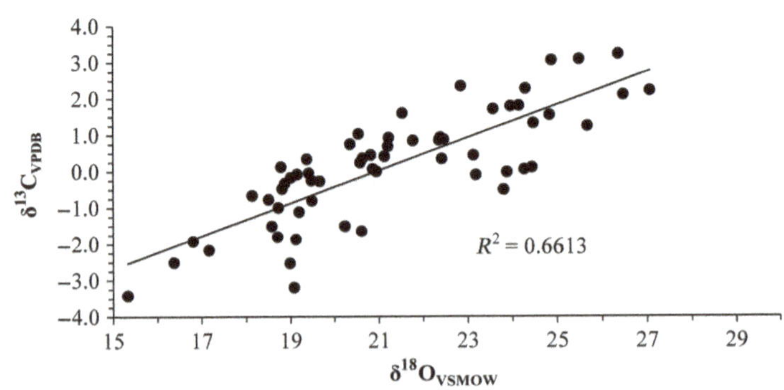

FIG. 10.19 Scatter plot of $\delta^{13}C_{VPDB}$ vs $\delta^{18}O_{VSMOW}$ from PVC 15.

types occurring in core PVC 15 are similar to those described in previous investigations of Lake Bonneville marl (Spencer et al., 1984; Forester, 1987; Thompson et al., 1990; Oviatt, 1991, 2015b; Oviatt et al., 1994b).

In order to constrain the age of Unit IV, four ^{14}C ages were determined from this unit in core PVC 22 at 293, 307, 312, and 320 cm (Fig. 10.20;

FIG. 10.20 The ^{14}C sample locations from core PVC 22.

TABLE 10.3 Radiocarbon Ages from Unit IV of Pilot Valley Cores

Sample #	Lab Code	^{14}C Age	^{14}C 1s	CALIB Mid	CALIB Error	Material	Method
1	PVC22-1012-8	15,835	40	17,137	148	Algal marl	AMS
2	PVC22-1012-22	19,320	50	21,317	241	Algal marl	AMS
3	PVC22-1012-27	20,972	56	23,377	205	Algal marl	AMS
4	PVC22-1012-35	22,433	62	24,813	303	Algal marl	AMS
5	Pilot Valley 09-6	22,923	53	25,324	182	Algal marl	AMS

Table 10.3). These represent the top, two intermediate, and bottom horizons within Unit IV. Ages for these samples from the bottom to the top of the unit are 25.1–24.5, 23.6–23.2, 21.6–21.1, and 17.3–17.0 cal ka BP. An additional sample from near the base of Unit IV was obtained from an auger hole a few miles north of PVC 22. Its age of 25.5–25.1 cal ka BP is similar to the oldest measured age from PVC 22.

10.5 DISCUSSION

10.5.1 Unit V

Aragonite ooids are interpreted to have been actively forming throughout most of the Holocene in the Great Salt Lake in the shallow nearshore in hypersaline water (Eardley, 1938; Carozzi, 1962; Kahle, 1974; Halley, 1977; Pedone and Norgauer, 2002). The $\delta^{18}O$ composition of the carbonate minerals in Unit V, averaging of 26.6‰, is consistent with formation in highly evaporated saline water similar to that in the Great Salt Lake (Langbein, 1961; Arnow and Stephens, 1990) (Fig. 10.18). Using the modern-day Great Salt Lake as an analog for ooid formation in the Pilot Valley embayment, the oolitic sand with >90% aragonite in Unit V represents deposition in a shallow, saline lake that pre-dates Lake Bonneville.

10.5.2 Unit IV

The abundance of detrital material at the boundary between Units IV and V is not well understood. Oviatt et al. (1990) correlated Unit IV equivalent layers with the transgressive Stansbury oscillation. The Stansbury oscillation was a time when temperatures were cooler and with less precipitation than during later Lake Bonneville stages (Madsen et al., 2001; Nelson et al., 2005). A possible explanation for the detrital content in interval 309 cm is that sediments derived from weathered rocks in the surrounding mountains were transported only occurred during large and infrequent storm events. In addition, based on ages from McGuire (2014) and this work, a large gap in time occurred between Units IV and V.

The $\delta^{18}O$ variations of primary carbonate minerals of about 5‰ (Fig. 10.19) suggest that water temperature, and most likely precipitation, fluctuated during Unit IV time. Such fluctuations are consistent with fluctuations of the Lake Bonneville level during the initial transgressive phase, but are insufficient to indicate whether water temperature and precipitation fluctuations occurred during the Stansbury oscillation (Oviatt et al., 1990, 1992). Unit IV $\delta^{18}O$ compositions are lower than those in Unit V, but higher than those in Unit III, suggesting less saline conditions than those during Unit V time and more saline conditions than those during Unit III time. Nelson et al. (2005) found a similar pattern in Pilot Valley shoreline tufa $\delta^{18}O$ where high $\delta^{18}O$ was correlated with increased evaporation and low $\delta^{18}O$ was correlated with decreased evaporation.

The $\delta^{13}C$ values associated with Unit IV (Fig. 10.18) are about 2‰ more positive than those associated with the upper part of Unit III and are associated with the core intervals containing greater amounts of organic carbon. Both the positive $\delta^{13}C$ and abundance of organic material, particularly near the base of Unit IV (Fig. 10.17) are attributed to high organic productivity in the lake.

Based on investigations of outcrops in the vicinity of Bear River City, Utah in the northeastern part of the Bonneville basin, Oviatt et al. (1990) postulated that algae from nearshore floating algal mats was responsible for the elevated organic carbon contents of the sediment equivalent to the Pilot Valley embayment Unit IV. They further concluded that the mats floated on the lake surface and eventually settled to the lake bottom where they mixed with lake-bottom sediment. Oviatt et al. (1990) also suggested as an alternate hypothesis that the organic carbon was derived from benthic algae living in the photic zone on the bottom of the lake during the low stand of the Stansbury oscillation. However, they dismiss this idea, citing that the algal-laminated marl in core C of Spencer et al. (1984), the equivalent of Pilot Valley Unit IV, would have been under nearly 100 m of water at the lowest point in the oscillation and may have been below the photic zone. However, if the algae were present during the initial transgressive phase of Lake Bonneville, the water level would have been much lower than during the Stansbury oscillation.

The spatial extent of the algal-laminated mud in all Unit IV cores suggests that the elevated organic material was not from floating algal mats. They would have had to be so abundant as to deposit over the entire aerial extent of the Pilot Valley embayment multiple times. This seems unlikely given the size of the embayment and the prevailing winds from the west. In the Pilot Valley embayment, the hypothesis that these were benthic algae living on the lake bottom is more likely. The elevation of Pilot Valley averages 1295 m asl. The algal-laminated marl in Pilot Valley is about three meters beneath the surface, placing it around 1292 m asl. Assuming the depth of the lake was less than during Stansbury time, and using an elevation of 1360 m asl for the average lake level during Stansbury time, the depth of the lake in Pilot Valley embayment at the time of deposition of the algal-laminated mud would have been about 70 m. This is sufficiently shallow to permit light penetration and support photosynthesis in the photic zone.

The primary Unit IV ostracod is *Limnocythere staplini* which is consistent with the findings of Spencer et al. (1984), Thompson et al. (1990), and Oviatt (1991, 2015b). *L. staplini* is generally more tolerant of saline water than many other ostracods which may indicate a relatively shallow lake. However, the presence of ostracods, both calcite and aragonite phases, and lower $\delta^{18}O$ values compared to Unit V indicate early Lake Bonneville was deeper than the pre-Bonneville lakes where ooids were forming. In the upper parts of Unit IV *Fabaeformiscandona* sp. is present. *F.* sp. is less tolerant of high salinity and the transition from *L. staplini* to *F.* sp. indicates rising lake levels and decreasing salinity.

This study determined 4 ages from the top, middle and bottom of the algal-laminated marl showing a succession of ages. Unit IV ^{14}C ages of samples from core PVC 22 bracket the timing of deposition of the algal-laminated mud in Unit IV from 25.1 to 17.0 cal ka BP. The younger ages, based on their stratigraphic position in Unit IV, cannot be correct because they conflict with the chronology of Lake Bonneville (Oviatt, 2015a). Furthermore, the ages are scattered (they span many thousands of years). Published ages for the algal-laminated mud from elsewhere in the basin show similar enigmatic results. Thompson et al. (1990) showed ages which were inverted with respect to stratigraphic position.

It seems that relying on radiocarbon ages from the algal-laminated mud as an indicator for the age of this Unit is tenuous at best. This may call into question many of the radiocarbon ages which have been performed on this material elsewhere in the basin. It may be better, though less precise, to rely on stratigraphic position as a relative indicator of age.

10.5.3 Unit III

Unit III is comprised of massive to weakly bedded and mottled olive-gray colored marl. Ostracod fauna consisted of benthic varieties which were intolerant of high-salinity levels and the relative absence of laminations and bedding are indicative of bioturbation (Spencer et al., 1984). TOC is relatively constant with an average of 6.17% indicating relatively stable lake organic productivity (Fig. 10.17).

Carbonate mineralogy of the sediments indicate that water chemistry was relatively stable (Fig. 10.14). An exception to this can be located at depths between 237 and 249 cm in PVC 15 where several distinct white colored laminations with high aragonite concentrations occur. Smear slides of sediment samples from this interval contain abundant aragonite crystals, and XRD analysis shows aragonite concentrations abruptly increase from about 8% to more than 20% and then decline just as steeply back to around 5% (Fig. 10.15). Strontium also increases from ~450 ppm to a high of ~780 ppm and back to ~360 ppm (Fig. 10.16). Calcium exhibits the same general trend of increased and then decreased concentrations but magnesium remains consistent at around 3%. The $\delta^{18}O$ compositions are depleted in ^{18}O relative to the overlying and underlying units, suggesting a deeper water environment (Fig. 10.18), thus Unit III represents conditions of open water in a deep lake.

The cause of the deposition of the sandy layer in Unit III at 167 cm is not well understood. It could have been the result of a major storm event that washed coarse-grained material into the lake. The thickness of the sand bed generally decreases away from the Silver Island Mountains and the Pilot Range, and the layer is missing in some interior playa cores. Further work is required to better understand the origin, and potential cause, of the sandy layer.

10.5.4 Unit II

Unit II marks a distinct transition in core PVC 15, expressed as abrupt changes in marl color, mineralogy, ostracod fauna, element chemistry, TIC, TOC, $\delta^{18}O$, and $\delta^{13}C$ (Table 10.2; Figs. 10.13–10.18). This suggests a major shift in climate and/or water balance within the lake.

Ostracod fauna change as well. *Cytherissa lacustris*, not occurring at this point in deeper parts of the core, appears just after the transition from Unit III to Unit II and seems to have preferred cold, deep, freshwater bodies. The appearance of *C. lacustris* has been correlated with the Provo level of Lake Bonneville in other parts of the Bonneville basin (Oviatt, 2015b).

The changes recorded between Unit III and Unit II likely represent the Bonneville flood. The drop in water level after the flood, and after the lake stabilized, is the Provo level. An expected change in water chemistry likely would have occurred with such a large loss of lake water volume. However, additional work would need to be done to prove if this is the case or not. Obtaining ^{14}C ages from bulk organic or inorganic carbon at the transition would prove helpful in determining the timing of events around this transition.

10.5.5 Unit I

Unit I exhibits a change in lithology from principally clay to silty/sandy clay and then to silt and sand. Unit I is slightly laminated to bedded at the contact with Unit II indicating water level was deep enough to deposit laminated sediments. At 94 cm occurs an apparent unconformity. The bedding below this depth is slightly inclined and is juxtaposed below a 14 cm thick sand bed. It is possible the inclined bedding is an artifact of coring, however, the placement of the bedding against the sand indicates an erosional surface. Rod-shaped ooids are contained in the sand bed, though not in large amounts, and are present to a depth of 85 cm, at which point the sediment changes to silty clay. It is not clear if the ooids present in this interval were forming in the lake at this time or if they were reworked from older deposits. At 33 cm occurs a thin ash bed that may have its origin from the prehistoric Mount Mazama eruption in present-day Oregon.

Mineralogy does not change dramatically at this transition though silicates do rise a few weight percent with drops of a couple of weight percent in calcite and aragonite. Dolomite rises about 1.5%. Clays show a marked decrease from 18.5% to 11.5%. Strontium exhibits a rise that continues to the middle part of the unit. Salinity-resistant ostracods *Limnocythere ceriotuberosa* and *L. sappaensis* are present only at the very base of the unit with reworked valves occurring above 92 cm depth. Ostracod valves disappear entirely by 72 cm depth.

The above data indicate the lake continued to regress from the Provo level and possibly receded entirely from Pilot Valley embayment followed by a small transgression. Lower water levels decreased the elevation at which the shore zone occurs. Within the shore zone energy levels are higher, which would more easily move coarse material and deposit it as sand beds. The presence of salinity-tolerant ostracods also suggests lower water level and increased salinity. It is unlikely the lake stabilized at this level for long, given the decreased amount of ooids contained in Unit I compared to Unit V. The upper part of Unit I at a depth of about 41 cm was comprised principally of silt and sand and was devoid of any ostracods. It also appears to lack any stratigraphic features indicating deposition in water. It is likely that sediments from this depth upward are post-Lake Bonneville. Additionally, the hardpan in the northeast part of the playa, along with abundant salt crystals, indicate complete regression of the lake from the Pilot Valley embayment with only occasional flooding to support deposition of these features.

Unit I is interpreted at representing the latest regressive phase of Lake Bonneville on though modern Holocene time, including the Gilbert highstand with possible evidence showing the lake regressing, transgressing, and regressing again in Pilot Valley. The elevation of the Gilbert level was ~1300 m asl which would make the lake about 5 m deep in Pilot Valley. Much of the upper part of Unit 1 is post-Lake Bonneville and has fluvial and eolian origins.

10.6 CONCLUSIONS

Five pre-Lake Bonneville, Lake Bonneville, and post-lake Bonneville stratigraphic units have been identified in the shallow subsurface of the Pilot Valley embayment. From oldest to youngest the units are:

a. Unit V: deposited in a pre-Lake Bonneville shallow, saline lake that was no more than a few meters deep. The unit consists of oolitic sand deposited about 32,000 years BP.
b. Unit IV: the basal Lake Bonneville sequence consisting of algal-laminated mud that was deposited during the transgressive phase of Lake Bonneville prior to the Stansbury oscillation. In Pilot Valley, the algal-rich laminations most likely formed as benthic algal mats within the photic zone. Radiocarbon ages for the algal-laminated mud are inconclusive, especially the younger ages. It is likely that future attempts to use organic material from the algal-laminated mud will have the same general results.
c. Unit III: a massive to weakly bedded and mottled marl deposited in a deep freshwater lake.
d. Unit II: a transition from olive-gray to yellow-gray marl accompanied by abrupt changes in mineralogy, chemistry, and stable isotopes. The base of this unit preserves changes in water chemistry associated with the Bonneville Flood.
e. Unit I: a coarsening upward sequence from clayey marl to silty and sandy sediments represents the latest regressive phase of Lake Bonneville through modern time and the deposition of Holocene sediments. The Holocene sediments include fluvial and eolian deposits, as well as a thin ash bed (Mazama ash). This unit incorporates the Gilbert episode through a shallow, saline post-Lake Bonneville lake, as indicated by the presence of ooids as well as laminations in the lower parts of this unit. The presence of a probable unconformity indicates the lake may have dried up completely before filling during the Gilbert episode. This unit also preserves the transition to a playa environment as illustrated by the presence of euhedral halite from the hardpan.

ACKNOWLEDGMENTS

The authors wish to thank those who made this research possible. Charles "Jack" Oviatt provided valuable insight during the initial stages of this project as well as many helpful comments when preparing this chapter. In addition, those who reviewed this chapter offered very helpful comments. This research was made possible with funding provided by the BYU Department of Geological Sciences.

APPENDIX A XRF RESULTS FROM PVC 15

Depth	Major Elements (Oxides, wt%)											
(cm)	SiO_2	TiO_2	Al_2O_3	Fe_2O_3	*MnO*	*MgO*	*CaO*	Na_2O	*K2O*	P_2O_5	*Sum*	*LOI*[a]	*Total*
6	48.91	0.44	9.57	2.95	0.06	7.94	9.63	1.46	2.17	0.21	83.34	16.54	99.88
11	50.11	0.44	9.86	3.02	0.06	7.58	8.98	1.67	2.23	0.21	84.15	15.74	99.88
16	49.23	0.42	9.25	2.67	0.06	6.32	11.03	1.41	2.15	0.20	82.75	17.13	99.88
21	48.01	0.39	9.37	2.85	0.06	6.21	10.86	1.41	2.20	0.20	81.56	18.32	99.88
26	51.25	0.46	10.08	2.94	0.06	7.31	8.60	1.59	2.28	0.20	84.77	15.11	99.88
31	52.78	0.42	10.47	2.94	0.07	5.99	8.73	2.16	2.36	0.20	86.11	13.76	99.87
36	45.76	0.42	8.85	2.80	0.06	7.06	12.24	1.34	2.13	0.21	80.88	19.01	99.88
41	46.17	0.40	9.07	2.96	0.06	7.13	11.82	1.25	2.21	0.22	81.28	18.60	99.88
72	34.98	0.23	5.31	1.25	0.03	5.65	20.66	1.10	1.26	0.15	70.63	29.18	99.82
77	36.10	0.29	5.84	1.69	0.04	8.23	17.74	1.61	1.31	0.17	73.02	26.83	99.85
82	39.16	0.32	6.57	1.98	0.04	6.91	16.39	2.17	1.49	0.17	75.20	24.65	99.86
87	35.70	0.28	4.96	1.16	0.03	4.62	21.62	1.04	1.22	0.18	70.82	28.98	99.80
92	37.67	0.29	5.61	1.53	0.04	4.73	20.09	1.20	1.37	0.17	72.69	27.13	99.82
97	42.48	0.33	6.48	1.70	0.04	4.38	17.53	1.17	1.59	0.16	75.87	23.97	99.83
102	41.15	0.32	7.16	2.24	0.05	4.32	17.26	1.34	1.67	0.17	75.66	24.19	99.85
107	42.56	0.35	7.36	2.36	0.05	3.77	16.79	1.42	1.73	0.17	76.54	23.32	99.86

Continued

Depth	Major Elements (Oxides, wt%)												
(cm)	SiO_2	TiO_2	Al_2O_3	Fe_2O_3	*MnO*	*MgO*	*CaO*	Na_2O	*K2O*	P_2O_5	*Sum*	*LOI*	*Total*
112	43.43	0.34	7.03	2.18	0.04	3.87	16.82	1.36	1.65	0.17	76.89	22.96	99.85
114	37.55	0.31	6.86	2.54	0.05	2.81	20.16	1.66	1.55	0.19	73.67	26.15	99.82
119	34.19	0.29	6.29	2.25	0.04	2.52	21.94	1.45	1.42	0.19	70.57	29.23	99.80
124	32.55	0.22	5.49	2.19	0.04	2.23	22.77	1.94	1.24	0.17	68.84	30.95	99.80
129	31.55	0.22	4.95	2.23	0.05	1.96	24.24	1.40	1.13	0.16	67.90	31.85	99.75
134	29.87	0.22	4.83	1.94	0.11	1.89	24.89	1.36	1.11	0.13	66.35	33.29	99.63
139	43.96	0.31	8.14	2.42	0.10	2.40	16.39	1.81	1.95	0.16	77.63	22.14	99.78
144	48.18	0.41	9.24	2.98	0.05	2.68	12.72	2.10	2.11	0.19	80.66	19.18	99.83
149	46.72	0.41	9.13	2.67	0.05	2.67	13.89	1.72	2.09	0.19	79.53	20.32	99.85
154	50.51	0.40	9.74	3.24	0.04	2.71	11.24	1.92	2.25	0.20	82.26	17.43	99.69
159	51.34	0.45	10.17	3.21	0.04	2.77	10.81	1.83	2.29	0.21	83.12	16.70	99.82
160	49.09	0.40	9.36	2.81	0.04	2.70	12.71	1.71	2.10	0.20	81.11	18.55	99.66
165	41.14	0.29	7.06	1.78	0.04	3.04	17.51	1.25	1.71	0.15	73.98	24.93	98.91
170	47.56	0.42	8.85	2.34	0.04	2.87	13.73	1.57	2.11	0.18	79.67	20.11	99.78
175	48.26	0.42	8.99	2.61	0.06	2.98	13.43	1.70	2.13	0.18	80.74	18.98	99.73
180	45.33	0.40	8.57	2.36	0.06	3.26	14.90	1.53	2.10	0.19	78.69	21.19	99.87
185	48.45	0.41	9.07	2.88	0.06	3.23	13.64	1.50	2.19	0.19	81.64	18.20	99.84

190	46.52	0.37	8.77	2.98	0.07	3.34	14.59	1.56	2.10	0.19	80.49	19.34	99.83
195	47.46	0.42	8.87	2.93	0.06	3.44	14.02	1.54	2.10	0.20	81.04	18.82	99.86
200	49.36	0.43	9.22	2.85	0.07	3.18	13.05	1.50	2.15	0.20	82.00	17.87	99.87
205	48.52	0.39	9.06	2.67	0.07	3.25	13.43	1.53	2.12	0.20	81.25	18.63	99.88
210	50.07	0.41	9.39	2.75	0.06	3.21	12.86	1.51	2.18	0.20	82.65	17.22	99.87
211	48.49	0.40	9.09	2.88	0.07	3.48	13.48	1.51	2.15	0.20	81.74	18.13	99.87
221	52.44	0.44	9.84	3.42	0.07	3.08	11.19	1.76	2.27	0.20	84.71	15.16	99.87
226	49.03	0.40	9.28	2.66	0.06	3.15	13.10	1.67	2.14	0.20	81.70	18.16	99.86
231	50.70	0.43	9.61	2.87	0.07	3.21	13.42	1.73	2.22	0.20	84.46	15.40	99.86
236	42.16	0.34	8.19	3.11	0.07	2.97	17.07	1.33	1.89	0.19	77.31	22.52	99.83
241	44.15	0.41	8.51	2.70	0.06	2.85	17.17	1.60	1.94	0.19	79.58	20.24	99.83
246	42.15	0.41	8.25	2.67	0.06	2.72	18.00	1.50	1.89	0.20	77.84	21.98	99.82
251	44.74	0.42	8.76	2.66	0.06	2.70	16.05	1.71	2.00	0.19	79.29	20.53	99.83
256	47.76	0.47	9.87	3.03	0.06	3.01	13.47	1.98	2.23	0.20	82.07	17.78	99.85
261	50.45	0.48	10.44	3.24	0.06	3.21	11.94	1.86	2.37	0.20	84.23	15.61	99.85
266	40.27	0.44	8.52	3.02	0.05	2.38	16.89	1.60	1.97	0.19	75.33	24.48	99.81
269	39.72	0.40	8.19	2.61	0.05	2.47	18.34	1.52	1.85	0.18	75.32	24.50	99.82
274	42.58	0.42	8.71	2.70	0.05	2.60	16.90	1.51	1.97	0.19	77.62	22.20	99.82
284	44.87	0.43	9.38	2.99	0.06	3.13	15.46	1.39	2.14	0.20	80.06	19.78	99.84

Continued

Depth (cm)	Major Elements (Oxides, wt%)												
	SiO_2	TiO_2	Al_2O_3	Fe_2O_3	MnO	MgO	CaO	Na_2O	K2O	P_2O_5	Sum	LOI	Total
289	38.67	0.39	8.23	2.72	0.06	3.14	18.81	1.38	1.91	0.19	75.50	24.32	99.82
294	44.87	0.41	8.94	2.91	0.07	3.15	15.70	1.91	2.07	0.20	80.23	19.61	99.84
299	36.49	0.34	7.76	2.54	0.07	2.79	19.94	1.50	1.80	0.18	73.41	26.39	99.81
304	37.86	0.40	7.97	2.62	0.07	2.85	19.98	1.55	1.87	0.18	75.34	24.47	99.81
309	58.23	0.49	10.97	2.50	0.06	3.54	7.62	2.20	2.92	0.22	88.75	11.05	99.80
314	18.82	0.16	3.50	0.84	0.02	1.65	31.98	0.92	0.88	0.10	58.87	40.88	99.75
319	8.37	0.08	1.68	0.34	0.01	0.93	36.94	0.58	0.41	0.06	49.39	50.35	99.74
324	7.48	0.04	1.51	0.30	0.02	0.90	37.35	0.56	0.36	0.05	48.56	51.17	99.74

[a]*LOI: loss on ignition.*

Depth (cm)	Trace Elements (ppm)															
	Ba	Cr	Cu	Ga	Nb	Ni	Pb	Y	Rb	Sc	Sr	Th	U	V	Zn	Zr
6	573	50	23	15	13	23	17	24	126	17	370	11	6	77	88	159
11	548	48	23	15	13	24	16	24	120	14	333	11	5	81	86	166
16	608	47	21	15	13	25	16	24	117	19	379	12	5	76	81	174
21	563	48	23	15	13	24	16	24	120	19	387	11	5	78	88	165
26	593	47	23	15	13	22	18	24	120	16	331	13	6	74	90	175
31	624	44	22	15	12	22	17	24	112	15	349	11	5	70	85	184

36	567	52	23	15	13	24	16	24	127	20	403	13	6	82	89	153
41	568	55	25	15	14	24	16	24	135	18	372	13	3	86	94	154
72	622	34	13	12	11	13	11	17	58	11	838	11	1	52	43	171
77	558	38	17	12	10	18	11	18	65	19	705	11	7	70	58	145
82	538	44	17	13	11	19	11	20	73	19	667	12	5	66	64	158
87	720	33	12	12	12	12	12	18	54	10	946	15	4	50	41	248
92	691	35	15	13	12	15	11	19	64	13	852	13	4	57	51	212
97	674	40	15	13	11	15	12	20	71	13	755	14	6	56	55	219
102	534	51	19	14	11	19	13	22	83	20	680	13	5	67	69	182
107	524	53	20	14	12	21	14	21	86	20	649	14	6	73	71	162
112	558	49	23	14	12	21	13	21	83	19	688	13	4	76	71	186
114	553	47	21	15	10	20	16	22	83	23	1103	15	2	69	71	144
119	589	42	20	15	10	20	12	22	76	21	1229	14	4	69	70	149
124	551	38	18	14	9	18	11	20	67	20	1305	13	3	62	65	133
129	748	34	21	14	8	18	13	20	61	21	1604	16	6	61	60	130
134	843	33	24	14	7	21	16	24	61	21	2997	19	3	63	60	156
139	634	50	25	15	13	25	17	24	101	21	533	14	4	96	86	155
144	640	63	25	15	14	25	15	25	105	21	377	15	5	87	87	175
149	641	54	24	15	14	22	13	25	99	20	359	12	5	80	78	175

Continued

Depth (cm)	Trace Elements (ppm)															
	Ba	*Cr*	*Cu*	*Ga*	*Nb*	*Ni*	*Pb*	*Y*	*Rb*	*Sc*	*Sr*	*Th*	*U*	*V*	*Zn*	*Zr*
154	588	59	23	15	15	22	15	27	103	19	295	14	6	87	86	181
159	670	62	27	16	15	22	20	27	108	17	295	14	5	86	91	189
160	668	58	25	15	15	28	17	27	102	20	431	14	6	82	88	196
165	994	60	24	14	12	23	16	23	83	20	533	10	8	83	77	175
170	632	59	23	15	14	26	20	25	95	18	354	13	5	73	80	185
175	638	59	23	15	13	27	19	24	95	19	354	14	5	74	79	192
180	651	69	23	15	13	31	18	24	96	23	288	12	4	72	93	174
185	638	67	23	14	13	29	18	24	95	19	284	13	5	71	85	174
190	671	77	23	14	13	27	17	23	93	21	306	12	4	75	83	172
195	639	63	22	15	13	28	18	24	93	17	298	12	3	72	84	179
200	658	63	22	15	14	26	18	24	95	17	325	12	4	71	82	192
205	620	62	21	15	13	26	17	24	92	20	284	12	6	71	82	183
210	714	62	22	15	14	26	19	25	97	18	317	13	4	77	85	194
211	714	62	22	15	14	26	19	25	97	18	317	13	4	77	85	194
221	661	61	22	15	14	26	19	25	99	17	277	14	5	77	84	204
226	671	62	23	16	14	27	20	24	98	21	389	13	2	73	84	192
231	658	60	24	15	14	27	19	25	97	22	363	13	6	76	83	194

236	794	54	21	15	11	26	18	24	88	24	697	14	4	61	76	168
241	776	51	21	15	12	23	16	24	87	21	730	14	6	69	73	181
246	793	52	21	15	12	23	15	24	87	21	779	14	5	66	73	186
251	789	54	19	15	12	21	15	23	92	20	702	14	4	71	74	195
256	751	57	23	15	13	27	16	24	105	20	474	13	4	83	85	186
261	761	56	23	16	13	27	17	24	108	19	425	13	6	83	86	191
266	705	52	25	15	12	26	16	24	99	23	911	15	6	104	86	168
269	695	52	24	16	11	24	15	25	95	22	1006	15	6	85	83	165
274	705	52	24	15	12	24	16	24	99	20	874	14	6	89	86	174
284	819	60	23	16	15	24	17	25	108	22	603	13	3	94	89	183
289	779	52	26	16	14	25	14	24	105	22	859	16	5	84	84	168
294	635	54	25	15	13	23	19	24	112	21	703	16	6	87	89	165
299	678	44	20	15	13	20	16	24	100	22	1098	18	3	73	76	165
304	647	49	21	15	13	20	16	24	103	24	1057	17	2	71	76	168
309	941	47	18	15	21	18	19	22	127	13	491	21	4	70	64	262
314	614	14	11	13	9	5	9	17	53	3	2165	19	0	43	30	118
319	425	1	10	12	4	1	8	13	27	0	2581	20	0	26	15	79
324	388	3	8	12	4	0	7	13	25	0	2703	19	0	24	12	76

REFERENCES

Arnow, T., Stephens, D.W., 1990. Hydrologic Characteristics of the Great Salt Lake, Utah; 1847–1986. U.S. Geological Survey Water-Supply Paper, 32.

Benson, L., Lund, S.P., Smoot, J.P., Rhode, D.E., Spencer, R.J., Verosub, K.L., Lauderback, L.A., Johnson, C.A., Rye, R.O., Negrini, R.M., 2011. The rise and fall of Lake Bonneville between 45 and 10.5 ka. Quatern. Int. 235, 57–69. http://dx.doi.org/10.1016/j.quaint.2010.12.014.

Carling, G.T., Mayo, A.L., Tingey, D., Bruthans, J., 2012. Mechanisms, timing, and rates of arid region mountain front recharge. J. Hydrol. 428, 15–31. http://dx.doi.org/10.1016/j.jhydrol.2011.12.043.

Carozzi, A.V., 1962. Cerebroid oolites. Ill. State Acad. Sci. 55, 239–249.

Currey, D.R., 1980. Coastal geomorphology of Great Salt Lake and vicinity. In: Gwynn, J.W. (Ed.), Great Salt Lake: A Scientific, Historical, and Economic Overview, Utah Geological and Mineral Survey, Salt Lake City, UT, USA, pp. 69–82.

Currey, D.R., 1982. Lake Bonneville: selected features of relevance to neotectonic analysis: U.S. Geological Survey Open File Report 82-1070. .

Currey, D.R., 1990. Quaternary Palaeolakes in the evolution of Semidesert Basins, with special emphasis on Lake Bonneville and the Great-Basin, USA. Palaeogeogr. Palaeoclimatol. Palaeoecol. 76 (3–4), 189–214.

Eardley, A.L., 1938. Sediments of Great Salt Lake, Utah. Am. Assoc. Pet. Geol. Bull. 22, 1305–1411.

Eardley, A.J., Shuey, R.T., Gvosdetsky, V., Nash, W.P., Picard, M.D., Grey, D.C., Kukla, G.J., 1973. Lake cycles in the Bonneville Basin, Utah. Geol. Soc. Am. Bull. 84 (1), 211–215.

Eberl, D.D., 2003. Users Guide to RockJock—A Program for Determining Quantitative Mineralogy from Powder X-Ray Diffraction Data. Revised November 30, 2009 (Vol. 03-78, pp. 48), USGS, Reston, VA, USA.

Forester, R.M., 1987. Late Quaternary Paleoclimate Records from Lacustrine Ostracods. Geol. Soc. Am., Boulder, CO, USA.

Gilbert, G.K., 1890. Lake Bonneville. Monograph—US Geological Survey, 438.

Godsey, H.S., Currey, D.R., Chan, M.A., 2005. New evidence for an extended occupation of the Provo shoreline and implications for regional climate change, Pleistocene Lake Bonneville, Utah, USA. Quatern. Res. 63 (2), 212–223. http://dx.doi.org/10.1016/j.yqres.2005.01.002.

Godsey, H.S., Oviatt, C.G., Miller, D.M., Chan, M.A., 2011. Stratigraphy and chronology of offshore to nearshore deposits associated with the Provo shoreline, Pleistocene Lake Bonneville, Utah. Palaeogeogr. Palaeoclimatol. Palaeoecol. 310 (3–4), 442–450. http://dx.doi.org/10.1016/j.palaeo.2011.08.005.

Grey, D.C., Bennett, R., 1972. A preliminary sinological history of Great Salt Lake. In: The Great Salt Lake and Utah's Water Resources, Proceedings: First Annual Conference, Utah Section of the American Water Resources Association, Salt Lake City, pp. 3–18.

Halley, R.B., 1977. Ooid fabric and fracture in the Great Salt Lake and the geologic record. J. Sediment. Petrol. 47 (3), 1099–1120.

Heiri, O., Lotter, A.F., Lemcke, G., 2001. Loss on ignition as a method for estimating organic and carbonate content in sediments; reproducibility and comparability of results. J. Paleolimnol. 25 (1), 101–110.

Kahle, C.F., 1974. Ooids from Great Salt Lake, Utah, as an analogue for the genesis and diagenesis of ooids in marine limestones. J. Sediment. Petrol. 44 (1), 30–39.

Langbein, W.B., 1961. Salinity and Hydrology of Closed Lakes. U.S. Geological Survey Professional Paper, 20.

Louderback, L.A., Rhode, D.E., 2009. 15,000 years of vegetation change in the Bonneville basin: the Blue Lake pollen record. Quat. Sci. Rev. 28, 308–326.

Madsen, D.B., Rhode, D., Grayson, D.K., Broughton, J.M., Livingston, S.D., Hunt, J., Quade, J., Schmitt, D.N., Shaver III, M.W., 2001. Late quaternary environmental change in the Bonneville basin, western USA. Palaeogeogr. Palaeoclimatol. Palaeoecol. 167 (3–4), 243–271. http://dx.doi.org/10.1016/s0031-0182(00)00240-6.

McGuire, K.M., 2014. Comparative Sedimentology of Lake Bonneville and the Great Salt Lake. All theses and dissertations. Paper 4022, http://scholarsarchive.byu.edu/etd/4022.

Mehringer Jr., P.J., Nash, P., Fuller, R.H., 1971. A Holocene volcanic ash from northwestern Utah. Proc. Utah Acad. Sci. Arts Lett. 48, 46–51.

Miller, D.M., 1990a. Geologic Map of the Crater Island SW Quadrangle, Box Elder County, UT.

Miller, D.M., 1990b. Geologic Map of the Crater Island Quadrangle, Box Elder County, UT.

Miller, D.M., 1993. Geologic map of the Crater Island NW Quadrangle, Box Elder County, Utah. Utah, Map 145, Geological and Mineral Survey, Salt Lake City, UT (pp. 13, 12 sheets-13, 12 sheets).

Morrison, R.B., 1965. Lake Bonneville; Quaternary Stratigraphy of Eastern Jordan Valley, South of Salt Lake City, Utah. U.S. Geological Survey Professional Paper (1044-9612), 80.

Mueller, G., Foerstner, U., 1972. Formation and diagenesis of inorganic Ca-Mg carbonates in the lacustrine environment. Naturwissenschaften 59 (4), 158–164.

Murchison, S.B., 1989. Fluctuation History of the Great Salt Lake, Utah, During the Last 13,000 Years. Ph.D dissertation, University of Utah, Salt Lake City, UT.

Nelson, S.T., Wood, M.J., Mayo, A.L., Tingey, D.G., Eggett, D., 2005. Shoreline tufa and tufaglomerate from Pleistocene Lake Bonneville, Utah, USA: stable isotopic and mineralogical records of lake conditions, processes, and climate. J. Quatern. Sci. 20 (1), 3–19. http://dx.doi.org/10.1002/jqs.889.

O'Connor, J., 2016. Chapter 6. The Bonneville Flood—A veritable débâcle. In: Oviatt, C.G., Shroder Jr., J.F. (Eds.), Lake Bonneville: A Scientific Update, vol. 20. Elsevier. pp. 105–126.

O'Connor, J.E., 1993. Hydrology, hydraulics, and geomorphology of the Bonneville flood. Special paper, Geol. Soc. Am. 274, 83.

Oviatt, C.G., 1987. Lake Bonneville stratigraphy at the Old River Bed, Utah. Am. J. Sci. 287 (4), 383–398.

Oviatt, C.G., 1991. Stratigraphy of Lake Bonneville deposits along Grouse Creek, northwestern Utah: Open-File Report—U.S. Geological Survey, 22. 22 sheets.

Oviatt, C.G., 1997. Lake Bonneville fluctuations and global climate change. Geology (Boulder) 25 (2), 155–158.

Oviatt, C.G., 2014. The Gilbert Episode in the Great Salt Lake Basin. Geological Survey Miscellaneous Publication 14-3, 20 p.

Oviatt, C.G., 2015a. Chronology of Lake Bonneville, 30,000 to 10,000 yr B.P. Quat. Sci. Rev. 110, 166–171.

Oviatt, C.G., 2015b. Ostracods in Pleistocene Lake Bonneville, eastern Great Basin, North America. Hydrobiologia. 11 p. http://dx.doi.org/10.1007/s10750-015-2483-y.

Oviatt, C.G., Currey, D.R., Miller, D.M., 1990. Age and paleoclimatic significance of the Stansbury shoreline of Lake Bonneville, Northeastern Great Basin. Quatern. Res. 33 (3), 291–305. http://dx.doi.org/10.1016/0033-5894(90)90057-r.

Oviatt, C.G., Currey, D.R., Sack, D., 1992. Radiocarbon chronology of Lake Bonneville, Eastern Great Basin, USA. Palaeogeogr. Palaeoclimatol. Palaeoecol. 99 (3–4), 225–241. http://dx.doi.org/10.1016/0031-0182(92)90017-y.

Oviatt, C.G., Habiger, G.D., Hay, J.E., 1994a. Variation in the composition of Lake Bonneville marl a potential key to lake-level fluctuations and paleoclimate. J. Paleolimnol. 11 (1), 19.

Oviatt, C.G., McCoy, W.D., Nash, W.P., 1994b. Sequence stratigraphy of lacustrine deposits; a quaternary example from the Bonneville Basin, Utah. with Suppl. Data 9402, Geol. Soc. Am. Bull. 106 (1), 133–144.

Oviatt, C.G., Thompson, R.S., Kaufman, D.S., Bright, J., Forester, R.M., 1999. Reinterpretation of the Burmester Core, Bonneville Basin, Utah. Quatern. Res. 52 (2), 180–184. http://dx.doi.org/10.1006/qres.1999.2058.

Pedone, V.A., Norgauer, C.H., 2002. Petrology and Geochemistry of Recent Ooids from the Great Salt Lake, Utah. Great Salt Lake an Overview of Change. Utah Department of Natural Resources Special Publication, Salt Lake City, UT, pp. 33–41.

Pedone, V., Oviatt, C.G., 2016. Chapter 16. Water chemistry changes over time and space in Lake Bonneville during the post-stansbury transgression. In: Oviatt, C.G., Shroder Jr., J.F. (Eds.), Lake Bonneville: A Scientific Update, vol. 20. Elsevier. pp. 442–461.

Reimer, P.J., Bard, E., Bayliss, A., Beck, J.W., Blackwell, P.G., Ramsey, C.B., Buck, C.E., Cheng, H., Edwards, R.L., Friedrich, M., Grootes, P.M., Guilderson, T.P., Haflidason, H., Hajdas, I., Hatté, C., Heaton, T.J., Hoffmann, D.L., Hogg, A.G., Hughen, K.A., Kaiser, K.F., Kromer, B., Manning, S.W., Niu, M., Reimer, R.W., Richards, D.A., Scott, E.M., Southon, J.R., Staff, R.A., Turney, C.S.M., van der Plicht, J., 2013. IntCal13 and Marine13 radiocarbon age calibration curves 0–50,000 years cal BP. Radiocarbon 55, 1869–1887.

Schnurrenberger, D., Russell, J., Kelts, K., 2003. Classification of lacustrine sediments based on sedimentary components. J. Paleolimnol. 29 (2), 141–154.

Spencer, R.J., Baedecker, M.J., Eugster, H.P., Forester, R.M., Goldhaber, M.B., Jones, B.F., Kelts, K., Mckenzie, J., Madsen, D.B., Rettig, S.L., Rubin, M., Bowser, C.J., 1984. Great Salt Lake, and precursors, Utah; the last 30,000 years. Contrib. Mineral. Petrol. 86 (4), 321–334.

Swart, P.K., Burns, S.J., Leder, J.J., 1991. Fractionation of the stable isotopes of oxygen and carbon in carbon dioxide during the reaction of calcite with phosphoric acid as a function of temperature and technique. Chem. Geol. Isotope Geosci. Sect. 86 (2), 89–96.

Thompson, R.S., Toolin, L.J., Forester, R.M., Spencer, R.J., 1990. Accelerator-mass spectrometer (AMS) radiocarbon dating of Pleistocene lake sediments in the Great Basin, Netherlands (NLD). Palaeogeogr. Palaeoclimatol. Palaeoecol. 77, 301–313.

Thompson, R.S., Oviatt, C.G., Honke, J.S., McGeehin, J.P., 2016. Chapter 11. Late quaternary changes in lakes, vegetation, and climate in the Bonneville basin reconstructed from sediment cores from Great Salt Lake. In: Oviatt, C.G., Shroder Jr., J.F. (Eds.), Lake Bonneville: A Scientific Update, vol. 20. Elsevier. pp. 221–291.

Chapter 11

Late Quaternary Changes in Lakes, Vegetation, and Climate in the Bonneville Basin Reconstructed from Sediment Cores from Great Salt Lake

R.S. Thompson*, C.G. Oviatt†, J.S. Honke* and J.P. McGeehin‡
**U.S. Geological Survey, MS980, Denver Federal Center, Denver, CO, United States*
†Kansas State University, Manhattan, KS, United States
‡U.S. Geological Survey, MS926A, 12201 Sunrise Valley Drive, Reston, VA, United States

ABSTRACT

Sediment cores from Great Salt Lake (GSL) provide the basis for reconstructing changes in lakes, vegetation, and climate for the last ~40 cal ka. Initially, the coring site was covered by a shallow saline lake and surrounded by *Artemisia* steppe or steppe-tundra under a cold and dry climate. As Lake Bonneville began to rise (from ~30 to 28 cal ka), *Pinus* and subalpine conifer pollen percentages increased and *Artemisia* declined, suggesting the onset of wetter conditions. Lake Bonneville oscillated near the Stansbury shoreline between ~26 and ~24 cal ka, rose to the Bonneville shoreline by ~18 cal ka, and then fell to the Provo shoreline, which it occupied until ~15 cal ka. Vegetation changed during this time span, albeit not always with the same direction or amplitude as the lake. The pollen percentages of *Pinus* and subalpine conifers were high from ~25 to 21.5 cal ka, indicating cool and moist conditions during the Stansbury oscillation and for much of the rise toward the Bonneville shoreline. *Pinus* percentages then decreased and *Artemisia* became codominant, suggesting drier and perhaps colder conditions from ~21 to ~15 cal ka, when Lake Bonneville was at or near its highest levels.

Lake Bonneville declined to a low level by ~13 cal ka, while *Pinus* pollen percentages increased, indicating that conditions remained cooler and moister than today. During the Younger Dryas interval, the brief Gilbert episode rise in lake level was followed by a shallow lake with a stratified water column. This lake rise occurred as *Pinus* pollen

Developments in Earth Surface Processes, Vol. 20. http://dx.doi.org/10.1016/B978-0-444-63590-7.00011-1

percentages were declining and those of *Artemisia* were rising (reflecting increasingly dry conditions), after which *Artemisia* pollen was at very high levels (suggesting cold and dry conditions) for a brief period.

Since ~10.6 cal ka lacustrine conditions have resembled those of present-day GSL. Pollen spectra for the period from ~10.6 to 7.2 cal ka have low levels of conifer pollen and high (for the Holocene) levels of desert and steppe taxa, suggesting generally dry conditions with maximum aridity occurring prior to the deposition of the Mazama tephra (~7.6 cal ka). After ~10.6 cal ka, *Juniperus* pollen percentages began to increase and by ~7.2 cal ka juniper woodlands were well established on lower mountain slopes. From ~7 to 4 cal ka, pollen percentages fluctuated near their mean values for the entire Holocene. The neopluvial (~4 to 2 cal ka) was the wettest part of the Holocene, with higher levels of *Juniperus* pollen and lower levels for steppe and desert taxa than in older Holocene sediments. Pollen percentages for the last ~2 cal ka are variable, but generally indicate a return to drier conditions.

Keywords: Lake Bonneville, Great Salt Lake, Great Basin, Palynology, Sedimentology, Paleoclimate

11.1 INTRODUCTION

The remarkable environmental and climatic history of the Bonneville basin has been the focus of detailed and insightful studies for over 125 years (eg, Gilbert, 1890; Broecker and Orr, 1958; Eardley, 1967; Scott et al., 1983; Currey et al., 1984; Spencer et al., 1984; Oviatt et al., 1992; Madsen et al., 2001; Benson et al., 2011; McGee et al., 2012; Reheis et al., 2014; Oviatt, 2015a). Many of these studies have been directed toward the shoreline evidence of the rise, highstands, and decline of Lake Bonneville, whereas others have focused on paleontological records from dry caves and rockshelters in the arid portions of the basin. Sediment cores covering millions of years have been examined to identify long-term changes, and more detailed studies have been done on sediments related to portions of the most recent major lake cycle (Lake Bonneville). Despite the great breadth of knowledge provided by all of these studies, many uncertainties remain to be resolved, particularly in regard to the environmental and climatic conditions associated with paleohydrological changes through the late Pleistocene and Holocene.

In this report, we attempt to reduce some of these uncertainties through the detailed study of the history of lacustrine and vegetation changes in the Bonneville basin, based on studies of sediment cores from the deepest part of Great Salt Lake (GSL), the largest remnant of Lake Bonneville. We reconstruct changes in lacustrine environments based on sedimentology, geochemistry, and micropaleontology; and changes in terrestrial vegetation based on pollen analysis. Our cores cover approximately the last 40,000 years and provide evidence of conditions prior to the rise of Lake Bonneville, through the lifecycle of this lake, and through the much more arid conditions of the Holocene.

11.1.1 Geographic Setting

The hydrologic Bonneville basin is an area of internal drainage that encompasses roughly 130,000 km^2, primarily in Utah, but also including parts of Nevada, Idaho, and Wyoming (Fig. 11.1). Much of the western two-thirds of this basin consists of broad arid flatlands at relatively low elevations (<1600 m), areas that were largely covered by Lake Bonneville during its deepwater phases. In the eastern third of the basin, the massive mountains of the Wasatch Range and

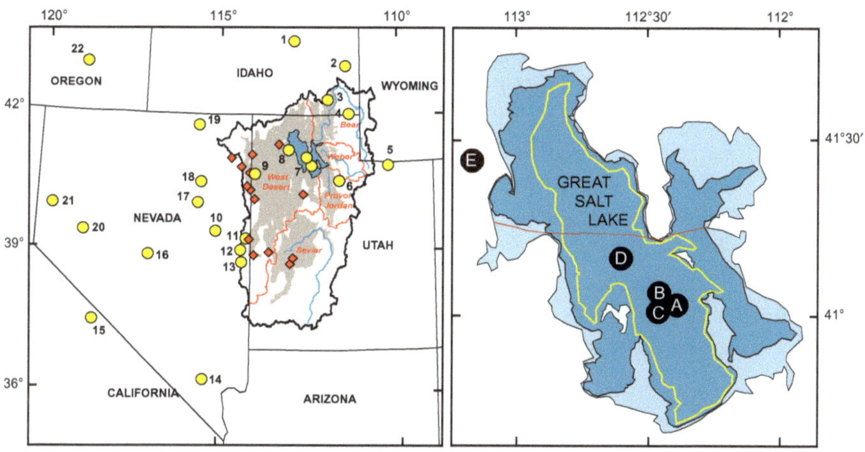

FIG. 11.1 Locations of sites discussed in the text. *Left panel*: Sites across the Great Basin and adjacent areas. *Heavy black line*, outline of the Bonneville basin. *Lines* within the Bonneville basin: *lighter red lines*, limits of subbasins (labeled in *heavy red italic font*); *blue lines*, present-day courses of rivers; *gray background*, extent of Lake Bonneville at its deepest level. *Dashed line* to the west of the Bonneville basin represents the outline of the remaining part of the Great Basin (as defined by internal drainage). Packrat middens are shown as *red diamonds*. Other study sites mentioned by name in the text are: (1) Idaho National Laboratory (Bright and Davis, 1982), (2) Grays Lake (Beiswenger, 1991), (3) Swan Lake (Bright, 1966) and Red Rock Pass, (4) Bear Lake (Doner, 2009; Moser and Kimball, 2009; Smoot and Rosenbaum, 2009), (5) Uinta Mountains pollen sites (Louderback et al., 2015), (6) Snowbird Bog (Madsen and Currey, 1979), (7) sediment cores from Great Salt Lake (see map on right), (8) Homestead Cave (Madsen et al., 2001) and Crescent Spring (Mehringer, 1985), (9) Blue Lake (Louderback and Rhode, 2009), (10) Stonehouse Meadow (Mensing et al., 2013), (11) Council Hall Cave (Thompson, 1984), (12) Stella Lake (Reinemann et al., 2009), (13) Mount Washington (LaMarche and Mooney, 1972), (14) Tule Springs (Mehringer, 1967), (15) White Mountains (LaMarche, 1973), (16) Gatecliff Shelter (Thompson and Kautz, 1983), (17) Ruby Marshes (Thompson, 1992), (18) Favre Lake (Wahl et al., 2015), (19) Mission Cross Bog (Thompson, 1984; Mensing et al., 2008), (20) Hidden Cave (Wigand and Mehringer, 1985), (21) Pyramid Lake (Mensing et al., 2004), and (22) Diamond Pond (Mehringer and Wigand, 1990). *Right panel*: Sites in or near modern Great Salt Lake (*light blue background* shows historical high levels (1283.8 m elevation) from 1986 to 1987 (Currey, 1980), *darker blue* the extent at the approximate mean historical elevation (1280 m), and *yellow line* the historical low elevation from 1963 (1277.5 m, Stephens, 1990)). *Red line* represents the railroad causeway across the lake. Key to pollen sites: (A) GSL96+, (B) GSL96-4, (C) Core C (Spencer et al., 1984), (D) GSL core of Mehringer (1985), and (E) Crescent Spring (Mehringer, 1985).

Uinta Mountains reach elevations above 3000 m, whereas less massive mountains are present on the perimeter of the basin to west (Fig. 11.1).

The Bonneville basin has a dry continental climate with most of the precipitation falling in the cooler months from frontal storms that bring moisture from the Pacific Ocean. The summer North American Monsoon does not have much effect in the region; thus summers are generally dry. Strong altitudinal gradients in both precipitation and temperature occur in the Bonneville basin, as well as geographic differences across the basin (with the west drier than the east, and the north cooler than the south). According to data available at the Western Regional Climate Center (http://www.wrcc.dri.edu) mean annual precipitation (ANNP) varies from ~125 mm in the low-elevation desert west of GSL to ~475 mm at Salt Lake City (at low-elevation southeast of the Lake) to more than 1000 mm in the higher parts of the Wasatch Range and Uinta Mountains to the east. Mean January temperatures (JANTs) range from approximately −4°C to −2°C in the deserts to from −12°C to −10°C at the highest elevations in the mountains on the eastern side of the basin, with mean January minima as cold as −10°C in the deserts and −21°C in the high mountains. For July, mean temperatures range from 24°C to 26°C in the low deserts to 8°C to 10°C in the highest mountains, and maximum temperatures of 35°C or more in the deserts and ~18°C at the highest elevations.

11.1.2 Past and Present Lakes in the Bonneville Basin

The history of lacustrine changes in the Bonneville basin was first investigated by Gilbert (1890) and has been the focus of intense study by many researchers since the 1950s, based on examination of outcrops, geomorphology, sediment cores, geochemistry, geophysics, radiometric dating, paleontology, paleoclimatic modeling, and many other techniques (examples include: Broecker and Orr, 1958; Broecker and Kaufman, 1965; Morrison and Frye, 1965; Eardley, 1967; Scott et al., 1983; Currey et al., 1984; Spencer et al., 1984; Currey, 1990; Oviatt et al., 1992, 2003; Hostetler et al., 1994; Sack, 1999; Madsen et al., 2001; Godsey et al., 2005, 2011; Broecker et al., 2010; Patrickson et al., 2010; Benson et al., 2011; McGee et al., 2012; Reheis et al., 2014; Oviatt, 2015a).

No evidence exists of large lakes in the Bonneville basin for tens of thousands of years prior to the rise of Lake Bonneville. This lake began rising from low levels, probably similar to modern GSL, approximately 30 cal ka (1000 calibrated radiocarbon years before present), and entered a long transgressive phase in the closed basin that was punctuated by numerous fluctuations and oscillations (Oviatt, 2015a; Fig. 11.2). It initially rose to altitudes close to the Stansbury shoreline (~1372 m, where it oscillated from ~26 to ~23 cal ka) and then rose to its highest level (the Bonneville shoreline, 1552 m) and began to overflow into the Snake River drainage near Red Rock Pass, Idaho (site #3 in Fig. 11.1). At about 18 cal ka, the overflow threshold

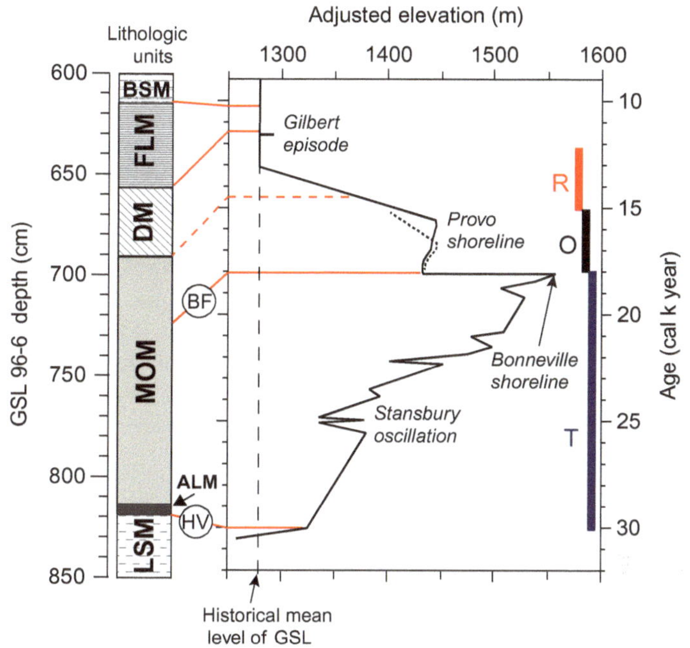

FIG. 11.2 Bonneville hydrograph vs GSL96+ lithologic units (for 600 to 850 cm depth); *red lines* show tie points between the two records (with the *dashed red line* indicating uncertainty regarding how far the lake had regressed when the transition from MOM to DM sediments occurred). Key to lithologic units: *ALM*, algal-laminated mud; *BF*, Bonneville flood; *BSM*, brine-shrimp mud; *DM*, dolomitic mud; *FLM*, finely laminated mud; *HV*, Hansel Valley ash; *LSM*, laminated sandy mud; *MOM*, mottled to massive ostracodal mud. Named shorelines and episodes of Lake Bonneville history shown in *italics*. Key to columns on right: *O*, overflowing; *R*, regressive phase; *T*, transgressive phase. The Mazama tephra is present at 538 cm depth (62 cm above the top of this diagram). The *solid black line* represents approximate lake-level changes, but it does not include the many fluctuations that must have occurred during the transgressive and regressive phases when the lake occupied a hydrographically closed basin. Two different possible chronologies are shown for variations in lake level during formation of the Provo shoreline (Godsey et al., 2011; Miller et al., 2013). *Modified from Oviatt, C.G., 2015a. Chronology of Lake Bonneville, 30,000 to 10,000 years B.P. Quat. Sci. Rev. 110, 166–171; Oviatt, C.G., 2015b. Ostracodes in Pleistocene Lake Bonneville, eastern Great Basin, North America. Hydrobiologia. Doi: 10.1007/s10750-015-2483-y.*

was catastrophically lowered by more than 100 m during the Bonneville flood (Gilbert, 1890; Malde, 1968; Currey, 1982; O'Connor, 1993; Miller et al., 2013; Oviatt, 2015a), which dropped the lake to a more stable threshold at Red Rock Pass. The subsequent Provo shoreline (1444 m) was formed during a period of open-basin hydrology that continued until ~15 cal ka (Miller et al., 2015; Oviatt, 2015a), after which the lake had ceased to overflow and began receding under closed-basin hydrology. Lake waters declined to near the modern level of GSL (1280 m) by 13 cal ka, after which there was a short-lived rise of approximately 15 m during the Gilbert episode (the last large lake in the Bonneville basin prior to Holocene GSL) ~11.6 cal ka. Since that time, the lake has apparently oscillated at elevations within or near the historical range for GSL.

GSL is the largest extant lake in the Bonneville hydrologic system and is fed primarily by the Bear, Weber, and Provo Rivers (Arnow and Stephens, 1990; see Fig. 11.1). It is a closed-basin lake in a desert setting and today hosts only brine shrimp (*Artemia franciscana*), bacteria, and other life forms that can tolerate hypersaline conditions. At its average historical level of ~1280 m, GSL is only ~10 m deep at its greatest depth. In response to climatic variations during the historical period (since AD 1847), GSL has fluctuated between a minimum depth of 7.6 m in 1963 (with a minimum areal extent of 2400 km^2) and a maximum depth of 14 m in 1986 (with a maximum extent of 5900 km^2) (Currey, 1980; Arnow and Stephens, 1990; Stephens, 1990; Fig. 11.1).

11.2 SEDIMENT CORES FROM GSL

Cores of sediments have been taken from GSL several times during the last 60 years, as part of the studies for the railroad causeway in the late 1950s (Eardley, 1967), for petroleum exploration in the 1970s (Davis, 1998; Davis and Moutoux, 1998; Kowalewska and Cohen, 1998), and for paleoclimatic and paleoseismic studies from the 1950s until 2000 (Eardley et al., 1957, 1973; Eardley and Gvosdetsky, 1960; Spencer et al., 1984; Oviatt et al., 1999; Balch et al., 2005). Our cores, taken in 1996 from the south arm of GSL (Fig. 11.1), complement and augment these studies by providing continuous, carefully described, and well-dated cores from the deepest and least faulted portion of the GSL basin. The 6-cm diameter cores were collected using a Kullenberg corer in 3-m core tubes. Core recovery was excellent except at the tops of core sections, where sediments were disturbed by reinsertion of the core barrel during subsequent drives. The cores were processed at the University of Minnesota, Limnological Research Center. Cores were split lengthwise, and one half was archived for future investigations and the other half was used for sampling and analysis. Samples were taken as contiguous 1-cm slices in the Holocene parts of the cores and as 0.5-cm contiguous slices in the Lake Bonneville and pre-Lake Bonneville parts of the cores.

Our primary sediment core[a] was taken in approximately 10 m of water depth at 41.017°N, 112.467°W. This core consists of an initial single drive of 2.155 m (GSL96-5) and, following difficulties with reentering that core

[a]For simplicity's sake for the rest of this paper, we use the label "GSL96+" to refer to the combination of the relatively short core (GSL96-5) and the much longer core (GSL96-6) that began at the depth where the shorter core ended. For comparative purposes in some sections of this report, we also refer to other sediment cores taken nearby (GSL96-4, 41.003°N, 112.385°W; core C from Spencer et al., 1984, 40.988°N, 112.449°W).

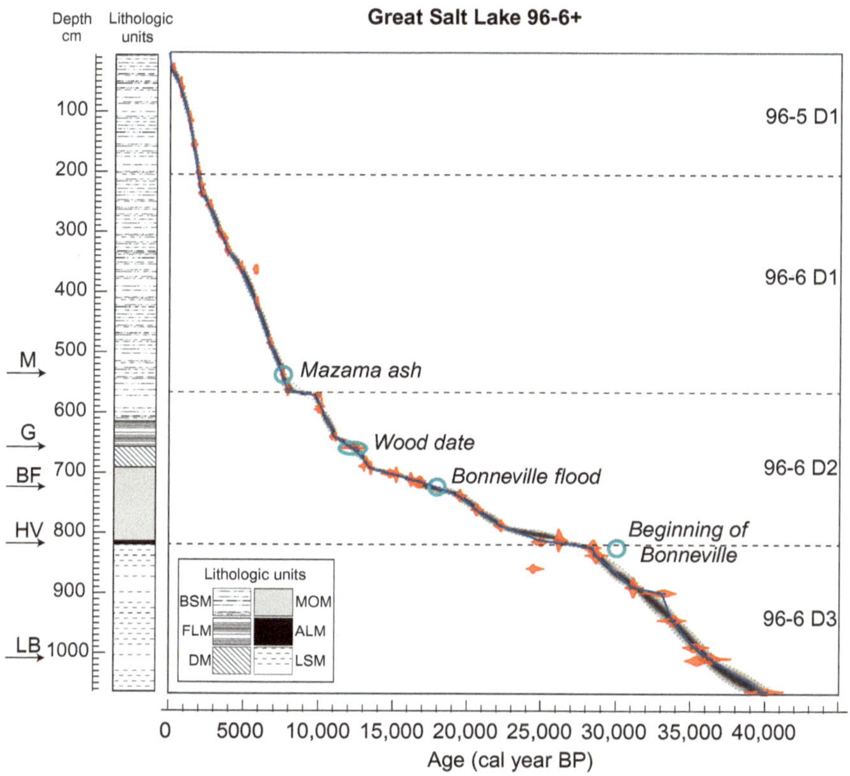

FIG. 11.3 Depth vs lithology and ages for the GSL96-6+ core. Bacon radiocarbon calibrations (Blaauw and Christen, 2011) are in *red* and probability trends in *black*. *Blue ellipses* indicate independent ages that are the basis for the −1800 year correction to sediment radiocarbon dates. *Thin blue line* shows the chronology for pollen samples derived by PSIMPOLL (Bennett, 2005) using linear interpolation. As discussed in the text, as much as a meter of sediment may not have been recovered between 96-6 Drives 1 and 2. *BF*, Bonneville flood; *G*, Gilbert episode; *HV*, Hansel Valley tephra; *LB*, lower basaltic ash; and *M*, Mazama ash. Lithologic unit abbreviations are defined in Table 11.1.

location, the coring rig was moved approximately 3 m, and the coring of GSL96-6 began at the approximate bottom depth of GSL96-5. GSL96-6 was taken in three drives (Fig. 11.3): Drive 1 from 211.4 to 567.5 cm depth below the sediment surface, Drive 2 from 568.5 to 819.0 cm depth, and Drive 3 from 819.2 to 1064.1 cm depth. As discussed later, an unknown length of sediment is apparently missing or disturbed between Drives 1 and 2. The uppermost portion of Drive 3 is disturbed and may also be missing a small depth range of sediments.

11.2.1 Lithologic Units

In this chapter, we focus on the lithologic units described for core GSL96+ (Table 11.1; Figs. 11.2–11.4), which was the primary source of pollen

TABLE 11.1 Lithologic Units in GSL Sediment Cores

Lithologic Unit	Depth Range (Thickness) (cm)	Core and Drive(s)	Description	Equivalent Units of Spencer et al. (1984)	Age Interpretation
BSM; brine-shrimp fecal pellet mud	0–614.5 [614.5]	96-5 D1, 96-6 D1 and top of D2	Light-to-medium gray, wavy-bedded to ripple-laminated, brine-shrimp pelletal aragonitic mud; includes abundant brine-shrimp cysts, small bodies of cyanobacteria, thin laminae of white pure aragonite mud. Mazama tephra present at 538.0 cm	I	Holocene Great Salt Lake
FLM; black, finely laminated mud	614.5–656.5 [42.0]	96-6 D2	Finely laminated organic-rich mud; laminations ~0.5 cm thick; abundant brine-shrimp cysts, few brine-shrimp pellets; pure black on initial exposure to air, oxidizes to medium gray to pale olive	II	Early Holocene (post-Gilbert shoreline)
DM; dolomitic mud	656.5–691.0 [34.5]	96-6 D2	Gray-to-light gray, dolomitic, ostracodal, color-banded mud; two inclined laminae of mud-coated ostracods in upper part	IIIa	Lake Bonneville; post-Provo shoreline regressive phase
MOM; mottled, ostracod-rich mud	691.0–819 [123.0]	96-6 D2	Medium-to-dark gray, ostracodal, massive to color-banded to vaguely laminated mud; light gray bed at 724.5 cm probably represents the Bonneville flood	IIIb, IIIc, IIId	Lake Bonneville; transgressive phase, Bonneville and Provo shoreline development
ALM; algal-laminated mud	814.0–819.0 [5.0]	Top of 96-6 D3	Medium gray, finely laminated mud; abundant algal filaments (present in other cores, but only evident in a small amount of disturbed sediment at the top of Drive 3 in 96-6). Hansel Valley basaltic ash is present here	IIIe	Lake Bonneville; early transgressive phase and development of Stansbury shoreline
LSM; laminated sandy mud	819.0–1065.6 [246.6]	96-6 D3	Medium-to-dark gray, fine-sandy, laminated mud; laminations <1 mm to ~1 cm thick. The lower basaltic ash is present at 1010.6 cm	IV	Pre-Bonneville shallow lake

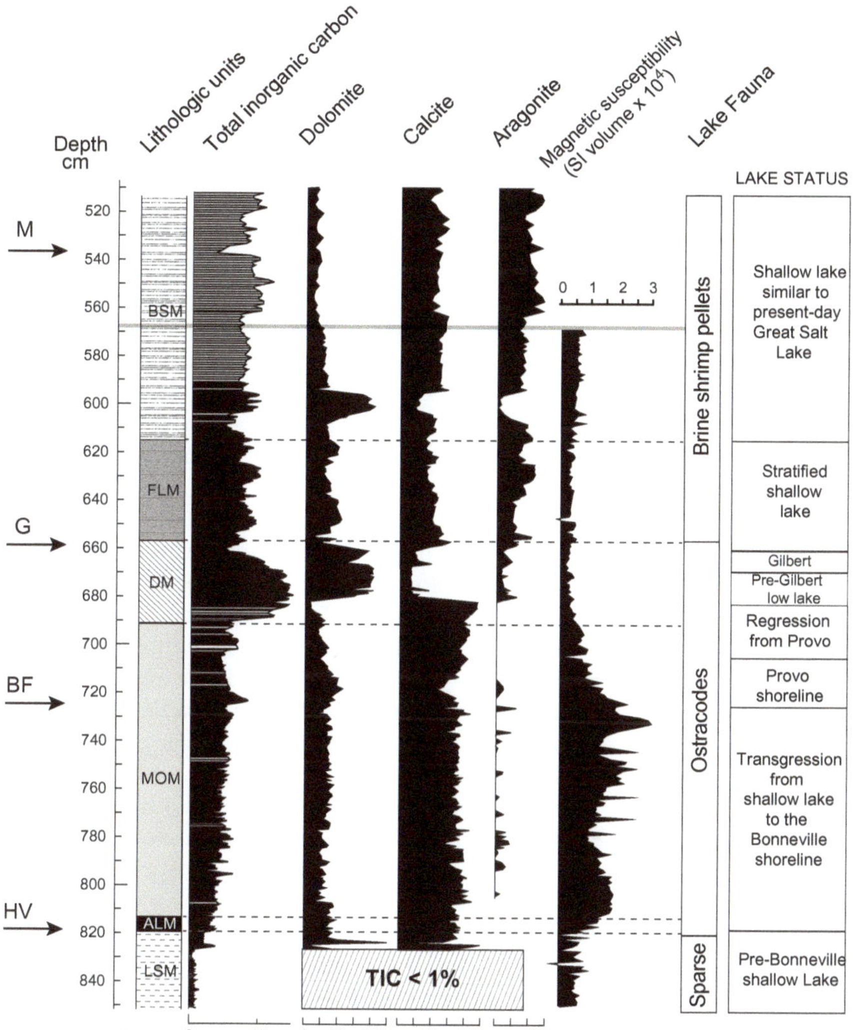

FIG. 11.4 From just below the rise of Lake Bonneville to just above the Mazama ash (850 to 510 cm depth), depth vs lithologic units, percent total inorganic carbon (TIC%), carbonate minerals (as percentages of total carbonate; the lowermost sediments have TIC% levels too low for meaningful calculations of these percentages), magnetic susceptibility, invertebrate paleontology (brine shrimp and ostracods), and inferred lake status. LSM below 820 cm has very low TIC% and sparse fossils of both brine shrimp and ostracods. *Gray line* through the BSM sediments marks where sediments may be disturbed or missing between Drives 1 and 2 of GSL96-6. *BF*, Bonneville flood; *G*, Gilbert episode; *HV*, Hansel Valley tephra; and *M*, Mazama ash. Lithologic unit abbreviations are defined in Table 11.1.

samples. The lowermost unit in core GSL96+ consists of laminated sandy mud (LSM; 1065.6 to 819.2-cm depth) in which both ostracods (indicative of relatively fresh or brackish water; Forester, 1987) and brine-shrimp (*A. franciscana*, indicative of saline water; Cole and Brown, 1967) cysts or fecal pellets are present. Neither are in high abundance, suggesting that the lake water, which probably fluctuated in chemistry through time, was saline but fresh enough to support some ostracods (*Limnocythere staplini*) at least part of the time. LSM sediments appear to represent a long period of relatively shallow-lake conditions prior to the expansion of Lake Bonneville. An abrupt contact at the top of the LSM is marked by the Hansel Valley (HV) basaltic ash, which was deposited when Lake Bonneville was at an altitude of least 1340 m (Fig. 11.2; 60 m above modern GSL; Oviatt et al., 1992; Miller et al., 2008), and in both the cores and outcrops it rests directly on sandy sediments deposited in shallow-lacustrine or marginal-lacustrine environments. Therefore, the lake must have transgressed very rapidly immediately prior to the eruption of the HV ash (Reheis et al., 2014).

Above the HV ash are laminated muds that grade upward into finely algal-laminated mud (ALM; in which algal filaments form horizontal bands within laminated mud; 819.2 to 814.2 cm depth). In core GSL96+ the ALM occurs at the boundary between two separate drives, and as a consequence this sedimentary unit appears to be truncated and disturbed when compared with its occurrence in other cores from GSL. The ALM lacks evidence of brine shrimp and ostracods increase in abundance upward (*L. staplini* grading upward to *L. staplini* plus *Fabaeformiscandona* aff. *caudata*; Thompson et al., 1990). The percentage of total inorganic carbon (TIC%) (Figs. 11.3 and 11.4) increases abruptly directly below the HV ash and continues to increase, with high variability, upward through the ALM and the overlying mottled to massive ostracodal mud (MOM). Higher levels of magnetic susceptibility in this record were found to be due to the dominance of greigite over pyrite in the sediments, which is interpreted as indicative of relatively deep water (Reynolds et al., 2008; Fig. 11.4), and the sharp rise in this parameter in the ALM provides further evidence of deepening lake waters.

The presence of benthic ostracods in ALM indicates that the lake-bottom waters were oxygenated, but the presence of the abundant algal filaments suggests that insufficient oxygen was available to decompose all the organic matter that settled to the bottom from the lake surface. The cause of this situation is uncertain, but ALM is a distinctive Bonneville facies that can be correlated between cores from GSL and low-altitude outcrops throughout the basin (Oviatt et al., 1990; Rey et al., 2016).

Above the ALM with a gradational contact (over a distance of 1 to 2 cm) is mottled to massive to vaguely laminated ostracodal mud (MOM, 814.5 to 691.0 cm depth). The MOM is medium to dark gray and contains thin laminae of white carbonate (aragonitic) irregularly distributed in its lower part (noted also by Spencer et al., 1984) and corresponds to the "white marl" of Gilbert

(1890) and Oviatt et al. (1994). Although we did not conduct a detailed examination of the ostracod faunas, they appear to change upward through the unit in the same general way as noted by Forester (1987), Thompson et al. (1990), and Oviatt (2015b). MOM is lithologically distinct from all other units (except for dolomitic mud [DM], with which it shares some characteristics), and it is characterized by slowly increasing TIC% up section, sustained dominance of calcite over dolomite, and the highest levels of magnetic susceptibility (and inferred deepest waters).

The presence of massive to mottled bedding, rather than distinct laminations, indicates that the lake bottom was well oxygenated (also indicated by the presence of ostracods) and that bottom-dwelling organisms probably kept the mud churned as it was slowly deposited in the deep water. Within the MOM, a spike in TIC% at 724.5 cm depth (Fig. 11.4) is followed by increased dolomite relative to calcite, a decline in magnetic susceptibility, and minor changes in the pollen percentage diagram (discussed later). We interpret these changes as indicative of changes within the lake basin caused by the Bonneville flood and its aftermath. Under this scenario, MOM sediments above 724.5 cm were deposited during the overflowing phase of Lake Bonneville and the development of the Provo shoreline.

DM (691.0 to 656.6 cm) overlies the MOM and is a relatively light-colored carbonate mud unit that contains ostracods and a high percentage of carbonate minerals. The overall increase in carbonate minerals, including dolomite, was noted by Spencer et al. (1984; their unit IIIa), who interpreted it (we believe correctly) as representing the rapid drawdown of Lake Bonneville from high, overflowing conditions. With the shift from MOM to DM sediments, a sharp upturn occurs in TIC%, along with a slight rise in calcite and decline in dolomite, and lower magnetic susceptibility (Fig. 11.4). Ostracods indicative of high alkalinity (*Lymnocythere sappaensis*; Oviatt, 2014, 2015b) appear in the upper part of the DM, and brine shrimp (indicating increased salinity) appear abruptly at the boundary between the DM and the overlying finely laminated mud (FLM) sediments. Mud-coated ostracods are present in the DM, which raises the possibility that reworked materials are present in this lithologic unit (potentially including reworked pollen grains). However, the mud-coated ostracods are a small fraction of the ostracods recovered from this unit, which may indicate that reworking is not a major factor in these sediments.

Light gray carbonate mud (DM) changes abruptly to darker gray (to black on fresh exposure) organic-rich FLM (656.5 to 614.5 cm depth) that lacks ostracods but contains abundant evidence of brine shrimp. This "sapropel" was probably deposited when the lake was very shallow and saline, but capped by a freshwater lens related to the discharge of groundwater that had been stored during the deep freshwater phases of Lake Bonneville (Oviatt et al., 2015). Stratigraphic and geochronologic data suggest that this unit postdates the Gilbert episode (Oviatt et al., 2015). At the base of the FLM, the percentages of calcite (and aragonite) increase and then fall, whereas the percentage of dolomite does the inverse. However, above that point in the FLM no large changes occur in TIC%, carbonate mineralogy, or magnetic susceptibility (Fig. 11.4). From the base of the FLM upward to the top of the sediment core, brine shrimp remains are present (cysts and fecal pellets) and ostracods are absent, indicating continuous hypersaline conditions through this part of the Holocene in GSL.

The uppermost lithologic unit in GSL96+ is light to medium gray carbonate mud dominated by brine-shrimp fecal pellets (brine-shrimp mud, BSM, 614.5 cm depth to the sediment surface). Thin laminae of white aragonitic mud are present and scattered irregularly through the unit and bedding is wavy laminated to faintly ripple laminated. The Mazama ash (7.6 cal ka, Zdanowicz et al., 1999) is present in BSM at 538.0 cm depth in GSL96+. Unit BSM is the thickest unit and represents relatively rapid sediment deposition in shallow highly saline lake waters, with inferred conditions similar to present-day GSL. As discussed in Section 11.2.3, radiocarbon dates provide evidence that suggests that ~1 m of sediment was not recovered from BSM sediments below the Mazama tephra at the junction of Drives 1 and 2 of GSL96+.

The presence of wavy bedding and ripple laminations throughout BSM suggests that the sediments were agitated during and after their deposition, a process that is not surprising in water that was generally less than 10 m deep. Consequences of bottom-sediment agitation were erosion in some places and deposition in others, and laterally discontinuous unconformities in the Holocene sediment of GSL. Differences in thickness of BSM, the presence or absence of the Mazama ash, and anomalies in the radiocarbon chronology at different locations in the lake indicate probable missing sediment in some cores (as discussed in Spencer et al., 1984). The Mazama ash is present in GSL96+, but radiocarbon ages suggest the possibility of an unconformity below the Mazama ash (see Section 11.6).

11.2.2 Correlations Between Core Sediments and Lake History

Stratigraphic comparisons establish a framework for relating core sediments to outcrop evidence of changing water levels in the succession of lakes before, during, and after Lake Bonneville. As noted in Section 11:2.1, several tie points occur between the sedimentary sequence in the GSL96+ core (and other cores from GSL) and the shoreline-based chronology of lake-level changes in the Bonneville basin (Figs. 11.2–11.4). These include (from the oldest to the youngest sediments):

(1) the occurrence of brine shrimp in LSM sediments, suggesting shallow pre-Bonneville conditions;
(2) the HV ash and rapid increase in TIC% that mark the beginning of the rise of Lake Bonneville;

(3) the occurrence of ostracods and absence of brine shrimp in MOM sediments, indicating relatively fresh and deeper water conditions;
(4) the Bonneville flood and its immediate aftermath in MOM sediments;
(5) the maximum in carbonate minerals in DM sediments that resulted from the drawdown of the lake from the Provo shoreline and consequent concentration of solutes in its waters; and
(6) the occurrence of brine-shrimp pellets and the absence of ostracods in FLM and BSM sediments, indicating hypersaline and shallow waters (which are continuous up to the sediments being deposited in modern GSL).

11.2.3 Chronology of GSL Sediment Cores

The primary source of numerical age estimates from the GSL cores is from 56 accelerator mass spectrometry (AMS) radiocarbon analyses of organic carbon from small sediment samples (Table 11.2). Minimally, all the samples were treated with an acid wash to remove carbonates (10% HCl solution at 70°C for 2–4 h). Seven of the samples received an acid–alkali–acid (AAA) treatment that additionally removed humic acids (10% HCl solution at 70°C for 2–4 h; 1 M NaOH solution at 70°C for 1–2 h; 10% HCl solution at 70°C for 1–2 h). Despite this different chemical treatment, the seven AAA-treated samples are in good stratigraphic agreement with the acid-only treated samples positioned above and below them when the 2σ calibrated age ranges are considered. All samples were combusted to CO_2 and reduced to graphite on an iron catalyst in the presence of H_2 at 575°C. The graphite was pressed into targets and sent to the Center for Accelerator Mass Spectrometry (CAMS) at the Lawrence Livermore National Laboratory (LLNL) in Livermore, California, for AMS radiocarbon dating. The radiocarbon ages were calibrated using CALIB 7.1 in conjunction with the IntCal13 data set (Stuiver and Reimer, 1993; Reimer et al., 2013).

As shown in Fig. 11.3, several sources of age estimates are independent of the radiocarbon ages on the sediments. These are: the HV ash and the rise of Lake Bonneville (~30 cal ka), the Bonneville flood (~18 cal ka), an AMS radiocarbon date on a small sample of wood from core GSL96+ with a calibrated median probability age of 12.4 cal ka (Table 11.2), the Mazama volcanic ash (7.6 cal ka, Zdanowicz et al., 1999), and the top of the sediment core (the lake bottom in AD 1996).

Radiocarbon dates from sediments in the GSL96 cores from the same depths as these independent ages are typically too old, and applying an offset of −1800 years to the radiocarbon dates brings the calibrated ages of those dates into approximate synchrony with calibrated ages of most of these tie points (Fig. 11.3; see further discussion in Oviatt et al., 2015). The correspondence is not as good for the comparison of the estimated age of the beginning of Lake Bonneville at ~819 cm depth in GSL96+ with the calibrated ages

TABLE 11.2 Radiocarbon Dates and Calibrated Ages from USGS 1996 Sediment Cores

Lab. Code	Core	Depth (cm)	^{14}C Age	Adjusted ^{14}C Age	S.D.	Lower 2σ bound	Median Prob.	Upper 2σ bound	Lith. Unit	Pollen Zone
WW-1383	96-3	341.0	12,260	10,460	60	12,111	12,386	12,562	DM	n/a
WW-1393	96-3	415.0	16,190	14,390	60	17,312	17,539	17,762	MOM	n/a
WW-1394	96-3	487.0	20,790	18,990	70	22,592	22,860	23,102	MOM	n/a
WW-2561	96-4	91.0	8550	6750	50	7514	7611	7680	BSM	n/a
WW-2562	96-4	94.5	8530	6730	40	7514	7597	7666	BSM	n/a
WW-1385	96-4	99.4	8660	6860	50	7592	7693	7795	BSM	n/a
WW-1386	96-4	291.5	12,270	10,470	50	12,136	12,419	12,561	DM	n/a
WW-1387	96-4	525.5	40,750	38,950	560	42,055	42,844	43,750	LSM	n/a
WW-2563	96-5	29.4	1890	90	50	10	118	273	BSM	E3
WW-2564	96-5	61.4	2600	800	40	673	718	784	BSM	E3
WW-2565	96-5	116.4	3220	1420	50	1270	1329	1408	BSM	E3
WW-2566	96-5	156.4	3480	1680	50	1417	1590	1711	BSM	E3
WW-2567	96-5	202.4	3730	1930	40	1742	1879	1988	BSM	E3
WW-2468	96-6	222.4	3850	2050	50	1896	2016	2137	BSM	E2/E3
WW-2469	96-6	236.4	3920	2120	60	1950	2103	2307	BSM	E2
WW-2193	96-6	256.4	4350	2550	50	2469	2623	2759	BSM	E2

WW-2470	96-6	301.4	4920	3120	50	3213	3332	3447	BSM	E2
WW-2471	96-6	312.1	5190	3390	50	3482	3637	3824	BSM	E2
WW-2198	96-6	331.1	5370	3570	40	3723	3871	3977	BSM	E2
WW-1865	96-6	360.1	6030	4230	50	4584	4747	4869	BSM	E1
WW-1718**	96-6	363.6	6820	5020	50	5653	5770	5900	BSM	E1
WW-1866	96-6	420.6	6910	5110	40	5746	5823	5929	BSM	E1
WW-2467	96-6	485.0	7710	5910	50	6637	6733	6881	BSM	E1
WW-1726**	96-6	538.1	8530	6730	40	7514	7597	7666	BSM	D2
WW-1388	96-6	563.0	8880	7080	60	7787	7902	8013	BSM	D2
WW-2472	96-6	572.5	10,610	8810	50	9671	9853	10,154	BSM	D2
WW-2192	96-6	590.5	10,810	9010	50	9933	10,195	10,249	BSM	D1
WW-1719**	96-6	595.5	10,730	8930	50	9904	10,046	10,214	BSM	D1
WW-2488	96-6	640.5	11,460	9660	40	10,793	11,095	11,199	FLM	C3
WW-2191	96-6	655.5	12,080	10,280	60	11,815	12,056	12,383	FLM	C2
WW-1389	96-6	657.0	12,340	10,540	60	12,237	12,506	12,688	DM	C1/C2
WW-4617*	96-6	659.5	10,600	n/a	270	11,503	12,410	13,060	DM	C1/C2
WW-1392	96-6	682.0	13,160	11,360	60	13,083	13,203	13,310	DM	B4
WW-1864	96-6	689.3	12,980	11,180	70	12,830	13,045	13,174	DM	B4
WW-2197	96-6	692.5	13,490	11,690	50	13,412	13,514	13,605	MOM	B4

Continued

TABLE 11.2 Radiocarbon Dates and Calibrated Ages from USGS 1996 Sediment Cores—Cont'd

Lab. Code	Core	Depth (cm)	^{14}C Age	Adjusted ^{14}C Age	S.D.	Lower 2σ bound	Median Prob.	Upper 2σ bound	Lith. Unit	Pollen Zone
WW-2489	96-6	701.5	14,320	12,520	40	14,434	14,807	15,090	MOM	B4
WW-2490	96-6	703.5	14,580	12,780	50	15,064	15,226	15,428	MOM	B4
WW-2491	96-6	711.5	15,280	13,480	50	16,026	16,223	16,440	MOM	B3
WW-2492	96-6	714.0	15,680	13,880	60	16,543	16,815	17,049	MOM	B3
WW-1382	96-6	717.0	15,570	13,770	60	16,381	16,646	16,918	MOM	B3
WW-2535	96-6	718.0	15,670	13,870	60	16,530	16,800	17,037	MOM	B3
WW-2195	96-6	738.5	17,980	16,180	70	19,277	19,530	19,754	MOM	B3
WW-2187	96-6	762.0	18,940	17,140	70	20,470	20,673	20,904	MOM	B3
WW-2188	96-6	788.0	20,180	18,380	80	21,976	22,263	22,446	MOM	B2
WW-1727**	96-6	812.0	23,810	22,010	70	26,000	26,201	26,457	MOM	B2
WW-1728**	96-6	815.0	22,490	20,690	90	24,550	24,937	25,252	MOM	B1
WW-1729**	96-6	824.2	26,210	24,410	130	28,105	28,457	28,754	LSM	A4
WW-2536	96-6	837.0	26,440	24,640	150	28,322	28,676	29,013	LSM	A4
WW-2537	96-6	859.2	22,200	20,400	90	24,225	24,508	24,917	LSM	A4
WW-2538	96-6	890.2	29,020	27,220	200	30,905	31,163	31,422	LSM	A4
WW-2539	96-6	899.6	30,750	28,950	300	32,157	33,105	33,790	LSM	A4

WW-2540	96-6	944.6	31,460	29,660	260	33,318	33,813	34,288	LSM	A4
WW-2568	96-6	989.6	33,220	31,420	250	34,769	35,300	35,872	LSM	A2
WW-1730**	96-6	1008.6	34,280	32,480	210	35,851	36,376	36,954	LSM	A1/A2
WW-1390	96-6	1011.6	33,380	31,580	230	34,936	35,459	36,012	LSM	A1
WW-1381	96-6	1063.1	37,140	35,340	370	39,005	39,914	40,797	LSM	A1

All samples were processed at the USGS Radiocarbon Dating Laboratory in Reston, Virginia. AMS radiocarbon dating was done at the Center for Accelerator Mass Spectrometry (CAMS), Lawrence Livermore National Laboratory (LLNL), Livermore, California. Radiocarbon dates are reported in ^{14}C year before present (year BP) with one-sigma errors, using the Libby half-life of 5568 year and an assumed $\delta^{13}C$ value of −25‰ following the conventions of Stuiver and Polach (1977). Median probabilities and two-sigma ranges for calendric ages are from CALIB 7.1 in conjunction with the IntCal13 data set (Stuiver and Reimer, 1993; Reimer et al., 2013). The single date in boldface (WW-4617*) is from one small piece of wood found in core sediment. All other analyses were conducted on organic carbon from sediment with 1800 year subtracted from the radiocarbon dates to account for the reservoir effect (see text for further explanation). All samples minimally received an acid treatment to remove carbonates. Seven samples received an AAA treatment and are indicated by a double asterisk (**). Some of these ages were previously published in Oviatt et al. (2015).

from sediment radiocarbon dates, and it is unclear whether the −1800 year offset is applicable in the lower parts of the core.

Based on the radiocarbon chronology, a hiatus, sediment disruption, or an error may have occurred in recording depths during the collection of the sediment core. The lowest radiocarbon date in Drive 1 of GSL96+ (WW-1388) has a median probability of 7.9 cal ka at a depth of 563.0 cm, and the uppermost radiocarbon date in Drive 2 (WW-2472) has a median probability of 9.9 cal ka at 572.5 cm (Table 11.2; Fig. 11.3). The apparent sediment accumulation rate between these two dates is 205.4 years per cm (1951 years/9.5 cm), which is more than 10 times slower than the rates based on the two lowest dates in Drive 1 (12.2 years per cm between WW-1726 and WW-1388) or the two uppermost dates in Drive 2 (19.0 years per cm between WW-2472 and WW-2192). If the latter two rates are applied to the ~2 cal ka difference between WW-1388 and WW-2472, the expected difference in depth between these two ages is approximately a meter (instead of 9.5 cm). For comparison, the stratigraphic difference between the top of FLM sediments and the Mazama tephra is 144.5 cm in core GSL96-4, where both of these stratigraphic markers occur within a single coring drive. This is almost twice the difference in depths between the top of FLM and Mazama in our depth scale for GSL96+, where this distance is 76.5 cm. Consequently, an unknown interval of sediments may not have been recovered when we began Drive 2. When we were describing the sediments of GSL96+, ~24 cm of disturbed sediment occurred at the top of Drive 2, which we believed to be younger sediment that had slumped down the core hole and fallen into the top of the coring tube. It now seems likely that these disturbed sediments represent some of the deposits that may not have been recovered from the top of Drive 2. Alternatively, this span of missing or disturbed sediments could represent an unconformity related to a period of very low lake level prior to the deposition of the Mazama tephra (Colman et al., 2002).

11.3 MODERN VEGETATION

Major aspects of the plant cover of the Bonneville basin can be seen in the "potential natural vegetation" (PNV) categories of Küchler (1964, 1985)[b] (Fig. 11.5), and in the following text we use these categories to characterize the distributions and climatic associations of vegetation types (arrayed from lower to higher elevations). The names of the Küchler PNV categories are underlined in the following text, where the key features of each PNV category are discussed. Distributions of plant species and communities in the Bonneville basin are generally arrayed in response to elevation; however, the relations between plant distributions and elevation are not totally linear.

[b]The Küchler PNV map is plotted on a relatively coarse scale and this results in some of the PNV groups appearing to extend farther downslope than they are commonly found (especially for the sagebrush shrub and sagebrush steppe, mountain mahogany—oak scrub, and montane and subalpine forests categories).

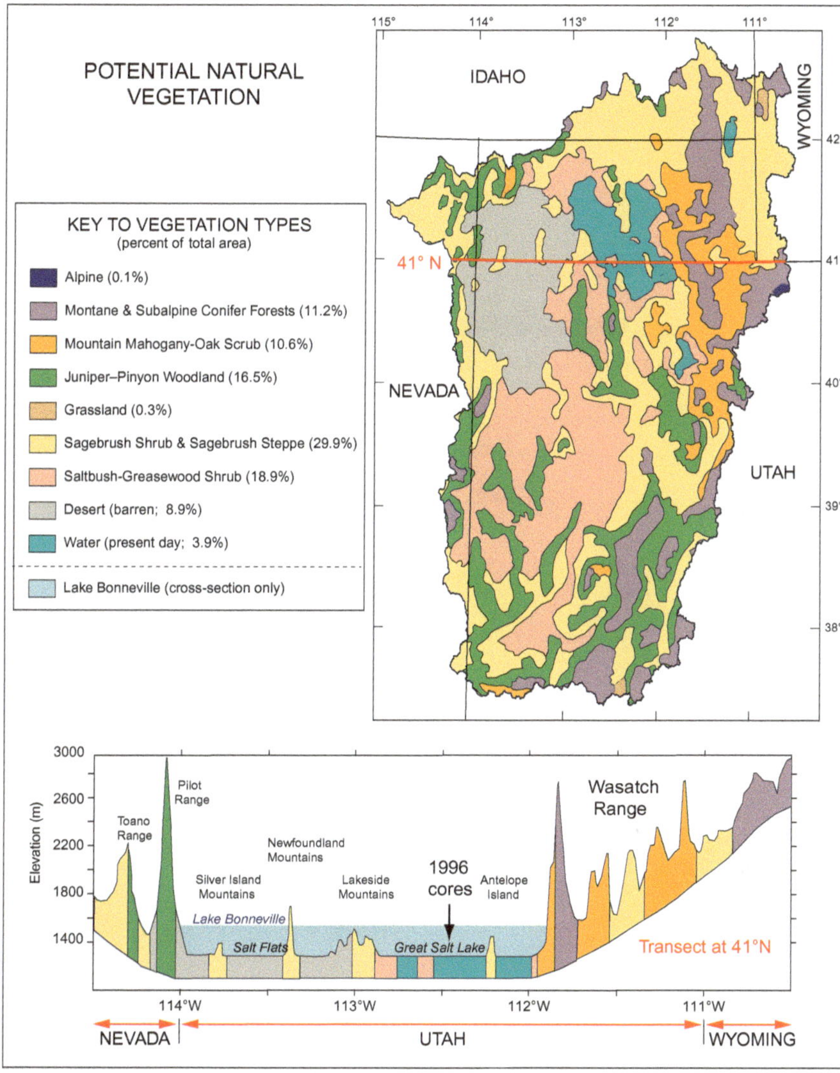

FIG. 11.5 Potential natural vegetation (PNV) of the Bonneville basin: *upper panel*: map by major PNV category, *lower panel*: transect at 41°N with ~200× vertical exaggeration. Key includes percentage of area by PNV category. *After Küchler, A.W., 1964. Potential Natural Vegetation of the Conterminous United States (map). American Geographical Society; Küchler, A.W., 1985. Potential Natural Vegetation. Reston, Virginia, National Atlas of the United States of America, Department of the Interior, U.S. Geological Survey.*

Fig. 11.6 illustrates the complexity of the relations between elevation and plant distributions; within a given elevation range, different PNV categories overlap, due in part to overlapping climatic tolerances (Fig. 11.7), but also due to effects of latitude, longitude, slope and exposure, substrate, and other factors.

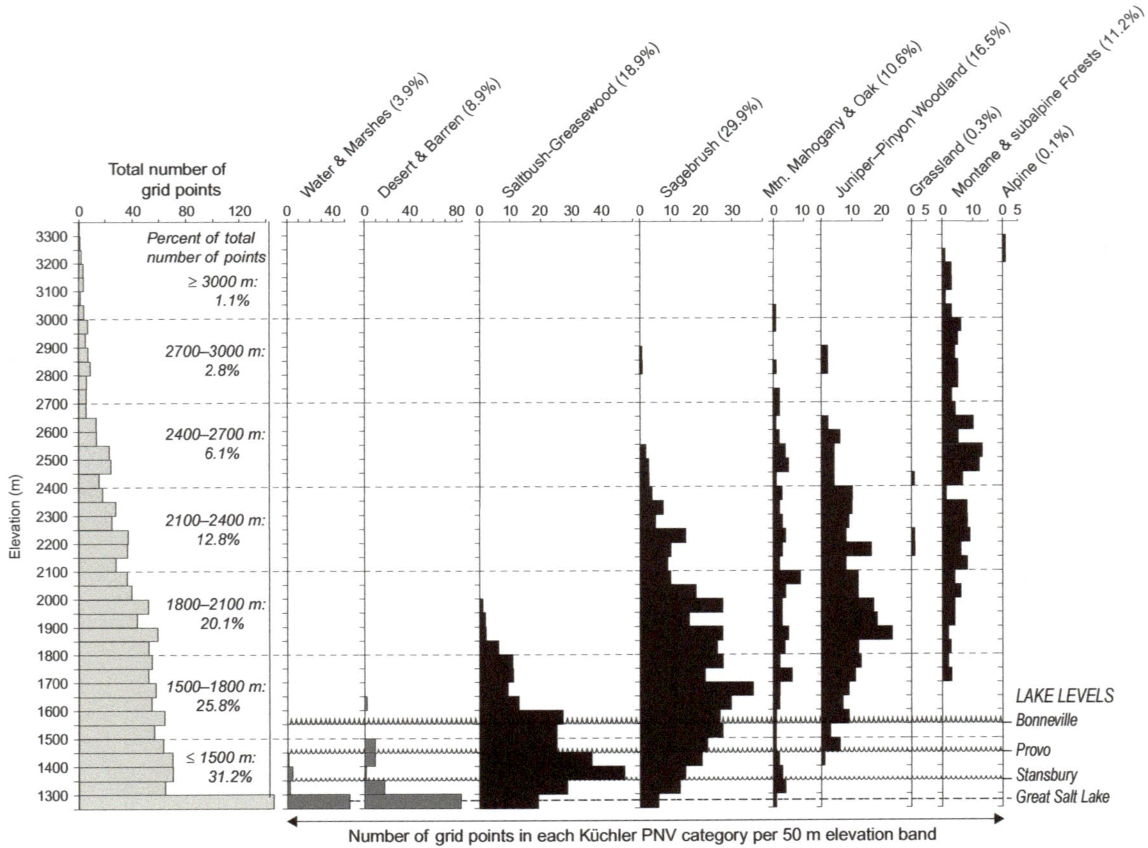

FIG. 11.6 Potential natural vegetation (PNV) of the Bonneville basin vs elevation and lake levels (as measured by number of 10-km grid points per 50 m of elevation). The *panel on the extreme left* shows the number of grid points in each 50 m interval of elevation (with percentages of the total number of grid points that fall within intervals of 300 m in elevation). The remaining panels show the number of grid points in each PNV categories that fall within each 50-m interval of elevation (note scale difference between the Water and Marshes and Desert and Barren PNV categories vs the remaining categories). Labels for PNV categories include percentage of total area by PNV category. *After Küchler, A.W., 1964. Potential Natural Vegetation of the Conterminous United States (map). American Geographical Society; Küchler, A.W., 1985. Potential Natural Vegetation. Reston, Virginia, National atlas of the United States of America, Department of the Interior, U.S. Geological Survey.*

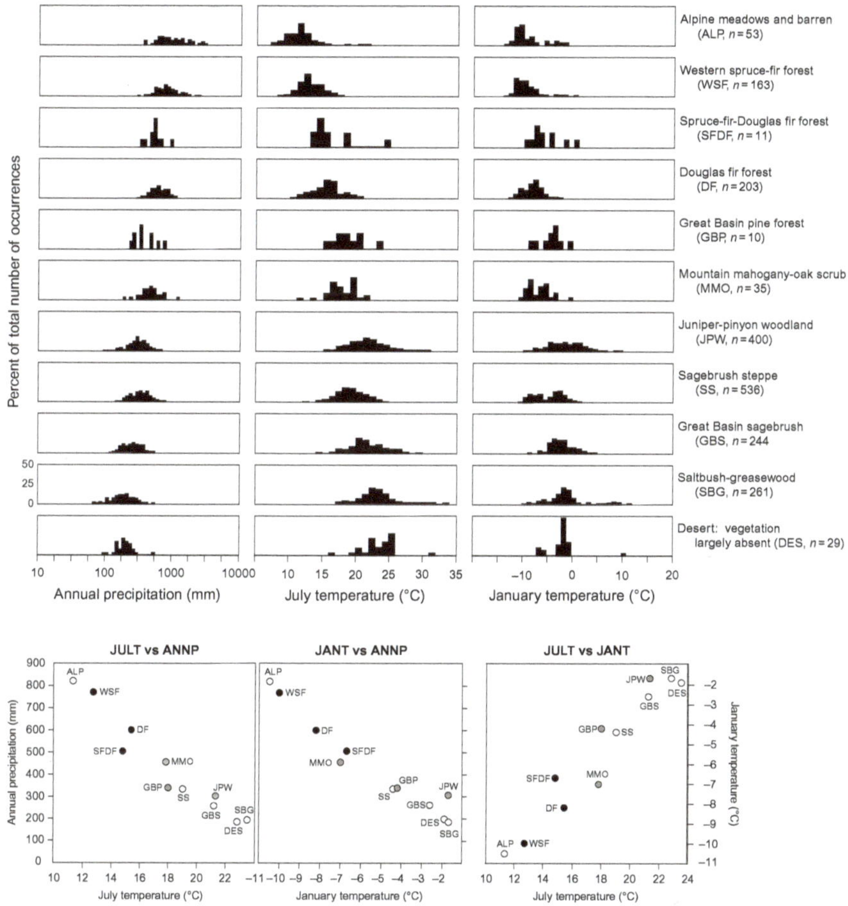

FIG. 11.7 Median climatic values for PNV categories based on vegetation and climatic data on a 25-km grid (Thompson et al., 2007). In Figs. 11.5 and 11.6, four PNV categories shown here (Western spruce-fir forest, spruce-fir–Douglas Fir, Douglas Fir, and Great Basin pine forest) are displayed as a single category titled "Subalpine and montane forests"; and the two sagebrush PNV categories shown here are displayed as "Sagebrush scrub and sagebrush steppe" (Fig. 11.5) or simply "Sagebrush" (Fig. 11.6).

11.3.1 Desert and Steppe

Desert, as defined by Küchler (1964, 1985), is largely barren of vegetation and occurs below ~1600 m on the GSL desert and other areas once covered by Lake Bonneville (Fig. 11.1). Desert overlaps with saltbush-greasewood shrub communities, which occur at elevations up to about 2000 m, and are dominated by plants that are tolerant of saline soils, including *Sarcobatus vermiculatus* and shrubs in the Amaranthaceae family (especially *Atriplex confertifolia*). These two PNV categories occupy the warmest and driest environments in the Bonneville basin (Fig. 11.7).

Shrub communities dominated by sagebrush (*Artemisia* spp.) occur on nonsaline soils at elevations up to ~2500 m. Great Basin sagebrush communities (Küchler, 1985) are present across the southern and central Bonneville basin, with *Artemisia tridentata* as the dominant species, along with other common species including *Artemisia nova*, *A. confertifolia*, and *Ephedra* spp. Sagebrush steppe occurs across the northern Bonneville basin and in the Wasatch Range and Uinta Mountains on its eastern flank. In this vegetation category, *A. tridentata* shares dominance with the grass *Pseudoroegneria spicata*, and other species of *Artemisia* shrubs are present (*A. nova*, *Artemisia arbuscula*) along with other grasses, forbs, and shrubs (including *Purshia tridentata*). Sagebrush PNV categories occur under generally cooler and wetter conditions than the desert and saltbush-greasewood categories (Fig. 11.7).

In addition to the PNV categories with sagebrush in their names, which occur across the Great Basin, sagebrush forms the matrix in which the taller trees and shrubs of juniper–pinyon woodland, Great Basin pine forest, and other communities occur (Billings, 1951). Accordingly, sagebrush is common, and generally abundant, across a large elevation and geographic range, from the upper limits of saltbush-greasewood shrub upslope to lower portions of forest communities.

11.3.2 Woodland and Scrub

Juniper–pinyon woodland occurs on lower mountain slopes, at elevations generally between ~1400 and 2700 m from the western slope of the Wasatch Range westward into the mountains of the Great Basin. This woodland is dominated by short-stature trees of *Juniperus osteosperma*, *Pinus edulis* (in the eastern part of the Bonneville basin), and *Pinus monophylla* (in the western part). These trees are generally widely spaced with intervening spaces hosting *A. tridentata*, shrubs, and small trees in the family Rosaceae (for example, *Cercocarpus*, *Cowania*, and *Purshia*), *Quercus*, grasses, and other shrubs. Due to persistent temperature inversions, throughout the year temperatures within this vegetation association may be warmer than those in both higher and lower elevation communities (Billings, 1954).

Mountain mahogany-oak scrub occurs at elevations that extend from below the juniper–pinyon woodland to slightly above it (Fig. 11.6); it is an important component of the vegetation of the Wasatch Range and also occurs in smaller patches (without *Quercus*) in the Great Basin mountain ranges to the west (Fig. 11.5). *Cercocarpus ledifolius*, *Quercus gambelii*, and *Acer grandidentatum* are dominant species in this PNV category, with the common occurrence of *Quercus undulata*, *Rhus trilobata*, and species of *Amelanchier*, *Cowania*, *Purshia*, and *Symphoricarpos*.

11.3.3 Montane and Subalpine Conifer Forests

The Küchler (1964, 1985) PNV categories of Douglas fir forest, western spruce-fir forest, spruce-fir–Douglas fir forest, and Great Basin pine forest are merged into montane and subalpine forests on Figs. 11.5 and 11.6. In the massive Uinta Mountains and Wasatch Range on the eastern side of the Bonneville basin (Fig. 11.4), these forests in the first three of these categories are relatively dense and include *Pseudotsuga menziesii*, *Populus tremuloides*, and species of *Pinus* (*P. contorta*, *P. ponderosa*), *Abies* (*A. concolor*, *A. lasiocarpa*), and *Picea* (*P. pungens*, *P. engelmannii*). With the exception of alpine vegetation (described later), the subalpine and montane forests (spruce-fir, "spruce-fir–Douglas fir," and "Douglas fir") are the PNV categories that occur at the highest elevations and live under the lowest temperatures and the highest precipitation levels (Fig. 11.7). In the drier mountain ranges south and west of GSL, the high-elevation Great Basin pine forest has *Pinus longaeva* and *Pinus flexilis* as the dominant species, along with *A. tridentata* and *C. ledifolius* (*Abies*, *Picea*, and *Pseudotsuga* are largely absent in these more arid habitats). The Great Basin pine forest is more open than forests on the eastern side of the Bonneville basin and occurs under generally warmer and drier conditions (Fig. 11.7).

11.3.4 Alpine Meadows and Barren

The Alpine meadows and barren PNV category occurs in small areas on the highest mountains and represents a treeless tundra dominated by grasses, sedges, willows, lichens, and many small shrubs and forbs. It is associated with the coldest temperatures and the highest levels of mean ANNP of any the PNV categories in the Bonneville basin.

11.3.5 Climate and Vegetation

The biplots and histograms in Fig. 11.7 illustrate both the strong relations between climatic variables and PNV categories but also show that substantial overlaps occur in the climatic conditions associated with the various categories (which is not surprising in that many species occur in more than one PNV and these communities interdigitate in their spatial and elevational distributions). For example, with the exception of Great Basin pine, conifer forest PNVs have median values for July temperature (JULT) between ~13°C and 16°C, between approximately −10°C and −6°C for JANT, and between ~500 and 900 mm for mean ANNP. Among the forest types (Fig. 11.7), western spruce-fir forest is associated with the coolest median temperatures and the wettest ANNP; spruce-fir and Douglas fir forests are intermediate between western spruce-fir and Great Basin Pine forests, the latter of which has substantially warmer and drier median tolerances (which overlap with those for Mountain mahogany-oak scrub and sagebrush steppe). Mountain mahogany-oak scrub is thus also intermediate between forest (on one hand) and woodland, steppe, and desert (on the other hand); its median values are cooler and wetter than those for woodland, steppe, and desert. Juniper–pinyon woodland and Great Basin sagebrush have similar summer temperatures, but this woodland requires warmer winters and higher ANNP levels than Great Basin sagebrush. The saltbush-greasewood and desert PNV categories occur under very similar climates, in the driest and warmest environments in the Bonneville basin.

11.4 VEGETATION HISTORY

11.4.1 Previous Palynological Studies

Davis (1998) and Davis and Moutoux (1998) reconstructed long-term vegetation history and paleohydrology from pollen assemblages covering the past 13.5 million yr (myr) from deep wells drilled in GSL by the Amoco Production Company. Desert and steppe taxa are present throughout the records, with *Sarcobatus* and *Ephedra* dominant in the late Miocene (prior to ~5 myr) and Amaranthaceae and *Artemisia* dominant in the Pliocene (~5 to 2.5 myr) and early Pleistocene (~2.5 Ma to ~750 ka). The pollen percentages of *Pinus* and *Artemisia* are much higher (and those of Amaranthaceae lower) during the late Pleistocene (<750 ka) than during these earlier time periods. Quasi-cyclic variations in pollen assemblages occur throughout the Amoco records, presumably reflecting glacial–interglacial cycles of varying intensity since the late Miocene. During the Pliocene and early Pleistocene, these cycles were dominated by *Artemisia* (glacials) vs Amaranthaceae (interglacials), whereas during the late Pleistocene the cycles were larger and expressed as *Picea*, *Abies*, and *Pseudotsuga* (glacials) vs *Juniperus*, *Ambrosia*-type, and *Sarcobatus* (interglacials). The pollen flora recovered from these wells is similar to that of the present-day throughout, with the exception of the presence of low levels of "Tertiary relicts" (*Ulmus*, *Carya*, and *Ostrya–Carpinus*) from the Miocene into the early Pleistocene (Davis, 1998; Davis and Moutoux, 1998). The changes seen in our pollen record represent the latest of the quasi-cyclic variations seen by these authors since the late Miocene in GSL.

In addition to the long records of Davis (1998), Davis and Moutoux (1998), and others described in Davis (2002), several previous studies have been done of late Quaternary pollen assemblages from cores taken in, or immediately adjacent to, GSL (Core C in Spencer et al., 1984; GSL core and Crescent Springs in Mehringer, 1985; see also the discussion in Rhode, 2016). The locations of these records are shown in Fig. 11.1, and they are discussed in detail later in comparison with the GSL96+ pollen profile.

11.4.2 Pollen Analysis

Sediment samples were processed for pollen analysis by using HCl, HF, and, when required, KOH (Faegri and Iversen, 1975). Percentage calculations were based on a terrestrial pollen sum that excluded algae, aquatic pollen, and spores. Sample residues were analyzed under 400–1000× magnification, and a minimum of 300 terrestrial pollen grains were counted from each sample. We analyzed 180 pollen samples from GSL96+ (including 5 samples that are from potentially disturbed early Holocene (EH) sediments) and 14 samples from GSL96-4. Pollen zones were determined using the CONISS program of Grimm (1987), as implemented by PSIMPOLL (Bennett, 2005). Due to differences in accumulation rates among the lithologic units, the mean stratigraphic differences in depth and time vary by pollen zone (Fig. 11.8). In pollen zone A (pre-Bonneville, LSM sediments) it is 6.6 cm and 317 years; in pollen zone B (ALM, MOM, DM sediments deposited in Lake Bonneville), it is 3.9 cm and 194 years; in pollen zone C (DM and FLM sediments; Pleistocene–Holocene transition), it is 3.7 cm and 150 years; and in the BSM sediments of the Holocene, it is 5.0 cm and 170 years for pollen zone D and 7.4 cm and 104 years for pollen zone E.

11.4.3 Pollen Flora

Eleven pollen types (*Pinus*, *Abies*, *Picea*, *Juniperus*, *Cercocarpus*-type and *Quercus*, *Artemisia*, *Ambrosia*-type, Asteraceae, Amaranthaceae, *Sarcobatus*, and Poaceae) compose ~90% or more of the terrestrial pollen sum for each pollen assemblage across the ~40 ka of the GSL96+ record (Fig. 11.8; Table 11.3). Although no large changes occur in the pollen flora through this record, the relative abundances of the various types change greatly through time. Furthermore, it is probable that not all of the plant species represented by a particular pollen taxon on the present-day landscape were present through the entire record (and it is possible that some species living in the Bonneville basin earlier are no longer present).

Many of these important pollen types include a variety of species from a range of habitats spanning large parts of the elevational gradient. For example, *Pinus* includes species present in the subalpine and montane forests of the mountains in and around the Bonneville basin, as well as pinyon pines of juniper–pinyon woodland of the lower mountain slopes. *Artemisia* also covers a broad elevational range—it is the dominant taxon in sagebrush communities, but is also an important understory element in the juniper–pinyon woodland, Great Basin pine forest, and some alpine communities. Asteraceae and Poaceae occur in nearly all of the vegetation communities in the Bonneville basin.

Other pollen types represent plants that, although widespread, are primarily indicative of particular vegetation communities. *Juniperus* is present in

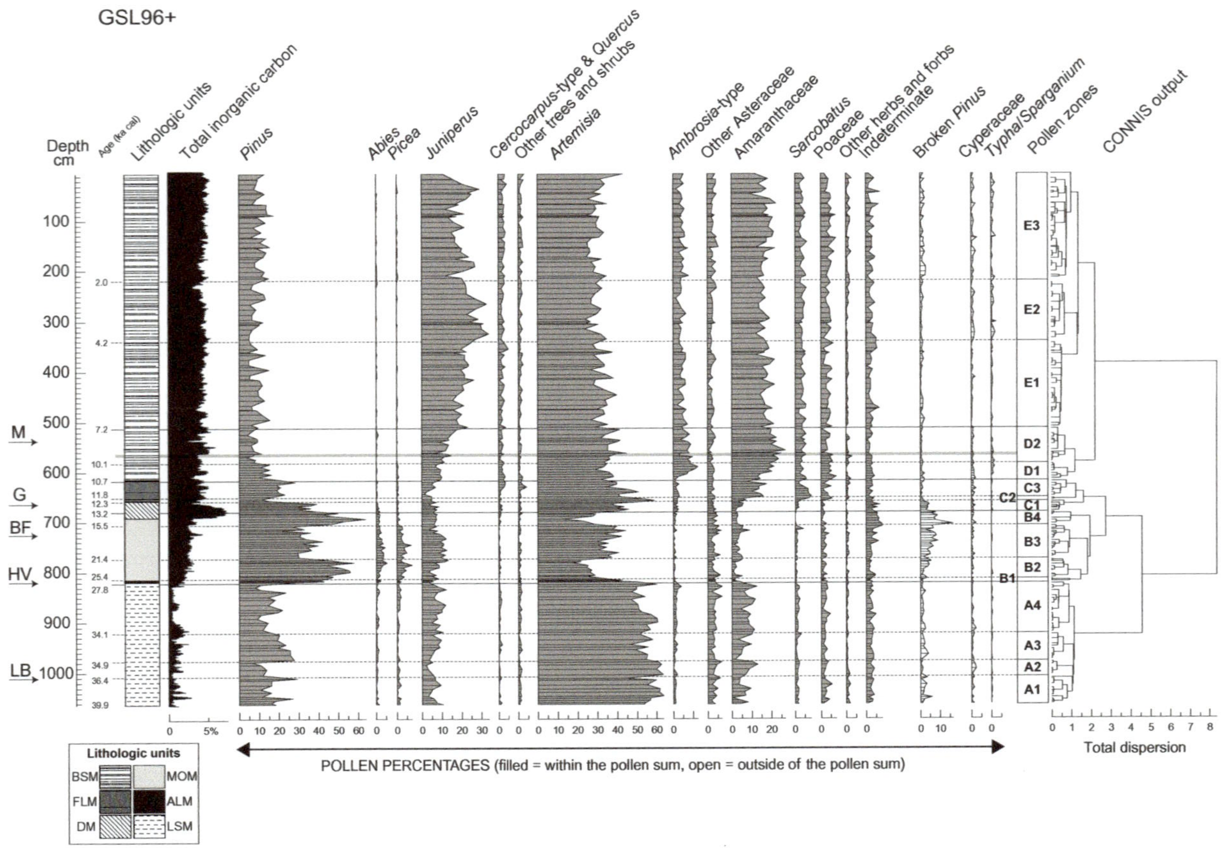

FIG. 11.8 For the entire length of the USGS GSL96+ sediment core, depth vs: approximate calibrated ages, lithologic units, TIC%, major pollen types, and pollen zones. *Solid gray line* through the BSM sediments marks where as much as a meter of sediment may not have been recovered between Drives 1 and 2 of GSL96-6. "Key to letter codes on *left side*: *BF*, Bonneville flood; *G*, Gilbert episode; *HV*, Hansel Valley tephra; *LB*, lower basaltic ash; and *M*, Mazama ash."

TABLE 11.3 Mean Pollen Percentages for Pollen Zones and Subzones

Pollen Zone	Lithologic Unit(s)	Depth Range (cm)	Estimated Age Range (cal ka)	Number of Samples	Pollen Grains Per Gram (×1000)	*Pinus*	*Abies*	*Picea*	*Juniperus*	*Cercocarpus*-Type and *Quercus*	Other Trees and Shrubs	*Artemisia*	*Ambrosia*-Type	Other Asteraceae	Amaranthaceae	*Sarcobatus*	Poaceae	Other Herbs and Forbs	Indeterminate	Broken *Pinus*	Cyperaceae	*Typha*/*Sparganium*
Average values for GSL96+																						
All	All	1062.6–5.4	39.9–0.0	*175*	*189*	*18.9*	*0.7*	*0.8*	*12.9*	*1.3*	*1.0*	*37.6*	*3.0*	*3.0*	*11.6*	*2.6*	*3.0*	*0.7*	*2.9*	*1.9*	*0.7*	*0.3*
Average values by pollen zone																						
E	BSM	510.0–5.4	7.1–0.0	68	188	9.8	0.1	0.1	21.1	1.8	0.2	30.4	4.5	2.6	16.7	3.6	4.0	0.6	2.8	1.0	0.7	0.6
D	*BSM*	614.5–515.0	10.6–7.2	20	300	9.9	0.1	0.2	10.9	1.7	0.8	36.3	6.8	2.3	19.9	4.0	4.0	0.9	2.3	0.7	0.9	0.6
C	*DM/FLM*	678.0–619.5	13.1–10.7	16	212	25.7	0.5	0.4	5.1	1.3	1.2	43.7	1.9	2.6	8.2	3.4	2.1	0.5	3.2	1.8	0.9	0.2
B	*ALM/ MOM/DM*	819.0–681.5	27.0–13.2	35	79	41.2	2.3	2.5	7.8	0.3	0.8	31.7	0.6	3.5	3.1	0.6	1.6	0.6	3.5	4.8	0.6	0.1
A	*LSM*	1062.6–824.2	39.9–28.5	36	161	16.6	0.5	0.8	6.9	0.4	0.9	55.1	1.2	3.6	7.2	1.3	2.3	0.7	2.7	1.5	0.7	0.1
Average values by pollen subzone																						
E3	*BSM*	214.9–5.4	2.0–0.0	31	208	11.1	0.1	0.2	18.9	2.1	0.2	29.8	4.5	3.0	16.9	3.6	4.3	0.7	2.7	1.3	0.9	0.6
E2	*BSM*	335.1–225.9	4.0–2.0	14	147	8.3	0.1	0.0	27.2	1.6	0.2	28.7	3.7	2.6	15.4	3.6	3.8	0.5	2.9	0.7	0.9	0.8
E1	*BSM*	510.0–345.1	7.1–4.3	23	187	8.9	0.1	0.1	20.4	1.5	0.2	32.4	4.9	2.1	17.2	3.8	3.6	0.5	2.9	0.9	0.4	0.3

Continued

TABLE 11.3 Mean Pollen Percentages for Pollen Zones and Subzones—Cont'd

Pollen Zone	Lithologic Unit(s)	Depth Range (cm)	Estimated Age Range (cal ka)	Number of Samples	Pollen Grains Per Gram (× 1000)	*Pinus*	*Abies*	*Picea*	*Juniperus*	*Cercocarpus*-Type and *Quercus*	Other Trees and Shrubs	*Artemisia*	*Ambrosia*-Type	Other Asteraceae	Amaranthaceae	*Sarcobatus*	Poaceae	Other Herbs and Forbs	Indeterminate	Broken *Pinus*	Cyperaceae	*Typha/Sparganium*
D2	*BSM*	579.5–515.0	10.0–7.2	13	301	7.6	0.2	0.0	12.1	1.5	0.7	37.4	6.6	2.1	21.5	4.1	3.5	0.9	2.0	0.6	0.7	0.5
D1	*BSM*	614.5–584.5	10.6–10.1	7	300	14.2	0.1	0.4	8.8	2.2	1.0	34.2	7.2	2.7	17.0	3.7	4.9	0.7	3.0	1.1	1.2	0.7
C3	*FLM*	648.5–619.5	11.6–10.7	7	342	20.9	0.3	0.4	4.3	1.7	1.7	41.7	2.1	2.5	13.2	5.2	2.8	0.9	2.4	0.7	1.0	0.1
C2	*FLM*	656.5–653.5	12.1–11.9	2	114	13.5	0.0	0.3	4.7	0.8	0.3	56.3	2.0	2.6	6.2	4.9	4.2	0.8	3.3	1.1	0.3	0.2
C1	*DM*	678.0–660.0	13.1–12.4	7	64	34.0	0.8	0.5	6.1	1.0	1.0	42.0	1.8	2.8	3.7	1.3	0.9	0.1	4.0	3.2	1.0	0.2
B4	*DM/MOM*	703.5–681.5	15.2–13.2	6	81	50.5	1.3	0.6	6.7	0.7	0.9	24.4	0.9	3.6	2.6	0.1	1.3	0.4	6.1	8.9	1.0	0.0
B3	*MOM*	770.5–708.5	21.2–15.9	15	45	32.4	2.6	3.1	9.1	0.2	0.8	36.3	0.5	3.7	4.0	1.0	1.9	0.7	3.6	5.4	0.5	0.1
B2	*MOM*	812.0–776.0	24.9–21.5	10	56	49.8	2.6	3.4	7.5	0.4	0.7	24.9	0.4	3.2	2.4	0.5	1.3	0.4	2.5	2.9	0.4	0.1
B1	*ALM*	819.0–815.0	27.0–25.8	4	245	38.6	1.7	1.1	4.9	0.0	0.8	42.1	0.6	3.4	2.0	0.4	1.8	0.6	2.1	1.4	0.4	0.0
A4	*LSM*	949.6–824.2	34.0–28.5	19	152	15.6	0.5	0.8	8.1	0.5	0.9	53.3	1.5	3.7	8.4	1.3	2.2	0.5	2.6	1.3	0.6	0.2
A3	*LSM*	974.6–954.6	34.8–34.1	3	242	26.3	0.9	0.6	4.9	0.3	0.7	52.9	0.9	2.5	3.7	0.7	2.0	0.5	3.3	2.4	0.5	0.1
A2	*LSM*	1004.6–979.6	36.1–35.0	4	109	10.6	0.3	0.5	3.9	0.2	0.9	61.5	0.7	4.3	9.6	1.6	2.3	0.7	2.8	0.6	1.2	0.1
A1	*LSM*	1062.6–1011.6	39.9–36.6	10	131	17.9	0.5	0.9	6.5	0.4	0.8	56.7	0.8	3.5	4.9	1.2	2.4	1.0	2.5	2.2	0.6	0.1

The italicized numbers represent the mean values for the entire length of the pollen record.

conifer forests, but only dominant in the juniper–pinyon woodland. Similarly, although plants represented by the "*Cercocarpus*-type and *Quercus*" (Davis, 2001) pollen category occur in juniper–pinyon woodland and in the lower reaches of the montane conifer forests, they are most abundant in the mountain mahogany-oak plant community. Finally, some of the pollen types can be fairly reliably associated with more narrowly defined habitat requirements. At the cool and moist end of the spectrum, *Picea* and *Abies* are largely restricted to middle and upper elevation conifer forests. Conversely, Amaranthaceae, *Sarcobatus*, and *Ambrosia*-type are associated largely with lower elevation saltbush-greasewood communities.

Three common (but minor) pollen types representing trees and shrubs are present at levels above 1% in at least one pollen sample in the GSL96-6 cores: *Alnus* (55 samples) and *Salix* (69 samples) are present in samples scattered through the stratigraphic column, and *Ephedra viridis*-type (100 samples) is scattered throughout but at higher levels above ~475 cm. Other pollen types related to woody plant species that are present at levels below 1% include: *Acer*, *Arceuthobium*, *Betula*, *Celtis*, *Corylus*, *Ephedra trifurca*-type, *Fraxinus*, *Populus*, *Pseudotsuga/Larix*, *Shepherdia/Eleaegnus*, *Symphoricarpos/Lonicera*, and *Tsuga heterophylla*. Two common (but minor) herbaceous pollen types (Rhamnaceae [47 samples] and *Eriogonum* [44 samples]) occur at levels higher than 1% in at least one pollen sample in the core. Both are scattered through the core, although Rhamnaceae grains are more common and abundant at 677.5 cm and above. Other herbaceous pollen types present at levels below 1% include: Apiaceae, *Campanula*, Caryophyllaceae, Brassicaceae, Fabaceae, *Galium*, Lamiaceae, Liguliflorae, Liliaceae, Onagraceae, *Phlox*, *Plantago*, *Polemonium*, Polygonaceae, Ranunculaceae, *Ribes*, Rosaceae, *Rumex*, *Sphaeralcea*-type, and *Thalictrum*.

Indeterminate pollen grains were included in the terrestrial pollen sum, and over the length of the pollen record the mean value for indeterminate pollen grains is 2.9% (Table 11.3). With one exception (pollen subzone B4, where indeterminate grains were 6.1% of the pollen sum; discussed later), the indeterminate pollen grains ranged between 2% and 4% across the subzones of the GSL96+ pollen diagram. We also tracked differences in the preservation of *Pinus* pollen grains through the record and found differences across lithologic units and pollen subzones (Table 11.3), with subzone B4 having the highest percentage of broken *Pinus* grains.

Aquatic/wetland pollen types, spores, and algae were counted for each of the pollen spectra, but are not included in the pollen sum for calculations of percentages. Aquatic and semiaquatic plants that occur in more than one sample include (with number of samples in parentheses): Cyperaceae (151), *Typha/Sparganium* (77), *Typha latifolia* (32), *Ruppia* (16), and *Myriophyllum* (9). No major stratigraphic patterns seem to appear for these pollen types, other than that *Typha/Sparganium* and *T. latifolia* are more abundant and more consistently present in BSM sediments than lower in the core. Among the algae, *Pediastrum* is present in 33 samples and *Botryococcus* in

36 samples, both largely scattered across the length of the record (although *Botryococcus* is fairly continuously present from ~740 to 840 cm depth).

11.4.4 Comparison of Modern Vegetation with Latest Holocene Pollen Spectra

The latest Holocene portion (~2 ka to present) of the GSL96+ pollen record is represented by subzone E3 (Table 11.3; Fig. 11.8). Comparison of the mean values of pollen types for this subzone with the percentage of cover by each of the Küchler (1964, 1985) PNV categories (Fig. 11.6) allows us to assess how well the individual PNV categories are represented in the pollen data. Collectively, major pollen types representing forest vegetation (*Abies*, *Picea*, *Populus*, *Pseudotsuga*, *Pinus* [also present in woodlands]) total ~11% of the mean pollen percentage for subzone E3, which is commensurate with the ~11% total PNV cover for the Bonneville basin. *Juniperus*, which lives primarily in woodlands, has a mean subzone E3 pollen percentage of 21%. This taxon, combined with some proportion of the *Pinus* category, aligns with a PNV coverage percentage of 17%. So, for the two conifer-dominated PNV categories, the E3 mean percentages correspond reasonably well with the total PNV cover percentages. In contrast, the other major tree-dominated PNV category of "mountain mahogany-oak scrub" has a total modern PNV cover percentage of 11%, but the *Cercocarpus*-type and *Quercus* pollen category (which largely aligns with this scrub vegetation) has a mean pollen percentage for subzone E3 of only ~2%. Steppe and desert communities (more local to GSL) are represented in subzone E3 in approximately the same percentages as in the PNV cover percentages. *Artemisia* is 30% in the pollen data, and sagebrush communities are 30% of the PNV cover. Amaranthaceae and *Sarcobatus* together are 21% of subzone E3, and the Saltbush-Greasewood PNV cover percentage is 19%. It may be coincidental, but with the exception of "mountain mahogany-oak scrub," the PNV categories are represented in the uppermost pollen zone of the GSL96+ core in approximately the same percentages as on the overall landscape of the Bonneville basin.

Although the general pattern of modern vegetation is reflected in pollen assemblages from the uppermost sediments of GSL, it should be noted that some taxa are important on the landscape but are greatly underrepresented in the pollen data. These include widely distributed tree species (such as *P. menziesii* and *P. tremuloides*) and shrubs (such as *Acer*, *Symphoricarpos*, and *Rhus*). In addition, pollen types such as "Other Asteraceae" and Poaceae that may include many species are distributed across the entire elevational gradient.

11.4.5 Interpretation of the Fossil Pollen Record from Lake Sediments in the Bonneville Basin

The Bonneville basin is an unusual, and to some degree difficult, setting for reconstructions of vegetation and climate changes from fossil pollen data. In particular, the huge changes in area and depth of lakes in this basin over the past 40,000 years have greatly affected the habitat available for plant species and the depositional environments in the lakes. As lakes changed depth and size, they covered or revealed large areas of habitat, and paleogeographic changes in basin morphology and river courses may have affected the geographic and elevational scope of the catchment that delivered pollen into these lakes. These changes may have included:

- Differing combinations of subbasins contributing to sediments in GSL through time (representing different elevational ranges and potentially different combinations of plant taxa). For example, changes in the course of the Bear River (Reheis et al., 2014; Pederson et al., 2016) may have added the upper portion of the Bear subbasin (Fig. 11.1) to the Bonneville basin during the rise of Lake Bonneville, which could have had the effect of increasing conifer pollen delivered into the lake.
- During the deepest lake phases the pollen record is probably dominated by river-borne palynomorphs from the large mountain masses to the east; during shallow water phases this input was probably considerably diminished and local input became more important.
- Rising and falling lake levels would have covered or exposed desert, saltbush-greasewood, and other vegetation communities. When the percentage by elevation range of the areal coverage of the Bonneville basin is compared with the elevations of the different stages of Lake Bonneville (Fig. 11.6), it can be seen that about a third of the Bonneville basin was covered by lake waters during the deepest phases. Today, this lowest elevation third of the area hosts desert and much of the saltbush-greasewood vegetation (as well as major parts of sagebrush, juniper–pinyon, and mountain mahogany-oak PNV categories; Fig. 11.6), and covering it with lake waters would alter the relative percentages of vegetation cover and their representation in pollen assemblages.

The collective effect of these changes on the pollen record may have been to heighten the apparent dichotomy between taxa representing cold and/or wet conditions (ie, montane and subalpine conifers) and those representing warmer and/or drier conditions (eg, *Artemisia*, Amaranthaceae, *Sarcobatus*, *Ambrosia*-type). The record may be further muddied during lake-level fluctuations during the post-Provo shoreline recession and subsequent shallow water phases (such as those that occurred during the historical period, as shown in Fig. 11.1). These fluctuations may have disturbed soft sediments of different ages along the margin and led to redeposition of older pollen grains.

An additional factor that may have further heightened this apparent dichotomy was that the elevational ranges of many species were lower during the late Pleistocene than they are today. As noted by Mehringer (1967), this could have greatly increased the areal coverage of higher-elevation plants, and because some of these are prolific pollen producers (especially

Pinus), they may have had a strong influence on the percentages of various pollen types in the pollen diagram. Similar effects may have occurred during the Holocene and during the time prior to Lake Bonneville, albeit of lesser amplitudes. For example, warmer growing-season temperatures and/or other factors may have caused upper treeline to rise, whereas moister conditions may have led to a decline in the lower limits of key plant taxa (such as *Juniperus*).

11.5 POLLEN STRATIGRAPHY OF GSL96+ (FIGS. 11.8 AND 11.9; TABLE 11.3)

11.5.1 Before Lake Bonneville

11.5.1.1 Pollen zone A (1062.6 to 824.2 cm depth; 39.9 to 28.5 cal ka; 36 samples)

Corresponds with lithologic unit LSM (which represents shallow-lake conditions before the rise of Lake Bonneville), and no indications occur of major changes in lake status or depth through the LSM (Fig. 11.4). *Artemisia* pollen is dominant throughout this pollen zone (mean = 55%, ranging from 53% to 62% across the four subzones), and the other important taxa here include: *Pinus* (mean = 17%, ranging from 11% to 26% across subzones), Amaranthaceae (mean = 7%, ranging from 4% to 10% across subzones), and *Juniperus* (mean = 7%, ranging from 4% to 8% across subzones). The pollen type "Other Asteraceae" is higher here than its average for the entire core; Amaranthaceae pollen is at its highest levels until the beginning of the Holocene; and *Ambrosia*-type, *Sarcobatus*, and Poaceae are higher than in the subsequent pollen zone B. The pollen of forest trees is at low levels through pollen zone A (*Abies* and *Picea* are each below 1%).

Pollen zone A has four subzones, all of which are dominated by *Artemisia* (Table 11.3). For zone A overall, *Artemisia* has strong negative correlations ($p < 0.05$) with *Pinus* ($r = -0.54$), *Juniperus* ($r = -0.52$), and "*Cercocarpus*-type and *Quercus*" ($r = -0.58$); and *Pinus* also has strong negative correlations with Amaranthaceae (−0.64). Relative to the mean pollen percentages for pollen zone A, the lowest subzone (A1: 1062.6 to 1011.6 cm depth; 39.9 to 36.6 cal ka; 10 samples) has relatively high values for *Pinus* and *Juniperus*, and lower values for *Artemisia* and Amaranthaceae. Subzone A2 (1004.6 to 979.6 cm; 36.1 to 35.0 cal ka; four samples) has the lowest percentages within zone A for *Pinus* and *Juniperus* and the highest percentages of *Artemisia* and Amaranthaceae. Subzone A3 (974.6 to 954.6 cm; 34.8 to 34.1 cal ka; three samples) has the highest *Pinus* and the lowest *Juniperus*, *Artemisia*, and Amaranthaceae percentages within zone A. The stratigraphically highest subzone (A4: 949.6 to 824.2 cm; 34.0 to 28.5 cal ka; 19 samples) has lower *Pinus* and *Artemisia*, high Amaranthaceae, and the highest *Juniperus* within zone A.

11.5.2 Lake Bonneville

11.5.2.1 Pollen Zone B (819.2 to 681.5 cm Depth; 27.0 to 12.2 cal ka, 35 Samples)

This pollen zone spans several lithologic units (ALM, MOM, and lower DM) and corresponds with the rise, deepwater phases, and regression of Lake Bonneville (Fig. 11.9). Overall, this pollen zone has the highest representations of coniferous forest trees (*Abies*, *Picea*, *Pinus*) and the lowest representations of steppe and desert plants (*Artemisia*, *Ambrosia*-type, Amaranthaceae, *Sarcobatus*). Four subzones occur, largely defined by the interplay between

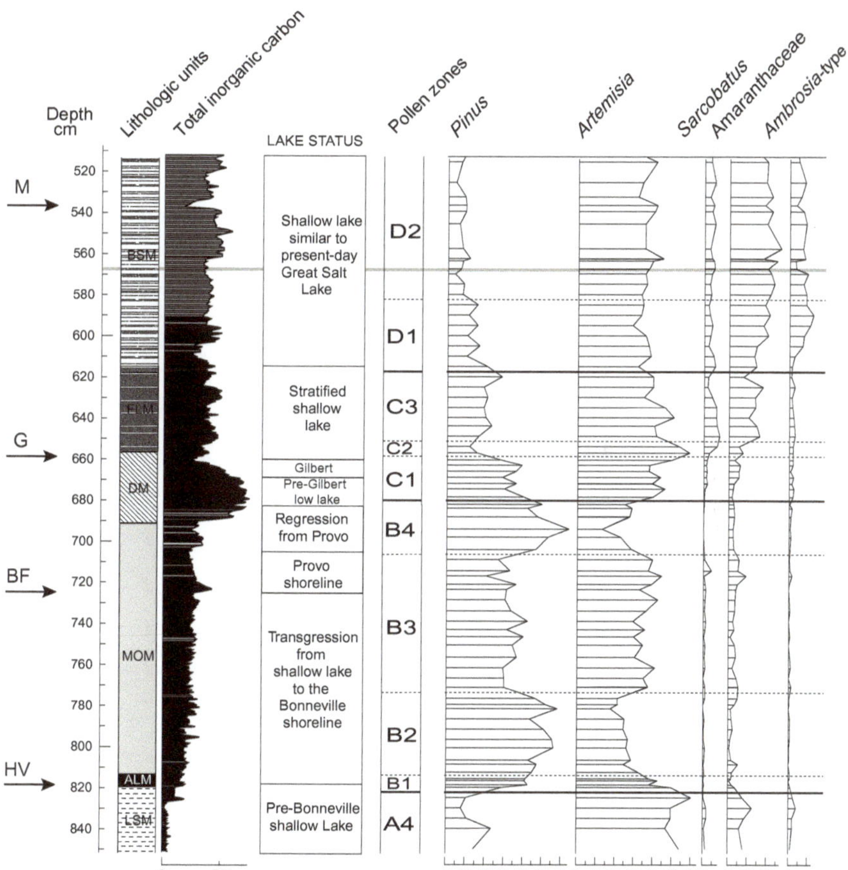

FIG. 11.9 From just below the rise of Lake Bonneville to just above the Mazama ash (850 to 510 cm depth), depth vs: inferred lake status (from Fig. 11.4), compared with pollen zones and selected pollen types. *Gray line* through the BSM sediments marks where as much as a meter of sediment may not have been recovered between Drives 1 and 2 of GSL96-6. *BF*, Bonneville flood; *G*, Gilbert episode; *HV*, Hansel Valley tephra; and *M*, Mazama ash.

Pinus and *Artemisia*, which are strongly negatively correlated ($r = -0.86$, $p < 0.01$) through this pollen zone (*Pinus* is also negatively correlated with Amaranthaceae; $r = -0.60$, $p < 0.01$). The first subzone (B1) has a rapid rise in *Pinus* and decline in *Artemisia* (B1), which is followed by a *Pinus* maximum/*Artemisia* minimum (subzone B2); then a return to moderate levels of *Artemisia* and lower percentages of *Pinus* (although the latter is still well above pre- and post-Lake Bonneville levels; subzone B3); and finally, a second maximum in *Pinus* and decline in *Artemisia* (subzone B4). The progression of changes in these subzones resembles the overall pattern seen in Madsen's pollen profile from GSL Core C (Fig. 11.11). As described later, these subzones correspond with the history of Lake Bonneville in some ways, but differ in others.

Pollen subzone B1 (819.0 to 815.0 cm depth; 27.0 to 25.8 cal ka, four samples, ALM sediments) is correlated with the initial rise of Lake Bonneville and is characterized by a sharp increase in the percentage of *Pinus* pollen, coupled with a decrease in *Artemisia*. *Picea* and *Abies* also increase in this subzone (although they remain below an average of their levels in subzones B2 and B3), and sharp declines occur in Amaranthaceae, *Sarcobatus*, and *Juniperus*. The diversion of the Bear River into the Bonneville basin (Reheis et al., 2014; Pederson et al., 2016) could have caused, in combination with climate change, the pollen spectra observed in this subzone.

Pollen subzone B1 is represented by only four samples and occurs at the break between Drives 2 and 3 of GSL96-6. A strong possibility exists that not all of this unit was recovered during coring. Fortunately, these sediments were recovered within a single drive of GSL96-4, which allows us to assess how the GSL96+ record may be affected (Fig. 11.10). The HV ash is present

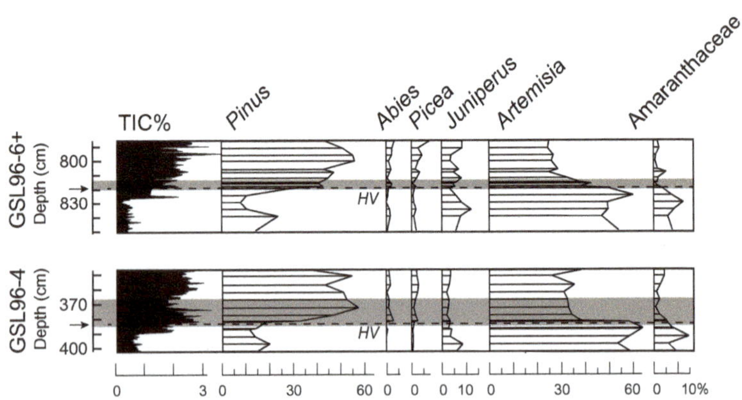

FIG. 11.10 The beginning of Lake Bonneville in sediment cores USGSL96-4 and USGSL96-6+. Depth vs the occurrence of ALM sediments (*gray band*), TIC%, Hansel Valley ash (HV; marked by the *arrow* and *dashed black line* in each core), and selected pollen types. This diagram shows that not much of the transition is missing from USGS96-6 at the core break between Drives 2 and 3.

in ALM sediments in both cores at a point just below the rapid rise in *Pinus* and decline in *Artemisia*. ALM sediments are considerably thicker in GSL96-4 than in GSL96+, suggesting that 10 cm or more of this unit may be missing from the latter core. However, the major features of the two pollen sequences are similar, suggesting little effect on the GSL96+ stratigraphy. In addition, Madsen's pollen diagram from GSL Core C (Spencer et al., 1984) shows a similarly abrupt increase in conifers and decline in *Artemisia* at about this level (Fig. 11.11).

Pollen subzone B2 (812.0 to 776.0 cm depth; 24.9 to 21.5 cal ka; 10 samples; MOM sediments) is approximately tied with the younger subzone B4 for having the greatest dominance by *Pinus* pollen (~50% for the mean for both subzones) and lowest levels of *Artemisia* pollen. However, subzone B2 also has the highest *Picea* pollen percentages (mean = 3.4%) and is tied with subzone B3 for the highest *Abies* pollen percentages (mean = 2.6%). Amaranthaceae and *Sarcobatus* pollen percentages are similar to those in the preceding pollen subzone B1, whereas *Juniperus* is higher. Through subzone B2, *Pinus* pollen ranges from 43% to 58%, with a rising, falling, then rising pattern. *Artemisia* pollen declines from 32% at the base of the subzone to 17–21% at the end. The highest levels of *Picea* and *Abies* pollen occur in sediments with estimated ages of 21.5–22 cal ka.

Pollen subzone B3 (770.5 to 708.5 cm depth; 21.2 to 15.8 cal ka; 15 samples; MOM sediments): Relative to the preceding and following subzones, B3 has much lower levels of *Pinus* pollen (mean = 32.4%) and higher levels of *Artemisia* (mean = 36.3%). *Picea* and *Abies* pollen percentages are similar to subzone B2, but significantly higher than in subzone B4. *Juniperus* is at its highest levels within pollen zone B (mean = 9.1%) and Amaranthaceae, *Sarcobatus*, and Poaceae are higher in subzone B3 than in either the B2 or B4 subzones. Through subzone B3, a pattern occurs of slightly higher *Artemisia* and lower *Pinus*, followed by a slight *Pinus* rise and *Artemisia* decline, and then another reversal toward the top of the subzone. This subzone has the second highest level of broken *Pinus* pollen grains and the third highest level of indeterminate pollen grains.

Pollen subzone B3 aligns with paleolimnological data (freshwater ostracods, calcite dominance in the sediment mineralogy) that indicate continued deep and freshwater conditions (Figs. 11.4 and 11.9). We correlate a pronounced rise in TIC%, an increase in dolomite relative to calcite, and a strong shift from high to declining levels of magnetic susceptibility (Reynolds et al., 2008) at 724.5 cm in GSL96+ with the Bonneville flood. Therefore, we think subzone B3 covers the final rise of the lake to the Bonneville shoreline, the Bonneville flood, and at least part of the time that the lake stood at the Provo shoreline. The dramatic changes in lake status through this pollen zone are not pronounced in the pollen data, and pollen spectra from subzone B3 do not appear to reflect conditions that were as cool and moist as those in subzone B2.

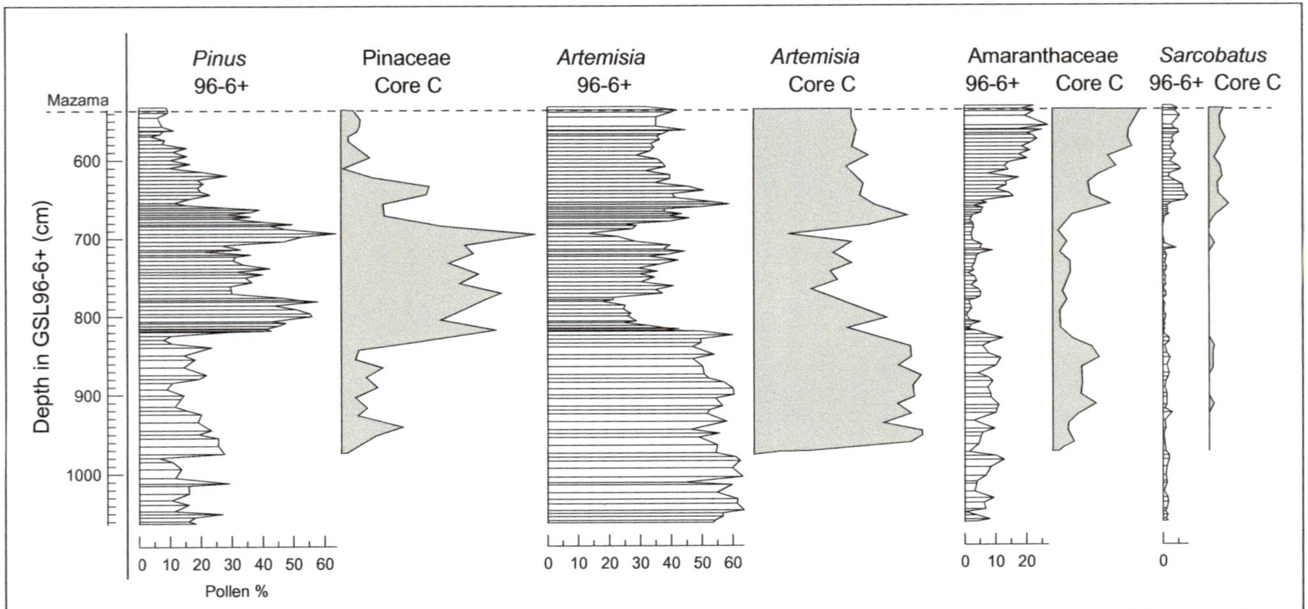

FIG. 11.11 Comparison of selected pollen types between the GSL96-6 record (from the Mazama tephra down to the base of the core) and D.B. Madsen's pollen diagram of the Lake Bonneville sediments in Core C (based on figure 8 in Spencer et al., 1984). To accommodate differences in analytical methods, *Pinus* in GSL96-6 is compared with Pinaceae in Core C.

Based on estimates from Currey et al. (1984), the rapid decline in lake level caused by the Bonneville flood—from the Bonneville shoreline to the Provo shoreline—would have exposed nearly 14,000 km^2 of land formerly covered by lake waters. This would have established new low-elevation habitats to be colonized by plant species tolerant of fine-grained sediments and perhaps saline soils. The pollen sequence above this level matches this pattern: the highest sustained levels of Amaranthaceae for pollen zone B occur in four samples from 720.5 to 708.5 cm (covering the estimated age of range of 17.5 to 15.8 cal ka) with a high of 9.2% at 716.5 cm (17.1 cal ka); and *Sarcobatus* pollen spikes (4.2%) in the sample from 714.0 cm (16.8 cal ka).

Pollen subzone B4 (703.5 to 681.5 cm depth; 15.2 to 13.2 cal ka; six samples, MOM and DM sediments): Similar to subzone B2, subzone B4 has the highest mean percentages of *Pinus* pollen and the lowest mean percentages of *Artemisia* for the entire record. However, subzone B4 has lower levels of *Picea* (3.4% for B2 vs 0.6% for B4), *Abies*, and *Juniperus*. The highest levels of broken *Pinus* pollen grains for the overall record occur in subzone B4, with highest for an individual sample at 16.7% at 703.5 cm (15.8 cal ka). For indeterminate pollen grains, the highest levels are in the two samples that span the B3/B4 boundary (~8%). The three lower samples of subzone B4 are in MOM sediments and here *Pinus* pollen rises from 47% to 64%, whereas *Artemisia* drops from 29% to 13%. The three upper samples of B4 are in DM sediments and here *Pinus* pollen drops to 42–50%, and *Artemisia* pollen rises to 25–29%. The end of subzone B4 occurs about a third of the way up from the base of DM sediments and aligns with the rise to the highest TIC% levels (Fig. 11.4), a strong shift from calcite to dolomite dominance, and the beginning of sustained presence of aragonite. On the basis of these indicators of lacustrine conditions, we infer the subzone B4 covers the regression from the Provo shoreline down to a shallow lake prior to the Gilbert episode (Fig. 11.9).

11.5.3 Pleistocene–Holocene Transition

11.5.3.1 Pollen Zone C (678.0 to 619.5 cm Depth; 13.1 to 10.7 cal ka, 16 Samples, DM and FLM Sediments)

This pollen zone begins in DM sediments with a sharp drop in TIC% levels (Fig. 11.4; subzone C1) and then continues upward to include all of the FLM sediments (subzones C2 and C3). These sediments are associated with a continuation of the shallow lake seen in subzone B4, a deepening of lake waters during the Gilbert episode, and a shallow stratified lake (represented by FLM sediments, Oviatt et al., 2015) during the earliest part of the Holocene (Fig. 11.9). Through these changes in lake status during pollen zone C, *Artemisia* rises to high levels similar to those in the pre-Lake Bonneville pollen zone A and then begins to decline; *Pinus* declines from the high levels of subzone B4 to the lowest levels in the pre-Holocene sediments and then rises; and Amaranthaceae and *Sarcobatus* have strong increases in their representations. For pollen zone C overall, *Pinus* has strong positive correlations with *Abies* ($r = 0.82$) and strong negative correlations with *Artemisia* ($r = -0.54$), Amaranthaceae ($r = -0.62$), *Sarcobatus* ($r = -0.74$), and Poaceae ($r = -0.62$). Consequently, as in the two lower pollen zones, variations in *Pinus* have strong influences on other taxa in the pollen diagram.

In pollen subzone C1 (678.0 to 660.0 cm depth; 13.1 to 12.4 cal ka; seven samples; DM sediments) an overall rise in *Artemisia* pollen (from 24% to 42%, Table 11.3) through this subzone is associated with a decline in *Pinus* pollen (from 51% to 34%) and a beginning of a trend toward increases in *Ambrosia*-type, Amaranthaceae, and *Sarcobatus*. Toward the top of subzone C1, *Pinus* pollen percentages decline, rise, and then drop again (and *Artemisia* pollen does the reverse). In the lake sediments represented by this subzone, dolomite is dominant over calcite, aragonite has its first sustained period of occurrence, and magnetic susceptibility is low (Fig. 11.4). A minimum in TIC% just below 660 cm depth in GSL96+ (Fig. 11.4) is interpreted as representing the freshest and deepest-water phase of the Gilbert episode (Oviatt, 2014), and thus subzone C1 represents vegetation changes that occurred during the Gilbert episode (Fig. 11.4).

Pollen subzone C2 (656.5 to 653.5 cm depth, 12.1 to 11.9 cal ka; two samples; FLM sediments) occurs at the base of the FLM lithologic unit, which reflects a strong shift in sedimentology and geochemistry from relatively freshwater late Pleistocene lake waters into hypersaline conditions where bacteria and brine shrimp were the dominant life forms. The two samples from this stratigraphically narrow subzone have the highest levels of *Artemisia* (mean = 56%) in the post-Bonneville portion of the record, along with very low levels of *Pinus* (mean = 14%). Compared with the underlying subzone C1, pronounced rises occur in pollen percentages for Amaranthaceae, *Sarcobatus*, and Poaceae (Figs. 11.8 and 11.9). Overall, the pollen spectra in subzone C2 resemble those of the pre-Bonneville pollen zone A and reflect cold and dry climatic conditions. As mentioned earlier, FLM sediments are interpreted as indicative of a shallow stratified lake under a dry climate (Oviatt et al., 2015); the pollen data from subzone C2 are compatible with this interpretation but also indicate that climatic conditions were relatively cold during some of the time represented by these sediments.

In pollen subzone C3 (648.5 to 619.5 cm depth; 11.6 to 10.7 cal ka; seven samples; FLM sediments), the mean level of *Artemisia* pollen percentages falls from the subzone C2 level (from 56% to 42%), whereas that of *Pinus* rises (from 14% to 21%). For both taxa the mean percentages remain above the levels seen in the Holocene sediments (Table 11.3, pollen zones D and E). However, in subzone C3 a very strong rise also occurs in Amaranthaceae pollen percentages (the average rising from 6% to 13%) and the highest mean levels of *Sarcobatus* (5.2%) for the GSL96+ pollen record. From subzone C3 upward to the present, the "*Cercocarpus*-type and *Quercus*" pollen taxon occurs at levels of 1.5% or greater, suggesting that the "mountain mahogany-oak scrub" vegetation community has existed in the Bonneville basin since the beginning of the Holocene.

11.5.4 The Holocene

The Holocene begins near or at the boundary between pollen subzones C2 and C3 (discussed earlier), but most of this epoch is represented by the BSM lithologic unit that covers the time since ~10.7 cal ka to the present in core GSL96+. BSM sediments are still being deposited in the modern lake, and so it appears that GSL has been a shallow hypersaline water body through all of the Holocene (unless there were times of near total desiccation, as discussed later). Sedimentation rates are much higher in BSM than in the underlying lithologic units. Although BSM sediments cover only ~27% of the time represented by GSL96+, this lithologic unit composes ~58% of the total length of this core. Holocene changes in lake level were apparently much smaller than during the period covered by Lake Bonneville, and it is difficult to interpret the apparently subtle, lake-level variations in the Holocene within the BSM deposits (due to high concentration of brine-shrimp fecal pellets, low variability in the levels of TIC% [Figs. 11.4 and 11.9], and little change in sedimentary structures). Changes within the pollen diagram for BSM sediments are also very subdued compared to those in Lake Bonneville deposits (Fig. 11.8); however, discernible patterns of changes occur through the Holocene. Within the BSM sediments, pollen zones D and E represent approximately the first third and latter two-thirds of the Holocene (with the boundary between the zones occurring around 7.1 and 7.2 cal ka). Compared to pollen zone E, zone D is characterized by somewhat higher mean pollen percentages of *Artemisia*, *Ambrosia*-type, Amaranthaceae, and *Sarcobatus*, and lower mean percentages of *Juniperus*. Differences among subzones in pollen zones D and E are less pronounced than in earlier pollen zones (Fig. 11.8; Table 11.3), but are evident in the standardized pollen values and pollen ratios shown in Fig. 11.12.

In BSM sediments (pollen zones D and E taken together), *Juniperus* ($p<0.01$) is significantly strongly negatively correlated with *Artemisia* ($r=-0.58$), *Ambrosia*-type ($r=-0.62$), and Amaranthaceae ($r=-0.60$). For the Holocene, *Juniperus* thus plays a similar role to that of *Pinus* during the late Pleistocene—that is, much of what happens in the pollen diagram for the Holocene is defined by the opposition between *Juniperus* and *Artemisia*, *Ambrosia*-type, and Amaranthaceae, which are from generally more arid environments than *Juniperus*.

11.5.4.1 Pollen Zone D (614.5 to 515.0 cm Depth; 10.6 to 7.2 cal ka; 13 Samples; BSM Sediments)

The BSM sediments of this pollen zone are the oldest in core GSL96+ that resemble sediments being deposited in present-day GSL. This EH pollen zone is dominated by *Artemisia* (mean = 37.1%) and other xerophytic taxa

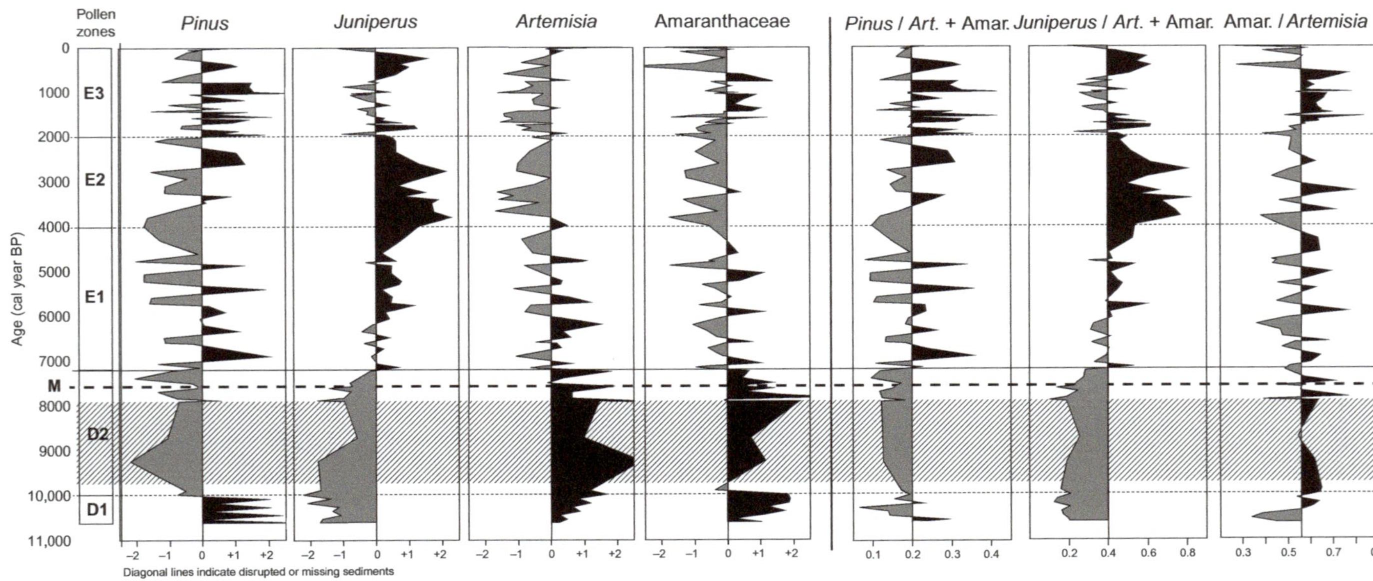

FIG. 11.12 Standardized pollen values and ratios between major pollen types through time for the BSM lithologic unit (pollen zones D and E taken together). The four profiles on the *left* present standardized values that are determined for each taxon by subtracting the mean (based on BSM samples) from a given value and then dividing by the standard deviation. The three profiles on the right present the ratios of major pollen types to each other (with the central axes for these profiles representing the mean ratio values for the BSM lithologic unit). Key to abbreviations: Amar., Amaranthaceae; *Art.*, Artemisia.

(including Amaranthaceae [21.2%], *Ambrosia*-type [7.2%], and *Sarcobatus* [3.8%]). The mean percentages of *Pinus* pollen are lower here than in the pollen zones related to Lake Bonneville and its precursors, but are essentially the same as those in the overlying pollen zone E. *Juniperus* pollen percentages are at higher levels than in underlying units (mean = 10.1%) and rising, but remain below the levels in overlying pollen zone E.

Pollen subzone D1 (614.5 to 584.5 cm, 10.6 to 10.1 cal ka, seven samples, BSM sediments): pollen taxa related to desert environments are high or rising in this subzone. Amaranthaceae rises from a mean of 13.2% in C3 to 17.0% here; *Sarcobatus* declines slightly from subzone C3, but remains relatively high (3.7%); and Poaceae is relatively high (4.9%). *Artemisia* is much lower in this subzone than in C3 (34.2% in D1 vs 41.7% in C3), as is *Pinus* (14.2% in D1 vs 20.9% in C3). *Abies* and *Picea* are both at very low levels, whereas *Juniperus* is higher in D1 than in C3 (8.8% in D1 vs 4.3% in C3), and this is part of a trend toward much higher juniper levels in younger Holocene units. Amaranthaceae and *Ambrosia*-type increase through this subzone from low levels to the highest in the GSL96+ record. Within subzone D1 a slight rise in TIC% occurs, associated with increased dolomite and decreased calcite and aragonite (Fig. 11.4). These lithologic changes occur during a sharp rise in *Ambrosia*-type (a pollen taxon that includes plants that grow on disturbed land surfaces), which could indicate a lowering of the level of GSL.

Pollen subzone D2 (579.5 to 515 cm; 10.0 to 7.2 cal ka, 13 samples; BSM sediments): the mean percentage for *Pinus* pollen for this subzone is the lowest for any subzone in the GSL96+ record, and *Abies* and *Picea* are also at very low levels. *Juniperus* continues its rapid rise in this subzone (8.8% in subzone D1, 12.1% in subzone D2, and 20.4% in subzone E1), with sustained percentages above ~15% beginning ~7.2 cal ka. Amaranthaceae pollen continues the high levels seen in pollen zone D1, *Artemisia* is at its highest level for the Holocene (37.4%), and *Sarcobatus*, *Ambrosia*-type, and Poaceae are at relatively high levels. The Mazama tephra is present at 538 cm depth. As discussed in Section 11.2.3, a gap seems to occur in the GSL96+ record for part or all of the time span from ~9.9 to 7.9 cal ka (Figs. 11.3, 11.4, and 11.9). This apparent hiatus could be due to sediments not accumulating during an episode of low lake level or may be due to missing or disturbed sediments due to problems in coring. The pollen spectra above and below this apparent hiatus fall within pollen subzone D2, and presumably pollen data from the missing interval would as well. Five additional samples were removed from the top of Drive 2 because sediments appear disturbed—these samples have pollen spectra similar to those above and below the stratigraphic range of probable unrecovered sediment (Fig. 11.13). Although complicated by the span of missing/disturbed sediments in D2, a sequence may occur of changing importance of Amaranthaceae vs *Artemisia* through D, with a peak in Amaranthaceae in pollen subzone D1; followed by lower levels of

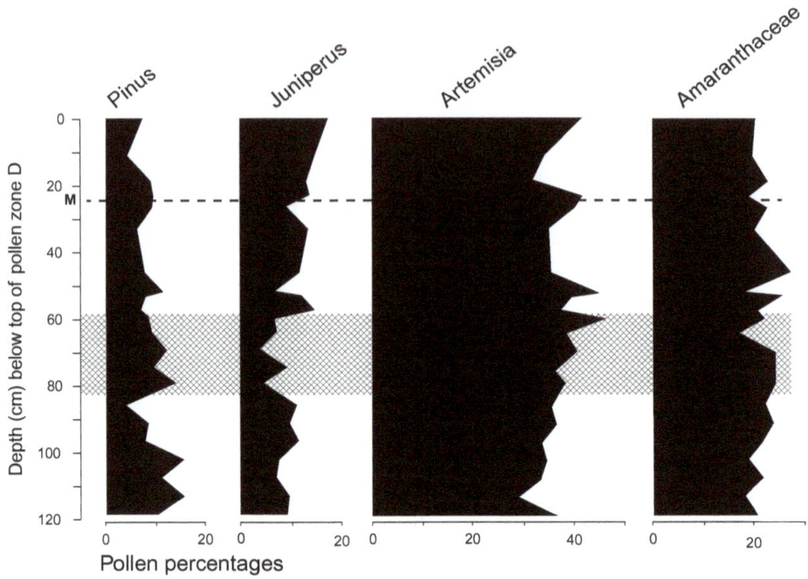

FIG. 11.13 Depiction of major pollen types for pollen zone D plotted by depth below the top of the zone with the samples from the "missing or disturbed" sediment interval (indicated by the *gray band*) inserted. The top of the *gray band* is also the top of Drive 2. Location of Mazama tephra indicated by *dotted line* and the letter M.

Amaranthaceae and higher *Artemisia*; then another peak in Amaranthaceae (as *Artemisia* declines); and then at the top of pollen subzone D2, a decline in Amaranthaceae (Fig. 11.12). This swapping back and forth in the relative importance of these steppe pollen taxa may reflect changes in the elevational and spatial extents of sagebrush vs saltbush-greasewood plant communities or may be artifacts of disturbances in the sediments through this interval.

11.5.4.2 Pollen Zone E (510.0 to 5.4 cm depth; 7.1 cal ka to near Present; 68 Samples; BSM Sediments Throughout)

This zone covers GSL through the middle and late Holocene (LH) and is characterized by low percentages of *Pinus*, the highest *Juniperus* in the record, and sustained presence of "*Cercocarpus*-type and *Quercus*" pollen (Table 11.3). *Artemisia* pollen is at lower percentages than in pollen zone D (mean of 30.4% in zone E vs 37.1% in zone D), as well as lower percentages for *Ambrosia*-type pollen (a mean of 4.5% in zone E vs 7.2% in zone D). Amaranthaceae pollen percentages are much higher than in pre-Holocene sediments, but lower than in zone D.

In pollen subzone E1 (510.0 to 345.1 cm, 7.1 to 4.3 cal ka; 23 samples) mean *Pinus* percentages are slightly higher than in pollen zone D, but still among the lowest in the entire record (Table 11.3); *Abies* and *Picea* are present only at trace levels, but *Juniperus* shows a strong rise from the underlying pollen zone (Table 11.3 and Fig. 11.12), indicating that woodland junipers were well established on the lower mountain slopes of the Bonneville basin by ~7 cal ka. In this subzone, *Artemisia* declines from subzone D2 levels, but has the highest mean percentages for pollen zone E. The overlying subzone E2 (335.1 to 225.9 cm, 4.0 to 2.0 cal ka; 14 samples) has the highest levels of *Juniperus* in the GSL96+ record, and the lowest mean levels of *Artemisia*, *Ambrosia*-type, and Amaranthaceae pollen for pollen zone E. As discussed later, the interpreted age and inferred paleoclimate fit within the neoglacial/neopluvial period (NP) that followed warmer/drier conditions of the early and middle Holocene (MH). The sediments below subzone E3 (214.9 to 5.4 cm, 2.0 cal ka to near present; 31 samples) were all within GSL96-6, whereas subzone E3 is entirely within GSL96-5. It has lower levels of *Juniperus* than in the underlying subzone, as well the highest percentages of *Pinus* in zone E. In addition, slight increases occur in mean values of "*Cercocarpus*-type and *Quercus*," *Artemisia*, *Ambrosia*-type, Amaranthaceae, and Poaceae relative to subzone E2 (Table 11.3). Within the last 1000 years, a rise and then decline in *Juniperus* (with the inverse happening with *Artemisia* and Amaranthaceae; Fig. 11.12) are plausibly correlated with the advance of mountain glaciers associated with the Little Ice Age (LIA) (Munroe et al., 2012).

11.6 DISCUSSION

Over the last 40,000 years in the Bonneville basin, a wide array has occurred of environmental responses to changes in climate, basin morphology, and perhaps other factors. Lakes, vegetation, and other proxies have different sensitivities to these changes; and as a result, the amplitude, timing, and duration of their responses have varied. During some periods, these proxies appear to respond in similar ways, whereas in other times divergent responses occur. These similarities and differences may provide evidence to the relative importance of lake-effect feedback to Lake Bonneville, groundwater buffering of lake conditions during dry episodes, and other phenomena. In this section, we review our reconstructed changes in lacustrine conditions and vegetation from sediment core data from GSL, compare these changes with other records from the Bonneville basin and surrounding region, and speculate on the paleoclimatic and other factors that may be associated with these lacustrine changes. A broad and deep pool of literature exists regarding paleoenvironmental and paleoclimatic changes in the Great Basin during the time represented in this study—we can only briefly refer to a small number of them in our comparisons with the GSL96+ record, and for greater understanding we refer the reader to the comprehensive and insightful review provided by Grayson (2011).

Most of the pollen analytical results discussed in this report are presented as percentages of the total number of terrestrial pollen grains in a given sample. Examination of the absolute number of pollen grains in individual samples (pollen concentrations, Table 11.3) also provides useful information, particularly in regard to the effects of changes in lake size and depth on the pollen record. Pollen concentrations are much lower in the deepwater phases of Lake Bonneville (pollen zone B, average concentration = 79 k) than in shallow water phases (average concentrations vary from 161 k for pollen zone A to 300 k for pollen zone D). In addition, the percentage data for some pollen taxa have negative correlations with pollen concentration (*Pinus*, $r = -0.30$; *Abies*, −0.29; *Picea*, −0.22), whereas others have positive correlations (Poaceae, $r = 0.36$; Amaranthaceae, 0.30; *Ambrosia*-type, 0.24; *Sarcobatus*, 0.23; "*Cercocarpus*-type and *Quercus*," 0.23). The taxa with negative correlations live primarily in the large mountain ranges on the eastern side of the Bonneville basin, and those with positive correlations generally occur in the desert and steppe environments nearer to GSL. This suggests that during the times when Lake Bonneville covered much of the basin floor, pollen from distant sources dominated the percentage diagram, but the overall amount of pollen reaching the core location was actually relatively low; conversely, when the lake was shallow, pollen from the local environment dominated and pollen concentrations were higher. (Two important taxa have no significant correlations with concentration: *Juniperus* ($r = 0.12$) and *Artemisia* (0.00).)

11.6.1 Conditions Before the Rise of Lake Bonneville

We interpret the basal lithologic unit of GSL96+ (LSM, Table 11.1) as representing a persistent shallow lake from at least 40 cal ka to sometime between ~30 and 28 cal ka. This unit is correlative with pollen zone A, which is strongly dominated by the pollen of *Artemisia*, and which has four subzones that largely are defined by differences in the percentages of *Pinus* (the second most abundant taxon). This combination of *Artemisia* dominance and relatively low and fluctuating levels of conifer pollen is also evident in pollen spectra in this age range from GSL Core C (Spencer et al., 1984, reproduced in part in Fig. 11.11) and Crescent Spring (Mehringer, 1985). We interpret these pollen spectra as representing widespread sagebrush steppe with small populations of subalpine pines in favored habitats scattered across the landscape. The climatic conditions associated with this vegetation would likely be cooler and wetter than today, based on the modern vegetation–climate relations shown in Fig. 11.7, although still quite dry compared with the later climates associated with Lake Bonneville. Amaranthaceae pollen is consistently present at significant levels through pollen zone A (although much lower than in Holocene sediments), also indicating a relatively dry climate.

Packrat (*Neotoma* spp.) middens from the northern Bonneville basin from the time period covered by pollen zone A contain *P. flexilis* and other plants that today live in dry montane to subalpine habitats in the Great Basin (Wigand and Rhode, 2002; Rhode, 2016). For this time period to the south,

Pinus is prominent in the Council Hall Cave pollen record in the Snake Range on the southwest margin of the Bonneville basin, and packrat middens from these and other mountains in the southern Bonneville basin contain the remains of xerophytic subalpine bristlecone pine (*P. longaeva*; Thompson and Mead, 1982; Wells, 1983).

These paleobotanical studies suggest that relatively dry climatic conditions were present prior to the rise of Lake Bonneville. The climate was also cooler than today, although it is difficult to estimate how much cooler, given the breadth of temperate ranges associated with sagebrush communities. However, studies of floristic isolates in the northern Sierra Nevada (Major and Bamberg, 1967) and of alpine phytogeography across the Great Basin (Billings, 1978) have been interpreted as indicating that cold dry steppe with alpine elements was present across much of this region during the Wisconsin glacial episode (marine oxygen isotope stages 4, 3, and 2). Sagebrush steppe or perhaps steppe-tundra was apparently spread across much of the central and northern parts of the interior western United States, from the Pacific Northwest (Barnosky et al., 1987) to southern Idaho (Davis, 1981; Beiswenger, 1991), east-central Nevada (Thompson, 1992), and the Yellowstone Plateau (Baker, 1976). Although the age of the establishment of this widespread sagebrush-dominated vegetation is poorly known, it appears to be present across much of the western interior states prior to the beginning of the GSL96+ pollen record.

11.6.2 The Rise of Lake Bonneville to the Bonneville and Provo Shorelines

The changes in lithology through ALM and MOM sediments (~30–28 to ~14 cal ka) are related to the rise, deep water, and initial drawdown phases of Lake Bonneville. In GSL96+ these sediments are covered by pollen zone B, which includes changes in the pollen diagram that resemble the changes in lake status in the initial part of the Lake Bonneville, but that seemingly diverge for portions of the deepwater phases of the lake. This period covers the advance, maximum, and initial retreat of the North American ice sheets; in the Bonneville basin, montane glaciers began to advance as early as ~32 cal ka in the headwaters of the Bear River in the Uinta Mountains, where they reached their maximum extent by ~24 cal ka (Laabs et al., 2007, 2009). Glaciers in other parts of these mountains remained at their full LGM extents until ~17.4 to ~16.8 cal ka (Munroe et al., 2006; Refsnider et al., 2008). Depending on the proximity of montane glaciers to Lake Bonneville (Laabs et al., 2011), their retreat began as early as 22 to 20 cal ka in the north and east Uinta Mountains (farther from Lake Bonneville) or as late as 18 to 16.5 cal ka in valleys of the south and west Uinta Mountains (closer to Lake Bonneville). Montane glaciers in the Wasatch Range that were very close to Lake Bonneville may have begun their retreats even later (15.7 ± 1.3 ka) and these observations suggest that the lake effect associated with this extremely large body of water had a strong influence on the persistence of nearby montane glaciers (Laabs et al., 2011).

In GSL96+, pollen subzone B1 covers the sharp upturn in the pollen of *Pinus* and other conifers that occurred during the rapid initial rise of Lake Bonneville (~27.0 to 25.8 cal ka), which, based on the core chronology, was later than the shoreline evidence suggests (Fig. 11.2). Subzone B2 represents the first sustained period of conifer dominance in the record (with relatively high levels of *Pinus*, *Picea*, and *Abies*, but not *Juniperus*) that aligns with part of the transgression of Lake Bonneville and the Stansbury oscillation (Fig. 11.2). The strong correspondence between rising lake waters and decreases in the proportions of *Artemisia* and other steppe elements in pollen spectra could be due, at least in part, to the effect of deepening waters covering all of the salt flats and substantial portions of the present-day habitats for saltbush-greasewood and sagebrush (Figs. 11.5 and 11.6).

Subzone B3 (~21.2 to 15.9 cal ka) has sustained relatively high levels of *Picea* and *Abies* pollen (perhaps related to the opening of high-elevation habitats for these taxa as glaciers retreated) and higher representations of *Juniperus* pollen than earlier, but also a major reduction in *Pinus* and increase in *Artemisia*. Interestingly, this apparent turn toward drier conditions in the vegetation record appears to correlate with the final transgression to the Bonneville shoreline, the Bonneville flood, the outflowing into the Snake River during the formation of the Provo shoreline, and the return to closed-basin hydrology after the lake waters receded below the Provo shoreline (Fig. 11.2). A similar sequence of conifer dominance followed by shared conifer-*Artemisia* dominance is present in the Core C pollen record (Fig. 11.11; Spencer et al., 1984), and Mehringer's (1985) core of Lake Bonneville sediments at Crescent Spring, indicating that this unexpected result is not confined to GSL96+.

Other pollen records in the region surrounding the Bonneville basin that cover the time represented by GSL96+ pollen subzones B1, B2, and B3 have relatively low stratigraphic resolution, are poorly dated or have problematic radiocarbon dates, and/or have hiatuses (eg, Council Hall Cave, Nevada [Thompson, 1984], Grays Lake, Idaho [Beiswenger, 1991], and the Ruby Marshes, Nevada [Thompson, 1992]). Consequently, it is not possible to make detailed comparisons between these records and the GSL96+ pollen profile. However, it can be said that these records resemble the GSL record in that pollen spectra of late Wisconsin age are dominated by *Artemisia* and *Pinus* (with *Picea* having significant representations at Council Hall Cave and Grays Lake), whereas the pollen percentages of many plant taxa that are important on the modern landscape are at low levels (including *Juniperus*, *Cercocarpus*-type and *Quercus*, Amaranthaceae, *Sarcobatus*, and Poaceae).

Although the GSL96+ pollen record provides a continuous record of vegetation change through the rise and highstands of Lake Bonneville, we cannot identify which species of plants were present, nor where they grew within the Bonneville basin. Fortunately, plant macrofossils from lacustrine sediments and packrat middens provide some of this information, albeit with sparse coverage in both geography and time. On the eastern side of the basin, where Lake Bonneville met the steep slopes of the Wasatch Range, plant macrofossils preserved in lagoonal and fine-grained alluvial deposits contain wood of *Picea*, along with *Abies* and/or *Juniperus* (Scott et al., 1983) with calibrated ages ranging from ~26.8 to 20.2 cal ka. In addition, a "matrix of clay and ostracod shells" from a shallow embayment of Lake Bonneville when it stood "well below the Provo stage" contains impressions of leaves and twigs of *A. lasiocarpa* (Cottam et al., 1959). Collectively, these plant remains indicate that elements of the modern spruce-fir forests were present on the moist lee side of Lake Bonneville as the lake was rising and montane glaciers were advancing.

Packrat middens provide coverage of vegetation changes in the more arid portions of the Bonneville basin to the west and south of Lake Bonneville during the time represented by GSL96+ pollen subzones B1, B2, and B3. Early in this time, in the northwestern part of the basin, *P. flexilis* grew with *P. engelmannii*, *Artemisia* spp., and shrubs that now live in upper montane settings, a combination of plants that suggests cool and moist conditions during the early rise of Lake Bonneville (~26.5 to 26 cal ka; Rhode, 2000a, 2016; Madsen et al., 2001; Wigand and Rhode, 2002). In the Snake Range of Nevada, in the southwestern sector of the Bonneville basin, packrat middens dated at 25.1 to 21.4 cal ka (Wells, 1983; Thompson, 1984; within the estimated time span of GSL96+ pollen subzone B2) contain *P. longaeva*, *P. flexilis*, *P. engelmannii*, and *Artemisia* spp. *P. longaeva* today lives in open woodlands and forests in relatively dry subalpine settings across the Great Basin, frequently interspersed with species of *Artemisia* (with substrate potentially influencing the amount and distribution of pines on the landscape; Thompson and Mead, 1982). *P. flexilis* and *P. engelmannii* can live in some of these settings but also have ranges extending northward into Canada.

By ~20.6 cal ka (near the beginning of the time represented by pollen subzone B3 in GSL96+), a packrat midden in the northwest Bonneville basin contains only low levels of *P. flexilis* and *P. engelmannii* and is dominated by *Artemisia* spp., with Poaceae, *Potentilla fruticosa*, and other shrubs, an assemblage that Madsen et al. (2001) interpret as signaling a shift to colder climatic conditions. Two middens dated at ~20.7 and 19.8 cal ka (so also of early pollen subzone B3 age) from the Snake Range have *P. longaeva*, *P. engelmannii*, and *Juniperus communis*, but lack *P. flexilis* (Wells, 1983; Thompson, 1984). No packrat middens have been discovered in the Bonneville basin for the middle part of the time represented by pollen subzone B3, with a gap between middens of 20.6 and 16.7 cal ka in the northwestern region of the basin and of 19.8 and 16.4 cal ka in the southwestern region. In both regions, packrat middens are continuously present through the rest of the Pleistocene (after 16.7 to 16.4 cal ka), which brings up the question: Was this gap caused by conditions inhospitable to packrats or by the preservation of packrat middens during time represented by the middle of pollen subzone B3?

By 16.7 cal ka in the northwest Bonneville basin, montane shrub vegetation was present with species of *Artemisia*, *Symphoricarpos*, and *Ribes*, along with *J. communis* and *P. fruticosa* (Rhode and Madsen, 1995; Rhode, 2016). Farther south, middens dated between 16.4 and 15.9 cal ka in the Deep Creek Mountains, Snake Valley, and Snake Range (Thompson, 1984) contain either *P. flexilis* or *P. longaeva*, indicating the presence of open subalpine forest or woodland late in pollen subzone B3 time.

Overall, the packrat midden record, although sparse, complements and enriches the pollen record for the period covered by GSL96+ pollen subzones B1, B2, and B3. In B2 and early B3 times, subalpine species of *Pinus* and *Picea* were present in both the northwest and southwest parts of the Bonneville basin, with a suggestion that conifers declined in importance in the northwest by ~20.6 cal ka, but persisted in the southwest. Although there are very few middens older than ~16.9 cal ka in the Bonneville basin, a gap in the midden record in the middle part of B3 time plausibly could be related to temperatures too cold for the survival of packrats. If true, then the lower *Pinus* and higher *Artemisia* pollen percentages in B3 could relate to expanded cold steppe or steppe-tundra. If this hypothesis is correct, then the increasing number of packrat middens beginning by ~16.7 to 16.4 cal ka (in the upper part of pollen subzone B3) would indicate warming conditions near the end of the time represented by the Provo shoreline (Fig. 11.2).

11.6.3 Regression from the Provo Shoreline and Return to Closed-Basin Hydrology

Lake Bonneville apparently overflowed for thousands of years during the formation of the Provo shoreline, dropped below its outlet threshold by about ~15 cal ka, and declined to a level at or below modern GSL by ~13 cal ka (Fig. 11.2). During this period, the Laurentide Ice Sheet was in full retreat, and montane glaciers in the Bonneville basin had receded far back from their maximal extents. In the GSL96+ pollen record, this interval is covered by pollen subzone B4 (~15.2 to 13.2 cal ka), which is characterized by a sharp rise in *Pinus*, coupled with reductions in *Picea*, *Abies*, *Juniperus*, and *Artemisia* (Figs. 11.8 and 11.9). This subzone has the highest percentages of indeterminate pollen grains and broken *Pinus* pollen grains in the record, perhaps related to more turbulence in the waters and/or rebedding of pollen grains from older deposits as the lake level declined. Although dating controls are inadequate for detailed correlations, pollen subzone B4 plausibly aligns with the last maximum in conifer pollen in Core C of Spencer et al. (1984; Fig. 11.11) and to relatively high levels of *Pinus* pollen in the lowest levels of the Swan Lake (Bright, 1966) and Blue Lake (Louderback and Rhode, 2009) records.

No macrofossil evidence exists on the wetter east side of the Bonneville basin for pollen subzone B4 time, but numerous packrat middens in this age range occur along the western side. These middens provide evidence of open woodland with *P. flexilis* in the north (at elevations from 1510 to 1800 m; Rhode and Madsen, 1995; Rhode, 2000b, 2016; Rhode et al., 2005), and *P. flexilis* and *P. longaeva* in the southwest (at elevations from 1670 to 2340 m; Wells, 1983; Thompson, 1984). Collectively, these midden assemblages reveal a broad area of open woodland dominated by subalpine pines, across a substantial elevational range, living in areas now hosting saltbush-greasewood, sagebrush, and juniper–pinyon woodland communities (Figs. 11.5 and 11.6). Rhode and Madsen (1995) estimated that growing-season temperatures would need to be ~6°C cooler than today for *P. flexilis* to survive at the lower elevation sites in this late Pleistocene woodland. The maximum in *Pinus* pollen in GSL96+ pollen subzone B4 is highly consistent with this vegetation reconstruction, and a strong correspondence occurs between the ages of these packrat middens and the temporal range of subzone B4 in our age model. However, the question remains of the composition of vegetation communities along the eastern side of the Bonneville basin—did the drier conditions associated with the drawdown of Lake Bonneville (and the loss of enhanced precipitation due to the lake effect on the lee side of Lake Bonneville) create conditions that led relatively xerophytic subalpine pines to cover these mountain slopes as well?

11.6.4 Pleistocene–Holocene Transition

Pollen zone C in GSL96+ (~13.1 to 10.7 cal ka; Figs. 11.8 and 11.9) includes the Younger Dryas (YD) (~12.9 to 11.5 cal ka; Carlson, 2013; a cold interval that is most strongly expressed in the North Atlantic Ocean and Europe), as well as the times immediately before and after this paleoclimatic episode. Evidence exists of climatic changes in western North America during the YD, including advances in glaciers in high mountains across the western United States (Menounos and Reasoner, 1997; Davis et al., 2009); rapid shifts in timberline (Reasoner and Jodry, 2000); cooler and/or wetter conditions recorded in speleothems (eg, Polyak et al., 2004; Vacco et al., 2005); and the widespread occurrence of "black mats" that are interpreted as reflecting a brief period of high groundwater discharge related to wetter conditions (Haynes, 2008; Pigati et al., 2009).

In the Bonneville basin, evidence exists of a rise in lake waters during the Gilbert episode, a wetter interval of short duration sandwiched between two periods of lower lake levels within the YD (Oviatt et al., 2005; Oviatt, 2014; Fig. 11.2). Faunal data from Homestead Cave, at 1406 m elevation near the present-day northwestern shoreline of GSL, indicate that freshwater fish living in Lake Bonneville experienced a catastrophic die-off at ~13.2 to 13.1 cal ka related to the final drawdown of the lake (in GSL96+ this is the estimated age of the boundary between pollen zones B and C). After a period of low lake levels, a fish population was reestablished (Broughton et al., 2000), presumably during the Gilbert event, and then died off completely sometime before ~12.4 to 12.3 cal ka (the approximate age of the boundary between pollen subzones C1 and C2).

In pollen subzone C1, the percentages of *Pinus* pollen are significantly reduced from those in the underlying subzone B4, but remain much higher than Holocene levels (pollen zones D and E). *Artemisia* pollen percentages are higher than those in either the older subzone B4 or the younger pollen zones of the Holocene. In the upper two samples of subzone C1 a sharp trend occurs toward declining *Pinus* pollen and rising *Artemisia* pollen (Fig. 11.9), leading into subzone C2, a short period following the Gilbert event with low levels of *Pinus* pollen and very high percentages of *Artemisia* pollen (Table 11.3). Hence, it appears that rise and fall of the lake waters associated with the Gilbert event were associated with a vegetation change that suggests increasingly drier conditions (Fig. 11.7). For a lake rise to occur during this increasing aridity, it follows that evaporation must have been declining, and thus temperatures must have been declining as well.

The boundaries between pollen subzones C1 and C2 and between lithologic units DM and FLM occur at the same stratigraphic position in GSL96+ (Fig. 11.9). The uppermost DM sediments were deposited in relatively freshwater during the Gilbert event, and the overlying FLM sediments contain the remains of brine shrimp and were deposited in shallower hypersaline waters associated with low precipitation and runoff (Oviatt et al., 2015). Subzone C3 (still within FLM sediments) represents a return to somewhat higher levels of *Pinus* and lower *Artemisia*, along with increased percentages of Amaranthaceae and *Sarcobatus*; this combination of changes could indicate temperatures warmer than in subzone C2, but still cooler than during the rest of the Holocene, coupled with a continued trend toward increasing aridity.

Madsen's pollen diagram for Core C from Spencer et al. (1984; Fig. 11.11) appears to have the main details of the changes occurring in pollen zone C in GSL96+, albeit at coarser resolution and with fewer age controls. In both Core C and the pollen profile from Swan Lake (Bright, 1966), a decline occurs in *Pinus* pollen (as in GSL96+ pollen subzone C1), followed by a short period of *Artemisia* dominance (as in subzone C2), and then a final minor increase in *Pinus* as Amaranthaceae and/or *Sarcobatus* rise (as in subzone C3), before settling into approximately mean Holocene pollen percentages. In the pollen profile from Blue Lake (Louderback and Rhode, 2009; Goebel et al., 2011), *Pinus* and *Artemisia* are codominants through the time leading into the YD, with *Artemisia* and then *Pinus* experiencing sharp declines toward the end of YD time (Louderback and Rhode, 2009), followed by a strong increase in Amaranthaceae (relative to *Artemisia*, Goebel et al., 2011) that indicates the establishment of shadscale and saltbush vegetation near the site. At high-elevation sites in the Uinta Mountains, on the eastern periphery of the Bonneville basin, Louderback et al. (2015) found strong shifts in vegetation, commensurate in age and direction of change with a YD fluctuation.

Plant macrofossils from packrat middens (Rhode and Madsen, 1995; Rhode, 2000b, 2016; Rhode and Louderback, 2015) indicate that the common occurrences of *P. flexilis* and *J. communis* at relatively low elevations ($\lesssim$1800 m) ended approximately the same time as the boundary between pollen zones B and C (~13.2 to 13.1 cal ka, although there is a single assemblage containing these species dated at ~12.9 cal ka). However, midden data indicate that these taxa, as well as *P. longaeva*, remained well below their modern lower elevation limits until approximately the age of the boundary between subzones C1 and C2 (~12.4 to 12.1 cal ka; Thompson, 1990; Rhode and Madsen, 1995; Rhode, 2000b). Concurrently, *A. confertifolia* became more abundant in the midden assemblages, reflecting a change from sagebrush to "sagebrush-shadscale scrub" (Rhode and Madsen, 1995).

Faunal records spanning the Pleistocene/Holocene transition in low-elevation sites in the saltbush-greasewood and desert communities west of GSL bolster the interpretations from the GSL96+ pollen record and from the packrat midden data. At Homestead Cave, changes in the small mammal fauna indicate a replacement of dense stands of sagebrush by saltbush-dominated vegetation between ~12.9 and 11.5 cal ka (Grayson, 2000, 2006), which is approximately the age range of pollen subzones C1 and C2 taken together. In the GSL96+ pollen diagram (Figs. 11.7 and 11.8), the first post-Lake Bonneville rise of Amaranthaceae (which includes *Atriplex* spp.) is strongly evident in pollen subzone C3 (~11.8 to 10.7 cal ka by our age model, so approximately the same as Grayson's age estimates for the sagebrush–saltbush transition). However, it should be noted that the transition to dominance by xerophytic small mammals appears to have occurred as late as ~10.1 to ~9.9 cal ka at the other sites in the western Bonneville basin (Schmitt et al., 2002; Schmitt and Lupo, 2012).

11.6.5 Early Holocene (~10.6 to 7.2 cal ka)

In GSL96+, much of the EH is represented by pollen zone D, and the sediments are entirely within the BSM lithologic unit, suggesting lacustrine conditions similar to those of today. These BSM sediments are fairly uniform through the Holocene and do not show any obvious correlation with the chronology of lake-level changes hypothesized by Murchison (1989). However, rivers and wetlands were much more extensive than today (at least in the western Bonneville basin) from ~12.9 to 9.8 ka cal (Oviatt et al., 2003), probably due to Sevier-basin overflow and to groundwater discharge. Based on the seismic studies of GSL sediments, Colman et al. (2002) hypothesized that a potential unconformity may be present at or immediately below the stratigraphic level where the Mazama tephra occurs. They interpreted this feature as indicative that GSL was at a very low level, or perhaps completely dry (a hypothesis that is difficult to evaluate with the GSL96+ record, given the possible occurrence of disturbed and/or missing sediments at the break between Drives 1 and 2 of core GSL96-6).

Two subzones occur in GSL96+ pollen zone D, with subzone D1 representing a relatively short period of time (~10.6 to 10.1 cal ka) during which *Pinus*, *Abies*, and *Picea* are at low levels, and *Juniperus* rises but is still well below the levels of the middle and late Holocene (Figs. 11.8 and 11.12). Among the steppe and desert pollen taxa, *Artemisia* is at higher levels than during most of the subsequent parts of the Holocene, whereas Amaranthaceae is high and rising, and *Ambrosia* type is at its highest percentage levels of the entire GSL96+ pollen record. Subzone D2 covers a longer span of time (~10.1 to 7.2 cal ka), although complications occur in the chronology for this period related to missing or disrupted sediments (discussed earlier). In this subzone, *Pinus* is at very low levels, whereas Amaranthaceae and *Artemisia* are at very high levels, suggesting that this is the driest part of the Holocene as seen in the pollen data.

For more than two decades debate has occurred on the nature of EH climates in this region, because seemingly contradictory evidence exists on EH environmental conditions and their paleoclimatic interpretations. Regrettably, the data are relatively sparse and many of the older studies have coarse resolution and inadequate dating. Some of the disagreements may relate to dating or semantic issues, whereas others seem to point to some unusual juxtapositions of cooler/wetter vs warmer/drier anomalies.

In the greater western United States, evidence occurs that montane glaciers were strongly retracted and (at least in some areas) treelines advanced to higher than modern levels very early in the Holocene (eg, Carrara, 2011, for the San Juan Mountains of southwestern Colorado). Yet, at lower elevations, pollen records, packrat middens, and vertebrate faunas suggest cooler-than-present summer temperatures. In addition, regional differences seem to have occurred, so that the EH appears relatively warm and dry in the Pacific Northwest, but relatively cool and moist in the Great Basin, Colorado Plateau, and other parts of the Southwest (Thompson et al., 1993). In other areas, such as the Yellowstone region of Wyoming and adjacent Idaho and Montana, EH climates exhibited considerable spatial heterogeneity related to how the higher-than-present summer solar radiation was expressed across complex topography and different seasonal precipitation regimes (Whitlock and Bartlein, 1993).

During the EH in the Bonneville basin and surrounding areas, cool and moist conditions (relative to the present) are indicated by a diversity of proxy data, including: faunal data from Homestead Cave (until ~9 cal ka; Madsen et al., 2001); the sizes of *Neotoma* fecal pellets across the northern Great Basin (until ~9 cal ka; Smith and Betancourt, 2003); the persistence of high levels of *Artemisia* in pollen spectra from relatively low elevations at the

Ruby Marshes (Thompson, 1992), Hidden Cave (Wigand and Mehringer, 1985), Gatecliff Shelter (Thompson and Kautz, 1983), and Tule Springs (Mehringer, 1967); the relatively high levels of *Picea* pollen and low levels of *Artemisia* pollen starting at ~9.8 cal ka at the high-elevation Snowbird Bog site in the Wasatch Range (Madsen and Currey, 1979); by a few scattered records of plants living below their modern lower limits in eastern and central Nevada until at least ~10 cal ka (Thompson and Hattori, 1983; Thompson, 1984, 1990); the persistence of mesophilic *Celtis reticulata* at sites where it no longer occurs at Homestead Cave and other sites in the Bonneville basin (declining after ~9.2 cal ka and rare by ~7.8 cal ka; Rhode and Louderback, 2015; Rhode, 2016) stable isotopes in speleothems on the northern (until at least ~9.5 cal ka; Lundeen et al., 2013) and western (until ~8.2 cal ka; Steponaitis et al., 2015) peripheries of the Bonneville basin; and, as discussed earlier, apparently high lake levels at Bear Lake.

In contrast, although the dating controls are sparse, pollen records from north of the Bonneville basin in southern Idaho and adjacent Montana show low levels of conifers and expanding steppe vegetation in the EH, suggesting increasing dry conditions. Relatively dry conditions are inferred from the pollen records from Lower Red Rock Lake, Montana (until ~7.1 cal ka, Mumma et al., 2012), Swan Lake, Idaho (until ~9.5 cal ka, Bright, 1966), and Grays Lake, Idaho (until ~8.2 cal ka, Beiswenger, 1991). Bright and Davis (1982) documented the expansion of shadscale vegetation through the EH on the eastern Snake River Plain of Idaho, culminating at ~7.8 cal ka. At Mission Cross Bog, in the Jarbidge Range of northeastern Nevada, immediately south of these sites in Idaho, pollen spectra dated between ~9.7 and 7.2 cal ka have low levels of *Abies* and *Pinus* pollen, coupled with high levels of Amaranthaceae, *Sarcobatus*, *Ambrosia*-type, and *Juniperus*, suggesting the warmest and driest conditions in this ~9.7 ka record (Thompson, 1984). Farther east, in the Uinta Mountains, pollen studies by Louderback et al. (2015) document a shift from *Picea–Artemisia* parkland to *P. contorta* forests at ~8.5 to 8.4 cal ka, which they interpreted as representing small changes in climate toward drier conditions with warmer winters.

Further evidence of relative aridity during at least part of the EH comes from the record from Blue Lake, on the western edge of the Bonneville basin, where *Pinus* and *Artemisia* pollen percentages decreased during the earliest Holocene, and Amaranthaceae pollen rose to its maximum percentages at ~8.2 cal ka (Louderback and Rhode, 2009). At Blue Lake, the wetlands were largely dried out from ~8.3 to 6.5 cal ka (Louderback and Rhode, 2009) and the ratio of Amaranthaceae to *Artemisia* pollen was at its highest levels for the entire pollen record during this interval (figure 4 in Goebel et al., 2011). Similarly, Stonehouse Meadow in Spring Valley in eastern Nevada was either absent or severely restricted in size from ~8.0 to ~7.5 cal ka (Mensing et al., 2013); and the oldest sediments recovered at Favre Lake in the northern Ruby Mountains have pollen assemblages indicative of relatively warm and dry conditions from ~7.6 to 5.8 cal ka, with the highest levels of Amaranthaceae and Sarcobatus pollen at ~7.2 cal ka (Wahl et al., 2015).

Yet a third pattern is interpreted for the record from Bear Lake, in the northeastern corner of the Bonneville basin, where studies of sediments, diatoms, and pollen suggest a dynamic history of changes in lake level through the EH. Although these indicators are not all in accord (Doner, 2009; Moser and Kimball, 2009; Smoot and Rosenbaum, 2009), a near consensus exists that the lake experienced a low stand from prior to 10 cal ka until ~9.2 cal ka or slightly later, followed by wetter conditions and higher lake levels until sometime between ~8.0 and 7.6 cal ka, and a return to drier conditions and lower lake levels after that time.

It is beyond the scope of this chapter to attempt to resolve the conundrum of whether the EH was relatively wet or dry in and near the Bonneville basin. However, it is worth noting that the interpretation that pollen zone D was the driest part of the GSL96+ record is in accord with some of the other proxy data, but in conflict with other interpretations.

11.6.6 Middle Holocene (~7.2 to 4 cal ka)

North of GSL, following the EH period of minimal glaciation, new glacial advances began as early as 6.9 to 5.6 cal ka (Clague et al., 2009) in southwestern Canada and 6.5 to 5.7 cal ka in Glacier National Park on the Canadian border in Montana (Munroe et al., 2012). In GSL96+, MH pollen subzone E1 (7.1 to 4.3 cal ka) appears to fit within this pattern, with pollen assemblages suggesting cooler/and or moister conditions after the aridity of pollen zone D2 (Fig. 11.12). *Pinus*, *Picea*, and *Abies* remain at low levels, whereas *Juniperus* is at significantly higher levels than in pollen subzone D2 (Table 11.3). However, relative to just pollen zone E, subzone E1 is characterized by mean percentage values of indicators of relative warmth and aridity (*Artemisia*, Amaranthaceae, *Sarcobatus*, and *Ambrosia*-type) that are slightly higher than in the other two subzones and "Other Asteraceae" and Poaceae values are slightly lower. Within subzone E1, standardized *Juniperus* scores are variable near the BSM mean until ~6.1 cal ka, and above the mean for the rest of the subzone (~6.0 to 4.3 cal ka, except for lower values at ~5.7 and 4.8 cal ka). Similarly, the ratio of *Juniperus* to "*Artemisia* and Amaranthaceae" in subzone E1 is higher after ~6.1 cal ka than before that time. In summary, pollen subzone E1 supports a shift to cooler and moister conditions after ~7.2 cal ka, following a long period of EH warmth and aridity. The climate cooled further after ~6.1 cal ka, leading incrementally to the cooler NP represented by subzone E2. As described later, this pattern resembles some of the other paleoclimatic reconstructions for the northeastern Great Basin and farther north, but is in opposition to many other sources of data from the Bonneville basin and sites to the west and south.

During the transition from EH to MH conditions, the general pattern is that sites that were previously cool and moist became warmer and drier, and the previously warm and dry sites became cooler and moister. On the western periphery of the Bonneville basin, Louderback and Rhode (2009) found signs of cooler and/or moister conditions beginning ~6.5 cal ka, following nearly 2000 years of relatively warm and dry conditions at Blue Lake. Here marshes replaced playas and grass meadows, and *Artemisia* increased relative to other, more xerophytic shrubs. As described earlier, pollen studies from north of GSL in southeastern Idaho also show a trend toward cooler and/or wetter conditions. These records have scant age controls and indicate different transition times for the onset of these conditions, ranging from ~9.5 cal to ~7.8 cal ka (Bright, 1966; Bright and Davis, 1982; Beiswenger, 1991).

In contrast, the Homestead Cave faunal record (Grayson, 2000), near the western shore of GSL, provides evidence of extreme aridity from ~8.3 to 5.1 cal ka that caused local extinctions and near-extinctions of animals adapted to wetter conditions and increases in animals adapted to very xeric conditions. On a broader scale, *Ochotona princeps*, a small mammal now restricted to a small number of high-elevation locales in the Great Basin, was spread across low elevations during the late Pleistocene, survived at lower-than-present sites through the EH, but was extirpated from these latter sites by the increased warmth of the MH (Grayson, 2005). Speleothem records suggest that increased aridity began as early as ~8.2 cal ka in the Snake Range (Steponaitis et al., 2015), and that conditions were warm and generally dry in the northern Wasatch Range from ~7.2 to ~4.1 cal ka (Lundeen et al., 2013).

Further support for a warm-dry MH in the eastern Great Basin is provided by the record from the Ruby Marshes, Nevada, west of the Bonneville basin. At this site, deeper-than-modern waters persisted until ~7.7 cal ka, and the early part of the MH was characterized by shallow waters, if not a complete drying out of the marsh complex (Thompson, 1992). This trend was coincident with an expansion of shadscale at the expense of sagebrush in the valleys (suggestive of drier conditions) and with the establishment of juniper–pinyon woodland in the adjacent mountains. This expansion of shadscale and reduction in sagebrush cover is seen at other sites farther west in the Great Basin (eg, Byrne et al., 1979; Wigand and Mehringer, 1985). The waters of the Ruby Marshes deepened after ~5.5 cal ka, as representation of *Artemisia* in the Ruby Marshes pollen diagram increased and that of Amaranthaceae decreased, indicating a return of somewhat cooler and/or moister climatic conditions. A similar, albeit more detailed, chronology is available from Pyramid Lake in western Nevada where Mensing et al. (2004) identified an extremely dry period for most of the time from ~7.6 to ~6.3 cal ka, with aridity decreasing from ~5.0 and 3.5 cal ka.

Many packrat midden studies (eg, Madsen, 1973; Thompson and Hattori, 1983; Thompson, 1984, 1990; Rhode, 2000b) provide evidence of the apparently rapid widespread establishment of juniper–pinyon woodlands across the central latitudes of the Great Basin between ~7.9 and 6.9 cal ka. Although this woodland establishment seems to indicate warmer conditions, *P. flexilis*, *Juniperus scopulorum*, and other montane plants persisted at sites below their modern elevational limits until ~7.0 cal ka (Thompson, 1990; Rhode, 2002b; Wigand and Rhode, 2002).

Studies of paleoclimatic proxy data from high-elevation lakes in the eastern Great Basin generally support the concept that the MH was warmer and drier than today, although these records do not extend back through much of the EH, so it is not possible to compare the relative warmth of the EH and MH. In addition, the timing of changes varies considerably among data sets. Pollen assemblages from Mission Cross Bog in far northeastern Nevada indicate that relatively warm conditions continued until ~4.6 cal ka, with the onset of cooler and moister conditions signaled by increases in *Abies* pollen between ~4.6 and 4.0 cal ka (Thompson, 1984). Farther south, at Favre Lake in the northern Ruby Mountains, relative warmth and aridity only persisted until ~5.8 cal ka, followed by cooler and wetter conditions associated with deeper waters from ~5.5 to 5.0 cal ka and ~4.2 to 3.9 cal ka (Wahl et al., 2015). Still farther south, at Stella Lake in the Snake Range on the southwestern edge of the Bonneville basin, chironomid-based estimates of July air temperatures suggest relatively cool summers from ~6.7 to 6.5 cal ka, followed by generally warm conditions until ~4.2 cal ka, with a short-lived cool episode ~6.2 cal ka and maximum Holocene warmth at ~5.4 cal ka (Reinemann et al., 2009). The hypothesis that MH summers were warmer than those of today is supported by indications that upper treeline was higher than today through much or all of the MH. In the Snake Range, LaMarche and Mooney (1972) found wood remnants dated between 4.5 and 2.2 cal ka at elevations 100 m or more higher than the modern upper treeline. On the western edge of the Great Basin, in the White Mountains of California, LaMarche (1973) found similar evidence of higher treelines beginning as early as ~7.5 cal ka and persisting to as recently as ~4.2 to 3.5 cal ka.

The GSL96+ record may be at or near the geographic transition between a region to the north where maximum Holocene warmth occurred during the EH, and a region to the west (across the Great Basin) where it occurred during the MH. It also may be unique in that it potentially integrates information across the range of environments and elevations present in the Bonneville basin. It may not be possible to determine why the GSL96+ record apparently aligns with the Blue Lake record and other pollen records from north of the Bonneville basin, but less well with the faunal data from Homestead Cave (which is located adjacent to GSL).

11.6.7 Late Holocene NP (~4.0 to 2.0 cal ka)

The onset of cooler temperatures in the LH is seen in glacial advances that occurred from Alaska south to the Rocky Mountains (Porter, 2007), starting as early as ~3.7 cal ka in the Rockies of Canada and the adjacent USA (Munroe et al., 2012). Notable neoglacial events are dated at ~3.3 to 2.7 cal ka and ~2.4 to 2.0 ka over this broad range (Porter, 2007). Faunal evidence from Homestead Cave indicates that during these cool phases GSL experienced an NP that roughly corresponded with the Neoglacial, with the waters of GSL being fresh enough to support fish. Peaks in the frequency of fish bones (*Gila atraria*) in the cave sediments suggest that around ~3.7 cal ka a GSL highstand occurred of undetermined duration (Broughton et al., 2000). Concurrently, cooler and wetter conditions are indicated by the post-MH return to Homestead Cave of *Neotoma cinerea*, *Reithrodontomys megalotis*, and aquatic avifauna; as well as shifts in the relative frequencies of *Dipodomys microps* vs *Dipodomys ordii* (Madsen et al., 2001).

In the GSL96+ pollen record, this return to cooler and moister conditions is reflected in pollen subzone E2, which is characterized by sustained high percentages of *Juniperus* pollen, largely offset by reductions in *Pinus* and in the steppe taxa *Artemisia*, *Ambrosia*-type, and Amaranthaceae (Figs. 11.8 and 11.12; Table 11.3). Within this subzone, the percentages of *Juniperus* pollen are above the Holocene mean for this taxon, with very high sustained values occurring ~3.8 to 3.4 cal ka and ~2.9 to 2.6 cal ka, followed by declining percentages that by ~2.0 cal ka approach the mean value for *Juniperus* during the Holocene (Fig. 11.12). The high levels of *Juniperus* pollen are almost certainly connected to a downslope NP expansion of the ranges of woodland junipers that occurred over a broad area of the northern Great Basin, from eastern Oregon (Wigand, 1987; Mehringer and Wigand, 1990; Wigand and Rhode, 2002) to the Bonneville basin. On the mountain slopes adjacent to GSL, the lower elevation limit of *J. osteosperma* was 50 to 100 m lower than at present between ~3.8 and ~3.1 cal ka (Rhode, 2000b; Madsen et al., 2001; Wigand and Rhode, 2002). In the Cricket Mountains of the southern Bonneville basin, this species also lived below its modern limits at ~2.3 cal ka (Madsen, 1984; Wigand and Rhode, 2002). Although the packrat midden evidence is sparse, the combination of the GSL96+ pollen data, which provide a quasi-continuous record of vegetation change, and the midden macrofossil data, which provide information on the amplitude and location of vegetation change, as well as the species involved, provides a fleshed out depiction of the ecological changes that occurred during subzone E2. This is analogous to the *Pinus* pollen maximum during GSL96+ pollen subzone B4, which appears to relate to the low-elevation occurrences of *P. flexilis* during the late Pleistocene.

Other pollen records in and near GSL have features that are commensurate with GSL96+ subzone E2. Mehringer's (1985) pollen profile from GSL shows consistently high percentages of *Juniperus* pollen between ~4 and 2 cal ka, and the pollen data from Crescent Spring (Mehringer, 1985) show a shift from relatively high Amaranthaceae during the period from ~5.0 to ~3.8 cal ka to lower Amaranthaceae and to higher levels of the more mesophytic *Artemisia* from ~3.8 to ~2.0 cal ka. At Blue Lake, Louderback and

Rhode (2009) identified an interval of cooler climate from ~4.4 to ~3.4 cal ka, with warmer conditions from ~3.4 to ~2.7 cal ka, and a return to cooler conditions from ~2.7 to ~1.5 cal ka. Although the timing does not precisely align, the sequence of changes inferred at Blue Lake is similar to the GSL96+ sequence through subzone E2, and, given the uncertainties involved in dating, they may actually be correlative. The pollen records from Swan Lake (north of GSL; Bright, 1966) and Snowbird Bog (at high elevation to the southeast of GSL; Madsen and Currey, 1979) both indicate relatively cool conditions for at least part of the NP, but age controls at both sites are inadequate for detailed correlations with the GSL96+ record.

At Stella Lake on the southwestern edge of the Bonneville basin, chironomid studies suggest that a cooling trend occurred in summer temperatures after ~5.4 cal ka, with the first cold episode occurring at ~3.8 cal ka and relatively cold conditions persisting until ~1.6 cal ka (Reinemann et al., 2009). To the northwest of GSL, at Mission Cross Bog, Thompson (1984) identified the period from ~4 to 2 cal ka as the coolest and wettest(?) part of the Holocene, as seen in the relatively high levels of *Abies*, *Pinus*, and Poaceae pollen and low levels of *Sarcobatus*, "*Cercocarpus*-type and *Quercus*," and *Juniperus* (at this high-elevation site, *Juniperus* indicates warmer and drier conditions). A subsequent higher resolution study (Norman, 2007; Mensing et al., 2008) of the past 4000 years provides more detail on drought periods not seen in the earlier study; however, a major conclusion remains the same, and the period from ~4 to ~2 cal ka was wetter than any time in the past 2000 year. Wetter conditions are also inferred from high percentages of *Pinus* pollen at Gatecliff Shelter (central Nevada) for the period from ~3.6 to ~2.9 cal ka (Thompson and Kautz, 1983).

However, not all records appear to fit the pattern of cooler and wetter conditions inferred for subzone E2. At Favre Lake (Wahl et al., 2015), pollen, sedimentological, and other data suggest a trend toward cooler and wetter conditions from ~5.8 to ~3.5 cal ka (with deeper water episodes from ~5.5 to ~5.0 cal ka and ~4.2 to ~3.9 cal ka), followed by drier and more variable climates from ~3.5 to ~1.8 cal ka (with relatively high levels of *Sarcobatus* and Amaranthaceae pollen indicating particularly dry conditions for ~3.5 to ~3.3 cal ka and ~2.0 to ~1.8 cal ka). Based on pollen and paleohydrological data, Mensing et al. (2008, 2013) identified several periods of major droughts during the LH across much of the Great Basin, starting as early as ~2.8 cal ka. These events are not clearly evident in the GSL96+ pollen record; however, GSL falls within the 40° to 42°N latitude range where Mensing et al. (2013) placed the boundary between relatively wet vs relatively dry climates for the LH.

Although it appears counterintuitive, the upper elevational limits of subalpine trees were higher than today (which are interpreted as indicating warmer-than-present summer temperatures) through the neopluvial time represented by GSL96+ subzone E2. In the Snake Range on the southwestern edge of the Bonneville basin, upper treeline was at least 100 m higher than today from

~4.5 to ~2.0 cal ka (LaMarche and Mooney, 1972). Although lower than in the MH, upper treelines in the White Mountains of California were higher than at present for most of the Holocene (LaMarche, 1973), with some reduction in summer temperatures beginning between ~3.5 to ~3.0 and ~2.5 cal ka (LaMarche, 1973).

The differences in inferred paleoclimates among sites in the Bonneville basin and surrounding regions are less pronounced in the NP, than in the EH or MH. Nevertheless, differences do occur in this area during the NP, suggesting that the transition from MH to LH conditions was not uniform in trend nor timing across this area. Based on the assumption that the past presence of *Juniperus* at elevations lower than it presently occurs is evidence of cooler and/or moister climates, pollen subzone E2 in GSL96+ fits the dominant pattern for low-elevation sites in this area, with a sustained period of higher effective moisture through the NP.

11.6.8 Late Holocene: The Last 2000 Years

Many paleoclimatic studies of the last 2000 years frame their results in terms of commonly employed terms such as the Medieval warm period and LIA. Regrettably, as discussed in Jones and Mann (2004) this practice can mask much of the variability seen through this time period, and an unfortunate human tendency exists to see what we already believe in our proxy records. In western North America, glaciers continued to advance and retreat through the last 2000 years (see Munroe et al. (2012) for a detailed discussion), with prominent periods of glaciation ~1.7 to ~1.3 cal ka ("the First Millennium" glaciation, FM) and ~0.7 to ~0.1 cal ka (the more extensive "LIA"). Within the Great Basin during this time, both sustained periods occurred of relatively high effective moisture (such as those associated with higher-than-historical lake levels in western Nevada from ~1.5 to ~1.3 cal ka and ~0.9 to ~0.7 cal ka; Adams, 2003), as well as major periods of drought (such as those identified by Mensing et al., 2008, as ending ~1.8, ~1.2, ~0.8, and ~0.6 cal ka). See Wigand (2006) and Grayson (2011) for more detailed discussions of the complexities of the last 2000 years of paleoclimatic variability in the Great Basin region.

After the period of cooler and moister conditions during the 2000 years of GSL96+ subzone E2, the succeeding (and final) subzone (E3) presents a more complex picture with considerably more apparent variability. The increased variability may in part be an artifact: we sampled the core at a relatively coarse interval of depth (7 to 9 cm on average across the three subzones of pollen zone E), and since the accumulation rate is much faster in E3 (and/or the sediments are less compacted) than in the lower subzones, the average temporal spacing between samples varies from ~139 years for E1, ~146 years for E2, and ~66 years for E3. Consequently, the greater variability in E3 may simply be a reflection of its higher sampling resolution.

Within pollen subzone E3, two episodes (~1.8 to 1.7 cal ka and ~0.6 to ~0.2 cal ka) occur of sustained high levels of *Juniperus* pollen that we interpret as representing relatively cool and/or wet climatic conditions. These episodes appear to align temporally with the FM and LIA glacial episodes, and fall outside of the drought periods identified by Mensing et al. (2008). Pollen spectra from the period between the *Juniperus* peaks have lower percentages of *Juniperus* and relatively high levels of Amaranthaceae, which together suggest relatively dry conditions. At the top of E3, *Juniperus* declines to its lowest levels in the subzone and *Artemisia* reaches its highest levels for pollen zone E overall. These changes can plausibly be due to warmer and drier conditions following the LIA, but may also reflect intensifying land use over the past century.

In general, vegetation represented by packrat middens dated to within the last 2 cal ka resembles the present-day native vegetation at the midden localities (Rhode, 2000b), although some middens contain *Juniperus* below its modern lower limits (Rhode, 2000b; Madsen et al., 2001). It appears that the lower limit of this taxon fluctuated through subzone E3 time, but it is not possible to align the high E3 *Juniperus* pollen episodes with the midden records. In addition, fish remains from Homestead Cave suggest a rise of GSL at ~1.1 cal ka (Madsen et al., 2001), which also falls outside of the high *Juniperus* episodes in subzone E3.

Many of the published pollen records in the Bonneville basin and surrounding region have low sampling resolution for the last 2000 years, and they also do not present a unified picture on paleoclimatic changes over this time. For example, the record from Swan Lake (Bright, 1966) has higher percentages of *Artemisia* and lower percentages of conifer pollen than earlier, suggesting warmer conditions than during the preceding 2000 years (and this is roughly similar to the differences between GSL96+ subzones E2 and E3). In contrast, the nearby Grays Lake pollen record (Beiswenger, 1991) has higher percentages of the pollen of montane conifers in the last 2000 years, and lower representation of *Juniperus* and steppe taxa, which seem to give the opposite signal to the Swan Lake record.

For the last 2000 years at Mission Cross Bog, Thompson (1984) found increased *Pinus* and *Ceanothus* and the lowest values in the record of Amaranthaceae, *Juniperus*, and Poaceae, with the interpretation that conditions were warmer than those of 4 to 2 cal ka, but cooler/wetter than prior to ~4.7 cal ka. A higher resolution study of the past 4 ka in this bog (Allan, 2003; Mensing et al., 2008) also found that the period from ~2 cal ka to the present was generally drier than, and had more extended droughts than, the prior 2000 years. The opposite interpretation was made for the Ruby Marshes (Thompson, 1992) where higher values of *Artemisia* relative to Amaranthaceae, together with apparently deeper waters, suggested cool and moist conditions over the past 500 years. Yet another pattern was identified at Favre Lake, where Wahl et al. (2015) found the last ~1.8 cal ka of the record to be more variable and wetter than earlier, but with lower lake levels occurring between ~1.0 and 0.75 cal ka.

In the Snake Range, the chironomid-based reconstruction from Stella Lake has relatively cold summer temperatures until ~1.6 cal ka, with the coldest temperatures occurring between ~1.8 and ~1.6 cal ka. Following this episode, a warming trend occurs, with the exception of a cold excursion at ~1.2 cal ka (Reinemann et al., 2009). At higher elevations of the Snake Range, LaMarche and Mooney (1972) found no evidence of higher-than-present upper treelines over the past 2000 years. However, in the White Mountains of California, treelines remained above present levels until ~0.9 cal ka and then declined to at least modern levels by ~0.5 cal ka (LaMarche, 1973). Similarly, in Colorado, higher-than-present upper treelines were present from prior to 2 cal ka until ~0.8 cal ka (Carrara and McGeehin, 2015).

To summarize, most paleoclimatic studies in and near the Bonneville basin are in agreement that the last 2000 years have been generally warmer and drier, and probably more variable, than the preceding 2000 years. The sequence of paleoclimatic changes that we interpret from pollen subzone E3 is commensurate with the pattern of glacial advances and retreats seen in the North America Cordillera and with the broad narrative of change for the Northern Hemisphere for the past 2000 years. However, it is difficult to make any detailed comparisons within the Bonneville basin region, given limitations of chronological controls, sampling resolutions, and apparently differing sensitivities and rates of responses for the various paleoclimatic proxy data.

11.7 SUMMARY AND CONCLUSIONS

The sediments recovered in GSL96+ contain an archive of the landscape history of the Bonneville basin covering the past ~40 cal ka. Over this span of time, the waters covering the core site varied in depth from less than 10 m to more than 300 m, and the amount of area covered by the different lake phases has varied from less than that of present-day GSL to most of western Utah and parts of adjacent states. Although vegetation cover also has changed over this period, the timing, amplitude, and direction of these changes were at times different from those inferred from lake-level changes. The paleolakes and vegetation were both responding to the large paleoclimatic changes related to the growth, still stands, and decline of continental-scale ice sheets. In addition, changes in lake level may have a direct effect on vegetation by covering or uncovering low-elevation habitats (which today host saltbush-greasewood, sagebrush, and other desert, steppe, and woodland communities). However, intrinsic aspects of each also occurred that conditioned their responses to climatic variations: for lakes these could include the lake effect on precipitation and feedback into the lake, and changes in drainage patterns in the Bonneville basin; and for vegetation, the direct effects of changing levels of carbon dioxide in the atmosphere, ecosystem dynamics, and disturbance regimes.

The GSL96+ core data indicate that for more than 10 cal ka prior to the rise of Lake Bonneville, a shallow saline lake was present at the coring site. The pollen profile (Fig. 11.8) shows strong dominance of *Artemisia* with fluctuating, but generally low levels of *Pinus*, *Juniperus*, and Amaranthaceae for this pre-Bonneville period. The inferred vegetation is a large expanse of sagebrush steppe or steppe-tundra, with scattered pines and junipers coming and going under climatic variations that were too minor to have much effect on lake levels. The climate was drier than during the subsequent Lake Bonneville period, but cooler than the Holocene.

At some point in the age range of ~30 to ~28 cal ka abrupt changes in both lithology and pollen percentages indicate a change toward wetter climatic conditions that led to the rise of Lake Bonneville and to a shift to conifer dominance in the pollen diagram. The sediments deposited from the beginning of the lake rise through the Stansbury oscillation (pollen subzones B1 and B2, ~27 to ~21.5 cal ka) have high percentages of *Pinus*, *Picea*, and *Abies* pollen; but as the lake continued to rise toward its highest levels (pollen subzone B3, ~21.2 to ~15.9 cal ka), *Pinus* percentages declined (although they remained well above pre-Bonneville or Holocene levels) and *Artemisia* resurged (which could indicate somewhat drier but cold conditions). During the final phases of Lake Bonneville (pollen subzone B4, ~15.2 to ~13.2 cal ka), an apparently counterintuitive increase in *Pinus* pollen (and concomitant decrease in the pollen of *Artemisia*) occurred as the lake regressed from the Provo shoreline. This rise in *Pinus* pollen correlates with the spread of relatively cold-tolerant and xerophytic *P. flexilis* (as seen in the packrat midden record) at relatively low levels around the declining lake.

During the YD (~12.9 to 11.5 cal ka, which falls within pollen zone C [~13.1 to ~10.7 cal ka]), after a period of lake levels at, or below, those of the present day, a final short-lived rise in lake waters occurred during the Gilbert episode. This rise was associated with higher-than-Holocene levels of *Pinus* pollen (declining at this time from earlier levels) and *Artemisia* (rising at this time); and low levels of Amaranthaceae, *Sarcobatus*, and *Ambrosia*-type. The subsequent decline of lake waters following the Gilbert episode to a shallow stratified lake is associated with a brief period of relatively cold and dry conditions indicated by high levels *Artemisia* pollen and low levels of conifer pollen (apparently still within the YD). The stratified lake continued into the earliest part of the Holocene, while *Artemisia* started to decline, *Pinus* experienced a minor rise, and a trend began toward increasing levels of pollen types representative of saltbush-greasewood and mountain mahogany-oak communities.

At ~10.6 cal ka, the stratification of the waters of the lake ceased, and the sediments being deposited changed from FLM to BSM, signaling conditions similar to the present-day hypersaline GSL. From that time forward, although no strong sedimentary evidence of changes in lake depth or size is apparent, the GSL96+ pollen record indicates continuing changes in climate. Pollen spectra from the lower section of BSM sediments (~ 10.6 to 7.2 cal ka) have the highest percentages of Amaranthaceae in the record, and relatively high levels of *Artemisia*, *Ambrosia*, and *Sarcobatus*, which we interpret as representative of very dry conditions. *Juniperus* pollen rises through this part of the Holocene, as *Pinus* declines to low levels.

The upper ~83% of the recovered BSM sediments were deposited in the last ~7.1 cal ka. The pollen changes through this period are subtle compared with the late Pleistocene changes, but do present a coherent picture of vegetation and climatic change that can be divided into three periods. First, between ~7.1 and ~4.3 cal ka, *Juniperus* pollen percentages rose to much higher levels than previously, *Pinus* remained low, and *Artemisia*, Amaranthaceae, and *Ambrosia*-type declined. From these changes, we infer that juniper woodlands were becoming established on the lower mountain slopes surrounding GSL and climatic conditions were wetter than the earlier part of the Holocene. Second, during the NP (~4 to ~2 cal ka), *Juniperus* pollen reached its maximum percentages in the in GSL96+ record, and *Artemisia*, *Ambrosia*-type, Amaranthaceae were at their lowest average levels for the Holocene. Packrat middens from this interval of time provide evidence that junipers grew at lower elevations than today and were closer to GSL. Relatively cool and moist conditions are inferred from the pollen and midden evidence. Finally, the pollen data for the last 2 cal ka are seemingly more variable than earlier in the Holocene. A general decline occurs in *Juniperus* pollen and small upticks in average values of *Artemisia*, *Ambrosia*-type, and Amaranthaceae, which suggests a return to drier conditions.

ACKNOWLEDGMENTS

We thank Kerry Kelts for his participation and insights in the early part of this project, and Don Currey, Don Grayson, Blair Jones, Dave Madsen, Dave Rhode, and Dave Miller for background information and helpful discussions over the years. Jenny Buchner (USGS), along with Dawn Graber and staff at the Limnological Research Center (University of Minnesota), provided help in both the field and laboratory; Rob Baskin provided equipment and field support; magnetic susceptibility analyses were carried out by Joe Rosenbaum, Rich Reynolds, and Andy Roberts; Rick Forester examined ostracod samples; Mike Perkins and Barbara Nash identified tephra samples; Gary Skipp, Carol Ann Chapman, and Wendell Jolly and other students at Kansas State University aided in sedimentological and mineralogical examinations; Richard Pelltier provided GIS support. Scott Minor, Dave Rhode, Janet Slate, and Laura Strickland provided very thoughtful and helpful reviews.

REFERENCES

Adams, K.D., 2003. Age and paleoclimatic significance of late Holocene lakes in the Carson Sink, NV, USA. Quat. Res. 60, 294–306.

Allan, M., 2003. A 2000 year record of vegetation and climate change in the Jarbidge Mountains of northeastern Nevada. M.S. Thesis, University of Nevada, Reno.

Arnow, T., Stephens, D.W., 1990. Hydrologic characteristics of the Great Salt Lake, Utah, 1847–1986. US Geological Survey water-supply paper [WSP number].

Baker, R.G., 1976. Late Quaternary vegetation history of the Yellowstone Basin, Wyoming. USGS Professional Paper 729-E.

Balch, D.P., Cohen, A.S., Schnurrenberger, D.W., Haskell, B.J., Valero Garces, B.L., Beck, J.W., Cheng, H., Edwards, R.L., 2005. Ecosystem and paleohydrological response to Quaternary climate change in the Bonneville Basin, Utah. Palaeogeogr. Palaeoclimatol. Palaeoecol. 221 (1), 99–122.

Barnosky, C.W., Anderson, P.M., Bartlein, P.J., 1987. The northwestern US during deglaciation; vegetational history and paleoclimatic implications. In: Ruddiman, W.F., Wright, H.E. (Eds.), North America and Adjacent Oceans During the Last Deglaciation. Geological Society of America, Boulder, pp. 289–321.

Beiswenger, J.M., 1991. Late quaternary vegetational history of Grays Lake, Idaho. Ecol. Monogr. 61, 165–182.

Bennett, K.D., 2005. Documentation for psimpoll 4.25 and pscomb 1.03: C programs for plotting pollen diagrams and analysing pollen data. Department of Earth Sciences, University of Uppsala.

Benson, L.V., Lund, S.P., Smoot, J.P., Rhode, D.E., Spencer, R.J., Verosub, K.L., Louderback, L.A., Johnson, C.A., Rye, R.O., Negrini, R.M., 2011. The rise and fall of Lake Bonneville between 45 and 10.5 ka. Quat. Int. 235 (1), 57–69.

Billings, W.D., 1951. Vegetational zonation in the Great Basin of western North America. In International Union Société Biologique Series B, 9: Les Bases Écologiques de la Regénération de la Végétation des Zones Arides, pp. 101–122.

Billings, W.D., 1954. Temperature inversions in the Pinyon-Juniper zone of a Nevada Mountain Range. Butler Univ. Bot. Studies 11, 112–118.

Billings, W.D., 1978. Alpine phytogeography across the Great Basin. Great Basin Nat. Memoirs 2, 105–117.

Blaauw, M., Christen, J.A., 2011. Flexible Paleoclimate age-depth models using an autoregressive gamma process. Bayesian Anal. 6, 457–474. http://dx.doi.org/10.1214/11-BA618.

Bright, R.C., 1966. Pollen and seed stratigraphy of Swan Lake, southeastern Idaho: its relation to regional vegetational history and to Lake Bonneville history. Tebiwa 9 (2), 1–47.

Bright, R.C., Davis, O.K., 1982. Quaternary paleoecology of the Idaho National Engineering Laboratory, Snake River Plain, Idaho. Am. Midl. Nat. 108, 21–33.

Broecker, W.S., Kaufman, A., 1965. Radiocarbon chronology of Lake Lahontan and Lake Bonneville II, Great Basin. Geol. Soc. Am. Bull. 76 (5), 537–566.

Broecker, W.S., Orr, P.C., 1958. Radiocarbon chronology of Lake Lahontan and Lake Bonneville. Geol. Soc. Am. Bull. 69 (8), 1009–1032.

Broecker, W.S., Mcgee, D., Adams, K.D., Cheng, H., Edwards, R.L., Oviatt, C.G., Quade, J., 2010. A great Basin-wide dry episode during the first half of the mystery interval? Quat. Sci. Rev. 28 (25–26), 2557–2563.

Broughton, J.M., Madsen, D.B., Quade, J., 2000. Fish remains from Homestead Cave and lake levels of the past 13,000 years in the Bonneville Basin. Quat. Res. 53 (3), 392–401.

Byrne, R., Busby, C., Heizer, R.F., 1979. The altithermal revisited: pollen evidence from Leonard Rockshelter. J. Calif. Gt. Basin Anthropol. 1, 280–294.

Carlson, A.E., 2013. The Younger Dryas climate event. In: Elias, S.A. (Ed.), The Encyclopedia of Quaternary Science, vol. 3. Elsevier, Amsterdam, pp. 126–134.

Carrara, P., 2011. Deglaciation and postglacial treeline fluctuation in the Northern San Juan Mountains, Colorado. U.S. Geological Survey Professional Paper 1782.

Carrara, P.E., McGeehin, J.P., 2015. Evidence of a higher late-Holocene treeline along the Continental Divide in central Colorado. The Holocene 25 (11), 1829–1837. http://dx.doi.org/10.1177/0959683615591353.

Clague, J.J., Menounos, B., Osborn, G., Luckman, B.H., Koch, J., 2009. Nomenclature and resolution in Holocene glacial chronologies. Quat. Sci. Rev. 28, 2231–2238. http://dx.doi.org/10.1016/j.quascirev.2008.11.016.

Cole, G.A., Brown, R.J., 1967. The chemistry of *Artemia* habitats. Ecology 48, 858–861.

Colman, S.M., Kelts, K.R., Dinter, D.A., 2002. Depositional history and neotectonics in Great Salt Lake, Utah, from high-resolution seismic stratigraphy. Sediment. Geol. 148 (1–2), 61–78.

Cottam, W.P., Tucker, J.M., Drobnick, R., 1959. Some clues to Great Basin postpluvial climates provided by oak distributions. Ecology 40, 361–377.

Currey, D.R., 1980. Coastal geomorphology of Great Salt Lake and vicinity. In: Gwynn, J.W. (Ed.), Great Salt Lake, a Scientific, Historical and Economic Overview. Utah Geol. Min. Surv. Bull., vol. 116, pp. 69–82.

Currey, D.R., 1982. Lake Bonneville: Selected features of relevance to neotectonic analysis. U.S. Geol. Surv. Open-File Rep. 82–1070, 30 pp. (and 1:500,000 map).

Currey, D.R., 1990. Quaternary palaeolakes in the evolution of semidesert basins, with special emphasis on Lake Bonneville and the Great Basin, USA. Palaeogeogr. Palaeoclimatol. Palaeoecol. 76 (3), 189–214.

Currey, D.R., Atwood, G., Mabey, D.R., Roy, J.S., Brown, K.D., 1984. Major Levels of Great Salt Lake and Lake Bonneville. Utah Geological and Mineral Survey, Salt Lake City, Utah.

Davis, O.K., 1981. Vegetation migration in southern Idaho during the Late Quaternary and Holocene. Ph.D. Dissertation, University of Minnesota, Minneapolis.

Davis, O.K., 1998. Palynological evidence for vegetation cycles in a 1.5 million year pollen record from the Great Salt Lake, Utah, USA. Palaeogeogr. Palaeoclimatol. Palaeoecol. 138, 175–185.

Davis, O.K., 2001. *Cercocarpus* type. http://www.geo.arizona.edu/palynology//pid00031.html.

Davis, O.K., 2002. Late Neogene environmental history of the northern Bonneville basin: a review of palynological studies. In: Hershler, R., Madsen, D.B., Currey, D.R. (Eds.), Great Basin Aquatic Systems History. Smithsonian Contributions to the Earth Sciences, vol. 33, pp. 295–307.

Davis, O.K., Moutoux, T.E., 1998. Tertiary and Quaternary vegetation history of the Great Salt Lake, Utah, USA. J. Paleolimnol. 19, 417–427.

Davis, P.T., Menounos, B., Osborn, G., 2009. Holocene and latest Pleistocene alpine glacier fluctuations: a global perspective. Quat. Sci. Rev. 28, 2021–2033.

Doner, L.A., 2009. A 19,000-year vegetation and climate record for Bear Lake, Utah and Idaho. Geol. Soc. Am. Spec. Pap. 450, 217–227.

Eardley, A.J., 1967. Bonneville chronology: correlation between the exposed stratigraphic record and the subsurface sedimentary succession. Geol. Soc. Am. Bull. 78 (7), 907–910.

Eardley, A.J., Gvosdetsky, V., 1960. Analysis of Pleistocene core from Great Salt Lake, Utah. Geol. Soc. Am. Bull. 71, 1323–1344.

Eardley, A.J., Gvosdetsky, V., Marsell, R.E., 1957. Hydrology of Lake Bonneville and sediments and soils of its basin. Geol. Soc. Am. Bull. 68, 1141–1202.

Eardley, A.J., Shuey, R.T., Gvosdetsky, V., Nash, W.P., Picard, M.D., Grey, D.C., Kukla, G.J., 1973. Lake cycles in the Bonneville basin, Utah. Geol. Soc. Am. Bull. 84, 211–216.

Faegri, K., Iversen, J., 1975. Textbook of Pollen Analysis, third ed. Blackwell Scientific, Oxford.

Forester, R.M., 1987. Late Quaternary paleoclimate records from lacustrine ostracodes. In: Ruddiman, W.F., Wright Jr., H.E. (Eds.), North America and Adjacent Oceans During

the Last Deglaciation. Geology of North America, vol. K-3. Geological Society of America, Boulder, CO, pp. 261–276.
Gilbert, G.K., 1890. Lake Bonneville. U.S. Geological Survey Monograph 1, 438 p.
Godsey, H.S., Currey, D.R., Chan, M.A., 2005. New evidence for an extended occupation of the Provo shoreline and implications for regional climate change, Pleistocene Lake Bonneville, Utah, USA. Quat. Res. 63 (2), 212–223.
Godsey, H.S., Oviatt, C.G., Miller, D.M., Chan, M.A., 2011. Stratigraphy and chronology of offshore to nearshore deposits associated with the Provo shoreline, Pleistocene Lake Bonneville, Utah. Palaeogeogr. Palaeoclimatol. Palaeoecol. 310 (3–4), 442–450.
Goebel, T., Hockett, B., Adams, K.D., Rhode, D., Graf, K., 2011. Climate, environment, and humans during the Younger Dryas, 12,900—11,600 calendar years ago. Quat. Int. 242, 479–501.
Grayson, D.K., 2000. Mammalian responses to middle Holocene climatic change in the Great Basin of the western United States. J. Biogeogr. 27, 181–192.
Grayson, D.K., 2005. A brief history of Great Basin pikas. J. Biogeogr. 32, 2103–2111.
Grayson, D.K., 2006. The Late Quaternary biogeographic histories of some Great Basin mammals (western USA). Quat. Sci. Rev. 25 (21–22), 2964–2991.
Grayson, D.K., 2011. The Great Basin: A Natural History. University of California Press, Berkeley, Los Angeles, London. 418 p.
Grimm, E.C., 1987. CONISS: a FORTRAN 77 program for stratigraphically constrained cluster analysis by the method of incremental sum of squares. Comput. Geosci. 13 (1), 13–35.
Haynes Jr., C.V., 2008. Younger Dryas "black mats" and the Rancholabrean termination in North America. Proc. Natl. Acad. Sci. U.S.A. 105, 6520–6525.
Hostetler, S.W., Giorgi, F., Bates, G.T., Bartlein, P.J., 1994. Lake-atmosphere feedbacks associated with paleolakes Bonneville and Lahontan. Science 668, 665–668.
Jones, P.D., Mann, M.E., 2004. Climate over past millennia. Rev. Geophys. 42, RG2002. http://dx.doi.org/10.1029/2003RG000143.
Kowalewska, A., Cohen, A.S., 1998. Reconstruction of paleoenvironments of the Great Salt Lake Basin during the late Cenozoic. J. Paleolimnol. 20 (4), 381–407.
Küchler, A.W., 1964. Potential Natural Vegetation of the Conterminous United States (Map). American Geographical Society, New York City.
Küchler, A.W., 1985. Potential Natural Vegetation. Reston, Virginia, National Atlas of the United States of America. Department of the Interior, U.S. Geological Survey, Reston, Virginia.
Laabs, B.J.C., Munroe, J.S., Rosenbaum, J.G., Refsnider, K.A., Mickelson, D.M., Singer, B.S., Caffee, M.W., 2007. Chronology of the last glacial maximum in the upper Bear River Basin, Utah. Arct. Antarct. Alp. Res. 39 (4), 537–548.
Laabs, B.J.C., Refsnider, K.A., Munroe, J.S., Mickelson, D.M., Applegate, P.J., Singer, B.S., Caffee, M.W., 2009. Latest Pleistocene glacial chronology of the Uinta Mountains: support for moisture-driven asynchrony of the last deglaciation. Quat. Sci. Rev. 28, 1171–1187.
Laabs, B.J.C., Marchetti, D.W., Munroe, J.S., Refsnider, K.A., Gosse, J.C., Lips, E.W., Becker, R.A., Mickelson, D.M., Singer, B.S., 2011. Chronology of latest Pleistocene mountain glaciation in the western Wasatch Mountains, Utah, USA. Quat. Res. 76, 272–284.
LaMarche, V.C., 1973. Holocene climatic variations inferred from treeline fluctuations in the White Mountains, California. Quat. Res. 3, 632–660.
LaMarche, V.C., Mooney, H.A., 1972. Recent climatic change and development of the bristlecone pine (P. longaeva Bailey) krummholz zone, Mt. Washington, Nevada. Arct. Alp. Res. 4, 61–72.
Louderback, L.A., Rhode, D.E., 2009. 15,000 Years of vegetation change in the Bonneville basin: the Blue Lake pollen record. Quat. Sci. Rev. 28, 308–326.

Louderback, L.A., Rhode, D., Madsen, D.B., Metcalf, M., 2015. Rapid vegetation shifts in the Uinta Mountains (Utah and Wyoming, USA) during the Late Pleistocene and Holocene. Palaeogeogr. Palaeoclimatol. Palaeoecol. 438, 327–343. http://dx.doi.org/10.1016/j.palaeo.2015.08.026.

Lundeen, Z., Brunelle, A., Burns, S.J., Polyak, V., Asmerom, Y., 2013. A speleothem record of Holocene paleoclimate from the northern Wasatch Mountains, southeast Idaho, USA. Quat. Int. 310, 83–95.

Madsen, D.B., 1973. Late Quaternary paleoecology in the Southeastern Great Basin. Ph.D. Dissertation, University of Missouri.

Madsen, D.B., 1984. Pollen, packrats, and pine along the IPP. In: Simms, S.R., Isgreen, M.C. (Eds.), Archaeological Excavations in the Sevier and Escalante Deserts, Western UtahUniversity of Utah Archaeological Center Reports of Investigations 83-12. University of Utah Archaeological Center, Salt Lake City, Utah, pp. 336–353.

Madsen, D.B., Currey, D.R., 1979. Late Quaternary glacial and vegetation changes, Little Cottonwood Canyon area, Wasatch Mountains, Utah. Quat. Res. 12 (2), 254–270.

Madsen, D.B., Rhode, D., Grayson, D.K., Broughton, J.M., Livingston, S.D., Hunt, J., Schmitt, D.N., Shaver, M.W., 2001. Late Quaternary environmental change in the Bonneville Basin, western USA. Palaeogeogr. Palaeoclimatol. Palaeoecol. 167, 243–271.

Major, J.A., Bamberg, S.A., 1967. Some cordilleran plants disjunct in the Sierra Nevada of California and their bearing on Pleistocene ecological conditions. In: Wright Jr., H.E., Osborn, W.H. (Eds.), Arctic and Alpine Environments. Indiana University Press, Bloomington, pp. 171–189.

Malde, H.E., 1968. The catastrophic late Pleistocene Bonneville flood in the Snake River Plain, Idaho. U.S. Geological Survey Professional Paper 596, 68 p.

McGee, D., Quade, J., Edwards, R.L., Broecker, W.S., Cheng, H., Reiners, P.W., Evenson, N., 2012. Lacustrine cave carbonates: novel archives of paleohydrologic change in the Bonneville Basin (Utah, USA). Earth Planet. Sci. Lett. 351–352, 182–194.

Mehringer Jr., P.J., 1967. Pollen analysis of the Tule Springs Site, Nevada. In: Wormington, H.M., Ellis, D. (Eds.), Pleistocene Studies in Southern Nevada. Nevada State Museum Anthropological Papers 13. Nevada State Museum. Carson City, Nevada, pp. 130–200.

Mehringer Jr., P.J., 1985. Late-Quaternary pollen records from the interior Pacific Northwest and northern Great Basin of the United States. In: Holloway, R.G., Bryant Jr., V.M. (Eds.), Pollen Records of Late-Quaternary North American Sediments. American Association of Stratigraphic Palynologists, Dallas, Texas, pp. 167–189.

Mehringer Jr., P.J., Wigand, P.E., 1990. Comparison of late Holocene environments from woodrat middens and pollen. In: Betancourt, J.L., Van Devender, T.R., Martin, P.S. (Eds.), Packrat Middens: The Last 40,000 Years of Biotic Change. University of Arizona Press, Tucson, pp. 294–325.

Menounos, B., Reasoner, M.A., 1997. Evidence for cirque glaciation in the Colorado Front range during the Younger Dryas chronozone. Quat. Res. 48, 38–47.

Mensing, S.A., Benson, L.V., Kashgarian, M., Lund, S., 2004. A Holocene pollen record of persistent droughts from Pyramid Lake, Nevada, USA. Quat. Res. 62, 29–38.

Mensing, S.A., Smith, J.M., Norman, K., Allan, M., 2008. Extended drought in the Great Basin of western North America in the last two millennia reconstructed from pollen records. Quat. Int. 188, 79–89.

Mensing, S.A., Sharpe, S.E., Tunno, I., Sada, D.W., Thomas, J.M., Starratt, S., Smith, J., 2013. The Late Holocene Dry Period: multiproxy evidence for an extended drought between 2800 and 1850 cal yr BP across the central Great Basin, USA. Quat. Sci. Rev. 78, 266–282.

Miller, D.M., Oviatt, C.G., Nash, B.P., 2008. Late Pleistocene Hansel Valley basaltic ash, northern Lake Bonneville, Utah, USA. Quat. Int. 178, 238–245.

Miller, D.M., Oviatt, C.G., McGeehin, J.P., 2013. Stratigraphy and chronology of Provo shoreline deposits and lake-level implications, late Pleistocene Lake Bonneville, eastern Great Basin, USA. Boreas 42, 342–361.

Miller, D.M., Wahl, D.B., McGeehin, J.P., Rosario, J., Oviatt, C.G., Anderson, L., Presnetsova, L., 2015. Limiting age for the Provo shoreline of Lake Bonneville, Quaternary International. 387, pp. 99–105.

Morrison, R.B., Frye, J.C., 1965. Correlation of the middle and late Quaternary successions of the Lake Lahontan, Lake Bonneville, Rocky Mountain (Wasatch Range), southern Great Plains, and eastern midwest areas. Mackay School of Mines, University of Nevada, Reno, Nevada.

Moser, K.A., Kimball, J.P., 2009. A 19,000-year record of hydrologic and climatic change inferred from diatoms from Bear Lake, Utah and Idaho. Geol. Soc. Am. Spec. Pap. 450, 229–246.

Mumma, S.A., Whitlock, C., Pierce, K., 2012. A 28,000 year history of vegetation and climate from Lower Red Rock Lake, Centennial Valley, Southwestern Montana, USA. Palaeogeogr. Palaeoclimatol. Palaeoecol. 326–328, 30–41.

Munroe, J.S., Laabs, B.J.C., Shakun, J.D., Singer, B.S., Mickelson, D.M., Refsnider, K.A., Caffee, M.W., 2006. Latest Pleistocene advance of alpine glaciers in the southwestern Uinta Mountains, Utah, USA: evidence for the influence of local moisture sources. Geology 34, 841–844.

Munroe, J.S., Crocker, T.A., Giesche, A.M., Rahlson, L.E., Duran, L.T., Bigl, M.F., Laabs, B.J.C., 2012. A lacustrine-based Neoglacial record for Glacier National Park, Montana, USA. Quat. Sci. Rev. 53, 39–54.

Murchison, S.B., 1989. Fluctuation history of the Great Salt Lake, Utah, during the last 13,000 years. Ph.D. Dissertation, University of Utah.

Norman, K., 2007. A high resolution re-examination of vegetation and climate change in the Jarbidge Mountains of Northeastern Nevada from 4000 to 2000 cal yr BP. M.S. Thesis, University of Nevada, Reno. 100 pp.

O'Connor, J.E., 1993. Hydrology, hydraulics, and geomorphology of the Bonneville flood. Geological Society of America Special Paper 274. 83 p.

Oviatt, C.G., 2014. The Gilbert Episode in the Great Salt Lake Basin, Utah. Utah Geological Survey Miscellaneous Publication 14-3. 20 p.

Oviatt, C.G., 2015a. Chronology of Lake Bonneville, 30,000 to 10,000 yr B.P. Quat. Sci. Rev. 110, 166–171.

Oviatt, C.G., 2015b. Ostracodes in Pleistocene Lake Bonneville, eastern Great Basin, North America. Hydrobiologia 1–11. http://dx.doi.org/10.1007/s10750-015-2483-y.

Oviatt, C.G., Currey, D.R., Miller, D.M., 1990. Age and paleoclimatic significance of the Stansbury shoreline of Lake Bonneville, northeastern Great Basin. Quat. Res. 33 (3), 291–305.

Oviatt, C.G., Currey, D.R., Sack, D., 1992. Radiocarbon chronology of Lake Bonneville, eastern Great Basin, USA. Palaeogeogr. Palaeoclimatol. Palaeoecol. 99 (3–4), 225–241.

Oviatt, C.G., Habiger, G.D., Hay, J.E., 1994. Variation in the composition of Lake Bonneville marl: a potential key to lake-level fluctuations and paleoclimate. J. Paleolimnol. 11 (1), 19–30.

Oviatt, C.G., Thompson, R.S., Kaufman, D.S., Bright, J., Forester, R.M., 1999. Reinterpretation of the Burmester core, Bonneville basin, Utah. Quat. Res. 52 (2), 180–184.

Oviatt, C.G., Madsen, D.B., Schmitt, D.N., 2003. Late Pleistocene and early Holocene rivers and wetlands in the Bonneville basin of western North America. Quat. Res. 60 (2), 200–210.

Oviatt, C.G., Miller, D.M., McGeehin, J.P., Zachary, C., Mahan, S., 2005. The Younger Dryas phase of Great Salt Lake, Utah, USA. Palaeogeogr. Palaeoclimatol. Palaeoecol. 219 (3–4), 263–284.

Oviatt, C.G., Madsen, D.B., Miller, D.M., Thompson, R.S., McGeehin, J.P., 2015. Early Holocene Great Salt Lake, USA. Quat. Res. 84, 57–68.

Patrickson, S.J., Sack, D., Brunelle, A.R., Moser, K.A., 2010. Late Pleistocene to early Holocene lake level and paleoclimate insights from Stansbury Island, Bonneville basin, Utah. Quat. Res. 73 (2), 237–246.

Pederson, J.L., Janecke, S.U., Reheis, M.C., Kaufman, D.S., Oaks Jr., R.Q., 2016. Chapter 2. The Bear River's History and Diversion: Constraints, Unsolved Problems, and Implications for the Lake Bonneville Record. In: Oviatt, C.G., Shroder Jr., J.F. (Eds.), Lake Bonneville: A Scientific Update, vol. 20. Elsevier. pp. 28–59.

Pigati, J.S., Bright, J.E., Shanahan, T.M., Mahan, S.A., 2009. Late Pleistocene paleohydrology near the boundary of the Sonoran and Chihuahuan deserts, southeastern Arizona, USA. Quat. Sci. Rev. 28, 286–300.

Polyak, V.J., Rasmussen, J.B.T., Asmerom, Y., 2004. Prolonged wet period in the southwestern United States through the Younger Dryas. Geology 32 (1), 5–8.

Porter, S.C., 2007. Neoglaciation in the American Cordilleras. In: Elias, S. (Ed.), Encyclopedia of Quaternary Science. Elsevier, Amsterdam, Netherlands, pp. 1133–1142.

Reasoner, M.A., Jodry, M.A., 2000. Rapid response of alpine timberline vegetation to the Younger Dryas climate oscillation in the Colorado Rocky Mountains, USA. Geology 28, 51–54.

Refsnider, K.A., Laabs, B.J.C., Plummer, M.A., Mickelson, D.M., Singer, B.S., Caffee, M.W., 2008. Last glacial maximum climate inferences from cosmogenic dating and glacier modeling of the western Uinta ice field, Uinta Mountains, Utah. Quat. Res. 69, 130–144.

Reheis, M.C., Adams, K.D., Oviatt, C.G., Bacon, S.N., 2014. Pluvial lakes in the western United States—a view from the outcrop. Quat. Sci. Rev. 97, 33–57.

Reimer, P.J., Bard, E., Bayliss, A., Beck, J.W., Blackwell, P.G., Ramsey, C.B., Buck, C.E., Cheng, H., Edwards, R.L., Friedrich, M., Grootes, P.M., Guilderson, T.P., Haflidason, H., Hajdas, I., Hatté, C., Heaton, T.J., Hoffmann, D.L., Hogg, A.G., Hughen, K.A., Kaiser, K.F., Kromer, B., Manning, S.W., Niu, M., Reimer, R.W., Richards, D.A., Scott, E.M., Southon, J.-R., Staff, R.A., Turney, C.S.M., van der Plicht, J., 2013. IntCal13 and Marine13 radiocarbon age calibration curves 0–50,000 years cal BP. Radiocarbon 55, 1869–1887.

Reinemann, S.A., Porinchu, D.F., Bloom, A.M., Mark, B.G., Box, J.E., 2009. A multi-proxy paleolimnological reconstruction of Holocene climate conditions in the Great Basin, United States. Quat. Res. 72, 347–358.

Rey, K.A., Mayo, A.L., Tingey, D.G., Nelson, S.T., 2016. Chapter 10. Late Pleistocene to Early Holocene Sedimentary History of the Lake Bonneville Pilot Valley Embayment, Utah-Nevada, USA. In: Oviatt, C.G., Shroder Jr., J.F. (Eds.), Lake Bonneville: A Scientific Update, vol. 20. Elsevier. pp. 184–220.

Reynolds, R.L., Rosenbaum, J.G., Thompson, R.S., 2008. Mineralogic causes of variations in magnetic susceptibility of Late Pleistocene and Holocene sediment from Great Salt Lake, Utah: U.S. Geological Survey Open-File Report 2008–1352. 17 p.

Rhode, D., 2000a. Middle and Late Wisconsin vegetation history in the Bonneville Basin. In: Madsen, D.B. (Ed.), Late Quaternary Paleoecology in the Bonneville Basin. Utah Geologic Survey Bulletin 130, pp. 137–148.

Rhode, D., 2000b. Holocene vegetation history in the Bonneville Basin. In: Madsen, D.B. (Ed.), Late Quaternary Paleoecology in the Bonneville Basin. Utah Geologic Survey Bulletin 130, pp. 149–163.

Rhode, D., 2016. Chapter 15. Quaternary Vegetation Changes in the Bonneville Basin. In: Oviatt, C.G., Shroder Jr., J.F. (Eds.), Lake Bonneville: A Scientific Update, vol. 20. Elsevier. pp. 420–441.

Rhode, D., Louderback, L.A., 2015. Bonneville basin paleoenvironments during the Pleistocene–Holocene transition. In: Madsen, D.B., Schmitt, D.N., Page, D. (Eds.), The Paleoarchaic Occupation of the Old River Bed Delta. University of Utah Anthropological Papers 128, Salt Lake City, pp. 22–29.

Rhode, D., Madsen, D.B., 1995. Late Wisconsin/early Holocene vegetation in the Bonneville Basin. Quat. Res. 44 (2), 246–256.

Rhode, D., Goebel, T., Graf, K.E., Hockett, B.S., Jones, K.T., Madsen, D.B., Oviatt, C.G., Schmitt, D.N., 2005. Latest Pleistocene–early Holocene human occupation and paleoenvironmental change in the Bonneville Basin, Utah–Nevada. In: Pederson, J., Dehler, C.M. (Eds.), Interior Western United States: Geological Society of America Field Guide 6. Geological Society of America, Boulder.

Sack, D., 1999. The composite nature of the Provo level of Lake Bonneville, Great Basin, western North America. Quat. Res. 52, 316–327.

Schmitt, D.N., Lupo, K.D., 2012. The Bonneville Estates Rockshelter rodent fauna and changes in late Pleistocene–middle Holocene climates and biogeography in the Northern Bonneville Basin, USA. Quat. Res. 78, 95–102.

Schmitt, D.N., Madsen, D.B., Lupo, K.D., 2002. Small-mammal data on early and middle Holocene climates and biotic communities in the Bonneville Basin, USA. Quat. Res. 58 (3), 255–260.

Scott, W.E., McCoy, W.D., Shroba, R.R., Rubin, M., 1983. Reinterpretation of the exposed record of the last two cycles of Lake Bonneville, western United States. Quat. Res. 20 (3), 261–285.

Smith, F.A., Betancourt, J.L., 2003. The effect of Holocene temperature fluctuations on the evolution and ecology of Neotoma (woodrats) in Idaho and northwestern Utah. Quat. Res. 59 (2), 160–171.

Smoot, J.P., Rosenbaum, J.G., 2009. Sedimentary constraints on late Quaternary lake-level fluctuations at Bear Lake, Utah and Idaho. Geol. Soc. Am. Spec. Pap. 450, 263–290.

Spencer, R.J., Baedecker, M.J., Eugster, H.P., Forester, R.M., Goldhaber, M.B., Jones, B.F., Kelts, K., McKenzie, J., Madsen, D.B., Rettig, S.L., Rubin, R., Bowser, C.J., 1984. Great Salt Lake and precursors, Utah: the last 30,000 years. Contrib. Mineral. Petrol. 86 (4), 321–334.

Stephens, D.W., 1990. Changes in lake levels, salinity and the biological community of Great Salt Lake (Utah, USA), 1847–1987. Hydrobiologia 197 (1), 139–146.

Steponaitis, E., Andrews, A., McGee, D., Quade, J., Hsieh, Y.-T., Broecker, W.S., Shuman, B.N., Burns, S.J., Cheng, H., 2015. Mid-Holocene drying of the U.S. Great Basin recorded in Nevada speleothems. Quat. Sci. Rev. 127, 174–185.

Stuiver, M., Polach, H.A., 1977. Discussion; reporting of C-14 data. Radiocarbon 19, 355–363.

Stuiver, M., Reimer, P.J., 1993. Extended ^{14}C database and revised CALIB radiocarbon calibration program. Radiocarbon 35, 215–230.

Thompson, R.S., 1984. Late Pleistocene and Holocene environments in the Great Basin. Ph.D. Dissertation, University of Arizona. 256 p.

Thompson, R.S., 1990. Late Quaternary vegetation and climate in the Great Basin. In: Betancourt, J.L., Van Devender, T.R., Martin, P.S. (Eds.), Packrat Middens: The Last 40,000 Years of Biotic Change. University of Arizona Press, Tucson, Arizona, pp. 200–239.

Thompson, R.S., 1992. Late Quaternary environments in Ruby Valley, Nevada. Quat. Res. 37, 1–15.

Thompson, R.S., Hattori, E.M., 1983. Packrat (*Neotoma*) middens from Gatecliff Shelter and Holocene migrations of woodland plants. In: Thomas, D.H. (Ed.), The Archaeology of Monitor Valley. 2. Gatecliff Shelter, Anthropological Papers of the American Museum of Natural History 59, pp. 157–167.

Thompson, R.S., Kautz, R.R., 1983. Pollen analysis. In: Thomas, D.H. (Ed.), The Archaeology of Monitor Valley. 2. Gatecliff Shelter, Anthropological Papers of the American Museum of Natural History 59, pp. 136–151.

Thompson, R.S., Mead, J.I., 1982. Late Quaternary environments and biogeography in the Great Basin. Quat. Res. 17, 39–55.

Thompson, R.S., Toolin, L.J., Forester, R.M., Spencer, R.J., 1990. Accelerator-mass spectrometer (AMS) radiocarbon dating of Pleistocene lake sediments in the Great Basin. Palaeogeogr. Palaeoclimatol. Palaeoecol. 78 (3–4), 301–313.

Thompson, R.S., Whitlock, C., Bartlein, P.J., Harrison, S.P., Spaulding, W.G., 1993. Climatic changes in the western United States since 18,000 yr B.P.. In: Wright Jr., H.E., Kutzbach, J.E., Webb, T.I.I.I., Ruddiman, W.F., Street-Perrott, F.A., Bartlein, P.J. (Eds.), Global Climates Since the Last Glacial Maximum. University of Minnesota Press, Minneapolis, Minnesota, pp. 468–513.

Thompson, R.S., Anderson, K.H., Pelltier, R.T., Shafer, S.L., Bartlein, P.J., 2007. Atlas of relations between climatic parameters and distributions of important trees and shrubs in North America—Ecoregions of North America. U.S. Geological Survey Professional Paper 1650-E.

Vacco, D.A., Clark, P.U., Mix, A.C., Cheng, H., Edwards, R.L., 2005. A speleothem record of Younger Dryas cooling, Klamath Mountains, Oregon, USA. Quat. Res. 64 (2), 249–256.

Wahl, D., Starratt, S., Anderson, L., Kusler, J., Fuller, C., Addison, J., Wan, E., 2015. Holocene environmental changes inferred from biological and sedimentological proxies in a high elevation Great Basin lake in the northern Ruby Mountains, Nevada, USA. Quat. Int. 387, 87–98.

Wells, P.V., 1983. Paleobiogeography of montane islands in the Great Basin since the last glaciopluvial. Ecol. Monogr. 53, 342–382.

Whitlock, C., Bartlein, P.J., 1993. Spatial variations of Holocene climatic change in the Yellowstone region. Quat. Res. 39, 231–238.

Wigand, P.E., 1987. Diamond Pond, Harney County, Oregon: vegetation history and water table in the eastern Oregon Desert. Great Basin Nat. 47, 427–458.

Wigand, P.E., 2006. Postglacial pollen Records of Southwestern North America. In: Elias, S.A. (Ed.), Encyclopedia of Quaternary Science, vols. 1–4. Elsevier, London, UK, pp. 2773–2783.

Wigand, P.E., Mehringer Jr., P.J., 1985. Pollen and seed analyses. In: Thomas, D.H. (Ed.), The Archaeology of Hidden Cave, Nevada. Anthropological Papers of the American Museum of Natural History, vol. 61, New York City, pp. 108–124.

Wigand, P.E., Rhode, D., 2002. Great Basin vegetation history and aquatic systems: the last 150,000 years. In: Hershler, R., Madsen, D.B., Currey, D.R. (Eds.), Smithsonian Contributions to Earth Sciences 33: Great Basin Aquatic Systems History. Smithsonian Institution Press, Washington, DC, pp. 309–367.

Zdanowicz, C.M., Zielinski, G.A., Germani, M.S., 1999. Mount Mazama eruption: calendrical age verified and atmospheric impact assessed. Geology 27 (7), 621–624.

Chapter 12

The Fishes of Lake Bonneville: Implications for Drainage History, Biogeography, and Lake Levels

J.M. Broughton* and G.R. Smith†
*University of Utah, Salt Lake City, UT, United States
†Museum of Zoology, University of Michigan, Ann Arbor, MI, United States

ABSTRACT

This chapter summarizes the paleontological and fish DNA evidence for the fishes that occupied the late Quaternary Bonneville basin and key biogeographic and hydrographic insights that emerge from analyses of those data. Fossils and mtDNA from modern fishes suggest nearly the entire assemblage of 21 species occurred in the region prior to the rise of Lake Bonneville. Ten sites have produced fish materials ($N > 16{,}100$ specimens) from the lake. The fauna was similar to that in modern Bear Lake, Utah–Idaho, and had two large salmonine top carnivores, three endemic whitefish zooplanktivores, three bottom or rocky shore dwelling sculpins, several minnows, a large lake sucker, a river sucker, and two mountain suckers. The collection from Homestead Cave, Utah, is the largest, and the richest late Quaternary fish assemblage from the basin. Fish bones from Stratum I of the cave provided $^{87}Sr/^{86}Sr$ values suggesting that they were derived from a low-elevation lake and change in fish size and taxonomic abundance suggest the fauna resulted from a series of die-offs resulting from increases in temperature and salinity. Radiocarbon dating suggests this occurred rapidly between 13.1 and 11.8 cal ka BP at the end of the regressive phase of Lake Bonneville. A recolonization of nearly the entire Lake Bonneville fish fauna occurred between 12.3 and 9.5 cal ka BP during the Gilbert episode, although the fauna is skewed to higher abundances of salinity- and temperature-tolerant taxa. No post-Gilbert early Holocene lake transgressions are suggested but peaks in *Gila atraria* frequencies in the upper strata may indicate Great Salt Lake reached highstands at ~3.6 and ~1.0 cal ka BP.

Keywords: Fishes, Lake Bonneville, Lake levels, Biogeography, Drainage history

Developments in Earth Surface Processes, Vol. 20. http://dx.doi.org/10.1016/B978-0-444-63590-7.00012-3

12.1 INTRODUCTION

> *No inconsiderable part of the progress that has been made in elucidating the problems of the Great Basin has come from the joint consideration of biotic and physiographic data. Even greater advances should ensue, when the biological and geological approaches are more definitely integrated.*
>
> (Hubbs and Miller, 1948, p. 120)

In their landmark synthesis of fish biogeography in the Great Basin, Hubbs and Miller envisioned a future where further synergy of ichthyological and geological analyses would lead to deeper insights into past hydrography and climates. For their part, Hubbs and Miller embraced the notion that the native fishes of the Great Basin occur only in waters that they could have reached through surface water connections so that the dispersal of fishes is closely tied to regional aquatic systems history. This may be considered the first principle of fish biogeography and succinctly paraphrased: Fish colonize wherever they can swim, survive, and reproduce. It follows that climatic inferences can be derived by documenting previous hydrographic connections from the geographic distributions of fishes and deducing the levels of ancient lake systems that would have allowed for such connections. Hubbs and Miller also explicitly recognized that such reconstructions could be more readily achieved "if we had more than the present dribble of paleontological data on the fishes" (Hubbs and Miller, 1948, p. 25). A richer fossil fish record would thus enhance our understanding of historical fish biogeography and ultimately lead to a greater understanding of regional paleohydrography and climate change.

Hubbs and Miller did not elaborate specifically on the nature of more "definitely integrated" analyses of biological and geological data, but it is clear that inferences on the nature of past climates can be derived from fossil or subfossil fish assemblages in ways beyond the biogeographic arguments that were their focus. Fluctuations in the size and depth of a lake can, of course, have dramatic effects on water chemistry and temperature and, in turn, profoundly influence the nature of its fish populations. In addition to controlling whether a lake can support fish at all, water temperature and salinity, for example, play critical roles in determining the species composition, growth characteristics, and the age and size structure of fish populations. As a result, high-resolution fossil fish sequences from Lake Bonneville could provide fine-scale details on lake-level oscillations to test and refine hypotheses based on other geological sources. And since perhaps a no-more-sensitive measure exists of regional climate than the size of closed basin lakes, paleofish faunas can potentially provide one of the more direct proxies of climate change.

In this chapter, we synthesize the available paleontological and fish DNA evidence for the fishes that occupied Lake Bonneville and summarize key biogeographic and hydrographic inferences and insights that emerge from those data. Of these collections, the well-dated ichthyofauna from Homestead Cave represents the largest, and the richest late Quaternary fish assemblage from the entire Bonneville basin, and the only stratified well-dated, terminal Pleistocene fish fauna of the region. This fauna has provided an unprecedented opportunity to examine the dynamics of Lake Bonneville's fish populations as they fluctuated toward the end of the Pleistocene and sheds light on both the nature and timing of fine-scale, climate-driven, lake-level oscillations, and late Quaternary fish biogeography in the Bonneville basin. With now considerably more than a dribble, we have begun to pursue the greater advances that Hubbs and Miller had presaged nearly 70 years ago.

12.2 MATERIALS AND METHODS: LAKE BONNEVILLE ICHTHYOFAUNAS

The fish remains presented and discussed here have been recovered and analyzed from 10 primary localities in the Bonneville basin (Fig. 12.1; Table 12.1). The materials derive from a wide range of depositional contexts and reflect fish sampled from various aquatic environments in the Bonneville basin over that past ~60,000 years, although the vast majority of materials have come from late Pleistocene through early Holocene sediments. We describe here each of these localities, their location, the nature of the deposits, the depositional origin of fish remains, collection methods, and dating. These data are presented in chronological order from oldest to youngest. An annotated list of the Bonneville drainage fishes is provided in Table 12.2.

12.2.1 Deweyville

Utah, Box Elder Co., elevation 1310 m (Fig. 12.1, locality 3). Fossils of *Catostomus ardens* and *Oncorhynchus clarkii* were collected by Oviatt (1984) from alluvium or marsh sediments along a canal bank below Cutler Dam at an elevation of 1310 m in pre-Bonneville sediments near the Bear River at Deweyville, about 16 km above Brigham City. The Cutler Dam Alloformation was deposited during Marine Oxygen Isotope Stage 4, dated at about 60 ka BP (Oviatt et al., 1987; Kaufman et al., 2001). The bones were recovered from a small sand lens in the lacustrine facies in a coarsening-upward sequence indicative of a regressive phase of the saline lake. Oviatt et al. (1987) suggested that the fish had been washed in from their fresh water fluvial habitat, upstream. The sequence, here about 15 m thick, is bounded above by the Fielding Geosol, which is, in turn, overlaid by the Bonneville Alloformation, and is exposed above a covered base of Section 12.1 (Oviatt et al., 1987). The materials are curated at the Museum of Paleontology, University of Michigan.

12.2.2 Hot Springs

Utah, Salt Lake Co., elevation 1350 m (Fig. 12.1, locality 5). A diverse fossil fish fauna and site details for the Hot Springs locality are reported in Smith

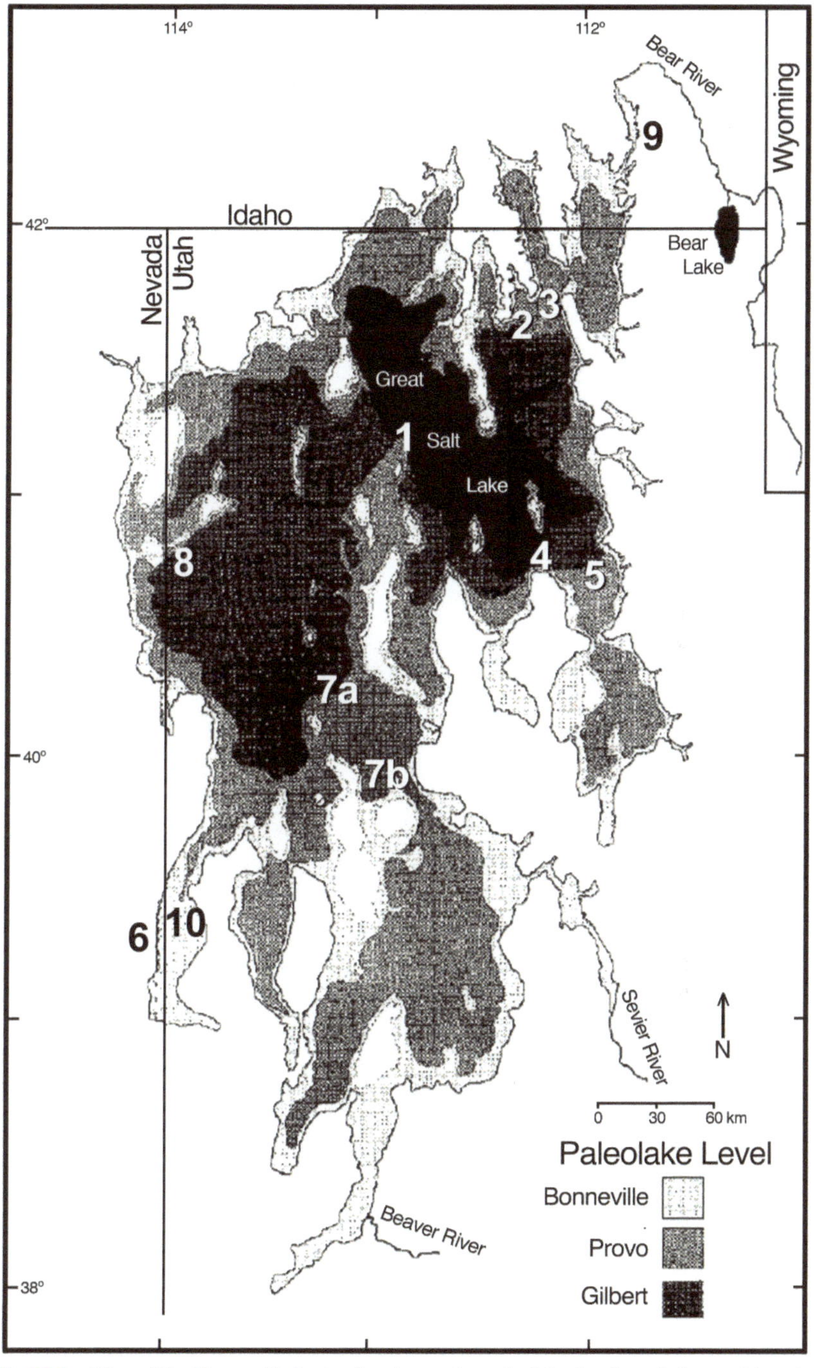

FIG. 12.1 Map of the Bonneville basin showing major paleolake levels and sites and locations discussed in the text. (1) Homestead Cave and Cathedral Cave, UT, (2) Public Shooting Grounds, (3) Deweyville, (4) Black Rock Canyon, (5) Hot Springs, (6) Smith Creek Cave and Cathedral Cave, NV, (7a and 7b) The Old River Bed, (8) Danger Cave, (9) Thatcher Basin, and (10) Crystal Ball Cave. *Adapted from Currey, D.R., 1990. Quaternary paleolakes in the evolution of semidesert basins, with emphasis on Lake Bonneville and the Great Basin, U.S.A. Palaeogeogr. Palaeoclimatol. Palaeoecol. 76, 189–214.*

TABLE 12.1 Numbers of Identified Fish Specimens per Taxon for Lake Bonneville Ichthyofaunas

Taxon	Site										Total
	Public Shoot. Grounds	*Deweyville*	*Black Rock Canyon*	*Hot Springs*	*Homestead Cave*	*Smith Ck. Cave*	*Cathedral Cave*	*Old River Bed*	*Danger Cave*	*Thatcher Basin*	
Cypriniformes					2245			34			2279
Cyprinidae					120			8			128
Gila atraria	+[a]		2	12	958	4		11			987
Richardsonius balteatus					1						1
Catostomus sp.					1343						1343
Catostomus ardens		1		11	179			+			191
Pantosteus virescens.					5						5
Chasmistes sp.										+	0
Chasmistes cf. *liorus*			+								0
Salmoninae					1022					+	1022
cf. *Salvelinus confluentus*					1						1

Oncorhynchus cf. *clarkii*					60						60
Oncorhynchus clarkii		29	30		2	27		+	+		88
Prosopium sp.					7536			+			7536
Prosopium cf. *spilonotus*					84						84
P. spilonotus			2	1	41						44
P. spilonotus/ abyssicola					198						198
P. abyssicola					8						8
P. gemmifer			10	507	778						1295
Cottus sp.					259		539				798
Cottus cf. *bairdii*					1						1
Cottus bairdii			2	4	6		3				15
C. extensus/ echinatus					14		5				19
Cottus extensus			9	55	5						69
Total		30	55	590	14,866	31	547	53			16,172

[a]*+ indicates the presence of a taxon, but without enumeration of elements.*

TABLE 12.2 Annotated List of Fishes of the Bonneville Drainage

Family Catostomidae, Suckers	*Chasmistes liorus,* June Sucker. Endemic to Utah Lake; represented by fossils (Miller and Smith, 1981) and DNA analyses of recent populations (Krabbenhoft, unpublished data). Restricted to lakes except during spawning runs up tributaries
	Catostomus ardens, Utah Sucker. Endemic to the Bonneville and upper Snake River drainages; represented by fossils and DNA analyses of recent fishes. Inhabits rivers and lakes
	Pantosteus platyrhynchus, Mountain Sucker. Endemic to the Bonneville and Upper Snake drainages; represented by DNA analyses of recent fishes. Usually inhabits moving water (Smith et al., 2013; Unmack et al., 2014)
	Pantosteus virescens, Green Sucker. Endemic to the Bonneville and Upper Snake drainages; represented by fossils and DNA analyses of recent fishes (Unmack et al., 2014). Usually inhabits moving water, but wave-swept littoral habitat was possible (Broughton, 2000b,c; Smith et al., 2013; Unmack et al., 2014)
Family Cyprinidae, Minnows	*Gila atraria,* Utah Chub. Endemic to Lake Bonneville and tributaries. Represented by Lake Bonneville fossils and DNA analyses of recent fishes (see Johnson and Belk (1999) for DNA analysis). The Snake and Bear River populations attributed to this species are a different form—*Gila domninus* (Cope)
	Iotichthys phlegethontis, Least Chub. Endemic to the Bonneville basin; represented by DNA analyses of recent fishes. No known fossils
	Richardsonius balteatus, Redside Shiner. Native to the Bonneville, Snake, Columbia, and peripheral drainages; represented in the Bonneville basin by a single fossil specimen (Broughton, 2000b) and DNA analyses of recent fishes (Dowling and Markle, unpublished data.)
	Snyderichthys copei, Leatherside Chub. Endemic to the Bonneville and Upper Snake drainages; represented by DNA analyses and historical information. Sometimes interpreted as part of *Lepidomeda* from the Colorado River drainage, with two species in the Bonneville basin (Johnson et al., 2004) based on DNA evidence, but morphology and osteology refute the interpretation based on DNA markers, indicating past DNA transfer by introgression. No fossil record

TABLE 12.2 Annotated List of Fishes of the Bonneville Drainage—Cont'd

	Rhinichthys cataractae, Longnose Dace. Widespread across northern North America. Found in the northern Bonneville basin. No fossil record. Represented by DNA analyses of recent phylogeographic information (Dowling, unpublished data)
	Rhinichthys osculus, Speckled Dace. Widespread across western North America. Represented by DNA analyses of recent fishes and 10 million year old fossils in the Snake and Columbia drainages. DNA indicates ancient presence in the pre-Bonneville basin and divergent northern and southern Bonneville populations (Smith et al., 2016; Dowling, unpublished data)
Family Salmonidae, Subfamily Coregoninae	*Prosopium williamsoni,* Rocky Mountain Whitefish. Native to the Colorado, Great Basin (including the Bear lake drainage), Columbia, and north to the McKenzie and Hudson Bay drainages; represented by DNA analyses of recent fishes and possibly fossil scales from the Old River Bed of Lake Bonneville
	Prosopium abyssicola, Bear Lake Whitefish. Endemic to the Bonneville basin, recent only in Bear Lake; represented by DNA analyses of recent fishes and Lake Bonneville fossils (Smith and Todd, 1984; Broughton, 2000b,c)
	Prosopium spilonotus, Bonneville Whitefish. Endemic to the Bonneville basin, recent only in Bear Lake; represented by DNA analyses of recent fishes and Lake Bonneville fossils (Smith and Todd, 1984; Broughton, 2000b,c)
	Prosopium gemmifer, Bonneville Cisco. Endemic to the Bonneville basin, recent only in Bear Lake; represented by DNA analyses of recent fishes and Lake Bonneville fossils (Smith and Todd, 1984; Broughton, 2000b,c)
Family Salmonidae, Subfamily Thymallinae	*Thymallus arcticus,* Arctic Grayling. Grayling were collected in the Bear River and Green River in 1834 according to the account of Nuttall and Townsend (Townsend, 1978, pp. 84, 89). Nuttall and Townsend were familiar with trout, grayling, and char (see later) because of Nuttall's earlier work on the natural history of the Great Lakes region, where they encountered grayling and char (Smith et al., 2002, p. 218)

Continued

TABLE 12.2 Annotated List of Fishes of the Bonneville Drainage—Cont'd

Family Salmonidae, Subfamily Salmoninae	*Salvelinus confluentus,* Bull Trout. Native to the Intermountain West; represented by fossils, but not extant currently in the Bonneville basin. Townsend (1978, p. 89) wrote about a collection made from the Bear River, 12 mi above Soda Springs, Idaho, on 6 Jul. 1834: "Trout, grayling, and a kind of char are very abundant here—the first very large." The Nuttall/Townsend record is supported by the fossil from Homestead Cave (see later) (Smith et al., 2002)
	Oncorhynchus clarkii utah, Bonneville Cutthroat Trout. Endemic to the Bonneville drainage, including the Bear River, Bear Lake, and Snake Valley, Nevada. Represented by fossils from Lake Bonneville and pre-Bonneville clays along the Bear River (see later) and DNA analyses of recent fishes
Family Cottidae	*Cottus bairdii semiscaber,* Bonneville Mottled Sculpin. The Mottled Sculpin is widespread in North America, with distinctive subspecies in the West. Represented by Lake Bonneville fossils and DNA analyses of recent fishes (Smith et al., 1968; Broughton, 2000b,c)
	Cottus extensus, Bear Lake Sculpin. Endemic to the Bonneville basin. Represented by Lake Bonneville fossils and DNA analyses of recent fishes, but recent only in Bear Lake (Smith et al., 1968; Broughton, 2000b,c)
	Cottus echinatus, Utah Lake Sculpin. Endemic to the Bonneville basin. Extinct with no fossil or DNA record
	Cottus beldingi, Paiute Sculpin. Native to the Columbia and Snake rivers, Bear River, western Lahontan, and part of the upper Colorado drainage. Fossil records in the Lahontan basin (Taylor and Smith, 1981)

See text and Smith et al. (2002) for geographic, stratigraphic, and evolutionary context.

et al. (1968). The fossils were collected by surface picking and screen washing several hundred pounds of matrix by K.F. Horn in Jun. 1963, and subsequently by Horn and G.R. Smith, and finally by C. Barbour. The fossils were in 3–5 cm thick lenses of weakly cemented, well-sorted sand in strata between layers of moderately cemented gravel, exposed on a steep face by excavation for gravel and fill material. The deposits, in the lower Draper Formation (Morrison, 1965), are below the calcareous tufa of the well-defined Stansbury lake level (1375 m). Larger pebbles of the Stansbury level locally are coated with crusts of calcite on their upper surfaces, whereas limestone pebbles are eroded on their under surfaces, indicating hot water ascending through the lacustrine sediments. It is speculated that the fish were attracted to the offshore hot springs by thermal-based, high productivity and that the elevated temperatures contributed to the death assemblage. At the time of collection, the hot springs of Wasatch Park were 210 m south of the area (Stokes et al., 1964; Smith et al., 1968). The fossils are deposited in collections of the Natural History Museum, University of Kansas, and the Museum of Paleontology, University of Michigan.

12.2.3 Cathedral Cave

Utah, Box Elder Co., elevation 1383 m (Fig. 12.1, locality 1). Cathedral Cave is a large limestone cavern located on Homestead Knoll a northwestern spur of the Lakeside Mountains just west of Great Salt Lake, ~10 km SW of Lakeside, Utah. The cave overlooks the expansive playa of Great Salt Lake Desert but its mouth is completely covered by a talus cone of tufa/breccia. The top of the depositional sequence of the cave floor is 1383 m, between two Stansbury level beaches of ~1405 and 1363 m. Investigations in 1995 led by Madsen (2000) involved excavation of a $1 \times 0.5\ m^2$ column sample with fill being passed in the field sequentially through 6.3 and 3.1 mm screens in addition to the collection of several bulk sediment samples. Eleven well-defined strata were evident in the column and two discrete sedimentary breaks divided the sequence into a lacustrine deposit (Zone 1) below and postlacustrine deposit (Zone 2) above. Fish remains, limited to several sculpin (*Cottus*) species, were confined to Stratum II, the first lacustrine deposit resting on limestone spall and bedrock (Stratum I) composed of laminated fine-grained lacustrine clays containing numerous ostracods and gastropods (Broughton, 2000a). The sculpin remains may owe their presence in these sediments to active burrowing or by settling out from natural mortality in the lake. A $15{,}310 \pm 60$ ^{14}C date was derived from pollen in an overlying stratum (Stratum IV). Fish remains and other fauna (see Madsen et al., 2001) are curated at the Natural History Museum of Utah.

12.2.4 Black Rock Canyon

Utah, Salt Lake Co., elevation 1423 m (Fig. 12.1, locality 4). J.H. Madsen, Jr., G.R. Smith, K.F. Horn, and A.D. Stock collected a diverse fish and gastropod fauna in November, 1964, from a fine- to medium-grained sand lens at an elevation of 1423 m. The site is below the Provo lake level, which is at an elevation of 1460 m in this area (Smith et al., 1968). Additional *Chasmistes* materials are reported in Miller and Smith (1981). Gastropod shells found with the fish were analyzed by M. Rubin, U.S.G.S. and yielded a radiocarbon age of $12{,}860 \pm 400$ ^{14}C years BP. This date and the elevation relative to the Provo terrace suggest assignment to the Upper Lacustrine Member of the Bonneville Formation of Morrison (1965). Associated gastropods, identified by Dwight Taylor, include *Fontelicella* sp., *Lithoglyphus hindsi*, *Lymnaea bonnevillensis*, *Physa* sp., and *Valvata humeralis*. The pelecypod, *Pisidium* sp. is also present. These materials are curated at the Natural History Museum of Utah and the Natural History Museum, University of Kansas.

12.2.5 Smith Creek Cave

Nevada, White Pine Co., elevation 1950 m (Fig. 12.1, locality 6). Smith Creek Cave is located in Smith Creek Canyon in the Snake Range on the Utah–Nevada border on the west shore of Lake Bonneville. Lake Bonneville reached an elevation of 1580 m, 4 km from the entrance of Smith Creek Canyon in Snake Valley (Mead et al., 1982). Although Smith Creek Cave has a long history of excavation and analysis, a small sample of fossil fish bones from three species was recovered from a single sedimentary layer, the Reddish-brown Silt Zone described by Bryan (1979). Bryan obtained a ^{14}C date of ~28 ka, but that may be inaccurate (Mead et al., 1982). The reddish-brown unit has abundant bird and mammal remains that could be the same age as fossils in Cathedral Cave, across the canyon from Smith Creek Cave. The reddish silt in the unit is interpreted to be either an eolian deposition or derived from erosion from water issuing from a nearby ceiling conduit: the abundant fossil material, including the fish bones, in the stratum are believed to have resulted from raptor activity (Mead et al., 1982). The chronology of the Reddish-brown silt unit is not well constrained, but a radiocarbon date of $12{,}600 \pm 170$ ^{14}C BP from an adjacent unit stratigraphically above is suggested to represent a minimum age for the deposit. Mead et al. (1982) reported *O. clarkii* and *Gila atraria* from the collection; recent work revealed specimens of *C. ardens*. The fish bones were collected in 1977 and 1978 by J. Mead, R.S. Thompson, and T.R. Van Devender (Mead et al., 1982). These specimens are curated at the Museum of Northern Arizona, Flagstaff.

12.2.6 Homestead Cave

Utah, Box Elder Co., elevation 1406 m (Fig. 12.1, locality 1). Homestead Cave (22BO763) is a wave-constructed cavern located in the Lakeside Mountains just west of Great Salt Lake, 10 km SW of Lakeside, Utah, in the northern Bonneville basin. The site is located on Homestead Knoll just ~850 m SW of Cathedral Cave. In 1993 and 1994, a 1×1 m^2 sample column was excavated to a depth of 3 m in the rear cave-floor sediments. The well-defined deposits contained 18 separate strata. Materials removed from the column were passed through a series of 6.4 and 3.2 mm screens in the lab. The deposits consisted primarily of degrading owl pellets, rich in small vertebrate remains, with little evidence of contamination by human foragers. A total of 93 radiocarbon dates spanning from 13.1 to 0.7 cal ka BP[1] have been obtained from a range of organic materials, including vertebrate bone (Madsen, 2000; Broughton et al., 2008; Smith and Betancourt, 2013; Terry and Novak, 2015). Overall, the chrono-stratigraphic coherency of the ^{14}C results suggests that the deposits were laid down sequentially over the past ~13.1 ka years, although a recently derived series of 56 dates from *Dipodomys* spp. (kangaroo rat) bone collagen expand both the temporal range and overlap of several strata in the deposit (Terry and Novak, 2015). Still, the relationship between stratum depth and mean calibrated age is highly significant ($r_s = 0.995$, $P < 0.0005$). A large and diverse fish fauna was concentrated in the lowest strata of the deposit (Strata I and II) with dates ranging from 13.1 to 9.5 cal ka BP. This period of time encompasses the very end of the regressive phase of Lake Bonneville, a postregressive lake near Great Salt Lake elevations, the Gilbert episode, and the first two early Holocene millennia of the hypersaline Great Salt Lake. A few late Holocene-aged strata also contain abundant fish remains. We follow Broughton (2000b) and report in the descriptive summary later the fish remains, by element, from 6.4 and 3.2 mm fractions of Stratum I and five noteworthy specimens present in higher strata of the cave. However, we present and draw on the entire sequence in our analyses of lake-level fluctuations. Analyses of the fish are described in Broughton (2000b,c) and Broughton et al. (2000, 2006); the Homestead Cave vertebrate fauna is housed at the Natural History Museum of Utah.

12.2.7 The Old River Bed

Utah, Juab, Toole, and Utah counties, elevation 1300 m (Fig. 12.1, localities 7a and 7b). In the regressive phase of Lake Bonneville when the lake fell below the threshold that separated the Sevier basin from the Great Salt Lake basin, a shallow lake in the Sevier basin overflowed at its north end, creating a river that carved a narrow, meandering valley on the floor of the basin—the Old River Bed. The lacustrine and fluvial geomorphology of this valley and the surrounding region have been recently investigated through geological assessments, including analyses of stratigraphy and dating of materials collected from backhoe pits and trenches (Oviatt et al., 2003; Madsen et al., 2015). Utah Chub (*G. atraria*) remains, (identified by JMB) dating between 12.9 and 9.9 cal ka BP, were present in several riverine "sand-channel" deposits derived from these trenches (42TO1161, 42TO1000, 42TO1182) and were reported briefly in Oviatt et al. (2003). We describe these materials (as Old River Bed I, locality 7a, Fig. 12.1), in more detail later. Earlier work in the region, that we designate here as Old River Bed II (locality 7b, Fig. 12.1), reported *C. ardens* bones from shut-off spit gravel deposits in a major

1. Notes: All dates within the radiocarbon time scale are provided in calibrated ka BP following CALIB 7.1 (Reimer et al., 2013; Stuiver et al., 2013) unless otherwise noted.

transgressive sequence at the Shutoff (Oviatt, 1984). *O. clarkii* bones and bones and scales of *Prosopium* sp. were recovered from nearby deep-water phase, white marl deposits of Lake Bonneville (Oviatt, 1984). The specimens from Old River Bed II are curated at the University of Michigan Museum of Paleontology; Old River Bed I materials are housed at the Department of Anthropology, University of Utah.

Several other localities have produced fish materials apparently derived from Lake Bonneville but refined identifications could not be made for them owing to the nondiagnostic elements represented (eg, vertebrae) or their state of preservation, or species level identifications are provided but elements are not enumerated. These include *O. clarkii* elements (identified by GRS) from basal deposits of Danger Cave, Tooele Co., Utah; small teleost vertebrae from Crystal Ball Cave, Utah, Millard Co., (Heaton, 1985), and Cathedral Cave, White Pine Co., Nevada (Jass, 2007); *G. atraria* materials from the Public Shooting Grounds, Box Elder Co., Utah (Murchison, 1989a); and *Chasmistes* sp. and salmonine fossils from late Pleistocene beds in the Thatcher Basin, Franklin Co., Idaho (Bright, 1963, 1967; Miller and Smith, 1981). We do not include these materials in Section 12.3 that provides the numbers of elements per taxon, but indicate their presence at these sites in Table 12.1 and draw on them in our biogeographic discussions.

12.3 RESULTS: SYSTEMATICS AND OSTEOLOGY

In this section, we describe the universe of fish materials that have been identified and quantified from the Lake Bonneville-aged deposits described earlier and discuss pertinent aspects of the distribution and habitat of the represented species and the biogeographic implications that follow from their occurrence at these localities. Our discussion of the taxa present and element representation is focused on species-level identifications but quantitative data on order, family and genus-level identifications are presented by site in Table 12.1. With few exceptions (noted later), the detailed osteological criteria used to make these identifications are provided in Smith et al. (1968), Miller and Smith (1981), and Broughton (2000b,c). Selected diagnostic fish elements from Hot Springs, Black Rock Canyon, and Homestead Cave are illustrated in Figs. 12.2–12.7.

Order Cypriniformes
Cyprinidae—Minnows
Gila atraria (Girard, 1856)
Utah Chub

Material—Homestead Cave: Frontal (40), ethmoid (4), infraorbital (2), vomer (6), parietal (29), pterotic (30), basioccipital (29), parasphenoid (2), articular (9), dentary (43), premaxilla (1), maxilla (11), posttemporal (7), quadrate (25), epihyal (14), ceratohyal (23), urohyal (8), pharyngeal (154), hyomandibular (34), opercle (49), preopercle (26), cleithrum (45), supracleithrum (3), pectoral fin spine (9), basipterygium (9), scapula (5), coracoid (3),

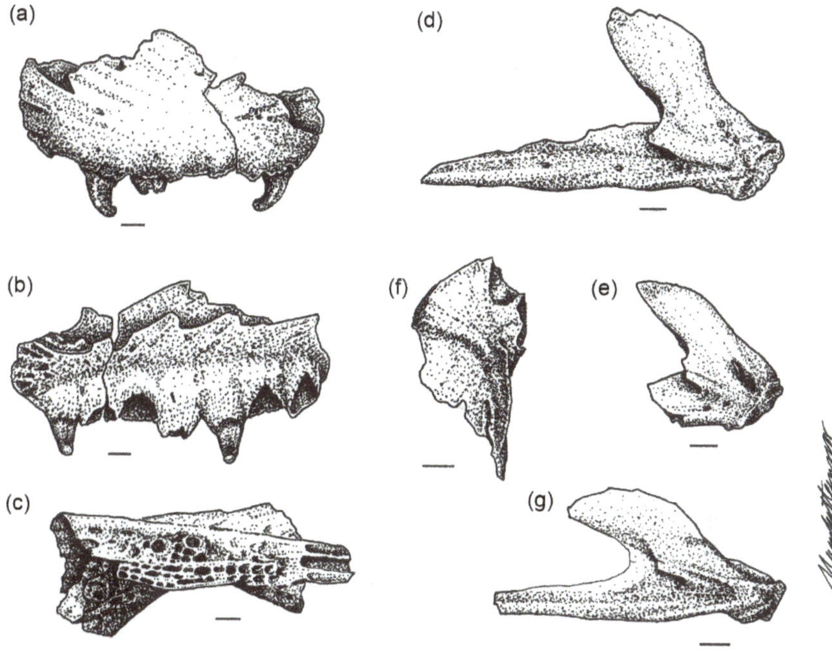

FIG. 12.2 Selected diagnostic salmonid elements from Homestead Cave. (a) Lateral view of right premaxilla of cf. *Salvelinus confluentus*; (b) mesial view of right premaxilla of cf. *S. confluentus* (same specimen as (a)); (c) basibranchial of *Oncorhynchus clarkii* (dorsal view); (d) left dentary of *Prosopium spilonotus*; (e) left dentary of *Prosopium abyssicola* (the ventral ramus is broken off); (f) left hyomandibular of *Prosopium gemmifer* (the posterolateral ridge is on the lower, right margin of the bone), and (g) left dentary of *P. gemmifer*. Scale: Bar length = 1 mm. *From Broughton, J.M., 2000b. Terminal Pleistocene fish remains from Homestead Cave, Utah, and implications for fish biogeography in the Bonneville basin. Copeia 2000, 645–656.*

supraorbital (1), supraoccipital (1), 1st vertebra (8), 2nd vertebra (16), 3rd vertebra (4), 4th vertebra (2), Weberian apparatus (8). Hot Springs: Pharyngeal (10), hyomandibular (1), opercle (1). Black Rock Canyon: Pharyngeal (1), Weberian suspensorium (1). Smith Creek Cave: Basioccipital (1), vertebra (3). Old River Bed I: Pharyngeal (10), dentary (1).

Richardsonius balteatus (Richardson, 1836)
Redside Shiner

Material—Homestead Cave: Pharyngeal (1).

Remarks—Six cyprinid species are native to the Bonneville basin: *G. atraria*, *Snyderichthys copei*, *Iotichthys phlegethontis*, *Rhinichthys cataractae*, *Rhinichthys osculus*, and *R. balteatus* (Table 12.2). With the exception of *G. atraria*, these fishes rarely attain total lengths over 150 mm. *G. atraria*, however, can reach lengths up to 559 mm but typically range from 127 to 305 mm (Sigler and Sigler, 1996). *Gila domninus* Cope, a close relative of *G. atraria*, lives in the Bear River, the Snake River above Shoshone Falls,

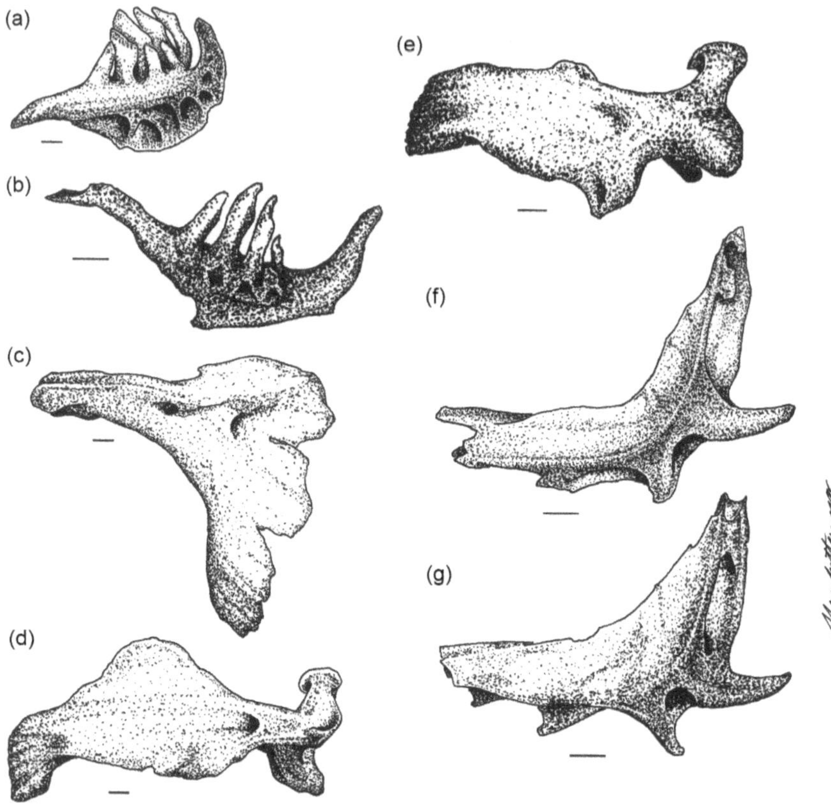

FIG. 12.3 Selected diagnostic cyprinid, catostomid, and cottid elements from Homestead Cave. (a) Left pharyngeal tooth of *Gila atraria*; (b) right pharyngeal tooth of *Richardsonius balteatus*; (c) left dentary of *Catostomus ardens*; (d) right maxilla of *C. ardens*; (e) right maxilla of *Pantosteus virescens*; (f) left preopercle of *Cottus bairdii*; and (g) left preopercle of *Cottus extensus*. Scale: Bar length = 1 mm. *From Broughton, J.M., 2000b. Terminal Pleistocene fish remains from Homestead Cave, Utah, and implications for fish biogeography in the Bonneville basin. Copeia 2000, 645–656.*

and the lower Wood River system, Idaho (Table 12.2). *G. atraria* has been reported from six of the Lake Bonneville-era ichthyofaunas and was apparently widespread in the lake at various levels.

G. atraria was identified from sand-channel deposits of the Old River Bed (I) dating from ~12.9 to 9.9 cal ka. The consistent presence of chubs in these contexts suggests that water flowed more-or-less continuously over this period (Oviatt et al., 2003). As we discuss later, changes in abundance of *G. atraria* across the Holocene sequence of deposits at Homestead Cave may reflect lake level variation of Great Salt Lake.

R. balteatus is widely distributed in northwestern North America (Lee et al., 1980) but the Homestead Cave specimen represents the only record of this species for Lake Bonneville. The presence of only a single pharyngeal

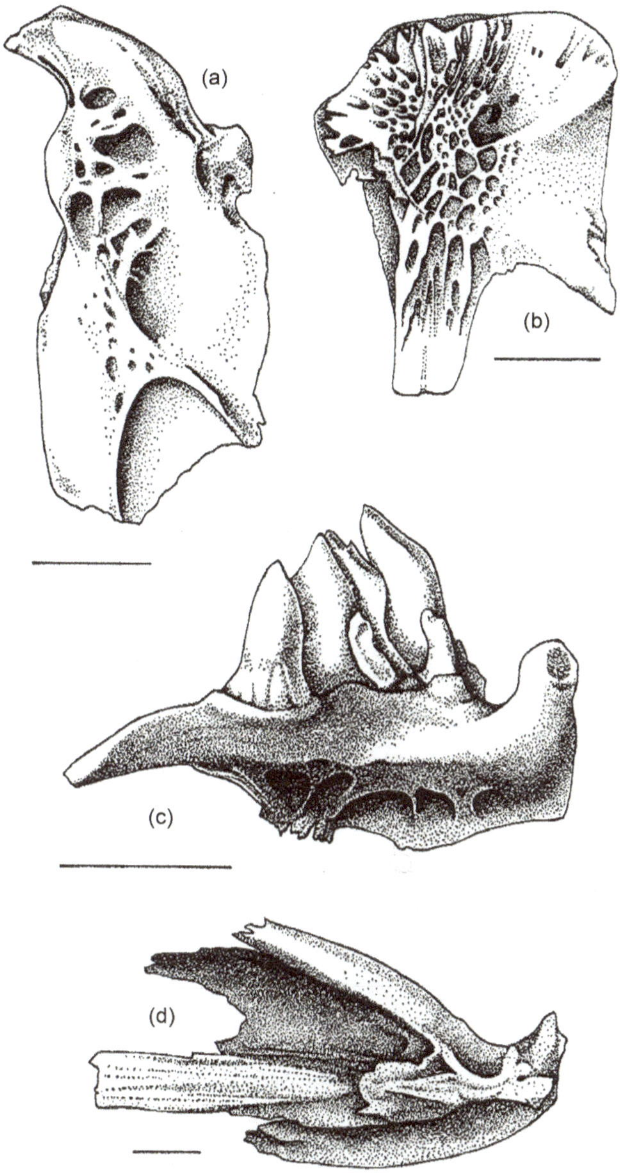

FIG. 12.4 Selected catostomid, cyprinid, and salmonine elements from Hot Springs and Black Rock Canyon. (a) Right hyomandibular of *Catostomus ardens* from the Hot Springs locality; (b) left hyomandibular of *Oncorhynchus clarkii*; (c) left pharyngeal arch of *Gila atraria*; and (d) right articular–angular of *O. clarkii* from the Black Rock Canyon locality. Scales for (a) and (c) = 5 mm; those for (b) and (d) = 10 mm. *From Smith, G.R., Stokes, W.L., Horn, K.F., 1968. Some late Pleistocene fishes of Lake Bonneville. Copeia 1968, 807–816.*

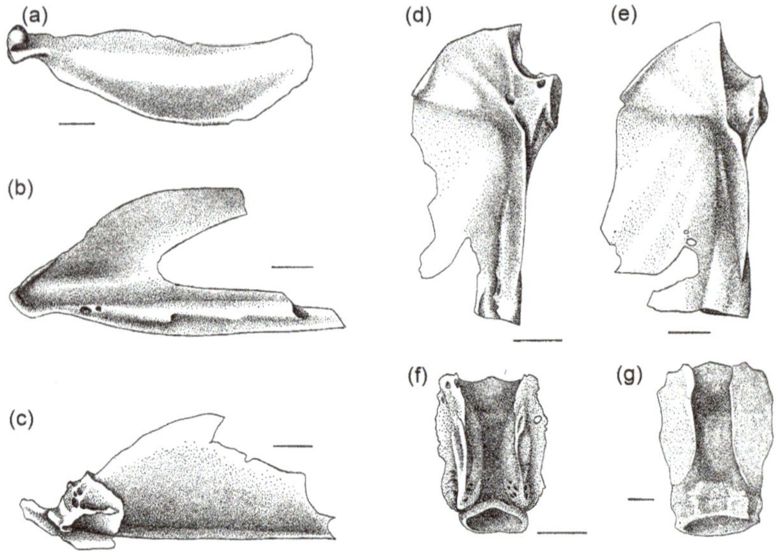

FIG. 12.5 Selected *Prosopium* elements from Hot Springs, Black Rock Canyon and modern Bear Lake. (a) Right maxilla, (b) left dentary, (c) left articular–angular, (d) left hyomandibular of fossil *Prosopium gemmifer* from the Hot Springs locality; (e) left hyomandibular of modern *P. gemmifer* from Bear Lake, Utah (KU 11748); ventral views of basioccipitals of (f) *P. gemmifer*, and (g) *P. spilonotus*, from the Black Rock Canyon locality. Scale = 1 mm. *From Smith, G.R., Stokes, W.L., Horn, K.F., 1968. Some late Pleistocene fishes of Lake Bonneville. Copeia 1968, 807–816.*

of the small *R. balteatus* may, in part, be function of recovery bias. Analysis of the as yet unexamined 1.6 mm fraction of the Homestead Cave fish materials may also document the presence of *S. copei*, *R. osculus* and *I. phlegethontis*, other widespread diminutive minnows of the Bonneville basin. Because *R. balteatus* specimens were previously unknown from Lake Bonneville faunas but were present in Pliocene Lake Idaho has led to the suggestion that this species entered the Bonneville basin from the Snake system as recently as about 14.0 cal ka (Sigler and Sigler, 1996, p. 24), but DNA evidence indicates that *Richardsonius* colonized the Bonneville basin from the Snake River drainage approximately 0.5 Ma (Houston, 2009).

Catostomidae—Suckers
Catostomus ardens (Jordan and Gilbert, 1881)
Utah Sucker

Material—Homestead Cave: Dentary (73), maxilla (74), premaxilla (16). Hot Springs: Ceratohyal (1), quadrate (2), hyomandibular (1), parietal (2), opercle (2), subopercle (1), maxilla (1), dentary (1). Deweyville: 2nd vertebra of Weberian apparatus (1). Smith Creek Cave: Cleithrum (1).

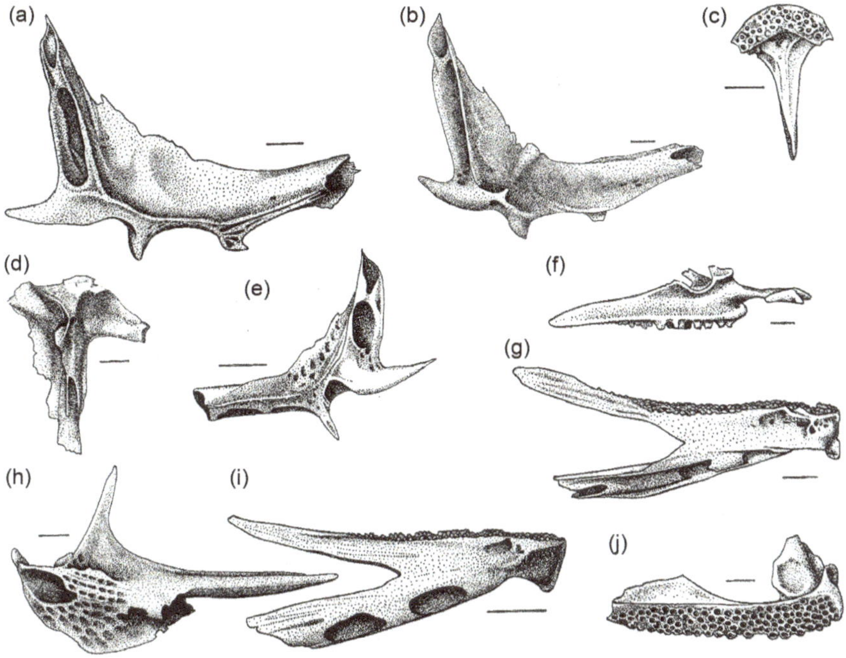

FIG. 12.6 Fossil *Cottus* from Black Rock Canyon and Hot Springs. (a, b, and i) are from the Black Rock Canyon locality. (b) is a specimen of *C. bairdii* (KU 16085); (a) and (i) represent *C. extensus* (KU 16089); the remainder of the specimens are from the Hot Springs locality and represent *C. extensus* (KU 16088) (a) and (b), right preopercles, old individuals; (c), prevomer; (d), left hyomandibular; (e), left preopercle, young individual; (f), right palatine; (g), right dentary, old individual; (h), right articular–angular; (i), right dentary, young individual; and (j), left premaxilla. Scale = 1 mm. *From Smith, G.R., Stokes, W.L., Horn, K.F., 1968. Some late Pleistocene fishes of Lake Bonneville. Copeia 1968, 807–816.*

Pantosteus virescens (Cope, 1872)
Green Sucker

Material—Homestead Cave: Dentary (1), Stratum VI; maxilla (3), Stratum IX; dentary (1), Stratum X.

Chasmistes cf. *liorus*
June Sucker

Material—Black Rock Canyon: Coracoid (1), dentary (1).

Remarks—Four catostomid species are native to the Bonneville basin: *Chasmistes liorus*, *C. ardens*, *Pantosteus virescens*, and *P. platyrhynchus* and all but the latter have now been securely identified from Lake Bonneville deposits. Both *C. liorus* and *C. ardens* can reach substantial sizes (>600 mm TL), whereas *P. virescens* and *P. platyrhynchus* reach maximum lengths of 457 and 305 mm, respectively (Sigler and Sigler, 1996). To judge from site occurrences and relative abundance, *C. ardens* was the most widespread and

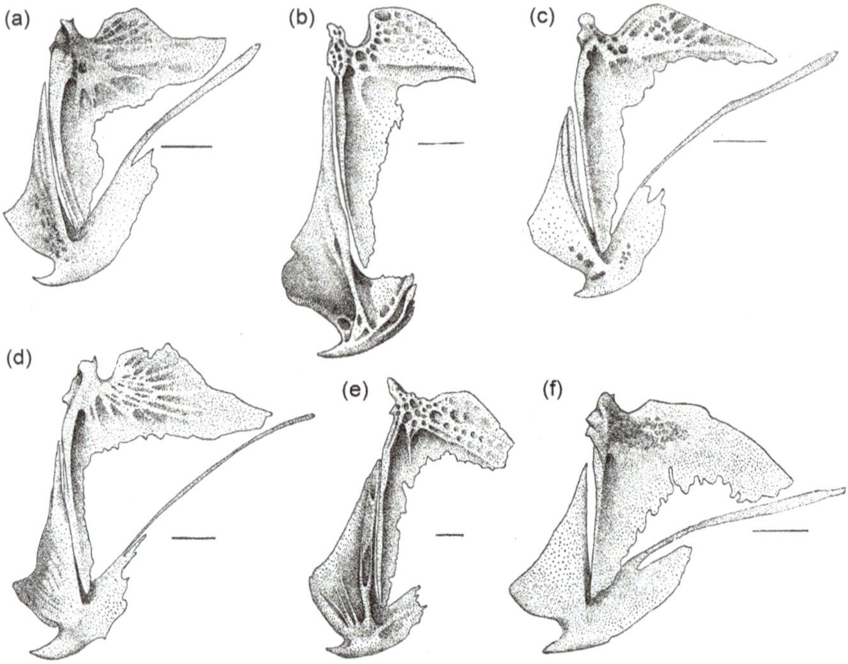

FIG. 12.7 Left opercles and subopercles of Great Basin sculpins. (a) *Cottus echinatus* (UMMZ 156794), Utah Lake, Utah; (b) *C. extensus* (opercle, KU 16089, Black Rock Canyon; subopercle, KU 16088, Hot Springs); (c) *C. extensus* (KU 11934), contemporary Bear Lake, Utah; (d) *C. bairdii* (KU 11940) contemporary Heiners Canyon, Utah; (e) *C. extensus* (KU 16086), Hot Springs locality; and (f) *C. beldingi* (KU 11933), contemporary Heiners Canyon, Utah. Scale = 1 mm. *From Smith, G.R., Stokes, W.L., Horn, K.F., 1968. Some late Pleistocene fishes of Lake Bonneville. Copeia 1968, 807–816.*

abundant sucker in the lake. It is also abundant in many late Holocene archaeological contexts in the basin (eg, Janetski and Smith, 2007).

Homestead Cave has provided the only record of *P. virescens* in Lake Bonneville and the only fossil record of the genus in the Great Basin. Today, two disjunct populations occur in swift and cool, mountain streams of the Bonneville basin: one in the upper Weber River drainage and one in the Bear River drainage. This present-distribution pattern led Smith (1966, p. 121) to suggest that the Weber River population must have colonized that system before the Bear and Weber Rivers were isolated by the desiccation of Lake Bonneville at the end of the Pleistocene. *P. virescens* probably occupied the Bonneville drainage prior to Late Pleistocene Lake Bonneville (Unmack et al., 2014), and was at least a transient along the stretch of shoreline between the Bear River and Weber Rivers, but evidence of their presence in the lake was lacking before the Homestead Cave record. The presence of *C. virescens* at Homestead Cave indicates not only that they occurred in Lake Bonneville but also that they ranged widely across the lake. Since *P. virescens* is a stream-adapted species, their presence also implies that populations of these fish had adapted to some aspect of the lacustrine environments of Lake Bonneville or suitable fluvial environments occurred near Homestead Cave in the past.

The distribution patterns of stream fishes in the Great Lakes region provide insight into how *P. virescens* may have adapted to lacustrine environments of Lake Bonneville. High-energy wave action along the shorelines of the Great Lakes creates substrate types and associated benthic algal and diatom communities that are analogous to those found in riffle sections of stream environments. As a result, some fish species that are primarily restricted to swift streams or rivers with rock or gravel substrates, also occur in certain high-energy shoreline settings of the Great Lakes (Hubbs and Lagler, 1958). Wave-swept shorelines clearly characterized Lake Bonneville and may have provided suitable habitat for *P. virescens* in those settings.

It is also possible that stream habitats may have existed near Homestead Cave in late Pleistocene times even though they do not occur there today. In this context, we note that riparian vegetation and avian taxa associated with such vegetation have been reported for the basal deposits of Homestead Cave and may reflect the presence of local fluvial habitats in the past (Madsen et al., 2001; Wolfe and Broughton, 2016). We also note that an ancient stream channel that would have drained much of the western margin of the Lakeside Mountains is located ~0.5 km from the mouth of the cave (Madsen, 2000).

C. liorus is a large, planktivorous lake sucker that is endemic to the shallow eutrophic waters of Utah Lake today, although it substantially introgressed with *C. ardens* in that setting (Miller and Smith, 1981). This sucker is rare in Lake Bonneville deposits. Miller and Smith (1981, p. 33) report a fragmentary coracoid and dentary referable to *Chasmistes* cf. *liorus* from the Black Rock Canyon locality. Hyomandibulae assigned to *Chasmistes* sp. were reported by Bright (1967) from Late Pleistocene sediments in the Thatcher Basin (Miller and Smith, 1981, p. 33). It is absent in the Homestead ichthyofauna that provided a substantial catostomid collection. The paucity of *C. liorus* is thus intriguing and may reflect a restricted spatial distribution of these fish in late Pleistocene Lake Bonneville. Perhaps they were restricted to warmer, shallower bays of the lake near major input rivers. *C. liorus* has been identified from late Holocene contexts adjacent to Utah Lake (Heaton and Smith, 2016).

Order Salmoniformes

Salmonidae—Trouts and Whitefish

cf. *Salvelinus confluentus* (Suckley, 1858)

Bull Trout

Material—Homestead Cave: Premaxilla (1 fragment).

Oncorhynchus cf. *clarkii* (Richardson, 1836)

Cutthroat Trout

Material—Homestead Cave: Vomer (15), basioccipital (2), maxilla (2), premaxilla (31), ceratohyal (1), glossohyal (9).

Oncorhynchus clarkii (Richardson, 1836)
Cutthroat Trout

Material—Deweyville: Vertebra (1), ceratohyal (1), prootic (1), pterotic (2), epiotic (2), cranial fragments (21), frontal (1). Homestead Cave: Basibranchial (2). Black Rock Canyon: Hyomandibular (1), supraethmoid (1), vomer (1), parasphenoid (1), articular–angular (2), quadrate (3), maxilla (1), preopercle (1), opercle (1), subopercle (1), ceratohyal (1), urohyal (1), cleithrum (1), and vertebra (13). Smith Creek Cave: Articular–angular (1), dentary (1), vertebra (25).

Remarks—Cutthroat Trout are present in a number of Lake Bonneville deposits and would have represented the dominant piscivore of the Lake Bonneville ecosystem, ecologically parallel to Lake Trout (*Salvelinus namaycush*) in the Great Lakes of eastern North America. Indeed, we observe that many of the salmonine elements from Homestead Cave (most likely representing *O. clarkii*) were very large and clearly represented very large individual fish. For example, many vertebrae exceeded 9.0 mm in maximum diameter, which based on vertebral measurements from known-sized reference specimens, would have represented live weights over 2.6 kg; some exceed 11.0 mm, suggesting weights over 6.2 kg (see Follett, 1980, p. 15; Follett, 1982; Fig. 12.3).

One the most surprising aspects of our recent work with Lake Bonneville fish faunas is the probable presence of Bull Trout (*S. confluentus*) in this system. A single premaxilla distinct from *O. clarkii* but exhibiting *Salvelinus* features was discovered in the Homestead Cave material and was examined by JMB and GRS (as well as K. Gobalet)—all concurred on the identification of *S. confluentus*. This specimen represents the first specimen record for the genus *Salvelinus* in the Great Basin during Quaternary times. A single historic account of char in the Bonneville basin was, however, provided in the early 19th century by Townsend (1978) whose account has also been considered "dubious" (Hass and McPhail, 1991, p. 2204). The Bonneville system may have been colonized by *S. confluentus* from the Snake River system when Lake Bonneville overflowed into the Snake approximately 18.0 cal ka BP. Alternatively, *S. confluentus* could have reached the Bonneville system sometime after approximately 50.0 cal ka BP when Lake Thatcher of southern Idaho, filled by the Bear River, began to spill southward into the Bonneville basin (Bouchard, 1997; Bouchard et al., 1998; but see also Bright, 1963, 1967). For a species as large as *S. confluentus*, the presence of only a single specimen at Homestead Cave suggests that even by ~13.1 cal ka BP they were uncommon in the Bonneville basin, or at least in Lake Bonneville. If *S. confluentus* was ever abundant in the Bonneville system, their decline appears to have begun before the Pleistocene–Holocene transition. The suggestion that *S. confluentus* was once a native species to cold waters of the Bonneville basin may have implications for the modern management of this species.

Prosopium spilonotus or *abyssicola* (Snyder, 1919)
Bonneville or Bear Lake Whitefish

Material—Homestead Cave: Opercle (80), prootic (45), exoccipital (29), dentary (17), quadrate (7), ceratohyal (4).

Prosopium cf. *spilonotus* (Snyder, 1919)
Bonneville Whitefish

Material—Homestead Cave: Dermethmoid (1), frontal (1), parietal (3), prootic (1), epiotic (2), maxilla (3), infraorbital (2), basioccipital (1), articular (5), quadrate (1), epihyal (3) ceratohyal (3), hyomandibular (11), preopercle (3), subopercle (3), supracleithrum (5), basipterygium (1), vertebra (40).

Prosopium spilonotus (Snyder, 1919)
Bonneville Whitefish

Material—Homestead Cave: Prootic (1), ceratohyal (6), opercle (4), dentary (23), quadrate (4). Hot Springs: Opercle (1). Black Rock: Basioccipital (1).

Prosopium abyssicola (Snyder, 1919)
Bear Lake Whitefish

Material—Homestead Cave: Dentary (8).

Prosopium gemmifer (Snyder, 1919)
Bonneville Cisco

Material—Homestead Cave: Articular (29), basioccipital (23), ceratohyal (10), dentary (302), maxilla (42), opercle (286), prootic (9), hyomandibular (58), parasphenoid (2). Hot Springs: Maxilla (63), dentary (54), articular-angular (63), quadrate (15), hyomandibular (62), preopercle (3), opercle (49), urohyal (40), posttemporal (18), frontal (13), supraoccipital (4), exoccipital (15), basioccipital (72), parasphenoid (10), prootic (17), hypural (9). Black Rock Canyon: Opercle (3), parietal (2), hyomandibular (1), articular-angular (2), maxilla (1), parasphenoid (1).

Remarks—Three of the four whitefish that occur today in the Bonneville drainage—*P. abyssicola*, *P. spilonotus*, and *P. gemmifer* (Fig. 12.8)—have been identified from Lake Bonneville ichthyofaunas. All three, now endemic to Bear Lake, apparently evolved from a *Prosopium williamsoni* ancestor, the more widespread and morphologically least-derived, western whitefish species (Smith and Todd, 1993). When and where the endemic Bonneville basin whitefish speciated remains uncertain, as we discuss later, although the documentation that they all lived in Lake Bonneville indicates that they have existed since the late Pleistocene.

Smith et al. (1968) identified differences in the shape of the hyomandibular between *P. gemmifer* from Lake Bonneville and those from modern Bear Lake. The fossil hyomandibulae of *P. gemmifer* from the Hot Springs locality

FIG. 12.8 Bonneville Cisco (*Prosopium gemmifer*) from Bear Lake, Utah–Idaho. A planktivorous whitefish, cisco were the most abundant fish species in Lake Bonneville.

differed from modern specimens in having a more perpendicular posterolateral ridge and a narrower gap between the dorsal and opercular condyles. The latter difference was evident in the ratio of the distance between the two condyles and the length of the dorsal condyle, which was substantially lower in the fossil form than in the sample of modern specimens from Bear Lake. Those differences were interpreted as evidence for the further specialization (ie, elongation) of the jaw structure in *P. gemmifer* over the last approximately 11.0 ka and were substantial enough to suggest that the Pleistocene and modern forms may even represent separate species (Smith et al., 1968; Smith, 1981). Table 12.3 provides the hyomandibular condylar ratios for the fossil and modern *P. gemmifer* materials provided by Smith et al. (1968) as well as those from the Homestead Cave and an additional modern sample from Bear Lake. The ratios are also provided for the sample of *P.* cf. *spilonotus* hyomandibulae recovered from Homestead Cave, as well as a modern sample of *P. spilonotus* from Bear Lake. These data show that the hyomandibular morphology of the two Lake Bonneville samples of *P. gemmifer*, Hot Springs and Homestead Cave, are similar, with ratios of 0.45 ± 0.017 and 0.49 ± 0.015, respectively. The ratios from the two modern samples of this species are nearly identical and are substantially higher (sample $A = 0.57 \pm 0.015$, sample $B = 0.56 \pm 0.010$) than the Pleistocene samples. The increase in this ratio documented between the fossil and modern samples of *P. gemmifer* is toward *P. spilonotus*. The sample of *P. gemmifer* from Homestead Cave is different from modern *P. gemmifer* and both *P. spilonotus* from Homestead Cave and modern *P. spilonotus* (Table 12.4). However, modern *P. gemmifer* does not differ from either the Homestead Cave sample of *P. spilonotus* or the modern sample of the latter. There is also no difference between the Homestead Cave and modern samples of *P. spilonotus* (Table 12.4). Insofar as these differences are reflecting

TABLE 12.3 Hyomandibular Condylar Ratios for Lake Bonneville and Modern *P. gemmifer* and *P. spilonotus*

Species	*N*	Mean	Range	SD	SE
P. gemmifer (Homestead Cave)	17	0.49	0.35–0.15	0.064	0.015
P. gemmifer (Hot Springs)	27	0.45	0.38–0.54	–	0.017
P. gemmifer (Modern A)[a]	22	0.57	0.51–0.65	–	0.015
P. gemmifer (Modern B)	23	0.56	0.48–0.65	0.047	0.01
P. cf. *spilonotus* (Homestead Cave)	11	0.53	0.46–0.61	0.05	0.015
P. spilonotus (Modern)	25	0.55	0.44–0.67	0.068	0.014

[a]*From Smith et al. (1968).*

TABLE 12.4 Test Statistics for Comparisons of Hyomandibular Condylar Ratios for the Lake Bonneville and Modern *Prosopium gemmifer* and *P. spilonotus*

Species (Sample) Comparison	*t* Statistic	*P*-Value
P. gemmifer (HC)/*P. gemmifer* (Modern sample B)	−4.13	<0.001
P. gemmifer (HC)/*P. spilonotus* (HC)	−1.96	=0.06
P. gemmifer (HC)/*P. spilonotus* (Modern)	−3.25	<0.01
P. gemmifer (Modern B)/*P. spilonotus* (HC)	−1.57	>0.10
P. gemmifer (Modern B)/*P. spilonotus* (Modern)	0.25	>0.50
P. spilonotus (Modern)/*P. spilonotus* (HC)	−1.01	>0.30

approximately 11,000 years of time, *P. gemmifer* has come to be phenotypically more similar to *P. spilonotus* over this interval; *P. spilonotus*, however, has not changed. Although the evolutionary mechanisms and significance of this are not clear, it is possible that introgression of morphological characters from *P. spilonotus* to *P. gemmifer* has occurred during their Holocene occupation of Bear Lake. Indeed, many examples of hybridization are known among coregonin populations including *Prosopium* populations from Bear Lake. Sigler and Miller (1963) reported three fish from Bear Lake they believed to be *P. gemmifer* × *P. spilonotus* hybrids. White (1974) also reported

"*P. gemmiferum*-like" fish from Bear Lake, suggesting they resulted from combinations of *P. gemmifer* and either *P. spilonotus* or *P. abyssicola*.

Order Scorpaeniformes
Cottidae—Sculpins
Cottus cf. *bairdii* (Girard, 1850)
Mottled Sculpin

Material—Homestead Cave: Preopercle (1).

Cottus bairdii (Girard, 1850)
Mottled Sculpin

Material—Homestead Cave: Preopercle (6). Hot Springs: Preopercle (3), subopercle (1). Black Rock Canyon: Preopercle (1), hyomandibular (1). Cathedral Cave: Preopercle (3).

Cottus extensus or *echinatus* (Bailey and Bond, 1963)
Bear Lake or Utah Lake Sculpin

Material—Homestead Cave: Preopercle (12). Cathedral Cave: Preopercle (5).

Cottus extensus (Bailey and Bond, 1963)
Bear Lake Sculpin

Material—Homestead Cave: Preopercle (5). Hot Springs: two partially complete, disarticulated individuals, including most of the skull bones and other bony elements, and assorted disarticulated bones including preopercle (7), hyomandibular (4), subopercle (4), opercle (2), palatine (5), articular–angular (12), dentary (8), premaxillae (11). Black Rock Canyon: Preopercle (4), opercle (1), articular–angular (1), dentary (3).

Remarks—Four species of sculpin are known from the Bonneville basin: *Cottus beldingi*, *C. bairdii*, *C. extensus*, and *C. echinatus*, a Utah Lake endemic that is now extinct. Of the four, *C. bairdii* and *C. extensus* have both been identified in several Lake Bonneville ichthyofaunas including Homestead Cave, Cathedral Cave (UT), Black Rock Canyon, and Hot Springs. The fact that *C. echinatus* has not yet been identified from Lake Bonneville deposits is noteworthy. Smith et al. (1968) suggested that *C. extensus* occurred in Lake Bonneville and Bear Lake during the late Pleistocene and underwent little evolution in the relatively cold and stable environment of Bear Lake during the Holocene. In contrast, they argued that in Utah Lake—the shallow, unstable low-elevation remnant of Lake Bonneville—*C. echinatus* diverged from a relictual *C. extensus* population over the last approximately 11,000 years. This suggestion was based on the historical distributions of these fishes as well as the fact that *C. extensus* was known from the fossil record of Lake Bonneville, but *C. echinatus*, as of the late 1960s, was not. This scenario is thus supported by the current pattern where both *C. echinatus* and *C. bairdii* are demonstrably present in the Lake Bonneville ichthyofauna but *C. echinatus* is not.

It is also noteworthy that the Cathedral Cave (UT) fish fauna is represented exclusively by sculpin and this appears to relate to the unique sampling mechanism involved in forming that deposit. Indeed, ostracod data, marl sediments, and a radiocarbon date of 15,310 ± 60 from an overlying stratum suggest that the material represents a deep-water phase deposit from the Bonneville or Provo level deposit that accumulated under 100–200 m of water (Oviatt, 2000). The presence of *C. extensus* is consistent with this suggestion, since these are the most abundant fish that occur in the greater depths of Bear Lake. These fish are also known to actively burrow into the soft marl sediments in the bottom of the lake (McConnell et al., 1957; Dalton et al., 1965; Sigler and Sigler, 1996). Although *C. bairdii* is a bottom dwelling, cold-water stream fish over most of its range (Sigler and Sigler, 1996), lake populations are also known for a form of this species in all of the Great Lakes of eastern North America (Keleher, 1952; Trautman, 1981, pp. 708–710).

12.4 THE LAKE BONNEVILLE ICHTHYOFAUNA

The fish fauna of Lake Bonneville can be understood from a combination of fossils and inference from the differentiation of recent fish DNA data. Assuming that the nearby tributary species were transients in estuaries and along the shore, the fauna comprised at least 21 species (Table 12.2) including two, large salmonine, top carnivores known from fossils; three endemic whitefish zooplanktivores, three known from fossils; three bottom or rocky shore dwelling sculpins, two known from fossils; several minnows, one abundantly represented by fossils; a large lake sucker known from fossils; a river sucker, abundantly represented by fossils; and two mountain suckers, one represented by fossils. Four small minnows and a small sculpin are inferred from their DNA diversity inside vs outside the basin to have been present in the drainage probably before the lake, but were too small to have left an easily detectable fossil record. The fauna of Bear Lake, a cold, high elevation, deep lake on the Utah–Idaho border, has 8 of the 21 species, and probably had two others, based on DNA, morphology, and historical accounts. All of the native species represented in Bear Lake are now known to also have occurred in Lake Bonneville. The small minnow, *R. osculus* surely occurred in all parts of the basin, probably since the Pliocene, based on ancient and diverged endemic DNA in all parts of the Bonneville basin, including Snake Valley in the west and Thousand Springs in northeast Nevada (Oakey et al., 2004). Patterns of relative abundance appear to be similar as well. The zooplanktivorous *P. gemmifer* (Fig. 12.8), for instance, was the most abundant fish in both Lake Bonneville and Bear Lake, along with Utah Chubs and Utah Suckers, whereas the large piscivorous cutthroat trout was as abundant as expected of the top carnivore in both settings.

On a broader continental scale, the Lake Bonneville fish assemblage also resembled the ecological composition of modern Lake Michigan. Although the two lakes are separated by over 13,000 years of time and ~2000 km of distance, they were similar in size, depth, regional air temperatures (Smith et al., 1968) and their short-lived histories. The Lake Michigan fauna is considerably richer however, reflecting the great diversity of its source fauna in the Mississippi, Missouri, Arctic, and eastern drainages (Bailey and Smith, 1981). The comparison of these two faunas is instructive relative to the importance of sources for colonization, ecological time, and evolutionary time. The Lake Michigan fauna has eight times the number of fish species as the Bonneville fish fauna, yet the lake has been in existence for less than 14,000 years (Hubbs and Lagler, 2004). Paleontological evidence (Shoshani and Smith, 1996) indicates that warm-water catfish and sunfish colonized the Great Lakes through warm, headwater-stream captures, prior to 11,000 BP, while cold-water fishes were colonizing through the cold, glacial outlet channels (Smith, 1985). The comparison shows that 10^3–10^4 years is enough ecological time to build up a fauna of scores of species if the source fauna is near. On the other hand, few species of fishes in Lake Michigan (coregonins) and few species in Lake Bonneville (coregonins and sculpins) have been suspected of having evolved in such a short time, and those examples may be questioned (Smith et al., 2010).

12.5 FISH INFERENCES REGARDING BONNEVILLE BASIN DRAINAGE HISTORY

Hydrographic drainages, if isolated as in the Great Basin, always have distinctive fish faunas. For example, only 4 of the 10 or 12 fish species in the Lahontan drainage occur among the 21 species in the Bonneville system. Differences between major geographic provinces are even greater: only 5 of the 21 species in the Bonneville drainage are native among the 35 species in the Colorado drainage. Therefore, species presence or absence in drainages is a useful source of information—most importantly, shared occupancy by the same species of fish in adjacent drainages is informative about past aquatic connections. On the basis of restriction of fishes to water (and lack of evidence for dispersal of fish eggs on birds' feet along flyways) past connections are assumed to be through lake spillovers, stream captures, and shared headwaters in upland meadows such as Two-Ocean Pass in the Tetons (Evermann, 1895) and Isa Lake in Yellowstone National Park.

Fish fossils and recent fish DNA both provide information concerning paleohydrology of the Great Basin. These data may be studied in the same way that stream-pebble petrology and source can be studied for provenance. Fish differ from stream pebbles in an important way, however—fish migrate long distances upstream just as frequently as they migrate downstream. Provided that the fish fossils are accurately identified and their host sediments are accurately dated by radiogenic or biostratigraphic ages, fish fossils provide information about where and when the species lived and died. Mitochondrial DNA differences between related fish can be used to estimate the date of the splitting event that created two populations if we can calibrate the genetic distances with fossil fish, their associated apomorphies, and their stratigraphic data (Smith et al., 2013).

Fossil fishes and DNA of recent fishes provide information about the timing and size of past outlets and tributaries to Lake Bonneville. Fish distributions help document precursor lakes, potential water balance, possible sources of sediment and salinity, and hydrographic connections that contributed to accumulation of its aquatic fauna. DNA of living fishes in the Great Basin, Snake River, and Colorado River is particularly useful in this regard because each member of the Bonneville fish fauna carries a record of its connection to relatives in or outside the basin and the timing of evolutionary separation from those relatives. This method depends on the principal that when a panmictic population is divided by a barrier, the daughter populations begin to accumulate DNA sequence differences at a quasi-constant rate (the molecular clock) that can be precisely measured (Broham and Penny, 2003).

12.5.1 Snake River to Great Basin Connections

Minnows (chubs, shiners, and dace) are the most common fishes in the Columbia-Snake and Bonneville drainages. Some of their mtDNA genes have been analyzed to determine evolutionary relationships within species and the genetic distances are rather small between drainages, but not as small as predicted if it is assumed that the Bonneville overflow at Red Rock Pass is the first connection. MtDNA sequence divergences between dace and chubs in the Bonneville and Snake Basins is greater than 1%, indicating that the two drainages were connected at times between 3.8 and 1.6 Ma assuming that the rate of substitution in these genes is about 1% per million years (Johnson and Belk, 1999; Dowling unpublished data).

Prior to that, the Salt Lake Formation was broadly connected to the western Snake River Plain, as indicated by shared fish lineages in the late Miocene Cache Valley Member and the Chalk Hills Formation in the Western Snake River Plain. Nine genera in four families of fishes are known from the Junction Hills local fauna of the Cache Valley member (McClellan and Smith, 2016): minnows (Cyprinidae) include pikeminnow (*Ptychocheilus*), chiselmouth chub (*Acrocheilus*), a snail crusher (*Mylocheilus* cf. *M. whitei*), hitch (*Lavinia*), and chub (*Gila*). Suckers (Catostomidae) include a relative of the Utah Sucker (*C. ardens*) and a lake sucker (*Chasmistes*). A bullhead catfish (*Ameiurus*) and a sunfish (*Archoplites*) complete the fish list. All of these except the *Catostomus*, a possible northern Utah endemic, have their closest relatives on the Western Snake River Plain (but species of *Catostomus* also occur in the MioPliocene of the Western Snake River Plain). This aquatic connection is important because it somehow circumvents the hypothetical hot spot bulge of the Twin Falls volcanic field (10.5–8.6) between the eastern and western Snake River Plains. None of these species and only three of the genera are in the Lahontan basin, casting doubt on the hypothesis that the upper Snake River flowed into the Humboldt (Beranek et al, 2006; Houston, 2009). The chiselmouth chub, *Gila* chub, *Catostomus*, and *Chasmistes*, persisted into the Pleistocene Bonneville basin, although the chiselmouth chub went extinct before the Holocene.

The Bonneville basin was also connected to the Virgin, Kanab, and Meadow Valley Wash drainages about 1.1 Ma according to evidence of morphological introgression and genetic distance measurements in dace (*Rhinichthys*) and mountain suckers (*Pantosteus*) from these drainages (Smith et al., 2002; Smith and Dowling, 2008; Unmack et al., 2014; Smith et al., 2016). The Speckled Dace (*R. osculus*) of the Sevier drainage is derived from the Virgin River, not the northern Bonneville basin, and populations of Speckled Dace and mountain suckers in Shoal Creek (Park Canyon Creek) tributary to the Sevier Drainage near Enterprise, Utah, show evidence of introgression with their counterparts in the nearby Meadow Valley Wash Drainage of eastern Nevada, possibly through the region near Modena. The modern irrigation diversions cited by Hubbs and Miller (1948, pp. 29–31) as possible sources of fish transfer across divides mentioned earlier were not responsible for the much older connections indicated by the mtDNA distances. Our DNA methods are not sensitive enough to detect differences evolved in 100–200 years.

Fishes in Snake Valley, at the western edge of the Bonneville basin between the Snake Range and the Deep Creek range, are the expected Bonneville species (Hubbs and Miller, 1948, p. 29) and include an additional form of *Rhinichthys*, formerly thought to be a distinct species (Miller, 1984; Smith et al., 2002). Observers have long struggled, however, to resolve the unusual morphological diversity among populations in Snake Valley. This puzzle was solved when Tom Dowling sequenced the cytochrome *b* gene, of the mitochondrial DNA of several populations and discovered two haplotypes one related to *Rhinichthys* of the Sevier and Virgin rivers and the other related to dace in the Lahontan Basin, but more than a million years older than other populations in that basin. The two mtDNA haplotypes were found in one population that showed evidence of having been mixed with other populations of dace in Snake Valley. The presence of the older dace haplotype as well as the presence of mountain suckers associated with this drainage (Smith, 1966) suggest a connection between the pre-Bonneville and Lahontan drainages as much as 3.3 Ma (Dowling unpublished data).

Dace in the ~1000 km^2 Thousand Springs drainage in northeast Nevada attest to an ancient and long-existent connection between the Bonneville and Snake River drainages. Dace from Rock Springs in the Thousand Springs drainage are part of one of the oldest lineages in the Snake River, according to mtDNA. The Rock Springs dace are morphologically similar to northern Bonneville dace, and Thousand Spring Creek can be traced to below the higher Bonneville shorelines (Hubbs and Miller, 1948, p. 29).

These connections offer insights into the antiquity of the Bonneville and pre-Bonneville fish fauna. Hubbs and Miller (1948) were tempted to explain many of the Bonneville fish species as arrivals through the outlet through Red Rock Pass. DNA data suggest that the entire fish fauna predates Lake Bonneville, but is not as old as the Cache Valley Member of the Salt Lake Formation.

The separation of Bonneville fishes into northern and southern drainage groups was first pointed out by Johnson et al. (2004) on the basis of differentiation in the Leatherside Chub, *S. copei*. Similarly, Smith and Dowling (2008) noted major DNA differences between northern and southern basin *R. osculus*. Unmack et al. (2014) followed with evidence in *P. platyrhynchus* suggestive of over 3 million years of differentiation that is consistent with evidence for water in the basin as early as 3 Ma (Oviatt, 1994).

12.5.2 Evolution of Prosopium Species in Lake Bonneville and Bear Lake

The six North American species of *Prosopium* exhibit profound morphological and geographic diversity. The Round Whitefish (*P. cylindraceum*) occurs from the Great Lakes and across most of Canada to Alaska, the Pygmy Whitefish (*P. coulterii*) occurs in the Rocky Mountains from Montana and Washington to Alaska and the deep waters of Lake Superior. The Rocky Mountain Whitefish (*P. williamsoni*) occurs in the Intermountain West, including the Bear River drainage, north to the Yukon Territory; the other three species occur in Bear Lake, Utah and Idaho, and existed in Lake Bonneville. This is a cold-water group with a special evolutionary history.

As the diversity of *Prosopium* at Hot Springs and Homestead Cave became known, including apparent Holocene morphological change in *P. gemmifer* as described earlier, the question of Lake Bonneville's contribution to the evolution of the Bear Lake species became important (Broughton, 2000b,c). It is tempting to assume that the larger Pleistocene lake was the original place of origin of at least some of the diversity. But Bear Lake is one of the oldest continuous lakes on the North American continent (Kaufman et al., 2009). Bear Lake also has the most direct connection to the long-lived, Pliocene Glenns Ferry Lake on the Western Snake River Plain. This Pliocene lake had one large and one small species of *Prosopium* for up to 2 million years (Stearley and Smith, 2016). Therefore, when the Bear River was tributary to the upper Snake River, it was in the same drainage as the Glenns Ferry Lake. Although the connection was hydrographic, the genetic communication among the fishes was not contemporaneous. Genes of lacustrine *Prosopium* were probably sequestered in fluvial *Prosopium*.

The inference that the Snake River drainage was an important site of *Prosopium* evolution, with six species in the Pliocene and Pleistocene, is a starting point for considering the role of Lake Bonneville, which was also an important part of the Snake River drainage at times. This investigation benefits from the Global Lakes Drilling program, which extracted long cores from the Great Salt Lake sediments (Dean et al., 2002) as well as Bear Lake (Kaufman et al., 2009). The question we ask of the comparison of these cores is: (1) What were the comparative histories of cold *Prosopium* habitat in the two basins? (2) What do these histories tell us about possible times of origin and times of extinction of the Bear Lake *Prosopium* species flock? (Smith and Todd, 1984).

All of the fossil and recent species, except the Round Whitefish, have a significant presence in the Snake-Columbia River drainage. The earliest two, *Prosopium prolixus* and *P.* sp., in the Pliocene, are in no ways "primitive," but are as advanced as any of the modern species. This implies that the group had a much longer history, probably in the Snake-Columbia river drainage. A corollary is that the morphologically divergent Bear Lake Whitefishes are possibly older than Pleistocene.

Bear Lake and Lake Bonneville were exposed to the same regional climatic and precipitation changes, with secondary features governed by local geomorphic and fluvial processes (Reheis et al., 2009). Lake-level fluctuations were actually greater in the Bonneville basin because of its confinement. Bear Lake did not dry out during the last 220,000 years (Kaufman et al., 2009), but the Bonneville basin accumulated salts and saline marsh sediments during dry episodes (Oviatt et al., 1999; Balch et al., 2005). Dry episodes in the Bear Lake basin, by contrast, interrupted the connections between Bear Lake and the Bear River (Kaufman et al., 2009). These two processes provide the most important controls on population dynamics and population genetics of *Prosopium* in the two connected systems. Optimal lake habitat was approximately continuous in Bear Lake, but not in Lake Bonneville, and opportunities for immigration between the lakes were best when both were fresh.

Balch et al. (2005) found four periods of marsh deposits when the Bonneville core site was above lake level and two periods under a deep lake. Deep water was indicated at ~170, 60, 45–35, and 25–15 ka. The Great Salt Lake core records low water levels at 215, 130, 105, and 45 ka. The Bear Lake core records low water levels in response to aridity at the first three of these dry intervals, but records a continuous lake, with a high stand at 47–39 ka, which delivered increased output of sediment to Lake Bonneville (Balch et al., 2005). These multi-millennial cycles surely selected for lacustrine adaptations, but with more serious extinctions and population bottlenecks in Lake Bonneville. Lake Bonneville was possibly a repeated dead end for *Prosopium* and other lacustrine fish evolution.

12.6 BONNEVILLE BASIN FISHES AND LATE QUATERNARY LAKE LEVELS

Although we have learned a great deal about the timing and size of past outlets and tributaries to Lake Bonneville from the collective sample of fish fossils and DNA of recent fishes, analysis of the large, well-dated fossil fish sequence from Homestead Cave provides an opportunity to understand the chronology and nature of change in the fauna as lake levels fluctuated over the last ~13,000 years. Most notably, this sequence provides a singular record on when fish were first eliminated from the lake toward the end of the Pleistocene and the timing and nature of several apparent recolonization events that occurred during the Holocene. We address these issues through analyses of: (1) the chronology of fish bone deposition in relation to reconstructions of lake level history; (2) strontium isotopes derived from the fish remains; (3) change in fish size; and (4) change in taxonomic composition in connection with taxon-specific temperature and salinity tolerances.

Recent geological research suggests that Lake Bonneville declined to elevations approaching those of modern Great Salt Lake by about 13.0 cal ka BP and remained at low altitudes for the next 1400 years. At about 11.6 cal ka BP, the lake transgressed again to a maximum altitude of 1295–1297 m, but regressed about 15 m to altitudes comparable to modern Great Salt Lake a very short time later (Oviatt, 2014, 2015a; Oviatt et al., 2015; Fig. 12.9). Analysis of sediments from the latter transgression, referred to as the Gilbert episode, suggest that the lake was not hypersaline at this time and did not contain brine

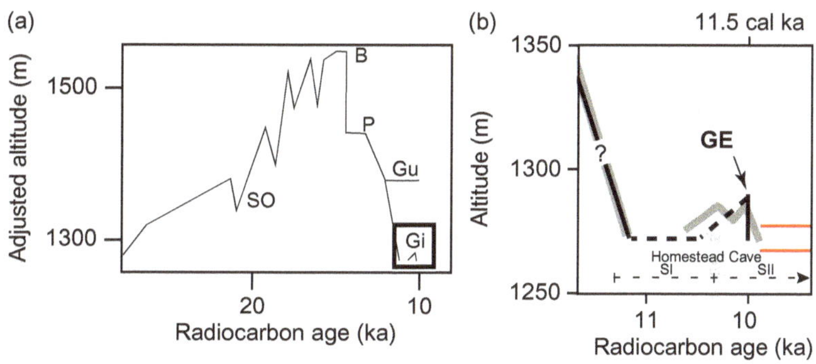

FIG. 12.9 Late Quaternary shoreline history of the Bonneville basin. (a) Generalized hydrograph for Lake Bonneville. *SO* = Stansbury oscillation; *B* = Bonneville shoreline; *P* = Provo shoreline; *Gu* = Gunnison shoreline; *Gi* = Gilbert shoreline; *T* = transgressive phase; *O* = overflowing phase; *R* = regressive phase. The *dark rectangle* shows the area enlarged in (b). (b) Enlargement of the part of (a) that shows the Gilbert episode. The *gray line* represents the interpretation of Oviatt et al. (2005); the *solid black line* is the reconstruction provided in Oviatt (2014); the *thick dashed line* reflects uncertainty in lake level chronology during the pre-Gilbert interval; and the two horizontal *red lines* younger than the Gilbert episode show the upper and lower altitude limits of Holocene Great Salt Lake fluctuations. The *thin dashed line* represents the period of deposition of the Homestead Cave fishes from Stratum I (SI) and Stratum II (SII). *GE* = Gilbert episode. *From Oviatt, C.G., 2014. The Gilbert episode in the Great Salt Lake Basin, Utah. Miscellaneous Publication 14-3, Utah Geological Survey. Panel (a): Modified after Oviatt, C.G., Miller, D.M., McGeehin, J.P., Zachary, C., Mahan, S., 2005. The Younger Dryas phase of Great Salt Lake, Utah, USA. Palaeogeogr. Palaeoclimatol. Palaeoecol. 219, 263–284; Reheis, M.C., Adams, K.D., Oviatt, C.G., Bacon, S.N., 2015. Pluvial lakes in the Western United States—a view from the outcrop. Quatern. Sci. Rev. 97, 33–57.*

shrimp (*Artemia franciscana*) and may have instead been brackish (Oviatt, 2014). Although previous work had suggested that the rich Stratum I fish fauna from Homestead Cave represented fish derived from the Gilbert episode lake (Broughton, 2000a,b; Broughton et al., 2000; Madsen et al., 2001), we propose an alternative interpretation here that is consistent with the current understanding of lake history and refinements in the Homestead Cave chronology.

12.6.1 The Homestead Cave Fish Sequence

Fish remains were heavily concentrated in Stratum I of the cave, although Stratum II and several higher units contain substantial numbers of fish as well (Fig. 12.10). Stratum I consists primarily of degraded organic material, including vast quantities of small vertebrate bone (Grayson, 1998; Madsen, 2000). Stratum I was removed as a single layer because the numerous micro-laminae evident in the unit could not be confidently separated. However, the laminated nature of the Stratum I sediments, together with a difference in more than a 1000 years in radiocarbon ages from the top to the bottom of the stratum (Table 12.5) suggest it is actually composed of multiple depositional events. To increase the stratigraphic resolution within Stratum I, sediment samples of the upper and lower halves of the stratum were collected from the profile and samples were also taken from the upper and lower 10 cm of Stratum I where it was thickest. Stratum Ia is the arbitrarily defined lower half of Stratum I, whereas Stratum Ib is the arbitrarily defined upper half. Five radiocarbon assays on individual *Neotoma* (woodrat) fecal pellets collected from the lowest 5 cm of Stratum Ia, directly overlying bedrock, are statistically indistinguishable and range from 13.1 to 13.0 cal ka BP. Three dates, also from fecal pellets, taken directly from the upper 5 cm of Stratum Ib

FIG. 12.10 View of the mouth Homestead Cave.

TABLE 12.5 Radiocarbon Dates from Homestead Cave (Age Estimates Are Listed Chronologically Within Stratigraphic Units)

Stratum	^{14}C Age (yr BP)	Cal Age (yr BP)[a]	Median Cal Age	Lab. No.	Material
XVII[b]	1020±40	820–1050	940	Beta101877	30 g artiodactyl pellets
XVI[b]	1200±50	990–1260	1130	Beta66940	Charcoal
XV[c]	767±41	690–760	700	AA72777	Bone collagen
XV[d]	980±40	800–960	870	CAMS70099	1 artiodactyl pellet
XV[d]	1790±40	1610–1820	1760	CAMS70100	1 artiodactyl pellet
XV[c]	1890±45	1720–1930	1830	AA72779	Bone collagen
XV[c]	2221±46	2130–2340	2230	AA72784	Bone collagen
XV[e]	2025±775	520–3860	2090	–	*Neotoma* fec. pellet(s)
XV[c]	2073±44	1930–2150	2040	AA72774	Bone collagen
XV[c]	2216±45	2130–2340	2230	AA72782	Bone collagen
XV[c]	2220±44	2140–2340	2230	AA72775	Bone collagen
XV[c]	2237±46	2150–2340	2230	AA72789	Bone collagen
XV[c]	2244±45	2150–2340	2240	AA72783	Bone collagen
XV[c]	2302±78	2120–2700	2320	AA69315	Bone collagen
XV[c]	2472±48	2370–2720	2560	AA69314	Bone collagen
XV[c]	2500±46	2380–2740	2580	AA72785	Bone collagen
XV[c]	2508±46	2380–2750	2590	AA72778	Bone collagen

Continued

TABLE 12.5 Radiocarbon Dates from Homestead Cave (Age Estimates Are Listed Chronologically Within Stratigraphic Units)—Cont'd

Stratum	^{14}C Age (yr BP)	Cal Age (yr BP)	Median Cal Age	Lab. No.	Material
XV[c]	2551±39	2490–2760	2630	AA72788	Bone collagen
XV[c]	2553±46	2490–2960	2630	AA72776	Bone collagen
XV[c]	2559±46	2490–2760	2640	AA72773	Bone collagen
XV[c]	2722±39	2750–2920	2820	AA72786	Bone collagen
XV[c]	2850±48	2850–3140	2970	AA72781	Bone collagen
XV[c]	3031±48	3080–3360	3230	AA72780	Bone collagen
XV[c]	4531±54	4580–5440	5170	AA72787	Bone collagen
XIV[b]	2850±50	2850–3080	2970	Beta103692	30 g artiodactyl pellets
XIII[b]	3480±40	3630–3880	3760	Beta101878	30 g artiodactyl pellets
XII[b]	3400±60	3550–3830	3650	Beta63179	Wood/charcoal
XI[d]	4450±40	4880–5290	5100	CAMS70101	1 artiodactyl pellet
XI[d]	4750±40	5330–5590	5510	CAMS70102	1 artiodactyl pellet
X[b]	5330±65	5950–6270	6110	AA14822	1 artiodactyl pellet
IX[d]	5400±40	6020–6290	6220	CAMS70103	1 artiodactyl pellet
IX[d]	6270±50	7020–7300	7200	CAMS70104	1 artiodactyl pellet
VIII[d]	6190±50	6960–7180	7090	CAMS70105	1 artiodactyl pellet

VIII[d]	6200±40	6990–7240	7090	CAMS70106	1 artiodactyl pellet
VII[c]	6160±35	6960–7160	7070	CAMS134591	Bone collagen
VII[d]	6160±85	6850–7260	7060	AA14824	1 artiodactyl pellet
VII[c]	6165±40	6950–7170	7070	CAMS134580	Bone collagen
VII[b]	6185±105	6760–7310	7080	AA14825	1 artiodactyl pellet
VII[c]	6245±40	7020–7260	7180	CAMS136730	Bone collagen
VII[c]	6275±35	7030–7280	7210	CAMS136729	Bone collagen
VII[c]	6345±35	7170–7410	7280	CAMS134586	Bone collagen
VII[c]	6345±40	7170–7410	7280	CAMS134583	Bone collagen
VII[c]	6355±40	7180–7420	7290	CAMS134581	Bone collagen
VII[c]	6370±50	7180–7420	7310	CAMS136733	Bone collagen
VII[c]	6395±40	7260–7420	7330	CAMS134589	Bone collagen
VII[c]	6435±40	7280–7430	7360	CAMS136732	Bone collagen
VII[c]	6440±35	7290–7430	7370	CAMS136731	Bone collagen
VII[c]	6455±35	7290–7430	7370	CAMS134588	Bone collagen
VII[c]	6505±35	7320–7480	7430	CAMS134590	Bone collagen
VII[c]	6625±35	7440–7550	7520	CAMS134587	Bone collagen
VII[c]	6625±45	7440–7580	7510	CAMS134584	Bone collagen
VII[c]	6675±35	7480–7610	7540	CAMS134582	Bone collagen

Continued

TABLE 12.5 Radiocarbon Dates from Homestead Cave (Age Estimates Are Listed Chronologically Within Stratigraphic Units)—Cont'd

Stratum	^{14}C Age (yr BP)	Cal Age (yr BP)	Median Cal Age	Lab. No.	Material
VII[c]	6685 ± 40	7480–7620	7550	CAMS136728	Bone collagen
VII[c]	6735 ± 40	7520–7670	7600	CAMS136734	Bone collagen
VII[c]	6830 ± 35	7590–7720	7590	CAMS134585	Bone collagen
VII[c]	7005 ± 40	7740–7940	7850	CAMS136735	Bone collagen
VI[b]	7120 ± 70	7790–8150	7950	AA14826	1 artiodactyl pellet
V[b]	8230 ± 70	9020–9410	9200	AA16810	1 artiodactyl pellet
IV[b]	6600 ± 40	7430–7570	7500	CAMS70108	1 artiodactyl pellet
IV[b]	8040 ± 50	8720–9030	8910	CAMS70107	1 artiodactyl pellet
IV[b]	8195 ± 85	9000–9420	9170	AA14823	1 artiodactyl pellet
III[d]	8450 ± 40	9420–9530	9480	CAMS71147	1 artiodactyl pellet
III[d]	9650 ± 40	10,880–11,190	9910	CAMS71148	1 artiodactyl pellet
II[c]	7075 ± 35	7840–7970	7900	CAMS134592	Bone collagen
II[c]	8480 ± 35	9460–9540	9500	CAMS134594	Bone collagen
II[b]	8520 ± 80	9400–9680	9510	AA14821	1 *Celtis* endocarp
II[c]	8615 ± 35	9530–9660	9560	CAMS141271	Bone collagen
II[e]	8675 ± 235	9120–10,270	9750	–	*Neotoma* fec. pellet(s)

II[b]	8790±90	9560–10,160	9840	AA14820	1 *Celtis* endocarp
II[b]	8830±240	9400–10,520	9910	Beta63438	*Celtis* endocarps
II[c]	9295±40	10,370–10,590	10,500	CAMS141278	Bone collagen
II[c]	9370±35	10,510–10,690	10,600	CAMS141279	Bone collagen
II[c]	9380±35	10,520–10,690	10,610	CAMS141280	Bone collagen
II[c]	9405±45	10,520–10,740	10,640	CAMS141274	Bone collagen
II[c]	9475±40	10,580–11,070	10,720	CAMS134595	Bone collagen
II[c]	9515±40	10,660–11,080	10,830	CAMS141272	Bone collagen
II[c]	9835±40	11,200–11,310	11,240	CAMS134597	Bone collagen
II[c]	10,165±40	11,700–12,310	11,860	CAMS141275	Bone collagen
II[c]	10,195±40	11,750–12,080	11,900	CAMS141276	Bone collagen
II[c]	10,200±40	11,750–12,080	11,900	CAMS141273	Bone collagen
II[c]	10,235±40	11,800–12,130	11,970	CAMS134596	Bone collagen
II[c]	10,255±35	11,830–12,140	12,010	CAMS141277	Bone collagen
II[c]	10,310±40	11,960–12,380	12,110	CAMS134593	Bone collagen
II[c]	10,395±40	12,070–12,420	12,270	CAMS134598	Bone collagen
Ib-5 cm[b,f]	10,160±85	11,400–12,120	11,810	AA14819	1 *Neotoma* fec. pellet
Ib-5 cm[e]	10,235±180	11,340–12,550	11,980	–	*Neotoma* fec. pellet(s)
Ib-5 cm[b]	10,350±80	11,940–12,440	12,210	AA14818	1 *Neotoma* fec. pellet

Continued

TABLE 12.5 Radiocarbon Dates from Homestead Cave (Age Estimates Are Listed Chronologically Within Stratigraphic Units)—Cont'd

Stratum	^{14}C Age (yr BP)	Cal Age (yr BP)	Median Cal Age	Lab. No.	Material
I-middle[b]	10,910±60	12,700–12,940	12,780	Beta72205	Bone collagen
Ia-5 cm[b]	11,065±105	12,730–13,090	12,950	AA14817	2 *Neotoma* fec. pellets
Ia-5 cm[e]	11,168±208	12,710–13,410	13,030	–	*Neotoma* fec. pellet(s)
Ia-5 cm[b]	11,181±85	12,810–13,210	13,140	AA16808	1 *Neotoma* fec. pellet
Ia-5 cm[b]	11,263±85	12,960–13,310	13,130	AA16809	1 *Neotoma* fec. pellet
Ia-5 cm[b]	11,270±135	12,820–13,390	13,130	AA14816	1 *Neotoma* fec. pellet

[a]*At two s.d. range calculated using CALIB 7.1.*
[b]*Dates from individual* Neotoma *or artiodactyl fecal pellets collected from the face of the mapped profile at approximately the middle of each stratigraphic unit (from Madsen, 2000).*
[c]*Dates from bone collagen from kangaroo rat* (Dipodomys *spp.) femora (from Terry and Novak, 2015).*
[d]*Dates obtained by Larry Benson from processed artiodactyl fecal pellets in an amalgam of all pellets from a particular stratigraphic unit (from Broughton et al., 2008).*
[e]*Dates were presumably on* Neotoma *fecal pellets obtained by Felisa Smith collected from processed pellets in an amalgam of all the pellets from a particular stratigraphic unit (from Smith and Betancourt, 2013).*
[f]*Dates for Stratum Ib and Stratum Ia were taken from the upper and lower 5 cm of the unit, respectively.*

provided a date range of 12.2–11.8 cal ka BP. The undisturbed microlaminae in Stratum I suggest the age of these fecal pellets can be used to date the associated faunal remains.

A substantial sample of fish was also derived from Stratum II at Homestead Cave and 21 radiocarbon dates for that unit (Terry and Novak, 2015) span the period from 12.3 to 9.5 cal ka BP (excluding an anomalous late date of 7.9 cal ka BP). That range encompasses the Gilbert episode that culminated at ~11.6 cal ka BP. Stratum II was excavated as a single unit without further sampling of materials from the profile as in Stratum I.

12.6.2 Summary and Deposition of the Homestead Cave Fishes

A total of 11 freshwater species, represented by 14,866 specimens, were identified from Homestead Cave (Table 12.6). Several lines of evidence suggest that owls were the agents that deposited the cave fauna: owl pellet cones were observed on the floor of the cave when the excavation began; the deposits contain numerous owl pellets; and many of the bones themselves, including fish bones from Stratum I, still have pellet material adhering to them. Evidence for human and carnivore involvement in the accumulation of the fauna is sparse, and no evidence exists for water-laid deposits (Grayson, 1998; Madsen, 2000). Moreover, quantitative analyses of skeletal part representation, bone fragmentation, and surface digestive damage of the Homestead Cave fish, in comparison to those variables derived from fish remains from modern owl pellets and many other depositional agents (humans, coyote [*Canis latrans*], diurnal raptors), strongly implicate large owls as the primary source of the fish materials in this site (Broughton et al., 2006).

Only two of the raptor taxa identified in the lower strata (I–III) of the deposit (Livingston, 2000; Wolfe and Broughton, 2016), routinely roost and nest in deep, open caves; Barn Owl (*Tyto alba*) and Great Horned Owl (*Bubo virginianus*). Although these owls mainly eat small mammals, they will scavenge fish carcasses and take fish exposed in very shallow water (Errington et al., 1940; Gallup, 1949; Bogiatto et al., 2006; Broughton et al., 2006). The maximum prey size that *B. virginianus*, the larger of the two species, can carry whole is ~1.5 kg (Marti, K., 1997, personal communication). As noted earlier, many of the fish specimens recovered from the cave come from very large individuals (>~2.6 kg estimated live weight; Broughton, 2000c). This evidence suggests that the owls did not take such fish whole, out of deep water, but instead scavenged them as parts. The maximum distance from roost to foraging areas reported for the owls is 5.6 km (Smith, 1971; Marti et al., 2005). Thus, aquatic habitats that supported fish, and that at least occasionally experienced substantial die-offs, likely existed within a radius of ~6 km or less from the cave during the periods of fish bone deposition.

TABLE 12.6 Numbers of Identified Fish Specimens per Taxon by Stratum, Homestead Cave[a]

Taxon	Stratum																					Total
	I	*Ia*	*Ib*	*I–II*	*II*	*III*	*IV*	*V*	*VI*	*VII*	*VIII*	*IX*	*X*	*XI*	*XII*	*XIII*	*XIV*	*XV*	*XVI*	*XVII*	*XVIII*	
Salmoninae	701	258	23	–	21	2	6	1	2	1	–	1	–	–	2	1	1	–	1	–	1	1022
cf. *Salvelinus confluentus*	1	0	–	–	–	–	–	–	–	–	–	–	–	–	–	–	–	–	–	–	–	1
Oncorhynchus cf. *clarkii*	40	20	–	–	–	–	–	–	–	–	–	–	–	–	–	–	–	–	–	–	–	60
Oncorhynchus clarkii	1	1	–	–	–	–	–	–	–	–	–	–	–	–	–	–	–	–	–	–	–	2
Prosopium sp.	945	5944	449	–	111	19	32	–	10	6	3	5	2	2	–	1	4	–	–	2	1	7536
Prosopium cf. *spilonotus*	79	5	–	–	–	–	–	–	–	–	–	–	–	–	–	–	–	–	–	–	–	84
Prosopium spilonotus	27	10	3	–	1	–	–	–	–	–	–	–	–	–	–	–	–	–	–	–	–	41
P. spilonotus/ abyssicola	93	87	2	–	14	–	1	–	–	–	–	–	–	–	1	–	–	–	–	–	–	198
Prosopium abyssicola	2	6	–	–	–	–	–	–	–	–	–	–	–	–	–	–	–	–	–	–	–	8
Prosopium gemmifer	225	499	37	–	15	1	–	–	–	1	–	–	–	–	–	–	–	–	–	–	–	778
Cypriniformes	1167	471	58	–	44	10	28	4	19	10	8	22	13	45	224	48	23	1	7	38	5	2245
Cyprinidae	34	35	3	–	3	3	5	–	1	1	–	1	–	2	14	4	2	–	–	12	–	120

Gila atraria	443	191	26	–	12	7	17	1	13	10	3	14	6	13	59	40	17	2	20	63	1	958
Richardsonius balteatus	–	–	–	1	–	–	–	–	–	–	–	–	–	–	–	–	–	–	–	–	–	1
Catostomus sp.	900	299	24	–	19	4	20	2	7	6	7	5	2	3	25	6	9	2	1	2	–	1343
Catostomus ardens	134	22	7	–	2	–	4	–	1	1	–	1	–	–	5	1	1	–	–	–	–	179
Pantosteus virescens	–	–	–	–	–	–	–	–	1	–	–	3	1	–	–	–	–	–	–	–	–	5
Cottus sp.	53	178	9	–	13	5	1	–	–	–	–	–	–	–	–	–	–	–	–	–	–	259
Cottus cf. *bairdii*	0	1	–	–	–	–	–	–	–	–	–	–	–	–	–	–	–	–	–	–	–	1
Cottus bairdii	2	4	–	–	–	–	–	–	–	–	–	–	–	–	–	–	–	–	–	–	–	6
C. extensus/ echinatus	3	9	–	–	2	–	–	–	–	–	–	–	–	–	–	–	–	–	–	–	–	14
Cottus extensus	0	5	–	–	–	–	–	–	–	–	–	–	–	–	–	–	–	–	–	–	–	5
Total	4850	8045	641	1	257	51	114	8	54	36	21	52	24	65	330	101	57	5	29	117	8	14,866

[a]*Specimens identified are from the 1/4″ and 1/8″ sample fractions from Homestead Cave; nomenclature follows Eschmeyer (2015). The osteological criteria used to identify these specimens are discussed in Broughton (2000b,c).*

12.6.3 Strontium Ratios, the Homestead Fishes, and Paleolake Levels

Strontium has four stable isotopes and the ratios of two them, ^{87}Sr and ^{86}Sr, are widely used as tracers in geological analyses. In the context of lake level analyses, it has been established that $^{87}Sr/^{86}Sr$ ratios of lake water are identical to the ratio of any carbonate (eg, tufa, shell) or apatite (eg, fish bone) growing in that water—the $^{87}Sr/^{86}Sr$ ratio of paleowaters can thus be reconstructed from analysis of these materials (Quade, 2000a,b; see also Hart et al., 2004). To reconstruct paleolake levels from $^{87}Sr/^{86}Sr$ ratios of fossil material requires either: (1) lake systems with multiple basins, each fed by rivers with unique $^{87}Sr/^{86}Sr$ ratios, as in Lake Bonneville; or (2) inflowing rivers in which the $^{87}Sr/^{86}Sr$ ratios change with recharge fluctuations. The main rivers in this setting are the Bear ($^{87}Sr/^{86}Sr=0.71926$) in the north and the Beaver/Sevier ($^{87}Sr/^{86}Sr=0.7074$) in the south (Jones and Faure, 1972; Bouchard, 1997). Analysis of tufas and shell from the high (Provo and Bonneville) shorelines shows that the $^{87}Sr/^{86}Sr$ ratios of paleolake waters in an integrated Lake Bonneville were 0.7114–0.7119, which represents some average of all values of rivers feeding into the lake (Quade, 2000a).

During the terminal Pleistocene when Lake Bonneville dropped below the Old River Bed threshold at 1390 m, it segmented and the $^{87}Sr/^{86}Sr$ ratio of at least some component water bodies shifted toward the value of local river inputs. This "splitting" event is clearly recorded in the $^{87}Sr/^{86}Sr$ ratios of carbonates associated with the smaller terminal Pleistocene lake in the southern basin. Shells from the southern basin filled by paleolake Gunnison have low ratios (0.7098), consistent with input from the Sevier/Beaver River. Tufa and shell from the northern basin collected from what was interpreted as the Gilbert shoreline give ratios of 0.7120–0.7144 (see Oviatt (2014) for discussion on the lack of evidence for a basin-wide Gilbert shoreline). Mass balance estimates from modern rivers suggest that these elevated ratios are not the result of isolation of the northern basin from the southern rivers (Bouchard, 1997). Instead, the $^{87}Sr/^{86}Sr$ ratio of the Bear River, the main source of Sr in the northern basin, may have shifted toward higher values as the lake dropped. Although the exact causes are unclear, the key point here is that lower Gilbert shoreline elevations (ie, $\sim$1300 m) in the northern basin yield consistently higher $^{87}Sr/^{86}Sr$ ratios (0.7120–0.7144) than those of the higher, integrated lake (0.7114–0.7119) (Quade, 2000a).

The lake level in which the Stratum I fish lived can be broadly estimated by the $^{87}Sr/^{86}Sr$ ratios of six Stratum I fish specimens. All the fish analyzed, including *C. ardens*, *G. atraria*, and *O. clarkii* returned $^{87}Sr/^{86}Sr$ ratios between 0.7125 and 0.7129 (Broughton et al., 2000; Quade, 2000b). These values are markedly distinct from those of high-shoreline tufas and shell (0.7114–0.7119), and are close to values from materials derived from Gilbert shoreline features. The presence of freshwater fish taxa with such high

$^{87}Sr/^{86}Sr$ ratios implies the presence of a relatively shallow but freshwater lake within the raptor foraging radius (~6 km) of the cave between 13.1 and 11.8 cal ka BP.

12.6.4 Change in Fish Size

The demographic structure of vertebrate populations is determined by age-specific patterns of fertility and mortality. Variation in these parameters can not only mechanically alter the distribution and abundance of individuals across age classes in a population, but can provide the selective context for the emergence and persistence of different life history strategies (eg, Gadgil and Bossert, 1970; Stearns, 1992; Charnov, 1993). In the context of intermountain fishes, Smith (1981) developed a specific model linking intraspecific variation in life-history strategies to local regimes of habitat stability and adult mortality. This model, consistent with general predictions from life history theory (eg, Roff, 1984; Charnov, 1993), has implications for change in fish-body size in populations that endure high rates of adult mortality:

> *Many intermountain minnows, suckers, and trouts are selected locally for large size by increased adult survival and consequent late reproduction in large habitats. Annual fluctuations that reduce habitats seasonally, thus causing heavy mortality, lead to persistence of phenotypes that reproduce early at the expense of later growth ... When adult mortality is low individuals leave more descendants by growing larger and producing more young over several seasons.*
>
> (Smith, 1981, pp. 126, 162)

Positive relationships between habitat size, as a measure of environmental stability, and maximum adult size have been documented in many western fish species (Smith, 1981), including several of the cypriniform taxa represented at Homestead Cave, namely, *G. atraria* and *P. virescens*. Insofar as the Homestead fish assemblage resulted from a progression of die-offs and elevated levels of adult mortality associated with ever increasing temperature and salinity we anticipate a shift to smaller-sized fishes across the period over which the lake lost its fish.

Vertebrae from Cypriniformes identified from Stratum Ia and Ib were selected for analysis of size change (Table 12.7). The maximum diameters of vertebral centra, as a measure of size, differ distinctively between Stratum Ia and Ib; the later deposit (Ib) has a smaller maximum size and exhibits a significantly smaller mean size than Stratum Ia (Mann–Whitney $U=8546.5$, $P<0.05$). Since the cypriniform vertebrae were not differentiated below the order level, it is possible that this decrease in size is reflecting species composition change—for instance, increasing proportional abundances of the smaller *G. atraria* relative to the larger *C. ardens*. However, no significant difference occurs in the relative abundance of these taxa between Stratum Ia and Ib ($X^2=1.09$, $P=0.30$).

TABLE 12.7 Vertebral Widths (mm) of the Stratum I Cypriniformes

Stratum	*N*	Mean[a]	Range	Standard Deviation
Ia	365	4.40	1.26–9.20	1.46
Ib	56	4.02	1.74–7.62	1.38

[a]*As a frame of reference, the mean vertebral widths of a modern 450 mm SL* Catostomus ardens *and a 202 mm SL* P. virescens *are 8.76 and 2.84 mm, respectively.*

Since fish size changed in the direction predicted for conditions of elevated adult mortality, the fauna appears to have accumulated over a period of time long enough for size change to occur, that is, over several episodes of substantial mortality, perhaps seasonally as the lake trended downwards.

12.6.5 Change in Taxonomic Abundances

Change in the relative abundance of fish taxa that can and cannot tolerate elevated water temperature or total dissolved solids, especially salinity, can also inform on lake-level fluctuations across the period represented by the Stratum I deposits. Of the taxa represented at the cave, *O. clarkii* and *G. atraria* are the most tolerant of higher salinities—up to ~19,000 mg/L total dissolved solids (Westenfelder et al., 1988; Nelson and Flickinger, 1992; Dickerson and Vinyard, 1999; Wright, 2009) and these taxa are significantly overrepresented in Stratum Ib, compared to Stratum Ia ($X^2=4.31$, $P<0.05$; Table 12.8). The fishes from the deposit that are tolerant of warmer (>~27°C) water temperatures, *C. ardens* and *G. atraria* (Beitinger et al., 2000; Broughton, 2000c; and references therein) are significantly overrepresented in Stratum Ib as well ($X^2=7.41$, $P\ll0.01$; Table 12.8). These changes are consistent with other data that suggest the assemblage resulted from a series of die-offs as salinity and temperature exceeded the tolerances of the various taxa making up the Lake Bonneville fauna.

12.6.6 Fish Die-Offs of Terminal Pleistocene Lake Bonneville: 13.1–11.8 cal ka BP

The fish materials from Homestead Cave are clearly concentrated in Stratum I (Table 12.4)—they account for over 26% of the vertebrate assemblage in this unit but comprise, in most cases, far less than 1% of the fauna in the higher strata. To further pinpoint the stratigraphic break in the cave deposits, after which fish remains became far less abundant, we examined the distribution of fish materials within Stratum I itself. In the bulk sediment sample taken from the lower 10 cm of the 30-cm-thick Stratum I, fish comprise

TABLE 12.8 **Numbers of Identified Specimens for Salinity- and Temperature-Tolerant and Intolerant Fish Taxa from Stratum I, Homestead Cave**

	Stratum		Stratum	
Taxon	*Ia*	%	*Ib*	%
Salinity tolerant				
Gila atraria	226	3.0	29	4.9
Salmoninae	279	3.7	23	3.9
Total	505	6.7	52	8.9
Salinity intolerant				
Prosopium	6551	86.5	491	84.2
Catostomus	321	4.2	31	5.3
Cottus	197	2.6	9	1.5
Total	7069	93.3	531	91.1
Temperature tolerant				
Gila atraria	226	3	29	5.0
Catostomus	321	4.2	31	5.3
Total	547	7.2	60	10.3
Temperature intolerant				
Prosopium	6551	86.5	491	84.2
Salmoninae	279	3.7	23	4.0
Cottus	197	2.6	9	1.5
Total	7027	92.8	523	89.7

23.5% (807 of 3434 specimens) of the entire vertebrate faunal assemblage. By contrast, fish comprise only 0.02% (10 of 3896 specimens) of the vertebrate materials from the upper 10 cm—a figure very similar, with several noteworthy exceptions, to the relative abundance of fishes throughout the rest of the column. The difference in the frequencies of fish and nonfish vertebrate specimens across these divisions is highly significant ($X^2=955.34$, $P \ll 0.0001$). Fish remains are essentially absent from the upper one-third of Stratum I and are confined to Stratum Ia and the lowest microlaminae of Stratum Ib. The die-offs of the Lake Bonneville fishes thus appear to have begun at about 13.1 cal ka BP but ended prior to ~11.8 cal ka BP.

Occupancy of a shallow-elevation lake (or lakes) perhaps near the Gilbert episode altitude or lower is suggested from $^{87}Sr/^{86}Sr$ ratios of the Stratum I fish, and the data on size and relative taxonomic abundance suggest the fauna resulted from a series of die-offs that occurred when the lake dropped near this level. Insofar as fish deposition began at 13.1 cal ka but stopped no later than 11.8 cal ka, the Stratum I fauna was most likely derived from Lake Bonneville during the final centuries of the regressive phase and from the postregressive phase when lake elevations were near those of modern Great Salt Lake. Most notably, it is at this time when the brackish tolerant ostracod species, *Limnocythere sappaensis*, first appears in the Lake Bonneville sequence and is indicative of lower lake levels and increases in water temperature and salinity (Oviatt, 2015b). Brine shrimp (*A. franciscana*) cysts and pellets that mark hypersaline conditions do not, however, appear until after the Gilbert episode during the earliest Holocene (11.5–10.2 ka BP).

12.6.7 Fish Recolonize the Gilbert Episode Lake

Although fish are largely absent from the upper portion of Stratum Ib, as noted earlier, suggesting fish were eliminated from the lake sometime before ~11.8 cal ka, fish are present again in the Stratum II deposit (NISP = 257), and comprise 3.0% of the vertebrate fauna from that unit. These materials thus appear to represent a return of fish to the lake basin. Recent research suggests that Gilbert episode lake levels peaked at 11.6 cal ka BP (Oviatt, 2014) and that date is bracketed by the 21 radiocarbon assays from Stratum II at Homestead Cave that range from 12.3 to 9.5 cal ka BP. Stratum II thus includes the Gilbert episode lake but also the first two millennia (11.5–9.5 cal ka BP) of the hypersaline Great Salt Lake.

Analyses of ostracodes from Gilbert episode sediments also suggest the lake may have been brackish, again based largely on the presence of *L. sappaensis*, although that taxon occurs in waters with a range of total dissolved solids between about 500 and 10,000 mg/L. Water at the very low end of this range (<~1000 mg/L is fresh water) would have been tolerable for the entire Lake Bonneville fish fauna and the more salinity-tolerant taxa (ie, *O. clarkii*, *G. atraria*) would have been able to readily withstand values at the high end (Nelson and Flickinger, 1992; Weber-Scannell and Duffy, 2007; Wright, 2009). Indeed, both *O. clarkii* and *Siphateles bicolor* (Tui Chub) survive in Walker Lake, Nevada, where total dissolved solids heavy in sodium have ranged from 13,000 and 19,000 mg/L since the 1990s. (Tui Chub is a close relative of Utah Chub with similar adaptations and histories.) Many other minnows readily tolerate water in excess of 4000 mg/L (Pimentel and Buckley, 1983; Nelson and Flickinger, 1992).

Although the sample size of fish bone from Stratum II is considerably smaller than that from Stratum I, the assemblage is nonetheless represented by 7 of the 11 species identified from the latter unit. This suggests that nearly the entire Lake Bonneville fish fauna may have recolonized the basin during the deposition of the Gilbert-aged sediments of Stratum II. It is possible, however, that the fish in Stratum II owe their presence to limited mixing with Stratum I.

The mixing issue can, however, be directly addressed through an examination of the relative taxonomic abundance of the Stratum I and II faunas. Insofar as taxa more resilient to high water temperature and salinity were able to establish larger populations during the interval including the Gilbert episode, its regressive phase, and the emergence of Great Salt Lake, the resulting Stratum II fauna should exhibit higher relative frequencies of these fishes. If, on the other hand, the fish materials from Stratum II were derived from mixing with Stratum I, we would not expect such differences.

Chi-square tests document significant increases in the relative abundance of temperature- and salinity-tolerant fish in Stratum II compared to both Stratum Ia and Ib (Tables 12.9 and 12.10). Moreover, a significant linear increase (using Cochran's test of linear trends; see Cannon, 2001) in the relative abundance of temperature- and salinity-tolerant fish specimens is indicated across the full sequence of Stratum Ia, Ib, and II (Table 12.10). These analyses strongly suggest that the Stratum II fish materials did not result from bioturbative mixing but were instead derived from the foraging behavior of owls using fish that resulted from die-offs of the Gilbert episode lake(s).

12.6.8 Lake Level Fluctuations of Great Salt Lake

The Holocene history of Great Salt Lake remains poorly understood, although recent summaries of the limited available data suggest the lake may not have exceeded much above historic period altitudes (about 1280 m) over the last ~10.2 ka years (Oviatt, 2014; Oviatt et al., 2015). Several lake transgressions during the Holocene have, however, been proposed based in part on inferred climatic conditions and limited core and shoreline data (eg, Currey et al., 1984; Murchison, 1989b; Madsen et al., 2001). Most recently, Patrickson et al. (2010) suggested an early Holocene highstand at 1360 m based on a 7.51 ± 0.04 ^{14}C ka BP (8.2–8.4 cal ka BP) date from Stansbury Island lacustrine sediments, although others interpret the deposit as colluvial (Oviatt et al., 2015). Late Holocene transgressions have also been proposed, most notably one between ~3.2 and 2.0 cal ka BP that may have transgressed to a highstand of 1287 m and flooded the Great Salt Lake Desert (Mehringer, 1985; McKenzie and Eberli, 1987; Murchison, 1989b). A transgression near ~1.0 cal ka BP has also been inferred from shoreline data (ie, the "Fremont beach") that suggest the lake reached an elevation of ~1283 m (Currey et al., 1984; Murchison, 1989b; Currey, 1990; Madsen et al., 2001); this episode may be associated with increased precipitation and is synchronous with the agriculturally based, Fremont cultural complex of the eastern Great Basin and Colorado Plateau (Madsen and Simms, 1998).

TABLE 12.9 Numbers of Identified Specimens for Salinity- and Temperature-Tolerant and Intolerant Fish Taxa from Stratum II, Homestead Cave

	Stratum	
Taxon	*II*	%
Salinity tolerant		
Gila atraria	15	7.1
Salmoninae	21	9.9
Total	36	16.9
Salinity intolerant		
Prosopium	141	66.2
Catostomus	21	9.9
Cottus	15	7.0
Total	177	83.1
Temperature tolerant		
Gila atraria	15	7.0
Catostomus	21	9.9
Total	36	16.9
Temperature intolerant		
Prosopium	141	66.2
Salmoninae	21	9.9
Cottus	15	7.0
Total	177	83.1

The *G. atraria* (Utah Chub) record from Homestead Cave provides an additional source of information on the Holocene history of Great Salt Lake (Broughton et al., 2000). Although *G. atraria* has never inhabited Great Salt Lake historically, this species is tolerant of relatively high salinity levels and warm-water temperatures and is the only member of the Lake Bonneville fauna that now occupies low-elevation creeks, springs, and marshes of the northern Bonneville basin. However, based on genetic analyses of modern populations of *G. atraria* in the region, Rosenfeld (1991, p. 89) suggested that during wetter periods of the Holocene, Great Salt Lake may have become sufficiently diluted to support this species. Since Great Salt Lake sediment cores

TABLE 12.10 Chi-Square Test Statistics for Comparisons of the Abundance of Temperature- and Salinity-Tolerant and Intolerant Taxa for Strata I and II, Homestead Cave

	Strata		
Taxonomic Abundance Groups	*Ia and II*	*Ib and II*	*Ia, Ib, and II*[a]
Temperature tol./intol.	28.02*	6.43***	31.87*
Salinity tol./intol.	33.56*	10.11**	31.90*

*Significant at $P<0.0005$.
**Significant at $P<0.001$.
***Significant at $P<0.01$.
[a]*Test statistic is for the linear trend component of the total chi-square (see Cannon, 2001).*

show no late Holocene gaps in deposition of brine shrimp, it is possible that fish moved within plumes of freshwater that entered the basin from input rivers to the east and floated west above the higher density hypersaline core of Great Salt Lake. Even if *G. atraria* could not have invaded Great Salt Lake during wetter cycles, the extent of local fresh water marshes should have varied with regional moisture and the elevation of the lake. In any case, change in the Holocene abundance of *G. atraria* near the cave should be linked to higher elevations of Great Salt Lake.

Materials representing the Lake Bonneville fish fauna occur in extremely low frequencies above Stratum II and likely owe their presence in those contexts to limited mixing. However, the *G. atraria* materials in these upper strata may be derived not only from limited mixing with Stratum I and II but also from Holocene-aged scavenging by owls. If owls deposited substantial numbers of *G. atraria* during Holocene transgression–regression cycles of Great Salt Lake, then major frequency peaks of *G. atraria* within the cave deposits should signal them. These data cannot be used to estimate the absolute elevations of the lake and may be more sensitive at registering highstands that were persistent over extended periods—that would allow sustained fish colonizations—rather than highly ephemeral (eg, seasonal) ones that may have even briefly reached considerable altitudes. In other words, high amplitude fluctuations even with high average depths would preclude fish occupation of the lake. The Holocene sequence of deposits at Homestead Cave (Strata III–XVIII) is constrained by 67 radiocarbon dates that span the period from 9.9 to 0.7 cal ka BP (Table 12.5).

G. atraria shows two main peaks in both absolute abundance and density (*G. atraria* NISP/L) in the Holocene strata at the cave, first within Stratum XII and second within Strata XVI and XVII (Table 12.6; Fig. 12.11). These strata provided radiocarbon dates of 3.6, and 1.1 and 0.9 cal ka BP, respectively. Insofar as the chub record is reflecting periods of sustained higher

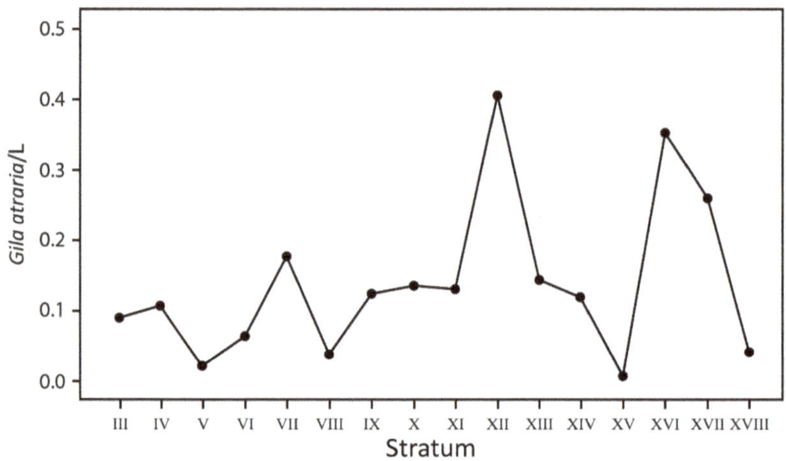

FIG. 12.11 Density (NISP/L) of Utah Chub (*Gila atraria*) by stratum (exclusive of Strata I and II) at Homestead Cave. Specimens identified as "Cyprinidae" are included with *Gila atraria* here because that species represents well over 99% of the identified cyprinids at Homestead Cave.

levels of Great Salt Lake, it suggests that such transgressions were limited to the late Holocene. Moreover, the dates associated with the chub spikes align with previous suggestions based on core and shoreline data for highstands between ~3.4 and ~2.0 cal ka, and again near 1.0 cal ka.

No suggestion occurs from the fish record, however, for a post-Gilbert early Holocene (8.4–8.2 cal ka BP) highstand of Great Salt Lake—strata IV through VI date to this period and chub densities are consistently low in these units. This conclusion was also reached by recent summaries of the shoreline, outcrop, and offshore deposition data and may have paleoclimatic implications (Oviatt et al., 2015). Although climate models and paleoclimatic proxy data from the Bonneville basin and adjacent areas indicate enhanced seasonality and warm and dry summer conditions during the early Holocene (eg, Broughton, et al., 2008; Doner, 2009; Moser and Kimball, 2009; Lundeen et al., 2013), many small mammal and plant macrofossil records have been read to suggest cooler and moister climate at this time (eg, Grayson, 1998, 2000, 2011; Rhode, 2000; Schmitt et al., 2002; Schmitt and Lupo, 2012). Oviatt et al. (2015) have recently proposed a reconciliation of these apparent inconsistencies. Specifically, they suggested that early Holocene climate was dry, but Lake Bonneville-derived groundwater—trapped in the fractured limestone of small interior basin ranges—was steadily discharged, feeding local plant communities at this time. The release of water stored in Homestead Knoll, for instance, may have been enough to produce localized, mesic refugia of early Holocene age for plant and small mammal communities, but insufficient to produce a sustained lake transgression.

Not only is this hypothesis consistent with the Homestead fish record that shows no indication of a post-Gilbert highstand during the early Holocene, but it accords well with our current understanding of Holocene variation in artiodactyl populations in the region. The primary taxa that occur in the Bonneville basin are mule deer (*Odocoileus hemionus*), pronghorn (*Antilocapra americana*), and bighorn sheep (*Ovis canadensis*), and it is unlikely that the restricted spatial extent of groundwater-fed vegetation in the early Holocene would have been enough to affect substantial improvements in the overall quality and quantity of forage to the wide-ranging artiodactyl herds. The high seasonality and dry climate of the early Holocene is, thus, anticipated to have depressed regional artiodactyl population densities.

In addition to a rich record of small vertebrates, Homestead Cave has also produced a singular high-resolution paleontological record of artiodactyl abundances in the region in the form of fecal pellets. Insofar as the number of fecal pellets deposited in the cave reflect the frequency of artiodactyl visits, and that this is an index of the relative abundance of those animals on the landscape, trends in fecal pellet densities (pellets/L) can be used to track variation in the regional artiodactyl population size.

The late Holocene spikes in chub abundances at Homestead Cave (eg, at 3.6 and 1.0 cal ka) are, in fact, closely aligned with dramatic increases in the density of artiodactyl fecal pellet abundances in the Homestead Cave sediments, the latter likely reflecting moisture-driven increases in regional artiodactyl populations at these times (Broughton et al., 2008). As with the density of chub bones, fecal-pellet densities are noticeably depressed in the early Holocene strata of the cave. Moreover, numerous Great Basin archeological faunas exhibit significant increases in the abundance of artiodactyl specimens at the times that chub bones and fecal pellets increase, suggesting that prehistoric human foragers took full advantage of increasing encounters with large game that resulted from ameliorating climate (Byers and Broughton, 2004; Broughton et al., 2008, 2011).

12.7 SUMMARY AND CONCLUSIONS

The study of Bonneville basin fishes with a comprehensive interpretive approach began with Hubbs and Miller in the mid-20th century. Although they made significant strides in our understanding of regional fish biogeography and paleohydrology, they recognized that a deeper integration with geology and a far richer fish paleontological record would lead to still greater advances. Still, neither we suspect could have envisioned that strontium derived from fish bones could be used to estimate the depth of the water the fish lived in or that DNA extracted from modern tissues could inform on species divergence times and places. They also surely could not have dreamed that generations of owls living in the heart of the Bonneville basin at Homestead Cave had already fortuitously assembled a relatively fine-grained, 13,000-year terminal Pleistocene and Holocene fish sequence that could provide a series of inferences on the dynamics of fish populations in relation to climate-based fluctuations in lake levels.

With these methodological advances, the substantial growth of the fish-fossil record, and increasing interdisciplinary work, our understanding of the dynamics of the late Quaternary ichthyofauna of the Bonneville basin has, indeed, grown. As we report here, the fish remains from 10 Bonneville basin sites dating to the late Pleistocene or early Holocene have now been studied, yielding a total 16,172 identified specimens representing 13 of the 21 species known for the drainage—these data show that the fauna was similar to the modern assemblage from Bear Lake, a cold, deep lake on the Utah–Idaho border.

The conjoined DNA and pre-Bonneville fossil fish records suggest extra-basinal connections that offer insights into the antiquity of the region's unique fish fauna. Although Hubbs and Miller (1948) were tempted to explain many of the Bonneville fish species as arrivals through the outlet at Red Rock Pass, the current data suggest that nearly the entire fish fauna predates Lake Bonneville, but is not as old as the Cache Valley Member of the Salt Lake Formation.

The Homestead Cave fish sequence has provided detailed insights on the nature and dynamics of the Bonneville basin ichthyofauna as lake levels fluctuated over the last ~13,000 years. The subfossil fish bones, accumulated by large owls, were concentrated in the lowest strata of the deposit and are represented by 11 of the 13 species that have been derived from Lake Bonneville sediments. The $^{87}Sr/^{86}Sr$ values from the Stratum I fish suggest they were derived from a low-elevation lake and a significant decrease in the size of cypriniform vertebrae documented across two divisions of Stratum I, suggests the fauna resulted from a process involving a series of die-offs. Taxa tolerant of higher salinities (*O. clarkii* and *G. atraria*) and temperatures (*C. ardens* and *G. atraria*) are more abundant in the upper division of Stratum I than in the lower section of this deposit. The terminal regressive and/or postregressive die-offs of the Lake Bonneville fish fauna appear to have occurred rapidly beginning at ~13.1 cal ka, and fish were eliminated from the lake no later than 11.8 cal ka. A return of nearly the entire Lake Bonneville fish fauna, although skewed to higher abundances of salinity- and temperature-tolerant taxa, occurs in the Stratum II deposits, with dates ranging from 12.3 to 9.5 and is most consistent with recolonization during the Gilbert episode.

No post-Gilbert highstands in the early Holocene are registered in the fauna, but frequency peaks of Utah Chub in the upper strata of the cave imply two prominent late Holocene transgressions of Great Salt Lake, one at ~3.6 cal ka and the other at ~1.0 cal ka. Both periods are associated with apparent expansions of artiodactyl herds in the Bonneville basin and the human hunting of those animals. The final moisture pulse at ~1.0 cal ka is coincident with the agriculturally based Fremont cultural complex of the eastern Great Basin and Colorado Plateau.

ACKNOWLEDGMENTS

We thank Virginia Butler, Charles Oviatt, and John Shroder for very helpful comments on the chapter; James Mead for help with the Smith Creek Cave fauna; Bryce Nielson and Scott Tolentino for providing fish reference specimens from Bear Lake; Marlene Lambert-Tempest for Figs. 12.2 and 12.3; Charles Oviatt for Fig. 12.9; Shawn Miller and Isaac Hart for assistance with Figs. 12.1 and 12.11, respectively; and David Madsen for all his work on the Homestead Cave and Cathedral Cave projects, including the recent compilation of the Homestead radiocarbon dates we present here in Table 12.5. Figs. 12.2–12.7 are reprinted here with permission from the American Society of Ichthyologists and Herpetologists. The Homestead Cave analysis was supported by grants from the Legacy Project, Department of Defense, and Hill Air Force Base.

REFERENCES

Bailey, R.M., Smith, G.R., 1981. Origin and geography of the fish fauna of the Laurentian Great Lakes Basin. Can. J. Fish. Aquat. Sci. 38, 1539–1561.

Balch, D.P., Cohen, A.S., Schnurrenberger, D.W., Haskell, B.J., Valero-Garcés, B.L., Beck, J.W., Cheng, H., Edwards, R.L., 2005. Ecosystem and paleohydrological response to Quaternary climate change in the Bonneville Basin, Utah. Palaeogeogr. Palaeoclimatol. Palaeoecol. 221, 99–122.

Beitinger, T.L., Bennett, W.A., McCauley, R.W., 2000. Temperature tolerances of North American freshwater fishes exposed to dynamic changes in temperature. Environ. Biol. Fishes 58, 237–275.

Beranek, L.P., Link, P.K., Fanning, C.M., 2006. Miocene to Holocene landscape evolution of the western Snake River Plain region Idaho: using the SHRIMP detrital zircon provenance record to track eastern migration of the Yellowstone hotspot. Geol. Soc. Am. Bull. 118, 1027–1050.

Bogiatto, R.J., Broughton, J.M., Cannon, V.I., Dalton, K., Arnold, S., 2006. Fish remains dominate barn owl pellets in northwestern Nevada. West. N. Am. Nat. 66, 395–396.

Bouchard, D.P., 1997. Quaternary Bear River Paleohydrogeography Reconstructed from the $^{87}Sr/^{86}Sr$ Composition of Lacustrine Fossils. Unpublished M.S. thesis, Utah State University, Logan, USA.

Bouchard, D.P., Kauffman, D.S., Hochberg, A., Quade, J., 1998. Quaternary history of the Thatcher Basin, Idaho, reconstructed from the $^{87}Sr/^{86}Sr$ and amino acid composition of lacustrine fossils: implications for the diversion of the Bear River into the Bonneville Basin. Palaeogeogr. Palaeoclimatol. Palaeoecol. 141, 95–114.

Bright, R.C., 1963. Pleistocene Lakes Thatcher and Bonneville, Southeastern Idaho. Ph.D. dissertation, University of Minnesota, Minneapolis, USA.

Bright, R.C., 1967. Late-Pleistocene stratigraphy in Thatcher basin, southeastern Idaho. Tebiwa 10, 1–7.

Broham, L., Penny, D., 2003. The modern molecular clock. Nat. Rev. Genet. 4, 216–224.

Broughton, J.M., 2000a. Cathedral Cave fishes. In: Madsen, D. (Ed.), Late Quaternary Paleoecology in the Bonneville Basin, pp. 134–135. Utah Geological Survey, Bulletin 130.

Broughton, J.M., 2000b. Terminal Pleistocene fish remains from Homestead Cave, Utah, and implications for fish biogeography in the Bonneville basin. Copeia 2000, 645–656.

Broughton, J.M., 2000c. The Homestead Cave ichthyofauna. In: Madsen, D. (Ed.), Late Quaternary Paleoecology in the Bonneville Basin, pp. 103–121. Utah Geological Survey, Bulletin 130.

Broughton, J.M., Madsen, D.B., Quade, J., 2000. Fish remains from Homestead Cave and lake levels of the past 13,000 years in the Bonneville Basin. Quatern. Res. 53, 392–401.

Broughton, J.M., Cannon, V.I., Bogiatto, R.J., Arnold, S., Dalton, K., 2006. The taphonomy of owl-deposited fish remains and the origin of the Homestead Cave ichthyofauna. J. Taphon. 4, 69–95.

Broughton, J.M., Byers, D., Bryson, R., Eckerle, W., Madsen, D., 2008. Did climatic seasonality control late Quaternary artiodactyls densities in western North America? Quatern. Sci. Rev. 37, 1916–1937.

Broughton, J.M., Cannon, M., Bayham, F., Byers, D., 2011. Prey body size and ranking in zooarchaeology: theory, empirical evidence, and applications from the northern Great Basin. Am. Antiq. 76, 403–428.

Bryan, A.L., 1979. Smith Creek Cave. In: Tuohy, D.R., Rendel, D.L. (Eds.), The Archaeology of Smith Creek Canyon, Eastern Nevada, 251–262. 394 pp. Nevada State Museum Anthroplogical Papers 17.

Byers, D.A., Broughton, J.M., 2004. Holocene environmental change, artiodactyl abundances and human hunting strategies in the Great Basin. Am. Antiq. 69 (2), 235–255.

Cannon, M., 2001. Archaeofaunal relative abundance, sample size, and statistical methods. J. Archaeol. Sci. 28, 185–195.

Charnov, E.L., 1993. Life History Invariants: Some Explorations of Symmetry in Evolutionary Ecology. Oxford University Press, Oxford.

Currey, D.R., 1990. Quaternary paleolakes in the evolution of semidesert basins, with emphasis on Lake Bonneville and the Great Basin, U.S.A. Palaeogeogr. Palaeoclimatol. Palaeoecol. 76, 189–214.

Currey, D.R., Atwood, G., Mabey, D.R., 1984. Major Levels of Great Salt Lake and Lake Bonneville. Utah Geological and Mineral Survey Map 73.

Dalton, R.A., Lee, T.H., Hesse, J.L., Helm, W.T., 1965. Distribution of sculpin, *Cottus extensus*, in Bear Lake, Utah–Idaho. Proc. Utah Acad. Sci. Arts Lett. 42, 70–73.

Dean, W., Rosenbaum, J., Haskell, B., Kelts, K., Schnurrenberger, D., ValeroGarcés, B., Cohen, A., Davis, O., Dinter, D., Nielson, D., 2002. Progress in global lake drilling holds potential for global change research: Eos. Trans. Am. Geophys. Union 83 (85), 90–91.

Dickerson, B.R., Vinyard, G.L., 1999. Effects of high levels of total dissolved solids in Walker Lake, Nevada, on survival and growth of Lahontan Cutthroat Trout. Trans. Am. Fish. Soc. 128, 507–515.

Doner, L.A., 2009. A 19,000-year vegetation and climate record for Bear Lake, Utah and Idaho. Paleoenvironments of Bear Lake, Utah and Idaho, and its catchment. In: Rosenbaum, J.G., Kaufman, D.S. (Eds.), Paleoenvironments of Bear Lake, Utah and Idaho, and Its Catchment, pp. 217–227. Geological Society of America Special Paper 450.

Errington, P., Hammerstrom, F., Hammerstrom Jr., F.N., 1940. The Great Horned Owl and its prey in north-central United States. Iowa Agric. Exp. Station Res. Bull. 277, 757–850.

Eschmeyer, W.N. (Ed.), 2015. Catalog of Fishes: Genera, Species, References. http://researcharchive.calacademy.org/research/ichthyology/catalog/fishcatmain.asp.

Evermann, B.W., 1895. Two-ocean pass. Popular Science Monthly 47, 175–186.

Follett, W.F., 1980. Fish remains from the Karlo Site (Ca-Las-7), Lassen County, California. J. Calif. Great Basin Anthropol. 2, 114–122.

Follett, W.F., 1982. An analysis of fish remains from ten archaeological sites at Falcon Hill, Washoe County, Nevada, with notes on fishing practices of the ethnographic Kuyuidikade Northern Paiute, pp. 181–203. Nevada State Museum Anthropological Papers 18.

Gadgil, M., Bossert, W., 1970. Life history consequences of natural selection. Am. Nat. 104, 1–24.

Gallup, F.N., 1949. Banding recoveries of *Tyto alba*. Bird Band. 20, 150.

Grayson, D.K., 1998. Moisture history and small mammal community richness during the latest Pleistocene and Holocene, northern Bonneville Basin, Utah. Quatern. Res. 49, 330–334.

Grayson, D.K., 2000. The Homestead Cave mammals. In: Madsen, D.B. (Ed.), Late Quaternary Paleoecology in the Bonneville Basin, pp. 67–89. Utah Geological Survey Bulletin 130.

Grayson, D.K., 2011. The Great Basin: A Natural Prehistory. University of California Press, Berkeley.

Hart, W.S., Quade, J., Madsen, D.B., Kaufman, D.S., Oviatt, C.G., 2004. The $^{87}Sr/^{86}Sr$ ratios of lacustrine carbonates and lake-level history of the Bonneville paleolake system. Bull. Geol. Soc. Am. 116, 1107–1119.

Hass, G.R., McPhail, J.D., 1991. Systematics and distributions of dolly varden (*Salvelinus malma*) and bull trout (*Salvelinus confluentus*) in North America. Can. J. Fish. Aquat. Sci. 48, 2191–2211.

Heaton, T., 1985. Quaternary paleontology and paleoecology of Crystal Ball Cave, Millard County, Utah: with emphasis on mammals and description of a new species of fossil skunk. Great Basin Nat. 45, 337–390.

Heaton, T., Smith, G.R., 2016. Lake suckers, *Chasmistes*, from Mouse Hole Cave, Utah County, Utah. West. N. Am. Nat. 76. In press.

Houston, D.D., 2009. Molecular Systematics and Phylogeography of the Genus *Richardsonius*. PhD dissertation, University of Nevada, Las Vegas.

Hubbs, C.L., Lagler, K.F., 1958. Fishes of the Great Lakes Region. University of Michigan Press, Ann Arbor.

Hubbs, C.L., Lagler, K.F., 2004. Fishes of the Great Lakes Region. University of Michigan Press, Ann Arbor. Revised by G.R. Smith.

Hubbs, C.L., Miller, R.R., 1948. The Great Basin, with Emphasis on Glacial and Postglacial Times. II. The Zoological Evidence, pp. 17–144. Bulletin of the University of Utah 38, Biological Series 10.

Janetski, J., Smith, G.C., 2007. Hunter–Gatherer Archaeology in Utah Valley. Museum of Peoples and Cultures, BYU, Occassional Papers No. 12, Provo, UT.

Jass, C.J., 2007. New Perspectives on Pleistocene Biochronology and Biotic Change in the East-Central Great Basin: An Examination of the Vertebrate Fauna from Cathedral Cave, Nevada. Ph.D. dissertation, University of Texas, Austin.

Johnson, J.B., Belk, M.C., 1999. Effects of predation on life-history evolution in Utah Chub (*Gila atraria*). Copeia 1999 (4), 948–957.

Johnson, J.B., Dowling, T.E., Belk, M.C., 2004. Neglected taxonomy of rare desert fishes: congruent evidence for two species of leatherside chub. Syst. Biol. 53 (6), 841–855.

Jones, L.M., Faure, G., 1972. Strontium isotope geochemistry of Great Salt Lake, Utah. Geol. Soc. Am. Bull. 83, 1875–1880.

Kaufman, D.S., Forman, S.L., Bright, J., 2001. Age of the Cutler Dam Alloformation (Late Pleistocene), Bonneville Basin, Utah. Quatern. Res. 56, 322–334.

Kaufman, D.S., Bright, J., Dean, W.E., Rosenbaum, J.G., Moser, K., Anderson, R.S., Colman, S.M., Heil Jr., C.W., Jiménez-Moreno, G., Reheis, M.C., Simmons, K.R., 2009. A quarter-million years of paleoenvironmental change at Bear Lake, Utah and Idaho. In: Rosenbaum, J.G., Kaufman, D.S. (Eds.), Paleoenvironments of Bear Lake, Utah and Idaho, and Its Catchment, pp. 311–351. Geological Society of America Special Paper 450.

Keleher, J.J., 1952. Notes on fishes collected from Lake Winnipeg region. Can. Field Nat. 66, 170–173.

Lee, D.S., Gilbert, C.R., Hocutt, C.H., Jenkins, R.E., McAllister, D.E., Stauffer Jr., J.R., 1980. Atlas of North American Freshwater Fishes. North Carolina State Museum of Natural History, Raleigh.

Livingston, S., 2000. The Homestead Cave avifauna. In: Madsen, (Ed.), Late Quaternary Paleoecology in the Bonneville Basin, pp. 91–102. Utah Geological Survey, Bulletin 130.

Lundeen, Z., Brunelle, A., Burns, S.J., Polyak, V., Asmerom, Y., 2013. A speleothem record of Holocene paleoclimate from the northern Wasatch Mountains, southeast Idaho, USA. Quatern. Int. 310, 83–95.

Madsen, D.B. (Ed.), 2000. Late Quaternary Paleoecology in the Bonneville Basin. Utah Geological Survey, Bulletin 130.

Madsen, D.B., Simms, S.R., 1998. The Fremont complex: a behavioral perspective. J. World Prehist. 12, 255–336.

Madsen, D.B., Rhode, D., Grayson, D.K., Broughton, J.M., Livingston, S.D., Hunt, J., Quade, J., Schmitt, D.N., Shaver III, M.W., 2001. Late Quaternary environmental change in the Bonneville Basin, western USA. Palaeogeogr. Palaeoclimatol. Palaeoecol. 167, 243–271.

Madsen, D.B., Schmitt, D., Page, D., 2015. The Paleoarchaic occupation of the Old River Bed Delta. University of Utah Press, Salt Lake City, USA. University of Utah Anthropological Papers 128.

Marti, C.D., Poole, A.F., Bevier, L.R., 2005. Barn Owl (*Tyto alba*). In: Poole, A. (Ed.), The Birds of North America Online. Cornell Lab of Ornithology, Ithaca.

McConnell, W.J., Clark, W.J., Sigler, W.F., 1957. Bear Lake: Its Fish and Fishing. Utah State Department of Fish and Game, Idaho Department of Fish and Game, Wildlife Management Department of Utah State Agricultural College, Logan.

McClellan, P., Smith, G. R., 2016. Fishes from late Miocene Cache Valley beds, Northern Utah, and Relationships to the Snake River Plain, In: Fink, W.L.,Carpenter, (Eds.), Late Miocene to Modern Fishes of the Snake River Plain and Northern Great Basin. University of Michigan Museum of Zoology Miscellaneous Publications, vol. 204. (in prep.).

McKenzie, J.A., Eberli, G.P., 1987. Indications for abrupt Holocene climatic change: late Holocene oxygen isotope stratigraphy of the Great Salt Lake, Utah. In: Berger, H.W., Labeyrie, L.D. (Eds.), Abrupt Climate Change. Reidel, Dordrecht, pp. 127–136.

Mead, J.I., Thompson, R.S., Van Devender, T.R., 1982. Late Wisconsinan and Holocene fauna from Smith Creek Canyon, Snake Range, Nevada. Trans. San Diego Soc. Nat. Hist. 20, 1–26.

Mehringer Jr., P.J., 1985. Late-Quaternary pollen records from the interior Pacific Northwest and northern Great Basin of the United States. In: Bryant, V.M. (Ed.), Pollen Records of Late-Quaternary North American Sediments. American Association of Stratigraphic Palynologists Foundation, Dallas, Texas, pp. 167–189.

Miller, R.R., 1984. *Rhinichthys deaconi*, a New Species of Dace (Pisces: Cyprinidae) from southern Nevada. 1–21. Occasional Papers of the University of Michigan Museum of Zoology 707.

Miller, R.R., Smith, G.R., 1981. Distribution and Evolution of *Chasmistes* (Pisces: Catostomidae) in Western North America. Occasional Papers of the Museum of Zoology University of Michigan 696.

Morrison, R.B., 1965. Lake Bonneville: Quaternary Stratigraphy of Eastern Jordan Valley, South of Salt Lake City, Utah. U.S. Geological Survey, Professional Paper No. 477.

Moser, K.A., Kimball, J.P., 2009. A 19,000-year record of hydrologic and climatic change inferred from diatoms from Bear Lake, Utah and Idaho. In: Rosenbaum, J.G., Kaufman, D.S. (Eds.), Paleoenvironments of Bear Lake, Utah and Idaho, and Its Catchment, pp. 229–246. Geological Society of America special papers 450.

Murchison, S.B., 1989a. Utah chub (*Gila atraria*) from the latest Pleistocene Gilbert shoreline, west of Corrine Utah. Great Basin Nat. 49, 131–133.

Murchison, S.B., 1989b. Fluctuation History of Great Salt Lake, Utah, During the Last 13,000 years. PhD dissertation, Department of Geography, University of Utah.

Nelson, S.M., Flickinger, S.A., 1992. Salinity tolerance of Colorado Squawfish, *Ptychocheilus lucius* (Pisces: Cyprinidae). Hydrobiologia 246, 165–168.

Oakey, D.D., Douglas, M.E., Douglas, M.R., 2004. Small fish in a large land-scape: diversification of *Rhinichthys osculus* (Cyprinidae) in western North America. Copeia 2004, 207–221.

Oviatt, C.G., 1984. Lake Bonneville Stratigraphy at the Old River Bed and Leamington, Utah. PhD dissertation, University of Utah, Salt Lake City.

Oviatt, C.G., 1994. Review of the Quaternary Geology of the Sevier and Black Rock Deserts. In: Blackett, R.E., Moore, J.N. (Eds.), Cenezoic Geology and Geothermal Systems of Southwestern Utah, 97–103. Utah Geological Association Publication 23.

Oviatt, C.G., 2000. Ostracodes. In: Madsen, D.B. (Ed.), Late Quaternary Paleoecology in the Bonneville Basin, pp. 132–133. Utah Geological Survey, Bulletin 130.

Oviatt, C.G., 2014. The Gilbert Episode in the Great Salt Lake Basin, Utah. Miscellaneous Publication 14-3, Utah Geological Survey.

Oviatt, C.G., 2015a. Chronology of Lake Bonneville, 30,000 to 10,000 yr B.P. Quatern. Sci. Rev. 110, 166–171.

Oviatt, C.G., 2015b. Ostracodes in Pleistocene Lake Bonneville, eastern Great Basin, North America. Hydrobiologia. 2015, 1–11.

Oviatt, C.G., McCoy, W.D., Reider, R.G., 1987. Evidence for a shallow early or middle Wisconsin-age lake in the Bonneville Basin, Utah. Quatern. Res. 27, 248–262.

Oviatt, C.G., Thompson, R.S., Kaufman, D.S., Bright, J., Forester, R.M., 1999. Reinterpretation of the Burmester core, Bonneville basin, Utah. Quatern. Res. 52, 180–184.

Oviatt, C.G., Madsen, D.B., Schmitt, D.N., 2003. Late Pleistocene and early Holocene rivers and wetlands in the Bonneville Basin of western North America. Quatern. Res. 60, 200–210.

Oviatt, C.G., Miller, D.M., McGeehin, J.P., Zachary, C., Mahan, S., 2005. The Younger Dryas phase of Great Salt Lake, Utah, USA. Palaeogeogr. Palaeoclimatol. Palaeoecol. 219, 263–284.

Oviatt, C.G., Madsen, D.B., Miller, D.M., Thompson, R.S., McGeehin, J.P., 2015. Early Holocene Great Salt Lake, USA. Quatern. Res. 84, 57–68.

Patrickson, S.J., Sack, D., Brunelle, A.R., Moser, K.A., 2010. Late Pleistocene to early Holocene lake level and paleoclimate insights from Stansbury Island, Bonneville basin, Utah. Quatern. Res. 73, 237–246.

Pimentel, R., Buckley, R.V., 1983. Concentrations of total dissolved solids preferred or avoided by endangered Colorado River fishes. Trans. Am. Fish. Soc. 112, 595–600.

Quade, J., 2000a. $^{87}Sr/^{86}Sr$ ratios and Lake Bonneville chronostratigraphy. In: Madsen, D.B. (Ed.), Late Quaternary Paleoecology in the Bonneville Basin, pp. 21–23. Utah Geological Survey, Bulletin 130.

Quade, J., 2000b. Strontium ratios and the origin of the early Homestead Cave biota. In: Madsen, D.B. (Ed.), Late Quaternary Paleoecology in the Bonneville Basin, pp. 44–46. Utah Geological Survey, Bulletin 130.

Reheis, M.C., Laabs, B.J.C., Kaufman, D.S., 2009. Geology and Geomorphology of Bear Lake Valley and Upper Bear River, Utah and Idaho. Special Paper 450 of the Geological Society of America.

Reimer, P.J., Bard, E., Bayliss, A., Beck, J.W., Blackwell, P.G., Ramsey, C.B., Grootes, P.M., Guilderson, T.P., Haflidason, H., Hajdas, I., Hatté, C., Heaton, T.J., Hoffmann, D.L., Hogg, A.G., Hughen, K.A., Kaiser, K.F., Kromer, B., Manning, S.W., Niu, M., Reimer, R.W.,

Richards, D.A., Scott, E.M., Southon, J.R., Staff, R.A., Turney, C.S.M., van der Plicht, J., 2013. IntCal13 and Marine13 radiocarbon age calibration curves 0–50,000 years cal BP. Radiocarbon 55, 1869–1887.

Rhode, D., 2000. Holocene vegetation history in the Bonneville Basin. In: Madsen, D.B. (Ed.), Late Quaternary Paleoecology in the Bonneville Basin, pp. 149–164. Utah Geological Survey Bulletin 130.

Roff, D.A., 1984. The evolution of life history parameters in teleosts. Can. J. Fish. Aquat. Sci. 41, 989–1000.

Rosenfeld, M.J., 1991. Systematic Studies of Members of the Genus *Gila* (Pisces: Cyprinidae) from the Great Basin and Colorado River: Protein Electrophoretic and Cytogenetic Variation. PhD dissertation, University of Utah, Salt Lake City, USA.

Schmitt, D.N., Lupo, K.D., 2012. The Bonneville Estates rockshelter rodent fauna and changes in late Pleistocene–middle Holocene climates and biogeography in the Northern Bonneville Basin, USA. Quatern. Res. 78 (1), 95–102.

Schmitt, D.N., Madsen, D.B., Lupo, K.D., 2002. Small-mammal data on early and middle Holocene climates and biotic communities in the Bonneville basin, USA. Quatern. Res. 58, 255–260.

Shoshani, J., Smith, G.R., 1996. Late Pleistocene fishes from the Shelton Mastodon site (Oakland County, Michigan) and their ecological context. Contrib. Univ. Mich. Mus. Paleontol. 29, 419–433.

Sigler, W.F., Miller, R.R., 1963. Fishes of Utah. Utah State Department of Fish and Game, Salt Lake City.

Sigler, W.F., Sigler, J.W., 1996. Fishes of Utah: A Natural History. University of Utah Press, Salt Lake City.

Smith, G.R., 1966. Distribution and evolution of North American catostomid fishes of the subgenus *Pantosteus*, genus *Catostomus*. Misc. Publ. Mus. Zool. Univ. Mich. 129, 1–132.

Smith, D.G., 1971. Population Dynamics, Habitat Selection, and Partitioning of Breeding Raptors in the Eastern Great Basin of Utah. PhD dissertation, Brigham Young University, Provo, UT.

Smith, G.R., 1981. Effects of habitat size on species richness and adult body size of desert fishes. In: Naiman, R.J., Soltz, D.L. (Eds.), Fishes in North American Deserts. John Wiley and Sons, New York, pp. 125–171.

Smith, C.L., 1985. The Inland Fishes of New York State. New York State Department of Environmental Conservation, Albany.

Smith, F.A., Betancourt, J.L., 2013. The effect of Holocene temperature fluctuations on the evolution and ecology of *Neotoma* (woodrats) in Idaho and northwestern Utah. Quatern. Res. 59, 160–171.

Smith, G.R., Dowling, T.E., 2008. Corrleating hydrographic events and divergence times of speckled dace (*Rhinichthys*: Teleostei: Cyprinidae) in the Colorado River drainage. In: Reheis, M., Hershler, R., Miller, D. (Eds.), Late Cenezoic Drainage History of the Southwestern Great Basin and Lower Colorado River Region: Geologic and Biotic Perspectives, pp. 301–317. Geological Society of America Special Paper 439.

Smith, G.R., Todd, T.N., 1984. Evolution of species flocks of fishes in north temperate lakes. In: Echelle, A., Kornfield, I. (Eds.), Evolution of Fish Species Flocks. University of Maine Press, Orono, pp. 45–68.

Smith, G.R., Todd, T.N., 1993. Morphological cladistic study of coregonine fishes. Pol. Arch. Hydrobiol. 39, 479–490.

Smith, G.R., Stokes, W.L., Horn, K.F., 1968. Some late Pleistocene fishes of Lake Bonneville. Copeia 1968, 807–816.

Smith, G.R., Chow, J., Unmack, P., Markle, D., Dowling, T.E., 2016. Evolution of the *Rhinichthys osculus* Complex (Teleostei, Cyprinidae) in Western North America. In: Fink, W.L., Carpenter, N.E. (Eds.), Late miocene to modern fishes of the Snake River Plain and Northern Great Basin. University of Michigan Museum of Zoology Miscellaneous Publications, vol. 204. (in review).

Smith, G.R., Dowling, T.E., Gobalet, K.W., Lugaski, T., Shiozawa, D.K., Evans, R.P., 2002. Biogeography and timing of evolutionary events among Great Basin fishes. In: Hershler, R., Currey, D.R., Madsen, D.B. (Eds.), Great Basin Aquatic Systems History, pp. 175–234. Smithsonian Contributions to Earth Sciences 33.

Smith, G.R., Badgley, C.E., Eiting, T.P., Larson, P., 2010. Species diversity gradients in relation to geological history in North American freshwater fishes. Evol. Ecol. Res. 12, 693–726.

Smith, G.R., Stewart, J.D., Carpenter, N.E., 2013. Fossil and recent mountain suckers and significance of introgression in catostomin fishes of western United States. Univ. Mich. Mus. Zool. Occas. Publ. 743, 1–59.

Stearley, R.F., Smith, G.R., 2016. Salmonid Fishes from Mio-Pliocene Lake Sediments in the Western Snake River Plain, Oregon and Idaho, and Great Basin Salmonid Diversity. Miscellaneous Publications of the Museum of Zoology, University of Michigan.

Stearns, S.C., 1992. The Evolution of Life Histories. Oxford University Press, Oxford.

Stokes, W.L., Smith, G.R., Horn, K.F., 1964. Fossil fishes from the Stansbury level of Lake Bonneville, Utah. Proc. Utah Acad. Sci. Arts Lett. 41, 87–88.

Stuiver, M., Reimer, P.J., Reimer, R.W., 2013. CALIB 7.1. 14CHRONO Center, Queen's University Belfast, Belfast. Available from: URL: http://calib.qub.ac.uk/calib/.

Taylor, D.W., Smith, G.R., 1981. Pliocene molluscs and fishes from northeastern California and northwestern Nevada. Contrib. Mus. Paleontol. Univ. Mich. 25 (18), 339–413.

Terry, R.C., Novak, M., 2015. Where does the time go?: Mixing and the depth-dependent distribution of fossil ages. Geology 43, 487–490.

Townsend, J.K., 1978. Narrative of a Journey Across the Rocky Mountains to the Columbia River. University of Nebraska Press, Lincoln.

Trautman, M.B., 1981. The Fishes of Ohio. Ohio State University, Columbus.

Unmack, P.J., Dowling, T.E., Laitinen, N.J., Secor, C.L., Mayden, R.L., Shiozawa, D.K., Smith, G.R., 2014. Influence of introgression and geological processes on phylogenetic relationships of western North American mountain suckers (*Pantosteus*, Catostomidae). PLoS One 9 (3), e90061. http://dx.doi.org/10.1371/journal.pone.0090061.

Weber-Scannell, P.K., Duffy, L.K., 2007. Effects of total dissolved solids on aquatic organisms: a review of literature and recommendation for salmonid species. Am. J. Environ. Sci. 3, 1–6.

Westenfelder, C., Birch, F.M., Baranowski, R.L., Rosenfeld, M.J., Shiozawa, D.K., Kablitz, C., 1988. Atrial natriuretic factor and salt adaptation in the teleost fish *Gila atraria*. Am. J. Physiol. 256, F1281–F1286.

White, R.G., 1974. Endemic Whitefishes of Bear Lake, Utah–Idaho. PhD dissertation, Utah State University, Logan.

Wolfe, A.L., Broughton, J.M., 2016. Chapter 14. Bonneville basin avifaunal change at the Pleistocene/Holocene transition: Evidence from Homestead Cave. In: Oviatt, C.G., Shroder Jr., J.F. (Eds.), Lake Bonneville: A Scientific Update, vol. 20. Elsevier. pp. 371–419.

Wright, K.A., 2009. Effects of increasing total dissolved solids on the Walker Lake, Nevada, fishery. Nat. Resour. Environ. Issues 15, 92. Article 16.

Chapter 13

Changes in Late Quaternary Mammalian Biogeography in the Bonneville Basin

D.N. Schmitt* and K.D. Lupo†
*SWCA Environmental Consultants, Salt Lake City, UT, United States
†Southern Methodist University, Dallas, TX, United States

ABSTRACT
Late Quaternary Bonneville basin mammal records chronicle the dynamic nature of regional biogeography and reveal marked transformations in taxonomic abundances and distributions that were primarily the result of changes in climate and biotic communities. During the late Pleistocene a number of species became extinct across the region, including mammoths (*Mammuthus columbi*) and other megafauna, while others persevered and continued to inhabit areas in and around the basin today. Mesic-adapted species in early Holocene low-elevation sites were thought to be present due to ample precipitation, but further examination of small-mammal assemblages suggests that regional moisture was largely a product of groundwater discharge. As middle Holocene desertification took hold, valley settings witnessed extirpations, declines in taxonomic richness, and increases in kangaroo rats (*Dipodomys* sp.), and other taxa well adapted to xeric habitats. The late Holocene (>4.5 cal ka BP) encompassed substantial shifts in climate that included moist events during which bison (*Bison bison*) and other artiodactyl populations expanded and mesic-adapted species returned to valley lowlands, especially ~2.0–0.5 cal ka BP.

Keywords: Mammalian biogeography, Climate change, Extinctions, Early Holocene, Rodents, Extirpations, Taxonomic richness

13.1 INTRODUCTION

Late Quaternary mammalian collections in and adjacent to the Bonneville basin (BB) are extensive and commonly rich. Given the number, context, and content of these assemblages, it is not surprising that most have been subject to detailed scrutiny and together with neighboring Great Basin assemblages provide one of the most extensive zoogeographic histories in the

Developments in Earth Surface Processes, Vol. 20. http://dx.doi.org/10.1016/B978-0-444-63590-7.00013-5

world. Although a host of paleontologists, archaeologists, and biologists have contributed to identifying and interpreting the regional record, the most notable investigations have been conducted by Donald Grayson and include assemblage analyses (eg, Grayson, 1988, 1990, 2000b), species-specific biogeographic histories (eg, Grayson and Madsen, 2000; Grayson, 2005, 2006a; Grayson and Fisher, 2009), and comprehensive syntheses (eg, Grayson, 1987, 1993, 2000a, 2006b, 2011, 2016). Drawing from these and other reports, this chapter presents an overview of mammalian presences, extinctions, and extirpations over the past ~40,000 years. The types of BB mammals and changes in their distributions are presented in four general time periods—late Pleistocene and early, middle, and late Holocene—focusing on now-extinct species that roamed the BB vicinity during the late Pleistocene, and extant lagomorphs, rodents, carnivores, and artiodactyls. Information on bats (Chiroptera) and shrews (Soricomorpha) are not presented and discussions of early and middle Holocene records largely center on small-mammals. All direct (^{14}C) and calibrated (cal) age estimates are presented in thousands of years (ka) before present. Calibrated (2-sigma) ages were calculated with the CALIB7.1 program (http:/calib.qub.ac.uk/calib/) in concert with the IntCal13 calibration curve (Stuiver and Reimer, 1993; Reimer et al., 2013).

13.2 LATE PLEISTOCENE EXTINCTIONS (>12.7 cal ka BP)

During the Bonneville lake cycle (~24–18 cal ka BP), the lake basin was surrounded by more than 90 species of terrestrial mammals that included both large and small species that occupy the region today, and mammals that became extinct during the declining years of the Pleistocene. Of the 39 genera of extinct late Pleistocene mammals in North America (Grayson, 2011, table 7.2) 19 have been identified in the BB or in nearby contexts. Included are the short-faced skunk (*Brachyprotoma obtusata*), sabertooth cat (*Smilodon fatalis*), flat-headed peccary (*Platygonus compressus*), helmeted muskox (*Bootherium bombifrons*), Columbian mammoth (*Mammuthus columbi*), and two species of ground sloth (*Megalonyx jeffersonii* and *Paramylodon harlani*) (Table 13.1). Records for a third species of ground sloth (*Nothrotheriops shastensis* [Shasta ground sloth]) are reported in neighboring Great Basin and Colorado Plateau settings (eg, Harrington, 1933; Mead and Agenbroad, 1992) and it is possible that this animal may have also once occupied the southern margin of the BB. The most commonly represented extinct mammals in the region consist primarily of camels (*Camelops hesternus*), horses (*Equus* spp.), and mammoths (*M. columbi*) (eg, Madsen et al., 1976; Gillette et al., 1999; Heaton, 1999; Hockett and Dillingham, 2004). Although late Pleistocene paleontological records suggest that their numbers were few, the abundances of these large-bodied animals in the BB mirror the overall proportional frequencies of now-extinct mammals across the continent, save

TABLE 13.1 **Extinct Late Pleistocene Mammals of the Bonneville Basin and Vicinity**

Order	Family	Genus/Species[a]	Common Name
Pilosa	Megalonychidae	*Megalonyx jeffersonii*	Jefferson's ground sloth
	Mylodontidae	*Paramylodon harlani*	Harlan's ground sloth
Carnivora	Felidae	cf. *Miracinonyx trumani*	American cheetah
		Panthera leo atrox[b]	American lion
		Smilodon fatalis	Sabertooth cat
	Canidae	*Canis dirus*[b]	Dire wolf
	Ursidae	*Arctodus simus*	Giant short-faced bear
	Musteliadae	*Brachyprotoma obtusata*	Short-faced skunk
Lagomorpha	Leporidae	*Aztlanolagus agilis*	Aztlán rabbit
Artiodactyla	Bovidae	*Bootherium bombifrons*	Harlan's (helmeted) muskox
		Oreamnos harringtoni[b]	Harrington's mountain goat
	Tayassuidae	*Platygonus compressus*	Flat-headed peccary
	Camelidae	*Camelops hesternus*	Yesterday's camel
		Hemiauchenia macrocephala	Large-headed llama
	Cervidae	*Navahoceros* cf. *fricki*	Mountain deer
	Antilocapridae	*Tetrameryx shuleri*	Shuler's pronghorn
Perissodactyla	Equidae	*Equus* cf. *occidentalis*	Western horse
		Equus cf. *conversidens*	Mexican ass
Proboscidea	Elephantidae	*Mammuthus columbi*	Columbian mammoth
	Mammutidae	*Mammut americanum*	American mastodon

[a]*Compiled from Miller (1976), Nelson and Madsen (1978), Miller (1979), Nelson and Madsen (1980), Heaton (1985), Miller (1987), Jefferson et al. (1994), Mead and Lawler (1994), Gillette and Miller (1999), Gillette et al. (1999), Heaton (1999), Madsen (2000a), McDonald et al. (2001), McDonald (2002), Hockett and Dillingham (2004), Grayson (2007), Barnett et al. (2009), and Grayson (2011, 2016).*
[b]*Extinct species of extant genera that occur elsewhere in the Americas.*

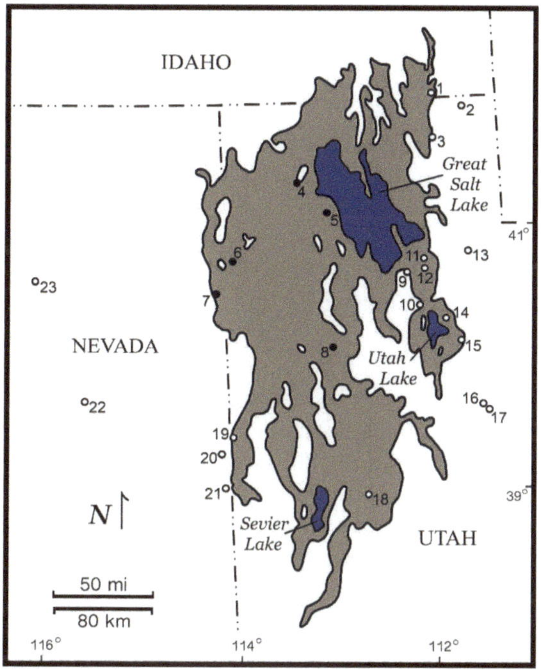

FIG. 13.1 Map of the Bonneville basin and vicinity showing the location of selected paleontological and archeological sites. *Open circles* mark locations containing extinct late Pleistocene mammals: (1) Franklin (McDonald, 2002), (2) Bear Lake mammoths (Gillette and Miller, 1999), (3) Logan Cemetery (Nelson and Madsen, 1978), (4) Hogup Cave (Aikens, 1970), (5) Homestead Cave (eg, Grayson and Madsen, 2000; Madsen, 2000b; Madsen et al., 2001), (6) Danger Cave (Jennings, 1957; Grayson, 1988), (7) Bonneville Estates Rockshelter (eg, Hockett, 2007; Schmitt and Lupo, 2012; Hockett, 2015), (8) Camels Back Cave (eg, Schmitt et al., 2002, 2004; Schmitt and Madsen, 2005), (9) Lark (Gillette and Miller, 1999), (10) Jordan Narrows (Gillette and Miller, 1999), (11) Huntsman (Gillette and Miller, 1999) and eastern Lake Bonneville (eg, Madsen et al., 1976; Nelson and Madsen, 1983), (12) Sandy (Madsen et al., 1976), (13) Silver Creek (Miller, 1976), (14) Orem (Gillette et al., 1999), (15) Spanish Fork (Gillette and Miller, 1999), (16) Sinkhole mastodon (Miller, 1987), (17) Huntington (eg, Gillette and Madsen, 1992; Madsen, 2000a), (18) Romer/Tabernacle Crater (eg, Romer, 1928; Nelson and Madsen, 1979), (19) Crystal Ball Cave (eg, Heaton, 1985), (20) Smith Creek Canyon (Miller, 1979; Mead et al., 1982) and Cathedral Cave (eg, Jass, 2009), (21) Snake Creek Burial Cave (eg, Mead and Mead, 1985; Heaton, 1987), (22) Sunshine Locality (eg, Huckleberry et al., 2001), and (23) Mineral Hill Cave (McGuire, 1978; Hockett and Dillingham, 2004).

for the numerous occurrences of mastodons (*Mammut americanum*) reported elsewhere in North America (eg, Grayson and Meltzer, 2003; Grayson, 2007).

Many remains of extinct mammals have been found along the margins of Lake Bonneville (Fig. 13.1) or are associated with purported regressive phase deposits (eg, Nelson and Madsen, 1978; Mead et al., 1982; Heaton, 1985; Gillette and Miller, 1999; McDonald, 2002), including a giant short-faced bear (*Arctodus simus*) discovered in a Provo level (~1475 m asl) gravel pit dated to 12.65±0.07 ^{14}C ka BP (14.71–15.28 cal ka BP) (Nelson and Madsen, 1983; Schubert, 2011) that corresponds with the beginning of the Provo level regression (Godsey et al., 2011). There are, however, some low-elevation records near the bottom of the basin and include a *M. columbi* tusk from the mouth of City Creek Canyon (1310 m asl) and the remains of a *M. columbi* discovered in nearby Sandy, Utah (1330 m asl) (Madsen et al., 1976). A sample of the City Creek tusk was dated to 14.15±0.80 ^{14}C ka BP (14.72–19.05 cal ka BP) which is too old for its context and clearly indicates that the specimen had been redeposited. Conversely, the oldest of three Sandy mammoth bone collagen samples returned an age estimate of 8.82±0.10 ^{14}C ka BP (9.59–10.18 cal ka BP) and this date is too recent. Because the Sandy *M. columbi* represents a cluster of semiarticulated elements and appears to have been in primary context, this low-elevation find warrants more reliable radiocarbon assay that would be valuable to understanding the timing of late Pleistocene extinctions and the regressive history of the lake.

Atop the Wasatch Plateau to the east, high elevation records (2740–2980 m asl) have been reported for *M. columbi* and *A. simus* at the Huntington site (Gillette and Madsen, 1992, 1993; Madsen, 2000a), and Miller (1987) documented *M. americanum* at the neighboring Sinkhole Mastodon site. In addition to the contextual significance of such high-elevation discoveries, analyses of associated paleobotanical remains, including the Huntington mammoth's last meal, have afforded detailed late Pleistocene vegetation records (Miller, 1987; Madsen, 2000a), and radiocarbon assay of the *M. columbi* and *A. simus* remains provide secure terminal Pleistocene dates. The mammoth amino acids returned an age estimate of 11.22±0.11 ^{14}C ka BP (12.81–13.39 cal ka BP) and the giant short-faced bear has been dated to 10.87±0.08 ^{14}C ka BP (12.67–12.97 cal ka BP) (Madsen, 2000a) and 10.98±0.04 ^{14}C ka BP (12.73–12.97 cal ka BP) (Schubert, 2011). These dates fall immediately after the most recent age estimates on extinct mammals in neighboring contexts, including *C. hesternus* (11.34±0.05 ^{14}C ka BP [13.09–13.29 cal ka BP]; Huckleberry et al., 2001), *P. compressus* (11.39±0.06 ^{14}C ka BP [13.10–13.36 cal ka BP]; McDonald, 2002), and doubtless mark the waning years of these species. Conversely, the youngest known dates for *Hemiauchenia macrocephala* (large-headed llama) and *Navahoceros* cf. *fricki* (mountain deer) in North America are 36.32±0.32 ^{14}C ka BP (40.24–41.59 cal ka BP) and 37.75±0.44 ^{14}C ka BP (41.43–42.67 cal ka BP), respectively. Both of these age estimates are on identified bones from west of the BB at Mineral Hill Cave (Fig. 13.1) (Hockett and Dillingham, 2004; Grayson and Meltzer, 2015, table 3) and it is possible that these taxa became extinct more than 20,000 years before the Pleistocene came to its end.

During the latest Pleistocene the now-extinct mammals shared the BB landscape with human foragers for several thousand years (Madsen, 2016), yet only two tentative artifact/extinct-mammal associations occur in the BB vicinity. Gillette and Madsen (1993) and Madsen (2000a) described possible cut marks on two *M. columbi* bones in association with Paleoindian artifacts at the Huntington site, but both the origin of the marks and artifact-mammoth association remain ambiguous. Similarly, Duke (2015) reported proboscidean protein residue on a Paleoindian spear point from the Old River Bed delta associated with stream channel (black mat) dates of 11.0–10.2 ^{14}C ka BP (~12.9–11.9 cal ka BP), but these results remain speculative due to the inconsistency of protein-residue analyses and lack of supporting evidence. Because no definitive associations of archaeological material occur with extinct mammal remains in the BB vicinity and, more importantly, the numbers of such occurrences across the Americas are so few, it has become clear that humans had little to do with the ultimate demise of these animals (Grayson and Meltzer, 2015; Grayson, 2016).

13.3 THE EARLY HOLOCENE (11.5–9.2 cal ka BP)

Table 13.2 presents extant lagomorphs and rodents, and Table 13.3 lists regional carnivores and artiodactyls that occupied the BB vicinity since the lake's latest highstand. Included are both small and large species that characteristically inhabit cool and moist settings in uplands surrounding the basin, especially its central and northern margins (eg, *Brachylagus idahoensis* [pygmy rabbit], *Neotoma cinerea* [bushy-tailed woodrat], *Tamias umbrinus* [Unita chipmunk], *Mustela erminea* [short-tailed weasel], and *Cervus canadensis* [wapiti or elk]). Also included are a number of mammals well adapted to the vast, arid valley bottoms and mountain foothills that comprised most BB environs over the last 9.5 cal ka BP (eg, *Lepus californicus* [black-tailed jackrabbit], *Dipodomys microps* [chisel-toothed kangaroo rat], *Neotoma lepida* [desert woodrat], *Vulpes macrotis* [kit fox], and *Antilocapra americana* [pronghorn]).

The terminal Pleistocene—early Holocene transition is marked by the final years of the Gilbert episode where the BB supported a lake, albeit briefly, that was higher than any other post-Bonneville lake transgression (eg, Oviatt et al., 2015). A rich series of archaeological mammal records dating to this time and extending throughout the Holocene have been investigated and together provide a detailed biogeographic history spanning the past 12.5 cal ka BP. The most comprehensive records have been recovered from excavations in stratified dry caves and shelters that consist of wave-cut caverns exposed during the lake's late Pleistocene regression, notably Hogup, Danger, Homestead, and Camels Back Caves, and Bonneville Estates Rockshelter (Fig. 13.1). Early Holocene assemblages in these sites contain the remains of small-mammals adapted to both low deserts and mesic settings, with the latter including *N. cinerea*, *Lepus townsendii* (white-tailed jackrabbit), and/or *Marmota flaviventris* (yellow-bellied marmot), each of which predominantly occurs today in boreal habitats on BB mountains (Table 13.2) (eg, Hall, 1946; Brown, 1971). The presence (and abundance) of these mesic-adapted mammals in

TABLE 13.2 Late Pleistocene–Modern Lagomorphs and Rodents in the Bonneville Basin and Neighboring Contexts

Order/Species[a]	Common Name	Order/Species[a]	Common Name
Lagomorpha		**Rodentia—Cont'd**	
Ochotona princeps	**Pika**	***Perognathus parvus***[b]	**Great Basin pocket mouse**
Brachylagus idahoensis[b]	**Pygmy rabbit**	***Chaetodipus formosus***[b]	**Long-tailed pocket mouse**
Lepus americanus	**Snowshoe hare**	*Microdipodops megacephalus*	Dark kangaroo mouse
Sylvilagus audubonii	Audubon's cottontail	*Dipodomys ordii*	Ord's kangaroo rat
Sylvilagus nuttalli[b]	**Nuttall's cottontail**	*Dipodomys microps*	Chisel-toothed kangaroo rat
Lepus townsendii[b]	**White-tailed jackrabbit**	***Castor canadensis***	**Beaver**
Lepus californicus	Black-tailed jackrabbit	*Reithrodontomys megalotis*[b]	Western harvest mouse
Rodentia		*Peromyscus crinitus*	Canyon mouse
Marmota flaviventris[b]	**Yellow-bellied marmot**	*Peromyscus maniculatus*	Deer mouse
Tamias minimus	Least chipmunk	*Peromyscus boylii*	Brush mouse
Tamias umbrinus	**Uinta chipmunk**	***Peromyscus truei***[b]	**Piñon mouse**
Tamias dorsalis	**Cliff chipmunk**	***Onychomys leucogaster***	Northern grasshopper mouse
Neotamias quadrivittatus	**Colorado chipmunk**	***Neotoma lepida***	Desert woodrat
Ammospermophilus leucurus	White-tailed antelope squirrel	***Neotoma cinerea***[b]	**Bushy-tailed woodrat**
Otospermophilus variegatus	**Rock squirrel**	***Clethrionomys gapperi***	**Southern red-backed vole**
Callospermophilus lateralis	Golden-mantled ground squirrel	***Ondatra zibethicus***	**Muskrat**

Urocitellus armatus	**Unita ground squirrel**	***Phenacomys intermedius***	**Heather vole**
Urocitellus mollis	Townsend's ground squirrel	***Microtus pennsylvanicus***	**Meadow vole**
Urocitellus beldingi	**Belding's ground squirrel**	***Microtus montanus***	**Montane vole**
Tamiasciurus hudsonicus	**Red squirrel**	*Microtus longicaudus*	Long-tailed vole
Glaucomys sabrinus	**Northern flying squirrel**	***Microtus richardsonii***	**Water vole**
Thomomys talpoides	**Northern pocket gopher**	***Lemmiscus curtatus***[b]	**Sage vole**
Thomomys bottae	Botta's pocket gopher	***Zapus princeps***	**Western jumping mouse**
Perognathus longimembris	Little pocket mouse	*Erethizon dorsatum*	Porcupine

[a]*Compiled from Hall (1946), Durrant (1952), Hall (1981), Zeveloff (1988), and Grayson (2006a, 2011). Boldface denotes species that largely occur in mesic contexts and/or upland settings along the margins of the Bonneville basin, notably the Wasatch Range (east), Raft River Mountains (northwest), and/or Deep Creek Mountains (west).*

[b]*Mesically adapted species recovered from early Holocene deposits at Homestead Cave (eg, Grayson, 1998, 2000b), Hogup Cave (Durrant, 1970), Danger Cave (Grayson, 1988), Bonneville Estates Rockshelter (Schmitt and Lupo, 2012), and/or Camels Back Cave (Schmitt et al., 2002; Schmitt and Lupo, 2005).* R. megalotis *are known to occur in some arid low-elevation settings but are most common in association with moist grasslands.*

TABLE 13.3 Late Pleistocene—Modern Carnivores and Artiodactyls in the Bonneville Basin and Neighboring Contexts

Order/Species[a]	Common Name
Carnivora	
Ursus americanus	**Black bear**
Canis latrans	Coyote
Canis lupus	Gray wolf
Vulpes vulpes	Red fox
Vulpes macrotis	Kit fox
Urocyon cinereoargenteus	Gray fox
Mustela erminea	**Short-tailed weasel**
Mustela frenata	Long-tailed weasel
Mustela vison	**Mink**
Martes americana	**Marten**
Martes americana nobilis[b]	**Noble marten**
Martes pennanti	**Fisher**
Gulo gulo	**Wolverine**
Taxidea taxus	Badger
Spilogale gracilis	Spotted skunk
Mephitis mephitis	Striped skunk
Lutra canadensis	**River otter**
Lynx canadensis	**Lynx**
Lynx rufus	Bobcat
Felis concolor	**Mountain lion**
Artiodactyla	
Cervus canadensis	**Wapiti**
Odocoileus hemionus	Mule deer
Alces americanus	**Moose**
Antilocapra americana	Pronghorn
Bison bison[c]	Bison
Ovis canadensis	Mountain sheep

[a]*Compiled from Hall (1946), Durrant (1952), Hall (1981), Zeveloff (1988), Grayson (2006b), and Grayson and Fisher (2009). Boldface denotes species that largely occur in mesic contexts and/or upland settings along the margins of the Bonneville basin, notably the Wasatch Range (east), Raft River Mountains (northwest), and/or Deep Creek Mountains (west).*
[b]*Great Basin records suggest that* M. a. nobilis *became extinct approximately 3000 years ago (eg, Mead and Mead, 1989; Grayson, 2011, 282).*
[c]*Bison were present prehistorically (eg, Lupo and Schmitt, 1997; Grayson, 2006b) and were extirpated from the region ~AD 1830 (Lupo, 1996).*

low-elevation contexts of early Holocene age led researchers to conclude reasonably that the period was cool and moist, and remained so until regional desertification took hold about 9.2 cal ka BP (eg, Grayson, 2000b, 2006b; Schmitt and Lupo, 2012). Until recently, it was assumed that this early Holocene moisture largely resulted from precipitation, but a recent study of BB shoreline and lakebed deposits (Oviatt et al., 2015) has led to a hypothesis that this moisture was the result of Lake Bonneville groundwater discharge and not rainfall. As such, valley bottoms during this period would have sustained groundwater-fed mesic oases longer than mountain foothills and alluvial fans, and a closer inspection of the changes in small-mammals from the early deposits at Bonneville Estates Rockshelter (BER), Homestead (HC), and Camels Back (CBC) caves indicates that this may have been the case.

To test this hypothesis we compared small-mammals from the mountain foothill context at BER with assemblages from the two lower elevation caves. BER is a massive cavern in the Lead Mine Hills of eastern Nevada along the Bonneville shoreline (1580 m asl; eg, Goebel et al., 2011; Hockett, 2015), and both HC and CBC are carved into low limestone ridges in valley bottoms some 200 m lower in elevation (see, eg, Madsen, 2000b; Madsen et al., 2001; Schmitt, 2004; Schmitt and Madsen, 2005). It is well known that in arid environments mammal species richness (measured by the number of species present) is positively correlated with moisture and productivity (eg, Abramsky and Rosenzweig, 1984; Grayson, 1998). Although sample sizes vary, differences between early Holocene rodent richness in the two low-elevation caves and BER (Fig. 13.2) suggest that the cave vicinities sustained moist environs from the beginning of the period until approximately 8.3 ^{14}C ka BP (9.3 cal ka BP), whereas mesic conditions in the BER uplands appear to have declined rapidly during the first few hundred years of the Holocene.

Early assemblages in each of these sites contained abundant *Neotoma* remains that reveal a similarly intriguing spatial and temporal trend. Fig. 13.3 plots the ratios of identified mesic-adapted *N. cinerea* specimens to those of both *N. cinerea* and *N. lepida* (a desert species) from late Pleistocene—early middle Holocene aggregates in the three sites. The BER record shows a marked decline in *N. cinerea* prior to 9.5 ^{14}C ka BP (~10.8 cal ka BP) (Schmitt and Lupo, 2012), whereas *N. cinerea* remained the dominant woodrat species at HC and CBC until approximately 8.5 ^{14}C ka BP (~9.5 cal ka BP) (Grayson, 2000b; Schmitt et al., 2002). Elsewhere in the BB, *N. cinerea* persisted well into the early Holocene at Danger (Grayson, 1988) and Hogup (Durrant, 1970) caves, both of which occur in low-elevation settings immediately overlooking the alkali mudflats of what is now the Great Salt Lake Desert. The woodrat records, like overall changes in rodent taxonomic richness, support the hypothesis that early Holocene moisture in the region resulted primarily from groundwater discharge (Oviatt et al., 2015), and that this discharge supported low-elevation moist

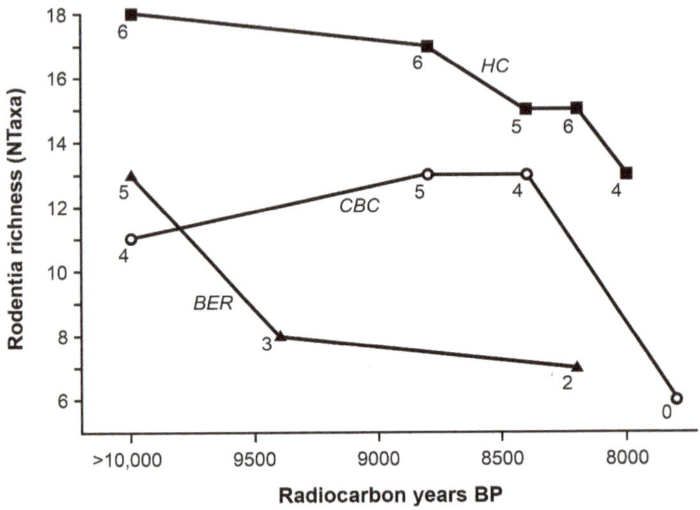

FIG. 13.2 Rodent species richness in the late Pleistocene (>10.0 ^{14}C ka BP)—early middle Holocene deposits at Bonneville Estates Rockshelter (*BER*) (Schmitt and Lupo, 2012), Camels Back Cave (*CBC*) (Schmitt et al., 2002; Schmitt and Lupo, 2005), and Homestead Cave (*HC*) (Grayson, 1998, 2000b). The numbers associated with each dated aggregate denote the number of mesic-adapted species present (see Table 13.2).

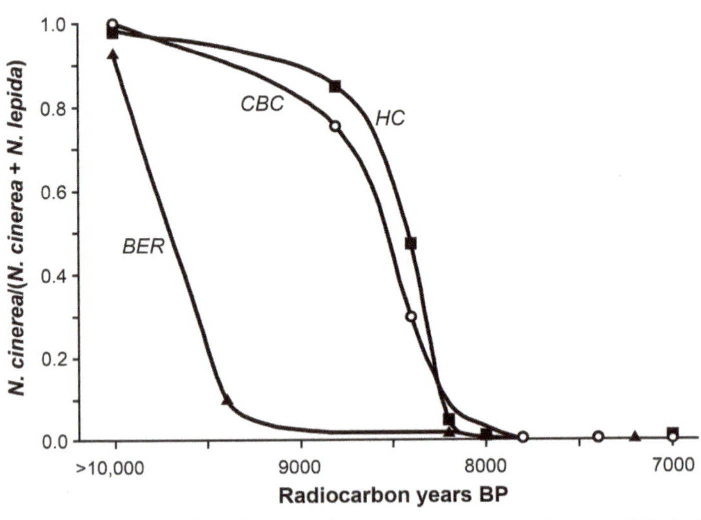

FIG. 13.3 The ratio of identified *Neotoma cinerea* specimens to *N. cinerea* and *N. lepida* specimens in the late Pleistocene (>10.0 ^{14}C ka BP)—early middle Holocene deposits at Bonneville Estates Rockshelter (*BER*) (Schmitt and Lupo, 2012, table 2), Camels Back Cave (*CBC*) (Schmitt et al., 2002, table 1), and Homestead Cave (*HC*) (Grayson, 2000b, table XV).

habitats for ~1500 years until just before the early Holocene came to an end. However, because the BER assemblage stands alone, robust confirmation awaits the recovery and analysis of additional late Pleistocene—early Holocene faunas from upland contexts.

13.4 MIDDLE HOLOCENE DESERTIFICATION (9.2–4.5 cal ka BP)

The onset of the middle Holocene was marked by rapid and severe desertification that lingered across most of the region for the next ~4000 years (eg, Madsen et al., 2001; Grayson, 2011, 242–259). With the loss of groundwater, increasing temperatures, and little precipitation, piedmont and valley bottom vegetation communities changed from mesic grass and sagebrush (*Artemisia* sp.) steppes to open desert scrub communities containing greasewood (*Sarcobatus vermiculatus*), saltbush (*Atriplex* spp.), and other xerophytic species (eg, Rhode, 2000). Mammalian responses to this warming and drying cycle in the low-elevation settings of the BB were considerable. For a number of species, especially small mesic-adapted taxa, this change was detrimental due to thermoregulatory constraints, the loss of primary foods, and/or a deficiency in vegetation cover that assisted in predator defense. Included were extirpations, conspicuous declines in small-mammal taxonomic richness, and the replacement of some mesic taxa by xeric-adapted relatives (Grayson, 1998, 2000a; Schmitt et al., 2002; Grayson, 2006b), as well as increases in heteromyid rodents that represent keystone species in arid environments (eg, Brown and Harney, 1993) and postulated struggling artiodactyl populations (Broughton et al., 2008).

Near HC, mesic-adapted *B. idahoensis*, *Chaetodipus formosus* (long-tailed pocket mouse), and *Perognathus parvus* (Great Basin pocket mouse) all but disappeared in the early middle Holocene, the numbers of harvest mice (*Reithrodontomys megalotis*) and especially microtine rodents declined (eg, Grayson, 2000b), and *N. lepida* supplanted *N. cinerea* shortly before 8.0 ^{14}C ka BP (~8.9 cal ka BP) (Fig. 13.3). In the CBC vicinity, *M. flaviventris*, *L. townsendii*, *B. idahoensis*, and *Peromyscus truei* (piñon mouse) became locally extinct, *N. cinerea* was replaced by *N. lepida* at about 8.0 ^{14}C ka BP, and heteromyid populations (especially *Dipodomys* sp.) expanded significantly and represented the most abundant rodent throughout the period (Schmitt and Lupo, 2005). In addition, cottontail (*Sylvilagus* sp.) populations radically declined while jackrabbit (*L.* cf. *californicus*) populations expanded in response to the increasingly open and arid environs near CBC (eg, Schmitt et al., 2004), BER (Hockett, 2015), and Danger Cave (Grayson, 1988).

13.5 THE LATE HOLOCENE (4.5 cal ka BP—PRESENT)

Shortly after the onset of the late Holocene, the Neopluvial period (or Neoglacial; ~ 4.0–2.5 cal ka BP) encompassed a major cool and moist event that brought forth several changes in BB small-mammal biogeography. Mesic-adapted *C. formosus* returned to the HC area after a more than 4000 years hiatus and the number of *R. megalotis* increased, a taxon characteristic of well-watered habitats (eg, Grayson, 2000b, 2011). At the same time, artiodactyl abundances increased at Danger Cave (Grayson, 1988) and CBC, and *P. parvus* returned to the valley bottom near CBC ~2.7 cal ka BP (Schmitt and Lupo, 2005).

During the latest Holocene (~2.5–0.5 cal ka BP) the BB witnessed another rather dramatic climatic event that included a brief warm period followed by an increase in moisture that included monsoonal storms (eg, Madsen et al., 2001; Wigand and Rhode, 2002). With respect to changes in regional mammal communities, this event was even more arresting than the Neopluvial episode that preceded it. For example, the HC area was recolonized by *N. cinerea* (Grayson, 2000b; see also Grayson and Madsen, 2000), the abundances of other mesic rodents near the cave increased, including *Lemmiscus curtatus* (sage vole), *P. parvus*, and *R. megalotis* (eg, Grayson, 2000b, 2006b), and *B. idahoensis* inhabited the Hogup Cave vicinity (Durrant, 1970), perhaps for the first time. Although few in number, both *L. curtatus* and *N. cinerea* returned to CBC, and the number of *Microdipodops megacephalus* (dark kangaroo mouse) remains in horizons dating between ca. 1.6 and 0.5 cal ka BP approach those from the cave's late Pleistocene deposits (Schmitt and Lupo, 2005).

Increase in summer precipitation during this period provided grasslands for *R. megalotis* and other mesic-adapted rodents as well as bison (*B. bison*) and other large herbivores whose populations expanded significantly, doubtless the result of the grassy plains that took root across many local piedmont and valley bottom settings. Most of the BB records for late Holocene *B. bison* have been recovered from open archaeological contexts and represent the densest concentrations of these mammals in the prehistoric Great Basin (eg, Lupo and Schmitt, 1997; Grayson, 2006a). These sites were occupied by Fremont period peoples who grew crops of maize, squash, and beans that also relied on summer storm incursions (eg, Madsen and Simms, 1998; Grayson, 2006a), and it appears that Fremont peoples often enjoyed their cultivated plant foods along with copious amounts of *B. bison* meat and marrow, especially between 1.5 and 0.5 cal ka BP. When this wet-summer interval ended, regional horticulture disappeared, populations of both *B. bison* and their human hunters decreased significantly, and *B. bison* eventually vanished about AD 1830 (Lupo, 1996). Given the number of Fremont sites containing abundant *B. bison* skeletal remains it is likely that human hunting pressure took its toll on some BB populations, but the perpetual decline in these large herbivores after ~0.6 cal ka BP suggests that climate change and the associated loss of moist grasslands affords a more plausible explanation (see Grayson, 2006a).

13.6 SUMMARY

Late Quaternary mammalian biogeography in the BB has been (and continues to be) shaped by a series of complex extinction, extirpation, and recolonization events (eg, Lawlor, 1998; Grayson and Madsen, 2000; Grayson, 2006b, 2011) and the timing of these events are convincingly tied to regional climatic change (eg, Grayson, 1998, 2000a; Madsen et al., 2001; Grayson, 2006a; Schmitt and Lupo, 2012). As temperatures rose and Lake Bonneville regressed during the declining years of the Pleistocene, a number of terrestrial mammals became extinct here and across the Americas. Those that persisted through the Holocene consist of a diverse array of generalists, species tethered to mesic habitats, and mammals that prefer open xeric valley bottoms and changes in their spatial and temporal distributions and/or abundances are similarly linked to climatic events. The numbers of mesic-adapted rodents in low-elevation contexts during the early Holocene was thought to be the consequence of precipitation, but BB valleys sustained these small-mammals significantly longer than upland settings and it appears that regional moisture was the product of groundwater discharge (Oviatt et al., 2015). As Middle Holocene desertification took hold, mesic-adapted species were extirpated from BB valleys, artiodactyl populations and overall taxonomic richness declined, and there were significant increases in mammals well adapted to low xeric settings, notably *L. californicus* and *Dipodomys* sp. The latest Holocene has encompassed a series of climatic changes that included a wet period enhanced by monsoonal storms that lasted over 1000 years. This major climatic transformation altered regional biotic communities and brought forth dramatic changes in mammalian biogeography, including the recolonization of some BB lowlands by *N cinerea*, *L. curtatus*, and other mesic-adapted species, and salient increases in *R. megalotis* and *B. bison*. As these summer storm incursions and associated moist grasslands waned ~0.6 cal ka BP, so too did the types and abundances of mammals that inhabited the vast valleys that comprise most BB environs today.

ACKNOWLEDGMENTS

We thank Jack Oviatt and Jack Shroder for inviting us to contribute to this volume and providing editorial comments; Bryan Hockett, R. Lee Lyman, and David Madsen for insights and comments on earlier drafts; and we gratefully acknowledge Donald Grayson for his voluminous and significant contributions on the ecology and biogeographic histories of Great Basin mammals.

REFERENCES

Abramsky, Z., Rosenzweig, M.L., 1984. Tilman's predicted productivity—diversity relationship shown by desert rodents. Nature 309, 150–151.

Aikens, C.M., 1970. Hogup Cave. University of Utah Press, Salt Lake City, UT (Anthropological Papers No. 93).

Barnett, R., et al., 2009. Phylogeography of lions (*Panthera leo* ssp.) reveals three distinct taxa and a late Pleistocene reduction in genetic diversity. Mol. Ecol. 18, 1668–1677.

Broughton, J.M., Beyers, D.A., Bryson, R.A., Eckerle, W., Madsen, D.B., 2008. Did climatic change control late Quaternary artiodactyl densities in western North America? Quat. Sci. Rev. 27, 1916–1937.

Brown, J.H., 1971. Mammals on mountaintops: nonequilibrium insular biogeography. Am. Nat. 105, 467–478.

Brown, J.H., Harney, B.A., 1993. Population and community ecology of heteromyid rodents in temperate habitats. In: Genoways, H.H., Brown, J.H. (Eds.), Biology of the Heteromyidae. American Society of Mammalogists, Provo, UT, pp. 618–651. (Special Publications No. 10).

Duke, D., 2015. Haskett spear weaponry and protein-residue evidence of Proboscidean hunting in the Great Salt Lake Desert, Utah. PaleoAmerica 1, 109–112.

Durrant, S.D., 1952. Mammals of Utah: Taxonomy and Distribution. Museum of Natural History, Lawrence (University of Kansas Special Publications 6).

Durrant, S.D., 1970. Faunal remains as indicators of neothermal climates at Hogup Cave. In: Aikens, C.M. (Ed.), Hogup Cave. University of Utah Press, Salt Lake City, UT, pp. 241–245 (Anthropological Papers No. 93).

Gillette, D.D., Madsen, D.B., 1992. The short-faced bear *Arctodus simus* from the late Quaternary in the Wasatch Mountains of central Utah. J. Vertebr. Paleontol. 12, 107–112.

Gillette, D.D., Madsen, D.B., 1993. The Columbian mammoth, *Mammuthus columbi*, from the Wasatch Mountains of central Utah. J. Paleontol. 67, 669–680.

Gillette, D.D., Miller, W.E., 1999. Catalogue of new Pleistocene mammalian sites and recovered fossils from Utah. In: Gillette, D.D. (Ed.), Vertebrate Paleontology in Utah: Miscellaneous Publication 99-1. Utah Geological Survey, Salt Lake City, UT, pp. 523–530.

Gillette, D.D., McDonald, H.G., Hayden, M.C., 1999. The first record of Jefferson's ground sloth, *Megalonyx jeffersonii*, in Utah. In: Gillette, D.D. (Ed.), Vertebrate Paleontology in Utah: Miscellaneous Publication 99-1. Utah Geological Survey, Salt Lake City, pp. 509–521.

Godsey, H.S., Oviatt, C.G., Miller, D.M., Chan, M.A., 2011. Stratigraphy and chronology of offshore to nearshore deposits associated with the Provo shoreline, Pleistocene Lake Bonneville, Utah, USA. Palaeogeogr. Palaeoclimatol. Palaeoecol. 310, 442–450.

Goebel, T., Hockett, B., Adams, K.D., Rhode, D., Graf, K., 2011. Climate, environment, and humans in North America's Great Basin during the Younger Dryas, 12,900–11,600 calendar years ago. Quat. Int. 242, 479–501.

Grayson, D.K., 1987. The biogeographic history of small mammals in the Great Basin: observations on the last 20,000 years. J. Mammal. 68, 359–375.

Grayson, D.K., 1988. Danger Cave, Last Supper Cave, and Hanging Rock Shelter: The Faunas. American Museum of Natural History, New York. (Anthropological Papers 66 (1)).

Grayson, D.K., 1990. The James Creek Shelter mammals. In: Elston, R.G., Budy, E.E. (Eds.), The Archaeology of James Creek Shelter. University of Utah Press, Salt Lake City, UT, pp. 87–98 (Anthropological Papers No. 115).

Grayson, D.K., 1993. The Deserts Past: A Natural Prehistory of the Great Basin. Smithsonian Institution Press, Washington, DC.

Grayson, D.K., 1998. Moisture history and small mammal community richness during the latest Pleistocene and Holocene, northern Bonneville basin, Utah. Quatern. Res. 49, 330–334.

Grayson, D.K., 2000a. Mammalian responses to middle Holocene climatic change in the Great Basin of the western United States. J. Biogeogr. 27, 181–192.

Grayson, D.K., 2000b. The Homestead Cave mammals. In: Madsen, D.B. (Ed.), Late Quaternary Paleoecology in the Bonneville Basin. Utah Geological Survey, Salt Lake City, UT, pp. 67–89 (Bulletin 130).

Grayson, D.K., 2005. A brief history of Great Basin pikas. J. Biogeogr. 32, 2101–2111.

Grayson, D.K., 2006a. Holocene bison in the Great Basin, western USA. Holocene 16, 913–925.

Grayson, D.K., 2006b. The late Quaternary biogeographic histories of some Great Basin mammals (western USA). Quat. Sci. Rev. 25, 2964–2991.

Grayson, D.K., 2007. Deciphering North American Pleistocene extinctions. J. Anthropol. Res. 63, 185–214.

Grayson, D.K., 2011. The Great Basin: A Natural Prehistory. University of California Press, Berkeley.

Grayson, D.K., 2016. Giant Sloths and Sabertooth Cats: Extinct Mammals and the Archaeology of the Ice Age Great Basin. University of Utah Press, Salt Lake City, UT.

Grayson, D.K., Fisher, J.L., 2009. Holocene Elk (*Cervus elaphus*) in the Great Basin. In: Hockett, B. (Ed.), Past, Present and Future Issues in Great Basin Archaeology: Papers in Honor of Don D. Fowler. Cultural Resource Series, vol. 20. Bureau of Land Management, Reno, Nevada, pp. 67–83.

Grayson, D.K., Madsen, D.B., 2000. Biogeographic implications of recent low-elevation recolonization by *Neotoma cinerea* in the Great Basin. J. Mammal. 81, 1100–1105.

Grayson, D.K., Meltzer, D.J., 2003. A requiem for North American overkill. J. Archaeol. Sci. 30, 585–593.

Grayson, D.K., Meltzer, D.J., 2015. Revising Paleoindian exploitation of extinct North American mammals. J. Archaeol. Sci. 56, 177–193.

Hall, E.R., 1946. Mammals of Nevada. University of California Press, Berkeley.

Hall, E.R., 1981, second ed. The Mammals of North America, vol. 1. John Wiley & Sons, New York.

Harrington, M.R., 1933. Gypsum Cave. Southwest Museum, Los Angeles (Papers No. 8).

Heaton, T.H., 1985. Quaternary paleontology and paleoecology of Crystal Ball Cave, Millard County, Utah: with emphasis on mammals and description of a new species of fossil skunk. Great Basin Nat. 45, 337–390.

Heaton, T.H., 1987. Initial investigation of vertebrate remains from Snake Creek Burial Cave, White Pine County, Nevada. Curr. Res. Pleist. 4, 107–109.

Heaton, T.H., 1999. Late Quaternary vertebrate history of the Great Basin. In: Gillette, D.D. (Ed.), Vertebrate Paleontology in Utah: Miscellaneous Publication 99-1. Utah Geological Survey, Salt Lake City, UT, pp. 501–507.

Hockett, B., 2007. Nutritional ecology of late Pleistocene to middle Holocene subsistence in the Great Basin: zooarchaeological evidence from Bonneville Estates Rockshelter. In: Graf, K.E., Schmitt, D.N. (Eds.), Paleoindian or Paleoarchaic? Great Basin Human Ecology at the Pleistocene-Holocene Transition. University of Utah Press, Salt Lake City, UT, pp. 201–230.

Hockett, B., 2015. The zooarchaeology of Bonneville Estates Rockshelter: 13,000 years of Great Basin hunting strategies. J. Archaeol. Sci. Rep. 2, 291–301.

Hockett, B., Dillingham, E., 2004. Paleontological Investigations at Mineral Hill Cave: Technical Report No. 18. Contributions to the Study of Cultural Resources. Bureau of Land Management, Reno, NV.

Huckleberry, G., Beck, C., Jones, G.T., Holmes, A., Cannon, M., Livingston, S.D., Broughton, J.M., 2001. Terminal Pleistocene/early Holocene environmental change at the Sunshine Locality, north-central Nevada, U.S.A. Quatern. Res. 55, 303–312.

Jass, C.N., 2009. Pleistocene lagomorphs from Cathedral Cave, Nevada. PaleoBios 29, 1–12.

Jefferson, G.T., Miller, W.E., Nelson, M.E., Madsen Jr., J.H., 1994. Catalogue of Late Quaternary Vertebrates from Utah: Technical Reports 9. Natural History Museum of Los Angeles County, Los Angeles.

Jennings, J.D., 1957. Danger Cave. University of Utah Press, Salt Lake City, UT (Anthropological Papers No. 27).

Lawlor, T.E., 1998. Biogeography of Great Basin mammals: paradigm lost? J. Mammal. 79, 1111–1130.

Lupo, K.D., 1996. The historical occurrence and demise of bison in northern Utah. Utah Hist. Quart. 64, 168–180.

Lupo, K.D., Schmitt, D.N., 1997. On Late Holocene variability in bison populations in the northeastern Great Basin. J. Calif. Gt. Basin Anthropol. 19, 50–69.

Madsen, D.B., 2000a. A high-elevation Allerød-Younger Dryas megafauna from the west-central Rocky Mountains. In: Madsen, D.B., Metcalf, M.D. (Eds.), Intermountain Archaeology. University of Utah Press, Salt Lake City, UT, pp. 100–115 (Anthropological Papers No. 122).

Madsen, D.B., 2000b. Late Quaternary Paleoecology in the Bonneville Basin. Utah Geological Survey, Salt Lake City, UT. Bulletin 130.

Madsen, D.B., 2016. Chapter 18. The early human occupation of the Bonneville basin. In: Oviatt, C.G., Shroder Jr., J.F. (Eds.), Lake Bonneville: A Scientific Update, vol. 20. Elsevier, pp. 504–525.

Madsen, D.B., Simms, S.R., 1998. The Fremont complex: a behavioral perspective. J. World Prehist. 12, 255–336.

Madsen, D.B., Currey, D.R., Madsen, J.H., 1976. Man, mammoth, and lake fluctuations in Utah. Utah Division of State History, Salt Lake City, UT (Antiquities Section Selected Papers 5).

Madsen, D.B., Rhode, D., Grayson, D.K., Broughton, J.M., Livingston, D.D., Hunt, J., Quade, J., Schmitt, D.N., Shaver III, M.W., 2001. Late Quaternary environmental change in the Bonneville Basin, western USA. Palaeogeogr. Palaeoclimatol. Palaeoecol. 167, 243–271.

McDonald, H.G., 2002. *Platygonus compressus* from Franklin County, Idaho, and a review of the genus in Idaho. In: Akersten, W.A., Thompson, M.E., Meldrum, D.J., Rapp, R.A., McDonald, H.G. (Eds.), And Whereas…Papers on the Vertebrate Paleontology of Idaho Honoring John A. White, vol. 2. Idaho Museum of Natural History, Pocatello, pp. 141–149 (Occasional Paper 37).

McDonald, K.A., Miller, W.E., Morris, T.H., 2001. Taphonomy and significance of Jefferson's ground sloth (Xenarthra: Megalonychidae) from Utah. West. N. Am. Naturalist 61, 64–77.

McGuire, K.R., 1978. Cave sites, faunal analysis, and big-game hunters of the Great Basin: a caution. Quatern. Res. 14, 263–268.

Mead, J.I., Agenbroad, L.D., 1992. Isotope dating of Pleistocene dung deposits from the Colorado Plateau, Arizona and Utah. Radiocarbon 34, 1–19.

Mead, J.I., Lawler, M.C., 1994. Skull, mandible, and metapodials of the extinct Harrington's mountain goat (*Oreamnos harringtoni*). J. Vertebr. Paleontol. 14, 562–576.

Mead, J.I., Mead, E.M., 1985. A natural trap for Pleistocene animals in Snake Valley, Utah. Curr. Res. Pleist. 2, 105–106.

Mead, E.I., Mead, J.I., 1989. Snake Creek Burial Cave and a review of the Quaternary mustelids of the Great Basin. Great Basin Nat. 49, 143–154.

Mead, J.I., Thompson, R.S., Van Devender, T.R., 1982. Late Wisconsinan and Holocene fauna from Smith Creek Canyon, Snake Range, Nevada. Trans. San Diego Soc. Nat. Hist. 20, 1–26.

Miller, W.E., 1976. Late Pleistocene vertebrates of the Silver Creek local fauna from north central Utah. Great Basin Nat. 36, 387–424.

Miller, S.J., 1979. The archaeological fauna of four sites in Smith Creek Canyon. In: Tuohy, D.R., Rendall, D.L. (Eds.), The Archaeology of Smith Creek Canyon, Eastern Nevada. Nevada State Museum, Carson City, pp. 272–329 (Anthropological Papers No. 17).

Miller, W.E., 1987. *Mammut americanum*, Utah's first record of the American mastodon. J. Paleontol. 61, 168–183.

Nelson, M.E., Madsen Jr., J.H., 1978. Late Pleistocene musk oxen from Utah. Trans. Kans. Acad. Sci. 81, 277–295.

Nelson, M.E., Madsen Jr., J.H., 1979. The Hay-Romer camel debate: fifty years later. Wyo. Univ. Contr. Geol. 18, 47–50.

Nelson, M.E., Madsen Jr., J.H., 1980. A summary of Pleistocene fossil vertebrate localities in the northern Bonneville Basin of Utah. In: Gwynn, J.W. (Ed.), Great Salt Lake: A Scientific, Historical, and Economic Overview. Utah Geological and Mineral Survey, Salt Lake City, UT, pp. 97–114 (Bulletin 116).

Nelson, M.E., Madsen Jr., J.H., 1983. A giant short-faced bear (*Arctodus simus*) from the Pleistocene of northern Utah. Trans. Kans. Acad. Sci. 86, 1–9.

Oviatt, C.G., Madsen, D.B., Miller, D.M., Thompson, R.S., McGeehin, J.P., 2015. Early Holocene Great Salt Lake, USA. Quatern. Res. 84, 57–68.

Reimer, P.J., Bard, E., Bayliss, A., Beck, J.W., Blackwell, P.G., Ramsey, C.B., Grootes, P.M., Guilderson, T.P., Haflidason, H., Hajdas, I., Hatté, C., Heaton, T.J., Hoffmann, D.L., Hogg, A.G., Hughen, K.A., Kaiser, K.F., Kromer, B., Manning, S.W., Niu, M., Reimer, R.W., Richards, D.A., Scott, E.M., Southon, J.R., Staff, R.A., Turney, C.S.M., van der Plicht, J., 2013. IntCal13 and Marine13 radiocarbon age calibration curves 0–50,000 years cal BP. Radiocarbon 55, 1869–1887.

Rhode, D., 2000. Holocene vegetation history in the Bonneville basin. In: Madsen, D.B. (Ed.), Late Quaternary Paleoecology in the Bonneville Basin. Utah Geological Survey, Salt Lake City, UT, pp. 149–163 (Bulletin 130).

Romer, A.S., 1928. A "fossil" camel recently living in Utah. Science 68, 19–20.

Schmitt, D.N., 2004. Ecological change in western Utah: comparisons between a late Holocene archaeological fauna and modern small-mammal surveys. In: Lyman, R.L., Cannon, K.P. (Eds.), Zooarchaeology and Conservation Biology. University of Utah Press, Salt Lake City, UT, pp. 178–192.

Schmitt, D.N., Lupo, K.D., 2005. The Camels Back Cave mammalian fauna. In: Schmitt, D.N., Madsen, D.B. (Eds.), Camels Back Cave. University of Utah Press, Salt Lake City, UT, pp. 136–176 (Anthropological Papers No. 125).

Schmitt, D.N., Lupo, K.D., 2012. The Bonneville Estates Rockshelter rodent fauna and changes in late Pleistocene-middle Holocene climates and biogeography in the northern Bonneville basin, USA. Quatern. Res. 78, 95–102.

Schmitt, D.N., Madsen, D.B., 2005. Camels Back Cave. University of Utah Press, Salt Lake City, UT (Anthropological Papers No. 125).

Schmitt, D.N., Madsen, D.B., Lupo, K.D., 2002. Small-mammal data on early and middle Holocene climates and biotic communities in the Bonneville Basin, USA. Quatern. Res. 58, 255–260.

Schmitt, D.N., Madsen, D.B., Lupo, K.D., 2004. The worst of times, the best of times: jackrabbit hunting by middle Holocene human foragers in the Bonneville basin of western North America. In: Mondini, M., Muñoz, S., Wickler, S. (Eds.), Colonization, Migration, and Marginal Areas: A Zooarchaeological Approach. Oxbow Books, Oxford, pp. 86–95.

Schubert, B.W., 2011. Late Quaternary chronology and extinction of North American giant short-faced bears (*Arctodus simus*). Quat. Int. 217, 188–194.

Stuiver, M., Reimer, P.J., 1993. Extended ^{14}C database and revised CALIB radiocarbon calibration program. Radiocarbon 35, 215–230. http://calib.qub.ac.uk/calib/calib.html. CALIB 7.1 accessed 22.7.15.

Wigand, P.E., Rhode, D., 2002. Great Basin vegetational history and aquatic systems: the last 150,000 years. In: Hershler, R., Madsen, D.B., Currey, D.R. (Eds.), Great Basin Aquatic Systems History: Smithsonian Contributions to Earth Sciences 33. Smithsonian Institution Press, Washington, D.C, pp. 309–367.

Zeveloff, S.I., 1988. Mammals of the Intermountain West. University of Utah Press, Salt Lake City, UT.

Chapter 14

Bonneville Basin Avifaunal Change at the Pleistocene/Holocene Transition: Evidence from Homestead Cave

A.L. Wolfe and J.M. Broughton
University of Utah, Salt Lake City, UT, United States

ABSTRACT

Although recent research of the faunal and floral remains of the Bonneville basin has increased our knowledge of biotic change at the end of the final regression of Lake Bonneville, our knowledge of avifaunal change during this period remains poorly understood. This study focuses on the bird remains from the lower strata (especially Stratum I) of Homestead Cave, the only site in the Bonneville basin that provides a substantial, well-dated avifaunal record dating to the terminal Pleistocene and early Holocene. Analysis of the entire Stratum I avifauna produced a sample of 4613 identified specimens; the fauna is dominated by aquatic taxa including ducks, grebes, gulls, and shorebirds. Terrestrial taxa are also well represented, most notably by Greater Sage-Grouse (*Centrocercus urophasianus*). We examine change in the taxonomic composition of the Bonneville basin aquatic avifauna as anticipated from previous research and find significant declines from 13.1 to 9.5 cal ka BP in piscivorous or deep-water taxa, such as diving ducks and *Aechmophorus* grebes. Our analysis captures the transition from the freshwater conditions associated with the terminal regressive phase of Lake Bonneville to the warmer hypersaline context that has characterized Great Salt Lake for much of the Holocene. We also document a dramatic loss of Greater Sage-Grouse across this period. This ~4000-year record of avifaunal change may prove useful in modern conservation efforts, most notably for that of Greater Sage-Grouse, as xeric conditions are forecasted to increase in the future.

Keywords: Lake Bonneville, Great Salt Lake, Avifauna, Greater Sage-Grouse, Aquatic birds

Developments in Earth Surface Processes, Vol. 20. http://dx.doi.org/10.1016/B978-0-444-63590-7.00014-7

14.1 INTRODUCTION

As the sun rose on a late October morning in 1849, Howard Stansbury embarked on his landmark scientific survey of Great Salt Lake and was immediately struck by the contrast between the stark and desolate landscape and the region's dazzling avifauna:

> *The Salt Lake, which lay about half a mile to the eastward, was covered by immense flocks of wild geese and ducks, [...] never did I behold any thing like the immense numbers here congregated together. Thousands of acres, as far as the eye could reach, seemed literally covered with them, presenting a scene of busy, animated cheerfulness, in most graceful contrast with the dreary, silent solitude by which we were immediately surrounded.*
>
> Stansbury (1852, p. 100)

Camped later on Gunnison Island, he was further impressed, and seemingly comforted, by other aspects of the bird life of Great Salt Lake:

> *Our friends the gulls and pelicans, had by no means decreased in numbers; the former filling the air with their interminable chattering, which continued the whole night, and formed a most striking contrast to the chilling and deathlike silence of the surrounding shores.*
>
> Stansbury (1852, p. 191)

Stansbury was surely unaware that just a year earlier the activity of his *friends* would secure them a deeply venerated position in local history:

> *The first I knew of the gulls, I heard their sharp cry ... there must have been thousands of them; their coming was like a great cloud; and when they passed between us and the sun a shadow covered the field. I could see the gulls settling for more than a mile around us. At first we thought that they also, were after the wheat and this fact added to our terror; but we soon discovered that they devoured only the crickets ... the gulls came every morning for about 3 weeks, when their mission was apparently ended. The previous crops were saved*
>
> Unknown Utah Pioneer, Jun. 9, 1848; Henderson (1931)

The large flocks of gulls, clearly derived from the Great Salt Lake colonies, saved the crops that sustained the Mormon pioneers during their first winter in the Salt Lake Valley. The California Gull became the Utah state bird and gull statues still stand in Temple Square, Salt Lake City, commemorating what to many was as a divine miracle. Although these observations and events took place over 150 years ago, the gulls from Great Salt Lake retain a unique place in state history and troves of visitors travel to the region year-round yearning to witness similar scenes of birds "as far as the eye could reach."

Indeed, Great Salt Lake harbors an unparalleled menagerie of aquatic bird life. The lake and the extensive wetlands along its east shore function as resting and staging areas for millions of migratory birds every year, including

FIG. 14.1 A flock of Franklin's Gulls (*Leucophaeus pipixcan*) on the shore of Great Salt Lake.

an abundance of grebes, marsh birds, shorebirds, and waterfowl, while the nine major islands that dot the lake are overtaken annually by colonial nesters seeking refuge from terrestrial predators. And although the lake today supports no fish, brine shrimp (*Artemia franciscana*) and brine flies (*Ephedra*) provide ample provisions for birds populating these aquatic habitats (Fig. 14.1).

The rich avifauna that inhabits Great Salt Lake today owes its existence to the unique aquatic systems history of the Bonneville basin, most notably the massive Pleistocene Lake Bonneville from which it derives. Over the last 30,000 years, the lake experienced numerous transgressions and regressions. The end of the Pleistocene marked the final regression of Lake Bonneville as it fell to elevations similar to modern Great Salt Lake by about 13.0 cal ka BP. It remained at low levels, becoming increasingly brackish, until the brief early Holocene Gilbert episode transgression at 11.6 cal ka BP. Thereafter, hypersaline conditions prevailed throughout the Holocene history of Great Salt Lake (Oviatt, 2014, 2015a; Oviatt et al., 2015).

Recent research focused on faunal and floral remains dating to this period (eg, Rhode and Madsen, 1995; Grayson, 1998, 2000a, 2000b, 2006; Broughton, 2000a, 2000b; Broughton et al., 2000, 2006; Livingston, 2000; Madsen et al., 2001; Lyman and O'Brien, 2005; Terry, 2007, 2008, 2010; Terry et al., 2011; Rowe and Terry, 2014; Terry and Rowe, 2015) has increased our knowledge of biotic change across the Pleistocene/Holocene transition in the Bonneville basin. The dynamics of the region's avifauna during this period, however, remain poorly understood due to a limited availability of avifaunal assemblages that date to this period. In fact, only three sites that have produced substantial vertebrate faunas—Danger Cave, Bonneville Estates Rockshelter, and Homestead Cave—were directly adjacent to the lake during the Pleistocene/Holocene transition (Fig. 14.2) and, of the three, Homestead Cave is the only site that has produced a substantial, fine-grained avifaunal record.

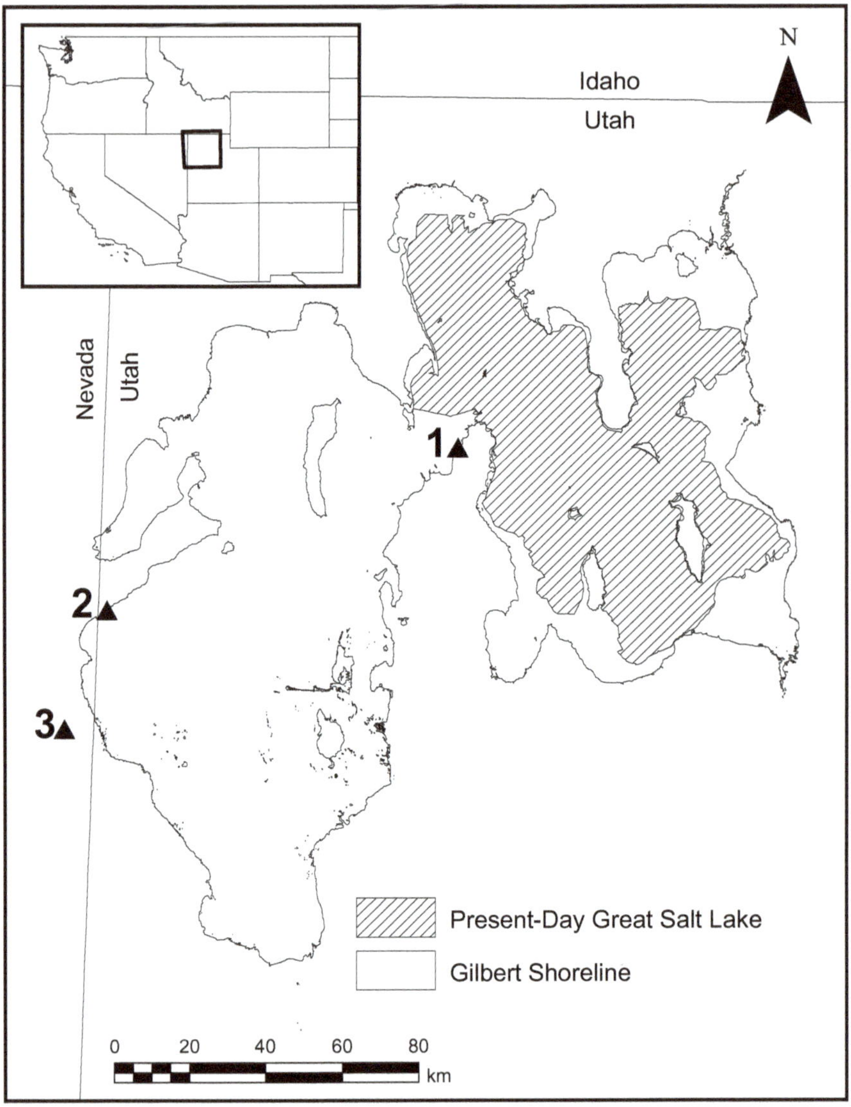

FIG. 14.2 Map of northern Utah showing the Gilbert shoreline as mapped by Currey (1982) (slightly higher in elevation than the highest elevation reached during the Gilbert episode; see Oviatt, 2014), and present-day Great Salt Lake, along with the location of sites discussed in the text. (1) Homestead Cave, (2) Danger Cave, (3) Bonneville Estates Rockshelter. *Figure prepared by Elic M. Weitzel. Data Sources: USGS Utah AGRC, Utah Geological Survey, US Census Bureau.*

The bird remains from Homestead Cave thus offer a unique opportunity to understand the nature of the Bonneville basin avifauna and how that fauna was altered in relation to fluctuations in lake levels and chemistry during the Pleistocene/Holocene transition. In addition, the wide range of other biotic materials analyzed from the Homestead deposits and the trends detected in these data allow for the generation of testable predictions of avifaunal change. Although previous work with the Homestead avifauna produced a sizable sample of identified specimens (N = 5862), including 39 species from 18 families (Livingston, 2000, Table XXIV), and allowed for provisional tests of change between major stratigraphic levels, only a portion of the materials were studied, and identifications were not provided for Strata Ia and Ib—the critical period that encompasses the end of the final Lake Bonneville regressive phase and the postregressive low-elevation lake.

This analysis is focused on the bird remains from the lower strata of Homestead Cave (Strata I–II) that date from 13.1 to 9.5 cal ka BP[a]. It will provide the most detailed description of the bird fauna that lived in and around the lake and track changes in the nature of the avifauna as the lake level fluctuated and became increasingly saline across the Pleistocene/Holocene transition.

14.2 MATERIALS AND METHODS

Homestead Cave is located a few kilometers southwest of Great Salt Lake in the Lakeside Mountains. Homestead Knoll is the low rocky promontory on the northwestern-most spur of these mountains, and Homestead Cave sits on the northern edge of the knoll at an elevation of 1406 m (Fig. 14.2). The cave is approximately 17 m deep and 4.5 m wide at the mouth (Madsen, 2000). The faunal materials from the cave were primarily deposited by raptors, as evidenced by the presence of owls roosting in the cave during excavation and the discovery of many complete and partially disintegrated owl pellets throughout the deposit. As we discuss in more detail later, we assume that the avifaunal remains found in this cave derive from birds captured or scavenged within the foraging radius of raptors and thus represent a sample of the local bird population.

A 1 m × 1 m column sample was excavated 2.7 m to bedrock in 18 well-defined strata, and all materials recovered were passed through a series of 6.3, 3.1, and 1.6 mm screens. A total of 93 radiocarbon dates spanning from 13.1 to 0.7 cal ka BP have since been obtained from a range of organic materials, including vertebrate bone (Table 14.1). Despite evidence of limited mixing, the relationship between stratum depth and mean calibrated age is highly significant ($r_s = 0.995$, $p < 0.0005$), suggesting that the strata were laid down sequentially (Madsen, 2000; Smith and Betancourt, 2003; Broughton et al., 2008; Terry and Novak, 2015).

This analysis will focus on the bird remains from Stratum I that accumulated between 13.1 and 11.8 cal ka BP, a period that encompasses the very

[a]All dates are given in calibrated ka BP following CALIB 7.1 (Reimer et al., 2013; Stuiver et al., 2013) unless otherwise noted.

TABLE 14.1 Radiocarbon Dates and Associated Calibrated Results from the Lower Strata of Homestead Cave, Listed Chronologically

Stratum	^{14}C Age (year BP)	Cal Age (year BP)[a]	Median Cal Age	Lab No.	Material
II[b]	7075±35	7840–7970	7900	CAMS134592	Bone collagen
II[b]	8480±35	9460–9540	9500	CAMS134594	Bone collagen
II[c]	8520±80	9400–9680	9510	AA 14821	1 Hackberry[d] endocarp
II[b]	8615±35	9530–9660	9560	CAMS141271	Bone collagen
II[e]	8675±235	9120–10,270	9750	–	Neotoma fecal pellet(s)
II[c]	8790±90	9560–10,160	9840	AA 14820	1 Hackberry endocarp
II[c]	8830±240	9400–10,520	9910	Beta 63438	Hackberry endocarps
II[b]	9295±40	10,370–10,590	10,500	CAMS141278	Bone collagen
II[b]	9370±35	10,510–10,690	10,600	CAMS141279	Bone collagen
II[b]	9380±35	10,520–10,690	10,610	CAMS141280	Bone collagen
II[b]	9405±45	10,520–10,740	10,640	CAMS141274	Bone collagen
II[b]	9475±40	10,580–11,070	10,720	CAMS134595	Bone collagen
II[b]	9515±40	10,660–11,080	10,830	CAMS141272	Bone collagen
II[b]	9835±40	11,200–11,310	11,240	CAMS134597	Bone collagen
II[b]	10,165±40	11,700–12,310	11,860	CAMS141275	Bone collagen
II[b]	10,195±40	11,750–12,080	11,900	CAMS141276	Bone collagen
II[b]	10,200±40	11,750–12,080	11,900	CAMS141273	Bone collagen
II[b]	10,235±40	11,800–12,130	11,970	CAMS134596	Bone collagen
II[b]	10,255±35	11,830–12,140	12,010	CAMS141277	Bone collagen
II[b]	10,310±40	11,960–12,380	12,110	CAMS134593	Bone collagen
II[b]	10,395±40	12,070–12,420	12,270	CAMS134598	Bone collagen
Ib[c] (upper 5 cm)	10,160±85	11,400–12,120	11,810	AA 14819	1 Neotoma fecal pellet
Ib[e] (upper 5 cm)	10,235±180	11,340–12,550	11,980	–	Neotoma fecal pellet(s)
Ib[c] (upper 5 cm)	10,350±80	11,940–12,440	12,210	AA 14818	1 Neotoma fecal pellet
I[c] (middle)	10,910±60	12,700–12,940	12,780	Beta 72205	Bone collagen
Ia[c] (lower 5 cm)	11,065±105	12,730–13,090	12,950	AA 14817	2 Neotoma fecal pellets
Ia[e] (lower 5 cm)	11,168±208	12,710–13,410	13,030	–	Neotoma fecal pellet(s)
Ia[c] (lower 5 cm)	11,181±85	12,810–13,210	13,140	AA 16808	1 Neotoma fecal pellet
Ia[c] (lower 5 cm)	11,263±85	12,960–13,310	13,130	AA 16809	1 Neotoma fecal pellet
Ia[c] (lower 5 cm)	11,270±135	12,820–13,390	13,130	AA 14816	1 Neotoma fecal pellet

[a] *Two sigma ranges; calculated using CALIB 7.1 (Reimer et al., 2013; Stuiver et al., 2013).*
[b] *Reported in Terry and Novak (2015).*
[c] *Reported in Madsen (2000).*
[d] Celtis reticulata.
[e] *Reported in Smith and Betancourt (2003).*

end of Lake Bonneville's regressive phase and the early postregressive phase when the lake approached elevations near modern Great Salt Lake (Oviatt, 2014). However, comparisons will also be made between the Stratum I collection and the avifauna from Stratum II, the latter of which has produced a total of 21 dates that span the early Holocene period from 12.3 to 9.5 cal ka BP (omitting an anomalous late date of 7.9 cal ka BP) (Table 14.1). Stratum II thus spans the period that includes the Gilbert episode (~11.6 cal ka BP) and the first two early Holocene millennia of the hypersaline Great Salt Lake.

During excavation of the column, Stratum I was removed as a single layer. However, the stratum did exhibit various microlaminae, and the dates obtained from the top and the bottom of the stratum differed by more than a 1000 years. Samples were thus subsequently taken from the profile of the top and the bottom of the stratum, creating the arbitrarily defined Stratum Ia (bottom half of Stratum I) and Stratum Ib (top half). Five dates were derived from Stratum Ia that ranged between 13.1 and 13.0 cal ka BP (11,270 $\pm$ 135 and 11,065 $\pm$ 105 ^{14}C year BP), whereas the three dates from Stratum Ib ranged between 12.2 and 11.8 cal ka BP (10,350 $\pm$ 80 and 10,160 $\pm$ 85 ^{14}C year BP) (Table 14.1). The avian materials from Stratum I thus consist of the sample from Stratum Ia, the sample from Stratum Ib, and the original Stratum I materials obtained from the main column sample.

Following a refinement of Livingston's (2000) identifications and additional identifications of previously unstudied material from the 3.1 and 1.6 mm screen samples, the final collective sample size for Stratum I is 4613 identified specimens. Taxonomic identifications are based on comparisons with reference material from the Natural History Museum of Utah. Identifications were not attempted for isolated vertebrae, ribs, pedal phalanges, carpals, and fragments that lacked substantial symphyses or articulation surfaces. A descriptive taxonomic summary of the represented elements by taxon and the diagnostic osteological characters used to identify the specimens are described in the Appendix; anatomical terminology follows Howard (1929), and taxonomic nomenclature follows AOU (2015).

14.3 TAXONOMIC COMPOSITION OF THE HOMESTEAD AVIFAUNA

The 4613 identified avian specimens from Stratum I of Homestead Cave represent 14 orders, 17 families, and at least 53 species (Table 14.2). As previous work had identified 954 avian specimens from the 6.1 mm sample from this stratum (Livingston, 2000), our analysis of the 3.1 and 1.6 mm fractions reported here adds 3659 additional specimens. Although the avifauna suggests raptors had sampled a range of local terrestrial habitats, and grouse, owls, hawks, goatsuckers, and perching birds are well represented in the assemblage, aquatic birds dominate (78.0% of all "aquatic" and "terrestrial" taxa) the assemblage. Among the aquatic taxa, waterfowl (Anatidae) are most

TABLE 14.2 Numbers of Identified Bird Specimens per Taxon by Stratum and Screen Size at Homestead Cave

	Provenience Unit						
	Stratum Ia		*Stratum Ib*		*Stratum I*		
Taxon	*(6.3 mm)*	*(1.6 mm)*	*(3.1 mm)*	*(1.6 mm)*	*(6.3 mm)*	*(3.1 mm)*	*Total*
Anserinae (small)					1		1
Anserinae (medium)					1		1
Anser albifrons		1			1		2
Branta canadensis					1		1
Anatinae	1	3		4	143	4	155
Anatinae (small)		1		1	19	2	23
Anatinae (large)	1	53	2	7	346	19	428
Anas sp.		2		1	66	2	71
A. platyrhynchos					3		3
Anas sp. (teal)					24	1	25
Aythya sp.	4	77	1	12	641	22	757
Bucephala albeola		1			20	1	22
B. clangula/islandica		1			3		4
Mergus sp.					3	1	4
Mergus merganser		3			7	2	12
Mergus serrator		5			6	1	12
Oxyura jamaicensis					1		1
Tetraoninae		1	1		3		5
Centrocercus urophasianus		6	1	5	84	15	111
Dendragapus obscurus					7		7
Tympanuchus phasianellus					2	2	4
Gavia immer					1		1
Podiceps auritus/nigricollis		11	1	7	39	12	70
P. auritus		13		6	108	14	141

Continued

TABLE 14.2 Numbers of Identified Bird Specimens per Taxon by Stratum and Screen Size at Homestead Cave—Cont'd

	Provenience Unit						
	Stratum Ia		*Stratum Ib*		*Stratum I*		
Taxon	*(6.3 mm)*	*(1.6 mm)*	*(3.1 mm)*	*(1.6 mm)*	*(6.3 mm)*	*(3.1 mm)*	*Total*
P. nigricollis		12		9	181	51	253
Aechmophorus occidentalis/clarkii		2		2	30	7	41
Phalacrocorax auritus		1			1		2
Accipitriformes		2			6	7	15
Cathartes aura					2		2
Circus cyaneus					2		2
Accipiter striatus/cooperii					9	2	11
A. striatus				2	2	2	6
A. cooperii					4		4
Buteo sp.		4		1	17		22
Rallus limicola					1	1	2
Porzana carolina		1		1	5	1	8
Gallinula galeata		1			2		3
Fulica americana					5		5
Pluvialis sp.				1	1		2
Charadrius vociferus					1		1
Actitis macularius		4		3	9	14	30
Tringa sp.		1		1	1		3
Tringa cf. *solitaria*					1		1
Numenius americanus						1	1
Calidris sp.		2			4	9	15
Phalaropus sp.		5		2	16	10	33
Laridae (small)		3		1	34	7	45
Laridae (large)		4		2	41	1	48

Continued

TABLE 14.2 Numbers of Identified Bird Specimens per Taxon by Stratum and Screen Size at Homestead Cave—Cont'd

	Provenience Unit						
	Stratum Ia		*Stratum Ib*		*Stratum I*		
Taxon	*(6.3 mm)*	*(1.6 mm)*	*(3.1 mm)*	*(1.6 mm)*	*(6.3 mm)*	*(3.1 mm)*	*Total*
Zenaida cf. *macroura*					2		2
Z. macroura					2		2
Strigiformes		2		5	17	4	28
Tyto alba					2		2
Megascops kennicottii					11	1	12
Bubo virginianus					1		1
Athene cunicularia					2	2	4
Asio sp.		2		6	108	5	121
A. otus				1	33		34
A. flammeus		1	1	2	43	2	49
Aegolius acadicus		1		1	10	1	13
Chordeiles cf. *minor*		1			4		5
Phalaenoptilus nuttallii				1	24	11	36
Sphyrapicus sp.		1			4	3	8
Colaptes auratus		1		4	18	4	27
Falco sparverius/columbarius		2		2	6		10
F. peregrinus/mexicanus					13		13
F. sparverius				2	35	12	49
F. columbarius					11	2	13
F. peregrinus				1	3		4
F. mexicanus					5		5
Passeriformes		288	2	157	492	820	1759
Total	6	519	9	250	2751	1078	4613

Stratum Ia consists of Field Specimen numbers 75, 93, and 94. Stratum Ib consists of Field Specimen numbers 76, 92, and 95. Stratum I consists of Field Specimen numbers 47, 63, and 64.

abundant (68.3% of aquatic bird specimens), followed by grebes (Podicipedidae, 22.7%), and much more distantly, by gulls (Laridae, 4.18%), shorebirds (Charadriidae and Scolopacidae, 3.86%), and rails and coots (Rallidae, 0.81%). It is also clear that within the aquatic avifauna, taxa-favoring deeper-water lacustrine contexts are most abundant. The majority of the duck specimens, for example, are diving ducks (eg, *Aythya*, *Mergus*, *Bucephala*, *Oxyura*; Fig. 14.3) that are more piscivorous and prefer deeper-water contexts relative to the more omnivorous dabbling duck species (*Anas*; Baldassarre, 2014). Similarly, the grebe assemblage includes a mix of taxa including Clark's or Western Grebe (*Aechmophorus clarkii/occidentalis*; Fig. 14.4) that are specialist piscivores (Table 14.2; LaPorte et al., 2013).

FIG. 14.3 Female Canvasback (*Aythya valisineria*).

FIG. 14.4 A pair of Western Grebes (*Aechmophorus occidentalis*) performing a "weed ceremony"—one of a number of complex courtship displays.

The avifauna from Stratum I—that again was deposited from the very end of Lake Bonneville's regressive phase through the early stages of the increasingly brackish postregressive lake—would thus, not surprisingly, appear distinctive from that of the hypersaline Great Salt Lake. The anatid fauna of the latter, for instance, although diverse, is dominated by dabbling ducks (Behle, 1958; Paul et al., 1999). The expansive mudflat and wetland habitats that surround the modern lake promote high densities of shorebird and marsh taxa that apparently occurred in much lower densities in the fauna of the terminal Pleistocene. Finally, the extensive series of islands in Great Salt Lake—most fully submerged in Lake Bonneville times—provide refuge today for extensive breeding colonies of Double-crested Cormorant (*Phalacrocorax auritus*), Great Blue Heron (*Ardea herodias*), California Gull (*Larus californicus*), and American White Pelican (*Pelecanus erythrorhynchos*)—birds that, as a group, are poorly represented in the Homestead fauna. Although modern Great Salt Lake is devoid of fish, these larger piscivores obtain fish resources from freshwater marshes and bays near freshwater inlet streams and rivers where fish occur (Behle, 1958).

As noted earlier, terrestrial birds that lived around the lake are also well represented in the assemblage and include a range of taxa from raptors to grouse to perching birds. The latter are, in fact, extremely abundant in the collection, although we have not yet made species-level identifications for these materials. Earlier preliminary analyses, however, suggest that Horned Larks (*Eremophila alpestris*) and waxwings (*Bombycilla*) are abundant in the collection (Livingston, 2000). Other well-represented terrestrial taxa include Greater Sage-Grouse (*Centrocercus urophasianus*), Long-eared and Short-eared Owls (*Asio otus* and *Asio flammeus*), Common Poorwills (*Phalaenoptilus nuttallii*), American Kestrels (*Falco sparverius*), and woodpeckers (Picidae), including Northern Flickers (*Colaptes auratus*) and sapsuckers (*Sphyrapicus*) (Table 14.2).

All of the terrestrial taxa present in the assemblage occur in the area today, with the exception of the forest-dwelling Dusky Grouse (*Dendragapus obscurus*), which is represented by seven specimens in the Stratum I fauna. The nearest records for Dusky Grouse in western Utah are from the Raft River and the Stansbury Ranges, which are located ~100 km north and south of Homestead Cave, respectively (Behle, 1985). Although the terrestrial vegetation around Homestead Cave historically consists of xeric desert shrub (eg, sagebrush [*Artemisia*], shadscale [*Atriplex*], rabbitbrush [*Chrysothamnus*], horsebrush [*Tetradymia*]), paleovegetation records from the Homestead region for the terminal Pleistocene suggest that limber pine and subalpine fir occurred along stream channels at slightly higher elevations. Indeed, the woodpecker materials also suggest the local presence of wooded habitats either in the form of scattered stands of coniferous forest or deciduous riparian communities that may have occurred within local washes that are too dry to support trees today (eg, Rhode and Madsen, 1995; Rhode, 2000; Madsen et al., 2001).

FIG. 14.5 Female Greater Sage-Grouse (*Centrocercus urophasianus*).

Finally, the relatively high frequency of Greater Sage-Grouse remains in the deposit is noteworthy, as this species was not identified in earlier work with the collection. Greater Sage-Grouse is indicative of dense sagebrush habitats (Fig. 14.5), and it is also well represented in the other terminal Pleistocene avifaunas from the Bonneville basin, namely, Danger Cave and Bonneville Estates Rockshelter (Fig. 14.2; Parmalee, 1988; Hockett, 2007). In the latter setting—which is largely an anthropogenic deposit—Greater Sage-Grouse is one of the best-represented vertebrate taxon in the collection, suggesting this bird was an important subsistence resource to early human foragers in the Bonneville basin, as they clearly were in many settings throughout the Holocene (Parmalee, 1970, 1980).

14.4 DEPOSITIONAL ORIGIN

As indicated earlier, the Homestead Cave assemblage appears to have been deposited almost entirely by raptors—especially owls—roosting in the cave. This is evidenced by the presence of owls roosting in the cave during the excavation, owl pellet residue adhering to many of the vertebrate remains, and, most importantly, detailed taphonomic analyses involving skeletal part representation and damage pattern on the Stratum I fish (Broughton et al., 2006) as well as the small mammals from the entire deposit (Terry, 2007).

Those analyses indicate that the fauna is consistent with an origin from large owls (eg, *Tyto alba* and *Bubo virginianus*) rather than other raptors or agents. In general, owls impart less damage and produce assemblages with more complete skeletal representation relative to other avian or carnivore agents (Broughton et al., 2006; Terry, 2007), and these features characterize the Homestead faunal deposits. Indeed, Terry (2007) documented that skeletal part frequency and fragmentation patterns indicate that large owls produced the entire Homestead sequence of small mammals and that changes in predator agents can be rejected as a source of temporal variation in taxonomic composition. In this context, we observe that none of the bird specimens studied here exhibit rodent gnawing, carnivore chewing, or human modifications (eg, cut marks, burning), and show excellent preservation with minimal fragmentation and what would appear to be complete skeletal part representation, although quantitative analyses of the latter have yet to be performed. Incidentally, we note that no change in the relative frequency of raptor families (Stratum Ia to Ib: $X^2=2.10$; $p=0.35$; Stratum I to II: $X^2=7.54$; $p=0.11$) or owl genera (Stratum Ia to Ib: $X^2=0.64$; $p=0.42$; Stratum I to II: $X^2=8.49$; $p=0.13$) exists across the Late Pleistocene/Early Holocene sequence of deposits that is our focus here.

Shifts though time in the taxonomic composition of the Homestead avifauna are thus likely to reflect real changes in the avian landscape and habitat composition across the depositional sequence at Homestead Cave, although some systematic bias against the deposition of very small birds and very large ones outside of the normal diet breadth of large owls undoubtedly exists.

14.5 TERMINAL PLEISTOCENE AND EARLY HOLOCENE CHANGE IN THE BONNEVILLE BASIN AVIFAUNA

In this section, we examine change in the taxonomic composition of the Bonneville basin avifauna as might be anticipated from terminal Pleistocene and early Holocene environmental change. We focus on change in the aquatic avifauna from Homestead Cave, as the assemblage is best represented by waterbirds, although we also explore trends in aspects of the terrestrial fauna. We draw largely upon previous analyses of the rich vertebrate fauna from Homestead Cave to derive expectations for change in the terminal Pleistocene and early Holocene avifauna. Although a wide range of data have been used to reconstruct environmental change and lake level dynamics in this setting (see Benson et al., 2011; Oviatt, 2014; Oviatt et al., 2015 and references therein), the Homestead vertebrate record provides the most analytically comparable data set to generate expectations for change in the avian fauna. We augment these data with paleovegetational records, largely derived from plant macrofossils retrieved from woodrat (*Neotoma*) middens.

Recent geologic research has revealed a more refined sequence of lake-level and lake-chemistry changes from the terminal Pleistocene into the early Holocene. Lake Bonneville regressed to elevations nearing those of modern Great Salt Lake by about 13.0 cal ka BP and remained at low elevations for the next 1400 years. The ostracod record suggests that this postregressive phase lake was likely becoming brackish, and remained so until after the Gilbert episode at about 11.6 cal ka BP. During this episode, the lake transgressed to a maximum elevation of 1295–1297 m, but quickly regressed ~15 m back to elevations similar to modern Great Salt Lake and became hypersaline, as indicated by the presence of brine shrimp (Oviatt, 2014, 2015a, 2015b; Oviatt et al. 2015; Thompson et al., 2016).

The Homestead Cave stratigraphy can easily be synced with this record of lake history. Stratum I (13.1–11.8 cal ka BP) was deposited during the terminal Pleistocene, spanning the end of the Lake Bonneville regressive phase and also including the postregressive low-elevation lake. Stratum Ia (13.1–13.0 cal ka BP) encompasses the very end of the regression, whereas Stratum Ib (12.2–11.8 cal ka BP) encompasses more of the low-elevation lake relative to Stratum Ia. Stratum II (12.3–9.5 cal ka BP) captures the brief Gilbert episode, and nearly two millennia (11.5–9.5 cal ka BP) of the early Holocene hypersaline Great Salt Lake.

The detailed fish record from Homestead Cave provides a useful framework for generating expectations for change in the aquatic avifauna, as it further informs on lake levels and aquatic environments during the terminal Pleistocene and early Holocene (Broughton, 2000a, 2000b; Broughton et al., 2000; Broughton and Smith, 2016). The fish remains are largely confined to the lowest stratum of Homestead Cave (Stratum I) and the fish analyses focused on changes across the lower and upper divisions of this unit—Strata Ia and Ib. Strontium analyses of fish bones from Stratum I are consistent with an origin from a low-elevation lake. A decrease in fish size along with an increase in species tolerant of higher salinities and/or water temperatures was documented from Stratum Ia (13.1–13.0 cal ka BP) to Stratum Ib (12.2–11.8 cal ka BP), suggesting a series of die-off events associated with increases in water temperature and salinity at the end of the regressive phase of Lake Bonneville. The early Holocene Stratum II deposits (12.3–9.5 cal ka BP) indicate that following these die-off events, fish recolonized the lake during the early Holocene Gilbert episode (~11.6 cal ka BP) and were characterized by a higher abundance of salinity and temperature tolerant taxa (Broughton, 2000a, 2000b; Broughton et al., 2000; Madsen et al., 2001; Broughton and Smith, 2016).

Analyses of both the small mammals from Homestead Cave (Grayson, 1998, 2000a, 2000b, 2006) and limited regional paleovegetation records (Rhode, 2000; Madsen et al., 2001; Rhode and Louderback, 2007) allow the derivation of expectations for temporal change in the terrestrial avifauna across the terminal Pleistocene and early Holocene deposits of the site. Although few northern Bonneville basin woodrat middens exist that date to the period represented by Stratum Ia (13.1–13.0 cal ka BP), what data do exist suggest that limber pine (*Pinus flexilis*) woodlands may have still existed at low elevations at this time. Needles, seeds, and cone parts are very abundant in woodrat middens dating to ~15.4–13.3 cal ka BP in the region and are commonly associated with a mix of shrubs in smaller quantities including sagebrush, snowberry (*Symphoricarpos*), prostrate juniper (*Juniperus horizontalis*), and soapberry (*Sapindus*) (Madsen et al., 2001). However, both the plant macrofossil record and small mammal fauna from Stratum I suggest that vegetation conditions were rapidly shifting to open, cold desert steppe dominated by dense sagebrush-grass communities. A similar environment characterized the period represented by Stratum II (12.3–9.5 cal ka BP), although conditions were growing increasingly xeric, especially toward the end of the period. Dense stands of big sagebrush were progressively thinned and increasingly mixed with other more xeric shrubs such as shadscale, greasewood (*Sarcobactus vermiculatus*), and horsebrush (Madsen et al., 2001).

These trends documented across the stratigraphic units at Homestead Cave allow for the derivation of expectations for the avifauna. We thus examine avifaunal change across two primary units of time. First, we compare the avifauna from the period associated with the terminal Pleistocene end of the Lake Bonneville regression—Stratum I (13.1–11.8 cal ka BP)—with that from the early Holocene Gilbert episode and subsequent hypersaline Great Salt Lake—Stratum II (12.3–9.5 cal ka BP). Then, at a finer level of chronological resolution covering the terminal Pleistocene itself, we examine change between Stratum Ia (13.1–13.0 cal ka BP) and Stratum Ib (12.2–11.8 cal ka BP). To provide more continuous coverage for the entire period examined (13.1–9.5 cal ka BP), we also analyze trends through time comparing the materials from Stratum Ia, Stratum Ib, and Stratum II. We expect similar trends across each period examined as lake conditions became increasingly saline and xeric vegetation expanded, with stronger trends evident across the analyses that extend over greater periods of time.

Finally, we note that the currently available sample of avifaunal identifications from the different provenience units described earlier from Homestead Cave represent materials retrieved with different screen mesh sizes. The avifauna from Stratum II, provided by Livingston (2000), represents only the 6.3 mm sample from that stratum. And although we present a substantial sample of 3.1 mm material from Stratum I here (Table 14.2), our analyses of change through time between Stratum I and Stratum II utilize only the 6.3 mm sample in order to ensure comparability. Our comparisons between Stratum Ia and Stratum Ib, however, represent the full suite of materials collected from 6.3, 3.1, 1.6 mm mesh screens. The absolute values of our relative abundance measures thus differ somewhat between these two sets of analyses, with some of that likely owing to differences in recovery methods. We discuss in more detail below possible effects of recovery methods on these and other aspects of the Homestead avifauna.

14.5.1 Change in the Aquatic Avifauna

We examine below three measures of relative abundance derived from the Homestead aquatic avifauna that should reflect variation in lake levels and associated aquatic conditions across the terminal Pleistocene and early Holocene of the Bonneville basin.

Deep- and shallow-water taxa. In order to examine broadscale change in the aquatic avifauna that should scale with changes in lake conditions, we divide the aquatic bird taxa first into two broad categories: (1) more carnivorous, diving taxa that occur in higher abundances in deeper-water habitats with abundant fish and aquatic invertebrates and (2) taxa that are more omnivorous, use higher proportions of aquatic vegetation, and are thus more abundant in brackish, shallow-water, mudflat, and marsh habitats. The deep-water taxa include diving ducks (*Aythya, Bucephala*, *Mergus*, and *Oxyura*), Western and Clark's Grebes (*Aechmophorus*), Double-crested Cormorants (*Phalacrocorax auritus*), and Common Loons (*Gavia immer*). The shallow-water taxa include geese (Anserinae), dabbling ducks (*Anas*), Eared and Horned Grebes (*Podiceps*), rails and coots (Rallidae), shorebirds (Charadriidae and Scolopacidae), and gulls (Laridae) (Behle, 1958; Behle and Perry, 1975; Storer and Nuechterlein, 1992; Cullen et al., 1999; LaPorte et al., 2013; Baldassarre, 2014; Dorr et al., 2014). Insofar as lake salinity and temperature gradually increased, fish died off, and marsh and mudflat habitats increased over time, we anticipate proportionate declines in deep-water birds.

Diving and dabbling ducks. Narrowing the taxonomic scope, we focus on change within the duck fauna and compare the relative frequencies of diving and dabbling ducks. Diving ducks are more piscivorous and take fish and aquatic invertebrates, as well as aquatic vegetation, through extended subsurface foraging bouts. Diving ducks thus occur in higher abundances in deep, freshwater, lacustrine contexts associated with substantial fish populations and include species in the genera *Aythya*, *Bucephala*, *Mergus*, and *Oxyura*. By contrast, dabbling ducks in the genus *Anas* do not use fish resources, feed principally on aquatic vegetation and invertebrates along the surface or by tipping into the water, and are more abundant in shallow-water ponds and lakes (Baldassarre, 2014). Insofar as lake salinity and temperature increased and fish populations were lost across the terminal Pleistocene and into the early Holocene, we anticipate a decline in diving ducks relative to dabbling ducks.

Aechmophorus *and* Podiceps *grebes*. Change in the relative abundance of *Aechmophorus* and *Podiceps* grebes should also track variation in lake conditions. Clark's and Western Grebes (*A. clarkii* and *A. occidentialis*, respectively) are both specialist piscivores where fish, acquired by extended subsurface diving, comprise between 81% and 100% of their diet (Storer and Nuechterlein, 1992; LaPorte et al., 2013). Eared and Horned Grebes (*Podiceps nigricollis* and *P. auritus*, respectively) are smaller, and although they also dive and take fish, they have broader diets that include higher proportions of plant material and aquatic invertebrates. Eared grebes, for instance, occur in extremely high densities in shallow, hypersaline lakes, such as Great Salt Lake, to molt and stage during fall migrations. Brine shrimp and brine flies comprise most of the diet in these settings (Cullen et al., 1999). Insofar as lake salinity and temperature increased over time and fish were ultimately replaced by brine shrimp, we expect a decline in *Aechmophorus* grebes relative to *Podiceps* grebes.

Trends in the aquatic avifauna: Results. For the comparisons in relative taxonomic abundance across the paired provenience units (eg, Stratum I vs. Stratum II, and Stratum Ia vs. Stratum Ib), we use chi-squared tests to statistically evaluate differences in the avifauna. To account for tests where marginal values are less than 5, *p*-values are simulated (R Core Team, 2015). For the series including Stratum Ia, Stratum Ib, and Stratum II we use Cochran's test of linear trends to evaluate the occurrence, strength, and significance of trends through time in the relative abundance of these taxa (see Cannon, 2001). Accompanying figures were produced in R (R Core Team, 2015).

On balance, the anticipated trends in the aquatic avifauna are revealed across the terminal Pleistocene and early Holocene deposits of Homestead Cave (Figs. 14.6–14.8; Tables 14.3 and 14.4). More specifically, significant declines in deep-water bird, diving duck, and *Aechmophorus* grebe abundances are indicated for each of the analyses comparing Stratum I and Stratum II, and for the assemblages representing the full sequence (Strata Ia, Ib, and II) (Figs. 14.7–14.8). We emphasize that the full sequence trends are not only statistically significant, but that their effect sizes (see Wolverton et al., 2014) are also each very strong—namely, the portion of the X^2 associated with the trends accounts for 100% (119.220/119.220), 93% (59.188/63.78), and 95% (5.962/6.259) of the total observed in the comparisons between the deep-water and shallow-water taxa, diving and dabbling ducks, and *Aechmophorus* and *Podiceps* grebes, respectively (Table 14.4).

Thus, a clear signal of increasing lake temperature and salinity is evident in the Homestead avifauna for the period that spans the transition from the terminal Pleistocene to the early Holocene. Trends within the terminal Pleistocene—between Stratum Ia and Ib—are, however, not as strong or consistent. Indeed, of the three relative abundance measures, only the most inclusive measure of deep-water and shallow-water birds declines significantly across Stratum Ia and Stratum Ib (Fig. 14.6; Table 14.4). These trends mirror the pattern documented with the Homestead fish remains that show some evidence of increasing water temperature and salinity across Stratum Ia and Stratum Ib, but with the most dramatic change occurring across the transition from the terminal Pleistocene to the early Holocene—between Stratum I and Stratum II. Further, this is consistent with the above geologic and ostracode evidence for a shift from brackish to hypersaline water conditions at the end of the Gilbert episode that occurs within the period represented by Stratum II.

These trends are clearly not a function of differences in recovery methods used to the collect the Homestead avifauna. For instance, significant trends were indicated in every case comparing materials from Stratum I and Stratum II where we have held recovery method constant by conducting those analyses with only materials collected with 6.3 mm screens (Fig. 14.7; Table 14.4). The full sequence analysis is more subject to recovery biases because 3.1 and 1.6 mm screen mesh was used to recover the Stratum Ia and Stratum Ib materials, but the identifications for the Stratum II assemblage are currently available only for materials that did not pass through 6.3 mm mesh. However, the taxa included

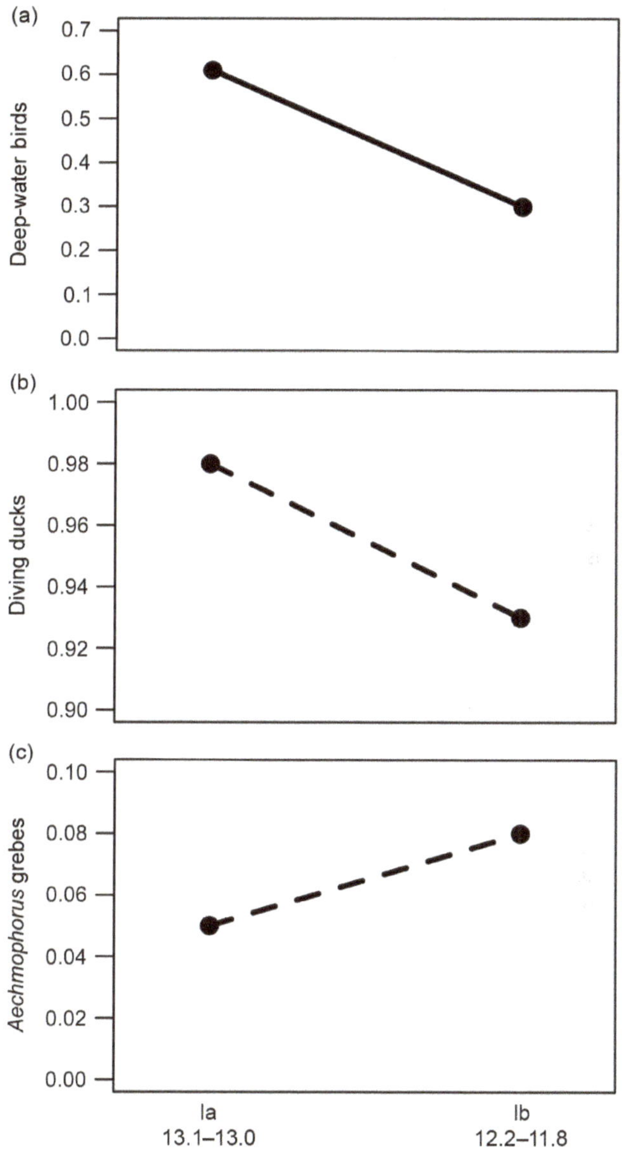

FIG. 14.6 Changes in the relative abundance of (a) deep-water birds, (b) diving ducks, and (c) *Aechmophorus* grebes during the terminal Pleistocene—from Stratum Ia to Stratum Ib. Tests significant at $p < 0.01$ are indicated by *solid lines*; *dashed lines* denote statistically insignificant trends.

in our analyses, with few exceptions, are of substantial size, so we anticipate that 6.3 mm mesh would have been sufficient to recover most elements from them. The exceptions include the smaller-sized rails and shorebirds (eg, *Rallus limicola*, *Porzana carolina*, *Calidris*, *Phaloropus*), for which we expect many

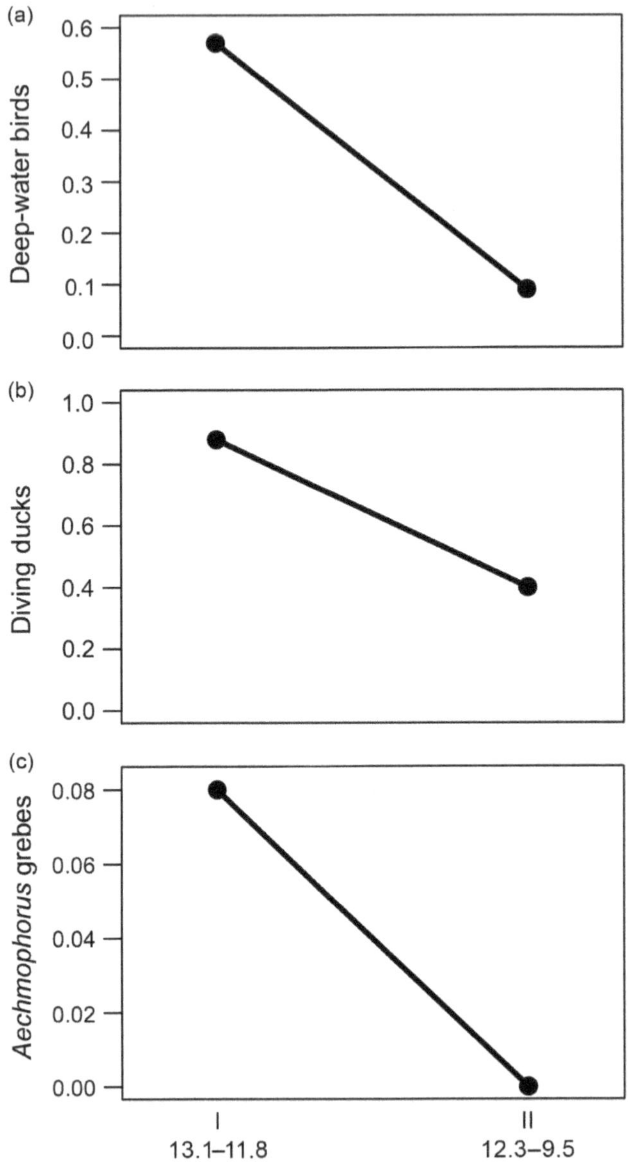

FIG. 14.7 Changes in the relative abundance of (a) deep-water birds, (b) diving ducks, and (c) *Aechmophorus* grebes from the terminal Pleistocene to the early Holocene—Stratum I to Stratum II. Tests significant at $p < 0.01$ are indicated by *solid lines*; *dashed lines* denote statistically insignificant trends.

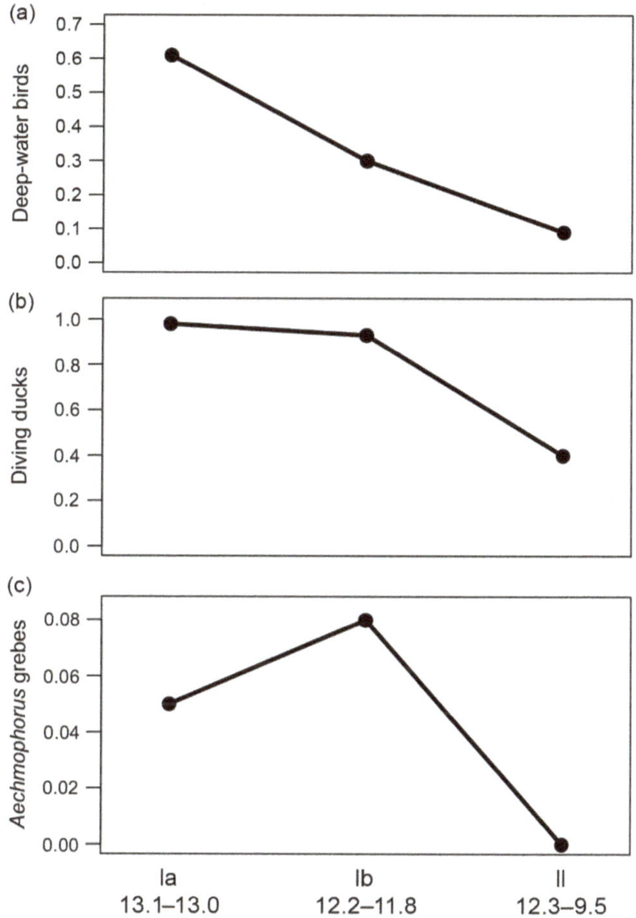

FIG. 14.8 Changes in the relative abundance of (a) deep-water birds, (b) diving ducks, and (c) *Aechmophorus* grebes through time during the terminal Pleistocene and into the early Holocene—from Stratum Ia, to Stratum Ib, to Stratum II. Tests significant at $p < 0.01$ are indicated by *solid lines*; *dashed lines* denote statistically insignificant trends.

elements would pass through 6.3 mm mesh. Still, the trend we document is that these taxa are more abundant in the Stratum II collection compared to Stratum Ia and Stratum Ib (Table 14.3)—the apparent increase in these small-sized shallow-water taxa is thus apparent despite a bias that would have reduced their recovery rate in Stratum II relative to Stratum Ia and Stratum Ib.

14.5.2 Change in the Terrestrial Avifauna: Greater Sage-Grouse

Greater Sage-Grouse do not occur around Homestead Cave today, but the historic range of this bird included the Lakeside Range earlier in the 20th century

TABLE 14.3 Numbers of Identified Specimens for Each Aquatic Taxonomic Group Used in the Tests of Diachronic Change

	Stratum			
Taxonomic Group	*Ia*	*Ib*	*I*	*II*
Deep-water birds	94	15	717	19
Shallow-water birds	60	35	546	204
Diving ducks	91	13	685	17
Dabbling ducks	2	1	93	25
Aechmophorus grebes	2	2	30	0
Podiceps grebes	36	23	328	125

The Stratum I data include only the 6.3 mm sample. Data from Strata Ia, Ib, and I derive from the analysis described earlier.
Data from Stratum II derive from Livingston, S.D., 2000. The Homestead Cave avifauna. In: Madsen, D.B. (Ed.), Late Quaternary Paleoecology in the Bonneville Basin. Bulletin, vol. 130. Utah Geological Survey, Salt Lake City, pp. 91–102.

TABLE 14.4 Summary of X^2 Values for the Tests of Diachronic Change of Aquatic Taxa

	Stratum		
Taxonomic Comparison	*Ia–Ib*	*I–II*	*Ia–Ib–II*
Deep/shallow-water birds	14.61[a]	176.51[a]	119.22[a] (119.22[a])
Diving/dabbling ducks	1.11	73.20[a]	59.19[a] (63.78[a])
Aechmophorus/Podiceps grebes	0.19	11.17[a]	5.96[b] (6.26[b])

The Stratum I data include only the 6.3 mm sample. Data from Strata Ia, Ib, and I derive from the analysis described earlier.
Values for Strata Ia–Ib–II represent the X^2_{trend}, and the values in parentheses represent the total X^2 values.
[a]*significant at* $p < 0.05$.
[b]*significant at* $p < 0.005$.
Data from Stratum II derive from Livingston, S.D., 2000. The Homestead Cave avifauna. In: Madsen, D.B. (Ed.), Late Quaternary Paleoecology in the Bonneville Basin. Bulletin, vol. 130. Utah Geological Survey, Salt Lake City, pp. 91–102.

(Schroeder et al., 2004). Additionally, as noted earlier, Greater Sage-Grouse is well represented in the Homestead avifauna during the late Pleistocene and early Holocene. Our analysis of the lower strata of Homestead Cave offers a 4000-year record of Greater Sage-Grouse in the Bonneville basin—a sequence that captures more climatic and environmental variability than other modern, short-term studies which may prove to be a better representation of the magnitude of change we might expect with future climate change, and could provide valuable information for the modern conservation of this bird. For these reasons, we focus our analysis of trends in Homestead terrestrial avifauna on Greater Sage-Grouse.

Insofar as thick stands of big sagebrush had become established near Homestead at the beginning of the sequence, we anticipate this taxon to be well represented in the Stratum I fauna. It is possible, however, that limber-pine woodlands may have occurred near the site as well, especially during the time represented by earliest deposits at Homestead (ie, Stratum Ia, dating to 13.1 cal ka BP)—a factor that would have constrained populations of sage-grouse. Indeed, the presence of Dusky Grouse and woodpeckers in the Stratum I collection is consistent with the presence of forest habitat within the foraging radius of the owls that deposited the assemblage. Nevertheless, across the transition to the early Holocene (Stratum II), big-sagebrush stands were progressively thinned and increasingly mixed with other more xeric shrubs, such as shadscale, greasewood, and horsebrush, to judge from both paleobotanical records and the small mammal record from Homestead Cave (Grayson, 2000b; Madsen et al., 2001). A decline in sage-grouse abundance is thus clearly anticipated across this transition.

Following the approach we took with the aquatic taxa above, we examine trends in sage-grouse abundance relative to other terrestrial avian taxa across the Homestead deposits and present chi-squared tests for those comparisons (Fig. 14.9; Tables 14.5 and 14.6). These analyses show, most clearly, that sage-grouse decline significantly in abundance from Stratum I to Stratum II (Fig. 14.9b; Table 14.6) as predicted. Indeed, sage-grouse represent nearly 10% of the terrestrial avifauna in Stratum I but are entirely absent from the Stratum II collection. (Livingston (2000) listed five phasianid specimens from Stratum II—our examination of those specimens indicates that they are all Sharp-tailed Grouse (*Tympanuchus phasianellus*)). The suggestive, but statistically insignificant, increase in sage-grouse abundance from Stratum Ia to Stratum Ib (Fig. 14.9a; Table 14.6) is intriguing and may reflect a subtle decline in limber-pine woodland from 13.1 to 11.8 cal ka BP in this setting.

Hockett (2007) and Parmalee (1988) have documented a similar abundance of sage-grouse in deposits dating to the terminal Pleistocene with a subsequent loss in the early Holocene, dated most precisely at 9.9 cal ka BP from the Bonneville Estates Rockshelter on the western margin of the Bonneville basin (Fig. 14.2). Sage-grouse then return to that area during the middle Holocene after 7.0 cal ka BP. Nearby at Danger Cave, a single specimen

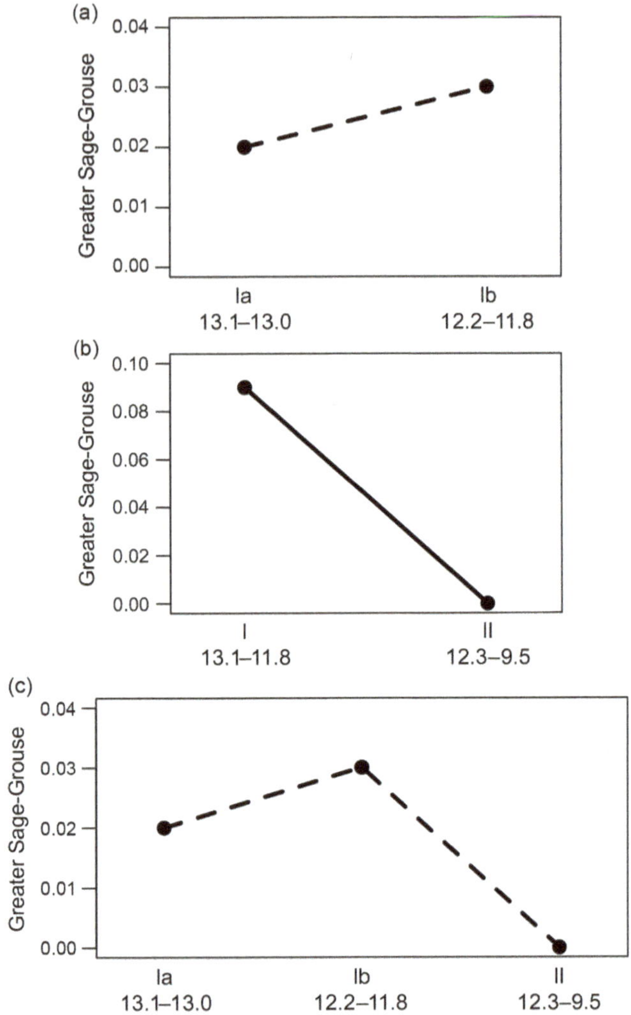

FIG. 14.9 Changes in the relative abundance of Greater Sage-Grouse from (a) Stratum Ia to Ib (during the terminal Pleistocene), (b) Stratum I to II (from the terminal Pleistocene to the early Holocene), and (c) Stratum Ia, to Stratum Ib, to Stratum II (during the terminal Pleistocene and into the early Holocene). Tests significant at $p < 0.01$ are indicated by *solid lines*; *dashed lines* denote statistically insignificant trends.

from Stratum V, with dates ranging from 5.6 to 1.8 cal ka BP, may also suggest a late Holocene return of sage-grouse in that setting. Further analyses of the Homestead avifaunal sequence above Stratum II will shed light on Holocene climate-based shifts in the local presence and abundance of Greater Sage-Grouse in the Lakeside Range and should provide data of interest to conservation biologists.

TABLE 14.5 Numbers of Identified Specimens for Greater Sage-Grouse and Other Terrestrial Taxa Used in the Tests of Diachronic Change of Greater Sage-Grouse

	Stratum			
Taxonomic Group	*Ia*	*Ib*	*I*	*II*
Greater Sage-Grouse	6	6	84	0
Terrestrial taxa	305	188	897	206

The Stratum I data include only the 6.3 mm sample. Data from Strata Ia, Ib, and I derive from the analysis described earlier.
Data from Stratum II derive from Livingston, S.D., 2000. The Homestead Cave avifauna. In: Madsen, D.B. (Ed.), Late Quaternary Paleoecology in the Bonneville Basin. Bulletin, vol. 130. Utah Geological Survey, Salt Lake City, pp. 91–102.

TABLE 14.6 Summary of X^2 Values for the Tests of Diachronic Change of Greater Sage-Grouse

	Stratum		
Taxonomic Comparison	*Ia–Ib*	*I–II*	*Ia–Ib–II*
Sage-Grouse/terrestrial taxa	0.70	18.98[a]	0.33 (22.36[a])

The Stratum I data include only the 6.3 mm sample. Data from Strata Ia, Ib, and I derive from the analysis described earlier.
Values for Strata Ia–Ib–II represent the X^2_{trend}, and the values in parentheses represent the total X^2 values.
[a]*significant at* $p < 0.005$.
Data from Stratum II derive from Livingston, S.D., 2000. The Homestead Cave avifauna. In: Madsen, D.B. (Ed.), Late Quaternary Paleoecology in the Bonneville Basin. Bulletin, vol. 130. Utah Geological Survey, Salt Lake City, pp. 91–102.

14.6 SUMMARY AND CONCLUSIONS

The basal strata of Homestead Cave have provided a singular fine-grained record of the avifauna that lived in and around Lake Bonneville over the period encompassing the very end of its final regressive phase, the rise and fall of the Gilbert episode lake, and the emergence of the hypersaline Great Salt Lake. The assemblage is represented by at least 53 species derived from a range of terrestrial and aquatic habitat types, although it is dominated by aquatic birds, with many favoring contexts associated with abundant fish resources—most notably diving ducks and grebes. Analyses of change in taxonomic abundances across the lower stratigraphic units (Strata I, Ia, Ib, and II) that span the period from 13.1 to 9.5 cal ka BP reveal change in both the aquatic and terrestrial avifauna. Within the aquatic avifauna, significant declines are registered in deep-water taxa, such as diving ducks and *Aechmophorus* grebes. Among the terrestrial avifauna, a dramatic loss of Greater Sage-Grouse is evident at the Pleistocene/early Holocene transition and a suggestive, but statistically insignificant, increase in the relative abundance of sage-grouse occurs across the terminal Pleistocene itself. The latter may signal a progressive loss of limber pine woodland, the initial presence of which is suggested by the occurrence of such forest taxa as Dusky Grouse and Northern Flicker. These changes are consistent with other records that suggest a dramatic decline in effective precipitation occurred over this period, with the most pronounced changes occurring at the transition between the terminal Pleistocene and the early Holocene.

Our analysis of the avifauna from the early strata of Homestead Cave has thus allowed us to capture the transition from freshwater Lake Bonneville to the hypersaline context that has characterized Great Salt Lake for much of the Holocene. We have shown that the plethora of clamorous gulls, pelicans, and cormorants that astonished and inspired Stansbury, and the myriad shorebirds and marshbirds that grace the shorelines and marshes of Great Salt Lake today, are a relatively recent condition that first began to form at the end of the Pleistocene. The dramatic final regression of Lake Bonneville resulted ultimately in the loss of both fish resources and an avifauna dominated by taxa geared toward piscivory. At the same time, wave-swept rocky shorelines were converted to productive marshes and expansive mudflats, creating habitat for a new and diverse assemblage of dabbling ducks, rails, and shorebirds. Although some islands within the lake would have existed at even the highest lake elevations, the final regression of the lake to nearly modern levels increased the number and extent of island contexts that ultimately favored the establishment of immense colonies of nesting birds that are an icon of Great Salt Lake today.

The abundance of Greater Sage-Grouse in the lower strata of Homestead Cave is particularly noteworthy, and we focused on this taxon due to the potential for our analysis to inform on modern sage-grouse conservation strategies. The plight of Greater Sage-Grouse has become one of the most prominent and controversial wildlife conservation issues in western North America today. The abundance and distributional range of Greater Sage-Grouse have declined dramatically since the mid-20th century in conjunction with the reduction of suitable sagebrush habitats as a result of anthropogenic impacts such as energy development, urban expansion, mining, military training, toxic waste disposal, grazing, wildfires, and the introduction of exotic species—notably cheatgrass (*Bromus tectorum*) (Connelly et al., 2004 and references therein; Schroeder et al., 2004; UDWR, 2009, 2013). These declines, likely exacerbated by climate change, have resulted in considerations of the federal listing of Greater-Sage Grouse under the Endangered Species Act and have motivated broadscale multimillion dollar, state- and federally sponsored management plans that have recently produced promising results (see UDWR, 2013, Appendix 8; WAFWA, 2015).

Although many have identified the value that palezoological data can provide for modern wildlife management and conservation strategies, and despite the numerous cases that have established those links (eg, Lyman, 1996, 2006, 2012; Lauwerier and Plug, 2004; Lyman and Cannon, 2004; Frazier, 2007; Dietl and Flessa, 2009; Hadly and Barnosky, 2009; Terry, 2010; Bovy, 2012; Wolverton and Lyman, 2012; Rick and Lockwood, 2013; Rowe and Terry, 2014), the applied use of such data has yet to become widespread among wildlife and conservation biologists. However, we stress that the tight linkage we have documented between climate of the Bonneville basin, vegetation change, and Greater Sage-Grouse population fluctuations should be of interest to conservation biologists and the local governing entities involved in developing management plans and legislation designed to protect these birds.

Documented at both Bonneville Estates Rockshelter and at Homestead Cave, Greater Sage-Grouse disappeared from both of these contexts in the Bonneville basin at the beginning of the Holocene when thick stands of big sage were opened and thinned as climatic conditions became more xeric. Sage-grouse returned to the area when climate ameliorated later in the Holocene, and they indeed were present in the Lakeside Range during historic times. However, the historic period has again witnessed dramatic range contractions due to habitat alteration of sagebrush communities that mirror the ones documented in the Homestead assemblage. Further analysis of the complete Homestead Cave avifaunal sequence should allow a more detailed evaluation of the population and range dynamics of Greater Sage-Grouse throughout the entire Holocene that may prove useful in conservation efforts under modeled future conditions that project increases in xeric conditions in this setting.

Although our focus here has been on the terminal Pleistocene and early Holocene avifauna of the Bonneville basin, we emphasize that significant insights will likely emerge from our continued work with the entire sequence from Homestead Cave. Such work should produce a high-resolution trans-Holocene avifaunal sequence that can be aligned against broadscale climatic systems, such as El Niño Southern Oscillation (ENSO), the Pacific Decadal Oscillation, and the North American Monsoon. Insofar as trends in the avifauna can be linked to such systems and we increase our ability to forecast how they will vary in the future, we can better anticipate future change in the Bonneville basin avifauna.

ACKNOWLEDGMENTS

We thank Glenna Nielsen-Grimm and Catie Webb from the Natural History Museum of Utah for access to the Homestead Cave materials. We also thank Elic M. Weitzel and Will

Rath for assistance with preparing figures. Finally, we thank Steve Wolverton, Charles G. Oviatt, and John F. Shroder Jr. for helpful comments on the chapter.

APPENDIX: DESCRIPTIVE TAXONOMIC SUMMARY

Order Anseriformes

Family Anatidae

Subfamily Anserinae (small)

Referred material: Humerus: 1 specimen.
Remarks: This specimen is similar in size to *Chen rossii* and *Branta hutchinsii minima*.

Subfamily Anserinae (medium)

Referred material: Distal phalanx: 1 specimen.
Remarks: This specimen is similar in size to *Chen caerulescens, Anser albifrons,* and medium-sized subspecies of *Branta canadensis*.

Anser albifrons (Scopoli, 1769)

Referred material: Quadrate, mandible: 2 specimens.
Remarks: (1) Quadrate: Displays a more deeply depressed orbital process than *Chen rossii,* a smaller foramen at the base of the otic process than in *Chen caerulescens,* and a more elongated mandibular articulation compared to *Branta canadensis*. (2) Mandible: The articular portion of the mandible is distinguished from *B. canadensis* by a more squared external process and a less rounded ventral border of the articular—it possesses a more triangular foramen medial to the postarticular process than in *C. caerulescens*.

Branta canadensis (Linnaeus, 1758)

Referred material: Quadrate: 1 specimen.
Remarks: This specimen is very large and is distinguished from *C. caerulescens* by a more obtuse angle between the otic and orbital processes and is distinguished from *A. albifrons* as described earlier.

Subfamily Anatinae

Referred material: Maxilla and premaxilla fragments: 155 specimens.
Remarks: These specimens are identified based on their distinctive porous appearance, but are too fragmented to identify further.

Subfamily Anatinae (small)

Referred material: Cranium, mandible, digit II phalanx I, scapula, coracoids, furcula, tibiotarsi, tarsometatarsi: 23 specimens.

Remarks: These specimens are similar in size to *Aix sponsa*, the *Anas* teals, *Bucephala albeola*, *Lophodytes cucullatus*, and *Oxyura jamaicensis* but are too fragmented to allow species- or genus-level identifications.

Subfamily Anatinae (large)

Referred material: A total of 428 specimens representing all main elements of the skeleton.
Remarks: These specimens may include any number of the larger duck species.

Anas sp. (large)

Referred material: Quadrates, mandibles, hyoids, ulnae, humeri, coracoids, femora, tibiotarsi, tarsometatarsi: 71 specimens.
Remarks: These specimens could represent any number of *Anas* species larger than the teals. Features used to distinguish *Anas* from other duck genera are as follows: (1) Quadrate: The orbital process is shallower than in *Aythya* and is distinguished from other genera by being more bulbous around the socket for the quadratojugal. (2) Mandible: Identified based on criteria from Broughton (2004). (3) Hyoids: Most similar to *Aythya* but are significantly wider. (4) Ulna: The proximal ulna possesses a shallower impression of the brachialis anticus and a more rounded external cotyla than in *Aythya*. Distinguished from *Bucephala* by a deeper humero-ulna depression and an external cotyla that extends further distally down the shaft. Distinguished from *Mergus* by a less rounded external cotyla. Distinguished from *A. sponsa* by a more excavated impression of brachialis anticus and a more squared proximal border of the external cotyla. (5) Humerus: The proximal humerus is identified based on criteria from Broughton (2004) and by a distinctive pneumatic foramen with bony struts that is shallower than that of *Mergus*. (6) Coracoid: Exhibits a distinctively deeper pneumatic foramina with a stronger anterior border compared to all other genera. (7) Femur: The proximal femur possesses a relatively smaller head and a trochanter that curves more medially relative to the other genera. The distal femur displays a shallower popliteal area than all other genera and is further distinguished from *Aythya* by a smaller internal condyle in the medial view with a smaller crest connecting it to the shaft, and a less proximally pointed external condyle. (8) Tibiotarsus: The proximal tibiotarsus possesses an outer cnemial crest that does not curve as inwardly as in *Aythya* and is distinguished from other genera by a more obtuse angle between the proximal articular surface and inner cnemial crest. The distal tibiotarsus exhibits a smaller tendinal groove than in *Aythya*, and is distinguished from other genera by external and internal condyles that are more even in thickness, and a condylar surface that extends further proximally on the posterior surface. (9) Tarsometatarsus: Significantly thinner for a given length compared to all other genera. Further distinguished from *Aythya* by a more shallow impression on the anterior surface beneath the cotylas.

Anas platyrhynchos (Linnaeus, 1758)

Referred material: Carpometacarpi, tarsometatarsus: 3 specimens.
Remarks: These specimens are significantly larger than all other *Anas*.

Anas sp. cf. *Anas discors* (Linnaeus, 1766)
or *Anas cyanoptera* (Vieillot, 1816)
or *Anas crecca* (Linnaeus, 1758)

Referred material: Mandibles, carpometacarpus, ulnae, coracoid, sternum, femora, tibiotarsi, tarsometatarsi: 25 specimens.
Remarks: The *Anas* teals are distinct from other *Anas* sp. based on their small size, but species-level identifications were not attempted.

Aythya sp.

Referred material: A total of 757 specimens representing all main elements of the skeleton.
Remarks: Features used to distinguish *Aythya* from other duck genera are as follows: (1) Cranium: Identified based on criteria from Broughton (2004), and by being generally more porous at the base. (2) Quadrate: Exhibits a deeper orbital process and more ovaloid mandibular articulations. (3) Mandible: The articular portion is identified based on criteria from Broughton (2004) and a stronger ridge extending distally and inferiorly from the external articular process. (4) Hyoid: Significantly thinner than *Anas*, with a more squared tip, and a flaring articulating portion. (5) Carpometacarpus: Distinguished from *Anas* by a thinner neck, and distinguished from other genera by a smaller and less flaring pollical facet and a smaller ligamental attachment of the pisiform process. (6) Ulna: The proximal ulna possesses a more squared external cotyla that extends further distally down the shaft and a more excavated humero-ulnar depression and impression of brachialis anticus. The distal ulna exhibits a thinner and shorter ligamental attachment and a more distally extending external condyle. (7) Humerus: The proximal humerus displays a straighter deltoid crest and a less rounded opening for the pneumatic foramen that lacks internal bony spicules. The distal humerus specimens are distinguished from *Anas* based on criteria from Broughton (2004) and by a fossa on the internal side that is closer to the entepicondylar prominence. Distinguished from other genera by a shallower olecranon fossa and by being less bulbous. (8) Coracoid: Possesses a more rounded and protruding furcular facet, a shallower pneumatic foramen that lacks a strong anterior border, and a flatter triosseal canal. (9) Sternum: Identified based on criteria from Broughton (2004). (10) Synsacrum: Identified based on criteria from Broughton (2004). (11) Femur: The proximal femur exhibits a relatively larger head than *Anas* but smaller than *Bucephala* and *Mergus*—the trochanteric ridge projects more externally and the iliac facet is larger. The distal femur possesses an external condyle that points more

proximally than dorsally, a deeper popliteal area and intercondylar fossa, and a larger internal condyle in the medial view that is connected to the shaft by an appreciable crest. (12) Tibiotarsus: The proximal tibiotarsus is distinguished from *Anas* as described earlier, and additionally by a more triangular inner cnemial crest and a larger outer cnemial crest that curves more into the shaft. The distal tibiotarsus possesses an external condyle that is thicker than the internal condyle, a condylar surface that extends less proximally on the posterior surface, and a stronger crest connecting the external condyle with the shaft. (13) Tarsometatarsus: Wider for its length, more even width throughout the shaft, longer calcaneal ridges of hypotarsus, and a very bulbous distal end.

Bucephala albeola (Linnaeus, 1758)

Referred material: Cranium, quadrates, hyoid, coracoid, femora, tibiotarsi, tarsometatarsi: 22 specimens.
Remarks: Distinguished from *Bucephala clangula* and *B. islandica* on the basis of their smaller size.

Bucephala clangula (Linnaeus, 1758)
or *Bucephala islandica* (Gmelin, 1789)

Referred material: Quadrates, femora: 4 specimens.
Remarks: Distinguished from *Anas* and *Aythya* as described earlier, and distinguished from other genera as follows: (1) Quadrate: Smaller and more bulbous body, smaller squamosal articulation, and a projection at the base of the otic process. (2) Femur: The proximal femur exhibits a smaller and rounder iliac facet and a relatively large and bulbous head.

Mergus merganser (Linnaeus, 1758)
or *Mergus serrator* (Linnaeus, 1758)

Referred material: Quadrate, mandible, hyoid, femur: 4 specimens.

Mergus merganser (Linnaeus, 1758)

Referred material: Quadrates, mandible, coracoids, femora, tibiotarsi: 12 specimens.

Mergus serrator (Linnaeus, 1758)

Referred material: Quadrates, mandibles, humerus, tibiotarsus, tarsometatarsi: 12 specimens.
Remarks: Very large and very small specimens are assigned to *M. serrator* and *M. merganser*, respectively. Species-level identifications were not attempted for intermediate-sized specimens. Features distinctive to *Mergus* are as follows: (1) Quadrate: Possesses two nearly parallel mandibular articulations. (2) Mandible: Identified based on criteria from Broughton (2004). (3) Hyoid: Distinctively

thin and triangular. (4) Humerus: The distal humerus is distinguished from *Aythya* as described earlier and from *Anas* by a distinctively deeper olecranon fossa. It is most similar to *Bucephala* but is distinguished from the latter by a more internally rotated attachment of anterior articular ligament and by a larger surface between the olecranon fossa and the internal condyle. *A. sponsa* also lacks this larger surface. (5) Coracoid: Distinguished from *Bucephala, Aythya,* and *Anas* by a larger and more flattened furcular facet. Distinguished from *A. sponsa* by a flatter furcular facet. It is too large to represent any of the other Anatinae species. (6) Femur: Distinguished from *Anas* and *Aythya* as described earlier. Too large to represent *O. jamaicensis, L. cucullatus,* or *A. sponsa.* Distinguished from *Bucephala* by a thicker shaft (notably on the lateral proximal end) and by a longer ridge medial to the popliteal area. (7) Tibiotarsus: The proximal tibiotarsus is distinguished from *Anas* and *Aythya* as described earlier. Distinguished from *Bucephala* by a less pointed corner of the outer cnemial crest. It is too large to represent *O. jamaicensis, L. cucullatus,* and *A. sponsa.* (8) Tarsometatarsus: Too thin to represent *Aythya* and is distinguished from *Anas* by a shorter innermost calcaneal ridge, a deeper anterior impression beneath the cotylas, and a thinner outer distal edge.

Oxyura jamaicensis (Gmelin, 1789)

Referred material: Tarsometatarsus: 1 specimen.
Remarks: This specimen is too thick to represent the small *Anas* sp., *L. cucullatus*, or *B. albeola*. It is too small to represent *Aythya* or the large *Anas* sp. It is distinguished from *A. sponsa* by being shorter and thicker, and by possessing a broader anterior impression.

Order Galliformes
Family Phasianidae
Subfamily Tetraoninae

Referred material: Distal phalanges: 5 specimens.

Centrocercus urophasianus (Bonaparte, 1827)

Referred material: A total of 111 specimens representing all main elements of the skeleton.
Remarks: These specimens are larger than other Galliformes including *D. obscurus*—they are further distinguished from the latter as follows: (1) Cranium: The cranium specimen is represented by the frontals and is wider with a shallower groove than in *D. obscurus.* (2) Quadrate: Displays a shorter orbital process and a less pronounced projection on the otic process. (3) Mandible: The articular portion of the mandible exhibits a thicker internal articular process and a deeper groove between the articular surface and the internal articular process. The symphysis portion of the mandible possesses a row of foramina on the

proximal inferior surface of the symphysis. (4) Digit II Phalanx I: Distinguished by a proportionately larger metacarpal facet and a more squared-off digital facet. (5) Carpometacarpus: The distal carpometacarpus specimen is distinguished by a larger and more triangular facet for digit II. (6) Radius: The distal radius exhibits a shallower ulnar depression and a more pointed scapho-lunar facet. (7) Ulna: The proximal ulna specimen possesses a more squared-off distal end of the internal cotyla and, from a medial view, a rounder and smoother internal edge of the articular surface. (8) Humerus: The proximal humerus specimens are distinguished by a larger and deeper pneumatic foramen. The distal humerus specimen is distinguished by a more prominent space between the external condyle and the ectepicondylar prominence. (9) Scapula: Possesses a thicker acromion and a larger glenoid facet. (10) Coracoid: Distinguished by a more pointed brachial tuberosity, a more ventrally protruding furcular facet, and a wider glenoid facet. The distal coracoid specimen is distinguished by a thicker sternocoracoidal process with a less distinct ridge, and a more pointed process lateral to the sternal facet. (11) Furculum: Identified by its unique crest along the symphysis. (12) Pygostyle: Exhibits a thinner base and a larger articulation surface dorsal to the vertebral articulation. (13) Femur: The distal femur specimen is distinguished by a less dorsally extending external condyle and by the shape of the anterior end of the ridge on the ventral side of the internal condyle, which ends abruptly in an angle rather than sloping into the shaft as it does in *D. obscurus*. (14) Tibiotarsus: The proximal tibiotarsus specimens display a smaller distance between the inner and outer cnemial crests, and by a more laterally placed nutrient foramen. The distal tibiotarsus exhibits a less ventrally extending internal condyle, a more rounded external condyle, and a narrower anterior intercondylar fossa.

Dendragapus obscurus (Say, 1822)

Referred material: Mandibles, carpometacarpus, pygostyle, tibiotarsi: 7 specimens.
Remarks: These specimens are smaller than *C. urophasianus* and larger than *T. phasianellus* and all other grouse that occur in the region.

Tympanuchus phasianellus (Linnaeus, 1758)

Referred material: Quadrates, digit II phalanx I, carpometacarpus: 4 specimens.
Remarks: (1) Quadrate: Distinguished from *D. obscurus* and *C. urophasianus* by a more elongated and curved mandibular articulation. Distinguished from *Bonasa umbellus* and *Falcipennis canadansis* by a thinner otic process that tapers more drastically toward the tip. (2) Digit II phalanx I: Too small to represent the larger grouse species and is distinguished from *F. canadensis* and *B. umbellus* by a smaller projection beneath the metacarpal facet, and relatively smaller facets. (3) Carpometacarpus: Distinguished from smaller

individuals of *D. obscurus* and *C. urophasianus* by a thinner process of metacarpal I and a smaller pollical facet. Distinguished from *F. canadensis* and *B. umbellus* by a thicker process of metacarpal I.

Order Gaviiformes
Family Gaviidae
Gavia immer (Brunnich, 1764)

Referred material: Quadrate: 1 specimen.
Remarks: Only one species of loon inhabits this area, and *G. immer* is easily distinguished from other loons on the basis of its large size.

Order Podicipediformes
Family Podicipedidae
Podiceps auritus (Linnaeus, 1758)
or *Podiceps nigricollis* (Brehm, 1831)

Referred material: Quadrates, digit II phalanx I, scapulae, coracoid, tarsometatarsi: 70 specimens.
Remarks: These specimens are either too fragmented to identify to species or intermediate in size between the two species.

Podiceps auritus (Linnaeus, 1758)

Referred material: A total of 141 specimens representing all main elements of the skeleton.

Podiceps nigricollis (Brehm, 1831)

Referred material: A total of 253 specimens representing all main elements of the skeleton.
Remarks: Most specimens are identifiable to either *P. auritus* or *P. nigricollis* based on their large or small size, respectively. Cranium, carpometacarpus, and tibiotarsus specimens are further identified based on criteria from Broughton (2004).

Aechmophorus occidentalis (Lawrence, 1858)
or *Aechmophorus clarkii* (Lawrence, 1858)

Referred material: Quadrates, mandibles, digit II phalanx I, carpometacarpi, radius, scapula, femur, tarsometatarsus: 41 specimens.
Remarks: In addition to the criteria presented in Howard (1929) and Broughton (2004), the following features are used to distinguish *Aechmophorus* from *Podiceps grisegena*: (1) Quadrate: Distinguished from *P. grisegena* by the shape of the pterygoid articulation—*Aechmophorus* quadrates possess one articulation surface whereas those of *P. grisegena* possess two separate articulation surfaces.

(2) Mandible: The articular portion of the mandible is distinguished from *P. grisegena* by a smaller and more pointed internal articular process and a larger and more rounded articulation surface. (3) Digit II phalanx I: Longer and thinner, and the metacarpal facet is more rounded.

Order Suliformes
Family Phalacrocoracidae
Phalacrocorax auritus (Lesson, 1831)

Referred material: Quadrate, pygostyle: 2 specimens.
Remarks: Only one species of cormorant, *Phalacrocorax auritus,* occurs in inland settings in western North America.

Order Accipitriformes

Referred material: Distal phalanges: 15 specimens.

Family Cathartidae
Cathartes aura (Linnaeus, 1758)

Referred material: Femora: 2 specimens.
Remarks: The proximal femur specimen is identified based on the distinctive rounded iliac facet. The distal femur specimen is distinguished from other Accipitriformes by its unique flat and broad internal condyle (in *Falco* the internal condyle is flat, but it narrows anteriorly), and it lacks projections on either the fibular condyle (as in *Buteo* and *Accipiter*) or the external condyle (as in *Falco*).

Family Accipitridae
Circus cyaneus (Linnaeus, 1766)

Referred material: Coracoid, tibiotarsus: 2 specimens.
Remarks: These specimens are most similar to *Buteo* and are identified based on the following criteria: (1) Coracoid: Distinguished from *Buteo* on the basis of smaller size, a smaller foramen beneath the furcular facet, and the lack of a procoracoid foramen. (2) Tibiotarsus: The proximal tibiotarsus is smaller than *Buteo* and possesses a more pointed external articular surface.

Accipiter striatus (Viellot, 1807)
or *Accipiter cooperii* (Bonaparte, 1928)

Referred material: Digit II phalanx I, carpometacarpi, ulna, humeri, scapula, coracoids, sternum: 11 specimens.
Remarks: These specimens were intermediate in size between *A. striatus* and *A. cooperii*.

Accipiter striatus (Vieillot, 1807)

Referred material: Carpometacarpus, ulna, humerus, coracoid, tibiotarsus, tarsometatarsus: 6 specimens.
Remarks: These specimens are too small to represent *A. cooperii.*

Accipiter cooperii (Bonaparte, 1828)

Referred material: Mandible, femora, tibiotarsus: 4 specimens.
Remarks: These specimens are too large to represent *A. striatus.*

Buteo sp.

Referred material: Quadrates, mandible, digit II phalanx I, carpometacarpi, humeri, sternum, pygostyles, tibiotarsi, tarsometatarsus, distal phalanges: 22 specimens.
Remarks: Four *Buteo* species occur in this area, and they range in size from smallest to largest as follows: *B. swainsoni, B. lagopus, B. jamaicensis,* and *B. regalis.* Due to the size overlap within the *Buteo* species and the limited comparative skeletons available, no species-level identifications were attempted.

Order Gruiformes
Family Rallidae
Rallus limicola (Vieillot, 1819)

Referred material: Tibiotarsus, tarsometatarsus: 2 specimens.
Remarks: These specimens are distinguished from *P. carolina* as follows: (1) Tibiotarsus: Exhibits a larger and more ovaloid external articular surface and a more excavated area beneath the external articular surface. (2) Tarsometatarsus: Exhibits a deeper ventral depression beneath the cotyla, a rounder internal cotyla, and more prominent ridges extending the length of the bone.

Porzana carolina (Linnaeus, 1758)

Referred material: Humeri, coracoid, sternae, femur, tarsometatarsi: 8 specimens.
Remarks: These specimens are distinguished from *R. limicola* as follows: (1) Humerus: Possesses a deeper pneumatic foramen and a deltoid crest that is more perpendicular to the shaft. (2) Coracoid: Exhibits a more triangular glenoid facet, and a more pointed ventral end of the furcular facet. (3) Sternum: Possesses more rounded ventral lips of coracoidal surface, and taller dorsal lips of coracoidal surface. (4) Femur: Exhibits a more rounded internal condyle and a smaller external condyle. (5) Tarsometatarsus: See above.

Gallinula galeata (Lichtenstein, 1818)

Referred material: Ulnae, humerus: 3 specimens.

Remarks: These specimens are too small to be *F. americana* and larger than *P. carolina* and *R. limicola.*

Fulica americana (Gmelin, 1789)

Referred material: Ulna, humerus, synsacrum, tarsometatarsi: 5 specimens.
Remarks: These specimens are too large to represent *P. carolina, R. limicola,* or *G. galeata.*

Order Charadriiformes
Family Charadriidae
Subfamily Charadriinae
Pluvialis sp.

Referred material: Femur, tarsometatarsus: 2 specimens.
Remarks: (1) Femur: The distal femur specimen is distinguished from similarly sized *Himantopus mexicanus* and *Tringa* by a deeper and more medially projecting popliteal area, and is further distinguished from *H. mexicanus* by a larger and more pointed external condyle. (2) Tarsometatarsus: Most similar to *Tringa* but distinguished from the latter by a thicker shaft for its length.

Charadrius vociferus (Linnaeus, 1758)

Referred material: Femur: 1 specimen.
Remarks: The proximal femur specimen is distinguished from *Limnodromus* by a less curved lateral edge of the iliac facet and from *Calidris* by a thicker iliac facet—it is too small to represent *Pluvialis*, but too large for all other *Charadrius* species.

Family Scolopacidae
Subfamily Scolopacinae
Actitis macularius (Linnaeus, 1766)

Referred material: Digit II phalanx I, carpometacarpi, radius, humeri, coracoids, tibiotarsus, tarsometatarsi: 30 specimens.
Remarks: A. macularius, Calidris, and *Phalaropus* elements are all very similar in size and form. They are distinguished from one another as follows: (1) Digit II phalanx I: *A. macularius* distinguished from *Phalaropus* by a shorter and more decurved process over the digital facet, and distinguished from *Calidris* by a more pointed metacarpal facet. (2) Carpometacarpus: The process of metacarpal I is thinner and forms a more acute angle with the carpal trochlea in *Calidris* and *A. macularius,* and compared to *Calidris*, *A. macularius* exhibits a thicker neck and a smaller internal carpal trochlea relative to the external carpal trochlea. (3) Radius: The distal radius exhibits a relatively higher elevation around the tendinal groove in *A. macularius* than in *Calidris*, and

a relatively thicker shaft than in *Phalaropus*. (4) Humerus: The proximal humerus of *A. macularius* possesses a more continuous bicipital crest, whereas *Calidris* and *Phalaropus* both exhibit prominent notches on the bicipital crest just distal to the ligamental furrow, and *Phalaropus* possesses a deeper and more lobed pneumatic fossa than *A. macularius* and *Calidris*. The distal humerus is distinguished by the shape and orientation of the ectepicondylar prominence. In *A. macularius* it is thinner and projects more distally and laterally—in both *Calidris* and *Phalaropus* it is thicker and more triangular. The ectepicondylar prominence is at an obtuse angle and projects more palmarly in *Phalaropus*, whereas in *Calidris* it is oriented more acutely with the shaft and projects laterally. (5) Coracoid: *A. macularius* possesses a relatively thinner shaft than both *Phalaropus* and *Calidris*, and *Calidris* is distinguished by a thinner and more proximally curved sternocoracoidal process. (6) Tibiotarsus: The distal tibiotarsus specimen of *A. macularius* is distinguished from *Phalaropus* and *Calidris* by a more rounded external condyle. (7) Tarsometatarsus: *Phalaropus* and *Calidris* possess tarsometatarsi with thinner shafts than *A. macularius*.

Tringa sp.

Referred material: Digit II phalanx I, ulna, tarsometatarsus: 3 specimens.
Remarks: (1) Digit II phalanx I: Distinguished from other similarly sized Charadriiformes by a more pointed metacarpal facet. (2) Ulna: The proximal ulna exhibits a deeper indentation between the olecranon and the external cotyla than in charadriids, exhibits a more proximal-facing external cotyla than in *Limnodromus*, and exhibits a more ridged proximal palmar surface than in *Calidris*. (3) Tarsometatarsus: The distal portion displays a less protruding trochlea for digit 3 than in charadriids, and a more evenly wide shaft compared to *Limnodromus*.

Tringa cf. *T. solitaria* (Wilson, 1813)

Referred material: Femur: 1 specimen.
Remarks: The specimen is distinguished from *Actitis* by a straighter shaft and smaller head, from *Phalaropus* by a more curved shaft, from *Calidris* by the lack of a prominent ridge on the ventral edge of the iliac facet, and from *Charadrius* by a shallower popliteal area. This specimen is too small to represent *Tringa flavipes*, *T. melanoleuca*, or *T. semipalmata*.

Numenius americanus (Bechstein, 1812)

Referred material: Mandible: 1 specimen.
Remarks: The large size of this specimen distinguishes it from *N. phaeopus* and *Limosa fedoa*. It is further distinguished from *L. fedoa* by a thinner internal articular process of the mandible.

Calidris sp.

Referred material: Carpometacarpi, humeri, coracoid, femur, tarsometatarsi: 15 specimens.
Remarks: Distinguished from *A. macularius* and *Phalaropus* by the criteria above and, for the femur, a thicker distal shaft than *A. macularius* and a smaller iliac facet than *Phalaropus*.

Phalaropus sp.

Referred material: Carpometacarpi, ulnae, humeri, coracoids, sternae, femur, tibiotarsi, tarsometatarsi: 33 specimens.
Remarks: Criteria for distinguishing from *A. macularius* and *Calidris* (as above) and for additional elements as follows: (1) Ulna: Shorter olecranon than *A. macularius* and a thinner external edge with a more internally curved olecranon than *Calidris*. (2) Sternum: More steeply sloping coracoidal sulcus. (3) Femur: Proximal end distinguished from *Calidris* by a more proximally extending trochanter, and distinguished from *A. macularius* by a thinner neck and smaller iliac facet. (4) Tarsometatarsi: *Phalaropus* tarsometatarsi are less flared proximally than *Calidris*.

Family Laridae (small)

Referred material: A total of 45 specimens representing all main elements of the skeleton.
Remarks: Small species include specimens similar in size to *Larus delawerensis, Leucophaeus pipixcan,* and *Chroicocephalus philadelphia*. Because of the extensive overlap in osteological features of gulls, no further species-level identifications were made.

Family Laridae (large)

Referred material: A total of 48 specimens representing all main elements of the skeleton.
Remarks: Large species include specimens similar in size to *L. californicus, L. argentatus,* and *L. thayeri*. Because of the extensive overlap in osteological features of gulls, no further species-level identifications were made.

Order Columbiformes
Family Columbidae (large)
Zenaida cf. *Z. macroura* (Linnaeus, 1758)

Referred material: Digit II phalanx I, ulna: 2 specimens.
Remarks: These specimens are too small to represent *Patagioenas fasciata*. They are larger than the comparative *Z. macroura* specimens but are

morphologically more similar to *Z. macroura* than *Z. asiatica*. Specifically, the process protruding from the metacarpal facet of the digit II phalanx I specimen is less bulbous than in *Z. asiatica*. The proximal ulna specimen exhibits a flatter internal border of the articulating surface (from olecranon along the edge of the internal cotyla) than in *Z. asiatica*.

Zenaida macroura (Linnaeus, 1758)

Referred material: Digit II phalanx I, humerus: 2 specimens.
Remarks: These specimens are too small to represent *Z. asiatica*—the proximal humerus is further distinguished from *Z. asiatica* by a relatively shallow pneumatic foramen and a more internally curved deltoid crest.

Order Strigiformes

Referred material: Coracoids, distal phalanges: 28 specimens.

Family Tytonidae
Tyto alba (Scopoli, 1769)

Referred material: Coracoid, femur: 2 specimens.
Remarks: T. alba elements are most similar to larger *Asio* sp. and are distinguished from the latter as follows: (1) Coracoid: The sternal-coracoidal process is larger relative to the sternal facet edge, and the sternal end exhibits a pronounced ridge on the ventral lateral portion and a more pronounced process at the margin of the sternal facet. (2) Femur: The distal end is larger with greater excavation in the popliteal area.

Family Strigidae
Megascops kennicottii (Elliot, 1867)

Referred material: Carpometacarpi, ulnae, humeri, femur, tibiotarsi, tarsometatarsus: 12 specimens.
Remarks: These specimens are distinguished from other small strigids by the following criteria: (1) Carpometacarpus: Too large to represent *Aegolius acadicus* and is distinct from *Athene cunicularia* by a longer neck between the head and the shaft and a wider intermetacarpal space. (2) Ulna: Too large to represent *A. acadicus* or *Psiloscops flammeolus*. The proximal end is distinguished from *A. cunicularia* by the shape of the bicipital attachment, which is thinner and more linear, and by the internal cotyla, which is shallower and less circular. The external cotyla is broader and curves into the shaft, whereas in *A. cunicularia* it projects more laterally. The distal end is distinguished from *A. cunicularia* by the shape of the external condyle: in *M. kennecottii* it slopes gently into the shaft, and in *A. cunicularia* it projects out more abruptly. The trochlea of the external condyle is larger and more squared in *M. kennecottii*. (3) Humerus: The distal end is identified based on criteria from Broughton (2004), and by a more pronounced ridge proximal to the ectepicondyle. (4)

Femur: The proximal end is too large to represent *A. acadicus* and distinguished from *A. cunicularia* by a more pronounced ridge on the anterior surface of the neck connecting the head and trochanter. Less obvious is the deeper sulcus medial to the obturator ridge and a shorter distal extension of the trochanteric ridge. (5) Tibiotarsus: The proximal end is larger than *A. acadicus* and distinguished from *A. cunicularia* by a rotular crest that is straighter, an outer cnemial crest that is thinner and more squared off, and a deeper impression on the surface between the external articular surface and the outer cnemial crest. (6) Tarsometatarsus: The proximal end is larger than *A. acadicus* and too wide and bowed proximally to represent *A. cunicularia.*

Bubo virginianus (Gmelin, 1788)

Referred material: Tarsometatarsus: 1 specimen.
Remarks: Distinguished from *Strix nebulosa* and *Bubo scandiacus* based on criteria in Howard (1929).

Athene cunicularia (Molina, 1782)

Referred material: Ulnae, femora: 4 specimens.
Remarks: These specimens are too large to represent *A. acadicus* or *P. flammeolus.* They are distinguished from other small strigids as follows: (1) Ulna: The proximal end is distinguished from *M. kennecottii* as described earlier; distal end is distinguished from *M. kennecottii* by a more pointed ligamental attachment of the carpal tuberosity. (2) Femur: The distal end is distinguished from *M. kennecottii* by a broader and flatter internal condyle and a deeper fossa superior to the fibular condyle.

Asio otus (Linnaeus, 1758)
or *Asio flammeus* (Pontoppidan, 1763)

Referred material: A total of 121 specimens representing all main elements of the skeleton.
Remarks: The elements of *Asio* are most similar to those of *T. alba* but can be distinguished based on size or other criteria from Howard (1929) as well as the criteria described later. The specimens identified to *Asio* sp. are either intermediate in size between the two *Asio* species or lack the landmarks required to derive the measurements that separate these taxa (see Emslie, 1982 and below). (1) Cranium: The fragment from the base of the skull is distinguished by shallower foramina for cerebral nerves lateral and anterior to the foramen magnum. (2) Quadrate: The mandibular articulation of the quadrate is wider and does not constrict along the midline as it does in *T. alba*. Furthermore, the otic process continues past the process that projects from the pneumatic foramen, whereas the otic process ends just at the projecting process in *T. alba*. (3) Premaxilla: Shorter and less decurved, and the anterior portion of the ventral groove does not contain a slight cleft as it

does in *T. alba*. (4) Mandible: The articular portion of the mandible is easily distinguished from *T. alba* as it displays a significantly more elongated internal articular process. The anterior portion is also easily distinguished by a wider symphysis. (5) Digit II phalanx I: Easily distinguished as the spine on it does not extend to the metacarpal facet as it does in *T. alba*. (6) Carpometacarpus: Displays a less prominent ridge distal to the tuberosity of metacarpal II. In contrast, *T. alba* exhibits a distinct ridge that approaches a pronounced groove that runs along the distal shaft. On the proximal end, *Asio* lacks the small groove on the distal surface of the process for metacarpal I that is seen in *T. alba*. Furthermore, the iliac facet does not extend as far anteriorly with the trochanter as it does in *T. alba*. (7) Radius: The distal end is distinguished by a smaller ligamental prominence that is less rounded and depressed. The proximal end is distinguished by a more angular head (in *T. alba* it is more circular) and by a smaller, more triangular capital tuberosity. (8) Ulna: The proximal end is distinguished by a shorter, less projecting external cotyla, with a less pronounced distal outline. The carpal tuberosity on the distal end is less pointed and projecting. (9) Scapula: In *T. alba* the glenoid facet is ovaloid, whereas in *Asio* a slight cleft is present on the superior edge. (10) Coracoid: The sternal-coracoidal process is smaller relative to the sternal facet edge. The head portion of the coracoid is distinguished by a larger foramen beneath the procoracoid. (11) Sternum: Exhibits a significantly more pronounced ventral manubrial spine, asymmetrical lips of the coracoidal surface, and a smaller gap at the midline along the dorsal lip of the coracoidal sulcus. (12) Femur: From a lateral view, the proximal end is more bulbous and ovaloid, whereas in *T. alba* it possesses a slight notch on the anterior border. The distal end exhibits a less excavated popliteal area. (13) Tibiotarsus: The condyles project less anteriorly and the tendinal groove is deeper, especially toward the internal condyle. (14) Tarsometatarsus: This element is substantially smaller and shorter in *Asio* than in *T. alba*.

Asio otus (Linnaeus, 1758)

Referred material: Carpometacarpi, radii, ulnae, humeri, fermora, tibiotarsi, tarsometatarsi: 34 specimens.
Remarks: A. otus elements are distinguished from those of *A. flammeus* based on size differences described in Emslie (1982).

Asio flammeus (Pontoppidan, 1763)

Referred material: Carpometacarpi, radii, ulnae, humeri, fermora, tibiotarsi, tarsometatarsi: 49 specimens.
Remarks: A. flammeus elements are distinguished from those of *A. otus* based on size differences described in Emslie (1982).

Aegolius acadicus (Gmelin, 1788)

Referred material: Carpometacarpi, ulna, tibiotarsi, tarsometatarsi: 13 specimens.
Remarks: These specimens are too small to represent *M. kennecottii* and *A. cunicularia* and too large to represent *P. flammeolus.*

Order Caprimulgiformes
Family Caprimulgidae
Subfamily Chordeilinae
Chordeiles cf. *minor* (Forster, 1771)

Referred material: Ulnae, humeri: 5 specimens.
Remarks: Common Nighthawk (*C. minor*) is extremely abundant in the northern Bonneville basin today as a summer migrant, whereas the northern margin of the historic range of Lesser Nighthawk (*C. acuttipennis*) does not extend north of the Mojave Desert in southwestern Utah. Our assignment of all *Chordeiles* specimens to *C.* cf. *minor* is based purely on modern abundance and geographic grounds. *Chordeiles* elements are fairly distinct but are distinguished from *P. nuttallii* based on their larger size and other criteria as follows: (1) Ulna: Distinguished from *P. nuttallii* by a relatively thicker shaft and less flaring proximal and distal ends. (2) Humerus: The distal end is distinguished from *P. nuttallii* by a deeper brachial depression and a thicker external condyle.

Subfamily Caprimulginae
Phalaenoptilus nuttallii (Ridgway, 1880)

Referred material: Quadrate, digit II phalanx I, carpometacarpi, ulnae, humeri, coracoids, sternae, tarsometatarsi: 36 specimens.
Remarks: These specimens are smaller than *Chordeiles* and are further distinguished by the criteria that follow: (1) Quadrate: Displays a thinner groove on the internal margin of the otic process. (2) Digit II phalanx I: The digital facet is less rounded, and the base is more concave. (3) Carpometacarpus: The process for metacarpal I is relatively small and less projecting. (4) Ulna: See above. (5) Humerus: Distal end, as above; the proximal end exhibits a smaller deltoid crest that is less triangular in shape, and a relatively short internal tuberosity. (6) Coracoid: Bears a more rounded furcular facet. (7) Sternum: Lacks a foramen on the manubrium and exhibits thinner coracoidal sulci. (8) Tarsometatarsus: The shaft is longer and thinner.

Order Piciformes
Family Picidae
Subfamily Picinae
Sphyrapicus sp.

Referred material: Mandible, carpometacarpus, ulna, humeri: 8 specimens.
Remarks: These specimens are most similar to *Picoides* and are distinguished from that taxon based on the following criteria: (1) Mandible: The articular portion displays a relatively thinner internal articular process that extends more internally than dorsally. (2) Carpometacarpus: The proximal end exhibits a more pointed proximal border of the intermetacarpal space and a smaller ligamental attachment of the pisiform process with deeper fovea proximal and distal to the process. (3) Ulna: The external condyle is more squared. (4) Humerus: The proximal end exhibits a smaller and less extending head and a more curved deltoid crest. The distal end exhibits a more pointed proximal end of the external condyle and a deeper and narrower brachial depression.

Colaptes auratus (Linnaeus, 1758)

Referred material: Quadrates, mandibles, ulnae, humeri, scapula, sternae, femora, tibiotarsi, tarsometatarsi, distal phalanx: 27 specimens.
Remarks: These specimens are too large to represent any of the other picids.

Order Falconiformes
Family Falconidae
Subfamily Falconinae
Falco sparverius (Linnaeus, 1758)
or *Falco columbarius* (Linnaeus, 1758)

Referred material: Quadrates, carpometacarpus, distal phalanges: 10 specimens.
Remarks: These specimens are either too fragmented or are within the range of size overlap of the two small *Falco* species.

Falco sparverius (Linnaeus, 1758)

Referred material: Digit II phalanx I, carpometacarpi, ulnae, humeri, scapula, coracoids, femora, tibiotarsi, tarsometatarsi: 49 specimens.
Remarks: These specimens are too small to represent *F. columbarius.*

Falco columbarius (Linnaeus, 1758)

Referred material: Mandible, digit II phalanx I, carpometacarpus, radii, ulna, humeri, femur, tibiotarsi, tarsometatarsus: 13 specimens.
Remarks: These specimens are too large to represent *F. sparverius.*

Falco peregrinus (Tunstall, 1771)
or *Falco mexicanus* (Schlegel, 1850)

Referred material: Mandibles, premaxillae, digit II phalanx I, radii, humerus, coracoid, femur, tibiotarsus, tarsometatarsus: 13 specimens.
Remarks: These specimens are either too fragmented or are within the range of size overlap of the two large *Falco* species.

Falco peregrinus (Tunstall, 1771)

Referred material: Mandibles: 4 specimens.
Remarks: The mandible is represented only by the articular, and it differs from *F. mexicanus* as follows: the depression on the superior surface is shallow and relatively small, the foramen on the internal articular process is larger and more ovaloid, and the dorsal surface is narrower.

Falco mexicanus (Schlegel, 1850)

Referred material: Mandible, ulna, tibiotarsi, tarsometatarsus: 5 specimens.
Remarks: Distinguished from *F. peregrinus* by the following: (1) Mandible: See above. (2) Ulna: The distal end is smaller with a shallower depression proximal to the external condyle. (3) Tibiotarsus: The proximal end is distinguished based on criteria from Broughton (2004). (4) Tarsometatarsus: The proximal shaft is thicker with a more pronounced ridge extending from the proximal medial edge to the posterior metatarsal groove.

Order Passeriformes

Referred material: A total of 1758 specimens representing all main elements of the skeleton.
Remarks: Taxonomic identifications below the order level were not attempted.

REFERENCES

American Ornithologists' Union, 2015. Checklist of North and Middle American Birds. Available from, http://www.checklist.aou.org.

Baldassarre, G., 2014. Ducks, Geese, and Swans of North America. John Hopkins University Press, Baltimore.

Behle, W.H., 1958. The Bird Life of Great Salt Lake. The University of Utah Press, Salt Lake City.

Behle, W.H., 1985. Utah Birds: Geographic Distribution and Systematics. Utah Museum of Natural History, University of Utah, Salt Lake City, Utah.

Behle, W.H., Perry, M.L., 1975. Utah Birds: Guide, Check-List and Occurrence Charts. Utah Museum of Natural History, University of Utah, Salt Lake City, Utah.

Benson, L.V., Lund, S.P., Smoot, J.P., Rhode, D.E., Spencer, R.J., Verosub, K.L., Louderback, L.A., Johnson, C.A., Rye, R.O., Negrini, R.M., 2011. The rise and fall of Lake Bonneville between 45 and 10.5 ka. Quat. Int. 235, 57–69.

Bovy, K.M., 2012. Zooarchaeological evidence for sandhill crane (*Grus Canadensis*) breeding in northwestern Washington State. In: Wolverton, S., Lyman, R.L. (Eds.), Conservation Biology and Applied Zooarchaeology. University of Arizona Press, Tucson, pp. 23–41.

Broughton, J.M., 2000a. The Homestead cave ichthyofauna. In: Madsen, D.B. (Ed.), Late Quaternary Paleoecology in the Bonneville Basin, Utah Geological Survey Bulletin, vol. 130, pp. 103–121.

Broughton, J.M., 2000b. Terminal Pleistocene fish remains from Homestead Cave, Utah, and implications for fish biogeography in the Bonneville basin. Copeia 3, 645–656.

Broughton, J.M., 2004. Prehistoric human impacts on California birds: evidence from the Emeryville Shellmound avifauna. Ornithological Monographs, vol. 56. American Ornithologists' Union, Washington, DC.

Broughton, J.M., Smith, G.R., 2016. Chapter 12. The fishes of Lake Bonneville: Implications for drainage history, biogeography and lake levels. In: Oviatt, C.G., Shroder Jr., J.F. (Eds.), Lake Bonneville: A Scientific Update, vol. 20. Elsevier, pp. 292–351.

Broughton, J.M., Madsen, D.B., Quade, J., 2000. Fish remains from Homestead Cave and lake levels of the past 13,000 years in the Bonneville Basin. Quat. Res. 53, 392–401.

Broughton, J.M., Cannon, V.I., Arnold, S., Bogiatto, R.J., Dalton, K., 2006. The taphonomy of owl-deposited fish remains and the origin of the Homestead Cave ichthyofauna. J. Taphonomy 4, 69–95.

Broughton, J.M., Byers, D.A., Bryson, R.A., Eckerle, W., Madsen, D.B., 2008. Did climatic seasonality control late Quaternary artiodactyl densities in western North America? Quat. Sci. Rev. 27, 1916–1937.

Cannon, M.D., 2001. Archaeofaunal relative abundance, sample size, and statistical methods. J. Archaeol. Sci. 28, 185–195.

Connelly, J.W., Knick, S.T., Schroeder, M.A., Stiver, S.J., 2004. Conservation Assessment of Greater Sage-Grouse and Sagebrush Habitats. Western Association of Fish and Wildlife Agencies, Cheyenne, Wyoming, USA.

Core Team, R., 2015. R: A Language and Environment for Statistical Computing. R Foundation for Statistical Computing, Vienna, Austria. Available from: http://www.R-project.org.

Cullen, S.A., Jehl Jr., J.A., Nuechterlein, G.L., 1999. Eared Grebe (*Podiceps nigricollis*). In: Poole, A., Gill, F. (Eds.), The Birds of North America, vol. 433. The Birds of North America, Inc., Philadelphia, pp. 1–28.

Currey, D.R., 1982. Lake Bonneville: Selected Features of Relevance to Neotectonic Analysis. U.S: Geological Survey Open-File Report 82-1070.

Dietl, G.P., Flessa, K.W., 2009. Conservation palaeobiology: using the past to manage for the future. In: The Palaeontology Society Papers, vol. 15. Paleontological Society, Lubbock.

Dorr, B.S., Hatch, J.J., Weseloh, D.V., 2014. Double-crested Cormorant (*Phalacrocorax auritus*). In: Poole, A. (Ed.), The Birds of North America Online. Cornell Lab of Ornithology, Ithaca. Available from: http://www.bna.birds.cornell.edu/bna/species/441.

Emslie, S.D., 1982. Osteological identification of Long-eared and Short-eared owls. Am. Antiq. 47, 155–157.

Frazier, J., 2007. Sustainable use of wildlife: the view from archaeozoology. J. Nat. Conserv. 15, 163–173.

Grayson, D.K., 1998. Moisture history and small mammal community richness during the latest Pleistocene and Holocene, northern Bonneville Basin, Utah. Quat. Res. 49, 330–334.

Grayson, D.K., 2000a. Mammalian responses to Middle Holocene climatic change in the Great Basin of the western United States. J. Biogeogr. 27, 181–192.

Grayson, D.K., 2000b. The Homestead Cave mammals. In: Madsen, D.B. (Ed.), Late Quaternary Paleoecology in the Bonneville Basin. Bulletin, vol. 130. Utah Geological Survey, Salt Lake City, pp. 67–89.

Grayson, D.K., 2006. The Late Quaternary biogeographic histories of some Great Basin mammals (western USA). Quat. Sci. Rev. 25, 2964–2991.

Hadly, E.A., Barnosky, A.D., 2009. Vertebrate fossils and the future of conservation biology. In: Dietl, G.P., Flessa, K.W. (Eds.), Conservation Paleobiology: Using the Past to Manage for the Future. The Paleontological Society Papers, vol. 15. Paleontological Society, Lubbock, pp. 39–59.

Henderson, W.W., 1931. Crickets and grasshoppers in Utah. Utah Agric. Exp. Stn. 96, 1–38.

Hockett, B., 2007. Nutritional ecology of late Pleistocene to middle Holocene subsistence in the Great Basin: zooarchaeological evidence from Bonneville Estates Rockshelter. In: Graf, K.E., Schmitt, D.N. (Eds.), Paleoindian or Paleoarchaic? Great Basin human ecology at the Pleistocene-Holocene transition. The University of Utah Press, Salt Lake City, pp. 204–230.

Howard, H., 1929. The Avifauna of the Emeryville Shellmound, vol. 32 University of California Publications in Zoology, Berkeley.

LaPorte, N., Storer, R.W., Nuechterlein, G.L., 2013. Western Grebe (*Aechmophorus occidentalis*). In: Poole, A. (Ed.), The Birds of North America Online. Cornell Lab of Ornithology, Ithaca. Available from: http://www.bna.birds.cornell.edu/bna/species/026a.

Lauwerier, R.C.G.M., Plug, I., 2004. The Future from the Past: Archaeozoology in Wildlife Conservation and Heritage Management. Oxbow Books, Oxford.

Livingston, S.D., 2000. The Homestead Cave avifauna. In: Madsen, D.B. (Ed.), Late Quaternary Paleoecology in the Bonneville Basin. Bulletin, vol. 130. Utah Geological Survey, Salt Lake City, pp. 91–102.

Lyman, R.L., 1996. Applied zooarchaeology: The relevance of faunal analysis to wildlife management. World Archaeol. 28, 110–125.

Lyman, R.L., 2006. Paleozoology in the service of conservation biology. Evol. Anthropol. 15, 11–19.

Lyman, R.L., 2012. Biodiversity, paleozoology, and conservation biology. In: Louys, J. (Ed.), Paleontology in Ecology and Conservation. Springer-Verlag, Berlin, pp. 147–169.

Lyman, R.L., Cannon, K.P., 2004. Zooarchaeology and Conservation Biology. University of Utah Press, Salt Lake City.

Lyman, R.L., O'Brien, M.J., 2005. Within-taxon morphological diversity in late-Quaternary Neotoma as a paleoenvironmental indicator, Bonneville Basin, northwestern Utah, USA. Quat. Res. 63, 274–282.

Madsen, D.B., 2000. Late Quaternary Paleoecology in the Bonneville Basin. Bulletin 130, Utah Geological Survey, Salt Lake City.

Madsen, D.B., Rhode, D., Grayson, D.K., Broughton, J.M., Livingston, S.D., Hunt, J., Quade, J., Schmitt, D.N., Shaver, M.W., 2001. Late Quaternary environmental change in the Bonneville basin, Western USA. Palaeogeogr. Palaeoclimatol. Palaeoecol. 167, 243–271.

Oviatt, C.G., 2014. The Gilbert Episode in the Great Salt Lake Basin, Utah. Utah Geological Survey. Miscellaneous Publication 14-3.

Oviatt, C.G., 2015a. Chronology of Lake Bonneville, 30,000 to 10,000 yr B.P. Quat. Sci. Rev. 110, 166–171.

Oviatt, C.G., 2015b. Ostracodes in Pleistocene Lake Bonneville, eastern Great Basin, North America. Hydrobiologia. 1–11. http://dx.doi.org/10.1007/s10750-015-2483-y.

Oviatt, C.G., Madsen, D.B., Miller, D.M., Thompson, R.S., McGeehin, J.P., 2015. Early Holocene Great Salt Lake, USA. Quat. Res. 84, 57–68.

Parmalee, P.W., 1970. Birds from Hogup Cave. In: Aikens, C.M. (Ed.), Hogup Cave. University of Utah Anthropological Papers, vol. 93, pp. 263–266.

Parmalee, P.W., 1980. Utilization of Birds by the Archaic and Fremont Cultural Groups of Utah. In: Campbell, K.E. (Ed.), Papers in Avian Paleontology Honoring Hildegarde Howard, Contributions in Science, Natural History Museum of Los Angeles County, Vol. 330v, pp. 237–250.

Parmalee, P.W., 1988. Avian remains from Danger Cave. In: Grayson, D.K. (Ed.), Danger Cave, Last Supper Cave, and Hanging Rock Shelter: the faunas. Anthropological papers of the AMNH, vol. 66, pp. 37–43.

Paul, D.S., Annand, E.M., Flory, J., 1999. Great Salt Lake Waterbird Survey 1997 and 1998 Seasons. Publication No. 99–19, Great Salt Lake Ecosystem Project, Utah Division of Wildlife Resources and Waterbird Survey Cooperators, Salt Lake City.

Reimer, P.J., Bard, E., Bayliss, A., Beck, J.W., Blackwell, P.G., Ramsey, C.B., Grootes, P.M., Guilderson, T.P., Haflidason, H., Hajdas, I., Hatté, C., Heaton, T.J., Hoffmann, D.L., Hogg, A.G., Hughen, K.A., Kaiser, K.F., Kromer, B., Manning, S.W., Niu, M., Reimer, R.W., Richards, D.A., Scott, E.M., Southon, J.R., Staff, R.A., Turney, C.S.M., van der Plicht, J., 2013. IntCal13 and Marine13 radiocarbon age calibration curves 0–50,000 Years cal BP. Radiocarbon 55, 1869–1887.

Rhode, D., 2000. Holocene vegetation history in the Bonneville Basin. In: Madsen, D.B. (Ed.), Late Quaternary Paleoecology in the Bonneville Basin. Bulletin, vol. 130. Utah Geological Survey, Salt Lake City, pp. 149–164.

Rhode, D., Louderback, L.A., 2007. Dietary plant use in the Bonneville basin during the terminal Pleistocene/early Holocene transition. In: Graf, K.E., Schmitt, D.N. (Eds.), Palaeoindian or Paleoarchaic? Great Basin human ecology at the Pleistocene-Holocene transition. The University of Utah Press, Salt Lake City, pp. 231–247.

Rhode, D., Madsen, D.B., 1995. Late Wisconsin/early Holocene vegetation in the Bonneville basin. Quat. Res. 44, 246–256.

Rick, T.C., Lockwood, R., 2013. Integrating paleobiology, archeology, and history to inform biological conservation. Conserv. Biol. 27, 45–54.

Rowe, R.J., Terry, R.C., 2014. Small mammal responses to environmental change: integrating past and present dynamics. J. Mammal. 95, 1157–1174.

Schroeder, M.A., Aldridge, C.L., Apa, A.D., Bohne, J.R., Braun, C.E., Bunnell, S.D., Connelly, J.W., Deibert, P.A., Gardner, S.C., Hilliard, M.A., Kobriger, G.D., McAdam, S.M., McCarthy, C.W., McCarthy, J.J., Mitchell, D.L., Rickerson, E.V., Stiver, S.J., 2004. Distribution of sage-grouse in North America. Condor 106, 363–376.

Smith, F.A., Betancourt, J.L., 2003. The effect of Holocene temperature fluctuations on the evolution and ecology of *Neotoma* (woodrats) in Idaho and northwestern Utah. Quat. Res. 59, 160–171.

Stansbury, H., 1852. Exploration and survey of the valley of the Great Salt Lake of Utah, including a reconnaissance of a new route through the Rocky Mountains.

Storer, R.W., Nuechterlein, G.L., 1992. Western Grebe (*Aechmophorus occidentalis*) and Clark's Grebe (*A. clarkii*). In: Poole, A., Gill, F. (Eds.), The Birds of North America, vol. 26. The Academy of Natural Sciences/The American Ornithologists' Union, Philadelphia/Washington, D.C, pp. 1–24.

Stuiver, M., Reimer, P.J., Reimer, R.W., 2013. CALIB 7.1. ^{14}CHRONO Center. Queen's University Belfast, Belfast. Available from: http://calib.qub.ac.uk/calib/.

Terry, R.C., 2007. Inferring predator identity from skeletal damage of small-mammal prey remains. Evol. Ecol. Res. 9, 199–219.

Terry, R.C., 2008. Raptors, rodents, and paleoecology: recovering ecological baselines from Great Basin caves. Ph.D. dissertation, University of Chicago, Chicago.

Terry, R.C., 2010. The dead do not lie: using skeletal remains for rapid assessment of historical small-mammal community baselines. Proc. R. Soc. Lond. B 277, 1193–1201.

Terry, R.C., Novak, M., 2015. Where does the time go?: mixing and the depth-dependent distribution of fossil ages. Geology 43, 487–490.

Terry, R.C., Rowe, R.J., 2015. Energy flow and functional compensation in Great Basin small mammals under natural and anthropogenic environmental change. Proc. Natl. Acad. Sci. 112, 9656–9661.

Terry, R.C., Li, C., Hadly, E.A., 2011. Predicting small-mammal responses to climatic warming: autoecology, geographic range, and the Holocene fossil record. Glob. Chang. Biol. 17, 3019–3034.

Thompson, R.S., Oviatt, C.G., Honke, J.S., McGeehin, J.P., 2016. Chapter 11. Late Quaternary changes in lakes, vegetation, and climate in the Bonneville basin reconstructed from sediment cores from Great Salt Lake. In: Oviatt, C.G., Shroder Jr., J.F. (Eds.), Lake Bonneville: A Scientific Update, vol. 20. Elsevier, pp. 221–291.

United States Census Bureau. Available at: URL: http://www.census.gov.

United States Geological Survey Utah Automated Geographic Reference Center (USGS Utah AGRC). Available from: URL: http://www.gis.utah.gov.

Utah Division of Wildlife Resources (UDWR), 2009. Utah Greater Sage-grouse management plan. Publication 09–17, Utah Department of Natural Resources, Division of Wildlife Resources, Salt Lake City.

Utah Division of Wildlife Resources (UDWR), 2013. Conservation Plan for Greater Sage-Grouse in Utah. Utah Department of Natural Resources, Division of Wildlife Resources, Salt Lake City.

Utah Geological Survey. Available at: URL: http://www.geology.utah.gov.

Western Association of Fish and Wildlife Agencies (WAFWA), 2015. Greater Sage-Grouse Population Trends: An Analysis of Lek Count Databases 1965–2015: Report from the Western Sage and Columbian Sharp-tailed Grouse Technical Committee. Western Association of Fish and Wildlife Agencies, Cheyenne.

Wolverton, S., Lyman, R.L., 2012. Conservation Biology and Applied Zooarchaeology. University of Arizona Press, Tucson.

Wolverton, S.J., Dombrosky, J., Lyman, R.L., 2014. Practical significance: ordinal scale data and effect size in zooarchaeology. Int. J. Osteoarchaeol. 26, 255–265. Available from: http://www.dx.doi.org/10.1002/oa.2416.

Chapter 15

Quaternary Vegetation Changes in the Bonneville Basin

D. Rhode
Desert Research Institute, Reno, NV, United States

ABSTRACT
Lake history and vegetation history complement each other in identifying key climate drivers of these interacting systems; for that reason comparing the two histories in the Bonneville basin has long held considerable research interest. Vegetation and constituent taxa respond to fluctuations in temperature, precipitation, seasonality, and other factors in a complex and mutually interactive interplay. Further complicating the interpretation of past vegetation patterns in relation to climatic factors are elements of considerable local climatic heterogeneity, differences in responses and spatiotemporal averaging of different proxy paleovegetation data sources, and chronometric uncertainties. Given these potential pitfalls, the fact that paleolake history–vegetation history comparisons work as well as they do is rather remarkable. Accommodating for complicating factors, long-term vegetation and hydrologic histories line up quite well through much of the Quaternary, and during the last glacial cycle in particular, on both orbital and millennial scales of climatic flux.

Keywords: Quaternary, Lake Bonneville, Vegetation history, Bonneville basin, Climate history

15.1 INTRODUCTION

Major fluctuations in temperature and precipitation that dramatically affected lake histories in the Bonneville basin during the Quaternary also shaped terrestrial vegetation distribution and composition. The Bonneville basin therefore provides a useful setting to explore linkages and discordances between long-term vegetation patterns and regional hydrology (Madsen, 1985; Madsen et al., 2001; Wigand and Rhode, 2002; Balch et al., 2005). This chapter reviews the long-term history of major Bonneville basin vegetation patterns in the Quaternary during the Bonneville lake cycle and early Holocene Great Salt Lake ($\gtrsim$9.0 cal ka BP).

Developments in Earth Surface Processes, Vol. 20. http://dx.doi.org/10.1016/B978-0-444-63590-7.00015-9

15.2 PALEOVEGETATION INFORMATION SOURCES

Key information about paleovegetation patterns comes from several main sources: (a) pollen from sediment cores and other localities; (b) plant macrofossils derived from packrat middens, archaeological sites, and other deposits; and (c) historic information obtainable from modern plant distributions and phylogeographic patterns. Use of ancient DNA preserved in sedimentary contexts has not yet been applied in the Bonneville basin, but will undoubtedly become an important information source in future work.

Pollen sequences derived from lake and bog sediments are a primary means to obtain paleovegetation information in the Bonneville basin (Table S1). Many of these come from basin-lacustrine deposits (Spencer et al., 1984; Louderback and Rhode, 2009; Eckerle et al., 2013; Thompson et al., 2016) that in some sediment cores extend as far back as the late Miocene (Davis, 2002). Pollen records have also been developed from smaller lakes and bogs, both lowland and montane (Madsen and Currey, 1979; Louderback and Rhode, 2009; Lundeen, 2012; Rhode and Louderback, 2013), and from archaeological and paleontological sites (Kelso, 1970; Dalley, 1976; Thompson, 1984).

Plant macrofossil deposits are common in the Bonneville basin, mainly in the form of woodrat (*Neotoma*) middens (Fig. 15.1). Approximately 150 midden samples have been dated and described from the Bonneville basin (Fig. 15.2, Table S2), more than half of them older than ~9.0 cal ka BP (Wells, 1976, 1983; Madsen, 1984; Thompson, 1984, 1990; Rhode and Madsen, 1995; Rhode, 2000a,b, 2013, 2014). Archaeological sites containing well-preserved plant remains provide a rich paleovegetation record spanning

FIG. 15.1 Bushy-tailed woodrat (*Neotoma cinerea*) reposing on a large midden deposit. The hard amberrat rind (dried urine mixed with fecal pellets and dust) can protect and preserve identifiable plant and animal remains encased in it for millennia, so long as it remains dry.

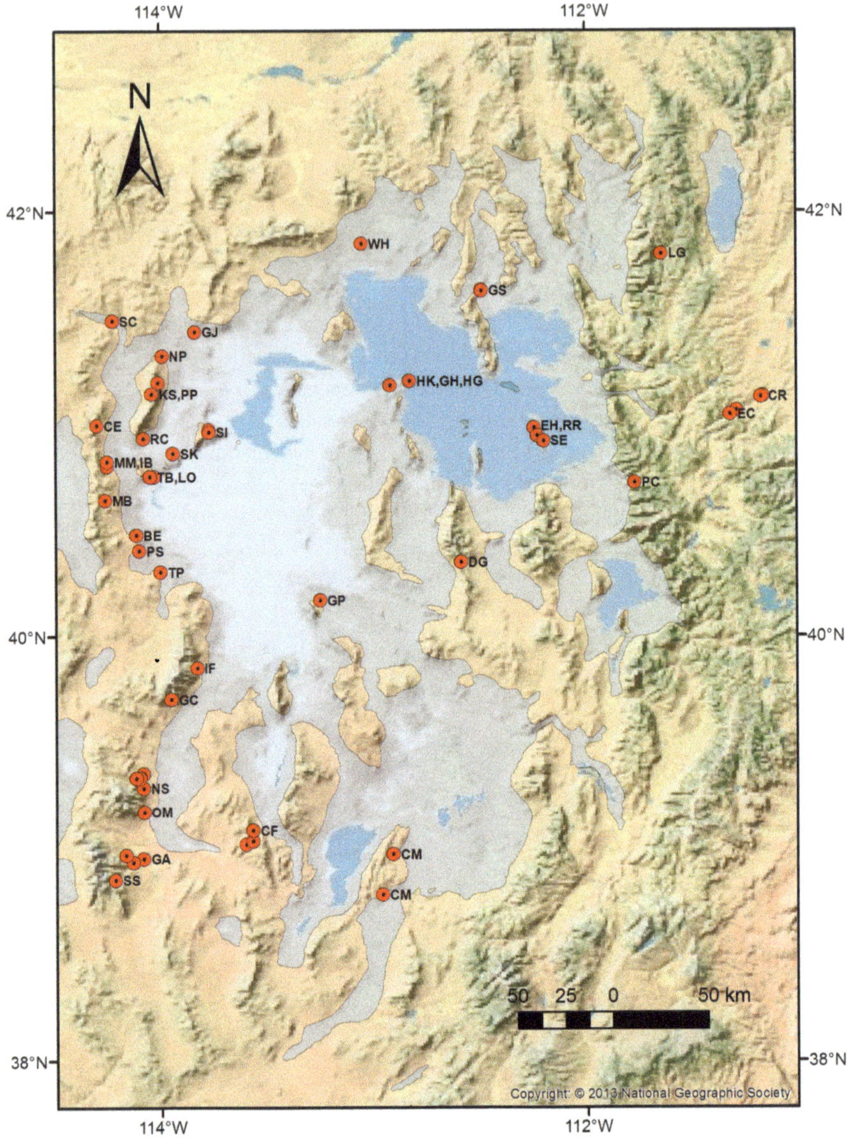

FIG. 15.2 Locations of dated woodrat middens from the Bonneville basin (see Table S2). Approximately maximum areal extent of Lake Bonneville (and other nearby Pleistocene lakes) is shown in translucent *gray*. Midden locations, marked by *dotted circles*, include BE (Bonneville Estates), CE (Collar and Elbow), CF (Confusion Range), CM (Cricket Mountains), CR (Castle Rock), DG (Devils Gate), EC (Echo Canyon), EH (Elephant Head), GA (Garrison), GC (Granite Canyon), GH (Gillespie Hills), GP (Granite Peak), GS (Golden Spike), HG (HAG), HK (Homestead Knoll), IB (Itchy Ban), IF (Indian Farm Canyon), KS (Killian Spring), LG (Logan Canyon), LO (Leppy Overhang), MB (Morgan Basin), MM (Marblehead Mine), NP (North Pilot Range), NS (Northern Snake Range, with several localities), OM (Old Man Cave), PC (Parley's Canyon), PP (Patterson Pass), PS (Pristine Shelter), RC (Raven Cave), RR (Red Rocks), SC (Schoolhouse Canyon), SE (Sentry Cave), SI (Silver Island Canyon), SK (Skunkbush Cave), SS (Southern Snake Range, with several localities), TB (Triple Barrel Cave), TP (Twin Peaks), and WH (Wildcat Hills).

the Holocene and extending into the latest Pleistocene (Table S3). Plant macrofossils and subfossils are also occasionally found in peats, organic mats, delta sediments, and other paleontological locales (Table S3).

Modern biogeographic patterns can yield important historic insights into past plant distributions. Cottam et al. (1959), for example, investigated scrub oak hybrids between Gambel oak (*Quercus gambelii*) and Sonoran scrub oak (*Quercus turbinella*) in the central Wasatch Range, concluding that the southerly scrub oak must have lived much further north than it does today, as a result of mid-Holocene warming (see also Nielson and Wullstein, 1983, 1985). Similarly, studies of small disjunct groves of singleleaf pinyon pine (*Pinus monophylla*), Colorado piñon (*Pinus edulis*), and their hybrids in the northern Wasatch Range led Lanner et al. (1972) to postulate these trees as relicts of more continuous conjoint past distributions. Dating these biogeographic puzzles is problematic without a fossil record, but they point to scenarios ripe for further investigation using pollen sequences and/or middens. The current explosion of phylogeographic research using next-generation DNA sequencing and improved methods of both modern and ancient DNA capture, replication, and analysis, coupled with internal DNA-based age estimation algorithms, promises to yield many more biogeographic puzzles and patterns to investigate. Ancient DNA from lake sediments and plant remains preserved in rodent middens and archaeological sites should allow fossil evidence to be incorporated in these phylogeographies. In turn, population and migration histories proposed using fossil evidence may be tested using molecular phylogeographic methods (Mitton et al., 2000; Hadly et al., 2004; Gugger and Sugita, 2010; Wilson and Pitts, 2012).

With these various data sets available, the following sections summarize main trends in Bonneville basin paleovegetation during the Quaternary period.

15.3 PRE-BONNEVILLE PLEISTOCENE VEGETATION

Records predating the Bonneville lake cycle include several long sediment cores taken from the Great Salt Lake, Sevier basin, and Bear Lake (Thompson et al., 1995; Davis, 1998, 1999, 2002; Davis and Moutoux, 1998; Oviatt et al., 1999; Balch et al., 2005; Jiménez-Moreno et al., 2007), some extending beyond the Quaternary. Pollen sequences obtained from some of these long cores demonstrate the regional dominance of xerophytic vegetation composed principally of various saltbushes (*Atriplex* and other shrubby members of the Amaranthaceae), greasewood (*Sarcobatus*), sagebrush (*Artemisia*), and pine (*Pinus*) since the late Miocene (Davis, 2002; Thompson, 1991; cf. also Davis and Ellis, 2010). Davis (2002:302) identified "three noteworthy events at ~3.7, 2.6, and 0.75 Ma [that] punctuated the gradual evolution of the [intermountain] desert flora." Of these, two pertain to the Quaternary. The 2.6 Ma event, at the beginning of the Pleistocene, shows increased abundance of wetland plants and lacustrine algae as shallow lakes and marshes filled the gradually coalescing Bonneville basin (cf. Kowalewska and Cohen, 1998). The 0.75 Ma event, most distinctive of the record, marks the dominance of sagebrush and pine and substantial decrease in saltbush, which Davis (2002:304) interpreted as "a permanent downward displacement of vegetation zones by ~500 m in elevation, in which greasewood steppe (with high percentages of Cheno-Ams [Amaranthaceae]) was replaced by sagebrush steppe in the intermountain basins" in the later Pleistocene. It also marks the start of open-lake conditions in the basin (Kowalewska and Cohen, 1998).

Several moderate- to deep-water lake episodes occurred in the last 750 ka, correlated with extensive northern hemisphere glacial intervals of ~100 ka cycle length (Davis, 2002:305; Balch et al., 2005). The first of these deep-water events was associated with oxygen isotope stage 16, ~620 ka (Oviatt et al., 1999; Nishizawa et al., 2012). Other lake cycles are associated with even-numbered isotope stages 14 through 2 (Lisiecki and Raymo, 2005), the last being the Bonneville lake cycle (though not all such cycles are well represented in core sediments; cf. Oviatt and Currey, 1987; Oviatt et al., 1999; Balch et al., 2005). These lacustrine cycles are matched by higher pollen contributions from montane conifers (pine, spruce, and fir) during glacial periods and from lowland shrubs and trees (sagebrush, saltbushes, greasewood, juniper) in the intervening interglacials (Davis and Moutoux, 1998). Wetland taxa are most abundant in the Great Salt Lake pollen records at the beginning of glacial stages, when lakes are fresh and shallow and pollen-producing wetlands are close to lake depocenters.

15.4 VEGETATION THROUGH THE LAKE BONNEVILLE CYCLE

Oxygen isotope stage 3 (OIS 3, ~57–29 cal ka BP; Voelker et al., 2002; Shackleton et al., 2004; Lisiecki and Raymo, 2005) was a period of mostly low lake levels prior to the onset of the Bonneville lake cycle (Balch et al., 2005; Benson et al., 2011; Oviatt, 2015; but cf. Nishizawa et al., 2013). Paleovegetation records are scarce but include pollen from the Great Salt Lake, Bear Lake, and Council Hall Cave (Martin and Mehringer, 1965; Mehringer, 1977; Spencer et al., 1984; Thompson, 1984; Thompson et al., 2016); a few packrat middens (Wells, 1983; Thompson, 1984; Wigand and Rhode, 2002); and isolated peat deposits or plant fragments in fluviolacustrine sediments (Scott et al., 1983; Oviatt et al., 1987). These records show that sagebrush dominated in intermontane valleys and piedmonts, with lesser amounts of saltbushes, greasewood, woody Asteraceae, and grass. Pine is consistently secondary to sagebrush in regional pollen records, with juniper and other conifers (spruce and fir) as minor or sporadic contributors. The pine pollen was undoubtedly derived from limber pine (*Pinus flexilis*) and bristlecone pine (*Pinus longaeva*), which are common constituents of woodrat middens dating to this period (Wells, 1983; Thompson, 1984; Wigand and Rhode, 2002). Bristlecone pine dominated in the southern Bonneville basin, while limber pine took its place further north. The juniper pollen may have come from Utah juniper (*Juniperus osteosperma*), which grew in the southern Bonneville basin before ~30 cal ka BP (Thompson, 1984), subalpine common juniper (*Juniperus communis*; Wells, 1983; Wigand and Rhode, 2002), and possibly even Rocky Mountain juniper (*Juniperus scopulorum*) though preglacial macrofossils of this last species are presently absent. Other plant associates include sagebrush, rabbitbrush (*Chrysothamnus*), snowberry (*Symphoricarpos*), currant (*Ribes*), and other mesophilic shrubs typical of high montane meadows today (Wells, 1983; Thompson, 1984). Subalpine conifers and dry-loving montane shrubs dominated over more mesophilic meadow vegetation during OIS 3. A climatic reconstruction of dry cool conditions based on vegetation proxy data coheres well with the shallow though occasionally expansive lake levels that characterized the Bonneville basin prior to ~30 cal ka BP (Benson et al., 2011; Oviatt, 2015). It does not cohere at all well with the postulated large OIS 3 lakes hypothesized by Nishizawa et al. (2013).

15.4.1 Early Bonneville Cycle

Lake Bonneville rose rapidly after ~30.7 cal ka BP, possibly in part from diversion of the upper Bear River into the Bonneville basin (Bouchard et al., 1998; Reheis et al., 2014; Oviatt, 2015). The period from ~30 to 24 cal ka BP is marked by fluctuating lake-level rise including the Stansbury oscillation to form a moderately deep freshwater lake of considerable surface area. Lake-sediment pollen records show a transition from a sagebrush/saltbush-covered landscape before ~30 cal ka BP to pine and spruce dominance afterward (Mehringer, 1977, 1985; Spencer et al., 1984; Thompson et al., 1990, 2016; Madsen et al., 2001). Scattered occurrences of subalpine conifer fossils in sediments on the east side of the Bonneville basin (Cottam et al., 1959; Scott et al., 1983) also demonstrate the presence of spruce/fir forests in the Wasatch front. Middens west of the lake basin show continued dominance of bristlecone pine in the southern part (<40°N), Engelmann spruce (*Picea engelmannii*) and limber pine further north (40°–41°N), and montane meadow plants contributing to both (Wells, 1983; Thompson, 1984; Wigand and Rhode, 2002). Significantly colder conditions are indicated, as Lake Bonneville began its great rise (Mering, 2015).

15.4.2 Middle Bonneville Cycle

From the Stansbury level, Lake Bonneville filled another ~235 m in depth to reach its highstand by ~18.3 cal ka BP (Oviatt, 2015). The rise to the highstand was by no means smooth and continuous; instead the lake underwent significant large-amplitude oscillations, in response to global-scale climate fluctuations (Oviatt, 1997; Patrickson et al., 2010; McGee et al., 2012; Nishizawa et al., 2013).

Pollen from Great Salt Lake Core C dating to the rising limb of the Bonneville hydrograph (Spencer et al., 1984; Thompson et al., 1990, 2016) shows dominance of spruce and pine during this interval; sagebrush and saltbush declined markedly as valley margins flooded. In the southern Bonneville basin, bristlecone pine forests also contained Engelmann spruce, Rocky Mountain juniper, limber pine, sagebrush, snowberry, common juniper, and other shrubs (Wells, 1983; Thompson, 1984; cf. Coats et al., 2008). Vegetation was sparser and more alpine in character further north. One midden sample from the Goshute Range dating ~20 cal ka BP suggests low sagebrush–grass communities with rare interspersed limber pine and Engelmann spruce grew as low as ~2000 m altitude in west central basin uplands (Wigand and Rhode, 2002). Evidence of subalpine fir macrofossils in lake sediments (Cottam et al., 1959) also indicate spruce–fir forests along the Wasatch Front.

15.4.3 Late Bonneville Cycle

After the Bonneville flood, the now-outflowing lake stabilized at the Provo shoreline level for a long time until it regressed sometime between ~16.4–14.6 cal ka BP (the exact timing of the regression remains unresolved; Benson et al., 2011; Godsey et al., 2011; McGee et al., 2012; Miller et al., 2013; Reheis et al., 2014). Smoot and Rosenbaum (2009:289), interpreting sediments from Bear Lake, suggest that a major drought occurred while Lake Bonneville overflowed. The available vegetation records, few and coarse-grained as they are, do not reveal any such drought; instead they substantiate cold conditions through the Provo-level overflow interval. The GSL 96-6 record of Thompson et al. (2016) suggests slightly higher sagebrush pollen and slightly lower pine pollen during this interval, with a high proportion of pine grains broken (a product of redeposition of marl exposed as the lake declined?). An increase in exposed land around the lake is also reflected in markedly higher values of Amaranthaceae and *Sarcobatus* pollen (Thompson et al., 2016). The Great Salt Lake Core C pollen record also shows abundant conifer presence until ~14.5 cal ka BP (Spencer et al., 1984; Thompson et al., 1990). Two midden samples from the southern Bonneville basin dating between ~18.2–16.4 cal ka BP contain limber pine, subalpine juniper, and montane shrubs (Thompson, 1984); three from the west basin further north contain subalpine meadow shrubs and herbs including common juniper, but no pine and only rare Engelmann spruce (Rhode, 2000a).

Many more middens date between ~16.4–14.6 cal ka BP; whether they were deposited when the lake stood at the Provo level or during its regression is uncertain. Middens from the southern Bonneville basin continue to show dominance by bristlecone pine and mesophilic montane shrubs (Thompson, 1984). Further north, middens located on the west central basin margins reflect a mosaic of limber pine woodlands mixed with sagebrush steppe, growing as low as 1475 m by ~15.6 cal ka BP (Rhode, 2000a). Limber pine forest mixed with Engelmann spruce and subalpine fir (*Abies lasiocarpa*) grew in nearby uplands (Madsen et al., 2001). The presence of dry and cold-adapted limber pine woodlands at low elevations suggests that average July temperatures were ~6°C cooler than today (Rhode and Madsen, 1995). These woodlands persisted until ~13.8 cal ka BP, well after Lake Bonneville had regressed from the Provo shoreline. Cold climate prevailed through the period as the lake regressed, indicating that the climate must have been quite dry for the lake to fall as rapidly as it apparently did. In this respect, it is worth noting that the lake was sufficiently large and cold to support the characteristic Bonneville fish fauna until at least ~13.2 cal ka BP (Broughton et al., 2000). Taken together, these vegetation records indicate continued dry and cool conditions (relative to the Holocene), as "the lake dropped rapidly, but not catastrophically, to levels comparable to those of modern [Great Salt Lake]" (Oviatt, 2015:167).

15.5 GREAT SALT LAKE CYCLE

The regression of Lake Bonneville marks the end of the Bonneville lake cycle and the beginning of the Great Salt Lake cycle (Oviatt and Currey, 1987; Oviatt et al., 1992; Oviatt, 2015). The post-Provo lake retreat opened up a vast area of former lakebed to the formation of marshes and playa-margin plant communities. Large wetlands formed in several places on the basin floor, as at the deltas of the Bear River and the Old River Bed (Oviatt et al., 2003, 2005; Madsen et al., 2015a,b), Fish Springs (Godsey et al., 2005), Blue Lake (Louderback and Rhode, 2009), and elsewhere.

The GSL 96-6 record of Thompson et al. (2016) places the post-Provo regression beginning at ~15.2 cal ka BP, beginning with very high values of pine and diminished values of sagebrush, spruce, fir, and juniper, followed by a sharp return of sagebrush by ~13.5 cal ka BP. Sediments from Great Salt Lake Core C suggest that sagebrush, greasewood, and saltbush increased rapidly after ~14.5 cal ka BP, at the expense of conifers (Spencer et al., 1984; Thompson et al., 1990). Pollen data from Swan Lake, located in Lake Bonneville's former overflow channel near Red Rock Pass, Idaho, show high values of conifers including limber pine, possibly lodgepole pine (*Pinus contorta*), spruce, and fir between ~13.9–12.9 cal ka BP (Bright, 1966). To the east in the Bear River Range, the Plan B Pond pollen sequence begins with high sagebrush, pine, spruce, and Asteraceae from ~13.8 cal ka BP with pine reaching a peak at ~12.9 cal ka BP, suggesting summer warming and drying (Lundeen, 2012). At Bear Lake, cold-tolerant sagebrush/shrub birch steppe-tundra was replaced by pine-dominated woodland between ~15–13.8 cal ka BP (Doner, 2009).

The Blue Lake pollen record shows high abundance of pine and sagebrush in lacustrine sediments (stratum 7) estimated to date ~15–11.8 cal ka BP (Louderback and Rhode, 2009; cf. Benson et al., 2011). These samples, though, come from relatively deep-water lake deposits (based on strontium isotope ratios; cf. Hart et al., 2004) and most likely reflect late Bonneville cycle vegetation, not postregression Great Salt Lake conditions. Lake level retreated below the Blue Lake area after the post-Provo regression, so sediments dating to the immediate aftermath of the regression may be unconformably absent (cf. Benson et al., 2013; Oviatt, 2014).

In mountain ranges in the southern Bonneville basin, bristlecone or limber pine alternated dominance on different sites, both accompanied by dry-loving as well as mesophilic montane trees and shrubs (Wells, 1983; Thompson, 1984, 1990). Midden records on the basin margins of the western central Bonneville basin show a replacement of lowland limber pine woodlands by sagebrush steppe including xerophytic shrubs including shadscale (*Atriplex confertifolia*), horsebrush (*Tetradymia*), and others. Even further north, on slopes fronting the northwest margin of the lake bed, middens dating ~14.2–13.6 cal ka BP contain the same desert shrubs but notably lack limber pine, which may not have ranged this far north (Rhode, 2013); Mehringer's (1985) hypothesis that largely treeless sagebrush–grass steppe prevailed in the northern Bonneville basin is here supported.

The vegetation transition from lower piedmont slopes dominated by limber pine woodlands at ~15.6 cal ka BP to their demise and replacement by sagebrush and saltbush-dominated communities after ~13.8 cal ka BP, occurred as the Bølling-Allerød (B/A) interstadial began (~14.7 cal ka BP), and during a time when drier conditions prevailed in much of the Great Basin and parts of the North American Southwest (Adams and Wesnousky, 1998; Wigand and Rhode, 2002; Asmerom et al., 2010; McGee et al., 2012; Polyak et al., 2012; Benson et al., 2013). The specific climatic changes that resulted in the demise of the low-elevation limber pine woodlands and the rise of sagebrush–saltbush scrub after ~13.9 cal ka BP remain uncertain, but a significant warming and drying trend during the B/A interval (the so-called Clovis Drought; Haynes, 1991) seems to account for the pattern of vegetation change, as well as the decline of the lake and the demise of its diverse native fish fauna beginning ~13.2 cal ka BP, just a few centuries before the Younger Dryas (YD) (Broughton and Smith, 2016).

15.5.1 The Younger Dryas Chronozone and Gilbert Episode

The YD stadial marked a millennium-long return to colder Northern Hemisphere temperatures at the close of the last deglaciation process, ~12.8–11.7 cal ka BP (Rasmussen et al., 2006; Steffensen et al., 2008). Climatic effects of the YD are geographically complex: areas close to the North Atlantic cooled dramatically but regions located farther away show reduced effects, and parts of North America actually warmed in response to increasing northern hemisphere insolation (Berger and Loutre, 1991; Shakun and Carlson, 2010).

In the Bonneville basin some YD records (especially those in uplands) indicate cooler and/or moister conditions at least part of the time, whereas others (mostly lowland records) indicate continuity of relatively dry conditions (Oviatt et al., 2005; Goebel et al., 2011; Lundeen et al., 2013; Rhode and Louderback, 2015; cf. Polyak et al., 2004). The Great Salt Lake apparently fluctuated around low levels near or below its modern level during much of the YD, suggesting continued dry conditions (Oviatt, 2014:17). The pollen and stratigraphic record from GSL 96-6 (Thompson et al., 2016) contains deposits representing the YD, with reduced pine values and very high values of sagebrush reflecting increasing aridity during the YD. The Great Salt Lake Core C record lacks dated YD age sediments (Spencer et al., 1984; Thompson et al., 1990; but cf. Oviatt, 2014:11; Thompson et al., 2016). At Swan Lake, the YD commenced with abundant conifers in the surrounding landscape, but they were soon replaced by sagebrush and grass, also possibly reflecting a drying trend (Bright, 1966). At Blue Lake, the carbonate-rich sediments of Strata 6, interpreted to date ~11.8–11.3 cal ka (Louderback and Rhode, 2009), may relate to part of the YD interval, perhaps including the Gilbert episode (Oviatt, 2014). It contains initially high but declining values of pine and sagebrush and rapidly rising Amaranthaceae, similar to Great Salt Lake Core C.

In the highlands of the Wasatch Range, subalpine sagebrush with mountain alder and birch prevailed over spruce and pine at Snowbird Bog, Little Cottonwood Canyon (Madsen and Currey, 1979). In nearby Silver Lake, Big Cottonwood Canyon, basal sediments dating to the YD are dominated by subalpine sagebrush and pine with low spruce (Rhode and Louderback, 2013). Further north in the Bear River Range, Lundeen's (2012) Plan B Pond pollen record shows elevated sagebrush over pine, slight increases in spruce and fir, and indices that suggest the early YD was cooler and moister than the centuries before or after.

The YD woodrat midden record shows that a mix of desert shrub and sagebrush–grass steppe continued to cover the piedmonts of the west-central basin valleys (Rhode, 2000b). Shadscale (*A. confertifolia*) had been a very minor component of the earlier limber pine woodland–sagebrush steppe mosaic, but it grew much more widely and abundantly in the YD. Middens collected by Wells (1983) in the southern Bonneville basin suggest a reversal in the fortunes of subalpine conifers: where limber pine had been most abundant before the YD, bristlecone pine came to dominate, possibly signaling moister YD conditions favoring the latter tree.

At the very end of the YD, the Great Salt Lake rose ~15 m and expanded to cover much of the western Great Salt Lake Desert up to ~1295–1297 m elevation, the so-called Gilbert shoreline (Eardley et al., 1957; Currey et al., 1984; Murchison, 1989a,b; Oviatt et al., 1992, 2005; Oviatt, 2014). The duration of this Gilbert-episode lake stand was apparently quite brief, on the order of decades, and its cause is uncertain. A similar short-duration spike in lake level may have occurred at Bear Lake (Smoot and Rosenbaum, 2009:288). Short-term fluctuations are observed in some vegetation records (eg, pine at Blue Lake), but whether these blips date to this Gilbert episode and reflect the same climatic drivers is not known and may be difficult to ascertain given the episode's apparent brevity. Thompson et al. (2016) suggested that the rising lake of the Gilbert episode, during a period of increasing aridity, must have arisen from a brief cold episode that reduced evaporation.

15.5.2 Early Holocene

The YD ended abruptly ~11.7 cal ka BP as deep oceanic circulation kicked into gear and tropical ocean waters flowed northward to warm the North Atlantic (Steffensen et al., 2008; Shakun et al., 2012). Northern Hemisphere summer solar insolation reached peak Holocene values 140% of modern, further heating continental interiors (Kutzbach and Guetter, 1986; Berger and Loutre, 1991). In the American Southwest, summer precipitation increased and temperatures may have cooled slightly (Polyak et al., 2004, 2012; Holmgren et al., 2009; Miller et al., 2010); the Pacific Northwest was drier than at present (Mehringer, 1985; Barnosky et al., 1987; Thompson et al., 1993). In the Bonneville basin, interseasonal temperature swings may have been strengthened (Broughton et al., 2008).

The decline of the Great Salt Lake following the Gilbert episode, after ~11.5 cal ka BP, allowed playa-margin marshlands to regrow in various parts of the Bonneville basin. At the Public Shooting Grounds locality, marshland sediments interpreted to date ~11.0–10.5 cal ka BP overlie near-shore sediments deposited during the Gilbert episode (Oviatt et al., 2005; Oviatt, 2014). At Locomotive Springs, wetland organics dating ~11.0–9.6 cal ka BP were deposited on Lake Bonneville marls below the highest level of the Gilbert episode (Oviatt et al., 2015). Near Wendover, organic-rich wetland muds overlying presumed Gilbert beach deposits date to ~10.8 cal ka BP (Murchison, 1989a; Oviatt, 2014). At Blue Lake, peat deposits also began to form by ~10.8 cal ka BP (Louderback and Rhode, 2009).

Regional pollen records show that sagebrush steppe persisted in lowlands, but that desert shrubs such as shadscale gained in abundance during the early Holocene (Bright, 1966; Madsen and Currey, 1979; Mehringer, 1985; Beiswenger, 1991; Louderback and Rhode, 2009). Pollen records from the Great Salt Lake core (Spencer et al., 1984; Eckerle et al., 2013; Thompson et al., 2016) document the continuing decline of conifer woodlands and expansion of xerophytic shrubs after ~11.0 cal ka BP. The pollen records from Swan Lake and Blue Lake highlight major transitions from sagebrush steppe to shadscale steppe ~10.2–9.5 cal ka BP (Bright, 1966; Louderback and Rhode, 2009). Likewise, the sparse early Holocene lowland midden record indicates that shadscale and sagebrush continued to dominate valley floors (Rhode, 2000b).

Oviatt et al. (2015) recently reviewed evidence for early Holocene Great Salt Lake levels between ~11.5–10.2 cal ka BP, and concluded that "there is no evidence that the GSL [Great Salt Lake] was higher than the modern lake during the early Holocene, with the low lake suggesting a warm and/or dry climate" (Oviatt et al., 2015:10; contra Currey et al., 1984; Murchison, 1989a; Patrickson et al., 2010). Some evidence has been reported for higher-than-modern early Holocene lake levels possibly after ~10.2 cal ka BP: for example, Elder (1992) used data from the Malad River drainage to suggest that Great Salt Lake rose to at least 1288 m before ~8.6 cal ka BP, and Oviatt et al. (2015:10) described undated ~1290 m shorelines at Locomotive Springs younger than ~11.0–9.6 cal ka BP.

The interpretation of generally low lake levels in much (if not all) of the early Holocene (Oviatt et al., 2015) contrasts with various proxy interpretations of cooler and/or wetter climatic conditions during that period relative to today (Grayson, 1998, 2000a,b, 2006, 2011; Rhode, 2000b; Madsen et al., 2001; Lundeen, 2012; Lundeen et al., 2013; Steponaitis et al., 2015). Proxy climate reconstructions derived from fossil records occurring in cave and rockshelter settings (ie, woodrat middens, small mammal faunas, and possibly speleothems) may characteristically reflect more mesic local microenvironments than do proxies derived from region-wide records such as lake-based pollen sequences, as noted above. Such differences may account for some of the variability in proxy climate interpretations (see discussion in Thompson et al., 2016). Oviatt et al. (2015) hypothesized that the early Holocene was cool but dry, and that enhanced groundwater discharge from lake waters stored in fractured limestone outcrops and valley slopes accounts for the persistence of mesophilic vegetation and small mammals (as well as the deposition of sapropels in Great Salt Lake sediments). How the Great Salt Lake could maintain an essentially modern level at a time when basin groundwater discharge was significantly enhanced and climate was cooler than modern implies that major streamflow inputs to the lake must have been much smaller than the modern situation, and that the climate was substantially drier than today.

Localized higher groundwater discharges in various areas around the basin evidently contributed to the development of large wetlands (eg, Old River Bed delta; Oviatt et al., 2003; Madsen et al., 2015a,b; Rhode and Louderback, 2015). The effects of higher surface-water discharge on shrub–grass communities in valleys and piedmonts across the region are not known. Mesophilic shrubs such as snowberry and netleaf hackberry (*Celtis reticulata*) persisted well into the early Holocene at Homestead Cave and elsewhere in the Bonneville basin, perhaps as a result of stored groundwater in rock outcrops as elsewhere previously suggested (Hunt et al., 2000; Rhode, 2000b; Wigand and Rhode, 2002:349). Oxygen isotopes in hackberry endocarps, a proxy measure of growing-season temperature (Jahren et al., 2001), suggest the early Holocene growing season was relatively cool around Homestead Cave at ~9.6 cal ka BP, when hackberry attained its peak abundance there (Rhode and Louderback, 2015). Hackberry endocarps at Homestead Cave declined markedly in numbers after ~9.2 cal ka BP and were rare by ~7.8 cal ka BP.

The Wasatch Range high country was a mosaic of high-elevation sagebrush communities and spruce–pine woodlands. At Silver Lake, pine (presumably limber pine) dominated through much of the early Holocene before ~9.0 cal ka BP, with one brief episode of moisture-loving spruce dominance about ~10.4 cal ka BP (Rhode and Louderback, 2013). At Snowbird Bog, conifer pollen increased dramatically relative to all other pollen shortly before ~8.9 cal ka BP, suggesting a rapid transition to a full spruce–pine forest that Madsen and Currey (1979) attributed to a sharp increase in temperature. At Plan B Pond, pine remained high and fairly constant through the early Holocene, sagebrush was dominant but declined through the early Holocene, spruce increased but both it and fir remained at low levels, and Asteraceae and Amaranthaceae increased. Overall, various indices show strongly warmer summer conditions (up to 2°C greater than today), but with possibly effectively wetter conditions due to larger snowpack (Lundeen, 2012).

Uplands in the western Bonneville basin harbored a mosaic of sagebrush–grass steppe, aspen (*Populus tremuloides*) groves, montane brush communities, and woodlands dominated by Rocky Mountain juniper. Limber pine retreated to protected canyons and upper slopes and ridges, mixed with Engelmann spruce and firs (both *Abies concolor* and *A. lasiocarpa*). In the southern Bonneville basin, woodlands were primarily a mix of Utah juniper and Rocky Mountain juniper on dry slopes, with bristlecone pine, limber pine, and common juniper on protected north-facing slopes (Thompson, 1984, 1990). Sagebrush and grass were common, and aspen, mountain mahogany (*Cercocarpus intricatus*), and chokecherry (*Prunus virginiana*) increased in abundance.

The warming and drying trend begun in the early Holocene resulted in significant changes in vegetation distribution, including both gradual shifts (eg, range expansions) as well as threshold effects (eg, desiccation of wetlands). With rising Holocene temperatures several important Great Basin tree taxa, including Rocky Mountain juniper, Utah juniper, and singleleaf pinyon pine, underwent substantial range shifts that in some cases have continued through the Holocene (Madsen and Rhode, 1990; Thompson, 1990; Rhode and Madsen, 1998; Rhode, 2000b; Cole et al., 2013; Weppner et al., 2013).

By ~8.3 cal ka BP, shadscale increasingly dominated valley floors through much of the northern Bonneville basin (Rhode, 2000b, 2013), though sagebrush habitats persisted in many places (cf. Schmitt and Madsen, 2005). This record of increased drying and a shift from sagebrush steppe to desert shrub communities during the early Holocene is mirrored by the mammalian faunal record from Homestead Cave and other sites (Grayson, 1998, 2000a,b, 2006; Schmitt et al., 2002; Schmitt and Lupo, 2012). Expansive valley-floor wetlands disappeared. The Blue Lake wetlands were in decline by ~9.0 cal ka BP and largely though not entirely desiccated by ~7.8 cal ka BP, when marshes were sharply diminished and replaced by patchy saltgrass (*Distichlis spicata*) meadows and playa margins dotted with greasewood and iodinebush (*Allenrolfea occidentalis*) (Louderback and Rhode, 2009). Other large wetland systems in the region declined or dried up completely between ~9.5–8.3 cal ka BP (Thompson, 1992; Wigand and Rhode, 2002; Oviatt et al., 2003; Rhode et al., 2005; Kiahtipes, 2009). The loss of key wetland habitats also resulted in significant shifts in human occupation and subsistence patterns in the region (Madsen, 2002; Rhode et al., 2006; Rhode, 2008; Madsen et al., 2015a,b; Madsen, 2016).

15.6 CONCLUSIONS

At orbital time scales, long-term vegetation history and hydrologic history in the Bonneville basin line up quite well through much of the late Quaternary, and during the last glacial cycle in particular (Oviatt, 1997; Davis, 2002; Jiménez-Moreno et al., 2007). Alternations of cold-mesic conifer forest formations vs more xerophilic shrublands correspond with major glacial–interglacial cycles and generally reflect shorter millennial-scale fluctuations as well, though these briefer episodes (eg, Dansgaard-Oeschger and Heinrich events) are often difficult to tease apart in vegetation records (Coats et al., 2008; Jiménez-Moreno et al., 2010).

The presence of shallow saline lakes during OIS 3 corresponds well with evidence for cold but dry climate favoring limber pine woodlands and steppe vegetation during this interval. Subalpine forests became increasingly abundant in montane settings as Lake Bonneville rose after ~30 cal ka BP. By the time the lake reached its highstand at ~18.3 cal ka BP, very cold conditions resulted in subalpine conifers dominating in the southern basin, while alpine sagebrush-dominated communities likely covered much of the north. Lake Bonneville's post-Provo regression coincides with the establishment of extensive wetlands on the newly exposed lakebed, the development of xerophytic shrubland communities at low elevations, and the expansion of cool- and dry-tolerant limber pine woodlands onto valley slopes and piedmonts in the central basin. The YD interval continues to be regionally variable and uncertain, with some records indicating continued dry conditions and others greater moisture, suggesting different responses along altitudinal and latitudinal gradients. The early Holocene also presents divergent proxy information related to climatic conditions driving the behavior of the Great Salt Lake and vegetation in the Bonneville basin (Oviatt et al., 2015). By ~9.0 cal ka BP, however, the accumulated effects of Holocene warming and drying trend brought these different proxies into better alignment and the early Holocene to a recognizable end.

Lake history and vegetation history potentially complement each other in identifying key climate drivers of both interacting systems. Vegetation formations, and their constituent taxa, respond to fluctuations in temperature, precipitation, effective moisture, seasonality, and atmospheric gas composition in complex, individualistic and mutually interactive ways such that "translating vegetation data to climate data is inexact because of the uncertainties generated by this interplay of variables" (Coats et al., 2008:335). Add to that complicating layers of spatial heterogeneity, spatiotemporal averaging of different proxies, and dating uncertainties, and it is somewhat amazing that lake history–vegetation history comparisons work as well as they do. Refinements in methods of biome reconstruction from paleovegetation data and inferring their climatic implications (Thompson and Anderson, 2000), greater chronometric control, new methods to obtain paleovegetation data such as ancient DNA of lacustrine sediments, and a growing base of paleovegetation data should significantly improve the comparisons of Bonneville basin vegetation histories with hydrologic histories in the future.

ACKNOWLEDGMENTS

My thanks to Jack Oviatt for the invitation to participate in this volume, to David Madsen for helpful editorial comments, and to David Madsen, Bryan Hockett, Ted Goebel, Mike Metcalf, Lisbeth Louderback, D. Craig Young, Monson Shaver III, Dave Schmitt, Bob Thompson, Rand Gruebel, and all the colleagues and students with whom I have had the pleasure to work in the Bonneville basin over the decades.

APPENDIX A SUPPORTING INFORMATION

Supplementary data associated with this article can be found in Tables S1–S3 in the online version (http://dx.doi.org/10.1016/B978-0-444-63590-7.00015-9).

REFERENCES

Adams, K.D., Wesnousky, S.G., 1998. Shoreline processes and the age of the Lake Lahontan highstand in the Jessup embayment, Nevada. Geol. Soc. Am. Bull. 110, 1318–1332.

Asmerom, Y., Polyak, V.J., Burns, S.J., 2010. Variable winter moisture in the southwestern United States linked to rapid glacial climate shifts. Nat. Geosci. 3, 114–117.

Balch, D.P., Cohen, A.S., Schnurrenberger, D.W., Haskell, B.J., Valero Garces, B.L., Beck, J.W., Cheng, H., Edwards, R.L., 2005. Ecosystem and paleohydrological response to Quaternary climate change in the Bonneville basin, Utah. Palaeogeogr. Palaeoclimatol. Palaeoecol. 221, 99–122.

Barnosky, C.W., Anderson, P.M., Bartlein, P.J., 1987. The northwestern U.S. during deglaciation: vegetational history and paleoclimatic implications. In: Ruddiman, W.F., Wright Jr., H.E. (Eds.), North America and Adjacent Oceans During the Last Deglaciation. The Geology of North America, vol. K-3. Geological Society of America, Boulder, CO, pp. 289–322.

Beiswenger, J.M., 1991. Late Quaternary vegetational history of Grays Lake, Idaho. Ecol. Monogr. 61, 165–182.

Benson, L.V., Lund, S.P., Smoot, J.P., Rhode, D.E., Spencer, R.J., Verosub, K.L., Louderback, L.A., Johnson, C.A., Rye, R.O., Negrini, R.M., 2011. The rise and fall of Lake Bonneville between 45 and 10.5 ka. Quat. Int. 235, 57–69.

Benson, L.V., Smoot, J.P., Lund, S.P., Mensing, S.A., Foit Jr., F.F., Rye, R.O., 2013. Insights from a synthesis of old and new climate-proxy data from the Pyramid and Winnemucca lake basins for the period 48 to 11.5 cal ka. Quat. Int. 310, 62–82.

Berger, A., Loutre, M.F., 1991. Insolation values for the climate of the last 10 million years. Quat. Sci. Rev. 10, 297–317.

Bouchard, D.P., Kaufman, D.S., Hochberg, A., Quade, J., 1998. Quaternary history of the Thatcher basin, Idaho, reconstructed from the $^{87}Sr/^{86}Sr$ and amino acid composition of lacustrine fossils: implications for the diversion of the Bear River into the Bonneville basin. Palaeogeogr. Palaeoclimatol. Palaeoecol. 141, 95–114.

Bright, R.C., 1966. Pollen and seed stratigraphy of Swan Lake, southeastern Idaho: its relation to regional vegetational history and to Lake Bonneville history. Tebiwa 9 (2), 1–47.

Broughton, J.M., Smith, G.R., 2016. Chapter 12. The fishes of Lake Bonneville: implications for drainage history, biogeography and lake levels. In: Oviatt, C.G., Shroder Jr., J.F. (Eds.), Lake Bonneville: A Scientific Update, vol. 20. Elsevier, pp. 292–351.

Broughton, J.M., Madsen, D.B., Quade, J., 2000. Fish remains from Homestead Cave and lake levels of the past 13,000 years in the Bonneville basin. Quat. Res. 53, 392–401.

Broughton, J.M., Byers, D.A., Bryson, R.A., Eckerle, W., Madsen, D.B., 2008. Did climatic seasonality control late Quaternary artiodactyl densities in western North America? Quat. Sci. Rev. 27, 1916–1937.

Coats, L., Cole, K.L., Mead, J.I., 2008. 50,000 years of vegetation and climate history on the Colorado Plateau, Utah and Arizona, USA. Quat. Res. 70, 322–338.

Cole, K.L., Fisher, J.F., Ironside, K., Mead, J.I., Koehler, P., 2013. The biogeographic histories of *Pinus edulis* and *Pinus monophylla*. Quat. Int. 310, 96–110.

Cottam, W.P., Tucker, J.M., Drobnick, R., 1959. Some clues to Great Basin postpluvial climates provided by oak distributions. Ecology 40 (3), 361–377.

Currey, D.R., Atwood, G., Mabey, D.R., 1984. Major Levels of Great Salt Lake and Lake Bonneville. Utah Geological and Mineral Survey, Map 73. Salt Lake City.

Dalley, G.F., 1976. Palynology of the Swallow Shelter deposits. In: Dalley, G.F. (Ed.), Swallow Shelter and Associated Sites, pp. 171–174. University of Utah Anthropological Papers 96, Salt Lake City.

Davis, O.K., 1998. Palynological evidence for vegetation cycles in a 1.5 million year pollen record from the Great Salt Lake, Utah, USA. Palaeogeogr. Palaeoclimatol. Palaeoecol. 138, 175–185.

Davis, O.K., 1999. Preliminary pollen analysis of Cenozoic sediments of the Great Salt Lake, Utah, U.S.A. In: Wrenn, J.H., Suc, J.-P., Leroy, S.A.G. (Eds.), The Pliocene: Time of Change. American Association of Stratigraphic Palynologists Foundation, Dallas, TX, pp. 227–240.

Davis, O.K., 2002. Late Neogene environmental history of the northern Bonneville basin: a review of Palynological studies. In: Hershler, R., Madsen, D.B., Currey, D.R. (Eds.), Great Basin Aquatic Systems History. Smithsonian Contributions to the Earth Sciences, vol. 33. Smithsonian Institution, Washington, DC, pp. 295–307.

Davis, O.K., Ellis, B., 2010. Early occurrence of sagebrush steppe, Miocene (12 MA) on the Snake River Plain. Rev. Palaeobot. Palynol. 160, 172–180.

Davis, O.K., Moutoux, T.E., 1998. Tertiary and Quaternary vegetation history of the Great Salt Lake. J. Paleolimnol. 19, 417–427.

Doner, L.A., 2009. A 19,000-year vegetation and climate record for Bear Lake, Utah and Idaho. In: Rosenbaum, J.G., Kaufman, D.S. (Eds.), Paleoenvironments of Bear Lake, Utah and Idaho, and Its Catchment, pp. 217–227. Geological Society of America Special Papers 450, Boulder.

Eardley, A.J., Gvosdetsky, V., Marsell, R.E., 1957. Hydrology of Lake Bonneville and sediments and soils of its basin. Geol. Soc. Am. Bull. 68, 1141–1201.

Eckerle, W., Brunelle, A., Macharia, A.N., Mayer, J., Petersen, K., Power, M., Rogers, O., Spencer, J., Taddie, S., 2013. Analysis of the GLAD 4 deepwater core from the Great Salt Lake: paleoenvironmental analysis and archaeological implications. In: Ruby Natural Gas Pipeline—Northern Utah, volume 3: Paleoenvironmental Studies, Alpine Archaeological Consultants, Inc., Montrose, CO.

Elder, A.S., 1992. The Paleoecology and Geomorphology of Holocene Deposits of the Southern Malad River, Box Elder County, Utah. MA thesis, Utah State University, Logan, UT.

Godsey, H.S., Currey, D.R., Chan, M.A., 2005. New evidence for an extended occupation of the Provo shoreline and implications for regional climate change, Pleistocene Lake Bonneville, Utah, USA. Quat. Res. 63, 212–223.

Godsey, H.S., Oviatt, C.G., Miller, D.M., Chan, M.A., 2011. Stratigraphy and chronology of offshore to near-shore deposits associated with the Provo shoreline, Pleistocene Lake Bonneville, Utah. Palaeogeogr. Palaeoclimatol. Palaeoecol. 310, 442–450.

Goebel, T., Hockett, B., Adams, K.D., Rhode, D., Graf, K., 2011. Climate, environment, and humans in North America's Great Basin during the Younger Dryas, 12,900-11,600 calendar years ago. Quat. Int. 242, 479–501.

Grayson, D.K., 1998. Moisture history and small mammal community richness during the latest Pleistocene and Holocene, northern Bonneville basin, Utah. Quat. Res. 49, 330–334.

Grayson, D.K., 2000a. The Homestead Cave mammals. In: Madsen, D.B. (Ed.), Late Quaternary Paleoecology in the Bonneville Basin. Utah Geological Survey Bulletin 130, pp. 67–89. Salt Lake City.

Grayson, D.K., 2000b. Mammalian responses to middle Holocene climatic change in the Great Basin of the western United States. J. Biogeogr. 27, 181–192.

Grayson, D.K., 2006. The late Quaternary biogeographic histories of some Great Basin mammals (western USA). Quat. Sci. Rev. 25, 2964–2991.

Grayson, D.K., 2011. The Great Basin: A Natural Prehistory. University of California Press, Berkeley. Revised and expanded edition.

Gugger, P.F., Sugita, S., 2010. Glacial populations and postglacial migration of Douglas-fir based on fossil pollen and macrofossil evidence. Quat. Sci. Rev. 29, 2052–2070.

Hadly, E.A., Ramakrishnan, U., Chan, Y.L., van Tuinen, M., O'Keefe, K., Spaeth, P.A., Conroy, C.J., 2004. Genetic response to climatic change: insights from ancient DNA and phylochronology. PLoS Biol. 2 (10), e290.

Hart, W.S., Quade, J., Madsen, D.B., Kaufman, D.S., Oviatt, C.G., 2004. The $^{87}Sr/^{86}Sr$ ratios of lacustrine carbonates and lake-level history of the Bonneville paleolake system. Geol. Soc. Am. Bull. 16, 1107–1119.

Haynes, C.V., 1991. Geoarchaeological and paleohydrological evidence for Clovis-age drought in North America and its bearing on extinction. Quat. Res. 35, 438–450.

Holmgren, C.A., Betancourt, J.L., Rylander, K.A., 2009. A long-term vegetation history of the Mojave-Colorado Desert ecotone at Joshua Tree National Park. J. Quat. Sci. 25, 222–236.

Hunt, J.M., Rhode, D., Madsen, D.B., 2000. Homestead Cave flora and non-vertebrate fauna. In: Madsen, D.B. (Ed.), Late Quaternary Paleoecology in the Bonneville Basin. Utah Geological Survey Bulletin 130, pp. 47–58. Salt Lake City.

Jahren, A.H., Amundsen, R., Kendall, C., Wigand, P.E., 2001. Paleoclimatic reconstruction using the correlation in $\delta\ ^{18}O$ of hackberry carbonate and environmental water, North America. Quat. Res. 56, 252–263.

Jiménez-Moreno, G., Anderson, R.S., Fawcett, P.J., 2007. Orbital-and millennial-scale vegetation and climate changes of the past 225 ka from Bear Lake, Utah–Idaho (USA). Quat. Sci. Rev. 26, 1713–1724.

Jiménez-Moreno, G., Anderson, R.S., Desprat, S., Grigg, L.D., Grimm, E.C., Heusser, L.E., Jacobs, B.F., López-Martínez, C., Whitlock, C.L., Willard, D.A., 2010. Millennial-scale variability during the last glacial in vegetation records from North America. Quat. Sci. Rev. 29, 2865–2881.

Kelso, G., 1970. Hogup Cave, Utah: comparative pollen analysis of human coprolites and cave fill. In: Aikens, C.M. (Ed.), Hogup Cave, pp. 251–262. University of Utah Anthropological Papers 93, Salt Lake City.

Kiahtipes, C.A., 2009. Fire in the Desert: Holocene Paleoenvironments in the Bonneville Basin. MA thesis, Anthropology, Washington State University, Pullman, WA.

Kowalewska, A., Cohen, A.S., 1998. Reconstruction of paleoenvironments of the Great Salt Lake basin during the late Cenozoic. J. Paleolimnol. 20, 381–407.

Kutzbach, J.E., Guetter, P.J., 1986. The influence of changing orbital parameters and surface boundary conditions on climate simulations for the past 18 000 years. J. Atmos. Sci. 43, 1726–1759.

Lanner, R.M., Hutchison, E.R., Hutchinson, E.R., 1972. Relict stands of pinyon hybrids in northern Utah. Great Basin Nat. 32, 171–175.

Lisiecki, L.E., Raymo, M.E., 2005. A Pliocene-Pleistocene stack of 57 globally distributed benthic $\delta^{18}O$ records. Paleoceanography 20, PA1003.

Louderback, L.A., Rhode, D., 2009. 15,000 years of vegetation change in the Bonneville basin: the Blue Lake pollen record. Quat. Sci. Rev. 28, 308–326.

Lundeen, Z.J., 2012. Paleoecological and Isotopic Records of Climate Change and Variability from Lakes and Speleothems, Bear River Range, Southeastern Idaho. PhD dissertation, Department of Geography, University of Utah, Salt Lake City.

Lundeen, Z., Brunelle, A., Burns, S.J., Polyak, P., Asmerom, Y., 2013. A speleothem record of Holocene paleoclimate from the northern Wasatch Mountains, southeast Idaho, USA. Quat. Int. 310, 83–95.

Madsen, D.B., 1984. Pollen, packrats, and pine along the IPP. In: Simms, S.R., Isgreen, M.C. (Eds.), Archaeological Excavations in the Sevier and Escalante Deserts, Western Utah, pp. 336–353. University of Utah Archaeological Center Reports of Investigations 83-12.

Madsen, D.B., 1985. Correlation between vegetative and lake level changes: the pollen evidence. In: Kay, P.A., Diaz, H.F. (Eds.), Problems of and Prospects for Predicting Great Salt Lake Levels. University of Utah Center for Public Affairs and Administration, Salt Lake City, pp. 54–62.

Madsen, D.B., 2002. Great Basin peoples and late Quaternary aquatic history. In: Hershler, R., Madsen, D.B., Currey, D.R. (Eds.), Great Basin Aquatic Systems History. Smithsonian Contributions to the Earth Sciences, vol. 33. Smithsonian Institution, Washington, DC, pp. 387–405.

Madsen, D.B., Currey, D.M., 1979. Late Quaternary glacial and vegetation changes, Little Cottonwood Canyon area, Wasatch Mountains, Utah. Quat. Res. 12, 254–270.

Madsen, D.B., Rhode, D., 1990. Early Holocene pinyon (*Pinus monophylla*) in the northeastern Great Basin. Quat. Res. 33, 94–101.

Madsen, D.B., Rhode, D., Grayson, D.K., Broughton, J.M., Livingston, S.D., Hunt, J., Quade, J., Schmitt, D.N., Shaver, M.W., 2001. Late Quaternary environmental change in the Bonneville basin, western USA. Palaeogeogr. Palaeoclimatol. Palaeoecol. 167, 243–271.

Madsen, D.B., Oviatt, C.G., Page, D., 2015a. Old River Bed delta geomorphology and chronology. In: Madsen, D.B., Schmitt, D.N., Page, D. (Eds.), The Paleoarchaic Occupation of the Old River Bed Delta, pp. 30–60. University of Utah Anthropological Papers 128, Salt Lake City.

Madsen, D.B., Schmitt, D.N., Page, D. (Eds.), 2015b. The Paleoarchaic Occupation of the Old River Bed Delta. University of Utah Anthropological Papers 128, Salt Lake City.

Madsen, D.B., 2016. Chapter 18. The early human occupation of the Bonneville basin. In: Oviatt, C.G., Shroder Jr., J.F. (Eds.), Lake Bonneville: A Scientific Update, vol. 20. Elsevier, pp. 504–525.

Martin, P.S., Mehringer Jr., P.J., 1965. Pleistocene pollen analysis and biogeography of the Southwest. In: Wright Jr., H.E., Frey, D.G. (Eds.), The Quaternary of the United States. Princeton University Press, Princeton, NJ, pp. 433–451.

McGee, D., Quade, J., Edwards, R.L., Broecker, W.S., Cheng, H., Reiners, P.W., Evenson, N., 2012. Lacustrine cave carbonates: novel archives of paleohydrologic change in the Bonneville basin (Utah, USA). Earth Planet. Sci. Lett. 351–352, 182–194.

Mehringer Jr., P.J., 1977. Great Basin late Quaternary environments and chronology. In: Fowler, D.D. (Ed.), Models and Great Basin Prehistory: A Symposium. Desert Research Institute Publications in Social Sciences, vol. 12. Desert Research Institute, Reno, NV, pp. 113–167.

Mehringer Jr., P.J., 1985. Late-Quaternary pollen records from the interior Pacific Northwest and northern Great Basin of the United States. In: Bryant Jr., V.M., Holloway, R.G. (Eds.), Pollen Records of Late-Quaternary North American Sediments. American Association of Stratigraphic Palynologists Foundation, pp. 167–189.

Mering, J.A., 2015. New Constraints on Water Temperature at Lake Bonneville from Carbonate Clumped Isotopes. MS thesis in Geochemistry, University of California, Los Angeles.

Miller, D.M., Schmidt, K.M., Mahan, S.A., McGeehin, J.P., Owen, L.A., Barron, J.A., Lehmkuhl, F., Löhrer, R., 2010. Holocene landscape response to seasonality of storms in the Mojave Desert. Quat. Int. 215, 45–61.

Miller, D.M., Oviatt, C.G., McGeehin, J.P., 2013. Stratigraphy and chronology of Provo shoreline deposits and lake-level implications, late Pleistocene Lake Bonneville, eastern Great Basin. Boreas 42, 342–361.

Mitton, J.B., Kreiser, B.R., Latta, R.G., 2000. Glacial refugia of limber pine (*Pinus flexilis* James) inferred from the population structure of mitochondrial DNA. Mol. Ecol. 9, 91–97.

Murchison, S.B., 1989a. Fluctuation History of Great Salt Lake, Utah, During the Last 13,000 years. Unpublished PhD dissertation, University of Utah, Salt Lake City.

Murchison, S.B., 1989b. Utah chub (*Gila atraria*) from the latest Pleistocene Gilbert shoreline, west of Corrine, Utah. Great Basin Nat. 49, 131–133.

Nielson, R.P., Wullstein, L.H., 1983. Biogeography of two southwest American oaks in relation to atmospheric dynamics. J. Biogeogr. 10, 275–297.

Nielson, R.P., Wullstein, L.H., 1985. Comparative drought physiology and biogeography of *Quercus gambelii* and *Quercus turbinella*. Am. Midl. Nat. 114, 259–271.

Nishizawa, S., Currrey, D.R., Brunelle, A., Sack, D., 2012. Bonneville basin shoreline records of a large lake during Marine Isotope Stage 16. Quat. Sci. Rev. 58, 1–6.

Nishizawa, S., Currey, D.R., Brunelle, A., Sack, D., 2013. Bonneville basin shoreline records of large lake intervals during Marine Isotope Stage 3 and the last glacial maximum. Palaeogeogr. Palaeoclimatol. Palaeoecol. 386, 374–391.

Oviatt, C.G., 1997. Lake Bonneville fluctuations and global climate change. Geology 25, 155–158.

Oviatt, C.G., 2014. The Gilbert Episode in the Great Salt Lake Basin, Utah. Miscellaneous Publication 14-3, Utah Geological Survey, Salt Lake City.

Oviatt, C.G., 2015. Chronology of Lake Bonneville, 30,000 to 10,000 yr B.P. Quat. Sci. Rev. 110, 166–171.

Oviatt, C.G., Currey, D.R., 1987. Pre-Bonneville Quaternary lakes in the Bonneville basin, Utah. In: Kopp, R.S., Cohenour, R.E. (Eds.), Cenozoic Geology of Western Utah: Sites for Precious Metal and Hydrocarbon Accumulations. Utah Geological Association Publication 16, Salt Lake City, pp. 257–264.

Oviatt, C.G., McCoy, W.D., Reider, R.G., 1987. Evidence for a shallow early or middle Wisconsin-age lake in the Bonneville basin, Utah. Quat. Res. 27, 248–262.

Oviatt, C.G., Currey, D.R., Sack, D., 1992. Radiocarbon chronology of Lake Bonneville, eastern Great Basin, USA. Paleogeogr. Paleoclimatol. Paleoecol. 99, 225–241.
Oviatt, C.G., Thompson, R.S., Kaufman, D.S., Bright, J., Forester, R.M., 1999. Reinterpretation of the Burmester core, Bonneville basin, Utah. Quat. Res. 52, 180–184.
Oviatt, C.G., Madsen, D.B., Schmitt, D.N., 2003. Late Pleistocene and early Holocene rivers and wetlands in the Bonneville basin of western North America. Quat. Res. 60, 200–210.
Oviatt, C.G., Miller, D.M., McGeehin, J.P., Zachary, C., Mahan, S., 2005. The Younger Dryas phase of Great Salt Lake, Utah, USA. Palaeogeogr. Palaeoclimatol. Palaeoecol. 219, 263–284.
Oviatt, C.G., Madsen, D.B., Miller, D.M., Thompson, R.S., McGeehin, J.P., 2015. Early Holocene Great Salt Lake, USA. Quat. Res. 84, 57–68.
Patrickson, S., Sack, D., Brunelle, A., Moser, K., 2010. Late Pleistocene to early Holocene lake level and paleoclimate insights from Stansbury Island, Bonneville basin, Utah. Quat. Res. 73, 237–246.
Polyak, V.J., Rasmussen, J.B., Asmerom, Y., 2004. Prolonged wet period in the southwestern United States through the Younger Dryas. Geology 32, 5–8.
Polyak, V.J., Asmerom, Y., Burns, S.J., Lachniet, M.S., 2012. Climatic backdrop to the terminal Pleistocene extinction of North American mammals. Geology 40, 1023–1026.
Rasmussen, S.O., Andersen, K.K., Svensson, A.M., Steffensen, J.P., Vinther, B.M., Calusen, H.B., Siggaard-Andersen, M.-L., Johnsen, S.J., Larsen, L.B., Dahl-Jensen, D., Bigler, M., Röthlisberger, R., Fischer, H., Goto-Azuma, K., Hansson, M.E., Ruth, U., 2006. A new Greenland ice core chronology for the last glacial termination. J. Geophys. Res. 111, D06102.
Reheis, M.C., Adams, K.D., Oviatt, C.G., Bacon, S.N., 2014. Pluvial lakes in the Great Basin of the western United States—a view from the outcrop. Quat. Sci. Rev. 97, 33–57.
Rhode, D., 2000a. Middle and late Wisconsin vegetation in the Bonneville basin. In: Madsen, D.B. (Ed.), Late Quaternary Paleoecology in the Bonneville Basin, Utah Geological Survey Bulletin, vol. 130, pp. 137–148. Salt Lake City.
Rhode, D., 2000b. Holocene vegetation history in the Bonneville basin. In: Madsen, D.B. (Ed.), Late Quaternary Paleoecology in the Bonneville Basin. Utah Geological Survey Bulletin. vol. 130, pp. 149–164. Salt Lake City.
Rhode, D., 2008. Dietary plant use by middle Holocene foragers in the Bonneville basin, western North America. Before Farming. [online version] 2008/3 article 2.
Rhode, D., 2013. 15,000 years of past environments along the Ruby Pipeline Route, Northern Utah. In: Ruby Natural Gas Pipeline—Northern Utah, Volume 3: Paleoenvironmental Studies, Alpine Archaeological Consultants, Inc., Montrose, CO, pp. 11–24.
Rhode, D., 2014. Woodrat Midden Paleovegetation Records from Around the Ruby Pipeline Route in Northern Nevada. Report prepared for Far Western Anthropological Research Group, Davis, CA.
Rhode, D., Louderback, L.A., 2013. Holocene vegetation history of the northern Wasatch Range: a pollen record from Silver Lake, Big Cottonwood Canyon, Salt Lake County, Utah. In: Rhode, D. (Ed.), 15,000 Years of Past Environments Along the Ruby Pipeline Route, Northern Utah. Ruby Natural Gas Pipeline—Northern Utah, Volume 3: Paleoenvironmental StudiesAlpine Archaeological Consultants, Inc., Montrose, CO, pp. 11–24
Rhode, D., Louderback, L.A., 2015. Bonneville basin paleoenvironments during the Pleistocene-Holocene transition. In: Madsen, D.B., Schmitt, D.N., Page, D. (Eds.), The Paleoarchaic Occupation of the Old River Bed Delta, pp. 22–29. University of Utah Anthropological Papers 128, Salt Lake City.
Rhode, D., Madsen, D.B., 1995. Early Holocene vegetation in the Bonneville basin. Quat. Res. 44, 246–256.

Rhode, D., Madsen, D.B., 1998. Pine nut use in the early Holocene and beyond: the Danger Cave archaeobotanical record. J. Archaeol. Sci. 25, 1199–1210.

Rhode, D., Goebel, T.E., Graf, K.E., Hockett, B.S., Jones, K.T., Madsen, D.B., Oviatt, C.G., Schmitt, D.N., 2005. Latest Pleistocene-early Holocene human occupation and paleoenvironmental change in the Bonneville basin, Utah-Nevada. In: Pederson, J., Dehler, C.M. (Eds.), Interior Western United States. Geological Society of America, Boulder, CO, pp. 211–230. Field Guide 6.

Rhode, D., Madsen, D.B., Jones, K.T., 2006. Antiquity of early Holocene small seed consumption and processing at Danger Cave, Utah, USA. Antiquity 80, 328–329.

Schmitt, Dave N., Lupo, K.D., 2012. The Bonneville Estates Rockshelter rodent fauna and changes in late Pleistocene-middle Holocene climates and biogeography in the northern Bonneville basin, USA. Quat. Res. 78, 95–102.

Schmitt, D.N., Madsen, D.B., 2005. Camels Back Cave. Anthropological Papers 125, University of Utah, Salt Lake City.

Schmitt, D.N., Madsen, D.B., Lupo, K.D., 2002. Small-mammal data on early and middle Holocene climates and biotic communities in the Bonneville basin, USA. Quat. Res. 58, 225–260.

Scott, W.E., McCoy, W.D., Shroba, R.R., Rubin, M., 1983. Reinterpretation of the exposed record of the last two cycles of Lake Bonneville, western United States. Quat. Res. 20, 261–285.

Shackleton, N., Fairbanks, R., Chiu, T.-C., Parrenin, F., 2004. Absolute calibration of the Greenland time scale: implications for Antarctic time scales and $\Delta^{14}C$. Quat. Sci. Rev. 23, 1513–1522.

Shakun, J.D., Carlson, A.E., 2010. A global perspective of last glacial maximum to Holocene climate change. Quat. Sci. Rev. 29, 1801–1816.

Shakun, J.D., Clark, P.U., He, F., Marcott, S.A., Mix, A.C., Liu, Z., Otto-Bliesner, B., Schmittner, A., Bard, E., 2012. Global warming preceded by increasing carbon dioxide concentrations during the last deglaciation. Nature 484, 49–55.

Smoot, J.P., Rosenbaum, J.G., 2009. Sedimentary constraints on late Quaternary lake-level fluctuations at Bear Lake, Utah and Idaho. In: Rosenbaum, J.G., Kaufman, D.S. (Eds.), Paleoenvironments of Bear Lake, Utah and Idaho, and its Catchment, pp. 263–290. Geological Society of America Special Paper 450, Boulder, CO.

Spencer, R.J., Baedecker, M.J., Eugster, H.P., Forester, R.M., Goldhaber, M.B., Jones, B.F., Kelts, K., McKenzie, J., Madsen, D.B., Rettig, S.L., Rubin, M., Bowser, C.J., 1984. Great Salt Lake and precursors, Utah: the last 30,000 years. Contrib. Mineral. Petrol. 86, 321–334.

Steffensen, J.P., Andersen, K.K., Bigler, M., Clausen, H.B., Dahl-Jenzen, D., Fischer, H., Goto-Azuma, K., Hansson, M., Johnsen, S.J., Jouzel, J., Masson-Delmotte, V., Popp, T., Rasmussen, S.O., Rothlisberger, R., Ruth, U., Stauffer, B., Siggaard-Andersen, M.-L., Sveinbjörnsdóttir, A.E., Svensson, A., White, J.W.C., 2008. High-resolution Greenland ice core data show abrupt climate change happens in few years. Science 312, 680–684.

Steponaitis, E., Andrews, A., McGee, D., Quade, J., Hsieh, Y.-T., Broecker, W.S., Shuman, B.N., Burns, S.J., Cheng, H., 2015. Mid-Holocene drying of the U.S. Great Basin recorded in Nevada speleothems. Quat. Sci. Rev. 127, 174–185.

Thompson, R.S., 1984. Late Pleistocene and Holocene Environments in the Great Basin. Unpublished PhD dissertation, University of Arizona, Tucson.

Thompson, R.S., 1990. Late Quaternary vegetation and climate in the Great Basin. In: Betancourt, J.L., van Devender, T.R., Martin, P.S. (Eds.), Packrat Middens—The Last 40,000 Years of Biotic Change. University of Arizona Press, Tucson, pp. 201–239.

Thompson, R.S., 1991. Pliocene environments and climates in the western United States. Quat. Sci. Rev. 10, 115–132.

Thompson, R.S., 1992. Late Quaternary environments in Ruby Valley, Nevada. Quat. Res. 37, 1–15.

Thompson, R.S., Anderson, K.H., 2000. Biomes of western North America at 18,000, 6000 and 0 ^{14}C yr bp reconstructed from pollen and packrat midden data. J. Biogeogr. 27, 555–584.

Thompson, R.S., Toolin, L.J., Forester, R.M., Spencer, R.J., 1990. Accelerator-mass spectrometer (AMS) radiocarbon dating of Pleistocene lake sediments in the Great Basin. Palaeogeogr. Palaeoclimatol. Palaeoecol. 78, 301–313.

Thompson, R.S., Whitlock, C., Bartlein, P.J., Harrison, S.P., Spaulding, W.G., 1993. Climatic changes in the western United States since 18,000 yr B.P.. In: Wright Jr., H.E., Kutzbach, J.E., Webb III, T., Ruddiman, W.F., Street-Perrott, F.A., Bartlein, P.J. (Eds.), Global Climates Since the Last Glacial Maximum. University of Minnesota Press, Minneapolis, pp. 468–513.

Thompson, R.S., Oviatt, C.G., Roberts, A.P., Buchner, J., Kelsy, R., Bracht, C., Forester, R.M., Bradbury, J.P., 1995. Stratigraphy, sedimentology, paleontology, and paleomagnetism of Pliocene-early Pleistocene lacustrine deposits in two cores from western Utah: U.S. Geological Survey open-file report 95-1.

Thompson, R.S., Oviatt, C.G., Honke, J.S., McGeehin, J.P., 2016. Chapter 11. Late Quaternary changes in lakes, vegetation, and climate in the Bonneville basin reconstructed from sediment cores from Great Salt Lake. In: Oviatt, C.G., Shroder Jr., J.F. (Eds.), Lake Bonneville: A Scientific Update, vol. 20. Elsevier, pp. 221–291.

Voelker, A., et al., 2002. Global distribution of centennial-scale records for Marine Isotope Stage (MIS) 3: a database. Quat. Sci. Rev. 21, 1185–1212.

Wells, P.V., 1976. Macrofossil analysis of wood rat (*Neotoma*) middens as a key to Quaternary vegetation history of arid America. Quat. Res. 6, 223–248.

Wells, P.V., 1983. Paleogeography of montane islands in the Great Basin since the last glaciopluvial. Ecol. Monogr. 53, 341–382.

Weppner, K.N., Pierce, J.L., Betancourt, J.L., 2013. Holocene fire occurrence and alluvial responses at the leading edge of pinyon-juniper migration in the northern Great Basin, USA. Quat. Res. 80, 143–157.

Wigand, P.E., Rhode, D., 2002. Great Basin vegetation history and aquatic systems: the last 150,000 years. In: Hershler, R., Madsen, D.B., Currey, D.R. (Eds.), Great Basin Aquatic Systems History. Smithsonian Contributions to the Earth Sciences, vol. 33. Smithsonian Institution, Washington, DC, pp. 309–368.

Wilson, J.S., Pitts, J.P., 2012. Identifying Pleistocene refugia in North American cold deserts using phylogeographic analyses and ecological niche modelling. Divers. Distrib. 18, 1139–1152.

Chapter 16

Water Chemistry Changes Over Time and Space in Lake Bonneville During the Post-Stansbury Transgression

V.A. Pedone* and C.G. Oviatt†
*California State University, Northridge, CA, United States
†Kansas State University, Manhattan, KS, United States

ABSTRACT
Carbonate abundance, mineralogy, and carbon-, oxygen-, and strontium-isotope compositions of marl deposited between ~24 and 18 cal ka differ significantly between locations in the Bonneville basin and over time at the same locations. The geochemistry of the lake in the Sevier subbasin evolved along a different pathway than the lake in the main body. Furthermore, variations in geochemistry within the main body show that shallow and deep-water masses developed and were not well mixed throughout the large basin. Advances in the understanding of the geochemistry of Lake Bonneville since the days of Gilbert have greatly increased our awareness of the hydrological complexity of this huge lake system.

Keywords: Lake Bonneville transgression, Total inorganic carbon, Carbonate mineralogy, Carbon and oxygen isotopes, Strontium isotopes

16.1 INTRODUCTION

The rise and fall of huge pluvial lakes in the western United States have been subjects of intense fascination and scientific study for more than a century (Russell, 1885; Gilbert, 1890). Present-day interest is largely motivated by determining how pluvial-lake history can contribute to understanding late Quaternary climate change, and many researchers seek to refine the geochronology and hydrologic history to improve correlation with the timing and magnitude of global climate change.

In this chapter, we examine the carbonate abundance; mineralogy; and carbon-, oxygen-, and strontium-isotope compositions of marl deposited

Developments in Earth Surface Processes, Vol. 20. http://dx.doi.org/10.1016/B978-0-444-63590-7.00016-0

during the transgressive phase of Lake Bonneville. These parameters are important tools in reconstructing spatial and temporal variability in lake size, water sources, and water mixing. The chapter serves as an example of how knowledge of the geochemistry of Lake Bonneville has grown since the days of Gilbert. The examples presented in this chapter help to illustrate that as large as Lake Bonneville was, it was not an ocean, and that chemical, physical, and biological factors were not uniform everywhere in the lake as the lake grew.

Lake Bonneville occupied a large area in what is now western Utah, eastern Nevada, and southern Idaho (Fig. 16.1). Lake history is divided into three phases: the transgressive, overflowing, and regressive phases (Fig. 16.2). In this chapter, the focus is on the post-Stansbury part of the transgressive phase, which occurred between about 24 and 18 cal ka, when the rising lake was between the Old River Bed threshold and the Bonneville shoreline. The lower chronological limit of 24 cal ka marks the time when the transgressing lake began to rise above the floor of the Sevier subbasin to form an integrated lake system. The upper chronological limit at 18 cal ka marks the Bonneville flood, which lowered the lake by about 125 m in less than a year (Oviatt, 2015a) and caused disruption of lake-bottom sediments and major changes in surface and bottom currents (Oviatt et al., 1994). Both time markers can be identified in cores and outcrops.

During the entire transgressive phase, Lake Bonneville occupied a hydrographically closed basin. Lake level generally rose during this time (Fig. 16.2), but it did undergo many oscillations and fluctuations, some of which have been identified in the stratigraphic record (Oviatt, 1997; Nelson and Jewell, 2015). The lake's maximum altitude (the Bonneville shoreline) was determined by the altitude of the saddle on the Marsh Creek alluvial fan at the Zenda threshold several km north of Red Rock Pass (Fig. 16.1), across which the lake overflowed to the Snake River drainage basin and initiated the Bonneville flood (Oviatt, 2015a; Shroder et al., 2016; Oviatt and Jewell, 2016).

16.2 MARL SEDIMENT

Carbonate material formed in Lake Bonneville consists of:

(1) shoreline tufas and carbonate-cemented beach gravels
(2) shoreline gastropods associated with shoreline sands and tufas
(3) endogenic carbonate minerals, primary precipitates formed in the surface waters that settle to the bottom in quiet, offshore areas
(4) benthic ostracods

Tufa and cemented beach gravels were best developed during periods of relatively stable lake levels at the Stansbury and Provo shorelines. Gastropod-rich

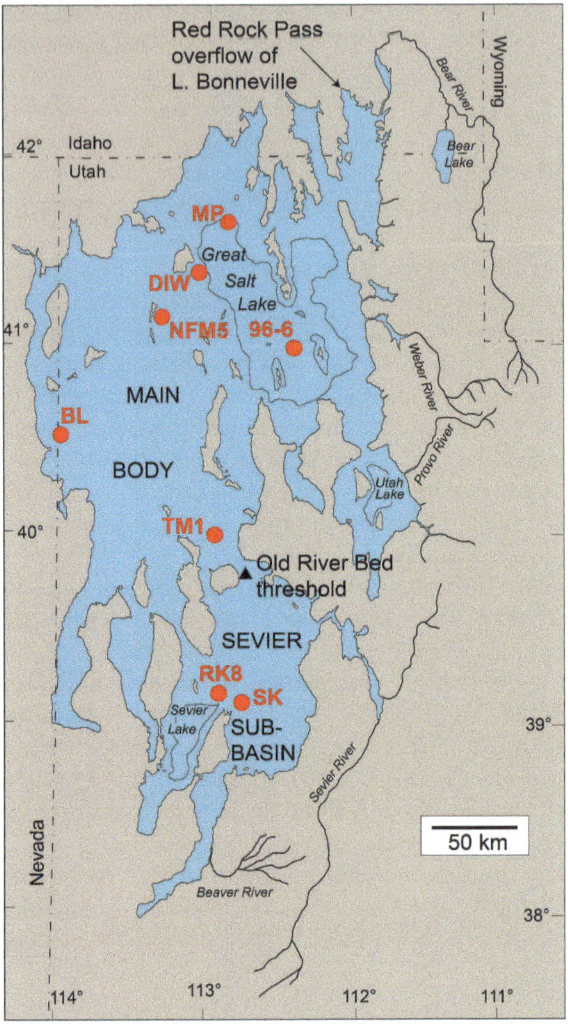

FIG. 16.1 Location of cores described in this paper. The Old River Bed threshold separates the main body from the Sevier subbasin. *Base map modified from Oviatt, C.G., 2015, Chronology of Lake Bonneville, 30,000 to 10,000 yr B.P. Quaternary Science Reviews 110, 166–171.*

sands occur at all altitudes above the Stansbury shoreline. Hence, mineralogical and geochemical studies of all these materials provide information about lake conditions at specific locations and points in time. These nearshore carbonates are also sources for ^{14}C ages (Oviatt, 2015a). In contrast, endogenic carbonate minerals formed continuously in surfaces waters and created marl deposits during all phases of Lake Bonneville history. Although shoreline deposits have the advantage of providing information on specific times and

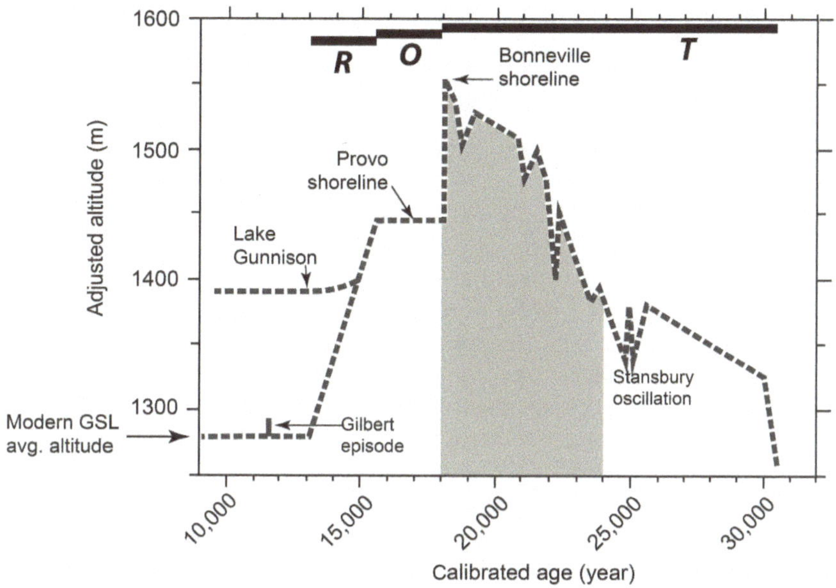

FIG. 16.2 Chronology of Lake Bonneville. The post-Stansbury transgression is shaded and is the focus of this study. *T*, transgressive phase; *O*, overflowing phase; *R*, regressive phase. *Modified from Oviatt, C.G., 2015, Chronology of Lake Bonneville, 30,000 to 10,000 yr B.P. Quaternary Science Reviews 110, 166–171.*

places, marl from cores can provide a higher resolution record of secular changes in lake geochemistry.

Analogs to endogenic carbonate formation in Lake Bonneville have been documented in a number of fresh- and brackish-water lakes (Kelts and Hsü, 1978 and references therein). The two most common triggers that lead to $CaCO_3$ supersaturation and precipitation are: (1) dissolved CO_2 removal by photosynthetic bacteria and plants in the near-surface photic zone; and (2) increase in temperature, which causes degassing of CO_2 from surface waters. Some lakes also experience episodic "whitings," in which an area of the lake forms a significant amount of endogenic carbonate over a few weeks and turns a milky color (Kelts and Hsü, 1978; Strong and Eadie, 1978; Galat and Jacobsen, 1985). Endogenic carbonate crystals range in size from 1 to 10 μm; exhibit subhedral to euhedral crystal forms; and include calcite, Mg-calcite, and aragonite (Kelts and Hsü, 1978; Galat and Jacobsen, 1985). These crystals slowly settle to the bottom with fine-grained silicate minerals to form marl. The calcite carapaces of benthic ostracods that live in the bottom muds are incorporated into the marl and are part of this sediment record. Bonneville marl samples contain endogenic aragonite and/or calcite that precipitated in the upper-water layer and ostracod carapaces that formed in the bottom-water layer.

16.3 RECONSTRUCTION OF THE HYDROGEOCHEMISTRY OF A CLOSED LAKE SYSTEM: TOOLS AND PROXIES

16.3.1 Carbonate Percentages and Mineralogy

From the time of Gilbert (1890) to the present, Lake Bonneville researchers have used carbonate content and mineralogy as general proxies for lake volume, particularly when the lake was a hydrologically closed system. Changes in the percentage of total inorganic carbon (%TIC), a measure of abundance of carbonate minerals, are interpreted as inversely proportional to lake volume, with less carbonate with increasing lake volume (eg, Haberzettl et al., 2005; Benson et al., 2011). Changes in carbonate mineralogy are also interpreted to be a function of lake volume. The sequence calcite to Mg-calcite to aragonite is interpreted to reflect increasing salinity, Mg^+/Ca^+, and HCO_3^- supply in the water brought about by decrease in lake volume through evaporative output (Kelts and Hsü, 1978; Given and Wilkinson, 1985; Oviatt et al., 1994; Benson et al., 2011). A return to calcite is interpreted as lake freshening and increase in volume owing to freshwater input. These generalities in a closed lake system assume that lake chemistry is a function of the balance of inflow, evaporation, and mineral precipitation and that the lake water is well mixed. In a large and complex lake system like Bonneville, geographic differences in parameters could result in heterogeneity in $CaCO_3$ precipitation in different areas of the lake at any given time.

16.3.2 Carbon and Oxygen Isotopes

C and O isotopes have been measured in Lake Bonneville marl since the 1980s (Spencer et al, 1984; Oviatt et al., 1994; Oviatt, 1997; Benson et al., 2011). Factors that affect the $\delta^{13}C$ and $\delta^{18}O$ values in lake water and how they are recorded in carbonate precipitates are complex and dependent on water budget, water sources, vertical and lateral water circulation, biologic productivity, reservoir effect over time, temperature, and abiotic and biotic kinetic effects (Kelts and Hsü, 1978; McKenzie, 1985; Gasse et al., 1987; Kelts and Talbot, 1989; Talbot, 1990; Benson, 1994; Benson et al., 1996, 2011; Li and Ku, 1997; Nelson et al., 2005). In hydrologically closed basins, $\delta^{18}O$ and $\delta^{13}C$ typically show positive covariance (Talbot, 1990 and references therein). Given all of the factors that influence C and O isotope measurements of lake carbonates over the large geographic area of Lake Bonneville, similarities in trends rather than correlation of specific values are likely the most meaningful.

16.3.3 Strontium Isotope Ratios

The $^{87}Sr/^{86}Sr$ ratio of the lake water represents the sum total of the $^{87}Sr/^{86}Sr$ of all input waters (streams, deep groundwater, and shallow groundwater), weighted by their Sr abundances. Sr readily substitutes for Ca in carbonate minerals. Because of the small mass difference between ^{87}Sr and ^{86}Sr, Sr isotopes are not fractionated in liquid–solid reactions or in biologically mediated mineral formation. Hence, the Sr-isotope composition is independent of mineralogy, temperature, and "vital effects," and the $^{87}Sr/^{86}Sr$ ratio of the carbonate is the same as that of the water from which it formed. Secular variation in $^{87}Sr/^{86}Sr$ in lake marl carbonate, therefore, can provide important information about water sources and lateral and vertical mixing of water masses (Pedone et al., 2000; Hart et al., 2004; Pedone and Oviatt, 2013).

16.4 METHODS

Eight cores were used for this study, two from the Sevier subbasin and six from the main body (Fig. 16.1). All were cores of outcrop exposures, except for GSL96-6, which was obtained from the south arm of the Great Salt Lake using a small raft as a coring platform (Thompson et al., 2016). Coring procedures for the Blue Lake core are described in Benson et al. (2011); reinterpretation of the position of the Bonneville flood in this core is given in Reheis et al. (2014). Table 16.1 shows the collector, sample interval, analyses performed on each core, and precision of laboratory analyses. Radiocarbon ages are calibrated using CALIB 7.0.4 (Stuiver and Reimer, 1993; Reimer et al., 2013). The time period of the post-Stansbury transgression is present in all cores. The lower limit of this time period (~24 cal ka) is identified in the cores based on the presence of the Pony Express ash (SK, TM1; Oviatt and Nash, 2014), radiocarbon ages (RK8, GSL96-6), paleomagnetic secular variation ages (BL 04-4), or changes in ostracod faunas (NFM5, DIW, MP; Oviatt, 2015b). The Bonneville flood (~18 cal ka; Oviatt et al., 1994; Oviatt, 2015a) is identified in cores and outcrops as an abrupt change in the marl, including bedding (from massive to laminated), grain size (from very fine to coarser grained, eg, clay to fine sand), %TIC, aragonite/calcite ratio, oxygen and carbon isotopes, and strontium-isotopic ratios.

In order to identify changes in lake chemistry, it is essential that analyses measure only carbonate minerals formed in the lake during its history (endogenic $CaCO_3$) and not detrital minerals from external sources or authigenic minerals formed later in the diagenetic history. Grain morphology and mineralogy were used to identify the origin of carbonate minerals in the marl. Point counts of 30 crystals per sample were made in the <20-μm fraction in 10 samples from NFM5 and 13 samples from RK8 using 1000× magnification on a scanning-electron microscope. Energy-dispersive X-ray analysis was used to distinguish calcite from dolomite and to identify silicate minerals. Detrital calcite and dolomite have irregular shapes with ragged margins and formed 8% and 1% of the 491 grains counted, respectively. Authigenic dolomite formed smooth-faced rhombs and formed 12% of grains counted. Samples that contain authigenic dolomite have the most "detrital" calcite, suggesting the lack of smooth faces and regular shape of the calcite might be the result

TABLE 16.1 Analytical Work Performed on Each Core Used by This Study

Core	Location	Collector and Data Analysis	%TIC	*A/C* Ratio	$\delta^{13}C$ and $\delta^{18}O$	$^{87}Sr/^{86}Sr$ Endogenic $CaCO_3$	$^{87}Sr/^{86}Sr$ Ostracodes
SK	Sevier basin	C. Oviatt	Sample thickness: 2 cm; sample interval: contiguous	Sample thickness: 2 cm; sample interval: contiguous			
RK8	Sevier basin				Sample thickness: 2 cm; sample interval: contiguous to 8 cm; 1-sigma precision ±0.1‰	Sample thickness: 2 cm; sample interval: variable; 2-sigma precision ±0.00001	Sample thickness: 2 cm; sample interval: variable; 2-sigma precision ±0.00001
TM1	Main body, south				Sample thickness: 2 cm; sample interval: contiguous; 1-sigma precision ±0.1‰		
NFM5	Main body, west central				Sample thickness: 2 cm; sample interval: variable; 1-sigma precision ±0.1‰		
DIW	Main body, north central						
MP	Main body, north						

GSL96-6	Main body, east central	R. Thompson, USGS, collector; C. Oviatt, analysis	Sample thickness: 0.5 cm; sample interval: contiguous	Sample thickness: 0.5 cm; sample interval: contiguous	Sample thickness: 0.5 cm; sample interval: contiguous to 8 cm. One-sigma precision: ±0.1‰	Sample thickness: 0.5 cm; sample interval; variable; 2-sigma precision ±0.00001	
BL04-4	Main body, western margin	Benson et al. (2011)	Sample thickness: 1 cm; sample interval: contiguous; 1-sigma precision ±0.1%	Sample thickness: 1 cm; sample interval: 5 cm	Sample thickness: 1 cm; sample interval: contiguous; 1-sigma precision ±0.04‰		

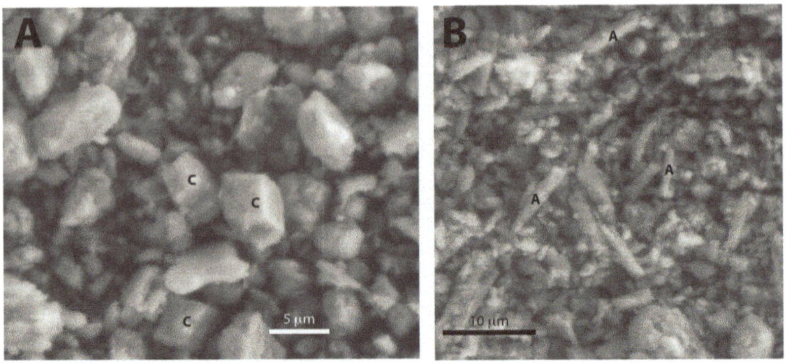

FIG. 16.3 (a) Scanning-electron microscope image of endogenic calcite in NMF5, identified by the rhombic crystal form with relatively smooth faces (labeled C in the image). (b) Scanning-electron microscope image of numerous endogenic aragonite crystals in RK8, identified by the square-ended needle crystal forms (labeled A in the image).

of partial dissolution during diagenesis. Smooth-faced rhombs of endogenic calcite and square-ended needles of endogenic aragonite formed the majority of grains counted (Fig. 16.3A and B). Aragonite can only be endogenic because this metastable mineral is absent in Paleozoic rocks that surround portions of the Bonneville basin. The point counts demonstrate that detrital carbonates will not influence geochemical analyses and that care must be taken not to dissolve dolomite during sample preparation. For example, for $^{87}Sr/^{86}Sr$ analyses, $CaCO_3$ in the samples was dissolved with 4% acetic acid so as not to dissolve dolomite (8% is typically used to dissolve dolomite).

16.5 RESULTS FOR THE POST-STANSBURY TRANSGRESSION

16.5.1 TIC and Mineralogy

The %TIC values in both cores from the Sevier subbasin, SK and RK8, during the post-Stansbury transgression are similar (6–8%) and are significantly higher than those in cores from the main body (Fig. 16.4). In contrast, secular variations in %TIC in the six cores in the main body are not similar. The average %TIC increases upward in GSL96-6 (2 to 2.5%) DIW (2 to 2.5%), and MP (2.5 to 3.5%), but decreases upward in TM1 (4 to 2.5%), BL04-4 (6 to 3%), and NFM5 (2.5 to 1.5%).

The post-Stansbury transgression in both cores from the Sevier subbasin, SK and RK8, are dominated throughout by aragonite (Fig. 16.4). However in the main body, the occurrence of aragonite in this interval is variable and differs between core locations. It is completely absent in MP, absent in upper 60 cm in DIW, absent in the upper 40 cm in NFM5, and absent in the upper 64 cm in BL04-4. Aragonite occurs sporadically throughout the post-Stansbury transgression in GSL96-6 in frequent small "spikes." In TM1,

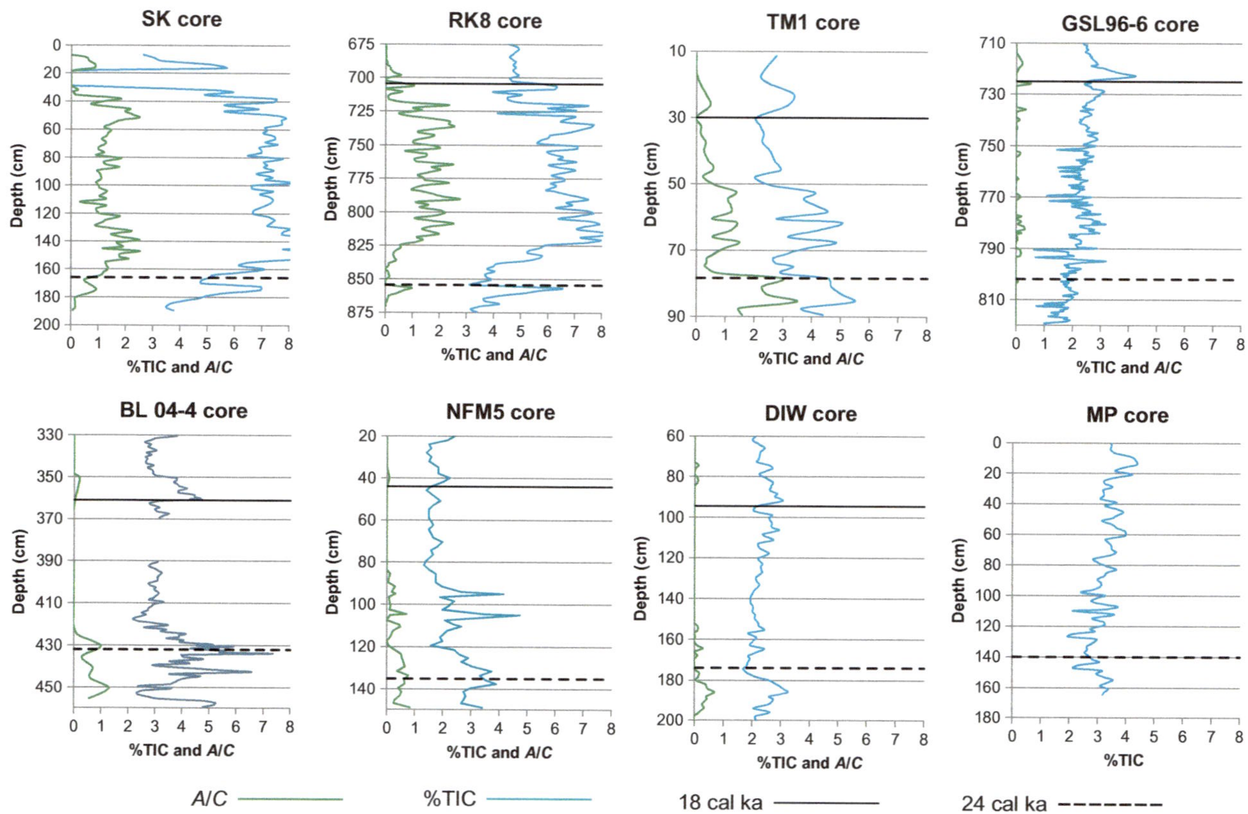

FIG. 16.4 The %TIC and *A/C* ratios in the eight cores shown in Fig. 16.1. Data are plotted in terms of depth because high-resolution age control in the cores is not available. Sedimentation rate is not uniform in these different areas of Lake Bonneville so depth intervals do not equate to equivalent time in the cores. The *bold black line* represents the Bonneville flood (~18 cal ka) in each core, and the *dashed bold black line* represents the approximate location of 24 cal ka in each core. The Bonneville flood is above the top of the SK and MP cores.

aragonite is present throughout the core, but decreases upwards in abundance. It should be noted that the short sampling interval in GSL96-6 compared to those in the other cores (Table 16.1) might be a factor in identifying aragonite because the peak height of the maximum intensity for aragonite is much lower than that of calcite, and aragonite could be lost in a large sampling interval over which small amounts of aragonite are present.

16.5.2 $\delta^{18}O$ and $\delta^{13}C$

The $\delta^{13}C$ and $\delta^{18}O$ values show strong positive covariance in RK8 ($R^2=0.85$), TM1 ($R^2=0.97$), BL04-4 ($R^2=0.89$), and GSL96-6 ($R^2=0.80$), as predicted in a hydrologically closed basin (Fig. 16.5). Data are sparse for NFM5, but they also show positive covariance.

In RK8 at 854 cm, the $\delta^{18}O$ and $\delta^{13}C$ values increase from −6‰ to 1.6‰, respectively, to relatively constant values of −5‰ and 4‰ at ~820 cm. This depth is 5 cm above the abrupt increase in %TIC and change from calcite to aragonite (Fig. 16.5). The $\delta^{18}O$ value of aragonite is ~1‰ higher than that of calcite formed at the same temperature from the same water (Swart, 2015), so the change in mineralogy can account for the increase in $\delta^{18}O$.

In TM, $\delta^{18}O$ and $\delta^{13}C$ increase upward. Between 78 and 53 cm, $\delta^{18}O$ values vary between −5.6‰ and −8.5‰, with a mean of −7.0‰. In the same interval, $\delta^{13}C$ values vary from 1.5‰ to 3.8‰, with a mean of 2.8‰. At 53 cm, the $\delta^{18}O$ and $\delta^{13}C$ values abruptly decrease. From this horizon to the Bonneville Flood horizon at 30 cm, the $\delta^{18}O$ varies between −7.0‰ and −9.6‰, mean of −8.8‰; and the $\delta^{13}C$ values vary between 0.3‰ and 1.9‰, mean of 1.1‰.

In BL04-4 between 432 and 406 cm, $\delta^{18}O$ and $\delta^{13}C$ values decrease from −6‰ to -9.8‰ and 3.3‰ to −0.4‰, respectively, and then are relatively constant at −9‰ and 0‰ up to 390 cm, where core was not recovered. Where core recovery resumed at 370 cm up to the Bonneville Flood horizon at 361 cm, $\delta^{18}O$ value was −7.8‰ and $\delta^{13}C$ value was 1‰.

In GSL96-6, $\delta^{18}O$ values in the post-Stansbury transgression are variable between −9.4‰ and −6.3‰ and show a slight overall decrease from average values of −7‰ to average values of −8.5‰. The $\delta^{13}C$ values in this interval vary between -1‰ and 2.3‰ and show no trends in values. Both isotopes show higher variability in the lower part of the transgression than in the upper part.

Overall in the post-Stansbury transgression interval, the $\delta^{18}O$ and $\delta^{13}C$ values in RK8 from the Sevier subbasin are higher by several permil and are less variable than those in the main body (Fig. 16.5). The shapes of the $\delta^{18}O$ and $\delta^{13}C$ curves in RK8 are distinctly different than those in the main body. The range in $\delta^{18}O$ values in the main body is similar in TM1, BL04-4, and GSL96-6 (~5.5‰ to −9.8‰). The range in $\delta^{13}C$ values is similar in BL04-4 and GSL96-6 (~−1‰ to 3‰), but is slightly higher in TM1 (0.5–4.5‰).

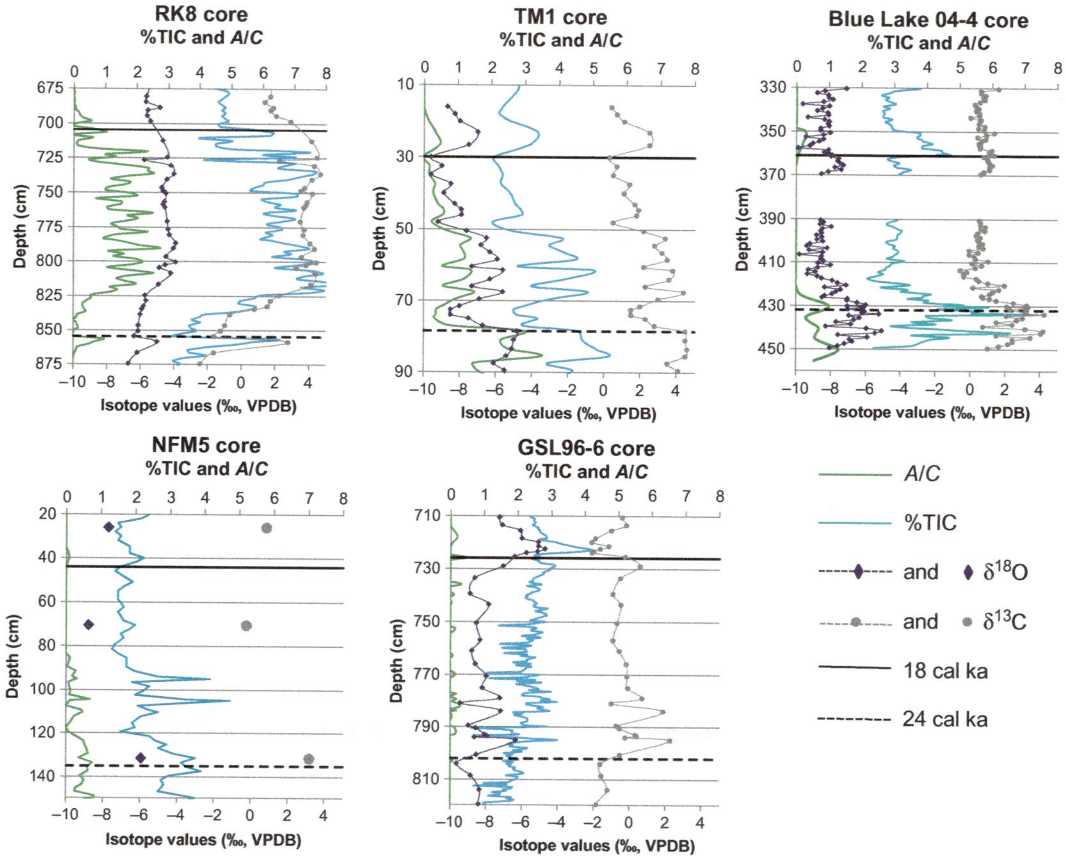

FIG. 16.5 The %TIC, *A*/*C* ratios, and $\delta^{18}O$ and $\delta^{13}C$ values in five cores. See Fig. 16.1 for locations of cores. The $\delta^{18}O$ in BL04-4, reported by Benson et al. (2011) relative to the VSMOW standard, was converted relative to the VPDB standard using the formula $\delta^{18}O_{VPDB} = (\delta^{18}O_{VSMOW} - 30.91)/1.03091$.

16.5.3 Sr Isotopes

Sr-isotope ratios were measured in endogenic carbonates in RK8, TM1, NFM5, and GSL96-6 (Table 16.1). Paired analyses of endogenic carbonate and ostracodes from the same sample were made wherever possible in RK8, TM1, and NFM5. Ostracods were not available from the samples obtained from the GSL96-6 core. The $^{87}Sr/^{86}Sr$ ratios in the marls between 24 and 18 ka show large changes over time, 0.71033–0.71185 in the main body and 0.71062–0.71156 in the Sevier subbasin (Fig. 16.6). Shifts in $^{87}Sr/^{86}Sr$ ratios

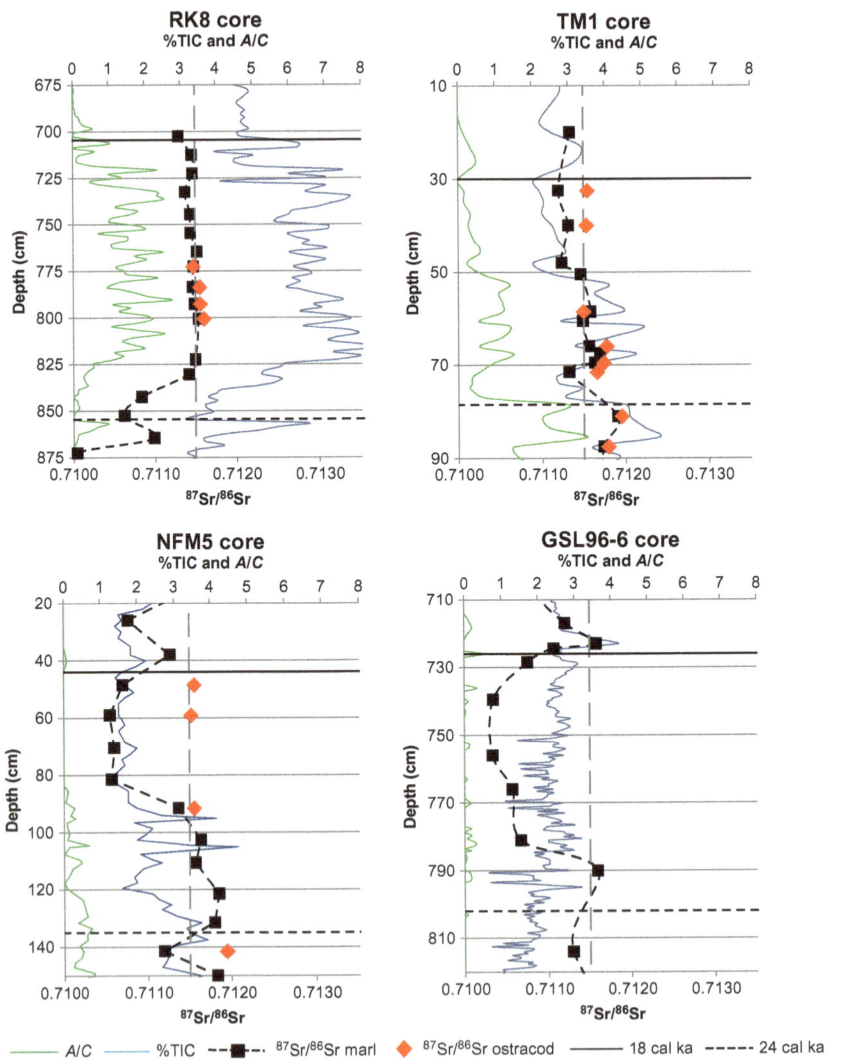

FIG. 16.6 Sr-isotope compositions of marl and ostracodes in four cores. See Fig. 16.1 for locations of cores.

>0.0001 mark major changes in lake hydrogeochemistry and are far above analytical uncertainties.

The Sr-isotopic composition in the RK8 core has a relatively simple history during the post-Stansbury transgression (Fig. 16.6). The $^{87}Sr/^{86}Sr$ ratios increase from 0.71062 at 853 cm to 0.71142 at 831 cm. The Sr-isotopic composition of marl in RK8 remains remarkably constant at this value (0.71148 ± 0.00005) throughout the remainder of the post-Stansbury transgression. Four samples yielded sufficient ostracods for analysis. The $^{87}Sr/^{86}Sr$ ratios of the ostracods are similar to those in the marl.

The Sr-isotopic composition history in the NFM5 core is different from that in RK8 (Fig. 16.5). In the early part of the post-Stansbury transgression from 135 to 90 cm, the Sr-values of the marl average 0.71165. At 90 cm, the values shift abruptly to significantly lower and relatively constant values between 0.71055 and 0.71070 for the remainder of the post-Stansbury transgression. In this interval, the $^{87}Sr/^{86}Sr$ ratios of the two ostracod samples are markedly more radiogenic than their paired marl samples, with similar values of ~0.71154.

The shape of the marl Sr-isotope curve in GSL96-6 is similar to that in NFM5, although values are generally lower (Fig. 16.5). In the later phase of the post-Stansbury transgression, values range from 0.71067 to 0.71034.

The marl Sr-isotope ratios in TM1 are generally intermediate in value between those of RK8 and NMF5 (Fig. 16.5). The shape of the Sr-isotope curve during the post-Stansbury transgression is similar to that of NFM5. Ostracods and marl have similar $^{87}Sr/^{86}Sr$ ratios during the early part of the post-Stansbury rise. However, where the marl $^{87}Sr/^{86}Sr$ ratios abrupt decrease, ostracod $^{87}Sr/^{86}Sr$ ratios do not. Between 48 and 32 cm, the $^{87}Sr/^{86}Sr$ ratios of ostracods are significantly higher than in the marl.

16.6 DISCUSSION

16.6.1 Sevier Subbasin

The differences in secular variation %TIC, *A/C*, $\delta^{18}O$, $\delta^{13}C$, and $^{87}Sr/^{86}Sr$ of the endogenic carbonate in the Sevier subbasin compared to those in the main body show clearly that the surface water of the Sevier subbasin was distinct from that of the main body throughout the post-Stansbury transgression (Figs. 16.4–16.6). The lake water produced significant amounts of aragonite throughout the post-Stansbury transgression. The $\delta^{18}O$ and $\delta^{13}C$ values of endogenic carbonate in RK8 exceeded those in main body, even after adjusting for mineral fractionation, indicating the surface water in the Sevier subbasin was likely slightly warmer with higher biological productivity than the main body. The $^{87}Sr/^{86}Sr$ ratios of the endogenic carbonate in the Sevier subbasin do not decrease as the lake expanded as do those in the main body.

The curves for %TIC, *A/C*, $\delta^{18}O$, $\delta^{13}C$, and marl $^{87}Sr/^{86}Sr$ have similar shapes in RK8. They show shifts at ~825 cm to relatively steady values that persisted throughout the post-Stansbury transgression. This similarity suggests that the processes that controlled formation of the abundance and types of carbonate minerals and the O-, C-, and Sr-isotope compositions of the water responded to the same change in hydrogeochemistry. Because the shapes of the curves in the Sevier subbasin are different than those in the main body, the hydrologic processes that drove secular change in the two subbasins must have been different. The increase in %TIC, specifically in the form of aragonite, as the lake expanded is opposite of what is expected in a hydrologically closed lake. Either the lake in the Sevier subbasin did not behave as a closed basin and/or a new source of water with significant total dissolved solids was introduced into the basin during the post-Stansbury transgression. The relatively constant values of $\delta^{18}O$, $\delta^{13}C$, and $^{87}Sr/^{86}Sr$ between 823 and 705 cm indicate that no major changes in lake hydrogeochemistry occurred during this interval.

The $^{87}Sr/^{86}Sr$ ratios of endogenic carbonate in the Sevier subbasin increase during the post-Stansbury transgression, whereas those in the main body decrease. Mixing of water between these subbasins cannot account for the changes, so another cause must be identified. Sr flux from springs exerts a major control on the $^{87}Sr/^{86}Sr$ ratio of the Great Salt Lake because springs have high Sr abundance (Pedone et al., 2000; Hart et al., 2004). Changes in the amount of spring discharge in each subbasin could result in the secular variation in $^{87}Sr/^{86}Sr$ ratios measured in the post-Stansbury transgressive marls. We are in the process of collecting and analyzing data from springs in the Sevier and Great Salt Lake subbasins. These data will be added to the Hart et al. (2004) database of rivers from the Sevier subbasin and rivers and streams in the main body in order to evaluate this hypothesis. Similar $^{87}Sr/^{86}Sr$ values of endogenic carbonate and ostracods between 800 cm and 772 cm in RK8 show that the water in the Sevier subbasin was vertically well mixed.

16.6.2 Main Body

In contrast to the Sevier subbasin, the differences in values and in secular variation in %TIC, *A/C*, $\delta^{18}O$, $\delta^{13}C$, and $^{87}Sr/^{86}Sr$ ratios of endogenic carbonate and of ostracodes in the cores from the main body indicate that circulation of surface water and deep water within this large main body was complex and not uniformly mixed during the post-Stansbury transgression. All main-body cores except GSL96-6 show a gradual decrease upward in aragonite, as expected in an expanding hydrologically closed lake. However, the disappearance of aragonite at different locations might not be time equivalent. Aragonite disappears at 21 cm above the start of the post-Stansbury rise in DIW and at 25 cm in BL04-4, but persists for 48 cm in NFM5. Although age equivalence of these thicknesses is not known, it seems likely that formation of aragonite persisted longer at NFM5 than at BL04-4 and DIW. Furthermore, the decrease in aragonite is not mirrored by a concurrent decrease in %TIC at all locations, as would be expected. The %TIC slightly increases upward in GSL96-6, DIW, and MP.

The ranges of values in $\delta^{18}O$ and $\delta^{13}C$ in BL04-4 and GSL96-6 are similar, although the secular variation in the stable isotope values, shown by the shapes of the curves, is not identical. This indicates that similar processes controlled the $\delta^{18}O$ and $\delta^{13}C_{DIC}$ of the water in these areas, but not always to the same extent at the same times during the post-Stansbury transgression. In the early portion of the post-Stansbury rise, the $\delta^{18}O$ and $\delta^{13}C$ values in TM1 are generally higher than those in BL04-4 and GSL96-6. In the later portion, they are similar. The higher values in %TIC and persistence of aragonite at this location throughout the post-Stansbury transgression indicates that processes occurred at TM1 that did not strongly influence other areas of the main body.

In the main body, the $^{87}Sr/^{86}Sr$ curves for endogenic carbonate have somewhat similar shapes in TM1, NFM5, and GSL96-6 in that they all show abrupt shifts to lower values in the upper portion of the transgression. The shift is much greater in NFM5 and GSL96-6 than in TM1. The shapes and values of the %TIC and *A/C* curves in these three cores differ. Hence, the hydrologic changes that produced secular change in Sr isotopes did so throughout the main body to different extents, but other, more local processes that influenced carbonate production were superposed on it.

Inflow from two sources most likely had the strongest influence on water circulation and lateral and vertical mixing of the main body: the Bear River and the Sevier subbasin (Fig. 16.1). Hart et al. (2004) estimate that the relative contribution to total discharge of the Bear River to the Great Salt Lake and Lake Bonneville was the same, but total runoff was much higher during Bonneville times. Oviatt et al. (1994) estimate that the ratio of surface-water inflow to evaporative output was two to three times higher in the small Sevier subbasin compared to that of the very large main body. The floor of the Sevier subbasin is also at higher elevation than that of the main body. Hence, water balance surpluses in the Sevier subbasin would have discharged into the main body. This effect would be greatest at relatively low lake levels and least at high lake levels.

TM1 is ideally located to examine the extent of mixing of waters between the main body and the Sevier subbasin. The %TIC, *A/C* ratios, and $^{87}Sr/^{86}Sr$ ratios of endogenic marl at TM1 are intermediate between those of RK8 and NFM5 throughout the post-Stansbury transgression. This demonstrates that waters from the Sevier subbasin did mix with those from the main body in this area. The decrease in %TIC, *A/C* ratios, and $^{87}Sr/^{86}Sr$ ratios of endogenic marl upward in the transgressive sequence of TM1 suggests that the influence of the Sevier subbasin water decreased, but did not end, as lake level rose.

Differences in %TIC and *A/C* in the cores in the central and northern parts of the main body, BL04-4, NFM5, GSL96-6, DIW, and MP, might reflect proximity to both Bear River discharge and northward flowing currents from the Sevier subbasin. For example, locations closest to the Bear River discharge, MP, DIW, and GSL96-6 do not show upward decreases in %TIC. Locations that contain aragonite in the early portion of the transgression might have received flow from the Sevier subbasin that reached less distant areas as lake level rose. The persistence of small amounts of aragonite in GSL96-6 might indicate episodic flow from the Sevier subbasin to this area, or from as yet unidentified water sources that would favor the precipitation of aragonite.

In TM1 and NFM5, the large separation of ostracod and marl $^{87}Sr/^{86}Sr$ ratios, coincident with the abrupt decreases in %TIC and *A/C*, indicate a major hydrologic change. In TM1, the $\delta^{18}O$ and $\delta^{13}C$ values also show abrupt decreases. The differences in $^{87}Sr/^{86}Sr$ ratios between the marl and ostracods indicate the absence of vertical mixing. Hydrologically, stratification is consistent with the low-oxygen waters interpreted by Thompson et al. (1990) from low ostracod abundance and diversity during the later portion of the post-Stansbury transgression in a core from the south arm of the Great Salt Lake obtained by Spencer et al. (1984). In GSL96-6, the $\delta^{18}O$ and $\delta^{13}C$ values show less variability a few centimeters above the abrupt decrease in $^{87}Sr/^{86}Sr$ of the endogenic carbonate, which might reflect development of a meromictic (nonoverturning) lake. Moreover, it is remarkable that the $^{87}Sr/^{86}Sr$ ratios of all ostracod samples in all cores during the post-Stansbury transgression have similar values of ~0.7115, which is also the same as the marl in the Sevier subbasin. Possibly, water from the Sevier subbasin continued to flow northward as a density-driven bottom current.

16.7 INTERPRETED HYDROGEOCHEMISTRY OF THE POST-STANSBURY TRANSGRESSION

As Lake Bonneville rose above the Old River Bed threshold following the Stansbury oscillations, water in the Sevier subbasin and main body did not mix to form a homogeneous water mass, but continued to follow separate hydrologic histories. Water in the Sevier subbasin evolved water chemistry that favored production of significant amounts of $CaCO_3$, particularly aragonite, and a relatively constant $^{87}Sr/^{86}Sr$ ratio of ~0.7115. Water in the lake in the Sevier subbasin was laterally and vertically well mixed during the post-Stansbury rise.

Water in the main body evolved along a different pathway. Overall, the carbonate saturation and the $^{87}Sr/^{86}Sr$ ratio decreased as lake level increased. The surface waters in the large main body were not homogeneous. Carbonate abundance and mineralogy in surface water seem to be related to the proximity to the inflow of the Bear River and/or south-to-north flow from the Sevier subbasin. During the later portion of the post-Stansbury transgression, $^{87}Sr/^{86}Sr$ ratios of endogenic carbonate were significantly lower than that those of contemporaneous ostracodes. This separation indicates that the lake in the main body ceased overturning and became vertically stratified. The bottom water had an $^{87}Sr/^{86}Sr$ ratio similar to water in the Sevier subbasin and might have been derived from a salinity-density-driven bottom current from the Sevier subbasin.

Mixing of water between the Sevier subbasin and main body occurred in the Old River Bed area and decreased, but did not end, as lake level rose. In the later portion of the post-Stansbury transgression, lake stratification extended into this area.

16.8 FUTURE RESEARCH

High spatial resolution analyses of carbonate mineralogy, C, O, and Sr isotope compositions in marl are important tools in reconstructing the hydrogeochemistry of Lake Bonneville. These parameters, however, do not provide an independent way to quantify lake size and must be correlated with the shoreline chronology. This remains problematic because of the lack of high temporal resolution and precise dating in marls.

Shoreline tufas and offshore endogenic carbonates may have significantly different carbonate mineralogy and geochemistry. For example, tufa and tufaglomerate near the Bonneville shoreline in the Silver Island Mountains is aragonite (Nelson et al., 2005), whereas endogenic carbonate from this time period (late transgressive phase) in the main body is dominated by calcite. Another puzzle is the difference in $^{87}Sr/^{86}Sr$ ratios between late transgressive-phase shoreline carbonates and endogenic carbonates. The $^{87}Sr/^{86}Sr$ measured by Hart et al. (2004) in tufa and mollusks collected near the Bonneville shoreline at six sites in the main body range from 0.71147 to 0.71184, significantly higher than those measured in contemporaneous marls in NFM5 and GSL96-6. Determination of the cause(s) of these differences is important in evaluating possible shoreline effects (such as groundwater inputs) and their relationship to overall lake geochemistry.

A system the size of Lake Bonneville must have had complex surface and bottom currents, perhaps even areas of upwelling driven by stream inflow, wind, and/or density differences. Future research in sedimentology and geochemistry will help to better define these processes in space and time. Modeling of the hydrogeochemistry, as for example Spencer et al. (1985) and Hart et al. (2004), has been and will be useful in formulating testable hypotheses about Lake Bonneville. Our current knowledge of the hydrogeochemistry of Lake Bonneville extends far beyond the knowledge base available in Gilbert's time, and we expect future research to take it much further.

ACKNOWLEDGMENTS

This research was partially funded by grants from Office of Research, CSU Northridge; ACS-PRF; U.S. Geological Survey; Utah Geological Survey; NSF GEO 0119936 to V.A.P.; and NSF EAR 9630047 and 1053129 to C.G.O. The authors thank Jay Quade for comments that led to improvements in the manuscript.

REFERENCES

Benson, L.V., 1994. Stable isotopes of oxygen and hydrogen in the Truckee River-Pyramid Lake surface-water system. 1. Data analysis and extraction of paleoclimatic information. Limnol. Oceanogr. 39, 344–355.

Benson, L.V., White, L.D., Rye, R., 1996. Carbonate deposition, Pyramid Lake Subbasin, Nevada, 4. Comparison of the stable isotope values of carbonate deposits (tufas) and the Lahontan lake-level record. Palaeogeogr. Palaeoclimatol. Palaeoecol. 122, 45–76.

Benson, L.V., Lund, S.P., Smoot, J.P., Rhode, D.E., Spencer, R.J., Verosub, K.L., Louderback, L.A., Johnson, C.A., Rye, R.O., Negrini, R.M., 2011. The rise and fall of Lake Bonneville between 45 and 10.5 ka. Quatern. Int. 235, 57–69.

Galat, D.L., Jacobsen, R.L., 1985. Recurrent aragonite precipitation in saline-alkaline Pyramid Lake, Nevada. Arch. Hydrobiol. 105, 137–159.

Gasse, F., Fontes, J.C., Plaziat, J.C., Carbonel, P., Kacmarska, I., De Deckker, P., Soulie-Marsche, I., Callot, Y., Dupeuble, P., 1987. Biological remains, geochemistry and stable isotopes for the reconstruction of environmental and hydrological changes in the Holocene lakes from North Sahara. Palaeogeogr. Palaeoclimatol. Palaeoecol. 60, 1–46.

Gilbert, G.K., 1890. Lake Bonneville. U.S. Geological Survey Monograph, vol. 1. 438 pp.

Given, R.K., Wilkinson, B.H., 1985. Kinetic control of morphology, composition, and mineralogy of abiotic sedimentary carbonates. J. Sediment. Petrol. 55, 109–119.

Haberzettl, T., Fey, M., Lücke, A., Maidana, N., Mayr, C., Ohlendorf, C., Schäbitz, F., Schleser, G.H., Wille, M., Zolitschka, B., 2005. Climatically induced lake level changes during the last two millennia as reflected in sediments of Laguna Potrok Aike, southern Patagonia (Santa Cruz, Argentina). J. Paleolimnol. 33, 283–302.

Hart, W., Quade, J., Madsen, D., Kaufman, D., Oviatt, C., 2004. The $^{87}Sr/^{86}Sr$ ratios of lacustrine carbonates and lake-level history of the Bonneville paleolake system. Geol. Soc. Am. Bull. 116, 1107–1119.

Kelts, K., Hsü, K.J., 1978. Freshwater carbonate sedimentation. In: Lerman, A. (Ed.), Lakes: Chemistry, Geology, Physics. Springer-Verlag, New York, pp. 295–323.

Kelts, K.R., Talbot, M.R., 1989. Lacustrine carbonates as geochemical archives of environmental change and biotic-abiotic interactions. In: Tilzer, M.M., Serruya, C. (Eds.), Ecological Structure and Function in large Lakes. Science and Technology Publ., Madison, WI, pp. 290–317.

Li, H.-C., Ku, T.-L., 1997. $\delta^{13}C$-$\delta^{18}O$ covariance as a paleohydrological indicator for closed-basin lakes. Palaeogeogr. Palaeoclimatol. Palaeoecol. 133, 69–80.

McKenzie, J.A., 1985. Carbon isotopes and productivity in the lacustrine and marine environment. In: Stumm, W. (Ed.), Chemical Processes in Lakes. Wiley, New York, pp. 99–118.

Nelson, D.T., Jewell, P.W., 2015. Transgressive stratigraphic record and possible oscillations of late Pleistocene Lake Bonneville, northern Hogup Mountains, Utah, U.S.A. Palaeogeography, Palaeoclimatology, Palaeoecology 432, 58–67.

Nelson, S.T., Wood, M.J., Mayo, A.L., Tingey, D.G., Eggett, D., 2005. Shoreline tufa and tufaglomerate from Pleistocene Lake Bonneville, Utah, USA: stable isotopic and mineralogical records of lake conditions, processes, and climate. J. Quatern. Sci. 20, 3–19.

Oviatt, C.G., 1997. Lake Bonneville fluctuations and global climate change. Geology 25, 155–158.

Oviatt, C.G., 2015a. Chronology of Lake Bonneville, 30,000 to 10,000 yr B.P. Quatern. Sci. Rev. 110, 166–171.

Oviatt, C.G., 2015b Ostracodes in Pleistocene Lake Bonneville, eastern Great Basin, North America. Special Issue of Hydrobiologia dedicated to R.M. Forester. http://dx.doi.org/10.1007/s10750-015-2483-y.

Oviatt, C.G., Jewell, P.W., 2016. Chapter 5. The Bonneville shoreline: Reconsidering Gilbert's interpretation. In: Oviatt, C.G., Shroder Jr, J.F. (Eds.), Lake Bonneville: A Scientific Update, vol. 20. Elsevier. pp. 88–104.

Oviatt, C.G., Nash, B.P., 2014. The Pony Express basaltic ash: a stratigraphic marker in late Pleistocene Lake Bonneville deposits, Utah. Utah Geological Survey Miscellaneous Publication 14-1, 10 pp.

Oviatt, C.G., Habiger, G.D., Hay, J.E., 1994. Variation in the composition of Lake Bonneville marl: a potential key to lake-level fluctuations and paleoclimate. J. Paleolimnol. 11, 11–30.

Pedone, V.A., Oviatt, C.G., 2013. South to north flow in Lake Bonneville: evidence from carbonate mineralogy and geochemistry. Geol. Soc. Am. Abstr. Programs 45, 552.

Pedone, V.A., Hemming, N.G., Kaufman, D.S., Quade, J., 2000. Negative covariance between lake volume and strontium isotope ratio in the Great Salt Lake, Utah. Geological Society of America, Annual Meeting, Abstracts with Programs, vol. 32. A-389.

Reheis, M.C., Adams, K.D., Oviatt, C.G., Bacon, S.N., 2014. Pluvial lakes in the western United States—a view from the outcrop. Quatern. Sci. Rev. 97, 33–57.

Reimer, P.J., Bard, E., Bayliss, A., Beck, J.W., Blackwell, P.G., Ramsey, C.B., Grootes, P.M., Guilderson, T.P., Haflidason, H., Hajdas, I., Hatté, C., Heaton, T.J., Hoffmann, D.L., Hogg, A.G., Hughen, K.A., Kaiser, K.F., Kromer, B., Manning, S.W., Niu, M., Reimer, R.W., Richards, D.A., Scott, E.M., Southon, J.R., Staff, R.A., Turney, C.S.M., van der Plicht, J., 2013. IntCal13 and Marine13 radiocarbon age calibration curves 0–50,000 Years cal BP. Radiocarbon 55, 1869–1887.

Russell, I.C., 1885. Geological History of Lake Lahontan, A Quaternary Lake of Northwestern Nevada. U.S. Geological Survey Monograph, vol. 11.

Shroder, J.F., Cornwell, K., Oviatt, C.G., Lowndes, T.C., 2016. Chapter 4. Landslides, Alluvial Fans, and Dam Failure at Red Rock Pass: The Outlet of Lake Bonneville. In: Oviatt, C.G., Shroder Jr., J.F. (Eds.), Lake Bonneville: A Scientific Update, vol. 20. Elsevier. pp. 75–87.

Spencer, R.J., Baedecker, M.J., Eugster, H.P., Forester, R.M., Goldhaber, M.B., Jones, B.F., Kelts, K., McKenzie, J., Madsen, D.B., Rettig, S.L., Rubin, M., Bowser, C.J., 1984. Great Salt Lake, and precursors, Utah: the last 30,000 years. Contrib. Mineral. Petrol. 86, 321–334.

Spencer, R.J., Eugster, H.P., Jones, B.F., 1985. Geochemistry of Great Salt Lake, Utah II: Pleistocene-Holocene evolution. Geochim. Cosmochim. Acta 49, 739–747.

Strong, A., Eadie, B.J., 1978. Satellite observations of calcium carbonate precipitation in the Great Lakes. Limnol. Oceanogr. 23, 877–887.

Stuiver, M., Reimer, P.J., 1993. Extended ^{14}C database and revised CALIB radiocarbon calibration program. Radiocarbon 35, 215–230. CALIB7.0, http://calib.qub.ac.uk/calib/calib.html.

Swart, P.K., 2015. The geochemistry of carbonate diagenesis: the past, present and future. Sedimentology 62, 1233–1304.

Talbot, M.R., 1990. A review of the palaeohydrological interpretation of carbon and oxygen isotopic ratios in primary lacustrine carbonates. Chem. Geol. (Isotope Geosciences Section) 80, 261–279.

Thompson, R.S., Toolin, L.J., Forester, R.M., Spencer, R.J., 1990. Accelerator-mass spectrometer (AMS) radiocarbon dating of Pleistocene lake sediments in the Great Basin. Palaeogeogr. Palaeoclimatol. Palaeoecol. 78, 301–313.

Thompson, R.S., Oviatt, C.G., Honke, J.S., McGeehin, J.P., 2016. Chapter 11. Late Quaternary Changes in Lakes, Vegetation, and Climate in the Bonneville Basin Reconstructed from Sediment Cores from Great Salt Lake. In: Oviatt, C.G., Shroder Jr., J.F. (Eds.), Lake Bonneville: A Scientific Update, vol. 20. Elsevier. pp. 221–291.

Chapter 17

Late Pleistocene Mountain Glaciation in the Lake Bonneville Basin

B.J.C. Laabs* and J.S. Munroe†
**North Dakota State University, Fargo, ND, United States*
†Middlebury College, Middlebury, VT, United States

ABSTRACT

The record of the last Pleistocene glaciation in the Lake Bonneville basin has been known for decades and has improved significantly due to expanded mapping and geologic dating efforts. Although ice occupied only a small portion of the basin and constituted a small component of the lake water budget, the pattern of glaciation in space and time can be combined with the stratigraphic record from Lake Bonneville to improve the understanding of regional climate change during and after the last glaciation. Mountain glaciers were most abundant in the eastern/Rocky Mountain sector of the basin, with the vast majority of ice volume concentrated in the Wasatch and Uinta Mountains. Here, the geomorphic record of the last glaciation is well preserved, stratigraphic observations of sediments of Lake Bonneville and valley glaciers are documented, and cosmogenic ^{10}Be exposure ages of several moraines based on new production models of in situ ^{10}Be are reported. Terminal moraines here, and in the western sector of the basin, were occupied near the time of the Bonneville highstand at 18 ka and subsequently abandoned while the lake continued to overflow. This constitutes a significant revision to initial reports of ^{10}Be exposure ages based on earlier production rate models, which suggest that ice retreat corresponded to the fall of Lake Bonneville from the Provo shoreline. Still, temporal correspondence of the Lake Bonneville highstand and glacier maxima support the previously proposed model that climate during the lake highstand was favorable for maxima of both glaciers and lakes. The onset of ice retreat while the lake overflowed, however, suggests warming in the basin at c.18 ka but with sufficient precipitation to sustain a positive water budget of Lake Bonneville.

Keywords: Lake Bonneville, Mountain glaciation, Cosmogenic exposure dating, Pleistocene climate change

Developments in Earth Surface Processes, Vol. 20. http://dx.doi.org/10.1016/B978-0-444-63590-7.00017-2

17.1 INTRODUCTION

During the last Pleistocene glaciation, mountain glaciers were present in 15 separate mountain ranges within the hydrologic basin of Lake Bonneville (Osborn and Bevis, 2001; Fig. 17.1). The vast majority of this total ice volume

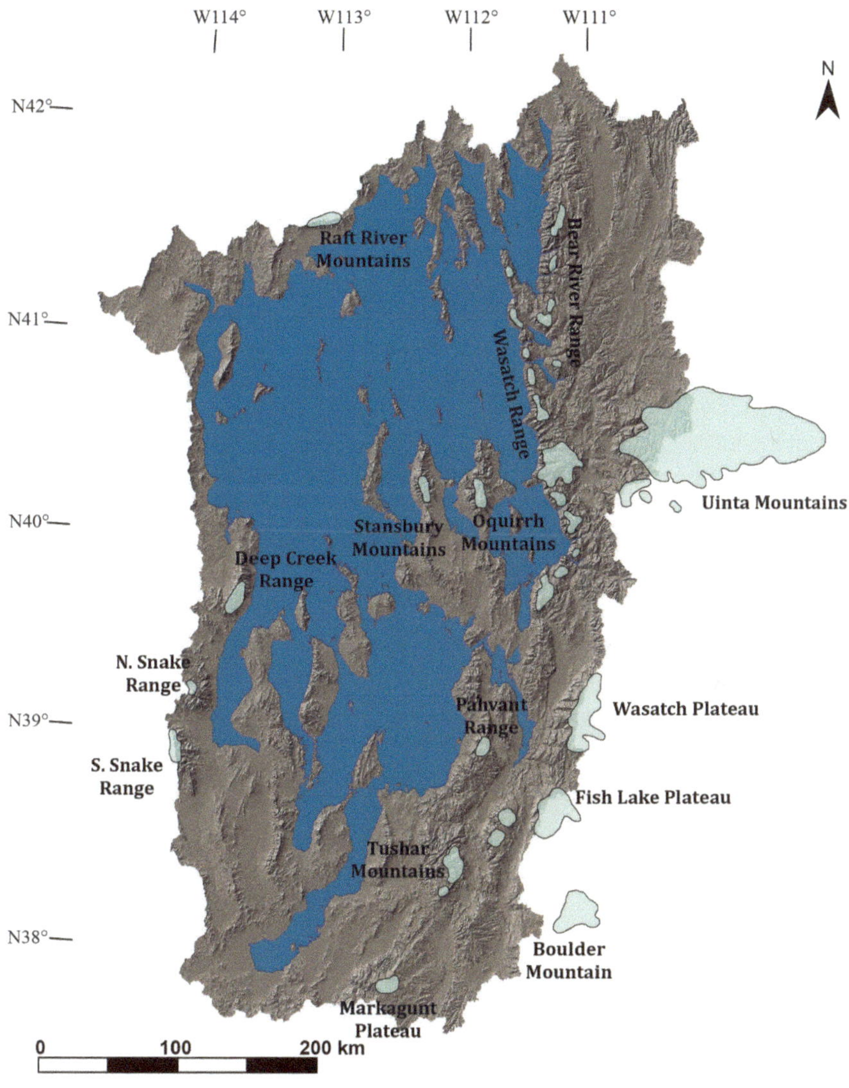

FIG. 17.1 Shaded-relief map of the Lake Bonneville basin produced from a mosaic of 30-m digital elevation models from the National Elevation Dataset (http://ned.usgs.gov). The maximum extent of Lake Bonneville is shown in *blue* (modified from http://gis.utah.gov) and Pleistocene mountain glacier systems in the basin are transparent *light blue*. *Modified from Biek, R., Willis, G., Ehler, B., 2010. Utah's Glacial Geology. Utah Geol. Surv. Notes, 42(3), 1–4.*

was concentrated in two ranges east of Lake Bonneville, the Wasatch and Uinta Mountains of northeastern Utah (Munroe and Laabs, 2009). These two ranges have been the focus of detailed glacial mapping efforts (eg, McCoy, 1977; Scott and Shroba, 1985; Personius and Scott, 1992; Laabs and Carson, 2005; Munroe, 2005; Refsnider et al., 2007; Laabs et al., 2011), which have supported reconstruction of glacier extent as well as the development of local cosmogenic ^{10}Be chronologies for the last glaciation (Munroe et al., 2006; Laabs et al., 2009; Laabs et al., 2011). These and other studies of the pattern of mountain glaciation in the Lake Bonneville basin, hereafter the "LBB" (eg, Munroe and Mickelson, 2002; Munroe et al., 2006), along with neighboring settings (Osborn and Bevis, 2001; Marchetti, 2007) have led to models for climatic and hydrologic conditions accompanying the glacier maxima, subsequent retreat, and the highstand of Lake Bonneville. Although mountain ice was likely a minor component of the total hydrologic budget of the lake (Fig. 17.2), documenting the spatial and temporal pattern of glaciation in the LBB provides a critical framework for understanding climatic changes during the last pluvial cycle.

Glacial deposits in the region surrounding Lake Bonneville were recognized by Gilbert (1890) and first documented by Atwood (1909) in the Wasatch and Uinta Mountains and by Blackwelder (1931) in northern ranges of the LBB. These studies interpreted the glacial deposits to represent at least two Pleistocene glaciations, with older glaciations being generally more extensive than later episodes. In the Wasatch and Uinta Mountains, the

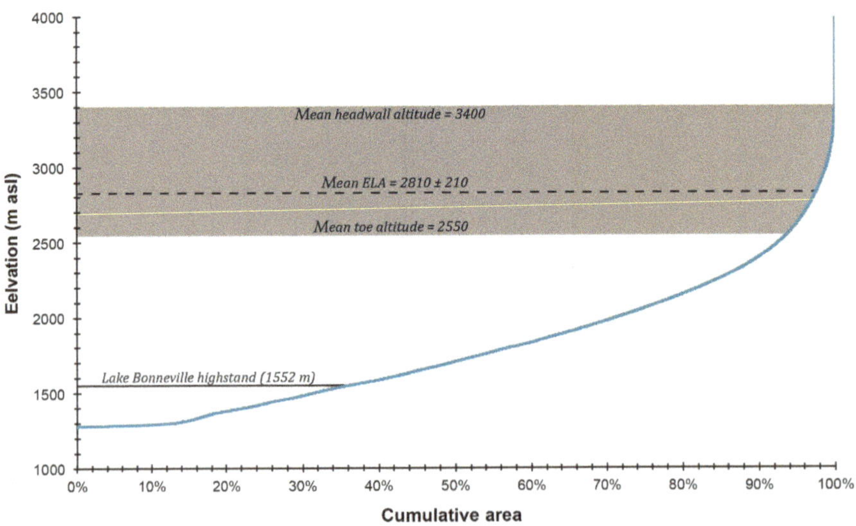

FIG. 17.2 Hypsometry of the Lake Bonneville basin. *Gray area* indicates the mean elevation range and equilibrium-line altitude (ELA) of reconstructed late Pleistocene glaciers listed in Table 17.2.

mapping of Atwood (1909) was succeeded by larger-scale mapping of glacial deposits by Bradley (1936), Richmond (1964), Osborn (1973), Scott and Shroba (1985), Bryant (1992), Personius and Scott (1992), Oviatt (1994), Munroe (2005), Laabs and Carson (2005), Refsnider et al. (2007), and Munroe and Laabs (2009) among others. Although glacial deposits are generally restricted to the mountain headwaters of the LBB, Madsen and Currey (1979), Scott (1988), and Godsey et al. (2005) documented stratigraphic relationships between glacial till and nearshore deposits of Lake Bonneville along the western front of the Wasatch Mountains near the mouths of Little Cottonwood and Bells Canyons. Although G.K. Gilbert's observations of morphological relationships of glacial and lacustrine deposits concluded correctly that glaciers advanced to the western front of the Wasatch Mountains prior to the Lake Bonneville highstand, some of the more recent studies reached a contrasting conclusion (as summarized by Scott, 1988).

Studies of glacial deposits in the Wasatch and Uinta Mountains, combined with mapping of glacial features in the smaller southeastern (Marchetti, 2007) and western ranges (Osborn and Bevis, 2001; and this study) of the LBB, have helped to reveal the pattern of mountain glaciation during the last glaciation. Despite the small glacial footprint relative to the size of Lake Bonneville (Figs. 17.1 and 17.2), the spatial distribution of ice provides a useful framework for understanding: (1) the range of possible temperature and precipitation changes accompanying the last glaciation in the LBB; and (2) the potential climatic effects that Lake Bonneville may have had on glacier mass balance in neighboring mountains. Previous work described the pattern of glaciation within and beyond the northern sector of the LBB in terms of glacier equilibrium-line altitudes (ELAs; Munroe and Mickelson, 2002; Munroe et al., 2006; Laabs et al., 2011). These studies reveal an intriguing west-to-east trend in ELAs across the northeastern Nevada and Utah (Fig. 17.3), such that ELAs decline nearly 400 m across Lake Bonneville from the Deep Creek–Stansbury–Oquirrh–Wasatch Ranges, and then rise sharply by 400–600 m across the western Uinta Mountains. This pattern is consistent with the hypothesis that Lake Bonneville affected glacier mass balance in downwind mountain ranges, but only if glacier maxima occurred when the lake was present.

In lieu of abundant and clear stratigraphic relationships between glacial and lacustrine deposits in the LBB, understanding the relative timing of moraine occupation and lake highstands relies chiefly on the comparison of radiocarbon-based chronologies of lacustrine deposits of Lake Bonneville (eg, Oviatt, 1997; Reheis et al., 2014; Oviatt, 2015) and cosmogenic chronologies of moraines (eg, Marchetti et al., 2005; Laabs et al., 2009, 2011; Marchetti et al., 2011). Such comparisons have been restricted (prior to this study) to glacial deposits in mountains east and downwind of Lake Bonneville. Broad application of cosmogenic ^{10}Be exposure dating of moraines in the Uinta (Munroe et al., 2006; Refsnider et al., 2008; Laabs et al., 2009)

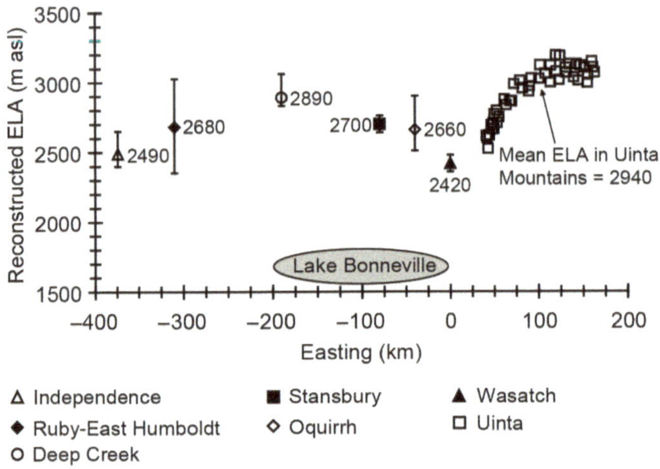

FIG. 17.3 Reconstructed glacier equilibrium-line altitudes (ELAs) across northeastern Nevada and northern Utah. Easting is relative to the eastern shoreline of Lake Bonneville near Salt Lake City, Utah. ELAs indicate the average of the toe-headwall altitude ratio and accumulation–area ratio, as defined in Table 17.1. ELAs for mountains in northeastern Nevada (Ruby–East Humboldt and Independence Ranges) and the eastern Uinta Mountains are from Laabs et al. (2011). *Modified from Laabs, B.J.C., Marchetti, D.W., Munroe, J.S., Refsnider, K.A., Gosse, J.C., Lips, E.W., Becker, R.A., Mickelson, D.M., Singer, B.S., 2011. Chronology of latest Pleistocene mountain glaciation in the western Wasatch Mountains, Utah, U.S.A. Quat. Res. 76, 272–284.*

and Wasatch Mountains (Laabs et al., 2011) has revealed apparent differences in the timing of local glacier maxima. Consistent with stratigraphic observations of Godsey et al. (2005), cosmogenic ^{10}Be exposure ages of moraines in the western Wasatch Mountains reported in these studies indicate that some glaciers persisted at, or readvanced to, their maximum length during the time when Lake Bonneville overflowed (Laabs et al., 2011). In the Uinta Mountains, reported exposure ages of moraines indicate that glaciers in southern and western valleys were at their terminal moraines while Lake Bonneville overflowed, whereas glacial retreat in northern and eastern valleys had begun by this time (Laabs et al., 2009). Although cosmogenic ^{10}Be exposure dating of glacial deposits has helped to establish a framework for reconstructing climate changes accompanying the last glaciation in the LBB, understanding the relative timing of glacier maxima and the rise and fall of Lake Bonneville has been complicated in part by uncertainties in the production of in situ ^{10}Be. A more reliable comparison is now possible due to a local calibration of the in situ production of ^{10}Be directly from geomorphic features constructed by Lake Bonneville (Lifton et al., 2015) and refinements to models of the temporal and spatial scaling of cosmic ray fluxes (Lifton et al., 2014).

The broad goal of this chapter is to describe the pattern of mountain glaciation in the LBB. This is achieved through compilation and review of the available mapping of glacial deposits and landforms in mountain ranges that hosted glaciers. Because the majority of glacier ice was concentrated in the Wasatch and Uinta Mountains, the spatial pattern of the last glaciation in these two regions is emphasized over other, smaller mountains. The temporal pattern of glaciation is reconsidered here through calculation of previously published and new cosmogenic ^{10}Be exposure ages of moraines in the LBB using the aforementioned refinements to the in situ production rate models by Lifton et al. (2014, 2015), and comparison of these ages to the chronology of Lake Bonneville as described by Oviatt (2015). Again, emphasis is given to the extensively glaciated Wasatch and Uinta Mountains, where the chronology of glacial deposits is known from cosmogenic ^{10}Be exposure dating of moraines. Lastly, previous and new inferences of the climatic and hydrologic relationship of Lake Bonneville and glaciers in neighboring mountains are discussed in the context of existing models of climate change in the southwestern United States during the last glaciation.

17.2 THE PATTERN OF PLEISTOCENE GLACIATION

This section describes the known distribution of glacial deposits and landforms in the glaciated mountain ranges of the LBB, based chiefly on recent 1:24,000-scale field mapping and interpretations of aerial photography. Reconstructions of the aerial and vertical extent of ice presented in numerous figures later are based on this mapping and represented in terms of ice area, volume, and ELA in Table 17.1. For specific information regarding the mapping of glacial deposits, the reader is referred to maps and other documents cited later.

17.2.1 The Wasatch Mountains

A product of late Cenozoic normal faulting at the junction of the Basin and Range and the Middle Rocky Mountains, the Wasatch Mountains are a north–south trending range forming a steep orographic barrier east of Lake Bonneville. The northern segment of the mountain range east of Salt Lake City and Provo features many of the highest summits of the range (eg, Mt. Timpanogos, 3582 m asl) and was the most heavily glaciated part of the range, with the largest glaciers occupying east-to-west flowing valleys (Fig. 17.4). Other segments of the range occupied by ice include a small number of valleys east of Ogden, valleys in the Bear River Range east of Logan (Reheis et al., 2005), and a small number of valleys east and southeast of Provo (Fig. 17.1). The clearest geomorphic indicators of Pleistocene glaciation are classic, mountain glacial–erosional landforms, such as cirques, arêtes, and broad U-shaped valleys (Atwood, 1909; Richmond, 1964). Throughout the Wasatch Mountains, glacial sediment is sparsely preserved within canyons (Richmond, 1964; McCoy, 1977), likely due to postglacial modification of glacial valleys. The best-preserved deposits of till and outwash are located

TABLE 17.1 Reconstructed Glacier Area, Volume, and Equilibrium-Line Altitudes in the Lake Bonneville Basin

Number	Mountains[a]	Drainage	Area (km^2)	Volume (km^3)[b]	ELA—Toe-Headwall Altitude Ratio[c] (m asl)	ELA—Accumulation-Area Ratio[d] (m asl)	Mean ELA (m asl)	1σ
1	South Snake	Blue Canyon	0.7	0.02	3133	3135	3134	1
2		Bald Mtn East	0.2	0.00	3110	3087	3098	16
3		Bald Mtn West	1.1	0.04	3063	3033	3048	22
4		Lehman Creek	10.8	1.12	2997	3083	3040	61
5		Baker Creek	5.5	0.42	3147	3080	3114	48
6		Snake Creek North	1.8	0.08	3114	3097	3105	12
7		Snake Creek Middle	2.2	0.11	3078	3018	3048	43
8		Snake Creek South	0.7	0.02	3162	3251	3207	63
9		Williams Canyon	1.0	0.04	3063	3056	3059	5
10		Big Canyon	0.9	0.03	3140	3110	3125	22
		Sum	*24.9*	*1.9*		*Rangewide mean ELA*	*3098*	*52*
11	Deep Creek[e]	Haystack Peak	1.7	0.08	3021	3170	3096	105
12		Red Cedar Creek	3.1	0.18	2951	3206	3079	181
13		Steves Creek North	0.6	0.02	2793	2804	2798	8
14		Steves Creek South	0.7	0.02	2834	2816	2825	13

15		Granite Creek	0.7	0.02	3078	3194	3136	82
		Sum	*6.8*	*0.3*		*Rangewide mean ELA*	*2987*	*161*
16	Stansbury[e]	Pass Canyon	0.4	0.01	2570	2606	2588	26
17		Little Pole Canyon	0.4	0.01	2692	2719	2705	19
18		North Willow	1.0	0.03	2488	2597	2542	77
19		Big Pole Canyon	0.8	0.02	2732	2658	2695	52
20		Mining fork	2.1	0.10	2441	2560	2500	85
21		Spring Canyon	0.6	0.02	2713	2719	2716	4
22		South Willow Canyon	5.2	0.39	2422	2606	2514	130
23		Big Creek Canyon	1.3	0.05	2692	2707	2699	10
		Sum	*11.8*	*0.6*		*Rangewide mean ELA*	*2620*	*93*
24	Fish Lake Plateau[f]	Jorgensen Creek North	0.5	0.01	3035	3020	3028	11
25		Jorgensen Creek South	0.9	0.03	3020	2980	3000	28
26		Seven Mile	3.5	0.22	3060	2950	3005	78
27		Rock Canyon	2.1	0.10	3165	3130	3148	25
28		Pelican Canyon	4.2	0.28	3030	3190	3110	113
29		Tasha Creek	11.3	1.20	3020	3130	3075	78
		Sum	*22.5*	*1.8*		*Rangewide mean ELA*	*3061*	*60*

Continued

TABLE 17.1 Reconstructed Glacier Area, Volume, and Equilibrium-Line Altitudes in the Lake Bonneville Basin—Cont'd

Number	Mountains	Drainage	Area (km^2)	Volume (km^3)	ELA—Toe-Headwall Altitude Ratio (m asl)	ELA—Accumulation-Area Ratio (m asl)	Mean ELA (m asl)	1σ
30	Oquirrh[e]	Settlement Canyon	1.0	0.03	2750	2804	2777	38
31		Lowe Canyon	0.7	0.02	2654	2658	2656	3
32		Jackson Hollow	0.5	0.01	2675	2755	2715	57
33		Jumpoff	2.2	0.11	2590	2606	2598	11
		Sum	*4.4*	*0.2*		*Rangewide mean ELA*	*2686*	*77*
34	Wasatch[e]	American Fork	20.5	2.86	2567	2560	2564	5
35		Little Cottonwood	42.2	8.35	2141	2606	2374	329
36		Bells	7.2	0.61	2221	2713	2467	348
37		Silver Lake	8.4	0.77	2610	2682	2646	51
38		Dry Creek	15.1	1.83	2226	2652	2439	301
		Sum	*93.3*	*14.43*		*Rangewide mean ELA*	*2498*	*107*
39	Uinta[e]	Heber Mountain	0.9	0.03	2403	2469	2436	47
40		Pinon Canyon	0.5	0.01	2610	2621	2616	8
41		Swifts Canyon	2.1	0.10	2520	2682	2601	115
42		Bear Basin	0.8	0.03	2524	2524	2524	0

43		Mill Fork	8.8	0.83	2570	2667	2619	69
44		South Fork	17.5	2.28	2610	2774	2692	116
45		Nobletts Creek	2.0	0.10	2680	2646	2663	24
46		Shingle Mil	1.6	0.07	2670	2743	2707	52
47		Shingle Mill East	0.3	0.01	2700	2633	2667	47
48		Bear Trap North	0.3	0.01	2770	2768	2769	1
49		Smith & Morehouse	125.0	41.83	2710	2880	2795	120
50		Slader Ridge West	0.4	0.01	2700	2719	2710	13
51		Broad Canyon	1.3	0.05	2800	2725	2763	53
52		Slader Creek	2.5	0.13	2770	2728	2749	30
53		Weber River	125.0	41.83	2780	2972	2876	136
54		Shingle Creek	16.6	2.11	2663	2902	2783	169
55		Coop Creek	7.8	0.69	2642	2793	2718	107
56		Yellow Pine Creek	5.6	0.43	2598	2744	2671	103
57		South Fork Weber	17.5	2.28	2774	2610	2692	116
58		West Fork Bear River	47.9	10.07	2788	2880	2834	65
59		Chalk Creek	5.0	0.36	2888	2840	2864	34
60		Provo River	132.7	45.72	2835	2950	2893	81

Continued

TABLE 17.1 Reconstructed Glacier Area, Volume, and Equilibrium-Line Altitudes in the Lake Bonneville Basin—Cont'd

Number	Mountains	Drainage	Area (km^2)	Volume (km^3)	ELA—Toe-Headwall Altitude Ratio (m asl)	ELA—Accumulation-Area Ratio (m asl)	Mean ELA (m asl)	1σ
61		North Fork Provo	59.5	13.88	2852	2950	2901	69
62		Hayden Fork Bear River	118.7	38.73	2948	3025	2987	54
63		East Fork Bear River	56.7	12.93	2988	3035	3012	33
		Sum	*757.0*	*214.5*		*Mean ELA of Western Uinta Icefield*	*2741*	*136*

Notes: List of glaciers is not exhaustive. Excludes ice in Big Cottonwood Canyon (Wasatch) and numerous small ice masses in valleys of the eastern sector of the basin where ice extents are poorly known.
[a]*Mountain ranges are listed from west to east.*
[b]*Based on area–volume scaling relationship for steady-state glaciers, where* $\log V = \gamma \log A + k$, *where V, volume, A, surface area,* γ, *1.375 (Bahr et al., 1997) or 1.56 (Radić et al., 2007).*
[c]*Toe-headwall altitude ratio = 0.4 (highest headwall altitude—toe altitude) + toe altitude.*
[d]*Assumed ratio accumulation–area ratio of 0.65.*
[e]*ELAs are from Laabs et al. (2011) and references therein.*
[f]*ELAs and ice area from Marchetti et al. (2011).*

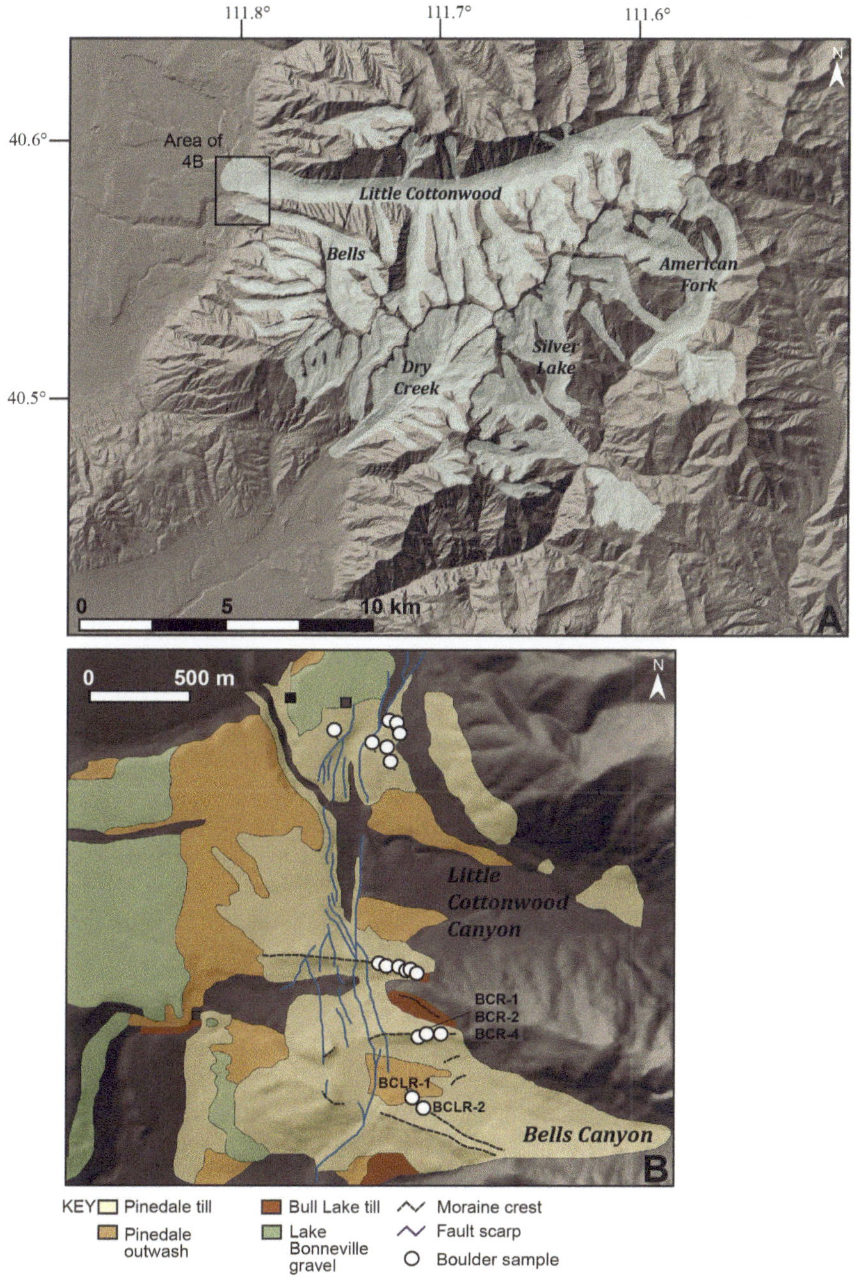

FIG. 17.4 (A) Shaded-relief map of a glaciated portion of the Wasatch Mountains near Salt Lake City, UT (produced from a mosaic of 30-m digital elevation models from the National Elevation Dataset, http://ned.usgs.gov). Pleistocene ice extents are transparent *light blue*. (B) Surficial geologic mapping of Personius and Scott (1992) with *shaded* relief. Boulder locations refer to samples of the terminal and left-lateral moraine at Little Cottonwood Canyon sampled for cosmogenic ^{10}Be exposure dating, and previously unpublished exposure ages of samples from Bells Canyon. *Boxes* indicate locations of outcrops where stratigraphic relationships of Lake Bonneville deposits and glacial till have been documented by Scott (1988; *black box*) and Godsey et al. (2005; *gray box*).

at or near the mountain front in a small number of glacial valleys, including Little Cottonwood, Bells, Dry Creek (Fig. 17.4), and Big Cottonwood Canyons (Quirk et al., 2015).

After the earliest documentation of glacial deposits in the western Wasatch Mountains by Gilbert (1890), Atwood (1909), and Ives (1950), the first 1:24,000-scale mapping of these deposits by Richmond (1964) subdivided Pleistocene till and outwash to represent at least two glacial episodes. The framework for these subdivisions followed terminology for glacial episodes coined by Blackwelder (1915) and later modified by Richmond (1957), wherein the last two episodes are the Bull Lake (penultimate) and Pinedale (last) Glaciations. These were correlated by Richmond (1965) to the Early and Late Wisconsin Glaciations, respectively, representing the last two periods of major global ice advance. The Pinedale Glaciation culminated in the Rocky Mountains and neighboring regions of the western United States during marine oxygen isotope stage 2 (26–14 ka; Lisiecki and Raymo, 2005), and the Bull Lake most likely during stage 6 (167–132 ka; Lisiecki and Raymo, 2005) or possibly later. Based largely on weathering characteristics of boulders and other sediment comprising moraines, Richmond (1964) concluded that deposits at the mouths of Little Cottonwood and Bells Canyons represent the Bull Lake Glaciation in the Wasatch Mountains and that ice did not reach the mountain front during the Pinedale Glaciation. Deposits in this area were remapped by McCoy (1977), who revised Richmond's interpretation and suggested that moraines at the mouth of Little Cottonwood and Bells Canyons were actually deposited during the last glaciation. This Pinedale age designation was confirmed by a reconsideration of soil formation in glacial and other Quaternary deposits near the mouth of Little Cottonwood Canyon (Scott and Shroba, 1985), and by radiocarbon dating of glacial features at the mountain front and within the canyons (Madsen and Currey, 1979).

The first radiocarbon age limits on glacial deposits led Madsen and Currey (1979) to conclude that glaciers in these two canyons began retreating from their terminal moraines prior to the Lake Bonneville highstand. This observation was supported by stratigraphic relationships of till and lacustrine nearshore gravel near the mouth of Little Cottonwood Canyon (Scott and Shroba, 1985; Scott, 1988). Here, till overlain by nearshore gravel displays a weakly developed soil, indicating a brief (centuries or millennia) period of exposure prior to burial by gravel deposited during the Lake Bonneville highstand. Conversely, Godsey et al. (2005) described interfingered till and shoreline deposits of the Lake Bonneville highstand in a separate exposure on the ice-distal side of the terminal moraine at Little Cottonwood Canyon (Fig. 17.4b), suggesting that the highstand and the local glacial maximum were synchronous.

Evaluating stratigraphic relationships of glacial and lacustrine sediment in the western Wasatch Mountains are complicated by the paucity of observations, and the fact that the exposures displaying clear stratigraphic relationships between till and lacustrine gravel are no longer visible due to urbanization and removal of deposits. Additionally, numerical ages of glacial deposits limit only the time when ice began retreating from moraine crests, not when ice first reached its maximum extent; therefore, ages of subsurface tills observed by Scott and Shroba (1985) and Godsey et al. (2005) in excavations at the mouths of Little Cottonwood and Bells Canyons could be significantly older than tills forming moraine crests dated by cosmogenic ^{10}Be (Laabs et al., 2011). When considered together, stratigraphic relationships between tills and lacustrine deposits of Lake Bonneville suggest that the termini of glaciers in Little Cottonwood and Bells Canyons occupied the mountain front more than once during the last Pleistocene glaciation. If so, glaciers likely extended to the mountain front both before and during the Lake Bonneville highstand at 18.0 ka.

Reconstructed ice extents in valleys where the locations of glacier termini are clearly delimited by moraines (Fig. 17.4) indicate that ELAs in the western Wasatch Mountains were the lowest in the LBB, ranging from 2374 to 2646 m (Table 17.1; Fig. 17.3). As discussed by Laabs et al. (2011), this pattern is not consistent with the distribution of modern precipitation across the northern sector of the LBB. In fact, winter precipitation at 2750 m, the regionally averaged ELA of the last glaciation, is greater in some ranges west of the Wasatch Mountains (eg, the Ruby and East Humboldt Mountains of northeastern Nevada) but the reconstructed ELA of the last glaciation there is higher. This suggests that, given a uniform temperature depression in mountains of the northern LBB, precipitation patterns differed from modern, possibly due to the presence of Lake Bonneville. This possibility is further discussed later.

17.2.2 The Uinta Mountains

The Uinta Mountains, an east–west-trending Laramide uplift in the Middle Rocky Mountains, are the only range in Utah with summits extending above 4000 m asl and were heavily glaciated during the Pleistocene. The range featured a large number of north- and south-flowing discrete valley glaciers as well as the western Uinta Icefield (Oviatt, 1994; Refsnider et al., 2008), which occupied the westernmost valleys of the Uinta Mountains and drained into Lake Bonneville via the Provo, Weber, and Bear Rivers (Refsnider et al., 2008; Munroe and Laabs, 2009; Fig. 17.5). The western Uinta Icefield was the largest single ice mass in the LBB, likely greater than all other ice masses combined in the basin (Table 17.1), even though the southeastern sector of it extended beyond the watershed divide and drained into the Colorado River system. Among glaciated areas of the western United States, the Uinta Mountains are widely known for their unique glacial geomorphology, prominently displayed along the summit ridge (eg, Hansen, 1975). Most glacial valleys in the Uinta Mountains feature broad, compound cirques formed in quartzite

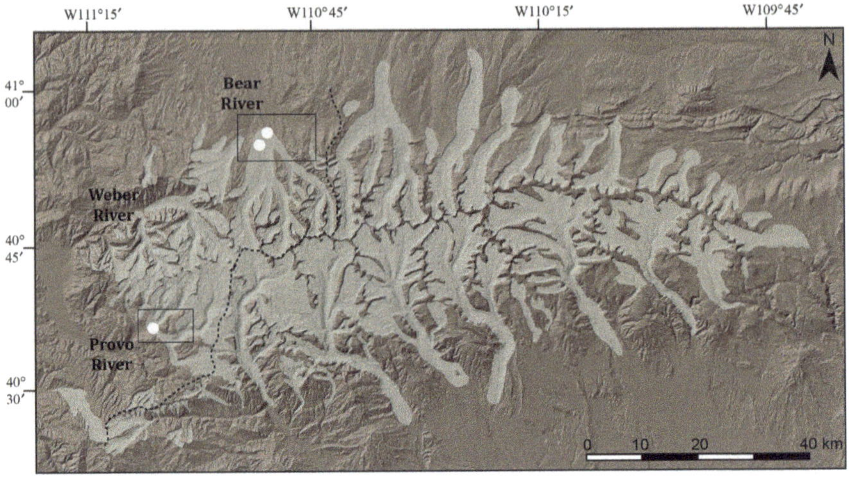

FIG. 17.5 Shaded-relief map of the glaciated portion of the Uinta Mountains (produced from a mosaic of 30-m digital elevation models from the National Elevation Dataset, http://ned.usgs.gov). Pleistocene ice extents are transparent *light blue* (from Munroe and Laabs, 2009). *White circles* indicate locations of moraines with cosmogenic ^{10}Be exposure ages. *Rectangles* indicate areas shown in Fig. 17.6.

and weakly metamorphosed sandstone that were occupied by laterally extensive icefields instead of valley glaciers (Munroe and Laabs, 2009). These icefields drained into one or more large U-shaped valleys that were occupied by valley glaciers extending to the range front. The Western Uinta Icefield is so-named because ice thicknesses were great enough to overwhelm the east–west-trending summit ridge, connecting ice masses on opposite sides of the range (Refsnider et al., 2007; Fig. 17.5).

The earliest mapping of glacial deposits in the Uinta Mountains by Atwood (1909) subdivided well-preserved moraines and outwash into two glacial episodes. These were later termed the Blacks Fork (penultimate) and Smiths Fork (last) Glaciations by Bradley (1936), after sets of well-preserved lateral and terminal moraine complexes in adjacent drainages of the northern Uinta Mountains. Mapping of glacial deposits in the southern Uinta Mountains by Osborn (1973) documented numerous moraine crests interpreted to represent multiple intervals of ice advance and retreat during the last and penultimate glaciations. More recent mapping by Munroe (2005) and Laabs and Carson (2005) subdivided moraines on the basis of soil development (based on soil development indices developed by Douglass, 2000), geometry (height and slope), frequency of large boulders, and morphological relationships with outwash terraces, confirming the extent of the last glaciation in the Uinta Mountains (Fig. 17.5).

Mapping of glacial deposits at 1:24,000-scale greatly expanded the understanding of the extent of the last glaciation in the headwaters of the Provo,

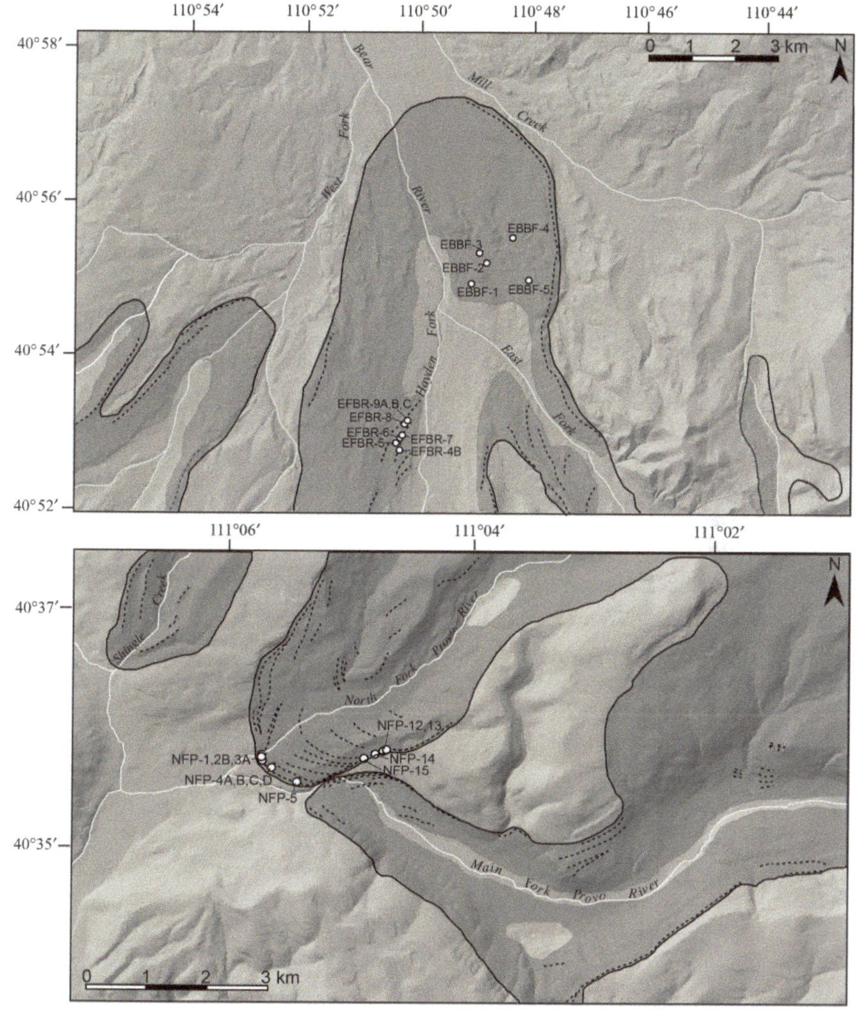

FIG. 17.6 Shaded-relief maps of the terminal moraine areas of the Bear River (top panel) and the Provo River (bottom panel) valleys. *Black lines* indicate the maximum ice extent. *Dashed lines* indicate moraine crests. *Darkened areas* indicate the extent of glacial till. Moraine boulder sample locations (*circles*) and *numbers* refer to samples shown in Table 17.2 and described in Laabs et al. (2009).

Weber, and Bear Rivers (Fig. 17.6), the latter being the largest river (in terms of drainage area) entering Lake Bonneville. In these and other valleys of the Uinta Mountains, glacial deposits are more widespread than in the Wasatch Mountains, due to a lesser degree of postglacial modification and generally larger glaciers. In all three valleys, moraines are well preserved, clearly .delimiting terminal ice positions during the last glaciation.

The Bear River valley was occupied by the largest sector of the Western Uinta Icefield and another, significantly smaller valley glacier. The icefield thickness exceeded 500 m in the largest troughs and formed an extensive piedmont lobe beyond the mountain front. A broad, high-relief, hummocky moraine that grades into an extensive outwash surface delimits the shape of the piedmont lobe. Up valley, recessional moraines are preserved in the East Fork and Hayden Fork of Bear River (Fig. 17.6a), indicating the terminal position of tributary glaciers during a stillstand or minor readvance of ice following construction of the piedmont moraine.

West of the Bear River valley, a smaller sector of the Western Uinta Icefield extended into the headwaters of the Weber River. Here, two large outlets of the Western Uinta Icefield occupied the Red Pine Creek tributary and the main valley of the Weber River, as indicated by well-preserved terminal moraines. A smaller outlet glacier terminated within a narrow canyon of the South Fork of Weber River (Fig. 17.5). Several small valley glaciers were present in high elevation tributary valleys of the Weber River, representing the westernmost extent of glaciation in the Uinta Mountains (Oviatt, 1994; Refsnider et al., 2007; Munroe and Laabs, 2009).

Across the westernmost summit of the Uinta Mountains are the headwaters of the Provo River, which featured the smallest sector of the Western Uinta Icefield (Refsnider et al., 2007). Most of this sector occupied the broad uplands at the heads of the North and South Forks of the Provo River, with smaller masses present in the Shingle, Coop, and Yellow Pine valleys (Munroe and Laabs, 2009). Detailed mapping by Refsnider et al. (2007) delimits the terminal position in the North Fork tributary during the last two glaciations, as well as the pattern of ice retreat up valley of the terminal moraine of the last glaciation. Glacial deposits are relatively sparse in the South Fork tributary, where only a few moraine segments are preserved above the confluence with the North Fork tributary and along Soapstone Creek. As a result, the up-valley extent of ice is less clear compared to other sectors of the western Uinta Icefield. The terminal position is clear, however, delimited by a set of small but well-preserved moraines near the terminal position in the North Fork tributary (Fig. 17.6).

Excellent preservation of glacial deposits throughout the Uinta Mountains, combined with 1:24,000-scale mapping in recent decades, afford reconstruction of glacier ELAs (Munroe et al., 2006). The sharp, west-to-east rise in glacier ELAs along the length of the Uinta Mountains (Fig. 17.3) has been interpreted to reflect a significant decline in precipitation along the range (Munroe and Mickelson, 2002; Munroe et al., 2006). This rise is too great to be explained by a west-to-east decline in precipitation due to simple orographic effects (Munroe and Mickelson, 2002; Laabs et al., 2006). Rather, the rise in ELAs may signify a declining impact of Lake Bonneville on glacier mass balance with distance away from the lake. This is further evidenced by the formation of the large western Uinta Icefield at the west end of the range and smaller, discrete valley glaciers at the east end (Munroe, 2005).

In addition to the reconstruction of ELAs for glaciers throughout the Uinta Mountains (Munroe et al., 2006; Table 17.1), the volume of ice in the Western Uinta Icefield can be approximated by applying area–volume scaling models for valley glaciers at steady state (Bahr et al., 1997; Radić et al., 2007; Table 17.1). This method was applied to glacial valleys where the planimetric surface area of the ice is known from mapping-based reconstructions. Although ice extents are not precisely known for several glacial valleys in the Wasatch Mountains, it is known for the largest valley glaciers in the range (Fig. 17.4). Volume calculations indicate that ice volume was an order of magnitude greater in the Bear, Weber, and Provo River sectors of the Western Uinta Icefield than in the entire Wasatch Mountains (Table 17.1). Furthermore, the total ice volume in the Uinta and Wasatch segments of the LBB (247 km^3) was approximately 2.9% of the volume of Lake Bonneville (8603 km^3), as determined by computing the volume below 1552 m in the 30-m digital elevation model of the LBB shown in Fig. 17.1. The total volume of ice in all 13 other glaciated mountains of the LBB is incompletely known, but based on surface area alone (Fig. 17.1) it comprises less than 5% of the total ice volume in the Wasatch Mountains. The total volume of ice in the LBB was therefore less than 3.5% of the volume of the lake, representing a minor component of the lake water budget.

17.2.3 Central and Southern Utah

Several mountains and parts of the Colorado Plateau draining into the southeastern sector of the LBB were glaciated, including the Wasatch, Sevier, Markagunt and Fish Lake Plateaus, and the Pahvant and Tushar Ranges (Osborn and Bevis, 2001; Marchetti, 2007; Fig. 17.1). The glacial record in these areas is briefly summarized here. For a more-detailed description of glacial deposits in these settings and reconstructions of ice extent, the reader is referred to the useful review by Marchetti (2007) and references therein.

The most extensively glaciated areas of central Utah were the Wasatch Plateau (Foley et al., 1986; Larsen, 1996), Boulder Mountain (Marchetti et al., 2005), and Fish Lake Plateau (Marchetti et al., 2011). Among these, the greatest volume of ice formed on the Wasatch Plateau just outside of the LBB divide, where east-flowing drainages display well-developed cirques and broad U-shaped valleys eroded into Cenozoic sedimentary rocks (Larsen, 1996). Moraines of the last glaciation are identified by their preservation and minimal weathering characteristics (Osborn and Bevis, 2001), delimiting terminal positions in most glacial valleys, and affording accurate reconstructions of ice extent. Fish Lake Plateau is a faulted upland composed of Tertiary volcanics that featured a small ice cap during the last glaciation (Marchetti, 2007;

Marchetti et al., 2011). Multiple moraine sets deposited by small cirque glaciers likely record ice advance during multiple Pleistocene glaciations (DeGraff and Gallegos, 1983). Moraines of the last glaciation include large boulders suitable for cosmogenic ^{3}He exposure dating, providing a key chronology of the last glaciation in central Utah (Marchetti et al., 2011). Age limits on the last glaciation from Fish Lake Plateau and from the nearby Boulder Mountain (Marchetti et al., 2005) are described later.

The largest glaciers outside of the Wasatch and Fish Lake Plateaus occupied two valleys draining the Markagunt Plateau (Mulvey et al., 1984; Osborn and Bevis, 2001). The glacier occupying Castle Creek attained a length of approximately 5 km, as delimited by a hummocky terminal moraine. More extensive ice in the vicinity of the summit of the Markagunt Plateau (Brian Head) is also possible, although glacial deposits and landforms are poorly preserved (if present at all) outside the Castle and Lowder Creek valleys (Osborn and Bevis, 2001).

The glacial record of the Sevier Plateau, the Pahvant Range, and the Tushar Mountains includes deposits of the last glaciation recognized in previous mapping in central Utah (summarized by Osborn and Bevis, 2001; Marchetti, 2007). Although glacial features are preserved in each of these settings, glacier ice was sparse and moraines are assigned to the last glaciation solely on the basis of preservation and surface weathering characteristics (eg, Bevis, 1995).

Marchetti (2007) summarized glacier ELAs for the Tushar and Pahvant Ranges and the Wasatch Plateau as estimated in previous studies. These range from approximately 2800 m asl in the Wasatch Plateau and Pahvant Range to nearly 3300 m asl in the Tushar Range (Marchetti, 2007 and references therein). The greater ELAs relative to the Wasatch Mountains and the mean ELA in the Uinta Mountains may be attributed to the lower latitude of glaciers in these settings, but may also reflect a stronger orographic effect in the Wasatch Mountains resulting in greater precipitation. Marchetti (2007) also attributes the variability of ELAs in central Utah to redistribution of snow by wind, such that wind-deposited snow Wasatch Plateau and Pahvant Range may have enhanced glacier mass balance and locally depressed the ELA.

17.2.4 Western Ranges

The western border of the LBB, consisting of generally smaller and fragmented mountains of the Basin and Range relative to the east, was sparsely glaciated. The South and North Snake Ranges and the Deep Creek Range featured small cirque and valley glaciers during the last glaciation (Osborn and Bevis, 2001). Glacial deposits in all three ranges were mapped by Bevis (1995) and subdivided based on weathering characteristics.

Among these three ranges, cirque and valley glaciers were most numerous in the South Snake Range, a rare setting where summit elevations in the Basin and

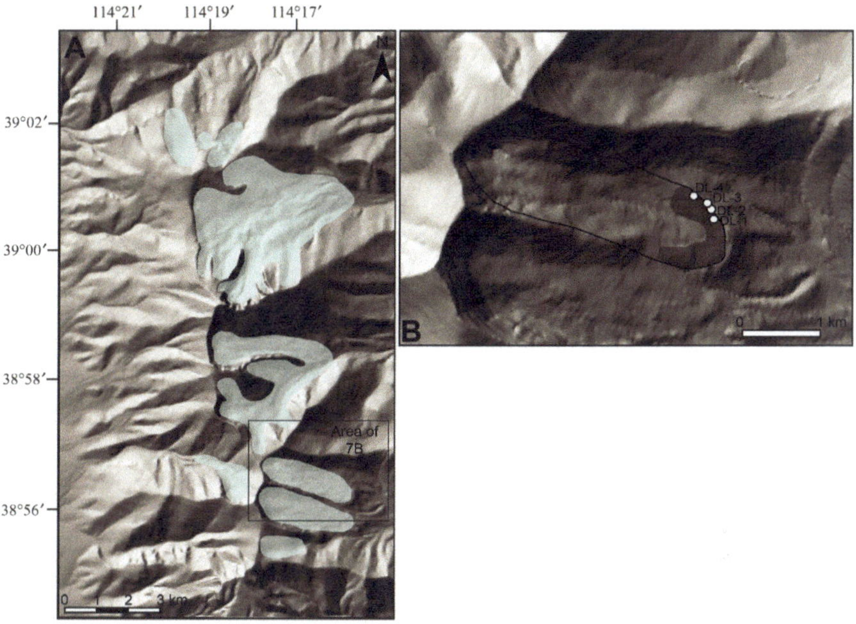

FIG. 17.7 Shaded-relief maps of (a) the glaciated portion of the South Snake Range, Nevada and (b) the northern tributary of the Snake Creek valley (produced from a mosaic of 10-m digital elevation models from http://epa.gov). Pleistocene ice extents are transparent *light blue* in (a) and outlined in *black* in (b). Darkened area in (b) is the terminal moraine. Moraine boulder sample locations (*circles*) and *numbers* refer to samples shown in Table 17.2.

Range exceed 4000 m asl (Fig. 17.7). At least 10 glacial valleys are identified in this range, each displaying well-developed cirques and U-shaped valleys noted in the early surveys of the Basin and Range (eg, Russell, 1885). Terminal moraines are preserved in some but not all glacial valleys. Piegat (1980) assigned moraines in several valleys to the Lamoille (penultimate) and Angel Lake (last) Glaciations, following terminology for Pleistocene glaciations in the Great Basin coined by Sharp (1938). The best-preserved moraine of the last glaciation is in the Snake Creek valley, forming a wide, hummocky moraine plateau in the vicinity of Dead Lake. This moraine features large boulders at the crest suitable for cosmogenic ^{10}Be exposure dating. Exposure ages of the moraine are given later. The largest valley glaciers in the South Snake Range occupied the Lehman and Baker Creek valleys, attaining maximum lengths of nearly 6 km (Osborn and Bevis, 2001).

The North Snake Range features four glacial valleys in the vicinity of the highest peak, Mt. Moriah (Piegat, 1980). Osborn and Bevis (2001) noted that a moraine deposited by a small cirque glacier displays surficial characteristics similar to that of moraines of the last glaciation elsewhere in the Great Basin. Piegat (1980) did not assign glacial deposits in the North Snake Range to a

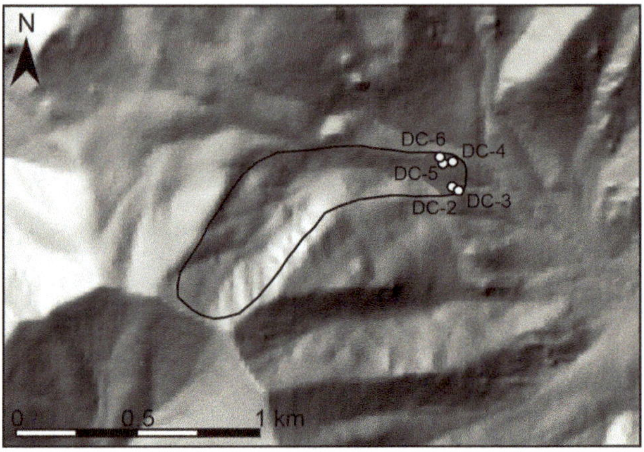

FIG. 17.8 Shaded-relief map of a portion of the Ibapah Peak, Deep Creek Mountains, digital elevation model (from http://epa.gov). Ice extent in a tributary of Granite Creek is outlined in *black*. *Darkened area* is the terminal moraine. Moraine boulder sample locations (*circles*) and *numbers* refer to samples shown in Table 17.2.

specific glaciation. Sparse preservation of moraines precludes an accurate reconstruction of ice extent and meaningful estimates of glacier ELAs.

Glacial deposits in the Deep Creek range, a remote range in western Utah more than 140 km north of the Snake Ranges, were identified by Bevis (1995) and Osborn and Bevis (2001). Bevis (1995) mapped moraines in four valleys and assigned terminal moraines to the last two Pleistocene glaciations based on preservation and weathering characteristics. A moraine in Granite Creek features large boulders suitable for cosmogenic ^{10}Be exposure dating (Fig. 17.8). Exposure ages of this moraine are given later.

At 2800–3000 m asl, ELAs of the South Snake and Deep Creek Ranges are similar to one another and to those reported for the central Utah ranges and plateaus, despite being on opposite sides of Lake Bonneville. This similarity suggests that Lake Bonneville did not enhance glacier mass balance in the central Utah ranges, other than possible local effects in the Wasatch Plateau and Pahvant Range.

17.2.5 Northern Ranges

Three small ranges in the north-central sector of the LBB, the Raft River, Stansbury and Oquirrh Ranges (Fig. 17.1), featured minor ice masses during the last glaciation. Glaciers in the Stansbury and Oquirrh Ranges were completely within the LBB. In contrast, only one small glacier in the Raft River Range flowed into Lake Bonneville, whereas other small ice masses in the range flowed northward and ultimately drained into the Snake River (Osborn and Bevis, 2001).

Glacial valleys appear in the northern segments of the Stansbury and Oquirrh Ranges, which were nearly surrounded by Lake Bonneville. These valleys are easily identified by well-developed cirques and U-shaped valleys in the Stansbury Range. Terminal and lateral moraines of the last glaciation are preserved in most valleys, indicating a maximum glacier length of 2.5–3 km in some east-flowing valleys (Osborn and Bevis, 2001). Conversely, glacial valleys are more difficult to identify in the Oquirrh Range due to weaker preservation of glacial erosional landforms and limited preservation of moraines. Here, glaciers were smaller in length and in number compared to the Stansbury Range (Osborn and Bevis, 2001).

In both ranges, glacial features are preserved well enough to enable reconstruction of ice extent and ELA in some valleys. ELAs in the two mountains follow the west-to-east decline across the northern sector of Lake Bonneville, displaying altitudes between those to the west and the Wasatch Mountains to the east (Table 17.1; Fig. 17.3).

17.3 CHRONOLOGY OF THE LAST GLACIATION

As described earlier, stratigraphic relationships between glacial and lacustrine deposits are limited to the western Wasatch Mountains, yet even there, previous stratigraphic models from the same glacial valley support different conclusions (Scott, 1988; Godsey et al., 2005). The lack of a complete stratigraphic record, combined with the limited number of numerical age limits on glacial deposits in the LBB, has for decades precluded a definitive understanding of the relative timing of glacier advance and retreat relative to the rise and fall of Lake Bonneville. This section presents chronologies of moraines in the basin based on cosmogenic ^{10}Be exposure dating of erratic boulders atop moraine crests. Most of the ages reported here are recalculated from previous work by the authors and their colleagues in the Wasatch and western Uinta Mountains (eg, Munroe et al., 2006; Laabs et al., 2009; Laabs et al., 2011). In addition, new cosmogenic ^{10}Be exposure ages from the western Wasatch, South Snake and Deep Creek Ranges are presented. The cosmogenic ^{3}He chronologies of glacial deposits in central Utah (Marchetti et al., 2005; Marchetti et al., 2011) are discussed later for comparison but are not recalculated. All of the available cosmogenic chronologies of terminal and recessional moraines are compared to the timing of high lake events in the LBB as described by Oviatt (2015).

Field and laboratory protocols involved in cosmogenic ^{10}Be exposure dating of moraine boulders in the Wasatch and Uinta Mountains are described in the original reports by Refsnider et al. (2008) and Laabs et al. (2009, 2011). Field and laboratory methods involved in cosmogenic ^{10}Be exposure dating of moraines in the western Wasatch, South Snake, and Deep Creek Ranges follow those of Laabs et al. (2013). Measurements of $^{10}Be/^{9}Be$ by accelerator mass spectrometry were done at the Purdue Rare Isotope Measurement Laboratory (PRIME Lab; Muzikar et al., 2003).

17.3.1 Calculations of Cosmogenic ^{10}Be Exposure Ages

Two recent reports of cosmogenic ^{10}Be exposure ages of moraines in the Wasatch and Uinta Mountains (Laabs et al., 2009, 2011) computed ages using a globally averaged, spallogenic-production rate, scaled for latitude and altitude using a model that assumes time-constant production of ^{10}Be (Balco et al., 2008). Since these reports were published, numerous additional calibrations of the spallogenic production of ^{10}Be have become available (as summarized by Borchers et al., 2016), nearly all of which yield a more precise and significantly lower reference production rate. Until recently, however, no surfaces with independent numerical ages were targeted for calibration of the ^{10}Be production rate in the western United States. That changed when Lifton et al. (2015, 2016) reported a calibrated ^{10}Be production rate model based on measurement of ^{10}Be concentrations in samples of two well-dated surfaces of Lake Bonneville deposits at Promontory Point in northern Utah. This model is useful for calculating the ^{10}Be production rate of late Pleistocene moraines in the LBB because of: (1) their proximity of the calibration site; and (2) the fact that the moraines and the surface used for the calibration both formed during the last glaciation.

In addition to this new, locally calibrated, production rate model, a separate study by Lifton et al. (2014) provides a new, physically based model for scaling the production of in situ ^{10}Be for latitude, altitude, and time. This model, termed the "LSD" (Lifton-Sato-Dunai) scaling scheme by Lifton et al. (2014), represents the most recent and robust consideration of the effects of geomagnitude latitude, atmospheric depth, and time on the spallogenic and muonic production of ^{10}Be. As discussed by Phillips et al. (2016), this model has the advantage of a stronger physical basis for computing the production of ^{10}Be and performs among the best of all available scaling models in a statistical analysis of known and predicted ages of calibration sites worldwide (Borchers et al., 2016). Given these advantages, and the risks of relying on a single calibration site, previously published and new cosmogenic ^{10}Be exposure ages reported here are recalculated using scaling factors generated by the LSD scaling model (Lifton et al., 2014) and a reference production rate of 4.0 ± 0.1 atoms g^{-1} $year^{-1}$ (after Shakun et al., 2015). As shown in the following paragraphs and Table 17.2, cosmogenic ^{10}Be exposure ages of moraines computed using the locally calibrated production model from Promontory Point and the LSD scaling factors yielded statistically overlapping exposure ages with central tendencies (mean and error-weighted mean ages) that differ by only hundreds of years. For consistency with Lifton et al. (2016), description of cosmogenic ^{10}Be exposure ages hereafter refers only to ages computed by the LSD scaling model. Recalculating cosmogenic ^{10}Be exposure ages in this way does not affect the distribution of ages about the central tendency. Rather, individual ages are uniformly ~10% older than initially reported and maintain their distribution. Because exposure ages from all but one moraine discussed later (the left-lateral segment of the moraine at

TABLE 17.2 ^{10}Be Data and Surface Exposure Ages for Moraines in the Lake Bonneville Basin

Moraine and Sample ID	Latitude (°N)	Longitude (°W)	Elevation (m asl)	Sample Thickness (cm)	Topographic Shielding Factor	Erosion Rate (cm/year)[a]	[^{10}Be] (Atoms g SiO_2^{-1})[b]	1σ	LSD Spall. Scaling Factor[c]	LSD Muon Scaling Factor[c]	LSD Exposure Age (ka)[d]	1σ External (kyr)[e]	Promontory Point-Lm Exposure Age (ka)[f]	1σ External (kyr)[e]
Wasatch—Little Cottonwood Canyon terminal														
02-UT-LCC-01	40.5823	−111.7973	1584	3	0.998	0.0003	2.54×10^5	8.21×10^3	3.282	1.409	**18.8**	**0.9**	**17.6**	**0.9**
02-UT-LCC-02	40.5815	−111.7938	1612	1.8	0.997	0.0003	2.46×10^5	6.53×10^3	3.343	1.417	**17.6**	**0.7**	**16.6**	**0.8**
02-UT-LCC-04	40.5828	−111.7937	1601	2	0.997	0.0003	2.48×10^5	6.32×10^3	3.317	1.414	**18.0**	**0.7**	**16.9**	**0.8**
02-UT-LCC-05	40.5828	−111.7934	1605	3	0.997	0.0003	2.61×10^5	6.56×10^3	3.337	1.415	**19.0**	**0.8**	**17.8**	**0.8**
02-UT-LCC-06	40.5817	−111.7948	1605	2	0.998	0.0003	2.44×10^5	6.13×10^3	3.325	1.415	**17.6**	**0.7**	**16.5**	**0.8**
02-UT-LCC-07	40.5823	−111.7931	1605	2	0.997	0.0003	2.43×10^5	6.22×10^3	3.325	1.415	**17.6**	**0.7**	**16.5**	**0.8**
02-UT-LCC-08	40.5808	−111.7935	1611	2	0.997	0.0003	2.70×10^5	7.04×10^3	3.357	1.417	**19.4**	**0.8**	**18.2**	**0.9**
											Mean ($n=7$, $\chi^2R=1.02$[g], 1σ) =**18.3±0.8**		Mean ($n=7$, $\chi^2R=0.69$, 1σ) =**17.2±0.7**	
Wasatch—Little Cottonwood Canyon lateral														
LCL-1	40.5705	−111.7941	1695	3	0.995	0.0003	1.98×10^5	6.74×10^3	3.534	1.443	**14.7**	**0.7**	**14.0**	**0.7**
LCL-2	40.5701	−111.7917	1714	1.5	0.992	0.0003	2.61×10^5	1.11×10^4	3.631	1.450	**18.9**	**1.1**	**18.1**	**1.1**
LCL-4	40.5704	−111.7938	1711	6	0.992	0.0003	4.07×10^5	1.92×10^4	3.748	1.451	**27.4**	**1.7**	**26.8**	**1.7**
LCL-5B	40.5702	−111.7919	1709	4	0.992	0.0003	1.62×10^5	5.58×10^3	3.559	1.446	**12.0**	**0.6**	**11.4**	**0.6**
LCL-6	40.5702	−111.7922	1706	2.5	0.995	0.0003	2.29×10^5	1.10×10^4	3.588	1.447	**16.7**	**1.0**	**16.1**	**1.0**
LCL-7	40.5702	−111.7925	1704	3.5	0.995	0.0003	3.21×10^5	1.47×10^4	3.685	1.448	**23.5**	**1.4**	**22.9**	**1.4**
LCL-9	40.5703	−111.7927	1699	2.5	0.996	0.0003	2.18×10^5	1.09×10^4	3.563	1.445	**16.0**	**1.0**	**15.3**	**1.0**
											Mean ($n=7$, $\chi^2R=32.43$, 1σ) =**18.5±5.3**		Mean ($n=7$, $\chi^2R=32.29$, 1σ) =**17.8±5.3**	

Continued

TABLE 17.2 ^{10}Be Data and Surface Exposure Ages for Moraines in the Lake Bonneville Basin—Cont'd

Moraine and Sample ID	Latitude (°N)	Longitude (°W)	Elevation (m asl)	Sample Thickness (cm)	Topographic Shielding Factor	Erosion Rate (cm/year)	[^{10}Be] (Atoms g SiO_2^{-1})	1σ	LSD Spall. Scaling Factor	LSD Muon Scaling Factor	LSD Exposure Age (ka)	1σ External (kyr)	Promontory Point-Lm Exposure Age (ka)	1σ External (kyr)
Wasatch—American Fork terminal														
AF-1	40.5200	−111.6197	2296	6	0.986	0	3.95×10^5	1.44×10^4	5.542	1.651	**17.2**	**0.8**	**16.6**	**0.9**
AF-2	40.5184	−111.6208	2312	3	0.994	0	3.91×10^5	1.69×10^4	5.596	1.656	**16.4**	**0.9**	**15.8**	**0.9**
AF-3	40.5177	−111.6208	2313	4	0.995	0	4.75×10^5	1.90×10^4	5.666	1.658	**19.7**	**1.0**	**19.2**	**1.0**
AF-4	40.5179	−111.6212	2312	2.5	0.993	0	4.53×10^5	1.23×10^4	5.634	1.657	**18.7**	**0.8**	**18.2**	**0.8**
AF-6	40.5193	−111.6206	2318	4	0.992	0	3.90×10^5	3.04×10^4	5.620	1.658	**16.4**	**1.4**	**15.9**	**1.4**
AF-7	40.5160	−111.6234	2285	4	0.992	0	3.38×10^5	1.52×10^4	5.453	1.645	**14.7**	**0.8**	**14.1**	**0.8**
AF-8	40.5168	−111.6236	2295	4.5	0.995	0	4.73×10^5	1.78×10^4	5.601	1.652	**19.9**	**1.0**	**19.4**	**1.0**
AF-9	40.5158	−111.6236	2283	4	0.994	0	3.92×10^5	1.13×10^4	5.486	1.646	**16.9**	**0.7**	**16.3**	**0.8**
AF-10	40.5185	−111.6223	2311	4.5	0.990	0	4.01×10^5	1.74×10^4	5.598	1.656	**17.0**	**0.9**	**16.5**	**0.9**
AF-11	40.5185	−111.6223	2311	4	0.990	0	3.93×10^5	7.97×10^3	5.595	1.656	**16.6**	**0.6**	**16.1**	**0.7**
											Mean (n=10, χ^2R=3.29, 1σ) =**17.4±1.6**		Mean (n=10, χ^2R=3.36, 1σ) =**16.8±1.6**	
Wasatch—Bells Canyon lateral														
BCR-1	40.56673	−111.78936	1803	4	0.9861	0	3.12×10^5	1.22×10^4	3.922	1.480	**20.6**	**0.8**	**20.0**	**0.8**
BCR-2	40.56674	−111.78958	1804	4	0.9859	0	3.81×10^5	1.31×10^4	3.990	1.482	**24.7**	**0.9**	**24.5**	**0.9**
BCR-4	40.56687	−111.79041	1800	4	0.9919	0	3.56×10^5	1.03×10^4	3.962	1.480	**23.1**	**0.8**	**22.8**	**0.8**
											Mean (n=3, χ^2R=6.08, 1σ) =**22.8±2.1**		Mean (n=3, χ^2R=7.37, 1σ) =**22.4±2.3**	

Wasatch—Bells Canyon recessional														
BCLR-1	40.56338	−111.79162	1741	4	0.987	0	2.42×10^5	9.07×10^3	3.6857	1.4585	**16.9**	**0.6**	**16.2**	**0.6**
BCLR-2	40.56338	−111.79162	1741	4	0.9889	0	2.58×10^5	7.62×10^3	3.6953	1.4588	**17.9**	**0.6**	**17.2**	**0.6**
											Mean ($n=2$, $\chi^2R=1.39$, 1σ) =**17.4±0.7**		Mean ($n=2$, $\chi^2R=1.39$, 1σ) =**16.7±0.7**	
Uinta—Bear River terminal														
EBBF-1	40.9178	−110.8210	2646	4.5	1	0	5.48×10^5	3.00×10^4	7.199	1.790	**17.7**	**1.2**	**17.7**	**1.2**
EBBF-2	40.9224	−110.8210	2593	6.5	1	0	5.83×10^5	2.86×10^4	7.017	1.770	**19.6**	**1.2**	**19.8**	**1.2**
EBBF-3	40.9241	−110.8210	2592	6.5	1	0	5.34×10^5	1.48×10^4	6.948	1.768	**18.2**	**0.8**	**18.2**	**0.8**
EBBF-4	40.9280	−110.8210	2578	3.5	1	0	5.70×10^5	3.11×10^4	6.915	1.764	**19.0**	**1.3**	**19.1**	**1.3**
EBBF-5	40.9181	−110.8210	2645	4.5	1	0	6.24×10^5	3.73×10^4	7.287	1.792	**19.9**	**1.4**	**20.1**	**1.4**
											Mean ($n=5$, $\chi^2R=0.65$, 1σ) =**18.9±0.9**		Mean ($n=5$, $\chi^2R=0.80$, 1σ) =**19.0±1.0**	
Uinta—Bear River lateral														
EFBR-1	40.8799	−110.7947	2757	4	1	0	5.18×10^5	5.09×10^4	7.711	1.834	**15.6**	**1.6**	**15.7**	**1.6**
EFBR-4A	40.8788	−110.8417	2640	4.5	1	0	7.29×10^5	1.34×10^5	7.355	1.792	**23.1**	**4.4**	**23.5**	**4.4**
EFBR-4B	40.8788	−110.8417	2640	4	1	0	6.46×10^5	2.47×10^4	7.286	1.790	**20.6**	**1.1**	**20.8**	**1.1**
EFBR-5	40.8805	−110.8429	2651	5	1	0	6.44×10^5	2.85×10^4	7.342	1.795	**20.5**	**1.2**	**20.8**	**1.2**
EFBR-7	40.8822	−110.8409	2646	3	1	0	6.41×10^5	5.97×10^4	7.298	1.792	**20.2**	**2.1**	**20.4**	**2.1**
EFBR-8	40.8847	−110.8403	2640	4.5	1	0	6.45×10^5	2.28×10^4	7.292	1.790	**20.6**	**1.1**	**20.9**	**1.1**
EFBR-9A	40.8855	−110.8394	2654	5	1	0	5.92×10^5	2.60×10^4	7.275	1.794	**19.0**	**1.1**	**19.1**	**1.1**
EFBR-9B	40.8855	−110.8394	2654	4.5	1	0	7.13×10^5	4.06×10^4	7.415	1.797	**22.4**	**1.6**	**22.8**	**1.6**
EFBR-9C	40.8855	−110.8394	2654	5	1	0	5.76×10^5	2.62×10^4	7.255	1.793	**18.5**	**1.1**	**18.6**	**1.1**
											Mean ($n=9$, $\chi^2R=1.74$, 1σ) =**20.1±2.2**		Mean ($n=9$, $\chi^2R=1.93$, 1σ) =**20.3±2.3**	

Continued

TABLE 17.2 ^{10}Be Data and Surface Exposure Ages for Moraines in the Lake Bonneville Basin—Cont'd

Moraine and Sample ID	Latitude (°N)	Longitude (°W)	Elevation (m asl)	Sample Thickness (cm)	Topographic Shielding Factor	Erosion Rate (cm/year)	[^{10}Be] (Atoms g SiO_2^{-1})	1σ	LSD Spall. Scaling Factor	LSD Muon Scaling Factor	LSD Exposure Age (ka)	1σ External (kyr)	Promontory Point-Lm Exposure Age (ka)	1σ External (kyr)
Uinta—North Fork Provo terminal														
NFP-1	40.5985	−111.0942	2319	5	1	0	3.22×10^5	1.56×10^4	5.597	1.659	**13.4**	**0.8**	**13.1**	**0.8**
NFP-2B	40.7023	−111.0945	2327	3	1	0	4.35×10^5	1.53×10^4	5.714	1.664	**17.5**	**0.9**	**17.2**	**0.9**
NFP-3A	40.7020	−111.0937	2321	6	1	0	4.81×10^5	5.16×10^4	5.747	1.663	**19.7**	**2.2**	**19.5**	**2.2**
NFP-4A	40.5965	−111.0927	2324	5	0.999	0	4.73×10^5	1.32×10^4	5.723	1.663	**19.3**	**0.9**	**19.0**	**0.9**
NFP-4B	40.5963	−111.0927	2324	2.5	1	0	3.91×10^5	1.94×10^4	5.659	1.661	**15.8**	**1.0**	**15.5**	**1.0**
NFP-4C	40.5963	−111.0927	2324	3	1	0	3.75×10^5	1.86×10^4	5.645	1.661	**15.2**	**0.9**	**15.0**	**0.9**
NFP-4D	40.5963	−111.0927	2324	3	0.999	0	5.33×10^5	2.55×10^4	5.798	1.665	**21.1**	**1.3**	**21.0**	**1.3**
NFP-5	40.5946	−111.0870	2346	3.5	1	0	4.39×10^5	1.18×10^4	5.771	1.670	**17.5**	**0.8**	**17.2**	**0.8**
											Mean ($n=8$, $\chi^2R=6.80$, 1σ) =**17.4±2.6**		Mean ($n=8$, $\chi^2R=6.95$, 1σ) =**17.2±2.6**	
Uinta—North Fork Provo lateral														
NFP-12	40.6006	−111.0719	2505	6	1	0	4.93×10^5	1.50×10^4	6.465	1.731	**17.9**	**0.9**	**17.8**	**0.9**
NFP-13	40.6007	−111.0718	2504	4.5	1	0	5.11×10^5	1.52×10^4	6.470	1.731	**18.4**	**0.9**	**18.2**	**0.9**
NFP-14	40.6003	−111.0733	2489	3	0.999	0	5.42×10^5	1.49×10^4	6.435	1.726	**19.4**	**0.9**	**19.3**	**0.9**
NFP-15	40.5996	−111.0753	2471	2.5	0.999	0	5.37×10^5	1.80×10^4	6.355	1.719	**19.3**	**1.0**	**19.2**	**1.0**
											Mean ($n=4$, $\chi^2R=0.62$, 1σ) =**18.8±0.7**		Mean ($n=4$, $\chi^2R=0.65$, 1σ) =**18.6±0.7**	

Deep Creek—Granite Creek terminal														
DC-02	39.8062	−113.9146	2661	6	0.982	0	3.22×10^5	1.29×10^4	6.881	1.783	*12.4*	*0.7*	*12.2*	*0.7*
DC-05	39.8081	−113.9160	2678	4.5	0.979	0	5.30×10^5	1.93×10^4	7.130	1.794	**19.5**	**1.0**	**19.4**	**1.0**
DC-06	39.8079	−113.9159	2675	6	0.979	0	5.33×10^5	2.00×10^4	7.137	1.794	**19.9**	**1.0**	**19.8**	**1.0**
DC-03	39.8064	−113.9150	2666	3	0.978	0	5.52×10^5	2.02×10^4	7.111	1.790	**20.2**	**1.0**	**20.1**	**1.0**
DC-04	39.8079	−113.9152	2668	5	0.977	0	8.88×10^5	1.45×10^5	7.433	1.800	*31.7*	*5.4*	*32.1*	*5.4*
										Mean (n=3, χ^2R=0.12, 1σ) =**19.9±0.4**		Mean (n=3, χ^2R=0.12, 1σ) =**19.8±0.4**		
South Snake Range—Dead Lake terminal														
DL-1	38.9355	−114.2724	2912	2.5	0.997	0	5.47×10^5	2.34×10^4	7.976	1.881	**17.4**	**1.0**	**17.2**	**1.0**
DL-2	38.9361	−114.2724	2911	3	0.997	0	6.30×10^5	2.17×10^4	8.097	1.884	**19.9**	**1.0**	**19.8**	**1.0**
DL-4	38.9364	−114.2731	2912	4.5	0.995	0	5.19×10^5	2.18×10^4	7.959	1.881	**16.9**	**0.9**	**16.6**	**0.9**
DL-6	38.9367	−114.2734	2908	2.5	0.995	0	4.37×10^5	1.75×10^4	7.855	1.877	*14.1*	*0.8*	*13.9*	*0.8*
										Mean (n=3, χ^2R=2.74, 1σ) =**18.1±1.6**		Mean (n=3, χ^2R=3.08, 1σ) =**17.9±1.7**		

Notes: All samples have a density of 2.65 g/cm^3. All ^{10}Be measurements were completed by accelerator mass spectrometer (AMS) at PRIME Lab (Purdue University). All samples were measured against AMS standard coded as "KNSTD" in the CRONUS Earth online exposure age calculator, except for the LCL (other than LCL-4), DC and DL samples, which were measured against Beryllium AMS standard Be-01-5-3 (Nishiizumi et al., 2007), which is coded "07KNSTD" in the CRONUS Earth exposure age calculator.
Ages and uncertainties are indicated in bold. Discarded ages are italicized.

[a]*See Laabs et al. (2011) for explanation of inferred erosion rates of moraine boulders in the Wasatch Mountains.*

[b]*Blank-corrected concentration.*

[c]*Computed from the scaling model of Lifton et al. (2014). LSD, Lifton-Sato-Dunai.*

[d]*Computed from the LSD scaling factors given in two preceding columns.*

[e]*Uncertainty of AMS measurement and production rate.*

[f]*Computed from the calibrated production rate at Promontory Point (Lifton et al., 2015) with "Lm" scaling by the CRONUS online exposure age calculator, version 2.2 (Balco et al., 2008).*

[g]*For all moraines, this value is computed from the surface exposure ages± the internal uncertainty.*

Little Cottonwood Canyon) are approximately normally distributed, we consider the mean age ($\pm 1\sigma$) to best represent the true age of the moraine. Competing effects of burial and prior exposure of erratic boulders may therefore cancel out in cases where the χ^2R statistic is greater than one, indicating that observed scatter exceeds that predicted by analytical uncertainty (Balco et al., 2009).

17.3.2 Cosmogenic ^{10}Be Chronologies of Moraines in the Western Uinta and Wasatch Mountains

Cosmogenic ^{10}Be exposure ages in the western Wasatch Mountains are from three moraine segments in two glacial valleys, Little Cottonwood and American Fork Canyons (Fig. 17.4). The two dated segments of the moraine at the mouth of Little Cottonwood, the right latero-frontal segment and the left-lateral segment, yield overlapping mean ages ($\pm 1\sigma$) of 18.3 ± 0.8 and 18.5 ± 5.3 ka, respectively. The geological or analytical reasons for the unusually broad scatter among exposure dates from the left-lateral segment (Table 17.2) are unclear. Cosmogenic ^{10}Be exposure ages of the terminal moraine in American Fork Canyon overlap with these ages, with a mean of 17.4 ± 1.6 ka. Together, ^{10}Be exposure ages from the western Wasatch Mountains indicate that overall retreat from these moraines was nearly synchronous.

Five new cosmogenic ^{10}Be exposure ages of moraines in Bells Canyon provide some useful insight to whether ice occupied the western front of the Wasatch Mountains during the Last Glacial Maximum, which was unclear from the original report of Laabs et al. (2011). Samples were collected from three erratic boulders atop the right-lateral sector of the terminal moraine and two from a recessional moraine less than 500 m upvalley. Both moraines represent times when ice occupied the western front of the Wasatch Mountains. Three cosmogenic ^{10}Be exposure dates from the right-lateral sector of the terminal moraine and two from the recessional moraine yield mean ages of 22.8 ± 2.8 and 17.4 ± 0.7 ka, respectively. Together, these ages indicate that ice occupied the mountain front during portions of the transgressive and overflowing phases of Lake Bonneville. The mean cosmogenic ^{10}Be exposure age of the recessional moraine overlaps with those of the right-lateral sector of the terminal moraine at Little Cottonwood Canyon and the terminal moraine in American Fork Canyon, further supporting the observation that final deglaciation began after the Lake Bonneville highstand but during the overflowing phase.

The timing of glaciation in the western Uinta Mountains is indicated by cosmogenic ^{10}Be exposure ages of moraines in the Bear River and North Fork Provo valleys. The ice-proximal sector of the terminal moraine on the piedmont of the Bear River valley has a mean cosmogenic ^{10}Be exposure age of 18.9 ± 0.9 ka and a recessional lateral moraine in the East Fork of Bear River has a mean exposure age of 20.1 ± 2.2 ka. Because these moraines delimit the ice margin in a recessional position, their cosmogenic ^{10}Be exposure ages represent minimum limits of the onset of overall ice retreat. Further limits on the start of ice retreat are available from radiocarbon ages of glacial flour identified in sediment cores of Bear Lake, downstream of the glaciated headwaters of the Bear River valley. These ages indicate a maximum in glacial-flour production at 20–19 ka, likely corresponding to the time of the glacier maximum or early ice retreat (Rosenbaum and Heil, 2009). Therefore, although the ^{10}Be exposure ages of the Bear River moraine segments overlap with those of moraines in the western Wasatch Mountains, the onset of ice retreat in Bear River likely occurred earlier.

In the North Fork Provo valley, cosmogenic ^{10}Be exposure ages of a terminal and left-lateral moraine indicate the onset of ice retreat at 17.4 ± 2.6 and 18.8 ± 0.7 ka, respectively. Overlapping ages between these two sectors of moraines are consistent with mapping of Refsnider et al. (2007), indicating that the two moraines are connected (Fig. 17.6). The possible reasons for the observed scatter among individual exposure ages from the terminal moraine are described in Refsnider et al. (2008). In any case, these exposure ages indicate that the onset of retreat of the North Fork Provo sector of the Western Uinta Icefield was in step with, or slightly later than, that of the Bear River sector and in step with the start of ice retreat in the western Wasatch Mountains.

17.3.3 Cosmogenic ^{10}Be Chronologies of Moraines in Western Ranges of the LBB

Cosmogenic ^{10}Be exposure ages of moraines in the western LBB are available from the Deep Creek and South Snake Ranges (Fig. 17.1; Table 17.2). In both settings, exposure ages of large erratic boulders atop a single-crested terminal moraine were determined, indicating the start of ice retreat in a single glacial valley (Table 17.2). In the South Snake Range, four cosmogenic ^{10}Be exposure ages have a mean age of 17.1 ± 2.4 ka, although the youngest of these does not overlap with the three oldest exposure ages within the measurement plus the production rate uncertainty (the latter is insignificant for comparing ages from the same moraine). Thus, the mean of the three oldest exposure ages, 18.1 ± 1.8 ka, provides a more accurate age of the moraine. In the Deep Creek Range, the oldest and youngest of five exposure ages do not overlap with any other ages in the population and are therefore considered outliers (Table 17.2). The mean of three overlapping exposure ages is 19.9 ± 0.4 ka. These ^{10}Be exposure ages are the first from the western LBB and are broadly consistent with those from the Wasatch and Uinta Mountains in the northeastern sector of the basin. Overall agreement among these ages (Fig. 17.9) suggests regional synchrony in the start of the last deglaciation.

17.3.4 The Pattern of the Last Glaciation in the LBB

Although chronological data are available for only a small number of glacial valleys in the LBB, the temporal pattern of the last glaciation provides a key

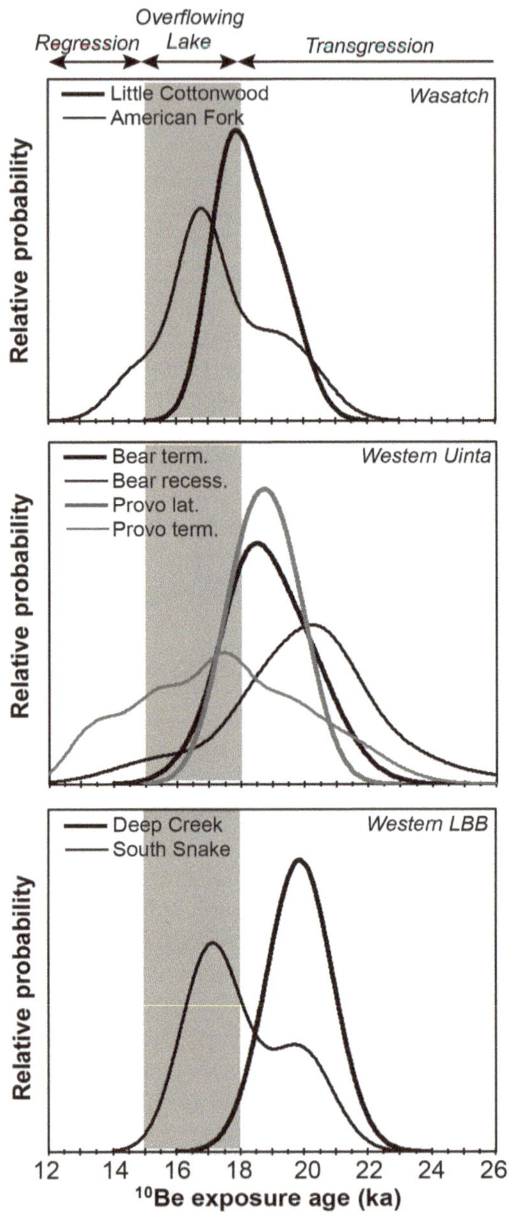

FIG. 17.9 Cumulative probably plots of cosmogenic ^{10}Be exposure ages of moraines in the Lake Bonneville basin relative to the transgressive (30–18 ka), overflowing (18–15 ka; *gray bar* on each plot), and regressive (15–12 ka) phases of Lake Bonneville, as defined by Oviatt (2015).

framework for understanding: (1) the relative timing of glaciation and the rise and fall of Lake Bonneville and (2) how climate changed during and after the last glaciation. The former is discussed here in light of the recalculated cosmogenic ^{10}Be exposure ages described earlier and additional age limits on glacial deposits in the LBB. The discussion is based on the chronology of Lake Bonneville as reviewed by Oviatt (2015), in which the transgressive (30–18 ka), overflowing (18–15 ka), and regressive (15–12.5 ka) phases of Lake Bonneville are identified largely on the basis of radiocarbon dating of shoreline deposits. This model reflects critical evaluation of a large number of radiocarbon dates and is broadly consistent with other recently reported chronologies of Lake Bonneville (eg, Benson et al., 2011; McGee et al., 2012; Miller et al., 2013).

The transgressive phase of Lake Bonneville began prior to, and ended after, the Last Glacial Maximum as defined by Clark et al. (2009) of 26.5–19.0 ka, a time when numerous mountain glaciers throughout the western United States attained their maximum extent (Shakun et al., 2015). In addition to cosmogenic ^{10}Be exposure ages of moraines in the western Uinta Mountains, Deep Creek and South Snake Ranges that fall within the latest part of the transgressive interval of the lake (Fig. 17.9), there is chronological and stratigraphic evidence indicating that glaciers in the western Wasatch Mountains attained their maxima prior to the overflowing phase of the lake. This includes stratigraphic observations of Scott et al. (1983) at the mouth of Little Cottonwood Canyon, where shoreline gravels of Lake Bonneville overlie weakly weathered till deposited during the last glaciation. Ice presence at the mountain front prior to the Lake Bonneville highstand is further evidenced by shoreline gravels overlying Pinedale age till comprising the distal slope of the moraine at Bells Canyon (Personius and Scott, 1992; Fig. 17.4).

Further support of glacier maxima during the transgressive phase of Lake Bonneville is yielded by cosmogenic ^{3}He exposure dating of moraines in central Utah (Marchetti et al., 2005, 2011). Terminal moraines of the last glaciation at Boulder Mountain (albeit just beyond the border of the LBB) and Fish Lake Plateau yield exposure ages corresponding to the Last Glacial Maximum, indicating that ice retreated from these moraines during the transgressive phase of Lake Bonneville.

Other terminal moraines in the LBB remained occupied, or were reoccupied, during the overflowing phase of the lake at 18–15 ka (Fig. 17.9). Cosmogenic ^{10}Be exposure ages of terminal moraines at Little Cottonwood and American Fork Canyons correspond to the time when Lake Bonneville overflowed, as does the mean age of the terminal moraine in the North Fork Provo River valley (Table 17.1; Fig. 17.9). The former ages are consistent with observations of interbedded till and lacustrine gravel near the mouth of Little Cottonwood Canyon of Godsey et al. (2005). East of the Provo River valley, a recessional moraine in the Uinta Mountains yields cosmogenic ^{10}Be exposure ages corresponding to the overflowing phase of Lake Bonneville, indicating that, in at least one glacial valley (Lake Fork Canyon), a readvance or stillstand occurred at 18 ka (Laabs et al., 2009). A readvance during the overflowing phase of Lake Bonneville is also indicated by cosmogenic ^{3}He exposure ages of recessional moraines at Boulder Mountain (Marchetti et al., 2005).

A key revision of the glacial chronology from earlier reports is that even the youngest terminal moraines in the LBB were abandoned near, and possibly before, the time of the Bonneville flood at 18 ka (Oviatt, 2015). Based on earlier ^{10}Be production rate models, Laabs et al. (2009, 2011) described the significance of the apparent synchrony of terminal moraine abandonment and the onset of the regressive phase of Lake Bonneville. The recalculated ^{10}Be exposure ages reported here indicate that terminal moraines were abandoned in advance of the fall from the Provo shoreline at or after 15 ka (Miller et al., 2013, 2015).

What remains unchanged from earlier reports is the strong likelihood that glaciers in the LBB attained their maxima (possibly multiple times) during the transgressive phase of Lake Bonneville (Laabs et al., 2011) and the Last Glacial Maximum. This is consistent with the earliest models of the relative timing of glacier maxima and the Lake Bonneville highstand (Gilbert, 1890; Atwood, 1909), and considerations of the asymmetry of moraine volumes at the mouths of Little Cottonwood and Bells Canyons (Laabs et al., 2011). Additional stratigraphic observations of Lake Bonneville sediment and till at the mouths of Little Cottonwood and Bells Canyons may help to clarify: (1) whether tills buried by the shoreline gravels were deposited during the last glaciation instead of during an earlier glacial episode; and (2) the precise timing of the onset of glacier maxima.

17.4 INFERRED CLIMATE DURING THE LAST GLACIATION

17.4.1 Regional Records of Climate Change and Inferences from the Glacial Record

Based on the available stratigraphic observations and numerical age limits for glacial deposits, the glacial record throughout the LBB indicates that glacier maxima occurred prior to the highstand of Lake Bonneville (ie, >18 ka) and that overall ice retreat began prior to or during the period when the lake overflowed at 18–15 ka (Fig. 17.9) with some local readvances. Glacier maxima during the transgressive phase of Lake Bonneville and the Last Glacial Maximum were undoubtedly driven by global- and regional-scale cooling in response to insolation forcing and minimal greenhouse gas concentrations (Clark et al., 2009) and is consistent in time with age controls on mountain glacier maxima elsewhere in the western United States (Phillips et al., 2009; Laabs et al., 2013; Shakun et al., 2015).

The fact that glaciers attained their maxima while Lake Bonneville was below its highstand elevation provides clues to understanding the magnitude of temperature and precipitation changes during the Last Glacial Maximum. Given that the length of mountain glaciers in continental interiors is largely controlled by temperature and radiation balance and that lakes are more sensitive to precipitation changes, climate in the LBB during the Last Glacial Maximum was likely cold and dry. This has been inferred in previous studies of glacial climate in the western United States (eg, Porter et al., 1983; Murray and Locke, 1989) and in the LBB (Kaufman, 2003), and is broadly consistent with atmosphere–ocean general circulation model simulations of the Last Glacial Maximum (eg, Bracconot et al., 2007). Amino-acid paleothermometry of Kaufman (2003) suggests that temperatures were 7–13°C colder than modern. Numerical modeling of the glacier in Little Cottonwood Canyon suggests that a temperature depression of 9°C or more was likely accompanied by less-than-modern precipitation (Laabs et al., 2006). A cold and drier-than-modern climate during the Last Glacial Maximum is consistent with the observation that Pleistocene lakes elsewhere in the Great Basin generally stood below their highstand elevations during the LGM (Munroe and Laabs, 2013), suggesting that precipitation levels were too low to sustain lakes at their highstands.

One possible interpretation of the correspondence of cosmogenic ^{10}Be exposure ages of some moraines and the start of the overflowing phase of Lake Bonneville (Fig. 17.9) is that climate in the LBB near the time of Bonneville highstand at 18 ka favored both glacier and lake maxima. If so, a cold and dry LGM climate may have transitioned into wetter conditions immediately post-LGM, favoring the expansion of lakes, and sustaining glaciers at or near their maxima. The retreat of the Laurentide Ice Sheet or hemisphere-scale effects of Heinrich Stadial 1 beginning after c.19 ka may have regionally affected atmospheric circulation in the western United States to permit the flow of southerly (Lyle et al., 2012) or westerly (Asmerom et al., 2010; Kirby et al., 2013; Oster et al., 2015), moisture-laden air into the Great Basin, regionally enhancing precipitation (Benson and Thompson, 1987; Licciardi et al., 2004; Thackray, 2008). Glacier modeling experiments and consideration of glacier ELAs in the northern LBB suggest that if temperature depression ranged from 6 to 9°C near the time of the Bonneville highstand, then precipitation was likely 1–2 × modern (Laabs et al., 2006; Leonard, 2007; Refsnider et al., 2008) or possibly greater (Laabs et al., 2015). If greater precipitation sustained mountain glaciers in the LBB at their maxima at c.18 ka, then this effect may have been short lived as indicated by the onset of overall ice retreat at 18 ka or shortly thereafter (Fig. 17.9) while Lake Bonneville continued to overflow until c.15 ka and numerous lakes elsewhere in the Great Basin attained their maxima at c.17 ka (Munroe and Laabs, 2013). Cosmogenic ^{10}Be exposure ages of some moraines beyond the LBB indicate that glaciers in the Great Basin and northern Rocky Mountains persisted at or readvanced to their maximum extent until 17–16 ka (Licciardi et al., 2004; Laabs et al., 2013), suggesting that, at least locally, climate remained favorable for lake and glacier maxima after Lake Bonneville began to overflow.

Further development of the chronology of glacial deposits, especially recessional moraines, within and beyond the LBB will help to understand how climate changed during the overflowing phase of Lake Bonneville. In any case, glaciers within the LBB and elsewhere in the western United States were greatly diminished or absent during the regressive phase of Lake Bonneville. The onset of ice retreat prior to the regressive phase of the lake suggests that, despite the strong possibility of enhanced precipitation while Lake Bonneville overflowed, glacier mass balance declined prior to regional-scale warming of the Bølling–Allerød interval (Stanford et al., 2006). Retreat was likely in response to warming driven by rising summer insolation and atmospheric CO_2 (Shakun et al., 2015).

17.4.2 Hydrological and Climatic Effects of Lake Bonneville on Mountain Glaciers

The potential impact of Lake Bonneville on glaciers in neighboring mountains has been discussed in numerous previous reports (eg, Zielinski and McCoy, 1987; Hostetler et al., 1994; Munroe and Mickelson, 2002; Munroe et al., 2006; Laabs et al., 2011). In light of the ^{10}Be exposure ages reported here and other chronological data pertaining to the relative timing of glacier maxima and the overflowing phase of Lake Bonneville, the potential hydrological and climatic linkages between the lake and glaciers are reconsidered here. Even in cases where glacier maxima occurred somewhat earlier (ie, at 20–19 ka) than the Lake Bonneville highstand, such as in the Bear River valley, Lake Bonneville was near its maximum extent during this time (Oviatt, 2015) and could have potentially affected glacier mass balance in neighboring, especially downwind, mountains.

The pattern of glacier ELAs in mountains across the northern LBB (Fig. 17.3) is especially intriguing given the correspondence of glacier maxima and the Lake Bonneville highstand at c.18 ka (Fig. 17.9). As noted earlier, the pattern of ELAs between the Ruby–East Humboldt Range in northeastern Nevada and the Wasatch Mountains does not follow the regional precipitation pattern at the mean ELA ($\sim$2750 m asl) for these mountains. If temperature depression relative to modern at $\sim$2750 m asl in this range of latitudes was approximately uniform (ie, a function of latitude) at 18 ka, then precipitation patterns likely differed from modern. Lake Bonneville may have been a significant, local moisture source, enhancing precipitation in the Stansbury, Oquirrh, Wasatch, and western Uinta Mountains and augmenting glacier mass balance. Lake-effect storms develop over Great Salt Lake today, depositing significant snowfall in the Wasatch and western Uinta Mountains (Carpenter, 1993), and may have been greater in magnitude when the lake was larger (Hostetler et al., 1994). As noted earlier, however, this effect was likely limited in duration, as indicated by the onset of overall ice retreat at 18 ka. The fact that Lake Bonneville continued to overflow until 15 ka indicates that the water budget in the LBB remained positive, but the onset of ice retreat suggests that climate was no longer favorable for glacier maxima. This may reflect warming at ~18 to 17 ka, but with sufficient precipitation to sustain an overflowing lake, or possibly a decline in local moisture availability for glaciers due to the 27% decline in the surface area of Lake Bonneville when it fell to the Provo shoreline (Currey et al., 1984). Given the greater sensitivity of mountain glaciers in continental interiors to temperature, ice retreat was most likely initiated by warming driven by rising summer insolation and rising atmospheric CO_2.

An alternative to the idea that Lake Bonneville enhanced precipitation delivery to glaciers in downwind mountains is that the corresponding highstand and glacier maxima represent a response to regional cooling. The Lake Bonneville highstand occurred shortly after a cooling pulse at 18.6 ka identified from a speleothem stable-isotope record from southern Nevada (Lachniet et al., 2011), suggesting that the final expansion of the lake to its highstand and corresponding occupation of terminal moraines in the LBB were driven chiefly by cooling rather than increased precipitation. This and other speleothem stable-isotope records from the southwestern United States (Asmerom et al., 2010; Wagner et al., 2010) indicate the onset of an effectively wetter climate at 18 ka or slightly later, suggesting a more complex model for temperature and precipitation changes that both sustained an overflowing Lake Bonneville and accompanied ice retreat in the LBB. Again, additional study of the pattern of ice retreat during the overflowing phase of Lake Bonneville would help to clarify the relative roles of temperature and precipitation in driving lake and glacier fluctuations during the period 18–15 ka.

17.5 SUMMARY AND CONCLUSIONS

The extent and chronology of the last glaciation in the LBB has significantly improved in recent decades due to widespread mapping efforts and the development of cosmogenic ^{10}Be and ^{3}He surface exposure dating of moraines. Mapping has enabled reconstruction of ice extents for most mountains, accounting for more than 95% of the surface area and volume of ice in the LBB during the last glaciation. The temporal correspondence of glacier maxima in the basin and the Lake Bonneville highstand is becoming clearer due to an improved understanding of the production rate of in situ ^{10}Be. Despite this improvement in the chronological framework, understanding changes in climate accompanying the later part of the transgressive phase of Lake Bonneville and the transition to the overflowing phase requires improved limits on either temperature or precipitation changes in the Great Basin. The pattern of the last glaciation as represented by mountain glacier ELAs across the northern LBB suggests enhanced precipitation in mountains engulfed by the lake and downwind of it. If such a precipitation pattern existed, it may have been limited to the brief duration of the Lake Bonneville highstand. Recent (eg, Lachniet et al., 2011) and emerging (eg, Oster et al., 2014) speleothem records for the Great Basin and neighboring regions combined with improved chronological limits on glacial deposits could help to improve our understanding of climate change during and after the last glaciation.

ACKNOWLEDGMENTS

New cosmogenic ^{10}Be exposure ages are based on measurement of $^{10}Be/^{9}Be$ by M. Caffee and others at PRIME Lab and were funded by NSF/EAR-0902472. The authors thank R. Becker, E. Carson, M. Devito, D. Douglass, P. Jewell, D. Koerner, D. Marchetti, D. Mickelson, J. Moore, N. Oprandy, J. Oviatt, B. Quirk, K. Refsnider, and numerous students and field assistants for thoughtful discussions of the glacial record of the LBB.

REFERENCES

Asmerom, Y., Polyak, V.J., Burns, S.J., 2010. Variable winter moisture in the southwestern United States linked to rapid glacial climate shifts. Nat. Geosci. 3, 114–117.

Atwood, W.W., 1909. Glaciation of the Uinta and Wasatch Mountains. U.S. Geological Survey Professional Paper, 61. 96 pp.

Bahr, D.B., Meier, M.F., Peckham, S.D., 1997. The physical basis of glacier volume–area scaling. J. Geophys. Res. 102 (B9), 20355–20362.

Balco, G., Briner, J., Finkel, R.C., Rayburn, J.A., Ridge, J.C., Schaefer, J.M., 2009. Regional beryllium-10 production rate calibration for northeastern North America. Quat. Geochronol. 4, 93–107.

Balco, G., Stone, J., Lifton, N., Dunai, T., 2008. A complete and easily accessible means of calculating surface exposure ages or erosion rates from ^{10}Be and ^{26}Al measurements. Quat. Geochronol. 3, 174–195.

Benson, L.V., Lund, S.P., Smoot, J.P., Rhode, D.E., Spencer, R.J., Verosub, K.L., Louderback, L.A., Johnson, C.A., Rye, R.O., Negrini, R.M., 2011. The rise and fall of Lake Bonneville between 45 and 10.5 ka. Quat. Int. 235, 57–69.

Benson, L.V., Thompson, R.S., 1987. The physical record of lakes in the Great Basin. In: Ruddiman, W.F., Wright Jr., H.E. (Eds.), North American and Adjacent Oceans During the Last Deglaciation. Geological Society of America, vol. K-3. The Geology of North America, Boulder, CO, pp. 241–260.

Bevis, K.A., 1995. Reconstruction of late Pleistocene paleoclimatic characteristics in the Great Basin and adjacent areas (Ph.D. thesis). Oregon State University, Corvallis. 278 pp.

Blackwelder, E., 1931. Pleistocene glaciation in the Sierra Nevada and Basin Ranges. Geol. Soc. Am. Bull. 42, 865–922.

Blackwelder, E., 1915. Post-Cretaceous history of the mountains of central-western Wyoming. J. Geol. 23, 97–340.

Borchers, B., Marrero, S., Balco, G., Caffee, M., Goehring, B., Lifton, N., Nishiizumi, K., Phillips, F., Schaefer, J., Stone, J., 2016. Geological calibration of spallation production rates in the CRONUS-earth project. Quat. Geochronol. 31, 188–198. http://dx.doi.org/10.1016/j.quageo.2015.01.009.

Bracconot, P., Otto-Bliesner, B., Harrison, S., Joussaume, S., Peterchmitt, J.-Y., Abe-Ouchi, A., Crucifix, M., Driesschaert, E., Fichefet, Th., Hewitt, C.D., Kageyama, M., Kitoh, A., Laîne, A., Loutre, M.-F., Marti, O., Merkel, U., Ramstein, G., Weber, S.L., Yu, Y.,

Zhao, Y., 2007. Results of PMIP2 coupled simulations of the Mid-Holocene and Last Glacial Maximum—part 1: experiments and large-scale features. Clim. Past 3, 261–277.

Bradley, W.A., 1936. Geomorphology of the north flank of the Uinta Mountains: U.S. Geological Survey Professional Paper 185-I. pp. 163–169.

Bryant, B., 1992. Geologic and Structure Maps of the Salt Lake City 1° × 2° Quadrangle, Utah and Wyoming: U.S. Geological Survey Miscellaneous Investigations Series Map I-1997. 1:125,000 scale.

Carpenter, D.M., 1993. The lake effect of the Great Salt Lake—overview and forecast problems. Weather Forecast. 8, 181–193.

Clark, P.U., Dyke, A.S., Shakun, J.D., Carlson, A.E., Clark, J., Wohlfarth, B., Mitrovica, J.X., Hostetler, S.W., McCabe, A.M., 2009. The Last Glacial Maximum. Science 325, 710–714.

Currey, D.R., Atwood, G., Mabey, D.R., 1984. Major levels of Great Salt Lake and Lake Bonneville. Utah Geological and Mineral Survey, Map 73. .

DeGraff, J.V., Gallegos, A.J., 1983. Relative dating of glacial deposits on the northern Sevier, Fish Lake, and UM Plateaus, central Utah: abstracts with programs. Geol. Soc. Am. 15 (5), 431.

Douglass, D.C., 2000. Glacial history of the west fork of Beaver Creek, Uinta Mountains, Utah. Unpublished M.S. thesis, University of Wisconsin, Madison. 64 pp.

Foley, L.L., Martin Jr., R.A., Sullivan, J.T., 1986. Seismotectonic study for Joes Valley, Scofield, and Huntington North Dams, Emergy County and Scofield Projects, Utah: U.S. Bureau of Reclamation Seismotectonic Report No. 86-7.

Gilbert, G.K., 1890. Lake Bonneville. U. S. Geological Survey Monograph, 1. 438 pp.

Godsey, H.S., Atwood, G., Lips, E., Miller, D.M., Milligan, M., Oviatt, C.G., 2005. Don R. Currey Memorial field trip to the shores of Pleistocene Lake Bonneville. In: Pederson, J., Dehler, C.M. (Eds.), Interior Western United States. Geological Society of America Field Trip Guide, vol. 6, pp. 419–448.

Hansen, W.R., 1975. The geologic story of the Uinta Mountains. U.S. Geol. Surv. Bull. 1291, 144.

Hostetler, S.W., Giorgi, F., Bates, G.T., Bartlein, P.J., 1994. Lake-atmosphere feedbacks associated with paleolakes Bonneville and Lahontan. Science 263, 665–668.

Ives, R.L., 1950. Glaciation in Little Cottonwood Canyon. Sci. Mon. 71, 105–117.

Kaufman, D.S., 2003. Amino acid paleothermometry of quaternary ostracodes from the Bonneville basin, Utah. Quat. Sci. Rev. 22, 899–914.

Kirby, M.E., Feakins, S.J., Bonuso, N., Fantozzi, J.M., Hiner, C.A., 2013. Latest Pleistocene to Holocene hydroclimates from Lake Elsinore, California. Quat. Sci. Rev. 76, 1–15.

Laabs, B.J.C., Amidon, W.A., Munroe, J.S., 2015. Combining Numerical Modeling of Pleistocene Glaciers and Lakes to Infer Climate Change During the Last Glaciation and Deglaciation in the Northern Great Basin, Utah and Nevada, U.S.A. In: Rosen, M.R., Cohen, A.S., Kirby, M., Gierlowski-Kordesch, K., Starratt, S.W., Valero Garces, B.L., Varekamp, J. (Eds.), Abstract Volume, Sixth International Limnogeology Congress, U.S., Geological Survey Open File Report 2015-1092, pp. 123–124.

Laabs, B.J.C., Munroe, J.S., Best, L.C., Caffee, M.W., 2013. Timing of the last glaciation and subsequent deglaciation in the Ruby Mountains, Great Basin, U.S.A. Earth Planet. Sci. Lett. 361, 16–25.

Laabs, B.J.C., Marchetti, D.W., Munroe, J.S., Refsnider, K.A., Gosse, J.C., Lips, E.W., Becker, R.A., Mickelson, D.M., Singer, B.S., 2011. Chronology of latest Pleistocene mountain glaciation in the western Wasatch Mountains, Utah, U.S.A. Quat. Res. 76, 272–284.

Laabs, B.J.C., Plummer, M.A., Mickelson, D.M., 2006. Climate during the last glacial maximum in the Wasatch and southern Uinta Mountains inferred from glacier modeling. Geomorphology 75, 300–317.

Laabs, B.J.C., Refsnider, K.A., Munroe, J.S., Mickelson, D.M., Applegate, P.A., Singer, B.S., Caffee, M.W., 2009. Latest Pleistocene glacial chronology of the Uinta Mountains: support for moisture-driven asynchrony of the last deglaciation. Quat. Sci. Rev. 28, 1171–1187.

Laabs, B.J.C., Carson, E.C., 2005. Glacial geology of the southern Uinta Mountains. In: Dehler, C.M., Pederson, J.L. (Eds.), Uinta Mountain Geology, vol. 33. Utah Geological Association Publication, Salt Lake City, Utah, pp. 235–253.

Lachniet, M.S., Asmerom, Y., Polyak, V., 2011. Deglacial paleoclimate in the southwestern United States: an abrupt 18.6 ka cold event and evidence for a North Atlantic forcing of Termination I. Quat. Sci. Rev. 30, 3803–3811.

Larsen, P.R., 1996. Glacial geomorphology of a portion of the Wasatch Plateau, Central Utah (Ph.D. dissertation). University of Utah, Salt Lake City. 186 pp.

Leonard, E., 2007. Modeled patterns of late Pleistocene glacier inception and growth in the Southern and Central Rocky Mountains, USA: sensitivity to climate change and paleoclimatic implications. Quat. Sci. Rev. 26, 2152–2166.

Licciardi, J.M., Clark, P.U., Brook, E.J., Elmore, D., Sharma, P., 2004. Variable responses of western U.S. glaciers during the last deglaciation. Geology 32, 81–84.

Lifton, N., Caffee, M., Finkel, R., Marrero, S., Nishiizumi, K., Phillips, F.M., Goehring, B., Gosse, J., Stone, J., Schaefer, J., Theriault, B., Jull, A.J.T., Fifield, K., 2015. In situ cosmogenic nuclide production rate calibration for the CRONUS-earth project from Lake Bonneville, Utah, shoreline features. Quat. Geochronol. 26, 56–69.

Lifton, N.A., Phillips, F.M., Cerling, T.E., 2016. Chapter 9. Using Lake Bonneville Features to Calibrate In Situ Cosmogenic Nuclide Production Rates. In: Oviatt, C.G., Shroder Jr., J.F. (Eds.), Lake Bonneville: A Scientific Update, vol. 20. Elsevier. pp. 165–183.

Lifton, N., Sato, T., Dunai, T.J., 2014. Scaling in situ cosmogenic nuclide production rates using analytical approximations to atmospheric cosmic-ray fluxes. Earth Planet. Sci. Lett. 386, 149–160.

Lisiecki, L.E., Raymo, M.E., 2005. A Pliocene–Pleistocene stack of 57 globally distributed benthic $\delta^{18}O$ records. Paleoceanography 20, PA1003. http://dx.doi.org/10.1029/2004PA001071.

Lyle, M., Heusser, L., Ravelo, C., Yamamoto, M., Barron, J., Diffenbaugh, N.S., Herbert, T., Andreasen, D., 2012. Out of the tropics: the Pacific, Great Basin lakes, and late Pleistocene water cycle in the western United States. Science 337, 1629–1633.

Madsen, D.B., Currey, D.R., 1979. Late quaternary glacial and vegetation changes, Little Cottonwood Canyon Area, Wasatch Mountains, Utah. Quat. Res. 12, 254–270.

Marchetti, D.W., Harris, M.S., Bailey, C.M., Cerling, T.E., Bergman, S., 2011. Timing of glaciation and last glacial maximum paleoclimate estimates from the Fish Lake Plateau, Utah. Quat. Res. 75, 183–195.

Marchetti, D.W., 2007. Pleistocene glaciations in central Utah—a review. In: Willis, G.C., Hylland, M.D., Clark, D.L., Chidsey Jr., T.C. (Eds.), Central Utah—Diverse Geology of a Dynamic Landscape, vol. 36. Utah Geological Survey Publication, Salt Lake City, Utah, pp. 197–203.

Marchetti, D.W., Cerling, T.E., Lips, E.W., 2005. A glacial chronology for the Fish Creek drainage of Boulder Mountain, USA. Quat. Res. 64, 263–271.

McCoy, W.D., 1977. A reinterpretation of certain aspects of the late Quaternary glacial history of Little Cottonwood Canyon, Wasatch Mountains, Utah. Unpublished M.A. thesis, University of Utah.

McGee, D., Quade, J., Edwards, R.L., Broecker, W.S., Cheng, H., Reiners, P.W., Evenson, N., 2012. Lacustrine cave carbonates: novel archives of paleohydrologic change in the Bonneville basin (Utah, U.S.A.). Earth Planet. Sci. Lett. 351–352, 182–194.

Miller, D.M., Wahl, D.B., McGeehin, J.P., Rosario, J., Oviatt, C.G., Anderson, L., Presnetsova, L., 2015. Limiting age for the Provo shoreline of Lake Bonneville. Quat. Int. 387, 99–105. http://dx.doi.org/10.1016/j.quaint.2015.01.001.

Miller, D.M., Oviatt, C.G., McGeehin, J.P., 2013. Stratigraphy and chronology of Provo shoreline deposits and lake-level implications, late Pleistocene Lake Bonneville, eastern Great Basin, USA. Boreas 42, 342–361.

Mulvey, W.E., Currey, D.R., Lindsay, L.M.W., 1984. Southernmost occurrences of later Pleistocene glaciation in Utah: Brian Head-Signey Peaks area, Markagunt Plateau. Encyclia 61, 97–101.

Munroe, J.S., Laabs, B.J.C., 2013. Temporal correspondence between pluvial lake highstands in the southwestern US and Heinrich Event 1. J. Quat. Sci. 28, 49–58.

Munroe, J.S., Laabs, B.J.C., 2009. Glacial geology of the Uinta Mountains area, Utah and Wyoming. DVD, 1 pl., scale 1:100,000, contains GIS data and text. Utah Geological Survey Miscellaneous Publications, Salt Lake City, Utah. ISBN: 1-55791-825-2. 09-4DM.

Munroe, J.S., Laabs, B.J.C., Shakun, J.D., Singer, B.S., Mickelson, D.M., Refsnider, K.A., Caffee, M.W., 2006. Latest Pleistocene advance of alpine glaciers in the southwestern Uinta Mountains, Utah, USA: evidence for the influence of local moisture sources. Geology 34, 841–844.

Munroe, J.S., 2005. Glacial geology of the northern Uinta Mountains. In: Dehler, C.M., Pederson, J.L. (Eds.), Uinta Mountain Geology, vol. 33. Utah Geological Association Publication, Salt Lake City, Utah, pp. 215–234.

Munroe, J., Mickelson, D., 2002. Last Glacial Maximum equilibrium-line altitudes and paleoclimate, northern Uinta Mountains, Utah, U.S.A. J. Glaciol. 48, 257–266.

Murray, D.R., Locke III, W.W., 1989. Dynamics of the late Pleistocene Big Timber Glacier, Crazy Mountains, Montana, U.S.A.. J. Glaciol. 35 (120), 183–190.

Muzikar, P., Elmore, D., Granger, D.E., 2003. Accelerator mass spectrometry in geologic research. Geol. Soc. Am. Bull. 115, 643–654.

Nishiizumi, K., Imamura, M., Caffee, M.W., Southon, J.R., Finkel, R.C., McAninch, J., 2007. Absolute calibration of ^{10}Be AMS standards. Nucl. Instrum. Meth. Phys. Res. B 258, 403–413.

Osborn, G., Bevis, K., 2001. Glaciation of the Great Basin of the western United States. Quat. Sci. Rev. 20, 1377–1410.

Osborn, G.D., 1973. Quaternary geology and geomorphology of the Uinta Basin and the south flank of the Uinta Mountains, Utah. Unpublished Ph.D. dissertation, University of California, Berkeley. 266 pp.

Oster, J.L., Ibarra, D.E., Winnick, M.J., Maher, K., 2015. Steering of western storms over western North America during the Last Glacial Maximum. Nat. Geosci. 8, 201–205. http://dx.doi.org/10.1038/NGEO2365.

Oster, J.L., Kelley, N.P., Montanez, I., 2014. Tracking regional and global teleconnections recorded by North American speleothem records. Abstracts with programs. Geol. Soc. Am. 46(6). 698.

Oviatt, C.G., 2015. Chronology of Lake Bonneville, 30,000 to 10,000 yr B.P. Quat. Sci. Rev. 110, 166–171.

Oviatt, C.G., 1997. Lake Bonneville fluctuations and global climate change. Geology 25 (2), 155–158.

Oviatt, C.G., 1994. Quaternary geologic map of the upper Weber River drainage basin, Summit County, Utah. Utah Geological Survey, Map 156, scale 1:50,000.

Personius, S.F., Scott, W.E., 1992. Surficial geologic map of the Sale Lake City segment and parts of adjacent segments of the Wasatch Fault Zone, Davis, Salt Lake, and Utah Counties, Utah: U.S. Geological Survey Miscellaneous Investigations Series Map I-2106, scale 1:50,000.

Phillips, F.M., Argento, D.C., Bourlès, D.L., Caffee, M.W., Dunai, T.J., Goehring, B., Gosse, J.C., Hudons, A.M., Jull, A.J.T., Kelly, M., Lifton, N., Marrero, S.M., Nishiizumi, K., Reedy, R.C., Stone, J.O.H., 2016. Where now? Reflections on future directions for cosmogenic nuclide research from the CRONUS Projects. Quat. Geochronol. 31, 155–159. http://dx.doi.org/10.1016/j.quageo.2015.04.010.

Phillips, F.M., Zreda, M., Plummer, M.A., Elmore, D., Clark, D.H., 2009. Glacial geology and chronology of Bishop Creek and vicinity, eastern Sierra Nevada, California. GSA Bull. 121, 1013–1023.

Piegat, J.J., 1980. Glacial geology of central Nevada. Unpublished M.S. thesis, Purdue University.

Porter, S.C., Pierce, K.L., Hamilton, T.D., 1983. Late Wisconsin mountain glaciation in the western United States. In: Porter, S.C. (Ed.), Late-Quaternary environments in the western United States. The Late Pleistocene, vol. 1. University of Minnesota Press, Minneapolis, MN, pp. 71–111.

Quirk, B.J., Moore, J., Laabs, B.J.C., Caffee, M.W., Plummer, M.A., 2015. Pleistocene Glacial Chronology and Paleoclimate of Big Cottonwood Canyon, Wasatch Range, Utah. Abstract EP53A-0997, Presented at the 2015 Fall Meeting, AGU, San Francisco, CA, 14–18 December.

Radić, V., Hock, R., Oerlemans, J., 2007. Volume-area scaling vs. flowline modeling in glacier volume projections. Ann. Glaciol. 46, 234–240.

Refsnider, K.A., Laabs, B.J.C., Plummer, M.A., Mickelson, D.M., Singer, B.S., Caffee, M.W., 2008. Last glacial maximum climate inferences from cosmogenic dating and glacier modeling of the western Uinta ice field, Uinta Mountains, Utah. Quat. Res. 69, 130–144.

Refsnider, K.A., Laabs, B.J.C., Mickelson, D.M., 2007. Glacial geology of the Provo River drainage, Uinta Mountains, U.S.A.. Arct. Antarct. Alp. Res. 39, 529–536.

Reheis, M.C., Adams, K.D., Oviatt, C.G., Bacon, S.N., 2014. Pluvial lakes in the Great Basin of the western United States—a view from the outcrop. Quat. Sci. Rev. 97, 33–57.

Reheis, M.C., Laabs, B.J.C., Forester, R.M., McGeehin, J.P., Kaufman, D.S., Bright, J., 2005. Surficial deposits in the Bear Lake basin. United States Geological Survey, Open-File Report 2005–1088. 30 pp.

Richmond, G., 1964. Glaciation of Little Cottonwood and Bells Canyons, Wasatch Mountains, Utah. U.S. Geological Survey Professional Paper 454-D, D1–D41.

Richmond, G., 1965. Glaciation in the Rocky Mountains. In: Wright, H.E., Frey, D.G. (Eds.), The Quaternary of the United States. Princeton University Press, Princeton, NJ, pp. 217–230.

Richmond, G., 1957. Three pre-Wisconsin glacial stages in the Rocky Mountain region. Geol. Soc. Am. Bull. 68, 239–262.

Rosenbaum, J.G., Heil Jr., C.W., 2009. The glacial/deglacial history of sedimentation in Bear Lake, Utah and Idaho. In: Rosenbaum, J.G., Kaufman, D.S. (Eds.), Paleoenvironments of Bear Lake, Utah and Idaho, and Its Catchment. Geological Society of America Special Paper 450, Boulder, CO, pp. 247–262.

Russell, I.C., 1885. Geological history of Lake Lahontan, a Quaternary lake of northwestern Nevada. U.S. Geological Survey Monograph 11. 288 pp.

Scott, W.E., 1988. Temporal relations of lacustrine and glacial events at Little Cottonwood and Bells canyons, Utah. In: Machette, M.N. (Ed.), In the Footsteps of G.K. Gilbert; Lake Bonneville and Neotectonics of the Eastern Basin and Range Province; Guidebook for Field Trip Twelve, Utah Geological Survey Miscellaneous Publication 88–1, 78–81, Salt Lake City, UT.

Scott, W.E., Shroba, R.R., 1985. Surficial geologic map of an area along the Wasatch Fault Zone in the Salt Lake Valley, Utah. U.S. Geological Survey Open-File Report 85-448. 19 pp.

Scott, W.E., McCoy, W.D., Shroba, R.R., Rubin, M., 1983. Reinterpretation of the exposed record of the last two cycles of Lake Bonneville, western United States. Quat. Res. 20, 261–285.

Shakun, J.D., Clark, P.U., He, F., Lifton, N.A., Liu, Z., Otto-Bliesner, B.L., 2015. Regional and global forcing of glacier retreat during the last deglaciation. Nat. Commun. 6, 8059. http://dx.doi.org/10.1038/ncomms9059.

Sharp, R.P., 1938. Pleistocene Glaciation in the Ruby-East Humboldt Range, northeastern Nevada. J. Geomorphol. 1, 296–323.

Stanford, J.D., Rohling, E.J., Hunter, S.E., Roberts, A.P., Rasmussen, S.O., Bard, E., McManus, J., Fairbanks, R.G., 2006. Timing of meltwater pulse 1a and climate responses to meltwater injections. Paleoceanography 21, PA4103.

Thackray, G.D., 2008. Varied climatic and topographic influences on late Plestocene mountain glaciation in the western United States. J. Quat. Sci. 23, 671–681.

Wagner, J.D.M., Cole, J.E., Beck, J.W., Patchett, P.J., Henderson, G.M., Barnett, H.R., 2010. Moisture variability in the southwestern United States linked to abrupt glacial climate change. Nat. Geosci. 3, 110–113.

Zielinski, G.A., McCoy, W.D., 1987. Paleoclimatic implications of the relationship between modern snowpack and late Pleistocene equilibrium-line altitudes in the mountains of the Great Basin, western U.S.A. Arct. Alp. Res. 19 (2), 127–134.

Chapter 18

The Early Human Occupation of the Bonneville Basin

D.B. Madsen
Texas State University, San Marcos, TX, United States

ABSTRACT
No direct evidence exists that people once occupied the shores of Lake Bonneville, but well-dated sites along coastal South America and overlooking a pluvial lake in the northwestern Great Basin suggest early foragers had reached the Bonneville Basin (BB) at least by ~14–15 cal ka. Dated hearths in cave sites overlooking the wetlands that developed during the regressive phase of Lake Bonneville indicate Paleoarchaic foragers, characterized by a variety of stemmed points and crescents, were definitely living in the BB by ~13.5–13.0 cal ka. Their subsistence and settlement apparently focused on extensive wetland ecosystems, where the highest number of Paleoarchaic sites is found. The largest of these wetlands, ~2600 km^2, formed at the terminal end of the Old River Bed (ORB) river that once flowed from Lake Gunnison in the Sevier basin north to the Great Salt Lake. More than 500 Paleoarchaic sites have been recorded in the ORB delta wetlands, and as many as 4–5000 may have been created during the ~2500 years or longer these wetlands were occupied. Geochemical sourcing of toolstone and intensive use of manufactured tools suggest these foragers stayed in the ORB wetlands for extended periods, perhaps moving about within the wetlands during the prolonged stays. When they did leave the wetlands they primarily moved about in a narrow ~400 km long north–south corridor within the BB. After the regression of Great Salt Lake from the Gilbert level by ~11 cal ka, the ORB wetlands were supported by groundwater flow supplied by stored Lake Bonneville water. Between ~9.5 and 9.0 cal ka this groundwater ceased to flow, the ORB wetlands dried up, BB foragers began to adapt to those changes, and the Paleoarchaic period effectively ended. These changes included a more intensive use of upland habitats, a seasonal reliance on pine nuts, and the development of grinding to process a variety of seeds. However, these Archaic foragers and later groups continued to seasonally occupy the many caves and rockshelters created by Lake Bonneville and the lake's impact on the people of the BB has never entirely disappeared.

Keywords: Paleoarchaic, Old River Bed delta, Western stemmed tradition, Caves and rockshelters, Archaic

Developments in Earth Surface Processes, Vol. 20. http://dx.doi.org/10.1016/B978-0-444-63590-7.00018-4

18.1 INTRODUCTION

It is becoming increasingly evident that the first people to enter the New World did so during the last high stand of Lake Bonneville, sometime between ~25 and 15 cal ka (Madsen, 2015; Raghavan et al., 2015) (all ages represent calendar years). As yet no direct evidence exists that these people occupied the shores of the lake during this period, but there is a good chance human foragers lived in the region at least during the period when it formed the Provo shoreline. The earliest known inhabitants of the Bonneville basin (BB) occupied the region beginning ~13 cal ka, well after the lake was at overflowing levels, but it is clear that even during its regressive phase the lake had a tremendous effect on the way people lived. Here, I review the early human occupation of the lake basin and the environmental impacts that Lake Bonneville had on the lifestyles of these early foragers.

18.2 THE INITIAL OCCUPATION OF THE AMERICAS DURING OR PRIOR TO THE BONNEVILLE HIGH STAND

It is now fairly well established that human foragers had spread throughout much of North and South America by 15–16 cal ka, much earlier than previously thought (Goebel et al., 2008; Meltzer, 2013; Madsen, 2015). Chronologically well-controlled excavations at two sites near the western margins of the two continents are largely responsible for this paradigmatic shift. One of these is the site of Monte Verde II in Chile, where dates on arrays of artifacts on house floors range from ~14.3 to 15.3 cal ka (Dillehay, 1997). The second is the Paisley Caves site in eastern Oregon, where dried human fecal remains in the stratified caves have been directly dated to as early as ~14.5 cal ka (Gilbert et al., 2008; Jenkins et al., 2012). Human DNA extracted from the fecal remains matches two of the five major mitochondrial haplogroups found in Native Americans. The ages of these sites are supported by an increasing host of sites of a similar age in both continents, such as Schaefer and Hebior located in southeastern Wisconsin (Overstreet and Kolb, 2003; Joyce, 2006) and Huaca Prieta in northern Peru (Dillehay et al., 2012). Reviews of these early sites are in Madsen (2015) and Meltzer (2009).

The Paisley Caves site is particularly important for understanding the early human occupation of the BB because the site is located within the Great Basin in a series of caves overlooking pluvial Lake Chewaucan. Summer and Albert lakes, at elevations of ~1260 and 1298 m, respectively, are the modern remnants of this Pleistocene lake. At its highest, Lake Chewaucan was at ~1378 m, and was only slightly below the ~1380 m elevation of the caves when they were first occupied (Licciardi, 2001; Negrini, 2002). These early Great Basin foragers appear to have been hunting extinct Pleistocene megafauna such as horses (*Equus* sp.) and camels (Camelidae), as well as mountain sheep (*Ovis* sp.) and other large game, but their diet was generally broad-based and included a variety of plants and smaller game animals found around the shores of the pluvial lake (Jenkins et al., 2012, 2013; Hockett and Jenkins, 2013). Given the presence of early Paleoamericans in the Great Basin by ~14.5–14.3 cal ka, together with the geomorphological and environmental similarity of the two lake basins at the time, it is highly likely that at least a small population of these early foragers was present in the BB when the lake occupied the Provo level. As yet, remains of these possible early occupations have not been found, but that may be because caves and rockshelters above the Provo level have rarely been investigated.

It is also possible that people were in the basin at even earlier times. A variety of sites occur throughout both North and South America that have been provisionally dated to between ~15 and 20 cal ka, but which are not yet widely accepted (Meltzer, 2013; Madsen, 2015). Yet, the sheer number of these sites suggests the New World was occupied before ~15 cal ka; and an increasing number of excavations now focused on this early period may soon conclusively demonstrate an earlier migration from the Eurasian continent. More importantly, it would have taken an unknown, but probably extended, period of time for the initial foragers to spread throughout the Americas. A number of mathematical models have been produced that estimate the rate at which human foragers would expand into previously unoccupied territory (eg, Bettinger and Young, 2004; Hazelwood and Steele, 2004; Bodner et al., 2012), and these models suggest that it would have taken the initial immigrants 1000 years at a minimum to spread throughout North America and at least 2000 years to spread throughout South America. The rate of expansion was probably slower than this, but even at these modest time estimates, the earliest peoples in the Americas must have been here by at least ~16.5 cal ka.

This is well within the last glacial maximum, when sea levels were 100–120 m lower and when the convergence of Laurentide and Cordilleran glaciation had closed off land-based entry points into North America (eg, Dyke, 2004; Clark et al., 2009). Since an ice-free corridor did not open until sometime after ~14.5 cal ka (Dyke, 2004; Clark et al., 2009), this entry point was most likely somewhere along the northwest coast of North America. Within the context of the North and South American continents as a whole, the BB is relatively close to this hypothetical northwest coast entry point, and thus people may have arrived in the basin soon after they first entered North America. This probable entry point, together with the location of the earliest well-dated New World sites along the Pacific coasts, has led to speculation that the earliest immigrants were boat-using foragers adapted to the resources of coastal estuaries and shorelines, and that they spread around the margins of the northern Pacific and south to the tip of South America before population expansion forced them to move inland (eg, Fladmark, 1979; Erlandson, 2002). If so, the earliest foragers to occupy the BB may not have done so until somewhat after the first people entered North America, but still prior to the Lake Bonneville regression. On the other hand, it is possible that the first immigrants to the New World entered North America well before the LGM and possibly even before Lake Bonneville began to transgress to its highest levels. The well-dated Yana site, located on coastal Siberia near the western margin of Beringia, was occupied as early as ~32 cal ka (Nikolskiy and Pitulko, 2013), and early land-based hunter-gatherers may have reached interior North America via the Bering land bridge as early as ~25 cal ka. Although an archeologically based, estimated time of arrival of ~15–25 cal ka closely matches estimates based on DNA and genetic clocks (O'Rourke and Raff, 2010; Raff et al., 2011; Reich et al., 2012) these estimates remain speculative, and it is not until ~13 cal ka that definitive evidence exists for the presence of early foragers in the BB.

18.3 THE PALEOARCHAIC IN THE GREAT BASIN

The earliest well-defined and widespread cultural tradition in western North America is characterized by a unique set of diagnostic lithic tools that together are known as the Western Stemmed Tradition (WST) (Bryan, 1980). Diagnostic forms include a variety of stemmed projectile points that vary geographically and chronologically, but which are generally consistent in their overall morphology and in their stems being laterally edge-ground (Fig. 18.1) (Beck and Jones, 2007). They are also characterized by a unique crescent-shaped biface that is thought to have been mounted perpendicular to the dart shaft (points out) and possibly used to hunt waterfowl (Beck and Jones, 1997; Jones and Beck, 1999), although this conjecture is by no means settled. These stemmed points and crescents, along with the rest of the lithic tool kit, constitute a lithic tradition that is very distinct from that of the largely contemporary Clovis tradition occurring in eastern North America (Beck and Jones, 2010, 2012, 2015a; Collins et al., 2013). Although Clovis points occur in a number of locations in the Great Basin (eg, at the Dietz site in Oregon (Pinson, 2011)), many of the fluted points occurring in western North America differ from Clovis points in a number ways and appear to have a different manufacturing technology than their eastern counterparts (Beck and Jones, 2012, 2015a,b; Collins et al., 2013). As with the rest of the Great Basin, these "western fluted" points are rare in the BB, but have been found at a number of locations (eg, Copeland and Fike, 1988; Davis and Brown, 1989; Oviatt et al., 2003).

In different forms and combinations, stemmed points and crescents are widely distributed in western North America, from the crest of the Rocky Mountains on the east to the Pacific coast on the west, and from Northern Baja, Mexico on the south, to British Columbia, Canada on the north (Fig. 18.2) (Madsen et al., 2015b). This distribution centers on the Great Basin and the Columbian Plateau, and it is in this central region where WST sites reach their greatest density and where the tradition is best defined. Sites

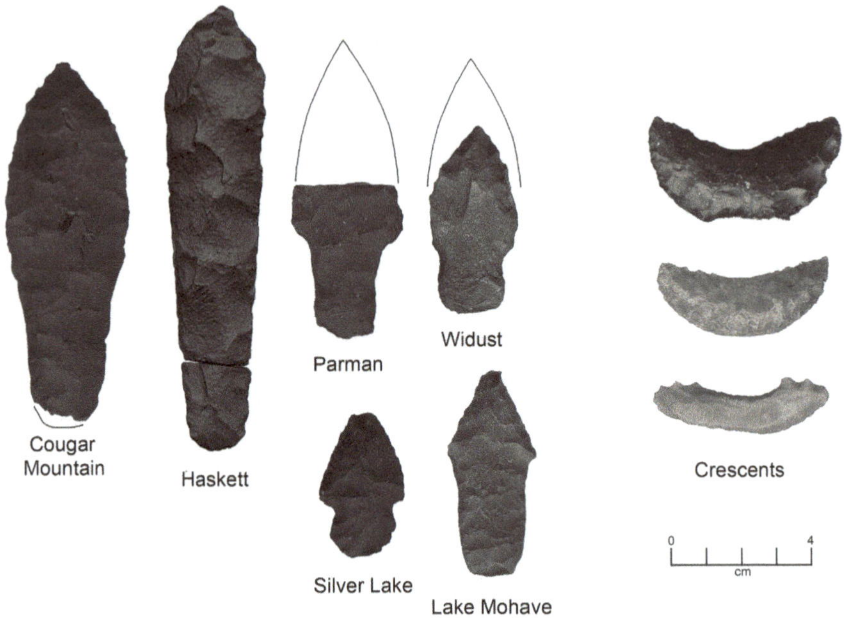

FIG. 18.1 Diagnostic Paleoarchaic stemmed point forms and crescents in western North America. The Cougar Mountain point comes from the Old River Bed delta. Image courtesy of Charlotte Beck.

are consistently found around the regressive shorelines of Great Basin pluvial lakes where extensive wetlands formed around the terminal distributaries of rivers and streams, and along Columbia River drainages north of the Great Basin where there was ready access to riverine resources (Bedwell, 1973; Beck and Jones, 1997). Most of these sites are surface sites, and the subsistence orientation of these early foragers is poorly known as a result. Based on the few sites where floral and faunal remains are present, subsistence was focused on a wide array of resources. This broad-based subsistence adaptation, together with a settlement orientation along marsh and riverine ecosystems and a lack of kill/butchering sites of such Pleistocene megafauna as mammoths and mastodons, has led to "Paleoarchaic" as a descriptive term for these early foragers (eg, Willig, 1988; Beck and Jones, 1997; Graf and Schmitt, 2007).

Because the vast majority of Paleoarchaic sites are surface sites, the chronological framework for the beginning of the tradition remains a subject of contention (eg, Goebel and Keene, 2014). The earliest well-dated site containing WST tool forms is the Paisley Caves site, where a number of stemmed point fragments are bracketed by ^{14}C dates of ~13.5–13.0 cal ka (Jenkins et al., 2013). At the Cooper's Ferry site, along the Salmon River in Idaho, stemmed points are associated with similar ^{14}C dates of ~13.3 cal ka (Davis et al., 2014; but see Goebel and Keene, 2014). At the Pilcher Creek site, also

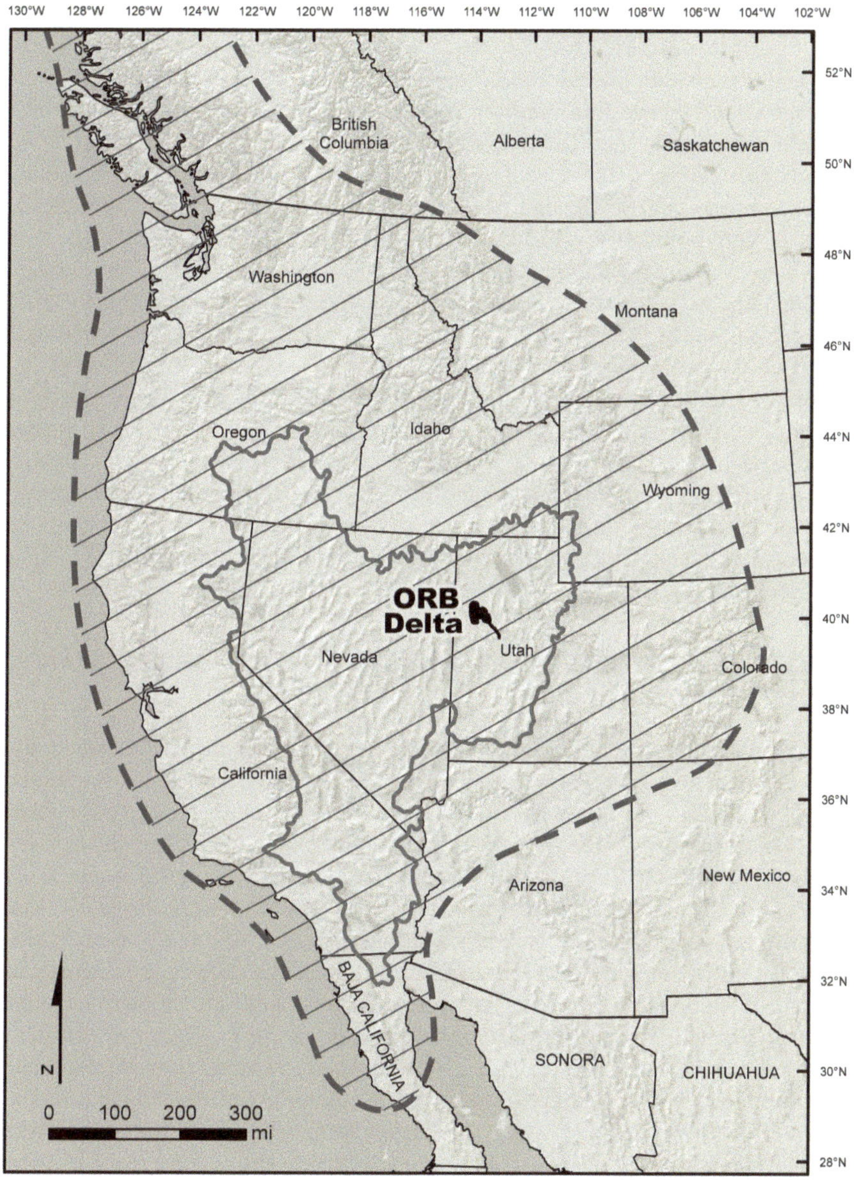

FIG. 18.2 Distribution of Western Stemmed Tradition diagnostic tool types in western North America. The Old River Bed (ORB) delta is where the greatest concentration of these Paleoarchaic tool types occur in the Bonneville Basin. *Modified from Madsen et al. (2015a).*

in Idaho, stemmed points were recovered in and on a Glacier Peak tephra dated to ~13.5 cal ka (Brauner, 1985; Kuehn et al., 2009). In the BB, the earliest Paleoarchaic dates are on charcoal from hearths in two cave sites on the western basin margin. At Smith Creek Cave in the foothills of the Snake

Range multiple hearths were initially dated to >13 cal ka (Bryan, 1979), but only one of these, dating to ~13.0 cal ka, has survived a stringent vetting process and is considered to be valid (Goebel et al., 2007). At Bonneville Estates, near the Utah/Nevada line south of Wendover, multiple hearths in the earliest cultural depositions date to between ~13.0 and 12.7 cal ka (Goebel et al., 2007; Graf, 2007).

The chronology for the end of the Paleoarchaic period is less contentious, but it is geographically variable, depending largely on when the large wetlands associated with pluvial Great Basin lakes finally dried up. In many of the smaller drainage basins, these marsh ecosystems were gone shortly after ~11 cal ka, but in the larger basins, particularly the BB, these large wetlands lasted until ~9.5 cal ka or even later (Jones and Beck, 2012; Madsen et al., 2015b). After this period, Great Basin foragers still used some of the now smaller wetlands at the ends of major rivers, but began to rely on a greater variety of habitats and developed the grinding of seeds of many upland grasses as a major supplement to their diet. It is the appearance of seed grinding that marks the end of the Paleoarchaic period and the beginning of the Archaic, but whether this transition was gradual (eg, Madsen, 2007; Jones and Beck, 2012) or rapid (eg, Pinson, 2007), remains a subject of debate.

18.4 THE IMPACT OF LAKE BONNEVILLE ON THE HUMAN ENVIRONMENT

Lake Bonneville began to regress from the Provo level by ~15 (Godsey et al., 2011), and within 500 years had regressed to the point where it had separated into two lakes in the northern, Great Salt Lake, and southern, Sevier, subbasins (Madsen et al., 2015a). By ~13.2 cal ka the lake had reached the modern levels of Great Salt Lake and had essentially ceased to exist (Madsen et al., 2001). Yet the lake continued to have a tremendous impact on the lives of BB peoples for thousands of years after its demise. Not the least of these impacts was the creation or enhancement of numerous caves and rockshelters formed in the soluble limestone of the basin-and-range mountains that were once islands within Lake Bonneville. Hundreds, possibly thousands, of these caves and rockshelters were created by solution and wave action during Lake Bonneville high stands, and are particularly numerous at the Bonneville and Provo lake levels, formed when the lake was threshold controlled and lake surface elevations were stable. Almost as soon as lake waters receded from the mouths of these caves they became home to the mobile foragers who lived in the basin and they continued to do so until historic times.

According to State Historical Preservation Office records for Utah, Idaho, and Nevada, nearly 200 BB caves and rockshelters have been formally recorded as archeological sites at elevations below 1550 m. From more than 30 years of personal experience working in the basin, I would estimate that at least two to three times this number remain to be identified. These caves were used for a variety of purposes ranging from storage to seed processing to temporary homes. Most of these caves were only occupied for short periods intermittently throughout the Holocene, but many, particularly those near permanent water sources, were occupied year after year as mobile foragers moved from place to place during a seasonal foraging round. In these cave sites, the accumulated residue of daily living left behind during these seasonal visits has resulted in stratified archeological deposits up to 4–5 m deep. The well-preserved organic remains in these dry caves, such as those in Danger Cave (Jennings, 1957) and Hogup Cave (Aikens, 1970) provide an astonishing record of daily life for periods of 12,000 or more, as well as climatic proxy records spanning the entire Holocene, such as those from Homestead Cave (eg, Madsen, 2000b). More than 30 of these caves have been tested or fully excavated and they provide a range of anthropological/archeological information that far exceeds that found at most open sites in the region (Fig. 18.3).

Perhaps the greatest impact Lake Bonneville had on the people who lived in its basin after its demise was the groundwater it left impounded in the fractured bedrock of the interior and surrounding mountain ranges. Until as late as ~9.5 cal ka this lake-derived groundwater fed large streams and springs that sustained extensive marsh and wetland ecosystems that lasted only until that stored groundwater was depleted (Madsen et al., 2015b; Oviatt et al., 2015). During the Younger Dryas, Great Salt Lake rebounded during the Gilbert episode, but by ~11 cal ka it had regressed to modern levels (Oviatt, 2014), and precipitation does not appear to have been greater than at present. Yet, many of the plants and animals that retreated to higher elevations later in the Holocene continued to persist in select low elevation locations, and it is likely that these refugia were supported by Lake Bonneville groundwater (Oviatt et al., 2015). These persistent BB wetlands allowed Paleoarchaic foragers to continue living largely as they had during previous millennia, unlike the rest of the Great Basin and Colorado Plateau, where the shift to an Archaic subsistence and settlement orientation had already started (Jones and Beck, 2014; Madsen et al., 2015a).

18.5 THE HUMAN OCCUPATION OF THE OLD RIVER BED DELTA AND OTHER BONNEVILLE BASIN WETLANDS

By far the largest of these wetland ecosystems formed in the delta of the Old River Bed (ORB) at the end of a river connecting the overflowing Sevier basin in the south with a lake in the Great Salt Lake basin to the north during the regressive phase of Lake Bonneville (Oviatt et al., 2003). The ORB delta was only a true delta during a brief period of the post-Provo regression and again for a brief time during the Gilbert episode (Oviatt, 2014; Madsen et al., 2015b). During the majority of their existence the wetlands consisted of a maze of interwoven distributary channels where the ORB river met the relatively flat

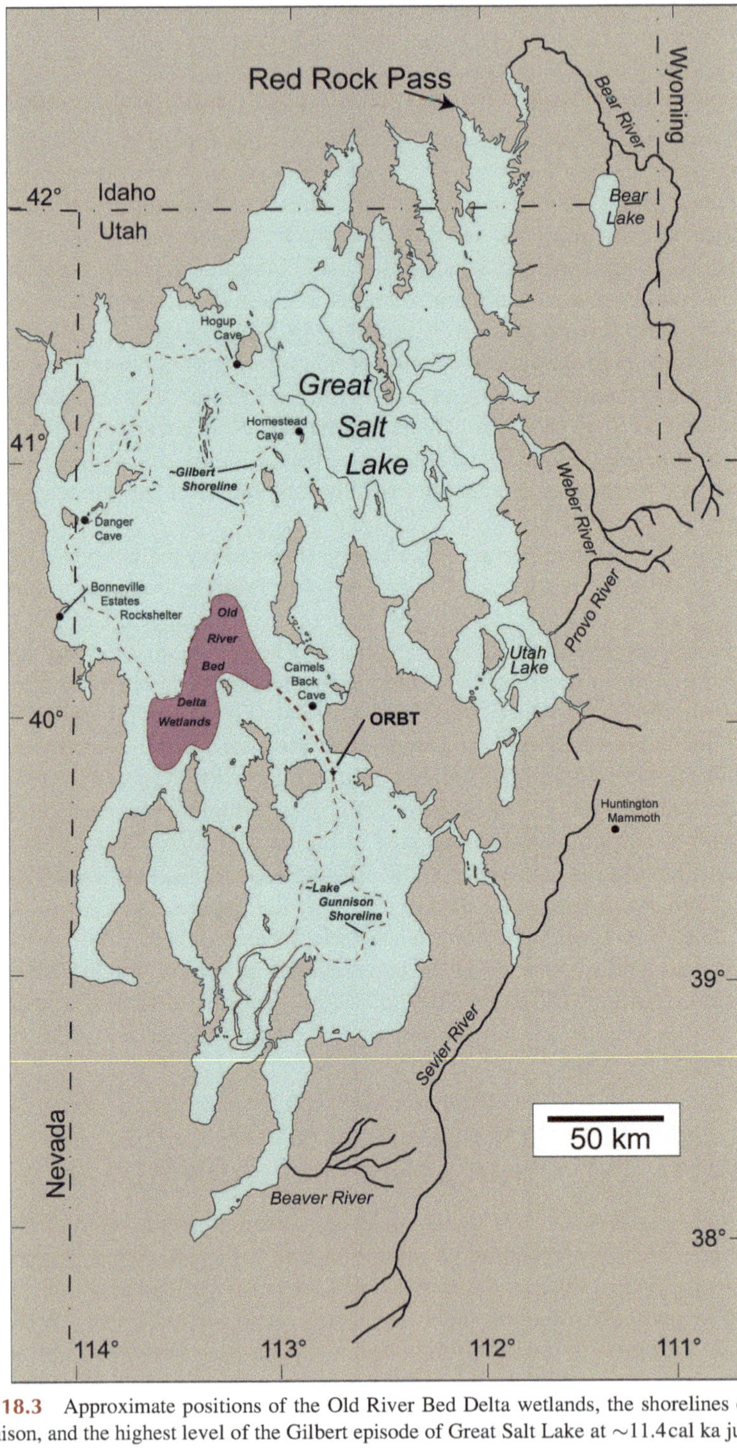

FIG. 18.3 Approximate positions of the Old River Bed Delta wetlands, the shorelines of Lake Gunnison, and the highest level of the Gilbert episode of Great Salt Lake at ~11.4 cal ka just prior to when Lake Gunnison ceased to overflow. Also shown are the locations of the Old River Bed threshold (ORBT) and the Old River Bed river connecting the two lakes (dashed line) and important archaeological sites mentioned in the text.

floor of the BB. This change in slope resulted in the creation of a highly meandering pattern of stream flow. At any one time these distributary channels ranged from those with flowing water, to abandoned channels with open water, to partially sediment-filled shallow water marshes, to completely filled channels where groundwater flow supported wetland meadows. Many of these channels are now topographically inverted, following the removal by deflation of the fine-grained Lake Bonneville lacustrine clays from around the coarser sands and gravels of the channel. Some of this deflation and channel inversion occurred during the formation of the ORB wetlands.

These differing channel forms supported a variety of wetland flora and fauna including fish, such as Utah chub (*Gila atraria*) and fresh-water mussels (*Anodonta* sp.) in channels with flowing water. Cattail (*Typha* sp.), cane (*Phragmites* spp.), and bulrush (*Scirpus* spp.) communities occurred in cut-off open water channels, while sedges (eg, *Schoenoplectus* sp.) and spikerush (eg, *Eleocharis* sp.) formed much of the wetland meadows. A large variety of waterfowl were present both as year-round and seasonal occupants, as were an equally large variety of small-game animals, probably including muskrats (*Ondatra* sp.) and rabbits (*Sylvilagus* sp.). The numbers and kinds of larger game is unknown, but based on modern Great Basin wetlands, likely included pronghorn (*Antilocapra* sp.), mule deer (*Odocoileus hemionus*), and seasonal elk (*Cervus* sp.).

Similar distributary stream channels originating in the Deep Creek and Fish Springs mountains merged with those fed by the ORB river, forming an ORB delta wetlands ecosystem that covered ~2600 km^2 (Fig. 18.4). In the proximal ORB delta area, constituting less than half of these overall wetlands, nearly 4000 linear km of paleochannels have been mapped and explored (Madsen et al., 2015b, p. 43). The total length of unmapped discontinuous channels that have been obscured by younger channel development and mudflat deflation between the mapped channels is likely three to four times this number, and for the wetlands as a whole the number is much higher. The chronology of channel formation in the ORB wetlands is controlled by 70 ^{14}C age estimates on plant remains and shells from 20 channels and on black mats not directly related to channels.

After Lake Bonneville separated into two major northern and southern basins, Lake Gunnison, in the Sevier basin, continued to overflow, while Great Salt Lake levels continued to fall. This regression in the northern basin appears to have been relatively rapid since the ORB distributary channels dating to this regressive period are relatively straight and suggest high-energy stream flow. These ORB streams may have emptied into a shallow playa lake in what is now the Great Salt Lake desert. After reaching its low point during the Bølling-Allerød, Great Salt Lake rebounded during the Gilbert episode to ~1295 m sometime between ~11.5 and 11.0 cal ka (Oviatt, 2014). During this time the ORB river formed a true delta and its distributary channels were carved by what was then a high-energy river carrying a significant sediment

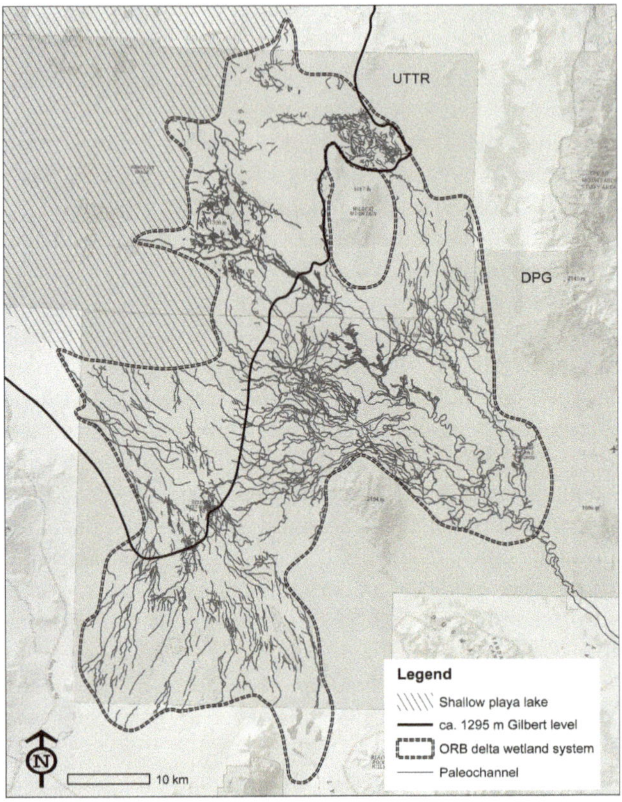

FIG. 18.4 Schematic plan view of the ORB delta wetlands system showing mapped contributory paleochannels originating from the Deep Creek Mountains and Fish Springs to the southwest, and the main ORB delta area to the north. A shallow playa lake seasonally covers the Great Salt Lake Desert north of the distributary system. The 1295m contour marks the elevation below which a Gilbert Lake formed ~11.3cal ka. The Utah Test and Training Range (UTTR) is a U.S. Air Force facility. Dugway Proving Ground (DPG) to the south is managed by the U.S. Army. *From Madsen et al. (2015a).*

load of sands and gravels from the Sevier basin to the delta margin. Stream flow and Great Salt Lake levels peaked shortly before ~11.4 cal ka, but slowed and regressed rapidly thereafter, and by ~11.3 cal ka, Lake Gunnison ceased to overflow. However, groundwater flow continued to feed open ORB distributary channels until sometime after ~9.8 cal ka. Dating is limited for the period ~9.8–9.3 cal ka, but sometime during this period even groundwater flow disappeared and the ORB wetlands reverted to a salt encrusted mudflat desert.

Human foragers appear to have been a part of the ORB delta wetlands throughout this 3000 year period. However, none of the archeological sites in the ORB are directly dated and the archeological chronology is controlled by the association of sites with individual well-dated channels. These associations begin definitively only ~12.5 cal ka because earlier channels have been obscured by wave action and sheet washing during the Gilbert episode, and because artifacts near the locations of these earlier channels have been scattered by those same erosional forces. As a result, human occupation of the delta wetlands can only be confirmed for a period between ~12.5 and 9.5 cal ka, but it is possible that occupation began by ~13 cal ka or even earlier. Buried sites in the wetlands are rare, and the large majority consists of lithic scatters of tools and manufacturing debris that lie on deflated surfaces. Most of these sites appear to be relatively intact, despite this deflation, with non-organic artifacts remaining in their horizontal positions as the fine-grained sediments around them were removed by wind erosion. This structural preservation has been enhanced by the ORB wetlands having been designated as a military reservation and protected from artifact collecting since the 1940s.

Sites in the delta wetlands were discovered and recorded by a combination of linear surveys conducted along mapped distributary channels and by block surveys. Intact sites are almost entirely restricted to the vicinity of channels, with only rare isolated artifacts occurring on the mudflats between channels. Slightly more than 500 Paleoarchaic sites have been formally recorded in the ORB delta, and they occur rather consistently about 1 km apart along the channels (Schmitt, 2015). By extrapolation to the nearly 4000 km of mapped channels in the ORB delta, at least that number of intact sites may occur in the wetlands and possibly many more along unmapped channels. A similar crude, area-based extrapolation from the number of sites in block surveys to the area of the ORB delta as a whole suggests perhaps that more than 5500 Paleoarchaic sites occur in the wetlands.

Not all of these sites were places where people lived. Analysis of kinds and numbers of tools and debitage on the sites suggests that the majority were special purpose sites where resources were collected and processed, and that only ~18% of the ~500 known sites are residential bases where people lived and from where they traveled to other locations to exploit the wetland resources of the ORB delta (Schmitt, 2015). Again, by a crude extrapolation, that means that somewhere between 600 and 900 such home-base residential sites may have occurred in the ORB wetlands. This seems a large number, but one that when applied to the period of occupation in the delta reduces to only one such site being created every 2–3 years.

Most of these residential base camps lie on areas of higher ground, such as point bars, natural levees, and older topographically inverted channels that formed "highways" along which it was possible to travel around in and out of the wetlands. Other extended sites occur along channel margins where the distribution of artifacts suggests foragers were exploiting resources in flowing and open-water distributaries. Still others occur within the channels, suggesting the exploitation of meadow or moist wetland resources in places where abandoned channels had largely filled in and where surface vegetation was supported by groundwater flow through the more permeable sands and gravels in the channel fill. Based on tool-to-debitage ratios, Schmitt (2015) estimated that in a sample of 100 of these 500+ Paleoarchaic sites, 18 functioned as camp sites, 20 were resource acquisition sites, 13 were sites where stone tools were fabricated, 7 were primarily resource processing localities, 5 were hunting sites, and 37 were less easily defined general/multiple task sites.

The plants and animals these early Paleoarchaic foragers were living on, and how and where they acquired them, is largely unknown since nearly all the sites from this period are surface sites with no organic preservation. However, in the BB, the ratio of upland Paleoarchaic sites to those in wetland settings is on the order of 1:100, suggesting their subsistence orientation focused almost entirely on marsh resources. One of the few Paleoarchaic sites with good preservation is Bonneville Estates, a Bonneville level rockshelter on the western margin of the BB. Fauna in the Paleoarchaic levels suggest a focus largely on small-game animals such as rabbits, hares (*Lepus* sp.), and sage grouse (*Centrocercus urophasianus*) (Hockett, 2015), and on a variety of plants (Rhode and Louderback, 2007). This plant use included the processing of seeds, but no evidence is known that these seeds were prepared for consumption by grinding, and the lack of any grinding stones at the >500 ORB Paleoarchaic sites suggests that such labor-intensive seed processing was not part of the Paleoarchaic subsistence strategy. Possible cut marks on a mammoth skeleton (*Mammuthus primigenius*), and a loose association with a stemmed point from the Huntington Canyon site on the eastern margin of the BB, suggest the possible rare hunting or scavenging of Pleistocene megafauna (Madsen, 2000a).

Based on the distribution of toolstone, primarily obsidian and other fine-grained volcanics, and to a lesser extent on lithic reduction strategies, it has been suggested that Paleoarchaic foragers were highly mobile, moving often from one wetland ecosystem to another across rather large foraging territories (eg, Jones et al., 2003, 2012). Much of this hypothesized mobility pattern was based on the geochemical sourcing of a limited number of obsidian tools, principally projectile points, at a limited number of sites, to particular obsidian outcrops. However, as more sites and more kinds of lithic tools and debitage have been geochemically analyzed, the estimated sizes of these foraging territories (or "conveyance zones" to use the most commonly employed term) have fallen dramatically (Jones et al., 2012). Similar sourcing studies, on both "points" and nonpoint artifacts, from a sample of ORB delta Paleoarchaic sites suggest that 95% of the volcanic toolstone comes from only three areas and that 76% of that toolstone comes from local sources around the ORB delta wetlands (Page and Duke, 2015). The other two sources are Browns Bench, on the Idaho/Nevada/Utah border just north of the BB (15%) and Black Rock/Wildhorse canyon in the Sevier basin on the southern margin of

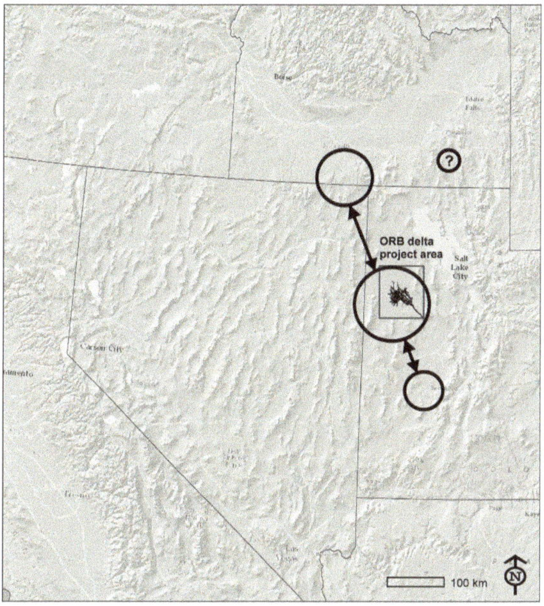

FIG. 18.5 The three major toolstone source areas likely visited by ORB foragers. The travel pattern was apparently consistent and equally limited throughout the Paleoarchaic occupation of the delta wetlands. The Malad obsidian source (small circle with the question mark), representing ~2% of the ORB delta collections, may be a product of either trade or rare direct procurement over the ~2500 years of occupation. *From Madsen et al. (2015a).*

the BB (5%). Assuming that toolstone distributions do represent Paleoarchaic foraging territories, an assumption that is largely untested, then the Paleoarchaic occupants of the ORB delta wetlands were moving around in a relatively narrow ~400 km long north–south corridor within the BB (Fig. 18.5), and spending most of their time in the delta itself (Madsen et al., 2015b). That does not mean that these early foragers lived in sedentary villages within the delta, rather, it appears that although they may have been living in the wetlands for much of the year, they were likely moving around from place to place within the wetlands ecosystem.

It is highly likely that other wetlands at the terminal ends of major rivers were also occupied by Paleoarchaic groups. Certainly an extensive wetland occurred at the ends of the Sevier and Beaver rivers in the Sevier Basin (Oviatt, 1988; Oviatt et al., 2003), and, although this wetland has not been as extensively studied archeologically, the Paleoarchaic occupation seems to have been just as intense (eg, Simms and Isgreen, 1984; Simms and Lindsay, 1989). Whether these were different groups, or the same people who occupied the ORB delta wetlands, remains unknown. It is also likely these or related groups lived in wetlands at the ends of the Bear, Provo, and Jordan rivers, but evidence for occupations is limited. This is because these wetlands, unlike the ORB delta and much of the Sevier basin wetlands, continued to form throughout the Holocene, and earlier sites are likely buried and/or have been subject to repeated scavenging for toolstone by later inhabitants (Beck and Jones, 2015b).

18.6 TRANSITION TO THE ARCHAIC

By sometime between ~9.5 and 9.0 cal ka, the wetlands supported by BB groundwater at the end of the ORB river, and likely those in the Sevier basin as well, had disappeared, as did the Paleoarchaic lifestyle these wetland ecosystems supported (Madsen et al., 2015b; Oviatt et al., 2015). They were replaced by unproductive saline mudflats that no longer attracted human foragers. Marsh areas still occurred at springs and at the ends of some of the larger rivers, but these wetlands were much smaller and more widely dispersed and could not support large numbers of people for extended periods. This reduction in groundwater flow, combined with a shift toward warmer and drier climatic conditions (Rhode and Louderback, 2015), led to a dramatic shift in floral and faunal communities (Rhode and Louderback, 2015; Schmitt and Lupo, 2016), and to a change in the subsistence and settlement strategies of BB foragers.

This change was not dramatic in any particular detail (Madsen, 2007). Archaic occupation patterns still consisted of small groups of hunter-gatherers moving around the landscape seasonally and living on a variety of resources collected from an assortment of habitats. What did change were the number and kinds of habitats later foragers occupied and the intensity with which they processed many resources for consumption. Smaller marshes and wetlands continued to be a focus of subsistence and settlement (Madsen and Kelly, 2008), but the generally smaller marshes led to shorter stays, and to the occupation and use of greater numbers and kinds of upland habitats (Jones and Beck, 1999, 2012). One of the more significant changes was a shift to a focus on the fall harvest of pine nuts, first limber pine (*Pinus flexilus*) and later pinyon pine (*P. monophylla*), once it had migrated into the BB during the early Holocene (Rhode and Madsen, 1998). Use of this eminently storable resource led to fall/winter occupations at favored locations in or near upland pinyon/juniper woodlands (Barlow and Metcalfe, 1996). Perhaps an even greater impact on Archaic lifestyles was a change in the way these people processed seeds. By ~9.6 cal ka (Rhode et al., 2006) BB foragers began to grind seeds using manos (hand stones) and metates (basal grinding stones) to break open the hard seed coats of heavy seed producers such as Great Basin wildrye (*Leymus cinereus*), Indian rice grass (*Achnatherum hymenoides*), and iodine bush/pickleweed (*Allenrolfea occidentalis*) to make them more digestible. The early mid-Holocene climatic shift and concomitant change in vegetation also led to changes in the animals that were targeted and in the way they were hunted. For example, work at Camels Back Cave (Schmitt et al., 2004) suggests there was a shift from a focus on rabbits to hares that probably involved a change from the individual hunting of rabbits to group net hunting of hares.

Together these small individual changes resulted in a more dramatic shift in the way people moved around the landscape. Stays were more prolonged at upland sites, and reduced at marsh sites, and the total number of moves in any one year likely increased as foragers more rapidly used up resources in the smaller resource patches. On the other hand, the size of a group's foraging territory likely decreased since the upland resources on which they now depended were more widely distributed than were wetland resources during the Paleoarchaic period, requiring less long-distance travel. Nevertheless, trade items suggest these smaller groups coalesced at times in particularly productive locations to socialize and exchange genes (eg, Hughes, 1994). These different foraging strategies led, in turn, to an evolution in lithic technology focused less on larger stemmed points and more on triangular notched projectile points used on atlatl darts (Jones and Beck, 2012). Together with ground stone artifacts, these notched projectile points are the diagnostic materials used to distinguish sites occupied during Archaic and later periods from those occupied during the Paleoarchaic.

18.7 CONCLUSIONS

Consensus is growing that the earliest Americans had spread throughout the western hemisphere by ~16 cal ka and possibly much earlier. The entry point for these immigrants was probably on the northwest coast of North America and by at least ~14.5 cal ka they had spread up the major drainages, such as the Columbia/Snake River system, into the northern Great Basin. Their population was small and thinly spread, and, although there is not yet any direct evidence that these early peoples were living near Lake Bonneville, it is highly likely. The paucity of sites dating to this period limits identification of any cultural diagnostics and precludes an understanding of how and where they lived.

The earliest widespread and easily recognized occupants of the Great Basin and the BB were Paleoarchaic foragers who occupied the region, beginning ~13.5–13.0 cal ka. These foragers can be readily identified by their stemmed projectile point forms, and the distribution of these points around shallow lakes, formed as Great Basin lakes regressed from their last glacial high stands, suggests their subsistence and settlement focus was on these wetland ecosystems. The earliest Paleoarchaic sites in the BB date to ~13 cal ka, and the largest concentration of these sites lies within a ~2600 km^2 wetland habitat at the terminal end of the Old River Bed. The Paleoarchaic occupation of these wetlands persisted until ~9.5–9.0 cal ka, much longer than elsewhere in the Great Basin, because the large wetland ecosystem was supported for its last ~1500 years by stored Lake Bonneville groundwater.

More than 500 Paleoarchaic sites have been officially recorded in these wetlands and as many as 4000–5000 may have been created during the

~2500–3000 years the wetlands were occupied. Less than 20% of these sites were camp sites, with the rest being special purpose sites used for such activities as resource acquisition and processing or tool manufacturing. Geochemical analyses of toolstone, together with evidence of extended tool use, scavenging, and resharpening, suggest stays within the ORB wetlands were prolonged and that when these Paleoarchaic foragers did move out of the wetlands they moved only along a narrow north–south zone within the Bonneville Basin. The subsistence focus is unknown because these sites are mostly surface sites with poor preservation, but from their location within the wetlands and from what little is known from nearby cave/rockshelter sites, these Paleoarchaic people relied on a broad array of lowland marsh plants and animals, with only rare trips to collect upland resources.

Once the ORB wetlands dried, the Paleoarchaic period came to an end, as did the major influence that Lake Bonneville had on people in the BB, and later foragers shifted their settlement/subsistence system to a larger mosaic of upland and valley bottom locations. This shift included the use of, and heavy reliance upon, pine nuts, and the grinding of a variety of seeds. During this Archaic period and later, foragers continued to use the caves and rockshelters created during the Bonneville period, and these caves are perhaps the most enduring aspect of Lake Bonneville's impact on human populations in the region.

ACKNOWLEDGMENTS

I am grateful to Jack Oviatt, David Rhode, Dave Schmitt, and Evelyn Seelinger for providing useful comments on an earlier draft of this chapter. I thank David Page for providing most of the artwork reproduced here and Arie Leeflang for assistance enumerating cave sites in the BB. Much of the work at the ORB delta reported here was supported by the US Army Dugway Proving Ground.

REFERENCES

Aikens, C.M., 1970. Hogup Cave. University of Utah Press, Salt Lake City. University of Utah Anthropological Papers 93.

Barlow, R.K., Metcalfe, D., 1996. Plant utility indices: two Great Basin examples. J. Archaeol. Sci. 23, 351–371.

Beck, C., Jones, G.T., 1997. The terminal Pleistocene/early Holocene archaeology of the Great Basin. J. World Prehist. 11, 161–236.

Beck, C., Jones, G.T., 2007. Early Paleoarchaic point morphology and chronology. In: Graf, K.E., Schmitt, D.N. (Eds.), Paleoarchaic or Paleoindian? Great Basin Human Ecology at the Pleistocene-Holocene Transition. University of Utah Press, Salt Lake City, pp. 23–41.

Beck, C., Jones, G.T., 2010. Clovis and western stemmed: population migration and the meeting of two technologies in the Intermountain West. Am. Antiq. 75, 81–116.

Beck, C., Jones, G.T., 2012. The Clovis-last hypothesis: investigating early lithic technology in the Intermountain West. In: Rhode, D. (Ed.), Meetings at the Margins: Prehistoric Cultural Interactions in the Intermountain West. University of Utah Press, Salt Lake City, pp. 23–46.

Beck, C., Jones, G.T., 2015a. A case of extinction in Paleoindian technology. In: Goodale, N., Andrefsky Jr., W. (Eds.), Lithic Technological Systems: Stone, Human Behavior, Evolution. Cambridge University Press, Cambridge, pp. 83–99.

Beck, C., Jones, G.T., 2015b. Lithic analysis. In: Madsen, D.B., Schmitt, D.N., Page, D. (Eds.), The Paleoarchaic Occupation of the Old River Bed Delta. University of Utah Press, Salt Lake City, pp. 97–208. University of Utah Anthropological Papers 128.

Bedwell, S.F., 1973. Fort Rock Basin: Prehistory and Environment. University of Oregon Books, Eugene.

Bettinger, R.L., Young, D.A., 2004. Hunter-gatherer population expansion in north Asia and in the Americas. In: Madsen, D.B. (Ed.), Entering America: Northeast Asia and Beringia before the Last Glacial Maximum. University of Utah Press, Salt Lake City, pp. 239–251.

Bodner, M., Perego, U.A., Huber, G., Fendt, L., Röck, A.W., Zimmermann, B., Olivieri, A., Gómez-Carballa, A., Lancioni, H., Angerhofer, N., Bobillo, M.C., Corach, D., Woodward, S.R., Salas, A., Achilli, A., Torroni, A., Bandelt, H.-J., Parson, W., 2012. Rapid coastal spread of First Americans: novel insights from South America's Southern Cone mitochondrial genomes. Genome Res. 22, 811–820.

Brauner, D.R., 1985. Early Human Occupation in the Uplands of the Southern Plateau: Archaeological Excavations at the Pilcher Creek Site (35UN147), Union County, Oregon: Report of the Department of Anthropology, Oregon State University, to the USDA Soil Conservation Service and the National Geographic Society, Washington, DC.

Bryan, A.L., 1979. Smith Creek Cave. In: Tuohy, D.R., Rendall, D.L. (Eds.), The Archaeology of Smith Creek Canyon, Eastern Nevada, pp. 162–251. Nevada State Museum Anthropological Papers 17, Carson City.

Bryan, A.L., 1980. The stemmed point tradition: an early technological tradition in western North America. In: Harten, L.B., Warren, C.N., Tuohy, D.R. (Eds.), Anthropological Papers in Memory of Earl H. Swanson. Idaho State Museum of Natural History, Pocatello, pp. 77–107.

Clark, P.U., Dyke, A.S., Shakun, J.D., Carlson, A.E., Clark, J., Wohlfarth, B., Mitrovica, J.X., Hostetler, S.W., McCabe, A.M., 2009. The Last Glacial Maximum. Science 325, 710–714.

Collins, M.B., Stanford, D.J., Lowery, D.L., Bradley, B.A., 2013. North America before Clovis: variance in temporal/spatial cultural patterns. In: Graf, K.E., Ketron, C.V., Waters, M.R. (Eds.), Paleoamerican Odyssey Center for the Study of the First Americans. Texas A&M University, College Station, Texas, pp. 521–540.

Copeland, J.M., Fike, R.E., 1988. Fluted projectile points in Utah. Utah Archaeol. 2, 5–28.

Davis, W.E., Brown, G.M., 1989. The Lime Ridge Clovis Site. Utah Archaeol. 2, 66–76.

Davis, L.G., Nyers, A.J., Willis, S.C., 2014. Context, provenance and technology of a Western Stemmed tradition from the Cooper's Ferry site, Idaho. Am. Antiq. 79, 596–615.

Dillehay, T.D., 1997. Monte Verde: A Late Pleistocene Settlement in Chile. Smithsonian Institution Press, Washington, D.C. vol. 2.

Dillehay, T.D., Bonavia, D., Goodbred Jr., S.L., Pino, M., Vásquez, V., Rosales Tham, T., 2012. A late Pleistocene human presence at Huaca Prieta, Peru, and early Pacific Coastal adaptations. Quat. Res. 77, 418–423.

Dyke, A.S., 2004. An outline of North American deglaciation with emphasis on central and northern Canada. In: Ehlers, J., Gibbard, P.L. (Eds.), Quaternary Glaciations: Extent and Chronology: Part 2: North America. Developments in Quaternary Sciences, vol. 2. Elsevier, Amsterdam, pp. 373–424.

Erlandson, J.M., 2002. Anatomically modern humans, maritime voyaging, and the Pleistocene colonization of the Americas. In: Jablonski, N.G. (Ed.), First Americans: the Pleistocene

Colonization of the New World. Memoirs of the California Academy of Science, vol. 27. California Academy of Science Press, Berkeley, pp. 59–92.

Fladmark, K.R., 1979. Routes: alternate migration corridors for early man in North America. Am. Antiq. 44, 55–69.

Gilbert, M.T.P., Jenkins, D.L., Götherstrom, A., Naveran, N., Sanchez, J.J., Hofreiter, M., Thomsen, P.H., Binladen, J., Higham, T.F.G., Yohe II, R.M., Parr, R., Scott Cummings, L., Willerslev, E., 2008. DNA from pre-Clovis human coprolites in Oregon, North America. Science 320, 786–789.

Godsey, H.S., Oviatt, C.G., Miller, D.M., Chan, M.A., 2011. Stratigraphy and Chronology of Offshore to Nearshore Deposits Associated with the Provo Shoreline, Pleistocene Lake Bonneville, Utah, USA. Palaeogeogr. Palaeoclimatol. Palaeoecol. 310, 442–450.

Goebel, T., Keene, J.L., 2014. Contextual and chronological hygiene and the earliest stemmed points of the intermountain west: making a long story short. In: Parezo, N.J., Janetski, J.C. (Eds.), Archaeology for All Times: Papers in Honor of Don D. Fowler. University of Utah Press, Salt Lake City, pp. 35–60.

Goebel, T., Graf, K., Hockett, B., Rhode, D., 2007. The Paleoindian occupations at Bonneville Estates Rockshelter, Danger Cave, and Smith Creek Cave (eastern Great Basin, U.S.A): interpreting their radiocarbon chronologies. In: Kornfield, M., Vasi'ev, S., Miotti, L. (Eds.), On Shelter's Ledge: Histories, Theories and Methods of Rockshelter Research. British Archaeological Reports, Oxford, pp. 147–161.

Goebel, T., Waters, M.R., O'Rourke, D.K., 2008. The late Pleistocene dispersal of modern humans in the Americas. Science 319, 1497–1502.

Graf, K.E., 2007. Stratigraphy and chronology of the Pleistocene to Holocene transition at Bonneville Estates Rockshelter, eastern Great Basin. In: Graf, K.E., Schmitt, D.N. (Eds.), Paleoarchaic or Paleoindian? Great Basin Human Ecology at the Pleistocene-Holocene Transition. University of Utah Press, Salt Lake City, pp. 82–104.

Graf, K.E., Schmitt, D.N. (Eds.), 2007. Paleoarchaic or Paleoindian? Great Basin Human Ecology at the Pleistocene-Holocene Transition. University of Utah Press, Salt Lake City.

Hazelwood, L., Steele, J., 2004. Spatial dynamics of human dispersals: constraints on modeling and archaeological validation. J. Archaeol. Sci. 31, 669–679.

Hockett, B., 2015. The zooarchaeology of Bonneville Estates Rockshelter: 13,000 years of Great Basin hunting strategies. J. Archaeol. Sci.: Reports 2, 291–301.

Hockett, B., Jenkins, D.L., 2013. Identifying stone tool cut marks and the pre-Clovis occupation of the Paisley Caves. Am. Antiq. 78, 762–778.

Hughes, R.E., 1994. Mosaic patterning in prehistoric California-Great Basin exchange. In: Baugh, T.G., Ericson, J.E. (Eds.), Prehistoric Exchange Systems in North America. Plenum Press, New York, pp. 363–383.

Jenkins, D.L., Davis, L.G., Stafford Jr., T.W., Campos, P.F., Hockett, B., Jones, G.T., Scott Cummings, L., Yost, C., Connolly, T.J., Yohe II, R.M., Gibbons, S.C., Raghaven, M., Rasmussen, M., Paijmans, J.L.A., Hofreiter, M., Kemp, B.M., Barta, J.L., Monroe, C., Gilbert, M.T.P., Willerslev, E., 2012. Clovis age Western Stemmed projectile points and human coprolites from the Paisley Caves. Science 337, 223–228.

Jenkins, D.L., Davis, L.G., Stafford Jr., T.W., Campos, P.F., Connolly, T.J., Scott Cummings, L., Hofreiter, M., Hockett, B., McDonough, K., Luthe, I., O'Grady, P.W., Reinhard, K.J., Swisher, M.E., White, F., Yates, B., Yohe II, R.M., Yost, C., Willerslev, E., 2013. Geochronology, archaeological context, and DNA at the Paisley Caves. In: Graf, K.E., Ketron, C.V., Waters, M.R. (Eds.), Paleoamerican Odyssey. Center for the Study of the First Americans, Texas A&M University, College Station, Texas, pp. 485–510.

Jennings, J.D., 1957. Danger Cave. University of Utah Press, Salt Lake City. University of Utah Anthropological Papers 27.

Jones, G.T., Beck, C., 1999. Paleoarchaic archaeology in the Great Basin. In: Beck, C. (Ed.), Models for the Millenium: Great Basin Anthropology Today. University of Utah Press, Salt Lake City, pp. 83–95.

Jones, G., Beck, C., 2012. The Emergence of the Desert Archaic in the Great Basin. In: Bousman, B.C., Vierra, B.J. (Eds.), From the Pleistocene to the Holocene: Human Organization and Cultural Transformations in Prehistoric North America, pp. 105–124. Texas A&M Anthropology Series No. 17, College Station.

Jones, G., Beck, C., 2014. Moving into the Mid-Holocene: the paleoarchaic/archaic transition in the Intermountain West. In: Parezo, N.J., Janetski, J.C. (Eds.), Archaeology in the Great Basin and Southwest: Papers in Honor of Don D. Fowler. University of Utah Press, Salt Lake City, pp. 61–84.

Jones, G.T., Beck, C., Jones, E.E., Hughes, R.E., 2003. Lithic source use and Paleoarchaic foraging territories in the Great Basin. Am. Antiq. 68, 5–38.

Jones, G.T., Fontes, L.M., Horowitz, R.A., Beck, C., Bailey, D.G., 2012. Reconsidering Paleoarchaic mobility in the central Great Basin. Am. Antiq. 77, 351–367.

Joyce, D.J., 2006. Chronology and new research on the Schaefer mammoth (*Mammuthus primigenius*) site, Kenosha County, Wisconsin, USA. Quat. Int. 142, 44–57.

Kuehn, S.C., Froese, D.G., Carrara, P.E., Foit Jr., F.F., Pearce, N.J.G., Rotheisler, P., 2009. Major- and trace-element characterization, expanded distribution, and a new chronology for the latest Pleistocene Glacier Peak tephras in western North America. Quat. Res. 71, 201–216.

Licciardi, J.M., 2001. Chronology of latest Pleistocene lake-level fluctuations in the pluvial Lake Chewaucan basin, Oregon, USA. J. Quat. Sci. 16, 545–553.

Madsen, D.B., 2000a. A high elevation Allerød-Younger Dryas megafauna from the west-central Rocky Mountains. In: Madsen, D.B., Metcalf, M.D. (Eds.), Intermountain Archaeology. University of Utah Press, Salt Lake City, pp. 100–115. University of Utah Anthropological Papers 122.

Madsen, D.B., 2000b. Late Quaternary Paleoecology in the Bonneville Basin. Utah Geological Survey Bulletin 130. Utah Geological Survey, Salt Lake City.

Madsen, D.B., 2007. The paleoarchaic to archaic transition in the Great Basin. In: Graf, K.E., Schmitt, D.N. (Eds.), Paleoarchaic or Paleoindian? Great Basin Human Ecology at the Pleistocene-Holocene Transition. University of Utah Press, Salt Lake City, pp. 3–20.

Madsen, D.B., 2015. A framework of the initial occupation of the Americas. PaleoAmerica 1 (3), 217–250.

Madsen, D.B., Kelly, R.L., 2008. The "Good Sweet Water" of Great Basin Marshes. In: Fowler, C.S., Fowler, D.D. (Eds.), The Great Basin: People and Place in Ancient Times. School for Advanced Research Press, Santa Fe, pp. 79–85.

Madsen, D.B., Rhode, D., Grayson, D.K., Broughton, J.M., Livingston, S.D., Hunt, J.M., Quade, J., Schmitt, D.M., Shaver III, M.W., 2001. Late quaternary environmental change in the Bonneville Basin, western U.S.A. Palaeogeogr. Palaeoclimatol. Palaeoecol. 167, 243–271.

Madsen, D.B., Schmitt, D.N., Page, D., 2015a. The Paleoarchaic Occupation of the Old River Bed Delta. University of Utah Press, Salt Lake City. University of Utah Anthropological Papers 128.

Madsen, D.B., Oviatt, C.G., Young, D.C., Page, D., 2015b. Old River Bed geomorphology and chronology. In: Madsen, D.B., Schmitt, D.N., Page, D. (Eds.), The Paleoarchaic Occupation of the Old River Bed Delta. University of Utah Press, Salt Lake City, pp. 30–60. University of Utah Anthropological Papers 128.

Meltzer, D.J., 2009. First Peoples in a New World: Colonizing Ice Age America. University of California Press, Berkeley.

Meltzer, D.J., 2013. The human colonization of the Americas: archaeology. In: Bellwood, P. (Ed.), The Encyclopedia of Global Human Migration. Prehistory, vol. 1. Wiley, New York, pp. 61–69.

Negrini, R.M., 2002. Pluvial lake sizes in the northwestern Great Basin throughout the Quaternary Period. In: Hershler, R., Madsen, D.B., Currey, D.R. (Eds.), Great Basin Aquatic Systems History. Smithsonian Contributions to the Earth Sciences, vol. 33. Smithsonian Institution Press, Washington D.C., pp. 11–52.

Nikolskiy, P., Pitulko, V., 2013. Evidence from the Yana Palaeolithic site, Arctic Siberia, yields clues to the riddle of mammoth hunting. J. Archaeol. Sci. 40, 4189–4197.

O'Rourke, D.H., Raff, J.A., 2010. The human genetic history of the Americas: the final frontier. Curr. Biol. 20, R202–R207.

Overstreet, D.F., Kolb, M.F., 2003. Geoarchaeological contexts for Late Pleistocene archaeological sites with human-modified woolly mammoth remains in southeastern Wisconsin, USA. Geoarchaeology 18, 91–114.

Oviatt, C.G., 1988. Late Pleistocene and Holocene lacustrine fluctuations in the Sevier Lake Basin, Utah, U.S.A. J. Paleolimnol. 1, 9–21.

Oviatt, C.G., 2014. The Gilbert Episode in the Great Salt Lake Basin, Utah. Utah Geological Survey, Salt Lake City.

Oviatt, C.G., Madsen, D.B., Schmitt, D.N., 2003. Late Pleistocene and early Holocene rivers and wetlands in the Bonneville basin of western North America. Quat. Res. 60, 200–210.

Oviatt, C.G., Madsen, D.B., Miller, D.M., Thompson, R.S., McGeehin, J.P., 2015. Early Holocene Great Salt Lake, USA. Quat. Res. 84, 57–68.

Page, D., Duke, D.G., 2015. Toolstone sourcing, lithic resource use, and paleoarchaic mobility in the western Bonneville Basin. In: Madsen, D.B., Schmitt, D.N., Page, D. (Eds.), The Paleoarchaic Occupation of the Old River Bed Delta. University of Utah Press, Salt Lake City, pp. 209–236. University of Utah Anthropological Papers 128.

Pinson, A.O., 2007. Artiodactyl use and adaptive discontinuity across the Paleoarchaic/Archaic transition in the northern Great Basin. In: Graf, K.E., Schmitt, D.N. (Eds.), Paleoarchaic or Paleoindian? Great Basin Human Ecology at the Pleistocene-Holocene Transition. University of Utah Press, Salt Lake City, pp. 187–203.

Pinson, A.O., 2011. The Clovis occupation of the Dietz site (35LK1529), Lake County, Oregon, and its bearing on the adaptive diversity of Clovis foragers. Am. Antiq. 76, 285–313.

Raff, J.A., Bolnick, D.A., Tackney, J., O'Rourke, D.H., 2011. Ancient DNA perspectives on American colonization and population history. Am. J. Phys. Anthropol. 146, 503–514.

Raghavan, M., Steinrücken, M., Harris, K., Schiffels, S., Rasmussen, S., DeGiorgio, M., Albrechtsen, A., Valdiosera, C., Ávila-Arcos, M.C., Malaspinas, A.-S., Eriksson, A., Moltke, I., Metspalu, M., Homburger, J.R., Wall, J., Cornejo, O.E., Moreno-Mayar, J.V., Korneliussen, T.S., Pierre, T., Rasmussen, M., Campos, P.F., De Barros Damgaard, P., Allentoft, M.E., Lindo, J., Metspalu, E., Rodríguez-Varela, R., Mansilla, J., Henrickson, C., Seguin-Orlando, A., Malmstöm, H., Stafford, T., Shringarpure, S.S., Moreno-Estrada, A., Karmin, M., Tambets, K., Bergström, A., Xue, Y., Warmuth, V., Friend, A.D., Singarayer, J., Valdes, P., Balloux, F., Leboreiro, I., Vera, J.L., Rangel-Villalobos, H., Pettener, D., Luiselli, D., Davis, L.G., Heyer, E., Zollikofer, C.P.E., Ponce De León, M.S., Smith, C.I., Grimes, V., Pike, K.A., Deal, M., Fuller, B.T., Arriaza, B., Standen, V., Luz, M.-F., Ricaut, F., Guidon, N., Osipova, L., Voevoda, M.I., Posukh, O.L., Balanovsky, O., Lavryashina, M., Bogunov, Y., Khusnutdinova, E., Gubina, M., Balanovska, E.,

Fedorova, S., Litvinov, S., Malyarchuk, B., Derenko, M., Mosher, M.J., Archer, D., Cybulski, J., Willerslev, E., 2015. Genomic evidence for the Pleistocene and recent population history of Native Americans. Science 349, 841–850.

Reich, D., Patterson, N., Campbell, D., Tandon, A., Mazieres, S., Ray, N., Parra, M.V., Rojas, W., Duque, C., Mesa, N., García, L.F., Triana, O., Blair, S., Maestre, A., Dib, J.C., Bravi, C.M., Bailliet, G., Corach, D., Hünemeier, T., Cátira Bortolini, M., Salzano, F.M., Petzl-Erler, M.L., Acuña-Alonzo, V., Aguilar-Salinas, C., Canizales-Quinteros, S., Tusié-Luna, T., Riba, L., Rodríguez-Cruz, M., Lopez-Alarcón, M., Coral-Vazquez, R., Canto-Cetina, T., Silva-Zolezzi, I., Fernandez-Lopez, J.C., Contreras, A.V., Jimenez-Sanchez, G., Gómez-Vázquez, M.J., Molina, J., Carracedo, A., Salas, A., Gallo, C., Poletti, G., Witonsky, D.B., Alkorta-Aranburu, G., Sukernik, R.I., Osipova, L., Fedorova, S.A., Vasquez, R., Villena, M., Moreau, C., Barrantes, R., Pauls, D., Excoffier, L., Bedoya, G., Rothhammer, F., Dugoujon, J.-M., Larrouy, G., Klitz, W., Labuda, D., Kidd, J., Kidd, K., Di Rienzo, A., Freimer, N.B., Price, A.L., Ruiz-Linares, A., 2012. Reconstructing native American population history. Nature 488, 370–374.

Rhode, D., Louderback, L.A., 2007. Dietary plant use in the Bonneville basin during the terminal Pleistocene/early Holocene transition. In: Graf, K.E., Schmitt, D.N. (Eds.), Paleoarchaic or Paleoindian? Great Basin Human Ecology at the Pleistocene-Holocene Transition. Great Basin Human Ecology at the Pleistocene-Holocene Transition. University of Utah Press, Salt Lake City, pp. 231–247.

Rhode, D., Louderback, L.A., 2015. Bonneville Basin environments during the Pleistocene-Holocene transition. In: Madsen, D.B., Schmitt, D.N., Page, D. (Eds.), The Paleoarchaic Occupation of the Old River Bed Delta. University of Utah Press, Salt Lake City, pp. 22–29. University of Utah Anthropological Papers 128.

Rhode, D., Madsen, D.B., 1998. Pine nut use in the early Holocene and beyond: the Danger Cave archaeobotanical record. J. Archaeol. Sci. 25, 1199–1210.

Rhode, D., Madsen, D.B., Jones, K.T., 2006. Antiquity of early Holocene small seed consumption and processing at Danger Cave, Utah, USA. Antiquity 80, 328–329.

Schmitt, D.N., 2015. Descriptions and classifications of Paleoarchaic sites in the proximal Old River Bed delta. In: Madsen, D.B., Schmitt, D.N., Page, D. (Eds.), The Paleoarchaic Occupation of the Old River Bed Delta. University of Utah Press, Salt Lake City, pp. 61–96. University of Utah Anthropological Papers 128.

Schmitt, D.N., Lupo, K.D., 2016. Chapter 13. Changes in Late Quaternary Mammalian Biogeography in the Bonneville Basin. In: Oviatt, C.G., Shroder Jr., J.F. (Eds.), Lake Bonneville: A Scientific Update, vol. 20. Elsevier. pp. 352–370.

Schmitt, D.N., Madsen, D.B., Lupo, K.D., 2004. The worst of times, the best of times: Jackrabbit hunting by Middle Holocene human foragers in the Bonneville Basin of western North America. In: Mondini, M., Muñoz, S., Wickler, S. (Eds.), Colonization, Migration and Marginal Areas: A Zooarchaeological Approach. Oxbow Books, Oxford, pp. 86–95.

Simms, S.R., Isgreen, M.C., 1984. Archaeological Investigations in the Sevier and Escalante Deserts, Western Utah. University of Utah Archaeological Center Reports of Investigations, Salt Lake City. pp. 83-12.

Simms, S.R., Lindsay, L.W., 1989. 42MD300, an early Holocene Site in the Sevier Desert. Utah Archaeol. 2, 56–66.

Willig, J.A., 1988. Paleo-Archaic adaptations and lakeside settlement patterns in the northern Alkali Basin, Oregon. In: Willig, J.A., Aikens Jr., C.M., Fagan, J.L. (Eds.), Early Human Occupation in Far Western North America: The Clovis-Archaic Interface, pp. 417–482. Nevada State Museum Anthropological Papers 21. Carson City.

Chapter 19

Imaging the Margins of Pleistocene Lake Deposits with High-Resolution Seismic Reflection in the Eastern Basin and Range: Pilot Valley, Utah (USA)

J.V. South, J.H. McBride, G.T. Carling, A.L. Mayo, D.G. Tingey, K.A. Rey and S.T. Nelson

Brigham Young University, Provo, UT, United States

ABSTRACT

A vast area of the northeastern Great Basin of the western USA was inundated by a succession of Pleistocene lakes, including Lake Bonneville. Playa-sediment deposition from these lakes onlapped onto alluvial fans that blanketed the slopes of adjacent mountain ranges to create prominent angular unconformities. Understanding these unconformities is useful for constraining the interpretation of the geologically recent tectonic evolution of the Basin and Range Province, as well as the interaction of lake sedimentation and alluvial fan development. The Pilot Valley playa, located just east of the Utah–Nevada border near Wendover, Utah, represents a remnant of these lakes. High-resolution seismic profiles have been acquired near the base of the bounding mountain ranges. The profiles reveal the stratigraphic relationships between Quaternary pluvial sediments as a shoreline depositional facies and the adjacent bounding fan deposits. On the western side of the basin, these profiles image subhorizontal playa sediments prograding over inclined alluvial fans. The boundary between the playa and fan sediments is marked by a prominent angular unconformity. Seismic images from the opposite side of the basin reveal a more heterogeneous structural and stratigraphic style, including down-to-the-basin normal faulting of shallow Paleozoic bedrock overlain by alluvial fan deposits, which are in turn onlapped by a thin veneer of playa sediments. The new geophysical images, when integrated with available geologic mapping, also aid in constraining how deep aquifers are locally recharged from an adjacent range. The results demonstrate the structural asymmetry of the range and playa system,

Developments in Earth Surface Processes, Vol. 20. http://dx.doi.org/10.1016/B978-0-444-63590-7.00019-6

consistent with a classic half-graben structure. Lastly, this study demonstrates the utility of the method of shallow seismic reflection to provide high-resolution subsurface images in the challenging environment of alluvial fan–playa geology.

Keywords: Seismic reflection, Lake Bonneville, Alluvial fans, Playa lakes

19.1 INTRODUCTION

Lake Bonneville in the eastern Great Basin, USA, is one of the world's best-studied Pleistocene lakes and represents one of a succession of pluvial inland lakes (Gilbert, 1890; Currey, 1990; Oviatt et al., 1999). The surface area of the lake has shrunk from approximately 51,000 km^2 (Gilbert, 1890; Mifflin and Wheat, 1979; Currey et al., 1984; Wambeam, 2001) to about one-eighth that amount in the remnant Great Salt Lake (Hostetler et al., 1994), leaving many areas of well-exposed shorelines and peripheral deposits available for study. The Pilot Valley alluvial fan–playa system, located just east of the Utah–Nevada border near Wendover, Utah, is a premier location for studying these ancient deposits. It is located at the west-central margin of the Great Salt Lake Desert in the northeastern part of the Basin and Range Province in northwestern Utah and northeastern Nevada (Fig. 19.1). Pilot Valley represents a typical basin within the greater Basin and Range Province. The predominant feature of the valley is a playa, with proportionally little alluvium at the surface between its bounding ranges (Fig. 19.2A and B). General scientific interest in the region began in the mid-1800s (Stansbury, 1852; Gilbert, 1890) and is reflected in the use and preservation of the nearby world-famous Bonneville Salt Flats and Speedway. As the need for accessing water for municipal and agricultural use becomes greater in the study area and western USA in general, it becomes increasingly important to understand the geologic subsurface along range fronts in areas such as Pilot Valley.

The purpose of this study is to reveal the stratigraphic relationship between playa-lake sediments and the underlying alluvial fan, which has clastics that have descended from the neighboring ranges. We use techniques of shallow, high-resolution seismic reflection to characterize the degree of shallow geological variability on opposite sides of, and along the playa. The interpretation of the seismic data is constrained by surface geological mapping. The seismic images furnish insights into the regional structure of the basin and its adjacent ranges. Our study also provides a stratigraphic context for interpreting a complex, freshwater–brine interface within the playa–fan system.

19.2 GEOLOGICAL SETTING

Pilot Valley is bounded on the west by the Pilot Range, where Pilot Peak is the highest point (3266 m above sea level (asl)) in the range (Fig. 19.1). The eastern boundary of the valley is marked by the Silver Island Range

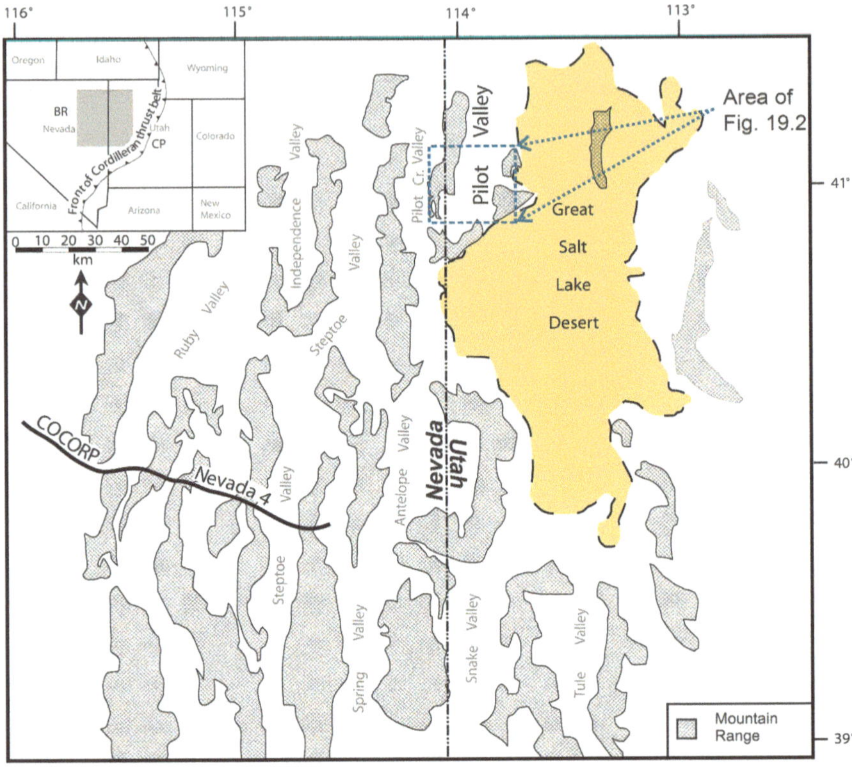

FIG. 19.1 Reference map centered over the northeastern Basin and Range Province (western USA) showing location of the study area and COCORP Nevada Profile 4 (Fig. 19.7). Other features of note are the Great Salt Lake Desert and Steptoe Valley. *BR* is Basin and Range; *CP* is Colorado Plateau.

and Crater Island mountains (Fig. 19.2A), where Graham Peak (2305 m asl) is the highest point (Fig. 19.3). The elevation of the Pilot Valley playa is slightly higher than the nearest elevation of the Great Salt Lake Desert (1295 vs 1285 m asl) and therefore has some surface drainage potential at the northeast edge of the valley where playa sediments slope toward the Desert. The valley is a fault-bounded graben, consistent with the general structure of the Basin and Range Province (Eaton, 1980). Previously, the valley has been described as a typical symmetric graben by Cook et al. (1964) with some *en echelon* normal faults on each side of the valley, having apparent equal or nearly equal displacements. More recently, however, geological mapping by Miller et al. (1990) and others shows a prominent range-front normal fault on the west side of the Pilot Valley basin, stepping down from Pilot Peak, with no or lesser faulting in the Crater Island and Silver Island Ranges on the east side of the basin. Prior seismic exploration east of the Silver Island and Crater Island mountains found that Cenozoic sediments unconformably overlie Paleozoic

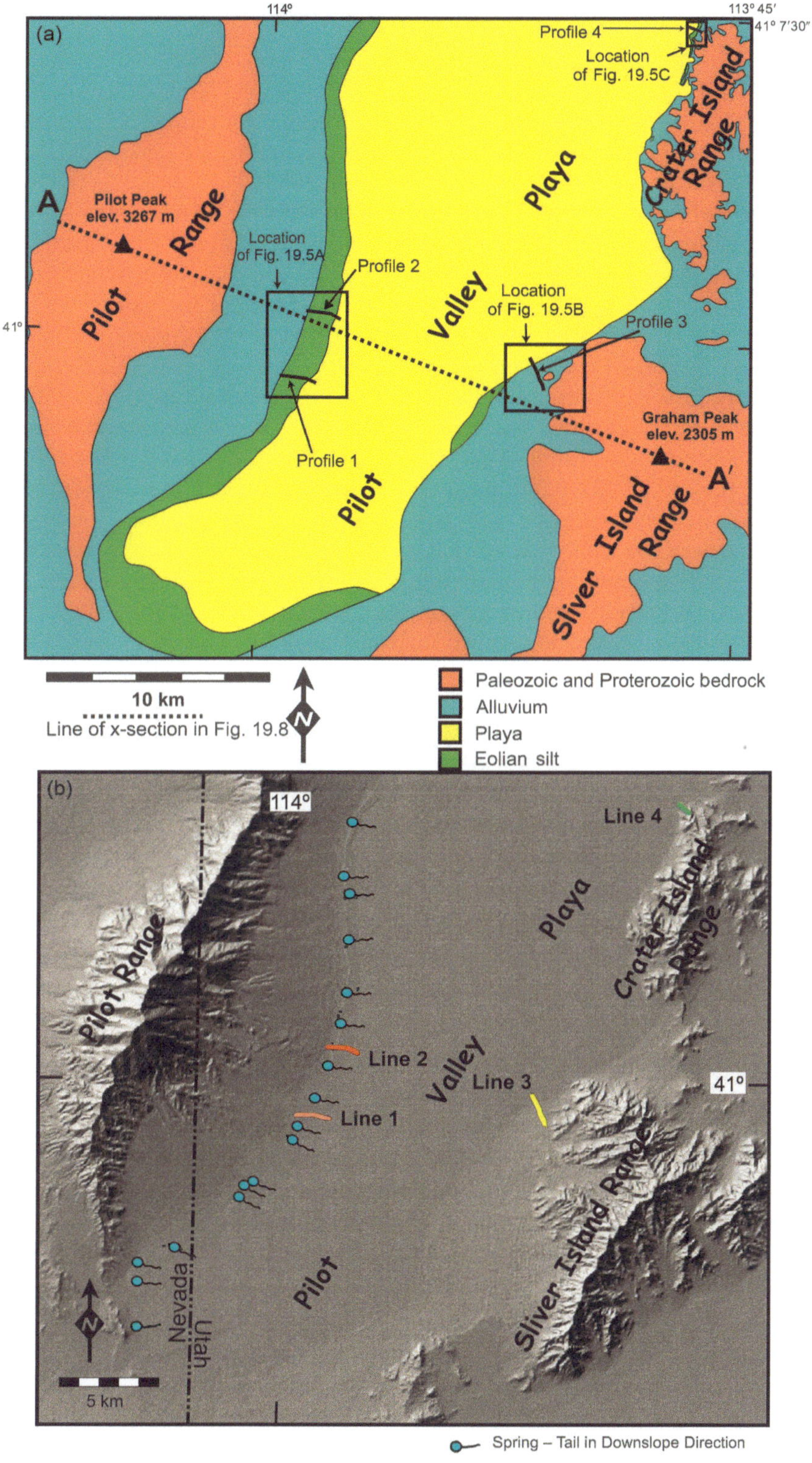

FIG. 19.2 See legend on opposite page.

rocks (Fitter, 1985; René et al., 1988). Other seismic reflection and refraction studies to the east of Pilot Valley suggested a complex and generally asymmetric basin structure (Murray, 1984; Forsyth, 1985).

The surface hydrology is very different on the west and east sides of the valley. On the west side, all streams stop at the playa margin but streams on the east side intermittently flood onto the playa and cut channels across it. Thus gravel overlies playa sediments on the east side of the valley, but not on the west side. This may result in more interfingering of playa sediments and alluvial gravels on the east side relative to the west side. Furthermore, the alluvial fans are steeper on the west side than the east side (Fig. 19.2B).

The Pilot Valley playa and the many visible ancient shorelines that rim it are evidence of lakes filling the region in times past, most notably Lake Bonneville (Bissell, 1968). A sediment core at Great Salt Lake in the Bonneville basin showed evidence for four large lakes since the Brunhes–Matuyama paleomagnetic boundary at 780 ka (Oviatt et al., 1999). The Bonneville, Provo, and Stansbury shorelines (Eardley et al., 1973; Currey, 1983, 1990) from Lake Bonneville are evident, if not continuous, around the valley. The Gilbert shoreline was also mapped in Pilot Valley by Currey (1983) but this is regarded with skepticism (Oviatt, 2014, 2015). A "Pilot Valley Shoreline" (Miller, 1990; Chapter 3) has been recognized at about 1305 m asl, which is slightly higher than the local Gilbert shoreline, as mapped by Currey (1983; ~1300 m asl) and has been observed at other locations around the Great Basin (Oviatt and Miller, 1997; Miller and Phelps, 2016).

Water is intermittently present on the playa, including during peak runoff and after extensive storm activity. The extent of the surface water varies with precipitation rates as well as with temperature and evaporation rates (Lines, 1979; Miller, 1990). Numerous investigations have been conducted to understand the behavior of groundwater in Pilot Valley, including those by Lines (1979), Mason et al. (1995), Mason and Kipp (1998), Petersen (1993), Duffy and Al-Hassan (1988), Fan et al. (1997), and Carling et al. (2012). Sampling and geochemical analyses (Tingey et al., 2002) show that the western margin of the playa has a wedge of higher-density groundwater overlying a lower-density groundwater zone. The prevailing explanation for the observed inversion is that the hydraulic head on the unconfined alluvial fan aquifer, above the valley floor, drives fresh water under the brine (Petersen, 1993). Whatever volume of water cannot be accommodated in the shallow sediments, discharges at the ground surface at a break in slope, where beach

---

FIG. 19.2 (a) Map showing generalized and highly simplified geology of Pilot Valley area and the location of seismic profiles and geological cross-section A–A′. The simplified map was created by the authors, but based partly on digital orthophoto quadrangles (USGS) and partly on maps by Miller (1985, 1990, 1993), Miller et al. (1990), Miller et al. (1993), and Miller and Lush (1994). (b) Digital elevation model (DEM) map showing locations of springs in Pilot Valley. Locations of seismic profiles are also shown.

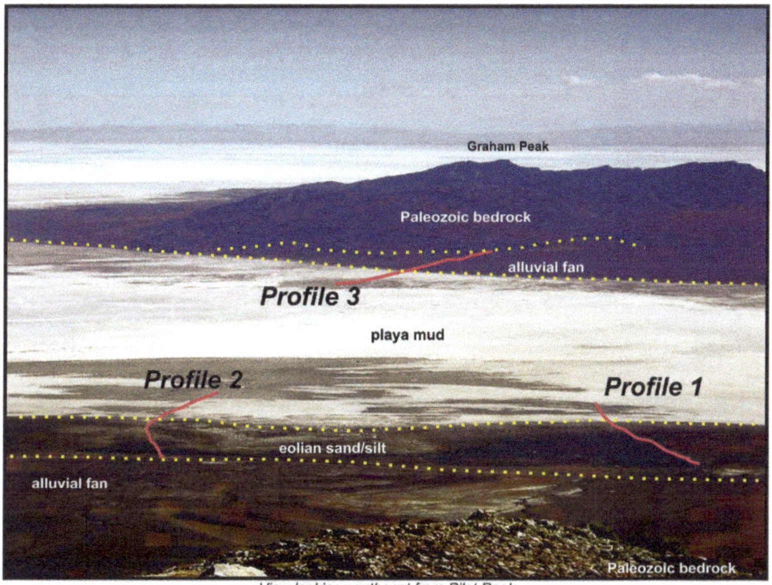

FIG. 19.3 Photo of Pilot Valley taken from the summit of Pilot Peak (elevation 3267 m asl). View is looking southeast. Approximate locations and orientations of the seismic Profiles 1–3 are shown. *Photo by J.V. South.*

gravels lie at the lower elevations of the surface alluvial deposits. Numerous fresh-to-saline water springs issue from the base of the steep alluvial fans along the Pilot Range (Fig. 19.2B). Thus, any subsurface data would contribute to the development of a hydrologic model of the margins of the basin.

19.3 METHODS

Due to the variable geology along the margin of the Pilot Valley playa, we pursued an experimental approach, testing different seismic methods. The principal seismic data were acquired as shallow-penetration, high-resolution compressional (P) wave reflection profiles (Fig. 19.4). We also performed a horizontally polarized shear (SH) wave survey along a portion of one of the P-wave surveys (Profile 2, Fig. 19.5A). The surveys are approximately perpendicular to the surface contact between playa lake sediments and the underlying sands and gravels of the alluvial fan so as to produce images of the playa margin, which, in some cases, crosses a line of springs (Fig. 19.2B). On the western side of the valley, two profiles (1 and 2) started on alluvial or eolian sediments, upslope from the playa, and continued out onto the playa. Profile 3 was acquired on the Silver Island (eastern) side of the playa, starting on exposed Paleozoic bedrock and descending onto the playa (Fig. 19.5B). Profile 4 was shorter and designed to examine the valley

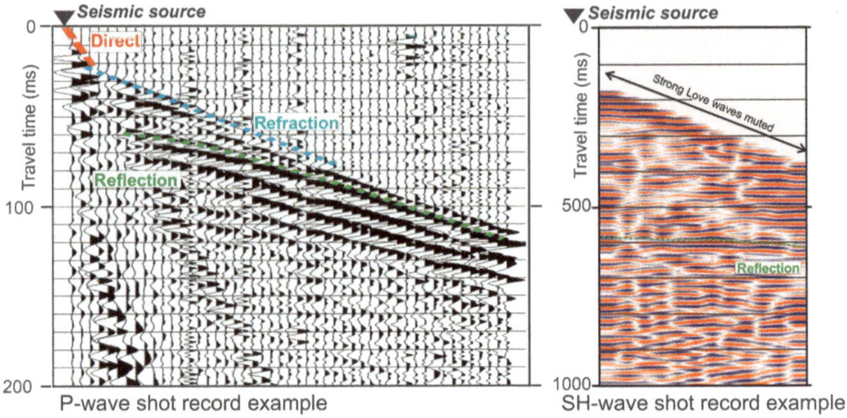

FIG. 19.4 Example of typical shot records, interpreted, from the 48-channel P-wave seismic data (Profile 4), *left*, and from the 12-channel SH-wave seismic data (Profile 2), *right*. P-wave data are shown as variable-area wiggle-trace; SH-wave data are shown as variable-density in order to enhance continuity.

margin near the northern outlet of Pilot Valley where the transition from bedrock to alluvial fan to playa is relatively condensed (Crater Island; Fig. 19.5C). Note that all seismic sections have been converted from time-to-depth using a simplified interval velocity derived from stacking velocities. All depths reported on the sections are with reference to the processing elevation datum (eg, a 50-m depth means 50 m below an elevation datum, not necessarily the ground surface).

19.3.1 P-Wave Seismic Data

The energy source for the P-wave profiles was a 45.4-kg accelerated elastic weight dropper mounted on an all-terrain vehicle, using an elastic band for acceleration, and powered by a 12-V hydraulic pump. All P-wave profiles were acquired with 72 28-Hz geophones in a common depth-point (CDP) "roll-along" mode (Sheriff and Geldart, 1995), providing for a nominal fold of cover of no less than 24 (Profile 4 used 96 geophones, yielding a nominal fold of cover of 48). For all P-wave reflection surveys, the first source–receiver offset, 3.0 m, was the same as the receiver and source spacing. Further explanation of the seismic technique can be found in Sheriff and Geldart (1995). Profiles 1–3 were each about 1.6 km long. The northernmost Profile 4 was about 580 m long. Data processing followed a routine series of steps, including 3D geometry assignment, muting of first-break arrivals (direct arrivals and refractions or diving waves), frequency band-pass filtering, CDP sorting, normal moveout (NMO) velocity correction, CDP stacking, migration (except for Profile 3), and time-to-depth conversion. Note that, due to the need to aggressively mute the first-break arrivals on the shot

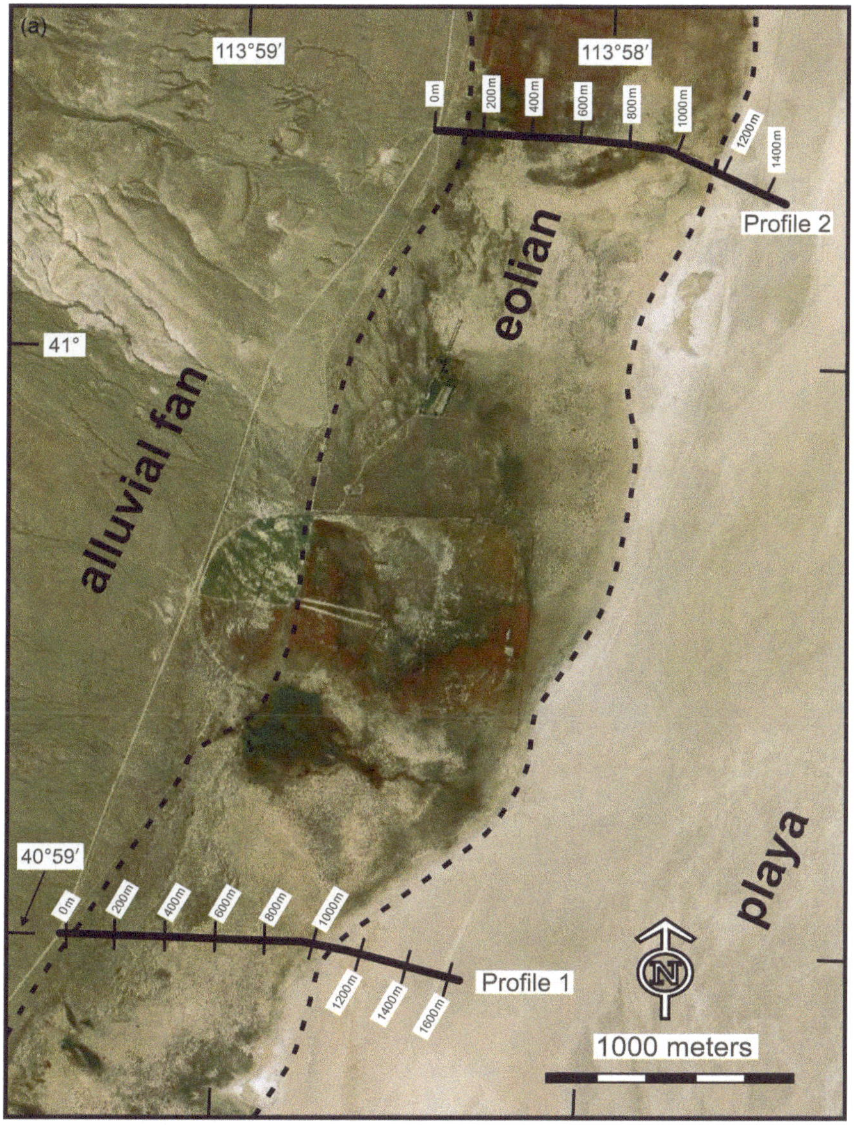

FIG. 19.5 (a) Aerial image (Map data: Google, DigitalGlobe) of the west side of Pilot Valley, showing the locations of seismic Profiles 1 and 2. The boundaries between the three sedimentary environments traversed by the seismic profiles are delineated by *dashed lines.* (b) Aerial image (Map data: Google, DigitalGlobe) of a portion of Silver Island Range where Profile 3 was acquired, with the seismic profile location and interpretations from field mapping by J.V. South. Faults mapped in the field and interpreted from the seismic profile are shown. Field observations were restricted to the area around the seismic profile and included kinematic indicators consistent with the two general fault strike orientations shown (NE-SW and NNW-SSE). (c) Aerial image (Map data: Google, DigitalGlobe) from the northeast part of Pilot Valley, west of the Crater Island Range, showing generalized geology and the location of seismic Profile 4. *Dashed lines* are generalized geologic contacts; *solid white lines* are faults. See Fig. 19.2A for the locations of the three maps.

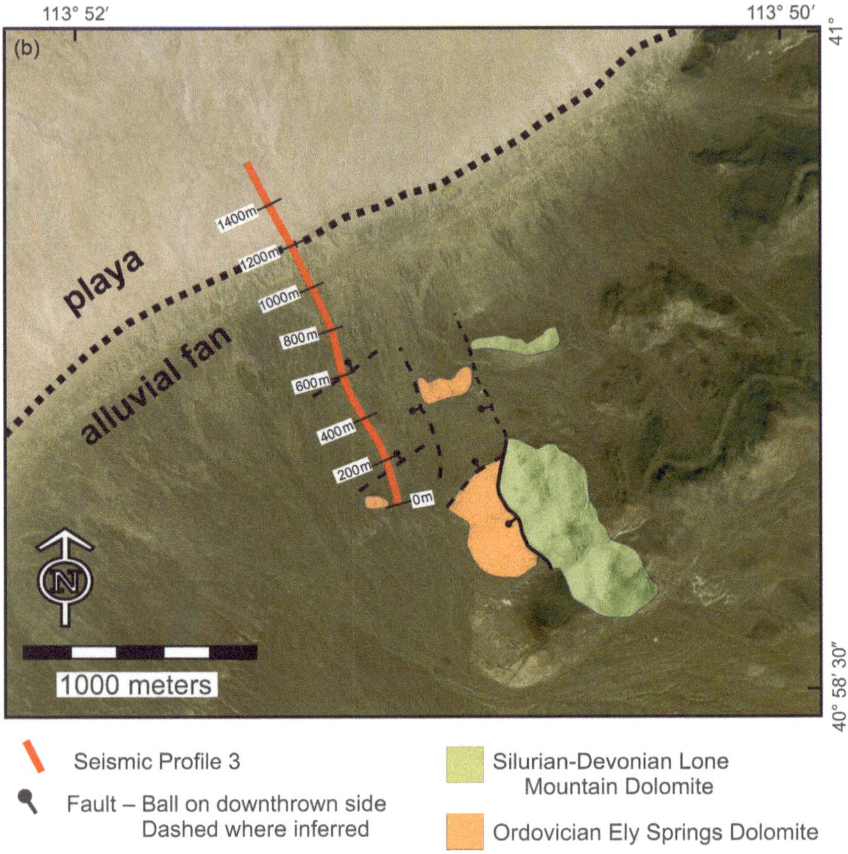

FIG. 19.5—CONT'D

records (Fig. 19.4), the uppermost part of the sections below the ground surface is blank (eg, Fig. 19.6). Based on the Rayleigh criterion, vertical resolution is between 1.0 and 3.1 m (using the applied band-pass filter limits and a typical P-wave velocity of 1000 m/s).

19.3.2 SH-Wave Seismic Data

The horizontally polarized shear-wave (SH-wave) seismic method was also employed for CDP profiling, but used a 12-channel land streamer (Pugin et al., 2004) pulled by a vehicle. The 12 14-Hz horizontal geophones were spaced at 1.52 m. The seismic source consisted of a 1-kg steel mallet, struck against a vertical metal plate coupled to the ground by the weight of the vehicle pulling the sleds. Our approach to SH-wave recording uses lower frequencies and requires a less energetic source compared to that used in P-wave

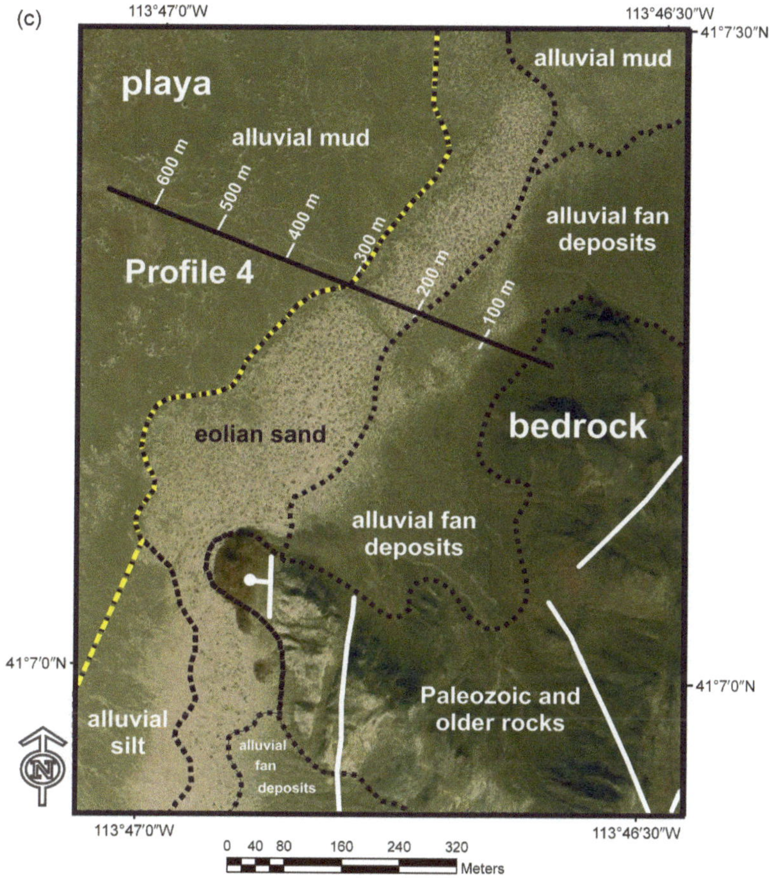

FIG. 19.5—CONT'D

surveys (Pugin et al., 2004). Although P-wave imaging is affected by the bulk modulus of the medium and thus affected by fluids, SH-wave imaging depends only on the density and shear modulus of the medium and thus is independent of water content, which was very shallow in our study area, and is independent of gas-charged layers. SH-wave imaging provides much higher resolution, relative to P-wave imaging, due to slower speed and shorter wavelength, which varies from one-half to one-third of the P-wave wavelength (Woolery et al., 1993; Pugin et al., 2004). The data processing steps are similar to those for the P-wave data, but include a step where the strong arrivals of Love waves (Fig. 19.4) must be muted (Pugin et al., 2004; McBride et al., 2010). Thus, the uppermost part of the section is blank (Fig. 19.6B). More details on the data-processing steps for SH-wave shallow imaging can be found in Woolery et al. (1993) and Pugin et al. (2004).

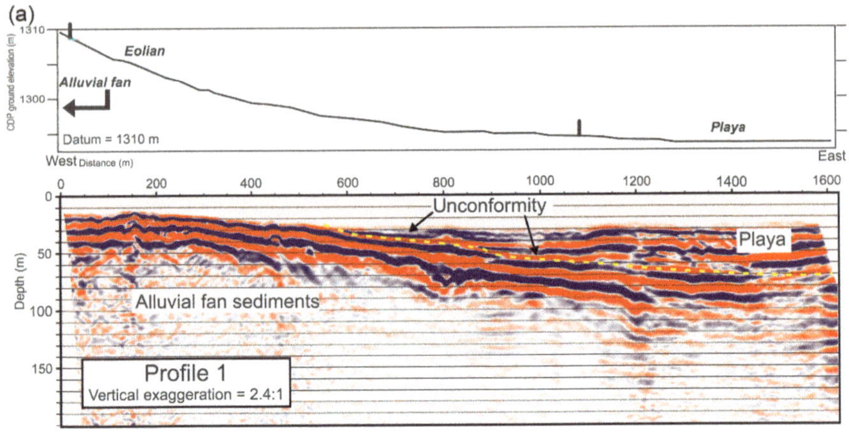

FIG. 19.6 (a) Stacked CDP seismic section of Profile 1. The angular unconformity (contact between alluvial and eolian–playa sediments) is highlighted by the *yellow dashed line*. Processing parameters include: Ormsby band-pass filter 40–80–240–400 Hz; automatic gain control 200-ms window; deconvolution (operator length = 120 ms; prediction distance = 20 ms); top and bottom mutes; refraction statics with a replacement velocity of 1000 m/s; normal moveout correction using 2D velocity function; 9-trace mix; migrated and depth conversion based on stacking velocities. Ground surface geologic contacts are shown. See Fig. 19.5A for profile location. (b) Same as for (a), but with an Ormsby band-pass filter of 70–100–240–400 Hz. Interpreted subsurface contacts shown by *yellow dashed line*. *Small double arrows* indicate poorly reflective layer thickening to the west, referred to in the text. Also shown is the test SH-wave seismic reflection profile for a coincident portion of the P-wave profile. The shear-wave velocity function derived from the MASW survey, used to establish the uppermost velocity structure of the area, is shown. (c) Same as for (b), but for Profile 3 on the east side of Pilot Valley (shown unmigrated due to complex structure). Apparently dipping arrivals on the western end of the section may represent guided wave contamination. See Fig. 19.5B for profile location. (d) Same as for (b), for Profile 4 on east side of valley, west of the Crater Island Range. See Fig. 19.5C for profile location.

A multichannel analysis of surface waves (MASW) survey conducted at this site guided the choice of velocity function used to convert the SH-wave profile from time-to-depth. The MASW technique relies on the inversion of Rayleigh-wave dispersion to derive a shear-wave velocity function (Park et al., 1999). Vertical resolution for the SH-wave section is between 0.9 and 1.7 m (using the applied band-pass filter limits and a typical S-wave velocity of 200 m/s).

19.4 GEOLOGY ALONG THE SEISMIC PROFILES

The environment for the seismic surveys is acoustically challenging. The basin is filled with thousands of meters of unconsolidated lacustrine and alluvial sediments (Berg et al., 1961), to depths well beyond that which can be imaged by high-resolution methods. Alluvial sediments are problematic for seismic surveys because they quickly attenuate the signal, which explains

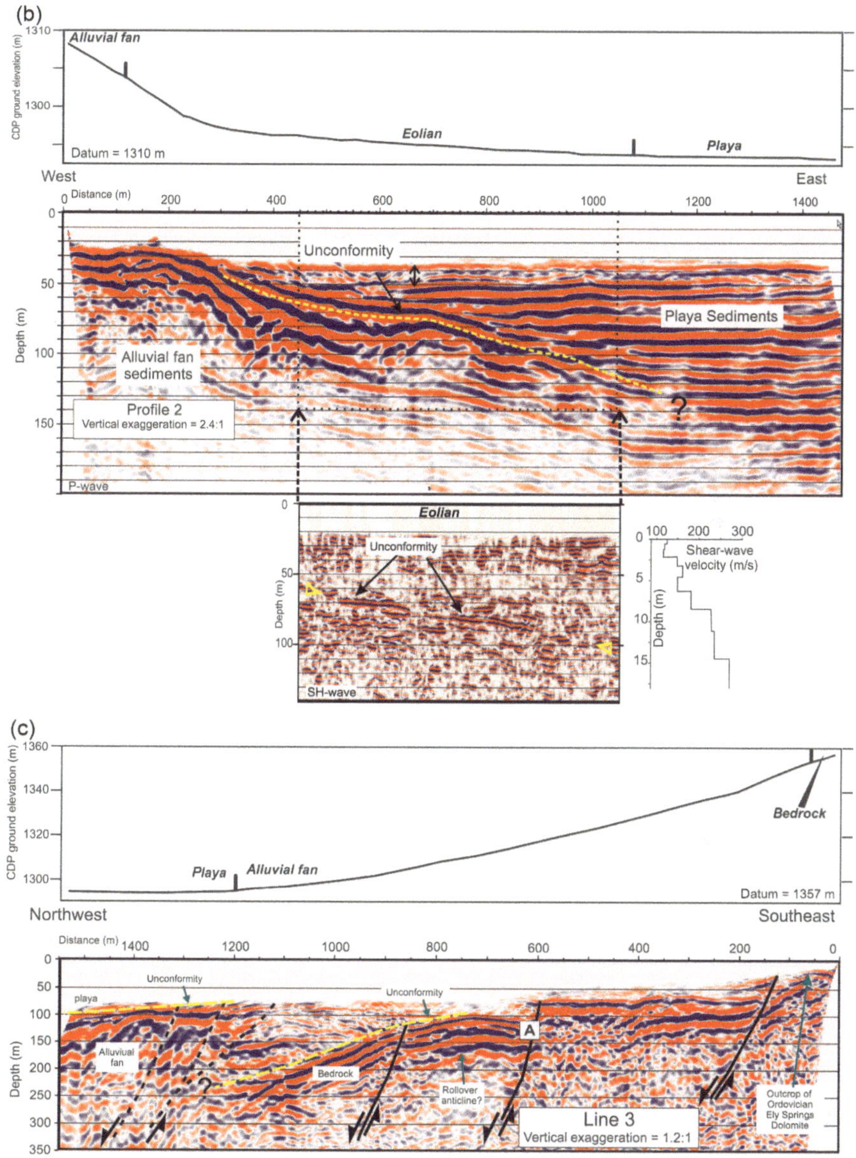

FIG. 19.6—CONT'D

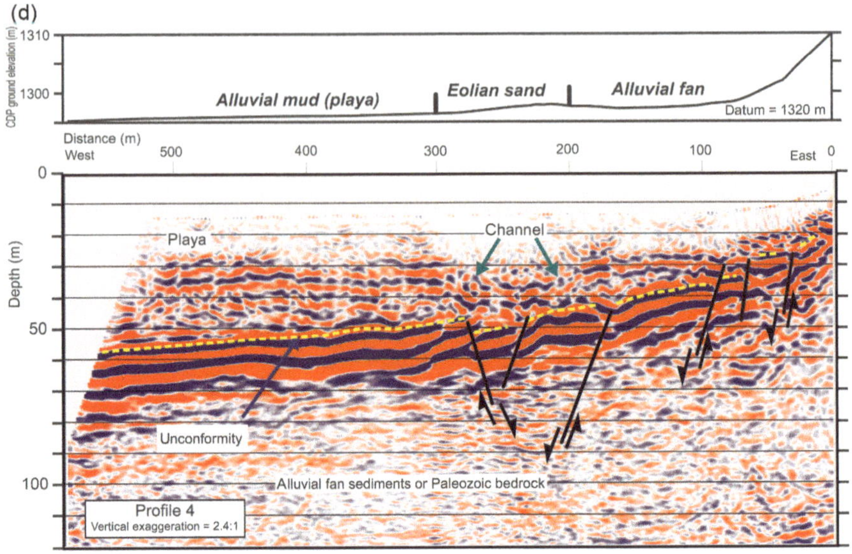

FIG. 19.6—CONT'D

the poor reflectivity below the alluvial fan unconformity. Bedrock is not clearly seen on any of the sections on the west side of Pilot Valley and only variably seen on the east side. The seismic-profile locations were chosen so as to be able to integrate geophysical interpretation with observations from the local surface geology.

Four different geological settings were studied (Profiles 1–4; Fig. 19.5): two on the western margin of the valley, one on the east side, and one near the northern outlet of the valley. The surveys traversed multiple, distinct geological surface environments: bedrock, alluvial fan, eolian and lacustrine silt, and playa muds. Generally speaking, the surface geology is predominantly poorly consolidated clastic sediments of Quaternary age (Fig. 19.2) (Miller, 1985, 1990, 1993; Miller et al., 1990, 1993; Miller and Lush, 1994).

The playa, which consists of mudflat and salt pan deposits, is the most prominent feature in the valley. Elevation changes very little across the playa, but the lowest point of the playa lies in the northwest portion of the valley (north of Profiles 1 and 2) at the surface contact with eolian dunes, where surface water last evaporates. Rimming much of the playa is a narrow band of eolian and lacustrine silt (described as Quaternary "surficial Lake Bonneville deposits" by Hintze et al. (2000)), which is continuous along the western margin of the playa, but relatively narrow and discontinuous along the eastern margin (Fig. 19.2).

The rocks exposed in the mountains are mostly late Proterozoic to early Paleozoic (Miller, 1990). The bedrock units located nearest the profiles on the west side in the Pilot Range (and therefore the most likely source for sediments in the alluvium on the west side) are predominantly the McCoy Creek Group. The McCoy Creek Group consists of late Proterozoic metamorphic rocks, primarily quartzite, but schist, marble, and phyllite are also present (Miller, 1990). On the east side of the valley, in the vicinity of seismic Profile 3, bedrock outcrops are typified by the Ordovician Ely Springs Dolomite. This highly resistant (to weathering), black dolomite is regionally a cliff former (Miller et al., 1990) (Fig. 19.5B). Some small outcrops of Devonian Guilmette Formation (limestone), Devonian Simonson Dolomite, Devonian-Silurian Lone Mountain Dolomite, and Ordovician Laketown Dolomite occur near the seismic profile (Schaeffer and Anderson, 1960; Hintze, 1988; Miller et al., 1990). Cobbles of Jurassic plutonic igneous rocks are also present, particularly on the nearby ancient beach terraces, which suggests considerable heterogeneity in the shallow subsurface. Further north along the eastern rim of the valley, the area of seismic Profile 4 includes Devonian limestone bedrock exposed in the northwestern Crater Island Range, onlapped by a succession of Quaternary alluvial fan, eolian, and playa sediments, analogous to the lateral succession of depositional environments on the west side of the valley, only narrower.

19.5 RESULTS AND INTERPRETATIONS

19.5.1 Western Margin of Pilot Valley: Profiles 1 and 2

Profiles 1 and 2 (Fig. 19.6A and B) begin on their western ends just at the eastern margin of the alluvial fan that descends from the Pilot Range (Figs. 19.2, 19.3, and 19.5A). The profiles then traverse the band of eolian and lacustrine silt, which rims the western margin of the valley, and then continue a short distance onto the flat playa surface (mapped as "surficial mud and salt flat deposits" by Hintze et al. (2000)). These two profiles are separated by a north–south distance of 3.6 km (on their western ends), are about the same length (between 1.5 and 1.6 km), span nearly identical elevation ranges (Profile 1, 1309–1294 m asl; Profile 2, 1308–1293 m asl), and traverse equivalent surface geologic environments (Fig. 19.6A and B).

Profile 2. Along Profile 2, the ground-surface contact between the depositional environments of the eolian silt and the alluvial fan occurs at about distance marker 150 m (ie, as measured along the profile, Fig. 19.6B) and the eolian silt and playa contact at distance marker 1150 m (Fig. 19.6B). A distinct, east-dipping angular unconformity is observed on the seismic profile, over which mostly horizontal playa sediments thicken eastward (Fig. 19.6B). The unconformity can be traced from a depth of 37 m below distance marker 325 m (western edge of profile), angling down-to-the-east to 100 m depth below distance marker 1050 m. The overall apparent dip of the unconformity is about 5 degrees to the east with minor variations. This angle is relatively congruent with the average dip of 5–6 degrees of the active alluvial fans immediately to the west. The overlying playa sediments have a very slight westward dip, particularly apparent between distance markers 650 and 1000 m at depths of 50–70 m (Fig. 19.6B). This is likely the result of the basin back-tilting toward the major, range-bounding fault (unnamed) further to the west (Miller, 1990). The sag in the unconformity between distance markers 300 and 700 m and the arch below about 200 m may be related to range-front tectonism. Between distance markers 300 and 1000 m, a very shallow and internally poorly reflective layer dips into the unconformity (ie, west) and thickens to the west to a maximum of 10–15 m (Fig. 19.6B). Strong multicyclic reflectivity can be seen in the cross section, particularly in the playa sediments. Cyclic temporal climatic changes may be responsible for the repetitive deposition expressed as well-developed reflectivity.

The original, predepth-converted seismic sections displayed numerous subparallel, coherent events arriving at later (eg, below the interpreted angular unconformity) travel times, which are likely multiple reflections from the unconformity (Fig. 19.6B). An autocorrelation was performed on the stacked travel-time data and further suggests that arrivals beyond the unconformity are likely to be mostly multiples and thus were not displayed or interpreted. Due to the relatively short receiver spread, the NMO sensitivity was insufficient to distinguish long-period multiple reflections from primary reflections. However, the strong multiple behavior nicely demonstrates the high acoustic impedance contrast across the angular unconformity separating playa and other lacustrine sediments from the underlying alluvial fan sediments.

The SH-wave seismic profile corresponds to the portion of the P-wave profile that extends over mostly horizontal playa–eolian sedimentary reflectors lapping onto the east-dipping angular unconformity (Fig. 19.6B). The MASW results indicate a shallow velocity variation from 125 m/s near the surface to 270 m/s at 18 m depth, consistent with the unconsolidated, low shear strength of the playa and lacustrine sediments at the site. The main aspect of the SH-wave reflection profile is the east-dipping reflection over a depth range of about 65–100 m. This arrival correlates exactly to the angular unconformity imaged on the P-wave section, but shows much greater detail and complexity, owing to the shorter wavelength of the signal, as discussed earlier. For example, the SH-wave profile shows two prominent gaps in the unconformity, which are accompanied by abrupt changes in its depth (below distance markers 660 and 920 m, Fig. 19.6B). Note how the SH-wave section provides greater detail, relative to the P-wave section, for the downward (toward the basin) bend in the unconformity reflector below distance marker 660 m, suggesting a down-to-the-basin offset in the reflector as opposed to only a gentle bend. The SH-wave profile was unable to image the stratigraphy above the dipping unconformity. Because the P-wave section above the unconformity is very reflective, the poor reflectivity on the SH-wave profile suggests that variations in bulk modulus are more important than variations in shear modulus.

Profile 1. Along Profile 1, the eolian silt and alluvial fan contact occurs at its west end, and the eolian silt and playa boundary at 1075 m (Fig. 19.6A). Profile 1 shows the unconformity (from distance marker 575 and depth 28 m to distance marker 1350 and 69 m depth), and flat playa–eolian sediments (above the unconformity) as observed in Profile 2. The unconformity has a gentler slope compared to Profile 2 (average dip along Profile 1 is 3 degrees compared with 5 degrees along Profile 2). The overlying lacustrine/playa sediments are thinner, perhaps as a result of the shallower dip of the unconformity. The unconformity can only be definitively traced to a depth of 70 m along Profile 2 compared to 125 m along Profile 1. Note also the arch in the unconformity on the western part of the Profile 1, which is broader than that on Profile 2.

19.5.2 Eastern Margin of Pilot Valley: Profiles 3 and 4

Profile 3. Profile 3 (Fig. 19.6C) begins on the northwestern flank of the Silver Island Range on the east side of Pilot Valley, near the western edge of bedrock outcrop. The profile traverses the alluvial fan, which is very narrow at this location (relative to the west side of the playa; Fig. 19.2A), and then continues onto the playa (Fig. 19.5B). In the vicinity of our profiles, the alluvial fans on the Silver Island Range do not slope as steeply (locally, 2–3 degrees) as do the fans on the Pilot Range (locally, 5–6 degrees). Profile 3, which is about the same length as Profiles 1 and 2, drops 61 m in elevation compared with 15 m in the profiles on the west side of the valley. Profile 3 is separated by a distance of 9.5 km from Profile 1 on the opposite side of the playa (between the western and eastern ends of the profiles, respectively). The geology for Profile 3 (Fig. 19.5B) is based on field mapping performed as part of the present study and on earlier geologic maps. The geology here differs significantly from that of Profiles 1 and 2 in that no deposits of eolian and lacustrine silt are crossed (Fig. 19.5B). The surface geology in the vicinity of Profile 3 is more complex. Profile 3 is more elevated above the playa (ground elevation range is 1357–1296 m asl), relative to Profiles 1 and 2.

Profile 3 (Fig. 19.6C) crossed the entire alluvial fan (from bedrock to playa). In contrast, Profiles 1 and 2 began several kilometers east of any outcrop in the Pilot Range. Profile 3 began only a few meters away from an outcrop of Ordovician Ely Springs Dolomite (Fig. 19.5B). A strong, shallow reflector appearing on the southeast end of the profile is interpreted as the sediment–Paleozoic bedrock interface, which can be traced westward to a depth of at least 200 m. The apparent dip of the bedrock reflector varies from near 0 degrees to about 20 degrees, which is considerably greater than the few degrees observed for the unconformity imaged on Profiles 1 and 2. The maximum bedrock dip is comparable to the nearby topographic slope of the

Paleozoic mountain front. The poor internal reflectivity of the section between the onset of the seismic data and the Paleozoic bedrock reflector is consistent with its interpretation as alluvial fan sediments. This is also consistent with a similar pattern for alluvial fan sediments lying over bedrock as described by McBride et al. (2010). Prominent reflector breaks on the section show that Paleozoic bedrock was vertically offset near the southeast end of the profile and becomes progressively more offset to the northwest (Fig. 19.6C). Several faults can be interpreted—the first major one appears at distance marker 610 m, which has an offset of approximately 35 m. Other faults can readily be interpreted. All the faults observed in the section have a generally westward dip.

The arch in the bedrock reflector (between 600 and 800 m) may be interpreted as a rollover anticline. The significance of such an interpretation would be that a rollover anticline suggests a listric fault bounding the arch on the east. Listric faulting may be related to west-northwest oriented crustal extension during the Jurassic (Miller and Allmendinger, 1991), which would not be surprising in the context of Basin and Range tectonism.

Only the last ~350 m of Profile 3 (1200–1550 m distance) extended over the playa (Fig. 19.6C). The unconformity that would be expected on the section at the playa–alluvial interface, and which would be similar to the one mentioned in the interpretation of Profiles 1 and 2, is difficult to detect in the seismic cross section due to its shallow depth. It appears at distance marker 1250 m, depth 80 m (below datum, not the ground surface), dipping to the northwest to distance marker 1525 m, depth 100 m, yielding an apparent dip of 4 degrees.

Profile 4. Profile 4 (Fig. 19.6D) began near Paleozoic bedrock outcrop on the northwestern flank of the Crater Island Range (Fig. 19.5C). The profile then extends across a small patch of alluvial fan, a narrow zone of eolian deposits, and then onto the mud playa (Fig. 19.6D). Unlike Profile 3, located 16.6 km (based on eastern ends of profiles) to the south, Profile 4 spans a relatively narrow elevation range (1310–1295 m asl) and does not traverse a significant extent of bedrock. Exposed Paleozoic bedrock in this part of the Crater Island Range is cut by normal faults that vary in strike from north to northeast, dipping toward the basin (Fig. 19.5C) (Miller et al., 1990).

The short seismic profile shows a strong, gently west-dipping (2 degrees, apparent) reflector overlain by weaker, mostly horizontal reflections with a maximum thickness of about 25 m. Below the strong reflector, the section is featureless. The strong reflector could be interpreted as the unconformity between playa and eolian sediments, above, and the typically nonreflective alluvial-fan sediments below (as on Profiles 1 and 2), which can be projected into the subsurface from outcrop at the eastern end of the profile. The strong reflector could also be interpreted as Paleozoic bedrock (as on Profile 3) although its relatively shallow dip does not compare well with the local bedrock topographic slope of 30 degrees. For this interpretation, the poorly reflective interval above the strong unconformity reflector would be relatively thin, alluvial-fan sediments overtopped by eolian and/or playa sediments. The material below the strong reflector would be the Permian Riepetown Sandstone, according to a cross section shown by Miller et al. (1990). The playa–eolian sediments show a well-expressed synformal reflector beneath distance 230 m. We interpret this feature to be a geologically recent channel that corresponds closely to the band of eolian sediments as shown mapped in Fig. 19.5C.

The interpretation of the reflectors on Profile 4 is analogous to that of Profiles 1 and 2, which show the alluvial fan–playa–eolian unconformity dipping basinward as it descends into the subsurface. Features of Profile 4 (Fig. 19.6D), not observed on the profiles along the Pilot Range (Profiles 1 and 2), are a series of normal faults, interpreted from small reflector offsets on the eastern half of the profile. These faults could be related to reactivation of (or developed in the same stress field as) normal faults mapped in the Paleozoic bedrock immediately to the east.

19.6 DISCUSSION

A primary feature of the Pilot Valley playa and adjoining mountain ranges is structural asymmetry. Our high-resolution seismic surveys provide a new context in which to examine the issue of asymmetry, as well as the along-strike and cross-basin variability in the shallow structure and stratigraphy of the basin. Berg et al. (1961) and Cook et al. (1964) conducted seismic refraction and gravity surveys, respectively, in Pilot Valley. Berg et al. (1961) modeled the playa with four seismic velocity layers (at 6 m (275 m/s), 20 m (860 m/s), 700 m (1980 m/s), and half-space (5570 m/s)). Cook et al. (1964) modeled the south end of the valley as a symmetric graben, containing Cenozoic basin fill up to a thickness of 1600 m. Although it is likely that the basin is neither a perfect half-graben nor a symmetric graben, it is probable that a high degree of asymmetry occurs. Berg et al.'s (1961) refraction survey ran along the axis of the valley (north–south) and thus did not shed light on the question of east–west basin asymmetry.

In Cook et al.'s (1964) gravity survey, a substantial fault was interpreted to exist only on the eastern edge of Pilot Valley, suggesting an asymmetric graben. Although our high-resolution seismic profiles do not cross any major range-front faults, they do indicate the variation in fine-scale faulting near the western and eastern edges of the playa. The disruption of shallow seismic reflectors on Profile 3 and, to a lesser extent, on Profile 4 (Fig. 19.6D), suggests that at least some shallow faulting has displaced bedrock along the east side of the valley, which is not observed to the same extent on the seismic profiles on the west side that straddle the playa boundary (ie, along Profiles 1 and 2). We do not observe significant faulting along the line of springs (Fig. 19.2B), as has been suggested as evidence for active faults along ranges in the Basin and Range province, including Pilot Valley (Hecker, 1993). This suggests that the springs do not discharge as a result of faulting. On the other hand, bedrock displacement on the east side is interpreted to be less significant than the major range-bounding faults on the west side (Miller, 1990), upslope from our seismic profiles, which substantiates the idea of an asymmetric graben with its moat on the west.

The asymmetry of the Pilot Valley basin is further emphasized by our observation that the strong unconformity reflector on the profiles of the western margin (Profiles 1 and 2, Fig. 19.6A and B) represents the top of the alluvial fan plunging beneath playa and eolian sediments that onlap the fan sediments, whereas on at least one of the profiles of the eastern margin (Profile 3, Fig. 19.6C), this reflector most likely represents the top of shallow Paleozoic bedrock that descends more or less gradually beneath the basin (ie, is not as greatly faulted and down-dropped as it would be along a range-front fault). The uncertainty in the interpretation of the shallow dipping reflector on Profile 4 (Fig. 19.6D) has implications for the regional structure of the northern margins of Pilot Valley. A shallow alluvial fan surface that dips gently into the center of the valley would suggest that Paleozoic bedrock drops off steeply perhaps due to major Basin and Range normal faulting associated with crustal extension in this location; however, such major faulting has not been observed along the east margin of the basin (Miller et al., 1990). Alternatively, a gently dipping bedrock reflector forming the shallow unconformity would suggest that Paleozoic bedrock continues at least partially across the northern part of the valley (ie, west of the Crater Island Range) with fewer faults and significantly less displacement than on the west side approaching Pilot Peak.

The shallow sedimentary layers imaged in our seismic sections above the interpreted unconformity represent basin fill from pluvial lakes that inundated the region throughout the Pleistocene and perhaps since the late Pliocene. The very uppermost layers of lacustrine sediment represent Lake Bonneville deposits (latest Pleistocene), although they are so thin (Rey, 2012) as to likely not be imaged in these seismic data. It is likely that the basin is dominantly filled by Quaternary and older playa sediments with subordinate lacustrine material (Thompson et al., 1995; Oviatt et al., 1999).

In order to provide a context for the shallow geophysical profiles in Pilot Valley, we have poststack reprocessed and reinterpreted the eastern portion of the COCORP (Consortium for Continental Reflection Profiling) deep seismic reflection profile (Hauser et al., 1987) (Fig. 19.7) in Steptoe Valley located 150 km to the southwest of the study area (Fig. 19.1). The valley, which is crossed by the eastern edge of the COCORP profile, furnishes a possible analog to the regional structure of Pilot Valley. Since Basin and Range structure tends to be repetitive along strike (Stewart, 1980), some comparisons can be drawn between Steptoe and Pilot Valleys. Although the intent of the COCORP project was to image the entire crust and the uppermost mantle,

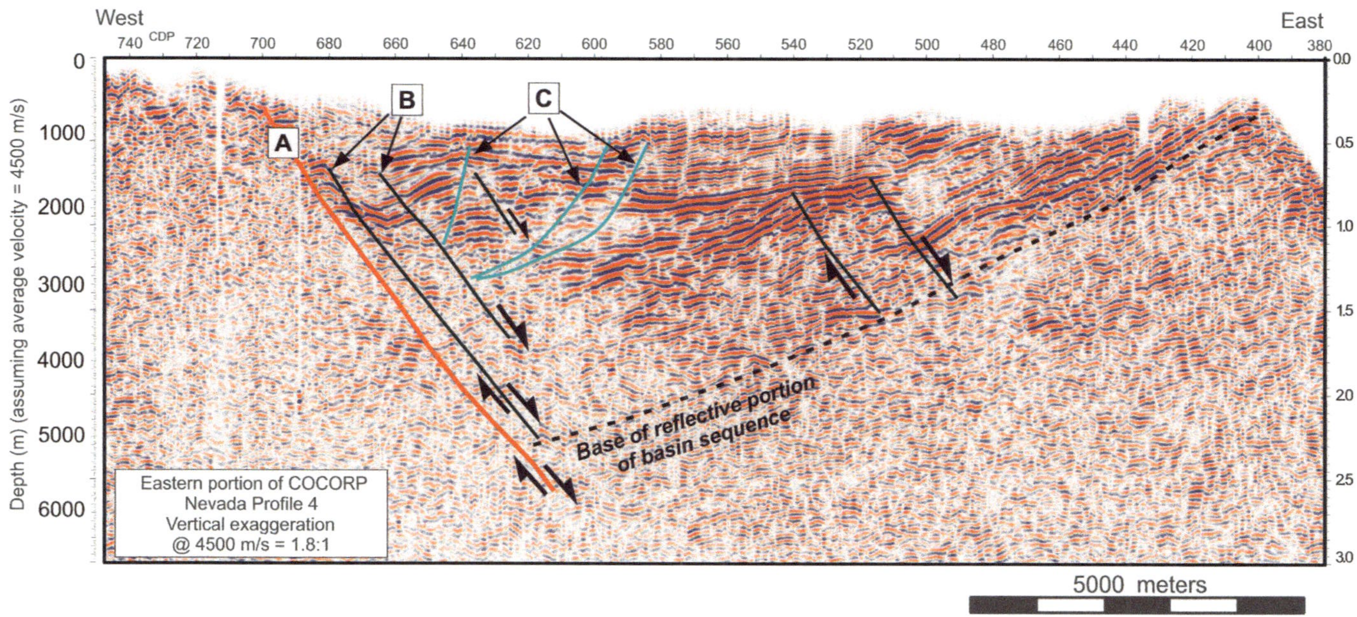

FIG. 19.7 Portion of COCORP Nevada Profile 4 deep seismic reflection profile over Steptoe Valley (see Fig. 19.1 for location). These data cover a much greater interval of the subsurface compared to our Pilot Valley sections (Fig. 19.6). The application of an Ormsby band-pass filter (7–26–32–34 Hz) enhanced and revealed more of the shallow tectonic features than originally observed. Prominent features of our interpretation are range-bounding fault (on the west—*red line A*), intrabasin faults (synthetic—*black lines B*, antithetic—*blue lines C*), and sediments dipping toward range-bounding fault. The SEGY-format data were accessed from www.geo.cornell.edu/geology/cocorp/COCORP.html.

the upper part of the section can be used to show what the structure of Pilot Valley might look like if a continuous, long-record seismic profile were available across its entire width and over the adjacent ranges.

As interpreted by Hauser et al. (1987), a major range-bounding fault occurs along the western edge of Steptoe Valley with an associated half-graben to the east. The asymmetry of the graben is apparent in the reprocessed seismic section shown in Fig. 19.7. The major range-bounding fault is interpreted on the west side of the valley (heavy red line, feature A), accompanied by west-dipping stratal reflectors between CDP 400 and 700. Other faults with an orientation synthetic to the range-bounding fault are also observed (features labeled B). Antithetic faults can also be interpreted (features labeled C). Using Steptoe Valley as a generalized analog, Pilot Valley may be interpreted as a half-graben with the main, range-bounding fault on the west side and infilled with basin sediments that might be expected to dip gently to the west, toward an east-dipping fault (Fig. 19.8). Perhaps the best evidence for this interpretation can be seen at the surface, where

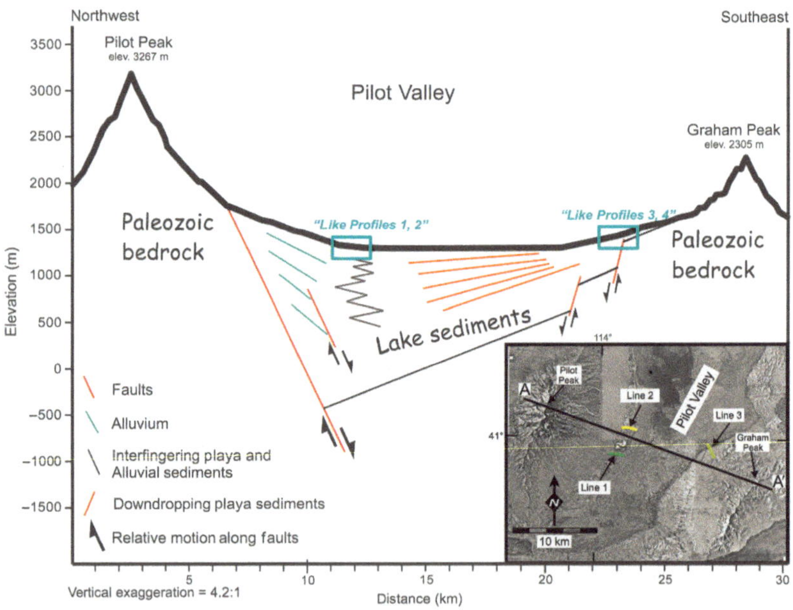

FIG. 19.8 Simplified conceptual cross section of Pilot Valley, showing the relative analogous locations of the high-resolution seismic profiles (see Fig. 19.2A for location). Elevations range from west of Pilot Peak over the summit, down the mountain front, across the sedimentary basin fill, up the slope of the Silver Island Range over Graham Peak, and to a point slightly east of the Graham Peak summit. *Inset map* shows locations of three seismic profiles and Pilot and Graham Peaks. *Based on Tingey, D.G., Mayo, A.L., Nelson, S.G., 2002. Recharge of a fresh water-brine groundwater system, Pilot Valley, Utah-Nevada. Geol. Soc. Am. Abst. Programs 34 (4), 56. Carling, G.T., Mayo, A.L., Tingey, D., Bruthans, J., 2012, Mechanisms, timing, and rates of arid region mountain front recharge. J. Hydrol. 428–429, 15–31.*

the alluvial fans have steeper slopes on the side of the valley with the range-bounding fault (west side) and are less steep on the opposite side (east) of the valley. Such an interpretation is supported by the fact that the lowest elevations in Pilot Valley follow a linear, north–south trend along the western edge of the playa.

19.7 CONCLUSION

Shallow, high-resolution seismic reflection profiles provide detailed images of the uppermost (<150 m) structure and stratigraphy of the Pilot Valley basin at the boundary between the playa and alluvial-fan sediments. This analysis represents one of the most comprehensive geophysical studies of a playa–alluvial fan margin. The most prominent features in the seismic profiles are a distinct angular unconformity on the west side of Pilot Valley and faulting of bedrock on the east side (Fig. 19.8). Although the alluvial fan sediments are acoustically opaque at the seismic frequencies used in our applications, the shallow playa–alluvial fan environment is ideal for acquisition of seismic data. The angular unconformity observed on the two seismic profiles from the west side (Fig. 19.5A) provides evidence of transgression of lakes over the alluvial fan, including Lake Bonneville and likely older ones (throughout the late Pliocene and Pleistocene). This transgression, which is most prominent on the west side, has locally overtaken and buried the progradation of alluvial fan sedimentation.

Our results demonstrate a significant amount of geological variation both across-strike (east to west) in Pilot Valley, as well as along strike (from south to north). Playa sediments onlap alluvial fan sediments on west side of Pilot Valley, along a line of springs (Fig. 19.2B); however, the absence of faults along the point of discharge suggests that the springs are unrelated to structural deformation. On the east side of the valley, multiple interpretations are possible for the west-dipping unconformity reflector, which is moderately to heavily faulted and nearer bedrock outcrops. We suggest that this unconformity on the valley's east side is, at least in places, a gently dipping Paleozoic bedrock surface, which further emphasizes the across-strike asymmetry of the basin.

The structure of Pilot Valley is believed to be an asymmetric graben similar to Steptoe Valley as illustrated by the COCORP deep seismic profile (Allmendinger et al., 1987; Hauser et al., 1987). This profile (Fig. 19.7) shows features that are analogous to what can be inferred for Pilot Valley, particularly intrabasin and range-bounding faults. The alluvial fans on the west side of Pilot Valley are steeper than those on the east side, consistent with the interpretation of major faulting on the west side of the basin. The seismic data show the angular unconformity atop plunging alluvial sediments with overlapping and nearly horizontal playa sediments on the west side of the valley. Where Profile 3 extends out onto the playa to a similar extent as Profiles 1 and 2, bedrock is visible in the section nearby at depths less than 300 m; however, on Profiles 1 and 2, bedrock must be much deeper.

ACKNOWLEDGMENTS

Many people helped acquire the seismic data. Specific gratitude is extended to Stephanie Ashley, Christopher Bexfield, Michelle Bushman, Nicole Cox, William Hokanson, Eleanor McBride, Maxwell Okure, Allan South, and Mary South. We are grateful for correspondence and insights from David M. Miller. Gratitude is expressed to the owners and operators of the TL Bar Ranch for allowing the use of their property for conducting much of the field work. This research was made possible through grants from the Geological Society of America and the BYU College of Physical and Mathematical Sciences. Generous software grants from Landmark Graphics Corporation (Halliburton) and IHS made the data processing and visualization possible. Reviews and edits of a previous version of the article by William J. Stephenson, Charles G. (Jack) Oviatt, and John F. Shroder, Jr. greatly improved the final version.

REFERENCES

Allmendinger, R.W., Hauge, T.A., Hauser, E.C., Potter, C.J., Klemperer, S.L., Nelson, K.D., Knuepfer, P., Oliver, J., 1987. Overview of the COCORP 40°N Transect, western United States: the fabric of an orogenic belt. Geol. Soc. Am. Bull. 98, 308–319.

Berg Jr., J.W., Cook, K.L., Narans Jr., H.D., Leamer, R.J., 1961. Seismic profiles in Northwestern Utah: Pilot Range and Grouse Creek Range area. J. Geophys. Res. 66 (4), 1255–1263.

Bissell, H.J., 1968. Bonneville—an ice-age lake. Brigham Young Univ. Geol. Stud. 15 (4), 66.

Carling, G.T., Mayo, A.L., Tingey, D., Bruthans, J., 2012. Mechanisms, timing, and rates of arid region mountain front recharge. J. Hydrol. 428–429, 15–31.

Cook, K.L., Halverson, M.O., Stepp, J.C., Berg, J.W., 1964. Regional gravity survey of the northern Great Salt Lake Desert and adjacent areas in Utah, Nevada, and Idaho. Geol. Soc. Am. Bull. 75, 715–740.

Currey, D.R., 1983. Lake Bonneville: selected features of relevance to neotectonic analysis: United States Geological Survey Series Open-File Report, 82-1070. 31 pp.

Currey, D.R., 1990. Quaternary palaeolakes in the evolution of semidesert basins, with special emphasis on Lake Bonneville and the Great Basin, U.S.A. Palaeogeogr. Palaeoclimatol. Palaeoecol. 76, 189–214.

Currey, D.R., Atwood, G., Mabey, D.R., 1984. Major Levels of Great Salt Lake and Lake Bonneville: Scale 1:750,000. Utah Geological and Mineral Survey Map 73.

Duffy, C.J., Al-Hassan, S., 1988. Groundwater circulation in a closed desert basin: topographical scaling and climatic forcing. Water Resour. Res. 24, 1675–1688.

Eardley, A.J., Shuey, R.T., Gvosdetsky, V., Nash, W.P., Dane Picard, M., Grey, D.C., Kukla, G.J., 1973. Lake cycles in the Bonneville Basin, Utah. Geol. Soc. Am. Bull. 84, 211–215.

Eaton, G.P., 1980. Geophysical and geological characteristics of the crust of the Basin and Range province. In: Burchfiel, B.C., Oliver, J.E., Silver, L.T. (Eds.), Continental Tectonics. National Academy of Sciences, Washington, DC, pp. 96–113.

Fan, Y., Duffy, C.J., Oliver Jr., D.S., 1997. Density-driven groundwater flow in closed desert basins: field investigations and numerical experiments. J. Hydrol. 196, 139–184.

Fitter, J.L., 1985. Seismic Investigation Near Silver Island in the Great Salt Lake Desert, Utah. M.S. thesis, University of Oklahoma. 89 pp.

Forsyth, P.M., 1985. Multicomponent Trace Analysis and Seismic Reflection Studies in the Great Salt Lake Desert, Utah. M.S. thesis, University of Oklahoma. 219 pp.

Gilbert, G.K., 1890. Lake Bonneville. U.S. Geological Survey Monograph 1. 438 pp.

Hauser, E., Potter, C., Hauge, T., Burgess, S., Burtch, S., Mutschler, J., Allmendinger, R., Brown, L., Kaufman, S., Oliver, J., 1987. Crustal structure of eastern Nevada from COCORP deep seismic reflection data. Geol. Soc. Am. Bull. 99, 833–844.

Hecker, S., 1993. Quaternary Tectonics of Utah with Emphasis on Earthquake-hazard Characterization. Utah Geological Survey Bulletin 127, 157 pp.

Hintze, L.F., 1988. Geologic history of Utah. Brigham Young University Geology Studies Special Publication 7. 202 pp.

Hintze, L.F., Willis, G.C., Laes, D.Y.M., Sprinkel, D.A., Brown, K.D., 2000. Digital Geologic Map of Utah. Utah Geological Survey, Salt Lake City, Utah. scale 1:500,000.

Hostetler, S.W., Giorgi, F., Bates, G.T., Bartlein, P.J., 1994. Lake-atmosphere feedbacks associated with Paleolakes Bonneville and Lahontan. Science 263, 665–668.

Lines, G.C., 1979. Hydrology and surface morphology of the Bonneville Salt Flats and Pilot Valley Playa, Utah. U.S. Geological Survey Water Supply Paper 2057.

Mason, J.L., Kipp Jr., K.L., 1998. Hydrology of the Bonneville Salt Flats, northwestern Utah, and simulation of ground-water flow and solute transport in the shallow-brine aquifer. U.S. Geological Survey Professional Paper 1585.

Mason, J.L., Brothers, W.C., Gerner, L.J., Muir, P.S., 1995. Selected hydrologic data for the Bonneville Salt Flats and Pilot Valley, western Utah, 1991–1993: U.S. Geological Survey Open File Report 95-104. .

McBride, J.H., Stephenson, W.J., Williams, R.A., Odum, J.K., Worley, D.M., South, J.V., Brinkerhoff, A.R., Keach, R.W., Okojie-Ayoro, A.O., 2010. Shallow subsurface structure of the Wasatch fault, Provo segment, Utah, from integrated compressional and shear-wave seismic reflection profiles with implications for fault structure and development. Geol. Soc. Am. Bull. 122, 1800–1814. http://dx.doi.org/10.1130/B30174.1.

Mifflin, M.D., Wheat, M.M., 1979. Pluvial lakes and estimated pluvial climates of Nevada. Nevada Bureau of Mines and Geology Bulletin 94, 58 pp.

Miller, D.M., 1985. Geologic map of the Lucin Quadrangle, Box Elder County, Utah. Utah Geological and Mineral Survey Map 78, 10 pp., scale 1:24,000.

Miller, D.M., 1990. Geologic map of the Crater Island SW Quadrangle, Box Elder County, Utah. Utah Geological and Mineral Survey Map 129, 8 pp., scale 1:24,000.

Miller, D.M., 1993. Geologic map of the Crater Island NW Quadrangle, Box Elder County, Utah. Utah Geological Survey Map 145, 13 pp., scale 1:24,000.

Miller, D.M., Allmendinger, R.W., 1991. Jurassic normal and strike-slip faults at Crater Island, northwestern Utah. Geol. Soc. Am. Bull. 103, 1239–1251.

Miller, D.M., Lush, A.P., 1994. Geologic map of the Pilot Peak Quadrangle, Box Elder County, Utah, and Elko County, Nevada. Utah Geological and Mineral Survey Map 160, 25 pp., scale 1:24,000.

Miller, D.M., Jordan, T.E., Allmendinger, R.W., 1990. Geologic map of the Crater Island Quadrangle, Box Elder County, Utah. Utah Geological and Mineral Survey Map 128, 16 pp., scale 1:24,000.

Miller, D.M., Lush, A.P., Schneyer, J.D., 1993. Geologic map of the Patterson Pass Quadrangle, Box Elder County, Utah and Elko County, Nevada. Utah Geological Survey Map 144, 20 pp., scale 1:24,000.

Miller, D.M., Phelps, G.A., 2016. Chapter 3. The Pilot Valley Shoreline: An Early Record of Lake Bonneville Dynamics. In: Oviatt, C.G., Shroder Jr., J.F. (Eds.), Lake Bonneville: A Scientific Update, vol. 20. Elsevier, pp. 60–74.

Murray, D.J., 1984. A Synergetic Study of Seismic Reflections and Critical Refractions in the Great Salt Lake Desert, Utah. M.S. thesis, University of Oklahoma. 109 pp.

Oviatt, C.G., 2014. The Gilbert episode in the Great Salt Lake basin, UT. Geological Survey Miscellaneous Publication 14-3. 20 pp.

Oviatt, C.G., 2015. Chronology of Lake Bonneville, 30,000 to 10,000 yr B.P. Quat. Sci. Rev. 110, 166–171.

Oviatt, C.G., Miller, D.M., 1997. New explorations along the northern shores of Lake Bonneville. In: Link, P.K., Kowallis, B.J. (Eds.), Mesozoic to Recent Geology of Utah. Brigham Young University Geology Studies, vol. 42-2, pp. 345–371.

Oviatt, C.G., Thompson, R.S., Kaufman, D.S., Bright, J., Forester, R.M., 1999. Reinterpretation of the Burmester Core, Bonneville Basin, Utah. Quat. Res. 52, 180–184.

Park, C.B., Miller, R.D., Xia, J., 1999. Multichannel analysis of surface waves MASW. Geophysics 64, 800–808.

Petersen, E.C., 1993. Geochemical Evolution of Groundwater with Implications for Groundwater Flow Patterns in the Pilot Valley Closed Basin, Utah and Nevada. M.S. thesis, Brigham Young University.

Pugin, A.J.M., Larson, T.H., Sargent, S.L., McBride, J.H., Bexfield, C.E., 2004. Near-surface mapping using SH-wave and P-wave seismic land streamer data acquisition in Illinois, U.S. Lead. Edge 23, 677–682.

René, R.M., Fittert, J.L., Murray, D.J., Walters, J.K., 1988. Reflection and refraction seismic studies in the Great Salt Lake Desert. Utah. Geophys. 53, 431–443.

Rey, K.A., 2012. Insights into the Early Transgressive History of Lake Bonneville from Stratigraphic Investigation of Pilot Valley Playa, UT/NV, USA. M.S. thesis, Brigham Young University. Paper 3803, 60 pp.

Schaeffer, F.E., Anderson, W.L., 1960. Geology of the Silver Island Mountains Box Elder and Tooele Counties, Utah and Elko County, Nevada. In: Schaeffer, F.E. (Ed.), Guidebook to the Geology of Utah. Utah Geological and Mineralogical Society, pp. 5–170.

Sheriff, R.E., Geldart, L.P., 1995. Exploration Seismology, second ed. Cambridge University Press, New York. 592 pp.

Stansbury, H., 1852. Exploration and survey of the valley of the Great Salt Lake of Utah, including a reconnaissance of a new route through the Rocky Mountains. Corps of U.S. Army Engineers, Washington. pp. 108–113.

Stewart, J.H., 1980. Regional tilt patterns of late Cenozoic basin-range fault blocks, western United States. Geol. Soc. Am. Bull., Pt. I 91, 460–464.

Thompson, R.S., Oviatt, C.G., Roberts, A.P., Buchner, J., Kelsey, R., Bracht, C., Forester, R.M., Bradbury, J.P., 1995. Stratigraphy, sedimentology, paleontology, and paleomagnetism of Pliocene-early Pleistocene lacustrine deposits in two cores from western Utah: U.S. Geological Survey, Report: OF 95-0001. p. 94.

Tingey, D.G., Mayo, A.L., Nelson, S.G., 2002. Recharge of a fresh water-brine groundwater system, Pilot Valley, Utah-Nevada. Geol. Soc. Am. Abst. Programs 34 (4), 56.

Wambeam, T.J., 2001. Modeling Lake Bonneville Basin Morphometry Using Digital Elevation Models. M.S. thesis, University of Utah, Salt Lake City. 60 pp.

Woolery, E.W., Street, R., Wang, Z., Harris, J.B., 1993. Near surface deformation in the New Madrid seismic zone as imaged by high-resolution SH-wave seismic methods. Geophys. Res. Lett. 20, 1615–1618.

Chapter 20

A Speleothem Record of Great Basin Paleoclimate: The Leviathan Chronology, Nevada

M.S. Lachniet

University of Nevada Las Vegas, Las Vegas, NV, United States

ABSTRACT

Speleothem studies from vadose-zone caves in the Great Basin have provided insight into the timing and controls of climatic change over the last 175,000 years. These studies integrate stable oxygen, carbon, and trace-element data to infer past temperature and wetness in the Great Basin, yielding new insights into the controls and timing on climate change on timescales ranging from centuries to orbital scales. This manuscript summarizes some of these records, with a focus on a long orbital-scale, oxygen-isotope reconstruction from Nevada called the Leviathan chronology, newly extended to 208,000 yr B2K (before year 2000). This speleothem-paleoclimate reconstruction is comprised of samples collected from Leviathan, Pinnacle, and Lehman Caves, Nevada, and shows orbital-scale variations in stable oxygen isotopes ($\delta^{18}O$) that closely follow the variations in Milankovitch-controlled summer insolation in the northern hemisphere. The data show that glacial-interglacial terminations arrive "on-time" as expected from Milankovitch variations, and that their duration was set by insolation. The strong match between Great Basin $\delta^{18}O$ values and insolation is also evident in minimum (cold) $\delta^{18}O$ values during marine isotope stages (MIS) 5d and 7, when insolation was low despite minimal ice volume. Holocene climate variation followed the orbital insolation trend, and agrees with Antevs' formulation of a distinct altithermal interval in the Great Basin. Surprisingly, the timing of $\delta^{18}O$ variations in the Leviathan chronology is markedly different to those originally published from Devils Hole, Nevada, suggesting that the phreatic origin of Devils Hole calcite may impart a nonclimatic forcing on calcite $\delta^{18}O$ values. This observation is confirmed by new Devils Hole cores that generally support the discoveries enabled by the Leviathan chronology, but which show a delayed and dampened $\delta^{18}O$ signal arising from dispersion in the regional carbonate aquifer.

Keywords: Speleothems, Great Basin, Leviathan Cave, Lehman Caves, Northern-hemisphere summer insolation, Precession, Marine isotope stage 5d, Interglacial

20.1 INTRODUCTION

20.1.1 Great Basin Paleoclimate in Context

The study of Great Basin (Fig. 20.1) paleoclimate has a long history extending to the 19th and early 20th-century investigations in the Lahontan, Bonneville, and Manly Basins (Russell, 1885; Gilbert, 1890; Blackwelder, 1933), where large pluvial lakes left behind prominent geomorphic evidence of their former extents. Initial pioneering radiocarbon and uranium-series dating of lake tufas and other lacustrine materials began to fix the timing of these lake extents to around the last 35,000 years (Broecker and Orr, 1958; Broecker and Kaufman, 1965; Benson et al., 1990). Following improved understanding of radiocarbon calibration into calendar years, a large compilation of radiocarbon ages and mapping efforts have begun to show a robust dating outline of pluvial lake extents and chronology (Adams et al., 2008; Reheis et al., 2014; Oviatt, 2015). These studies show the suborbital-scale expansion of pluvial lakes associated with an expanded Laurentide Ice Sheet and subsequent shrinking associated with the deglaciation. Prominent features include expanded extents around the last glacial maximum (LGM) in the Bonneville Basin (ca. 22–18 cal ka) and followed by a regression associated with sill down cutting at ca. 18 cal ka. In the Lahontan Basin, peak lake size was achieved between ca. 15 and 17 cal ka (Adams et al., 2008; Reheis et al., 2014), followed by shrinking then a small transgression around the Younger Dryas interval.

Despite evidence for pluvial lake expansions older than the last glacial cycle in the Great Basin (Ku et al., 1998; Reheis, 1999; Reheis et al.,

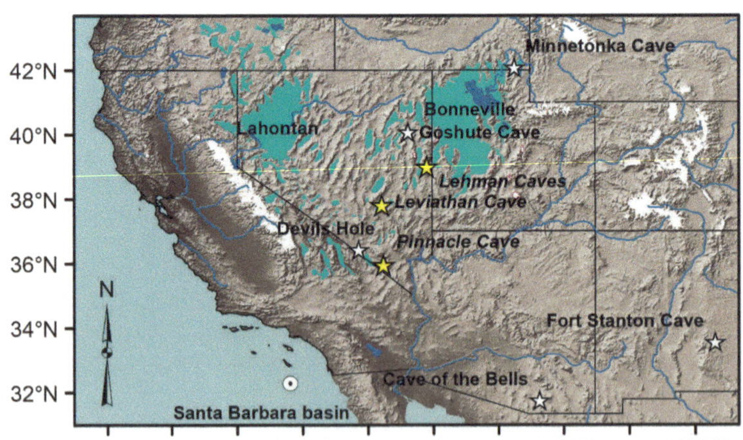

FIG. 20.1 Map of the Great Basin and cave locations mentioned in the text. *Light blue* and *white shading* indicates the extent of pluvial lakes and glaciers, respectively, around their maximum limits during MIS 2. *Blue* is the modern Great Salt Lake. *Yellow stars* indicate cave locations used in the Leviathan chronology.

2012), dating has been difficult to apply because U-series dates on tufas and other carbonates commonly exhibit open-system behavior, and many deposits are older than the radiocarbon limit of ca. 45 ka. Further, many pluvial lakes desiccate during warm periods, thus leaving only a fragmentary record of past climate conditions that is biased toward cool and wet conditions. As a result, the orbital-scale climate variations in the Great Basin have been less well understood than during the last glacial period.

20.1.2 Previous Speleothem Studies

An apparent resolution to this problem was the development of a 500,000-year climate record from Devils Hole, Nevada, which was based on uranium-series dated carbonate. The first pioneering work with stable isotope values (both $\delta^{18}O$ and $\delta^{13}C$) in cave calcite deposits was completed at Devils Hole, a tectonic window into a regional carbonate aquifer (Winograd et al., 1975, 1988, 1992; Coplen et al., 1994). The importance of the Devils Hole record was immediately recognized (Broecker, 1992), because it offered a window into the forcings and pacing of Great Basin paleoclimate on orbital timescales that were beyond the reach of the other terrestrial records. The Devils Hole $\delta^{18}O$ record documents a strong ice-age cyclicity and appears to show that Great Basin climate was out-of-step with the presumed Milankovitch forcings (Winograd et al., 2006), because the timing of $\delta^{18}O$ variations appear to precede corresponding changes in northern-hemisphere summer insolation (NHSI). The Devils Hole record is a phreatic (below the water table) mammillary calcite that grew carbonate from waters in the regional carbonate aquifer. The main discovery of the Leviathan chronology $\delta^{18}O$ data discussed below (Lachniet et al., 2014) is that the timing of Great Basin paleoclimatic change is indeed consistent with Milankovitch forcing, because $\delta^{18}O$ variations lag behind NHSI in vadose-zone speleothems, thus calling into question the interpretation or dating of the original (DH-11) Devils Hole record (Winograd et al., 2006).

Two valuable speleothem records are available for the greater southwestern United States (Fig. 20.1), those from Fort Stanton Cave, New Mexico (Asmerom et al., 2010) and Cave of the Bells, Arizona (Wagner et al., 2010). These records were important because they documented the atmospheric response to the abrupt climate changes also discovered in Greenland (Rasmussen et al., 2014) and the Santa Barbara Basin of offshore California (Hendy, 2010) called Dansgaard/Oeschger events. The most plausible interpretation of the $\delta^{18}O$ variations invoked changes in the winter-to-summer ratio of precipitation infiltrating into the cave, such that warm D/O events were associated with more summer infiltration, and vice versa for cold events. In these two records, the link to D/O events is very clear, but they contain little information about $\delta^{18}O$ variations linked to the Milankovitch variations associated with changes in precession and obliquity, as they are dominated

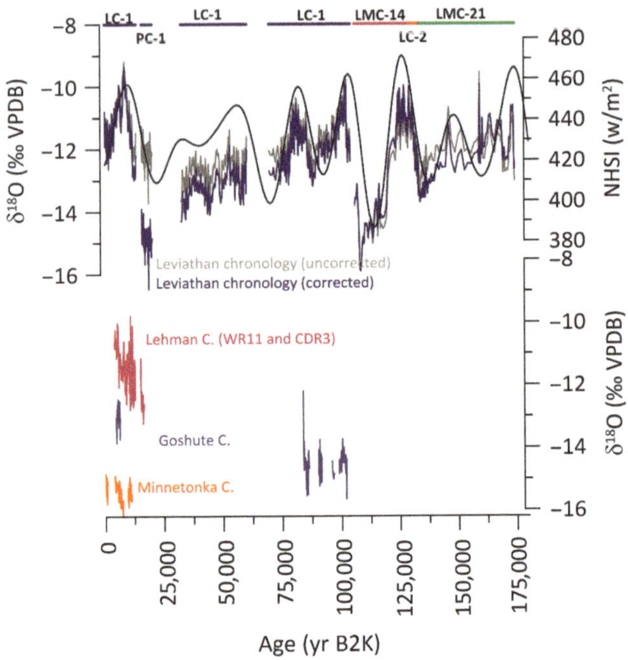

FIG. 20.2 Plot of stalagmite $\delta^{18}O$ time series from the Leviathan chronology (*blue* and *gray*, the latter uncorrected for changes in site water and ocean water $\delta^{18}O$ changes) and other caves in the Great Basin. *Bars* on top indicate growth intervals of Nevada stalagmites from Leviathan (LC-1), Pinnacle (PC-1), and Lehman Caves (LMC-14, LC-2, and LMC-21). Short and discontinuous time series from Lehman Caves (WR11 and CDR3), Goshute Cave, and Minnetonka Cave are shown at bottom for comparison to the Leviathan chronology.

by the millennial-scale, $\delta^{18}O$ oscillations. Additionally, these two records are located in regions that receive significant contributions of precipitation from the summer monsoon, whereas most of Nevada is dominated by winter precipitation. The first speleothem-based, climate reconstructions from Nevada, from Goshute Cave (Denniston et al., 2007) seemed to support the presence of millennial-scale, climate variability in the Great Basin (Fig. 20.2), but subsequent work described below provided a less clear picture. The Goshute Cave work also suggested that vadose-zone speleothems may be an alternative data source to test hypotheses of paleoclimatic change in the Great Basin.

In addition to Goshute Cave, paleoclimate studies began on age-dating and isotopic analyses of speleothems recovered from Lehman Caves, in Great Basin National Park, Nevada in 1999 by R. Denniston (Cornell College) and collaborators at the University of New Mexico. Initial efforts showed that many of the "dead and down" stalagmites recovered from the cave were beyond the limit of U-series dating, but that others presented suitable characteristics for dating (Denniston, 2000; Lachniet et al., 2014). Subsequent work recovered a stalagmite (LC-2) from Lehman Caves in 2004 that appeared to capture the transition out of the penultimate glacial maximum (PGM) (Shakun et al., 2011), a contention that was later confirmed by additional Lehman Caves isotopic and age-dating efforts from the 1999 samples (Lachniet et al., 2014) and more recent samples (Cross et al., 2015; Steponaitis et al., 2015). In Nevada, work began at Pinnacle Cave in 2007 with a stalagmite (PC-1) that showed full glacial conditions between ca. 20 and 15.6 ka (Lachniet et al., 2011). Both the Pinnacle Cave and Lehman Caves data presented conflicts of interpretation with the timing of $\delta^{18}O$ variations observed in the Devils Hole record, suggesting that one or more of these records may contain nonclimatic complications. A discontinuous record of winter climate was recovered from a stalagmite from Minnetonka Cave in the northern Wasatch Mountains of Idaho (Lundeen et al., 2013), which suggested that Holocene climate was marked by drier and warmer mid-Holocene conditions than early and late Holocene. Further work at Lehman Caves has provided stalagmite trace-element variations (Cross et al., 2015; Steponaitis et al., 2015) that further corroborate the broad details of the Minnetonka and Leviathan Cave climate histories, showing a broad interval of warm and dry conditions between ca. 8000 and 3000 yr BP.

To date, most of the cave speleothem records from the Great Basin have provided discontinuous snapshots of time, and thus have only begun to live up to their promise to provide continuous climate records back to ca. 500 ka, the approximate dating limit of the U-series technique. One reason for this is the difficult U-series dating characteristics of some Lehman Caves stalagmites, which may have experienced open-system conditions (Cross et al., 2015; Lachniet, unpublished data). Further, few stalagmites have been collected from growth position and under active drips for cave conservation reasons, thus disallowing rigorous tests for isotopic equilibrium in modern $CaCO_3$ that are required to evaluate how well the stalagmites record climate change (Lachniet, 2009, 2015). Additionally, some Lehman stalagmites are clearly precipitated out of oxygen isotopic disequilibrium, returning values that are implausible for reasonable cave temperatures and drip-water $\delta^{18}O$ values (Lachniet, unpublished data), suggesting that this cave presents challenges to paleoclimatologists for extraction of long and robust isotopic paleoclimate time series.

20.2 THE LEVIATHAN CHRONOLOGY

The longest and most continuous speleothem record (LC-1) recovered to date comes from Leviathan Cave, Nevada. Leviathan Cave is located in the Worthington Mountains of central Nevada, in what is now Basin and Range National Monument. The cave contains passages where temperature is constant at 8.2°C and relative humidity stays at 100%, and modern calcite appears to be in isotopic equilibrium with cave drip waters at the measured cave temperature (Lachniet et al., 2014). Cave drip waters, in turn, represent the mean average values of winter precipitation infiltration, thus making Leviathan Cave a proxy for winter precipitation $\delta^{18}O$ variations. Stalagmite LC-1 contains three main growth intervals separated by hiatuses: from modern to ca. 13.4 ka, from 32.3 to 60.2 ka, and from 69.6 to 103.8 ka. A small section of growth at the base of the stalagmite dates to around 120 ka, but appears to be highly discontinuous and thus not suitable for paleoclimate work. The dating of LC-1 is highly precise because of high ^{234}U concentrations in the stalagmite, with typical two σ age uncertainties of better than ±100 years for the entire Holocene, and ±200 to 500 years for most of the rest of the record (equivalent to better than $\pm0.5\%$). The presence of hiatuses in this sample during marine isotope stages 2 and 4 suggests that temperatures may have been too cold to support vigorous soil productivity over the cave at this high-altitude location (2240 m), a contention supported by evidence of periglacial sorted stone circles above the cave.

The Leviathan chronology is based on the compilation of carefully selected stalagmites from Leviathan (LC-1), Pinnacle (PC-1), and Lehman Caves (LMC-14, LMC-21, and LC-2), Nevada (Fig. 20.2). Sample selection was based on strong age-depth models without significant age inversions, which resulted in the exclusion of some Lehman Caves dead and down stalagmites that exhibit evidence for open-system behavior and postdepositional recrystallization. Together, the five stalagmites that comprise the Leviathan chronology cover 148.3 kyr of the last 174.1 ka (Fig. 20.2). A key time interval over Termination II is constrained by the published data from Lehman Cave stalagmite LC-2 (Shakun et al., 2011).

To create the composite chronology, all the time series were interpolated to a 50-year interval, and corrected for changes in the $\delta^{18}O$ of sea water according to methods in Lachniet et al. (2014), with a maximum adjustment of ~1.2‰ during maximum ice volume. Also, because the site-specific, drip-water and precipitation values of the three cave sites differ, the Pinnacle Cave and Lehman Caves records were adjusted by −1.73‰ and +0.36‰, respectively (Lachniet et al., 2014). The small adjustments for Lehman Cave have little influence on the timing or amplitude of $\delta^{18}O$ variations, but the relative values of peaks may differ if the isotopic gradient has changed over time. The combined corrections for changes in sea-water and site-water $\delta^{18}O$ variations are smaller than the observed intra-sample $\delta^{18}O$ variations (4–5‰). Most conservatively, the relative $\delta^{18}O$ variations in the Lehman Caves stalagmites between ca. 105 and 208 ka, and between 0 and 105 ka in the LC-1 stalagmite show variations that are unaffected by the site-water correction.

Additional samples from Lehman Caves are extending this record further back in time (Lachniet, unpublished data). The U-series and $\delta^{18}O$ data for one such stalagmite, LMC-23 from Lehman Cave, are shown in Table 20.1 and Fig. 20.3. The stalagmite grew during a short interval between 208 and 201 ka, and thus extends (discontinuously) the Leviathan chronology into

TABLE 20.1 U-Series Data for Stalagmite LMC-23

Depth (mm)	^{238}U (ng/g)		^{232}Th (pg/g)		$^{230}Th/^{232}Th$ Activity Ratio		$^{230}Th/^{238}U$ Activity Ratio		Measured $\delta^{234}U$ (‰)		Initial $\delta^{234}U$ (‰)		Uncorrected Age (yr BP)		Corrected Age[a] (yr BP)		Corrected Age (yr B2K)
10	198	0.12	558	±28	1410	72	1.30×10^{0}	1.58×10^{-03}	431	1.4	766	3	203,362	870	203,316	871	203,330
54	132	0.12	1447	±26	355	6	1.28×10^{0}	1.83×10^{-03}	404	1.4	723	3	206,931	998	206,746	1001	206,760
98	152	0.11	1693	±35	345	7	1.26×10^{0}	1.81×10^{-03}	386	1.4	693	3	207,959	1013	207,767	1016	207,781
135	233	0.17	1816	±34	497	9	1.27×10^{0}	1.63×10^{-03}	394	1.4	708	3	207,260	943	207,127	944	207,141
169	295	0.19	734	±27	1549	57	1.26×10^{0}	1.53×10^{-03}	387	1.4	696	3	208,196	923	208,153	923	208,167
204	183	0.14	671	±40	1613	95	1.93×10^{0}	2.74×10^{-03}	1023	2.0	1839	6	207,844	901	207,807	901	207,821

Years before present = yr BP, where present is AD 2011 or 2013. All errors are absolute 2σ. Subsample sizes range from 50 to 130 mg. $\delta^{234}U = ([^{234}U/^{238}U]\ \text{activity} - 1) \times 1000$. $[^{230}Th/^{238}U]\ \text{activity} = 1 - e^{-\lambda_{230}T} + (\delta^{234}U_{measured}/1000)[\lambda_{230}/(\lambda_{230} - \lambda_{234})](1 - e^{-(\lambda_{230}-\lambda_{234})T})$, where T is the age. Decay constants (λ) are 9.1577×10^{-6} year^{-1} for ^{230}Th, 2.8263×10^{-6} year^{-1} for ^{234}U, and 1.55125×10^{-10} year^{-1} for ^{238}U.

[a]*Corrected ages use an average crustal value for the initial $^{230}Th/^{232}Th$ atomic ratio = 4.4 ± 2.2 ppm.*

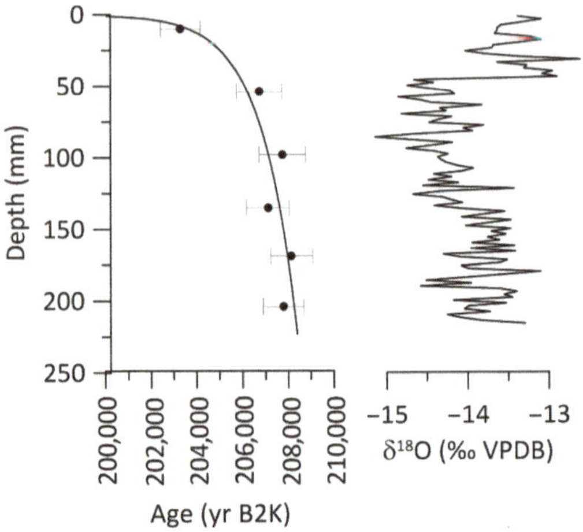

FIG. 20.3 U-series and measured $\delta^{18}O$ data for Lehman Caves stalagmite LMC-23 show growth over the 201–208 ka period during MIS 7. The age model is defined by the equation age = 1539.6 * ln(depth) + 200,074.

MIS 7, which is a key interval characterized by intermediate ice volume and CO_2 concentrations (Petit et al., 1999) but has several NHSI minima more pronounced than during MIS 2. The LMC-23 $\delta^{18}O$ data were corrected for changes in the $\delta^{18}O$ value of sea water and the difference in drip-water values between sites according to the method presented earlier (Lachniet et al., 2014).

Close inspection of Fig. 20.2 shows a strong correspondence between the Leviathan chronology and NHSI at 65°N over the last 175,000 years, an observation that will be discussed in more detail below. In contrast, the short records from Goshute, Minnetonka, and Lehman Caves (including the two stalagmites WMR11 and CDR3 for which data are publically available) do not as clearly fit into the orbital variation, suggesting they may have either different forcings or suffer from sample problems related to open-system behavior, recrystallization, or isotopic disequilibrium as has been observed in other Lehman Caves stalagmites (Lachniet, unpublished data). Additional long-term records are desirable to test possible hypotheses. For this reason, the remainder of this review focuses on the $\delta^{18}O$ signal from the Leviathan chronology. Additional insights into moisture availability derived from trace-element and carbon-isotope variations merit further attention (Cross et al., 2015; Steponaitis et al., 2015), because these data appear to replicate well in some cases and may be proxies for effective moisture over the cave site.

20.2.1 Stable Isotope Climate Record

The strength of the Leviathan chronology is that it provides evidence for climate variations on orbital timescales through MIS 7, thus providing a long record of changes in atmospheric circulation over the Great Basin. Because the $\delta^{18}O$ variations represent integrated histories of air masses at and upstream of the cave site, the Leviathan chronology's $\delta^{18}O$ variations are also relevant for the broader Western United States and for the northeastern subtropical Pacific Ocean, which is the dominant source of moisture reaching the Great Basin. Today, the $\delta^{18}O$ variations of atmospheric precipitation are strongly controlled by moisture source, with higher latitude moisture sources having lower $\delta^{18}O$ values, and lower latitude sources having higher $\delta^{18}O$ values (Friedman et al., 2002; Lachniet et al., 2014). Further, the moisture source is also strongly related to temperature, with higher latitude moisture sources delivering colder air masses and having lower $\delta^{18}O$ values. Thus, the $\delta^{18}O$ variations of the Leviathan chronology are considered a robust proxy for changes in winter atmospheric circulation at and upwind of the Great Basin.

The most notable feature of the Leviathan chronology is its strong precessional-scale, $\delta^{18}O$ cyclicity (Fig. 20.4). The $\delta^{18}O$ variations show a remarkable match to the shape of NHSI variations. Further, the $\delta^{18}O$ values lag peaks and troughs in the NHSI by a few thousand years, suggesting that Great Basin climate is responding to the classic Milankovitch forcing. In the broad sense, the Great Basin $\delta^{18}O$ variations mimic the changes in NHSI, and also show strong similarities with changes in global ice volume as indicated by benthic foraminifera $\delta^{18}O$ variations (Lisiecki and Raymo, 2005) (Fig. 20.4). These data are thus valuable for estimation of the duration and intensities of past temperature and moisture-source history for the Great Basin. Prominently, the data from three Lehman stalagmites (LC-2, LMC-14, and LMC-21) show that the penultimate interglacial period, corresponding to MIS 5e, arrived on time and was of short duration. Estimation of TII based on the time series shows a timing of ca. 131.8 ka, with a duration of 10.9 kyr (Lachniet et al., 2014). The on-time arrival of MIS 5e and its short duration conflict strongly with interpretations from Devils Hole (Winograd et al., 1997), which suggested the penultimate glacial was about 22 ka in duration. An update to the Devils Hole record (Moseley et al., 2016) using new calcite cores also inferred an unusually long interglacial period of 16.1 kyr, which is significantly longer than recorded in the vadose-zone stalagmites at Lehman Caves. The implication of this observation is discussed in more detail below.

Another notable feature of the Leviathan chronology is the return to cool, low-$\delta^{18}O$ conditions during MIS 5d, when $\delta^{18}O$ values were lower than during the preceding PGM of MIS 6. This result was surprising because the presence of the Laurentide Ice Sheet over much of northern North America during the PGM reflected significantly cooler temperatures than the following low-ice periods during MIS 5. One explanation for the unusual $\delta^{18}O$ pattern showing colder conditions in MIS 5d than in MIS 6 is the observation that obliquity was high during MIS 6, thus allowing warmer conditions relative to MIS 5d. The MIS 5d $\delta^{18}O$ low was interpreted to reflect a cold/high-

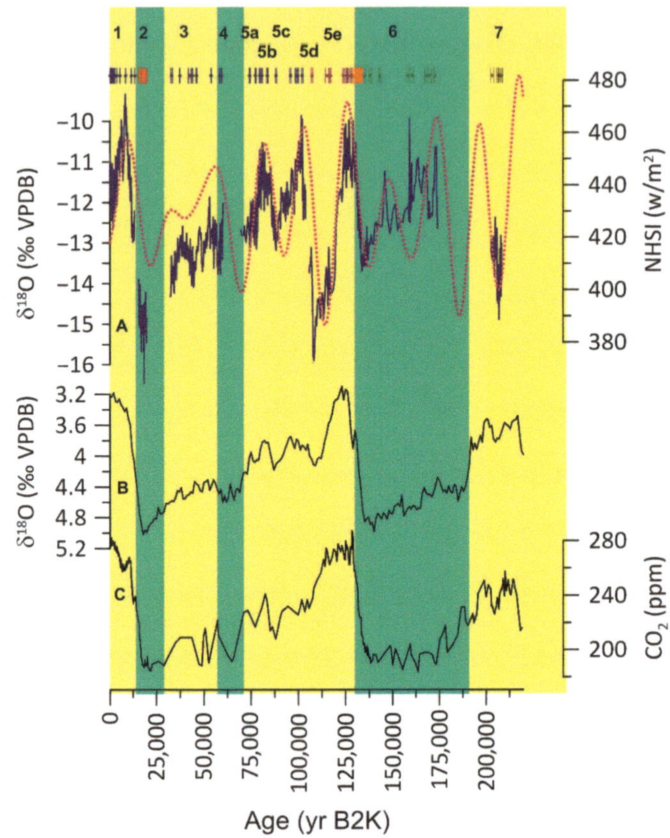

FIG. 20.4 Orbital scale paleoclimate records. A is the Leviathan chronology back to 208 ka (*blue line*) with 65°N summer insolation (*dotted magenta line*) and uranium-series age dates; B is the LR04 benthic marine $\delta^{18}O$ stack (Lisiecki and Raymo, 2005); and C is Antarctica CO_2 (Luethi et al., 2008). The Leviathan chronology $\delta^{18}O$ values have been corrected for changes in the $\delta^{18}O$ of sea water and the difference of drip-water values between sites.

latitude, winter moisture source associated with the prominent NHSI minimum, as it captures the relative magnitudes of the NHSI curve between MIS 6 and MIS 5a. Similarly, the new data from stalagmite LMC-23 show low $\delta^{18}O$ values during a prominent NHSI minimum during the 'interglacial' MIS 7. The MIS 7 $\delta^{18}O$ values, like those during MIS 5d, were lower than during MIS 6, despite lower ice volume at this time, and by implication a reduced extent of northern-hemisphere ice cover. The implication of these findings is that very cold conditions can be present in the Great Basin in the absence of significant ice-sheet cover and in the presence of intermediate to high CO_2 concentrations.

The Holocene $\delta^{18}O$ variations also strongly follow the NHSI trend. Following a clear cool event in the Younger Dryas characterized by relatively low $\delta^{18}O$ values, the $\delta^{18}O$ values at Leviathan then rise to peak at about 8.0 ka, some 3 ka after the NHSI peak, consistent with the lag of other $\delta^{18}O$ peaks and troughs behind NHSI over the past 210 ka. The $\delta^{18}O$ values then decrease to reach a local minimum at ca. 2.5 ka. The details of the Holocene record will be published separately, but it is worth noting that the Leviathan chronology supports the trip-partite division of Holocene climate first envisioned by Antevs (1938, 1948), with a clear mid-Holocene "altithermal" warm period between ca. 9 and 5 ka, transitioning into a cooler and wetter medi-thermal beginning around 4 ka.

20.3 DISCUSSION

The interpretation of the Nevada speleothem data indicates that changes in NHSI provoke changes in atmospheric circulation that are registered by $\delta^{18}O$ variability. The lag behind NHSI of $\delta^{18}O$ variations suggests that the forcing is indirect. Several possible forcings are associated with the Great Basin $\delta^{18}O$ variations. First, it is clear that CO_2 concentrations increase abruptly around the MIS 2/1 and MIS 6/5d transitions, which may explain some of the warming over Terminations I and II. However, Nevada $\delta^{18}O$ values drop steadily throughout the Holocene, whereas CO_2 remains high and even increases slightly over the last 8000 years (Ruddiman, 2003). Similarly, the strongly low $\delta^{18}O$ values during MIS 5d and MIS 7 happen during intervals with intermediate to high CO_2 values. Thus, the CO_2 changes may amplify a warming already in progress around ice volume terminations but are unlikely to be the source of the climate change, because they are decoupled during prominent intervals such as MIS 1 and 5d. A related hypothesis suffers from similar problems: the extent of the Laurentide Ice Sheet (LIS). The LIS retreated over the MIS 2/1 and MIS 6/5 transitions when temperatures in the Great Basin warmed (as inferred by increasing $\delta^{18}O$ values). However, decreasing $\delta^{18}O$ values from 8 ka to modern happened in the absence of any ice-sheet regrowth, and the prominent MIS 5d and MIS 7 minima also happened when ice sheets were small. Thus ice-sheet extent cannot be the primary driver of Great Basin $\delta^{18}O$ variations. The clear conclusion is that neither CO_2 nor ice-sheet extent were the sole or dominant controls on Great Basin paleoclimate over orbital timescales.

In contrast to the CO_2 and ice-sheet forcing, a more plausible forcing candidate is the NHSI-driven response of high-latitude sea ice and land-ice cover (Khodri et al., 2001, 2005). Robust and replicated proxy data for the extent of Arctic sea ice cover remain sparse. However, recent efforts have shown rapid and pronounced sea ice variations during MIS 5, with prominent maxima during 5e, c, and a separated by minima during 5b and d (Cronin et al., 2014). These variations mimic the Great Basin $\delta^{18}O$ record, and suggest that sea ice is an important contributor to forcing of atmospheric circulation changes in the Great Basin on precessional timescales. The observation of a strong link between precessional-scale NHSI variations and sea ice is supported by modeling observations of decreased sea ice during the mid-Holocene (Berger et al., 2013), when the Leviathan chronology registered warm and dry conditions (Lachniet et al., 2014). An additional forcing could be the extent of seasonal snow cover over the North American landmass, which should also respond sensitively to changes in NHSI. Increased snow cover appears to be linked to a stronger Aleutian low and a strengthening of the North Pacific storm tracks, each of which should be associated with delivery of higher latitude moisture (with low $\delta^{18}O$ values) to the Great Basin (Walland and Simmonds, 1996). A complete test of the high-latitude, snow-and-ice-cover hypothesis requires more robust and complete ice proxy records spanning from modern to at least MIS 7 to compare against the stalagmite $\delta^{18}O$ data.

The $\delta^{18}O$ data from the Great Basin vadose-zone speleothem also suggest a need to revisit interpretations of the $\delta^{18}O$ record in the phreatic speleothem of Devils Hole. Given that the vadose caves have very short response times to climate variation (on the timescale of seasons to a few years), and that the growth rates and hence U-series dating density is very high, the vadose-zone stalagmites are more likely to be representative of atmospheric $\delta^{18}O$ variations than Devils Hole. Devils Hole, being a window into a large regional carbonate aquifer (Winograd et al., 1975), receives waters from varying latitudes and ages that are mixed prior to arriving at Devils Hole and precipitating calcite. The lack of an "early" warming around ice volume Terminations I and II in the Leviathan, Pinnacle, and Lehman Caves data relative to the original (DH-11) Devils Hole further suggests that Devils Hole $\delta^{18}O$ variations are complicated by other nonclimatic effects. The problematic nature of Devils Hole calcite was confirmed by new dating efforts, which revealed systematic age discrepancies in cores collected at various elevations above the modern water table that were interpreted to reflect mobilization of thorium in the water column (Moseley et al., 2016). The result is inaccurate age estimates for calcite located significantly below the water table, as in the DH-11 core. Cores retrieved from near the water table (DH2-D) show timings of climate events that are more similar to the Leviathan chronology (Fig. 20.5), with some subtle but important differences. Importantly, the timing of peak $\delta^{18}O$ values during the penultimate interglacial (MIS 5e) are delayed by several thousand years in DH2-D (arriving at 129.5 ka in the Leviathan chronology and 125.9 ka in DH2-D). This delay is considerably longer than the <2000-year transit time inferred by (Winograd et al., 2006). Further, the length of the MIS 5e interglacial is too long (16.1 kyr) by about 5000 years relative to the 10.9 kyr duration in the Leviathan chronology. The Devils Hole data also exhibit dampened $\delta^{18}O$ variability relative to surface climate. The delay and dampening of the climate signal is attributable to the transit time and dispersion of isotopic signals as they traverse the regional carbonate aquifer feeding Devils Hole.

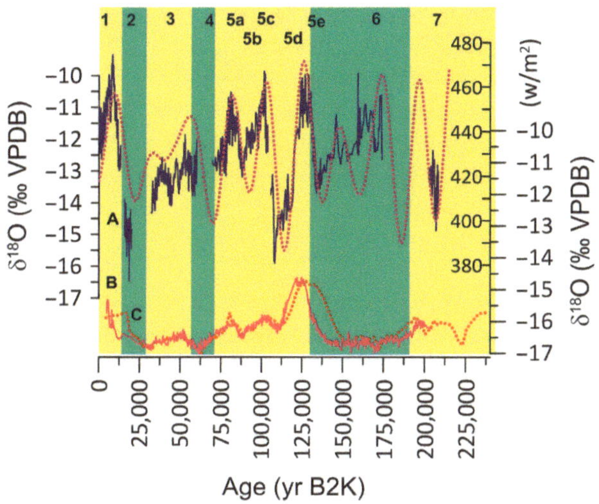

FIG. 20.5 Leviathan vs the Devils Hole record. A is the Leviathan chronology (*blue line*) with northern-hemisphere summer insolation at 65°N (*dotted magenta line*). Bottom is Devils Hole $\delta^{18}O$ record: DH2-D in *pink* (B) (Moseley et al., 2016) and DH-11 (Winograd et al., 2006) in *dotted red* (C). The *y*-axes for Leviathan and Devils Hole are scaled identically, showing the dampened isotopic signal in Devils Hole. *Top* indicates marine isotope stages and substages. The delayed and dampened $\delta^{18}O$ signal relative to surface climate is interpreted to reflect smoothing of the climate signal in the regional carbonate aquifer feeding Devils Hole.

Another intriguing observation of the Nevada data is the general lack of the pronounced millennial-scale climate variability in the $\delta^{18}O$ data—excepting the Younger Dryas cooling—that is so clearly evident in records from Greenland (Rasmussen et al., 2014), the Santa Barbara Basin (Hendy, 2010), Arizona (Wagner et al., 2010), and New Mexico (Asmerom et al., 2010). Suggestions of similar millennial-scale correlations to Greenland and the North Atlantic at Goshute and Lehman Caves await more continuous records (Denniston et al., 2007; Cross et al., 2015) for verification. One possibility for the general lack of millennial-scale $\delta^{18}O$ variability is that the summer monsoon results in essentially no recharge into Leviathan Cave, whereas a summer component is evident at Cave of the Bells and Fort Stanton Cave and imparts strong variations in the $\delta^{18}O$ of drip waters into the cave that are lacking at Leviathan. The presence of a Younger Dryas cool period is thus somewhat perplexing. Evidence from the Pacific Northwest (Oregon Caves) (Vacco et al., 2005), New Mexico (Asmerom et al., 2010), and Santa Barbara Basin (Hendy, 2010) offshore of southern California also exhibit a strong Younger Dryas, suggesting that this time interval had a widespread impact on climate in western North America. It is possible that the Younger Dryas had a significantly strong impact on winter-time $\delta^{18}O$ variations that were recorded in Leviathan Cave, whereas other millennial-scale climate events were not.

Finally, the relation between atmospheric circulation and pluvial lake extent in the Great Basin presents intriguing questions: did lake extents vary due to precession forcing? Were they associated with changes in moisture source or just changes in moisture amount? Were lake highstands during Heinrich events (Munroe and Laabs, 2013) associated with a change in precipitation $\delta^{18}O$ values? At present, it is only feasible to compare the Leviathan chronology $\delta^{18}O$ variations to Lake Bonneville shoreline altitudes in broad detail because of hiatuses in the speleothem record, but some interesting observation result nonetheless.

To make the comparison between lake levels, I used estimates from several sources: The Little Valley (MIS 6), Cutler Dam (MIS 4), and Holocene (MIS 1) curves are sourced from (Madsen, 2000), and appended to the MIS 3/2 stages of (Oviatt et al., 1992), which have been updated by (Oviatt, 2015). It is acknowledged that the timing of the older lake cycles is tentative, as they have not yet been securely dated by radiometric techniques. However, this comparison allows evaluation of $\delta^{18}O$ variations to estimates of lake extent on orbital timescales over the LGM and PGM. Full glacial conditions, as evidenced by low $\delta^{18}O$ values, coincide with the Bonneville shoreline and later Provo shoreline after sill down cutting, and rising $\delta^{18}O$ values coincide with the regression to low lake levels (Oviatt, 2015) over the last glacial cycle. During MIS 4, speleothem deposition ceased at the same time of a lake transgression, suggesting that effective moisture was not to blame for the speleothem hiatus. Evidence for lakes in the Bonneville Basin during MIS 5e is scant, and $\delta^{18}O$ values are intermediate to high during this time, suggesting that climate was either too warm or too dry to sustain them, while maintaining enough winter precipitation to infiltrate into Leviathan Cave and precipitate $CaCO_3$. High lake levels during MIS 6 (the Little Valley Lake cycle) coincide with low $\delta^{18}O$ values. Unfortunately, the key time interval constraining the Bonneville transgression between ca. 30 and 18 ka remains unconstrained in speleothem data. However, the strong link between NHSI and $\delta^{18}O$ values allows for estimation: a decrease in insolation between ca. 32 and 22 ka was likely associated with a decrease in $\delta^{18}O$ that coincided with both the expansion of the LIS and Lake Bonneville (Fig. 20.6).

In broad details, it can be concluded that lake extents appear to expand during periods of greater global ice volume and Laurentide Ice Sheet extent. Surprisingly, the pronounced lake expansions during Heinrich stadial 1 (Munroe and Laabs, 2013) are not associated with a pronounced change in Pinnacle Cave $\delta^{18}O$ values (Lachniet et al., 2011). A decrease in Pinnacle Cave $\delta^{13}C$ values, in contrast, at ca. 17 ka in southern Nevada suggests that soils became more productive, possibly in response to wetter conditions, a possibility supported by a speleothem $\delta^{13}C$ decrease at 16 ka from McLean's Cave on the west slope of the Sierra Nevada (Oster et al., 2015). Although not definitive yet, these data suggest that Heinrich stadial 1 may have been characterized by more frequent storms, but that no noticeable change occurred in

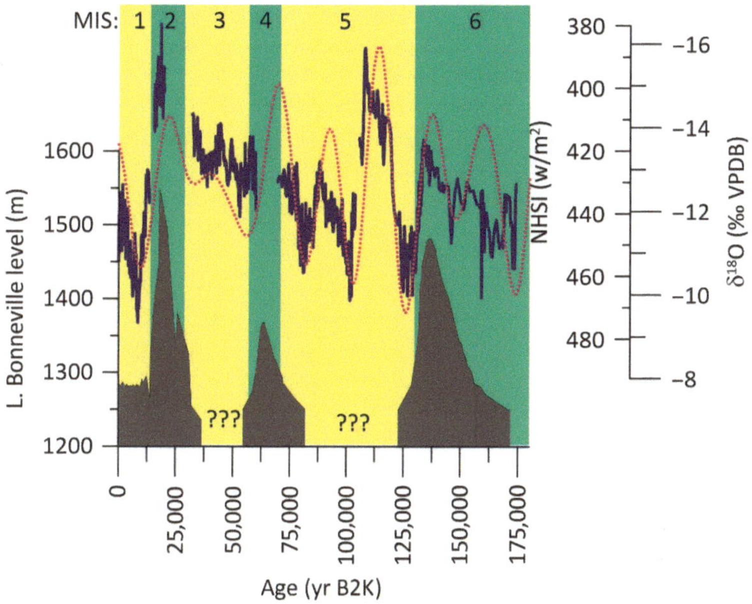

FIG. 20.6 Comparison of the Leviathan chronology on inverted $\delta^{18}O$ scale (cold/high-latitude moisture plots upward) and estimates of Lake Bonneville level over the past 175 ka (plotted in gray), see main text for references. A coincidence of low $\delta^{18}O$ and high lake levels during MIS 6 and MIS 2 suggests a dominant control of insolation and ice-sheet extent on Great Basin climate variations. The lake level history is poorly constrained over the intervals with "???".

moisture source. For this reason, it is suggestive that Nevada $\delta^{18}O$ values may not pick up some of the millennial-scale climate variability seen in the lakes and other speleothem records in the southwest. Further testing of this hypothesis for the Great Basin is currently underway using moisture-sensitive $\delta^{13}C$ values (Kirby et al., 2015) and trace-element ratios (Cross et al., 2015; Steponaitis et al., 2015).

20.4 CONCLUSIONS

The strongest conclusion from the Leviathan chronology $\delta^{18}O$ time series is that Great Basin paleoclimate is controlled by precessional-scale variations in summer insolation in the northern hemisphere. The likely forcing is indirect via sea- and land-ice forcing, as suggested by a 1–3 ka lag in $\delta^{18}O$ behind NHSI. The dominant forcing of atmospheric $\delta^{18}O$ variations is unlikely to be from CO_2 and ice-sheet extent, because the Leviathan $\delta^{18}O$ values document cold conditions during a time of minimal to intermediate ice-sheet extent and CO_2 concentrations, for example, during MIS 5d and MIS 7. The pronounced decrease in Leviathan $\delta^{18}O$ between 8 ka to modern time also cannot be explained by either CO_2 concentrations (which are stable or rising) or ice-sheet extent (which was continuously diminishing over the Holocene). But the Holocene decrease in $\delta^{18}O$ follows decreasing NHSI with a few thousand year lag, consistent with Milankovitch forcing. Finally, the clear discrepancies between the high-resolution (both in terms of dating density and $\delta^{18}O$ resolution) vadose zone Leviathan chronology and the low-resolution record from the phreatic Devils Hole, suggests that the Devils Hole $\delta^{18}O$ record contains strong nonclimatic fingerprints. The meaning of the Devils Hole $\delta^{18}O$ time series will thus require additional work and interpretations that include proxies for water-residence time and latitude of infiltration in the regional carbonate aquifer.

ACKNOWLEDGMENTS

This research was supported by NSF grants AGS-1405546 and AGS-1405557. We thank the Bureau of Land Management for the sampling permit for Leviathan Cave, and the Southern Nevada Grotto of the National Speleological Society for valuable collaboration.

REFERENCES

Adams, K.D., Goebel, T., Graf, K., Smith, G.M., Camp, A.J., Briggs, R.W., Rhode, D., 2008. Late Pleistocene and early Holocene lake-level fluctuations in the Lahontan basin, Nevada: implications for the distribution of archaeological sites. Geoarchaeology 23, 608–643.

Antevs, E., 1938. Climatic variations during the last glaciation in North America. Bull. Am. Meteorol. Soc. 19, 12–16.

Antevs, E., 1948. Climatic changes and pre-white man. Bull. Univ. Utah 38, 167–191.

Asmerom, Y., Polyak, V.J., Burns, S.J., 2010. Variable winter moisture in the southwestern United States linked to rapid glacial climate shifts. Nat. Geosci. 3, 114–117.

Benson, L.V., Currey, D.R., Dorn, R.I., Lajoie, K.R., Oviatt, C.G., Robinson, S.W., Smith, G.I., Stine, S., 1990. Chronology of expansion and contraction of four Great Basin lake systems during the past 35,000 years. Palaeogeogr. Palaeoclimatol. Palaeoecol. 78, 241–286.

Berger, M., Brandefelt, J., Nilsson, J., 2013. The sensitivity of the Arctic sea ice to orbitally induced insolation changes: a study of the mid-Holocene Paleoclimate Modelling Intercomparison Project 2 and 3 simulations. Clim Past 9, 969–982.

Blackwelder, E., 1933. Lake Manly: an extinct lake of Death Valley. Geogr. Rev. 23, 464–471.

Broecker, W.S., 1992. Upset for Milankovitch theory. Nature 359, 779–780.

Broecker, W.S., Kaufman, A., 1965. Radiocarbon chronology of Lake Lahontan and Lake Bonneville 2 Great Basin. Geol. Soc. Am. Bul. 76, 537–566.

Broecker, W.S., Orr, P.C., 1958. Radiocarbon Chronology of Lake Lahontan and Lake Bonneville. Geol. Soc. Am. Bull. 69, 1009–1032.

Coplen, T.B., Winograd, I.J., Landwehr, J.M., Riggs, A.C., 1994. 500,000-year stable carbon isotopic record from Devils Hole, Nevada. Science 263, 361–365.

Cronin, T., Marzen, R., DeNinno, L., 2014. Orbital-scale cyclostratigraphy and ice volume fluctuations from Arctic Ocean sediments. In: American Geophysical Union Fall Meeting, San Franciso, CA.

Cross, M., McGee, D., Broecker, W.S., Quade, J., Shakun, J.D., Cheng, H., Lu, Y., Edwards, R.L., 2015. Great Basin hydrology, paleoclimate, and connections with the North Atlantic: a speleothem stable isotope and trace element record from Lehman Caves, NV. Quatern. Sci. Rev. 127, 186–198.

Denniston, R.F., 2000. Mineralogic and stable isotopic characteristics of select Great Basin speleothems. Geological Society of America, 11.

Denniston, R., Asmerom, Y., Polyak, V.J., Dorale, J.A., Carpenter, S.J., Trodick, C., Hoye, B., Gonzalez, L.A., 2007. Synchronous millennial-scale climatic changes in the Great Basin and the North Atlantic during the last interglacial. Geology 35, 619–622.

Friedman, I., Harris, J.M., Smith, G.I., Johnson, C.A., 2002. Stable isotope composition of waters in the Great Basin, United States; 1, Air-mass trajectories. J. Geophys. Res. D Atmos. 107, 1–14.

Gilbert, G.K., 1890. Lake Bonneville. Government Printing Office, Washington. p. 438.

Hendy, I.L., 2010. The paleoclimatic response of the Southern Californian margin to the rapid climate change of the last 60 ka: a regional review. Quatern. Int. 215, 62–73.

Khodri, M., Leclainche, Y., Ramstein, G., Braconnot, P., Marti, O., Cortijo, E., 2001. Simulating the amplification of orbital forcing by ocean feedbacks in the last glaciation. Nature 410, 570–574.

Khodri, M., Cane, M.A., Kukla, G., Gavin, J., Braconnot, P., 2005. The impact of precession changes on the Arctic climate during the last interglacial-glacial transition. Earth Planet. Sci. Lett. 236, 285–304.

Kirby, M.E., Knell, E.J., Anderson, W.T., Lachniet, M.S., Palermo, J., Eeg, H., Lucero, R., Murrieta, R., Arevalo, A., Silveira, E., Hiner, C.A., 2015. Evidence for insolation and Pacific forcing of Late Glacial through Holocene climate in the Central Mojave Desert (Silver Lake, CA). Quatern. Res. 84, 174–186.

Ku, T.L., Luo, S.D., Lowenstein, T.K., Li, J.R., Spencer, R.J., 1998. U-series chronology of lacustrine deposits in Death Valley, California. Quatern. Res. 50, 261–275.

Lachniet, M.S., 2009. Climatic and environmental controls on speleothem oxygen isotope values. Quatern. Sci. Rev. 28, 412–432.

Lachniet, M.S., 2015. Are aragonite stalagmites reliable paleoclimate proxies? Tests for oxygen isotopic time series replication and equilibrium. Geol. Soc. Am. Bull. 127, 1521–1533.

Lachniet, M.S., Asmerom, Y., Polyak, V., 2011. Deglacial paleoclimate in the southwestern United States: an abrupt 18.6 ka cold event and evidence for a North Atlantic forcing of Termination I. Quatern. Sci. Rev. 30, 3803–3811.

Lachniet, M.S., Denniston, R.F., Asmerom, Y., Polyak, V.J., 2014. Orbital control of western North America atmospheric circulation and climate over two glacial cycles. Nat. Commun. http://dx.doi.org/10.1038/ncomms4805.

Lisiecki, L.E., Raymo, M.E., 2005. A Pliocene-Pleistocene stack of 57 globally distributed benthic $\delta^{18}O$ records. Paleoceanography 20, 1–17.

Luethi, D., Le Floch, M., Bereiter, B., Blunier, T., Barnola, J.-M., Siegenthaler, U., Raynaud, D., Jouzel, J., Fischer, H., Kawamura, K., Stocker, T.F., 2008. High-resolution carbon dioxide concentration record 650,000–800,000 years before present. Nature (London) 453, 379–382.

Lundeen, Z., Brunelle, A., Burns, S.J., Polyak, V., Asmerom, Y., 2013. A speleothem record of Holocene paleoclimate from the northern Wasatch Mountains, southeast Idaho, USA. Quatern. Int. 310, 83–95.

Madsen, D.B., 2000. Late Quaternary Paleoecology in the Bonneville Basin. Utah Geological Survey, Bulletin, vol. 130. p. 190.

Moseley, G.E., Edwards, R.L., Wendt, K.A., Cheng, H., Dublyansky, Y., Lu, Y., Boch, R., Spotl, C., 2016. Reconciliation of the Devils Hole climate record with orbital forcing. Science 351, 165–168.

Munroe, J.S., Laabs, B.J.C., 2013. Latest Pleistocene history of pluvial Lake Franklin, northeastern Nevada, USA. Geol. Soc. Am. Bull. 125, 322–342.

Oster, J.L., Montañez, I.P., Santare, L.R., Sharp, W.D., Wong, C., Cooper, K.M., 2015. Stalagmite records of hydroclimate in central California during termination 1. Quatern. Sci. Rev. 127, 199–214.

Oviatt, C.G., 2015. Chronology of Lake Bonneville, 30,000 to 10,000 yr BP. Quatern. Sci. Rev. 110, 166–171.

Oviatt, C.G., Currey, D.R., Sack, D., 1992. Radiocarbon chronology of Lake Bonneville, eastern Great Basin, USA. Palaeogeogr. Palaeoclimatol. Palaeoecol. 99, 225–241.

Petit, J.R., Jouzel, J., Raynaud, D., Barkov, N.I., Barnola, J.M., Basile, I., Bender, M., Chappellaz, J., Davis, M., Delaygue, G., Delmotte, M., Kotlyakov, V.M., Legrand, M., Lipenkov, V.Y., Lorius, C., Pepin, L., Ritz, C., Saltzman, E., Stievenard, M., 1999. Climate and atmospheric history of the past 420,000 years from the Vostok ice core, Antarctica. Nature 399, 429–436.

Rasmussen, S.O., Bigler, M., Blockley, S.P., Blunier, T., Buchardt, S.L., Clausen, H.B., Cvijanovic, I., Dahl-Jensen, D., Johnsen, S.J., Fischer, H., Gkinis, V., Guillevic, M., Hoek, W.Z., Lowe, J.J., Pedro, J.B., Popp, T., Seierstad, I.K., Steffensen, J.P., Svensson, A.M., Vallelonga, P., Vinther, B.M., Walker, M.J.C., Wheatley, J.J., Winstrup, M., 2014. A stratigraphic framework for abrupt climatic changes during the Last Glacial period based on three synchronized Greenland ice-core records: refining and extending the INTIMATE event stratigraphy. Quatern. Sci. Rev. 106, 14–28.

Reheis, M.C., 1999. Highest pluvial-lake shorelines and Pleistocene climate of the western Great Basin. Quatern. Res. (New York) 52, 196–205.

Reheis, M.C., Bright, J., Lund, S.P., Miller, D.M., Skipp, G., Fleck, R.J., 2012. A half-million-year record of paleoclimate from the Lake Manix Core, Mojave Desert, California. Palaeogeogr. Palaeoclimatol. Palaeoecol. 365, 11–37.

Reheis, M.C., Adams, K.D., Oviatt, C.G., Bacon, S.N., 2014. Pluvial lakes in the Great Basin of the western United States a view from the outcrop. Quatern. Sci. Rev. 97, 33–57.

Ruddiman, W.F., 2003. The anthropogenic greenhouse era began thousands of years ago. Clim. Change 61, 261–293.

Russell, I.C., 1885. Geological history of Lake Lahontan: a quaternary Lake of northwestern Nevada. Monograph of the United States Geological Survey, Government Printing Office, Washington. p. 448.

Shakun, J.D., Burns, S.J., Clark, P.U., Cheng, H., Edwards, R.L., 2011. Milankovitch-paced Termination II in a Nevada speleothem. Geophys. Res. Lett. 38, 1–6. http://dx.doi.org/10.1029/2011GL048560.

Steponaitis, E., Andrews, A., McGee, D., Quade, J., Hsieh, Y.-T., Broecker, W.S., Shuman, B.N., Burns, S.J., Cheng, H., 2015. Mid-Holocene drying of the U.S. Great Basin recorded in Nevada speleothems. Quatern. Sci. Rev. 127, 174–185.

Vacco, D.A., Clark, P.U., Mix, A.C., Cheng, H., Edward, R.L., 2005. A speleothem record of Younger Dryas cooling, Klamath Mountains, Oregon, USA. Quatern. Res. 64, 249–256.

Wagner, J.D.M., Cole, J.E., Beck, J.W., Patchett, P.J., Henderson, G.M., Barnett, H.R., 2010. Moisture variability in the southwestern United Sates linked to abrupt glacial climate change. Nat. Geosci. 3, 110–113.

Walland, D.J., Simmonds, I., 1996. Modelled atmospheric response to changes in Northern Hemisphere snow cover. Climate Dynam. 13, 25–34.

Winograd, I.J., Thordarson, W., U.S. Atomic Energy Commission, 1975. Hydrogeologic and Hydrochemical Framework, South-Central Great Basin, Nevada-California: With Special

Reference to the Nevada Test Site: Prepared on Behalf of the U.S. Atomic Energy Commission. U.S. Government Printing Office, Washington.

Winograd, I.J., Szabo, B.J., Coplen, T.B., Riggs, A.C., 1988. A 250,000-year climatic record from Great Basin vein calcite; implications for Milankovitch theory. Science 242, 1275–1280.

Winograd, I.J., Coplen, T.B., Landwehr, J.M., Riggs, A.C., Ludwig, K.R., Szabo, B.J., Kolesar, P.T., Revesz, K.M., 1992. Continuous 500,000-year climate record from vein calcite in Devils Hole, Nevada. Science 258, 255–260.

Winograd, I.J., Landwehr, J.M., Ludwig, K.R., Coplen, T.B., Riggs, A.C., 1997. Duration and structure of the past four interglaciations. Quatern. Res. (New York) 48, 141–154.

Winograd, I.J., Landwehr, J.M., Coplen, T.B., Sharp, W.D., Riggs, A.C., Ludwig, K.R., Kolesar, P.T., 2006. Devils Hole, Nevada, $d^{18}O$ record extended to the mid-Holocene. Quatern. Res. 66, 202–212.

Chapter 21

Pleistocene Lake Bonneville as an Analog for Extraterrestrial Lakes and Oceans

M.A. Chan*, P.W. Jewell*, T.J. Parker[†], J. Ormö[‡], C.H. Okubo[§] and G. Komatsu[¶]

*University of Utah, Salt Lake City, UT, United States

[†]Jet Propulsion Laboratory, California Institute of Technology, Pasadena, CA, United States

[‡]Centro de Astrobiología (CSIC-INTA), Torrejón de Ardoz, Spain

[§]U.S. Geological Survey, Tucson, AZ, United States

[¶]International Research School of Planetary Sciences, Università d'Annunzio, Pescara, Italy

ABSTRACT

Geomorphic confirmation for a putative ancient Mars ocean relies on analog comparisons of coastal-like features such as shoreline feature attributes and temporal scales of process formation. Pleistocene Lake Bonneville is one of the few large, geologically young, terrestrial lake systems that exemplify well-preserved shoreline characteristics that formed quickly, on the order of a thousand years or less. Studies of Lake Bonneville provide two essential analog considerations for interpreting shorelines on Mars: (1) morphological variations in expression depend on constructional vs erosional processes, and (2) shorelines are not always correlative at an equipotential elevation across a basin due to isostasy, heat flow, wave setup, fetch, and other factors. Although other large terrestrial lake systems display supporting evidence for geomorphic comparisons, Lake Bonneville encompasses the most integrated examples of preserved coastal features related to basin history, sediment supply, climate, and fetch, all within the context of a detailed hydrograph. These collective terrestrial lessons provide a framework to evaluate possible boundary conditions for ancient Mars hydrology and large water body environmental feedbacks. This knowledge of shoreline characteristics, processes, and environments can support explorations of habitable environments and guide future mission explorations.

Keywords: Lake Bonneville, Mars, Titan, Shorelines, Analogs, Extraterrestrial, Ocean

Developments in Earth Surface Processes, Vol. 20. http://dx.doi.org/10.1016/B978-0-444-63590-7.00021-4

21.1 INTRODUCTION

The purpose of this chapter is to detail important aspects of Pleistocene Lake Bonneville and its relevance to understanding lakes and oceans on Mars. Unlike spatially limited sediment cores and outcrops, shorelines are key mappable elements that are essential for estimating the total water mass and cyclical fluxes in lakes and oceans. The shorelines represent the edge of the water body, and thus allow volumetric estimates, and distinction between constructional and erosional features to enable interpretations of the geomorphic effectiveness of water and wind. Lake Bonneville (Gilbert, 1890) and its modern remnant, the Great Salt Lake, comprise a renowned geologic locality in the Basin and Range physiographic province of North America. Our study provides insight about the morphological development of shorelines, how long it takes to build prominent shorelines on Earth, and what processes can modify previously formed shoreline features. To evaluate extraterrestrial lake and ocean processes on Mars, we utilize a comparative geomorphic approach that examines the general geology, geography, geomorphology, and hydrodynamic modeling. The terrestrial analog of Lake Bonneville is thus critical to interpreting primary surface processes and subtle secondary modification of Martian shorelines.

21.1.1 Mars

Terminology for standing bodies of water on other planets has been problematic (eg, Ghatan and Zimbelman, 2006; Parker et al., 2010). The term ocean does have genetic and compositional (seawater) connotations based on its historical usage on Earth. However, we herein use the term ocean as a generic description for a widespread standing body of water that once existed on Mars (vs a smaller enclosed lake).

Currently, the Mars time scale is broadly categorized into three epochs (Noachian, Hesperian, and Amazonian), constrained mainly by crater counts, superposition and cross-cutting relationships, and surface mineralogy (eg, Phyllosian, Theiikian, Siderikian). Water played an important role on Mars (eg, Clifford, 1993; Baker, 2007), particularly in the early history of the planet (Carr, 1996; Malin and Edgett, 1999, 2000, 2003; Clifford and Parker, 2001; Jakosky and Phillips, 2001; Webb, 2004; Boyce et al., 2005). However, the spatiotemporal and geomorphic contexts of water-lain sediments and shorelines have not been well constrained or correlated across regions of the planet (eg, Squyres et al., 2004; Grotzinger et al., 2005; Bibring et al., 2007; Mustard et al., 2008).

One of the most prominent controversies in planetary geology involves the possibility of an ancient (no longer present) ocean on Mars. The Mars ocean hypothesis proposes that about a third of the surface of Mars was once covered by a vast ocean of liquid water, informally named "Oceanus Borealis," early in the planet's geologic history (Clifford and Parker, 2001), ~3.8 billion years ago. The inundated low-lying area is primarily located in the northern hemisphere; the northern plains region lies 4–5 km below Mars' datum. Published literature related to water and oceans on Mars dates back to the 1980s (eg, Lucchitta et al., 1986; Parker et al., 1989; Baker et al., 1991; Parker et al., 1993; Edgett and Malin, 1999; Head et al., 1999; Clifford and Parker, 2001; Parker and Currey, 2001; Fairén et al., 2003; Ormö et al., 2004; Boyce et al., 2005; Dohm et al., 2009; Parker et al., 2010; Cabrol and Grin, 2010; Lucchitta, 2010). Discussing the evidence for and against the Mars ocean hypothesis remains timely in the context of current missions to assess the habitability of the planet, and in ongoing efforts to guide exploration and surface lander studies.

21.1.2 Approach

Although recognizing the widely different environmental variables and time scales inherent in the past hydrospheres of the Earth vs other planets, features of Lake Bonneville represent the best baseline for establishing criteria for identifying evidence of past water bodies on Mars. Our ideas build on inputs from this terrestrial analog (Chan et al., 2011), which define three critical geomorphic and sedimentologic attributes of large water bodies useful for reappraising the Mars ocean hypothesis: (1) shoreline morphology, (2) correlative elevations, and (3) equipotential surfaces. Our discussions address features present in Mars' northern plains, which could define what was possibly once the largest water body on that planet. Although we cannot fully address other shoreline features in regions such as the Valles Marineris (eg, Lucchitta, 2010), Ma'adim Vallis (Irwin et al., 2004), and the Hellas and Argyre basins (Kargel and Strom, 1992; Parker, 1994; Ormö and Komatsu, 2003), we do include some reference to a few crater-lake features (Cabrol and Grin, 2002).

21.2 METHODOLOGY

In examining large bodies of water, it is important to consider issues of morphology such as: what is the nature of the geomorphic record left by a major standing body of water? What are the diagnostic attributes of a coastal zone? What is a shoreline? How are water body dimensions identified after the water is no longer present? What Earth analogs can provide insights for Mars?

Morphometric analyses at the meter-scale and larger are now possible, for both terrestrial shorelines and putative extraterrestrial analogs (Table 21.2). New high-resolution imagery and LiDAR (light detection and ranging) DEMs (digital elevation models) are now available for terrestrial shorelines. Imaging and derived DEMs from current Mars datasets (such as the high-resolution imaging science experiment (HiRISE) camera and context camera (CTX) on the Mars reconnaissance orbiter (MRO) have spatial resolutions that are comparable to the high-resolution terrestrial datasets. Other lower resolution Mars data (Table 21.1) such as Mars orbiter laser altimeter (MOLA) and high-resolution stereo camera (HRSC) provide the global coverage that is necessary to place the high-resolution data into a regional context. These datasets enable scaled comparisons of critical geomorphic attributes of terrestrial shorelines to hypothesized Martian shorezone features.

21.3 LAKE BONNEVILLE ANALOG AND DATA INTERPRETATION

In order to develop a terrestrial analog that can be valuable for interpreting extraterrestrial lakes or oceans, we focus on what we consider to be the best and most prominent analog: Pleistocene Lake Bonneville. Lake Bonneville was a terminal closed-basin lake and although this analog was notably smaller than the proposed Mars ocean, the only terrestrial analog to a Mars ocean that would not be smaller would be the Earth's ocean basins.

Characteristics of an appropriate terrestrial analog are

- *Lack of a strong tidal signal.* Given Mars' small moons and great distance from the sun, tidal forces are likely to have been negligible (Parker et al., 1989; Clifford and Parker, 2001). Likewise, Lake Bonneville was small enough that lunar and solar tidal forces were much less important than wind-driven surges.
- *Comparable geomorphic features.* A wide range of preserved landforms occur at Lake Bonneville (some showing cross-cutting relationships) that are similar to those observed on Mars (eg, shorelines, gullies, channels, deltas, fans). Geomorphic features that might unequivocally indicate an ancient Martian water body (eg, lacustrine or marine system) include beach ridges, shorelines (both erosional and constructional), bay-mouth barriers, and topography suggesting a former equipotential water surface.
- *Extensive literature base.* Observations and well-documented studies of Lake Bonneville extending back as far as a century are available to draw from for significant and summative conclusions.

We present interpretations of processes relevant to the descriptive attributes of: (1) coastal feature morphology, including "shorelines," and (2) correlative elevations and/or equipotential surfaces. Lake Bonneville is relatively young (Pleistocene); older systems are typically overprinted by tectonism, erosion, and/or burial by younger materials in Earth's rigorous hydrosphere, a situation that may not have existed on Mars for billions of years. Even relatively recently formed shorelines are easily altered by anthropogenic activities in populated areas. However, young lake systems offer precise resolution if they have been radiometrically dated and correlated with climate cycles (eg, Milankovitch orbital cycles), and they can yield maximum constraints

TABLE 21.1 Summary of Previously Collected Mars Datasets Relevant to Extraterrestrial Mapping and Analysis

Data Acronym	Program	Horizontal Resolution	Vertical Precision	Expected Use (DEMs)	Examples of Relevant Extraterrestrial Areas
MOLA (Mars)	Mars orbital laser altimeter	~500 m/pixel (gridded 128 pixels/degree), 150 m spatial	~30 cm after georeference to MOLA global & THEMIS (Mars Odyssey – thermal emission imaging system) regional data	Regional areas	– W. Deuteronilus (Arabia, Ismenius, Deuteronilus, Acidalia Levels) – Isidis (Arabia, Deuteronilus Levels) – Gusev to Gale craters (Elysium & Aeolis Levels)
HRSC (Mars)	High-resolution stereo camera	10–15 m/pixel, 75 m spatial	~10 m	Target areas	Local areas in basins above, for trend information (larger terraces)
CTX (Mars)	Context imager	6 m/pixel	Reference to HRSC	Target areas	Reference to HRSC above
HiRISE (Mars)	High-resolution imaging science experiment	30 cm/pixel	~1 m	Local areas	Deuteronilus (eastern Cydonia on flanks of Arabia Level, where images exist & can be processed)
MOC (Mars)	Mars orbital camera	1.5–12 m/pixel	Not applicable	Local areas	Supplementary to CTX, some HiRISE

DEM, digital elevation model.

regarding the temporal duration of surface-water storage. Finally, we synthesize specific Lake Bonneville attributes and its context of relevant paleoclimate settings, geologic processes, and controls. This section is followed by applications of the terrestrial concepts and their specific relevance for assessing the putative shorezone features in the region of Martian northern plains.

21.3.1 Setting

Pleistocene Lake Bonneville (Gilbert, 1890) was the largest rain-fed or pluvial lake (~50,000 sq km) in western North America during the Last Glacial Maximum (~20 kya). Lake Bonneville and its modern remnant, the Great Salt Lake, comprise an outstanding locality in the Basin and Range physiographic province, with remarkably well-preserved and exemplary geomorphic features, including prominent shorelines, wave-cut erosional terraces, spits and bay-mouth barriers, deltas (eg, the type locality of Gilbert deltas), gullies and rills, outburst channels, and playa lake features (eg, patterned grounds, eolian deposits, and evaporites). Lake Bonneville underwent numerous fluctuations due to climatic and basinal changes and has preserved several prominent shorelines named (oldest to youngest): Stansbury, Bonneville, Provo, and Gilbert (Currey et al., 1984). Several hundred radiocarbon dates constrain the chronology of the lake level history and establish its hydrograph (Fig. 21.1) (Oviatt, 2015). Thus, Lake Bonneville is a valuable dataset that is well suited to establish important sedimentary and geomorphic criteria that can be compared with Mars (Clifford and Parker, 2001; Irwin and Zimbelman, 2012). The shoreline history and facies analysis within the Lake Bonneville system have predictive capabilities where different sedimentary facies reflect paleoenvironmental conditions.

21.3.2 General Geomorphology

Lake Bonneville has been recognized as a viable Martian analog since the pioneering work of Tim Parker, Don Currey, and others in the 1980s (eg, Parker et al., 1989, 1993; Parker and Currey, 2001). The Lake Bonneville–Great Salt Lake system is among the best terrestrial examples for comparison with Mars shorelines for a number of reasons. The general geology, geochronology, and geomorphology of Lake Bonneville are easily accessible for developing specific criteria for recognition of shoreline features (eg, Currey et al., 1984; Scott, 1988; Oviatt et al., 1994). Lake Bonneville landforms include both erosional and constructional sedimentary features that developed over different timescales (Godsey et al., 2005, Jewell, 2007, 2010), ranging from gradual (eg, wave-cut shoreline terraces, lobate fan deltas) to the sudden or catastrophic (eg, outburst channels). The Bonneville basin is bounded by active normal faults that have produced a steep-sided water basin that closely approximates Mars morphologies, particularly craters (Cabrol and Grin, 2002). In the Great Basin

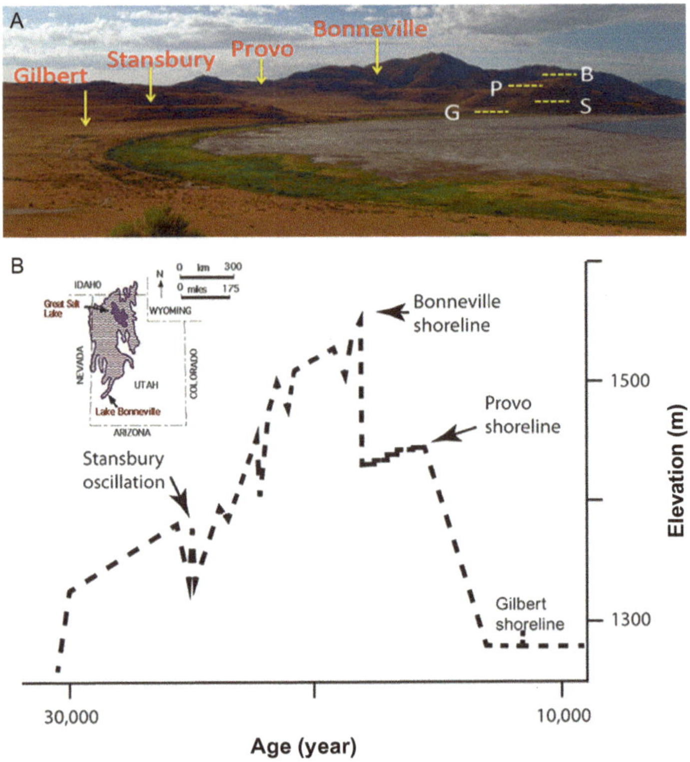

FIG. 21.1 Named shorelines of Lake Bonneville at Antelope Island State Park, Utah, with view of White Rock Bay, looking south (a). *Inset* locations (b) and the hydrograph of prominent shorelines showing the elevation vs relative age. Note: shoreline abbreviations in *upper right* of (a) correspond to first letter of each named shoreline. Gilbert shoreline (G) is generally accepted to be ~11,500–12,500 years ago. *Sources: Panels (b) From Oviatt, C.G., 2015. Chronology of Lake Bonneville, 30,000 to 10,000 yr B.P. Quat. Sci. Rev. 110, 166–171.*

of Utah, the rapid hydroclimatic change during the Pleistocene transitioned from wet to semiarid following the Last Glacial Maximum (Oviatt, 1997) and is comparable to the abrupt transition suggested to have occurred on Mars during its Late Noachian to Early Hesperian Epochs (~3.6–3.8 gya) (eg, Poulet et al., 2005; Bibring et al., 2006).

21.3.3 Shoreline Morphology and Expression

Shoreline morphology and expression are affected by many factors that are active during shoreline development and post-depositional modification. When water disappears from a basin, the area is likely to only preserve the last phases, obscuring an understanding of the older processes. The

Pleistocene Lake Bonneville analog is particularly instructive because it provides critical characteristics, temporal constraints and insight on the geomorphic evolution, hydroclimatic change, and landscape dynamics of large water bodies. Relative relationships may distinguish different events and the history of shoreline development. Prominent shorelines may indicate a longer period of water-body occupation or more extensive erosion, as compared to more subdued or discontinuous shorelines. Multiple scales of observation are generally helpful in identifying shorelines in aerial or satellite data, although interpretations may be limited by problems related to remote sensing (eg, image resolution). Linear shoreline features may be distinguishable in remote imagery with only 2 m horizontal resolution and 1 m vertical resolution, a scale that is presently available in areas covered by HiRISE images in some portions of Mars.

Slope may be an important factor in shoreline development and expression. If a shoreline has a shallow regional slope, the adjacent water body dissipates much of its energy through basal friction. Thus shallow slopes produce areally extensive, but subtle, shoreline features. Conversely, if the shoreline topography is steep (eg, block faulting in Lake Bonneville or impact crater walls on Mars), it is more likely to develop pronounced, erosional shorelines such as the "kalottberg" hills and washed-out cobble fields noted in other terrestrial large-lake systems (Linden, 2006). In areas of steep slope, wave energy encounters less bottom friction, and thus the water has more erosional capability. Parker and Currey (2001) discussed how wave energy absorption by gently sloping bottom topography, such as on alluvial fans in the western Bonneville basin, produced broader shore features that are easier to detect in satellite imagery than steeply sloping bottom topography along mountain fronts.

The terrestrial Lake Bonneville shoreline examples also show that threshold controls produce prominent shorelines (eg, Provo level). If lake level, and thus lateral shoreline placement, is climate-controlled and there is a spatially extensive accommodation zone, shorelines may vary locally in character (eg, transgressive shorelines that can have local gravel accumulations). Submerged or drowned shorelines may be transgressed or modified, creating a subdued expression (eg, the transgressive Stansbury level) that is not as prominent basin-wide or as pristine as younger/less modified shorelines. Topographic profiles for terrestrial shorelines may distinguish an erosional shoreline, or wave-cut terrace, from a depositional shoreline based on profile shape (eg, Parker et al., 1989).

21.3.4 Equipotential Surfaces

Some abandoned or remnant shorelines on Earth may approximate an equipotential surface of a constant elevation, but the majority seldom do. Even in Lake Bonneville, the distinctive shorelines are not always regionally continuous (eg, due to tilting, erosion, or expression). The Lake Bonneville shorelines occur at different elevations due to variable wave setup and fetch, isostatic adjustments due to loss of the water load, or tectonic activity. Recent measurements of storm surges in the modern Great Salt Lake of the Bonneville basin, recorded by the deposition of debris (flotsam) along the shorelines of Antelope Island indicate that water flood levels reached heights ~0.6–3.7 m amplitude (Atwood and Mabey, 2000; Atwood, 2006), and significant wave height (the average height of the upper one-third of waves) for storms of the 1980s was estimated as 1.5 m (Atwood, 2006). Further, isostatic rebound in the Bonneville basin results in higher shoreline elevations in the center of the basin than along the margins (up to 74 m difference; Currey, 1990), since the ~300 m+ of water load in the basin interior is no longer present (Gilbert, 1890; Bills et al., 2002). Similarly, aspects of differential thermal isostasy from Martian surface heat flow might account for elevational differences at the kilometer scale for paleoshorelines (Ruiz et al., 2004).

These examples demonstrate that it should not be at all surprising that mapping of geomorphic features in the Mars northern plains has revealed contacts or hypothesized equipotential surfaces that have been offset and/or occur at different elevations (eg, Parker et al., 1989; Baker et al., 1991; Parker et al., 1993, 2010; Head et al., 1999; Clifford and Parker, 2001; Parker and Currey, 2001; Dohm et al., 2009). Thus, the lack of lateral continuity or consistent elevation for Mars shorelines (Ghatan and Zimbelman, 2006) does not rule out the different levels as valid shoreline features. Instead, these differences will prove valuable in reconstructing the geomorphic landscape evolution. Any paucity of constructional landforms does not preclude the dominance of erosion in areas that are sediment-source limited. Therefore it is important to consider all landform attributes when evaluating evidence of Martian shorelines.

21.4 ANALOG APPLICATION TO MARS

In comparing terrestrial analogs to other planets, we acknowledge scaling issues as well as different boundary conditions such as atmospheric density, surface temperature, rock type, and surface gravity and gravitational constants. Moreover, one of the biggest issues on Mars is the lack of high-resolution temporal control. Fully addressing these boundary conditions is beyond the scope of this paper. Mars obviously had different orbital and environmental parameters than Earth (eg, axial tilt, gravity, atmospheric pressure, temperature, sustained presence of water and life, and so forth). Nevertheless, terrestrial lake examples are the best available baselines for investigation of extraterrestrial shorelines. Specific detailed descriptions and interpretations of Earth analogs form the best basis for comparing and evaluating the nature of oceans on Mars and the scale of features. Terrestrial lakes show the range of variability and provide a framework within which testable interpretations of Mars shoreline features can be made.

21.4.1 Current Science Regarding a Mars Ocean

A high priority objective in Mars science is to gain insight into the past distribution of water, including the occurrence of large water bodies, on the planet (eg, Carr, 1983, 2000). Recent discussions consider what phase (solid ice, liquid, gas vapor) of water interacted with the planetary surface (eg, Boynton et al., 2002; Wyatt et al., 2004), over what timescales the water bodies existed (eg, Clifford and Parker, 2001; Cabrol and Grin, 2002; Fairén et al., 2003; Madden et al., 2004), the importance of wet environments for life (eg, Mahaney et al., 2004; Squyres and Knoll, 2005; Knoll et al., 2005; Dohm et al., 2011), and the context of Martian climatic change (eg, Mustard et al., 2001; Kerr, 2007; Zuber, 2007).

The most extensive attempts to document Mars shorelines on a broad scale have been by Parker et al. (1989, 1993, 2010). The initial work relied on low-resolution Viking Orbiter imagery that is currently being augmented by higher resolution imagery from the more recent orbiters. A range of features has been identified and mapped as evidence of a former Mars ocean in the northern plains, including plains unit contacts and margins, terraces, and curvilinear ridges (Parker et al., 1989, 1993). Some of these features can be traced or correlated for hundreds of kilometers or more, and two of the contacts—the "gradational boundary" of Rossbacher (1985) and another contact known as the "interior plains boundary" (Parker et al., 1989)—were traced to closure within the basin; Parker et al. (1989) referred to these as Contacts 1 and 2, respectively. By assessing MOLA topographic data, Head et al. (1999) demonstrated that the elevation of Contact 1 varied significantly, calling its interpretation as an ocean "shoreline" into question, whereas the Contact 2 was interpreted as an equipotential surface, and it was used to generate the volume of a hypothetical northern ocean. The calculated volume was consistent with previous estimates on the amount of water that flowed through the Chryse outflow channels, as well as the volume of groundwater potentially available in megaregolith pore space. Head et al. (1999) further attributed the observed deviations in the altitude of Contact 2 (average elevation $\sim$4.7 km, with a standard deviation 560 m) to postformation modification and vertical movement, especially since most of the variation is observed in Elysium, Arabia, and around Tharsis, areas that likely experienced post-Contact 2 isostatic activity. Understanding these relationships of relative timing and postformational influences on the elevations of shoreline features is extremely important for assessing evidence both for and against a Martian ocean.

Parker et al. (2010) presented evidence of four primary shorelines in the west Deuteronilus Mensae region, where the original hypothesis was developed. These features, now termed "levels," exhibit distinctive morphologies and define contacts between regional surficial geomorphic units (Fig. 21.2). In addition to these four primary levels, a number of more subtle, less continuous levels were also identified at intermediate elevations from northern

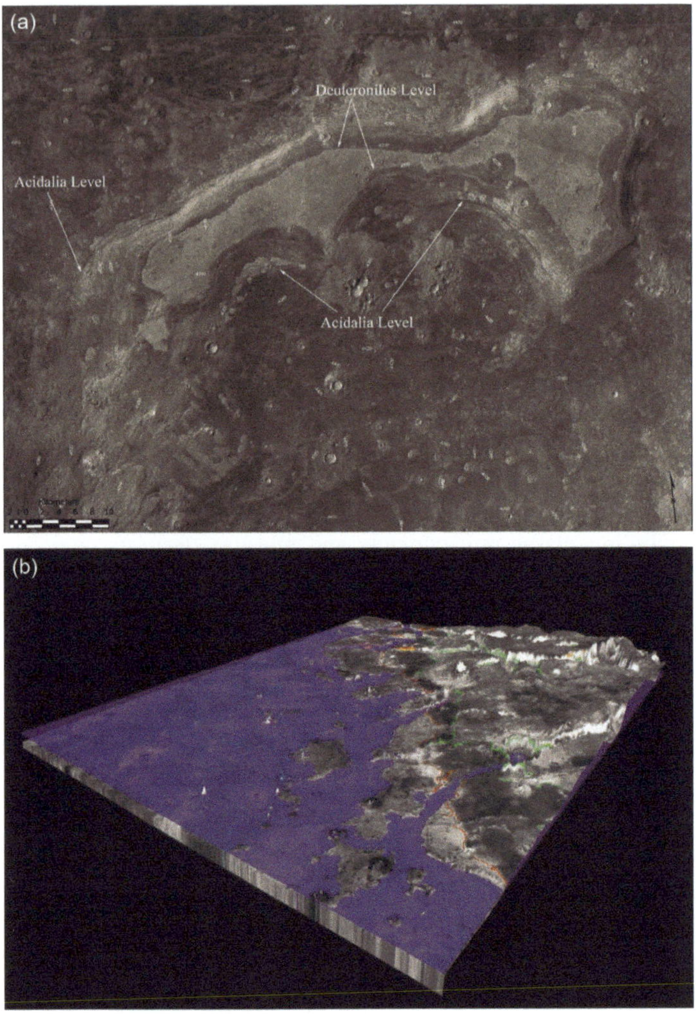

FIG. 21.2 Deuteronilus Mensae area of Mars depicting potential shorelines. (a) Eroded craters at Arabia Level from CTX image. *Arrows* indicate contact between dark "plains" surface and lighter elevated "highlands" surface, first identified in Viking Orbiter images. This contact appears to lie at an elevation of approximately −3800 m locally, and consists of an onlapping of dark material over the brighter material. (b) A diagrammatic 3D view (25× vertical exaggeration) flooded to about the median elevation of the Deuteronilus Level in the Deuteronilus Mensae (−3890 m) from left position 439719.688, −2548478.798 to 635468.600, −2781483.148 at right. *Panel (a) From Parker, T.J., Grant, J.A., Franklin, B.J., 2010. The Northern Plains: A Martian oceanic basin. In: Cabrol, N.A., Grin, E.A. (Eds.), Lakes on Mars. Elsevier, Amsterdam. ISBN 978-0-444-52854-4, Chapter 7.*

plains areas of east Acidalia Planitia and west Deuteronilus Mensae, Acheron Fossae, and the rim of the Isidis basin. Assuming these distinctive features are indeed shorelines, these suggest a history of eustatic changes, with different "still-stand" periods recorded at the "contacts" and more subtle dynamics of water storage reflected in the prominent levels. Putative Mars shorelines could be both constructional (depositional) and erosional, reflecting different hydrodynamic conditions. Erosional shorelines on Mars (eg, parts of Arabia Level) may be distinguished where slopes are steep. In places where there are gentler slopes, it is harder to identify terraces because they are not dramatic enough or may be more smoothed partly due to deposition. Shorelines appear to be depositional where they show thin onlap (eg, east Acidalia Planitia) or where thin debris flows exist (eg, parts of Deuteronilus Level) (Parker et al., 2010). Although Parker et al. (2010) believe imagery does suggest a large Martian ocean, the shorezone features were probably also heavily modified by ice and are even complicated or obscured by debris, lava flows or eolian dust, and debris.

A recent study details a global database of known deltaic deposits, valley networks, and present-day topography to test for the occurrence of an ocean on early Mars (Di Achille and Hynek, 2010). The consistent elevational distribution of ancient Mars deltas suggests that a planet-wide equipotential surface existed along the margins of the northern lowlands, supporting interpretations of a vast ocean that covered the northern hemisphere as recently as Hesperian time.

21.4.2 Visual Recognition of Shorezone Features

The active biosphere and hydrological realm on Earth means preserving evidence of shorelines is significantly different than on Mars. Implications from terrestrial studies—particularly in recent deposits—are instructive. For instance, aerial images of distinctive terrestrial shorelines initially look similar to those of Lake Bonneville (Ghatan and Zimbelman, 2006; Fig. 21.3B), but in reality they show a series of vegetated/nonvegetated patterns caused by differing soil moisture (Fig. 21.3). Obviously, vegetation cannot be used to elucidate similar features on Mars, but the underlying mineralogical cause for the variation in vegetation (ie, grain size related capacity of retaining moisture) could theoretically be detected with instruments such as CRISM. Shoreline feature detection by orbital imaging radar (synthetic aperture radar, SAR) has been conducted successfully on Earth in nonvegetated arid regions (eg, Komatsu et al., 2001). The shoreline features identifiable in the radar imagery are confirmed by optical remote sensing and field checking (Komatsu et al., 2007). This detection works because the radar signal is sensitive to grain sizes that affect different radar scattering capacities and the grain size variation characterizes surfaces of shoreline features such as beach ridges and spits. However, radar has not been employed nor is it currently planned for any future orbiter missions to Mars.

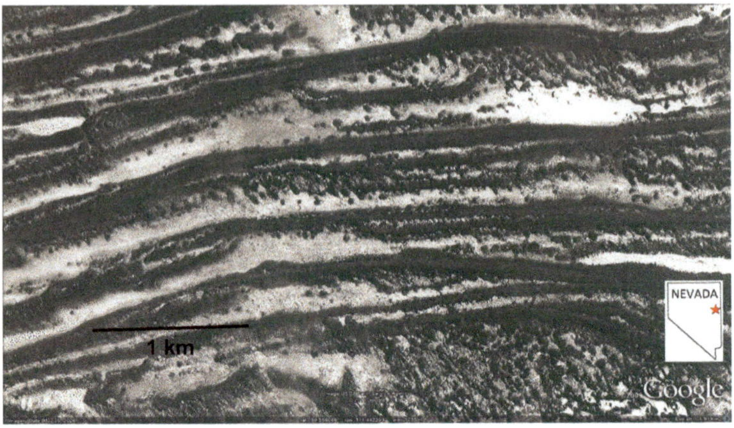

FIG. 21.3 Beach ridges from the Spring Valley area of Nevada (Ghatan and Zimbelman, 2006; their Fig. 3b) show the role of vegetation in defining shorezone features on Earth (39.6°N, 114.6° W, see *inset map*).

21.4.3 Digital Recognition of Shorezone Features

According to Ghatan and Zimbelman (2006), the paucity of Mars coastal ridges in MOLA profiles suggests a few possible scenarios: (1) no ocean existed related to the mapped "shorelines"; or (2) an ocean existed but it had conditions of little/no wave action, not enough sediment input to make ridges, water that readily froze and did not allow ridge development, or any constructed landforms were somehow eroded away. Although this is a reasonable argument, an important limitation of the MOLA data still exists. Although the vertical precision of the MOLA is on the order of 20–30 cm, the spatial resolution is hundreds of meters. Shot points are 150 m in diameter and may be 300 m apart along track or even up to 2–3 km apart across track at the equator. Therefore, unless coastal landforms on Mars are kilometers wide, they cannot be accurately resolved in MOLA data. Only regional slopes and breaks in slope (eg, changes in slope angle or shape across a shoreline) can be detected in MOLA data. The HRSC has better spatial resolution (typically 75 m postings), but less precise vertical resolution that depends on the pixel scale and clarity of the images, as well as noise.

By contrast on Earth, the recognition and analysis of lacustrine shorelines is critically dependent on high-resolution digital elevation data (Fig. 21.4). As pointed out by Parker et al. (2010), even the most prominent shorelines of a major lacustrine feature such as Lake Bonneville are difficult to discern in most satellite imagery. Some of the most significant work on terrestrial shorelines has relied on high-resolution (<10 m) DEM or GPS data (eg, Mattson and Bruhn, 2001; Pelletier et al., 2006). For standard satellite or aerial imagery on both Earth and Mars, sun angles are generally

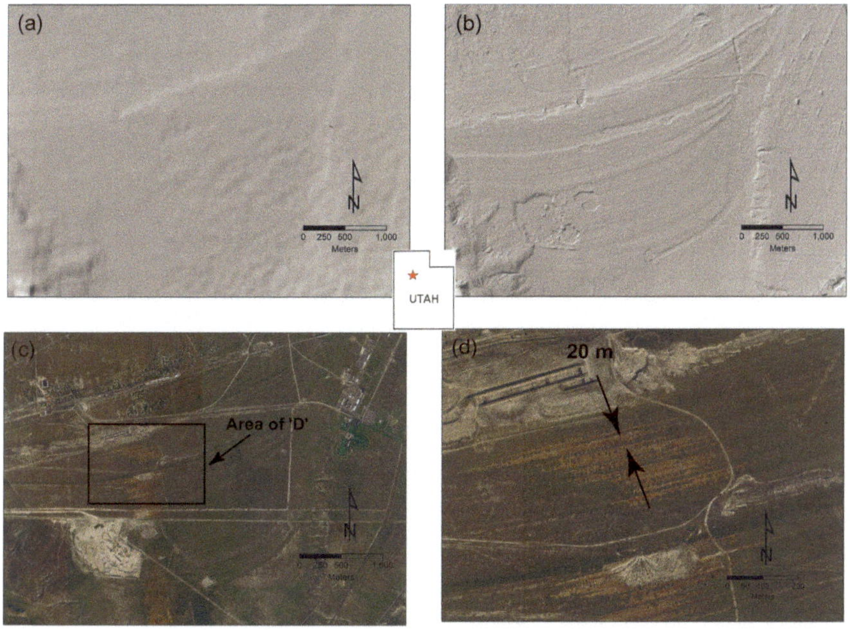

FIG. 21.4 Comparison showing the role of scale and vegetation in delineating shorezone features on Earth. All examples are from the Provo shoreline, Stockton Bar, Utah. (a) 30 m DEM hillshade. (b) 5 m DEM hillshade. (c) High-resolution (1 m pixel) photo. (d) Close-up showing scale of shorezone beach ridges and the role of vegetation in discerning the features on areal photography (40.46°N, 112.32°W, see *inset map*).

high, making it difficult to recognize subtle shoreline features. The power of digital-elevation data lies in the ability to overlay artificially low sun angles from a variety of directions on terrains, thereby increasing the chance to discern a possible shoreline.

21.4.4 Spatial Relationships

21.4.4.1 Mars

Identifying a shoreline is an interpretive process, particularly with remotely sensed images. In initial attempts to map and correlate shorelines in the northern plains of Mars (Parker et al., 1989, 1993, 2010; Parker and Currey, 2001), lateral continuity was one criterion used to identify potential shoreline features. Furthermore, a common opinion exists that shorelines on Mars should show a distinctive and recognizable shoreline profile, such as a bench and shelf-like topography in three-dimensions or along longitudinal profiles. In reality, studies of terrestrial shorelines demonstrate that this is only one of several criteria that should be used in conjunction with other evidence to invoke an interpretation of a standing body of water.

21.4.4.2 Lake Bonneville

Data from terrestrial analogs demonstrate that shoreline expressions are spatially complex. It is apparent that Lake Bonneville shorelines do not maintain a single morphologic expression along their entire length. Instead, the morphology of a shoreline varies along its length depending on local constructional and erosional processes, preservation potential, and exposure. Furthermore, identification of shorelines in remote imagery based solely on lateral-continuity criteria may lead to false conclusions. Shorelines on Earth are identified based on multiple criteria applied to field and high-resolution remote sensing data, with lateral continuity of a single geomorphic expression being a relatively minor criterion.

21.4.5 Temporal Relationships

21.4.5.1 Mars

An important difference between Earth and Mars is the duration of wet surface conditions. Obviously, Earth has had stable surface water for most of geologic time, possibly including as far back as the Hadean (Wilde et al., 2001). Mars, on the other hand, may have only had significant water bodies periodically (Table 21.1) under environmental conditions that differ from the current Mars surface (Baker et al., 1991; Carr, 2000; Carr and Head, 2003). Another point to consider is that, although liquid water might have been stable at the surface early on as a Mars ocean was accumulating water (via cometary impact, planetary outgassing, or both), the Mars shorelines younger than the Arabia Level and interior to it show more evidence of ice-shoving at their margins, suggesting that the ocean receded and progressively froze-over with time. It might even be that the ocean is still present but frozen solid at depth, covered by a few tens of centimeter of debris (Parker et al., 1989, 2010).

21.4.5.2 Lake Bonneville

One of the important lessons from Lake Bonneville is that water can accumulate and disappear relatively quickly, geologically speaking, on a time scale of thousands of years or less, and that the resulting geomorphic shorelines and related features can also form over short time scales. It is important to note that even in young Lake Bonneville deposits, distinctive terrestrial lake mineralogies, and geomorphic features are not necessarily detectable through remote imaging and are easily covered and obscured over the course of thousands to tens of thousands of years.

21.4.6 Shoreline Preservation

Preservation of small landforms on Mars is hampered by later impact cratering and eolian erosion that presents major challenges in their identification using orbital data sets (Irwin and Zimbelman, 2012). Thus, the potential to preserve small landforms on Mars as well as in Lake Bonneville depends strongly on age. Larger shoreline morphologies may remain intact on Mars, with little modification even after passage of a billion years or more. In the Deuteronilus Mensae region, the Arabia and basinward levels show little apparent modification, for example (Parker et al., 2010). But for older shorelines, which would have been subjected to more overprinting by subsequent geologic processes, this is probably not the case. Older and/or modified shorelines potentially associated with the Meridiani Planum deposits would have been eroded by continued fluvial activity, and thus would be more difficult to see in orbital imagery. Periglacial processes would also have degraded any shoreline features at higher latitudes.

Volatile introduction during Mars' history of Late Heavy Bombardment could have contributed water or enabled atmospheric pressures to sustain a large ocean-water body. However, shorelines that might have consequently developed may have been short-lived such that the water body regressed, sublimated, and froze not long after the end of the Late Heavy Bombardment (Parker et al., 2010).

On Earth, pristine and well-defined shorelines are typically young and can form quickly. A general shoreline may represent a longer lake-level stand, but individual ridges on a shoreline may represent exceptionally strong storms that are geologically very brief. More subtle shorelines may record a history of being either short-lived, or older, reworked or resurfaced, such as those modified by transgressions (ie, drowning by rising water levels). If there is little/no wave action and if water drainage occurred rapidly without pauses or standstills, it is also possible that few other modifications to shorelines and indeed all geomorphic features would have developed. Despite Earth's active hydrosphere, subtle shoreline and other geomorphic features are preserved for millions of years in the Antarctic Dry Valley and Atacama Desert (Matmon et al., 2009) where there is little weathering or erosion.

Given these Earth examples, the geomorphic expression of Mars shoreline features during its prominent watery phase may indicate very little subsequent reworking or weathering. Alternatively, Ghatan and Zimbelman (2006) attributed the paucity of coastal ridges in MOLA profiles to erosion (among multiple possibilities). This idea of eroded coastal ridges would require significant weathering and erosion to resurface a large area, but erosion was probably more rapid in the Noachian and into the Hesperian (Baker et al., 1992; Carr, 2006).

21.4.7 Prospects and Importance of Identifying Mars Shorelines

A significant aspect of the continuing controversy regarding the existence of a Mars ocean is the current lack of widespread and continuous high-resolution DEMs. With the Lake Bonneville analog as a guide, it is clear that additional detailed studies along the lines of Di Achille and Hynek (2010) are needed in greater quantity to link and correlate shorelines. Consequently, summative collections of studies and more detailed stratigraphic and environmental analysis of Martian sediments undertaken are necessary before the Mars ocean "controversy" can be fully resolved.

Developing a relative chronology of features formed by water is important for interpreting Mars' geologic history and constructing a possible ocean water budget. A plot of the relative age of Martian shorelines against elevation (ie, a hydrograph) can be useful for estimating the duration of water persistence at a specific shoreline. For example, broad shorelines with wide "shelf-like" morphologies indicate an extended period of water interaction and feature development (Godsey et al., 2005). Other extensive ridges parallel to the shorelines may indicate a slow recession of water. In a Mars hydrograph, each shoreline can be viewed as an indicator of water storage on the planet's surface, and each reflects a "stage" with a certain volume of water that can be estimated.

21.5 DISCUSSION

21.5.1 The Mars Ocean

Planetary scientists concur that Mars once possessed an active hydrosphere. Understanding the hydrologic processes leading to the formation of geomorphic features on Mars must necessarily rely on Earth systems for establishing certain boundary conditions (eg, size and longevity of the water body). Examination of terrestrial shorelines allows us to establish shorezone criteria for recognizing an ancient Mars ocean that in turn can elucidate hydroclimatic oscillations, periods of stable water, and shorezone development. Collective characterization of Earth ice-age lakes and new Mars data can provide input data to models addressing how Mars shorezones are similar and different to Earth, and how they might have evolved their geomorphic expression over time. A water-budget cycle of Mars through these shorezone models in turn could be transformative; the models provide comparative assessment relating global-scale environmental and climate change, and guide future Mars explorations to "follow the water" in the search for habitable environments for life.

The relative chronology of water features is important for interpreting Mars' geologic history and a possible ocean-water budget. All local shorelines should be linked to a base level, a datum that reflects the global water budget of the northern Martian lowlands. A Mars hydrograph is critical for developing more detailed interpretations of the relative duration of water persistence at a specific shoreline (ie, temporal minimum), bracketed or estimated from the terrestrial analogs.

Although some in the planetary community still debate whether Mars had an ocean (see summary in Ghatan and Zimbelman, 2006), the science is now at a critical juncture with high-resolution datasets (eg, HiRISE imagery and topography) that allow reevaluation of the geomorphic evidence for a past ocean. Inferred shorelines in the northern plains of Mars ("Mars ocean" of Parker et al., 1989; Baker et al., 1991; Head et al., 1999; Carr and Head, 2003) comprise some of the strongest evidence for surface water storage on the planet. The recent mapping of Di Achille et al. (2009) and Di Achille and Hynek (2010) demonstrates the need to reconsider the questions of what comprises geomorphic evidence for the Mars ocean and how to distinguish these from morphologically similar but genetically unrelated features.

Interpretation of shorelines on Mars could also have fundamental impacts on astrobiological arguments in search for present or past life on Mars. In a recent assessment of habitable environments on Mars, Dohm et al. (2011) stressed the importance of putative ancient oceans. Large bodies of water such as oceans could have provided the habitable environments, and the sinks for sediments preserving evidence of life (eg, Komatsu and Ori, 2000). Despite the differences and a lack of accurate temporal control, the relative geomorphic relationships of shorelines and landforms of the early Mars ocean offer important clues to geologic and hydrologic processes the late Noachian–Hesperian (approximate time of the putative Mars ocean).

Ultimately, the terrestrial analogs are critical for interpreting Mars facies, climate change, basinal changes, and history. Multiple analogs are useful to determine how some geologic processes dominate, and what the interplay might be. Knowledge of terrestrial paleoshorelines is useful to correlate and interpret the potential causes for variances from equipotential lines of Mars terraces and shoreline-like geomorphology. Explanations for inconsistent equipotential lines could include isostatic adjustments on Mars (Head et al., 1998, 1999), crustal heat flow (Ruiz et al., 2004), as well as other geologic processes that are decipherable in Earth analog examples.

21.5.2 Additional Terrestrial Analogs for Mars

A number of terrestrial settings might make suitable analogs for lakes and the Mars ocean in a fashion similar to that just outlined for Lake Bonneville. We briefly consider two additional Pleistocene settings particularly promising as analogs for extraterrestrial fluid bodies (Table 21.2).

21.5.2.1 Lake Agassiz

Lake Agassiz was the largest proglacial lake in North America during the Pleistocene (Upham, 1895). Colman et al. (1994) discuss major episodes of water discharge recorded within downstream bodies including the Lake Michigan basin. They further suggest that diversions of glacial meltwaters from the lake between the Mississippi, St. Lawrence and McKenzie river systems may have impacted and complicated widespread climatic variations that coincidently occurred during the Younger Dryas cooling event.

TABLE 21.2 Mars Water Body Features and Distinctive Features (Synthesis from Text Citations), Compared to Relevant and Diverse Analog Study Examples

Timing of Mars "Lakes or Oceans"	Scales or Conditions	Distinctive Features	Selected Relevant Analogs
1. Middle Noachian (Clifford and Parker, 2001)	Warm and wet? vs Cold and wet?	Extensive degradation of ancient impact basins and other large-scale geomorphic features	B, A, I
2. Late Noachian—Early Hesperian (Baker et al., 1991, 1992)	Warm (episodic?)	Episodic surface runoff for geologically significant periods to produce large valley networks and erosional terraces along lowland/upland boundary	A, I
3. Late Hesperian—Amazonian? (Tanaka et al., 2005; Rodriguez et al., 2015)	Episodic floods?	Major outflow channel formation, such as Circum-Chryse systems	B, A, I
Terrestrial Analog Features	**Scales or Conditions**	**Distinctive Features**	**Selected Relevant Analogs**
Isostatic adjustments	m to 100+ m	Elevations are not all the same equipotential	B, A, I
"Pristine" (major) shorelines	10s m to km	Prominent expression	B, A, I
"Degraded" shorelines	10s m to km	Degraded, modified expression	B
Constructional shorelines	10s m to km	Longshore drift	B
Spits, barriers, bars	10s m to 10s km	Longshore drift, lateral accretion	B, I
Erosional shorelines	10s m to km	Wave-cut terraces	B
Erosional landforms	10s m	Shoreline caves, coves, knobs, stacks	I
Beach ridges	m to 10s m	Water-level fluctuations	B, A, I
Glacial–fluvial	m to 100s m	Sculpted bedrock, poorly sorted tills; river channels & bars	A, I

TABLE 21.2 Mars Water Body Features and Distinctive Features (Synthesis from Text Citations), Compared to Relevant and Diverse Analog Study Examples—Cont'd

Terrestrial Analog Features	Scales or Conditions	Distinctive Features	Selected Relevant Analogs
Deltas	km	Fluvial gradients, progradational depocenter	B, I, A
Rills, Gullies	m to 10s m	Erosional features, sapping	B, A
Alluvial Fans	km	Basin geometry, radial fan apron, mass movement	B, A
Outburst channels	km	Catastrophic floods, megachannels, immense bars	B, A, I
Patterned ground, weathering	m to km	Desiccation, thermal expansion & contraction in frozen ground, evaporite growth surficial cracks	B, I, A
Eolian, dunes	m to km	Wind features, dune bedforms, morphology, cross-bedding, ripples, deflation, yardangs	B, A
Laminated sediments, basin fill	m to km	Varves, marl, mudstones, sandstones	B, A, I
"Kalottberg"	m to 100s m	Washed-out coarse clastic deposits	I
Deep basinal setting	km	Geometry (modern or ancient)	B, I
Sequence stratigraphy	m to 100s m	Sequential sediment packages	B, A

Distinctive Analog Characteristics Will Help Establish Necessary Criteria to Identify, Map, and Interpret Similar Features on Mars.
Pleistocene terrestrial analog systems: **B = Lake Bonneville (eg, also see Chan et al., 2011, for details and specific localities), A = Pleistocene Lake Agassiz, formed in the Hudson Bay drainage of central North America, I = stages of the developing Baltic Sea in particular the Pleistocene Baltic Ice Lake (BIL) in the present central Sweden area.

Distinctive erosional and depositional features related to outflow events can also be distinguished and might have relevance to similar events on Mars. Bathymetric profiles of Lake Agassiz in the north Superior basin show some steep and sudden drop-offs related to sudden discharge of high-velocity waters that would have eroded channels through the bay (Gary et al., 2011).

In contrast, depositional features include subaqueous fans that are thickest in the deepwater areas adjacent to the outlets (Gary et al., 2011) in a manner not unlike that of Mars.

21.5.2.2 Baltic Sea

The present-day Baltic Sea is a brackish water body located in Northern Europe, with a drainage basin approximately four times its modern surface area, and a morphometry influenced by glacial erosion and "drowned rivers" that empty into its two main bays: the Gulf of Finland and Gulf of Bothnia (Björck, 1995). The Baltic Sea (this term is herein used to include the Baltic Ice Lake (BIL) and its subsequent stages) has fluctuated in size, salinity, and degree of connectedness with nearby oceans since the end of the last (Weichselian) glaciation of the Pleistocene.

The Baltic Sea expresses a wide variety of both depositional and erosional shoreline features, such as beach ridges, spits, coastal caves, washed-out cobble fields, and "kalottberg." Some landforms may be of huge proportions, such as the spits of the Gulf of Gdansk. Small, but distinct beach ridges characteristically represent an individual storm event, indicating the geologically instantaneous formation of some shoreline features, whereas an accumulation of beach ridges commonly is associated with a certain stage (eg, BIL) in the development of the Baltic Sea. Similar to the Lake Bonneville example, variables affecting the isostatic adjustments of the Baltic Sea explain how contemporaneous shorelines are not always at consistent elevations across a basin. This is important to bear in mind when comparing with water-level changes in large water bodies on Mars (ie, the northern plains), where heat flow and isostatic adjustments of the crust (here by endogenous processes rather than glaciation) may have varied across different parts of the northern plains (cf. Head et al., 1999).

21.5.3 Lake Bonneville as an Analog for Titan

Mars is not the only extraterrestrial setting for which Lake Bonneville is an appropriate analog. Large bodies of liquid methane, ethane, more complex hydrocarbons, or some combination of these substances near the polar regions of Saturn's moon, Titan, have long been accepted (Stofan et al., 2007; Hayes et al., 2010). These are the only known, existing active fluid bodies on a rocky object outside of Earth. As such, the existence of lakes and seas on Titan is less controversial that those of the early geologic history of Mars.

Deployment of the Huygens probe to the surface of Titan, and multiple passes by the Cassini orbiter beginning in 2006, provide a wealth of tantalizing insights into this dynamic fluid system. Many features have similarities to Lake Bonneville. (1) Mass balances in the lakes are extremely dynamic due to Titan's tilted orbital plane relative to the Sun (26.7 degree) in conjunction with Saturn's orbital period (29.5 years) (eg, Lorenz and

Mitton, 2008). (2) Elaborate channel-features adjacent to the lakes and oceans have been identified (eg, Tomasko et al., 2005; Soderblom et al., 2007; Burr et al., 2009, 2013; Langhans et al., 2012). (3) Titan's lakes and seas are all closed basin systems. (4) Titan's lakes and oceans have extensive alluvial channels (up to 1 km wide) and deltas approximately the same size as those of Lake Bonneville have been identified (Wall et al., 2010). (5) Putative submerged shorelines associated with Ligeia Mare have been proposed (Lucas et al., 2014), meaning this feature has a history of fluid transgression. (6). A proposed maximum depth of 160 m of Ligeia Mare (Mastrogiuseppe et al., 2014) approximates Lake Bonneville's Provo shoreline period. (7) Approximately 4 m of lake recession has been documented in Ontario Lacus between 2005 and 2009 (Hayes et al., 2010) and shoreline retreat of 9–11 km (Turtle et al., 2011) a rate roughly analogous to wet/dry cycles in modern closed-basin lakes such as the Great Salt Lake on Earth (eg, Gwynn, 2002). (8) Tens of meters of oscillations are believed possible in response to orbital oscillations (Lucas et al., 2014), meaning Titan lakes and oceans have direct parallels to glacially driven lake cycles on Earth.

The drawback to fully utilizing Lake Bonneville as an analog to Titan lies in the current resolution of Titan data. To detect shorelines, future missions to Titan would need either: (1) a global SAR orbiter that could extend the coarse resolution that currently covers just over half of the planet; or (2) aerial platforms (balloon, blimp, or fixed wing glider or helicopter UAV—unmanned aerial vehicle) able to image at meter and even centimeter scales.

21.6 SUMMARY

Pleistocene Lake Bonneville is a premier large-lake example with spatial hydrograph and time controls as valuable contexts to compare to planetary shorezone features. The scales and nature of exposure and preservation of Bonneville shorezone features are spectacular even when compared to other important lake systems worldwide. Clearly, Lake Bonneville is a very strong terrestrial analog model to interpret and understand how water budgets interact with the landscape on other planetary bodies (eg, Mars and possibly Titan).

The collective study of terrestrial analogs applied to Mars shorelines can provide a relative-time global framework for extraterrestrial fluid budgets and climate. Although shorelines are not the only consideration for interpreting standing fluid bodies, they are one of the most noticeable and detectable features. The comparative work here establishes specific geomorphic and stratigraphic criteria to understand and interpret the Mars water cycle, and the sedimentary contexts for surficial facies. The three terrestrial analog examples shown here elucidate the interplay of climate, water source (rivers vs glaciers), geologic processes, tectonics, isostatic adjustments, and the sediment record. These examples of young terrestrial lakes show that many lake features can form quickly on the order of just thousands of years, or even at a single event (eg, beach ridges from strong storms). As a function of climate change, water can rapidly disappear, leaving the remnant geomorphic record of past shorelines and related features.

In the case of Mars, strong evidence exists that a marine or fluvio-lacustrine environment changed rapidly to an eolian/cryospheric (cold desert) environment, the remnants of which remain dynamically within the surface and subsurface of present-day Mars. Further examination of Mars shorelines and use of a systems approach to examine feedback relationships of climate, fluid budget, and the sedimentary record are all critical to understanding a planet's history. Understanding Mars' global water budget will be transformative, and will provide a comparative assessment relating planetary environmental and climate change and evolution.

ACKNOWLEDGMENTS

The work by M.A.C. and P.W.J. was supported by a small Rocky Mountain NASA Space Grant Consortium seed grant. The work by J.O. is supported by Grants AYA2008-03467/ESP and AYA2011-24780/ESP, AYA2012-39362-C02-01, and ESP2014-59789-P from the Spanish Ministry of Economy and Competitiveness. We greatly appreciate the input of Donald Barker and Kathleen Nicoll. We thank Bruce Bills for his constructive criticisms that helped us improve the manuscript. Any potential errors are the responsibility of the authors.

REFERENCES

Atwood, G., 2006. Shoreline Superelevation: Evidence of Coastal Processes of Great Salt Lake, Utah. Ph.D. dissertation, Department of Geography, University of Utah. 231 pp.

Atwood, G., Mabey, D.R., 2000. Shorelines of Antelope Island as evidence of fluctuations of the level of Great Salt Lake. In: King, J.K., Willis, G.C. (Eds.), The Geology of Antelope Island. Utah Geological Survey, Davis County, Utah, pp. 85–98.

Baker, V.R., 2007. Water cycling on Mars. Nature 446, 150–151.

Baker, V.R., Strom, R.G., Gulick, V.C., Kargel, J.S., Komatsu, G., Kale, V.S., 1991. Ancient oceans, ice sheets and the hydrological cycle on Mars. Nature 352, 589–594.

Baker, V.R., Carr, M.H., Gulick, V.C., Williams, C.R., Marley, M.S., 1992. Channels and valley networks. In: Kieffer, H.H., Jakosky, B.M., Snyder, C.W., Matthews, M.S. (Eds.), Mars. University of Arizona Press, Tucson, AZ, pp. 493–522.

Bibring, J.P., Langevin, Y., Mustard, J.F., Poulet, F., Arvidson, R., Gendrin, A., Gondet, B., Mangold, N., Pinet, P., Forget, F., Berthé, M., 2006. Global mineralogical and aqueous mars history derived from OMEGA/mars express data. Science 312, 400–404. http://dx.doi.org/10.1126/science.1122659.

Bibring, J.-P., Arvidson, R.E., Gendrin, A., Gondet, B., Langevin, Y., Le Mouelic, S., Mangold, N., Morris, R.V., Mustard, J.F., Poulet, F., Quantin, C., Sotin, C., 2007. Coupled ferric oxides and sulfates on the Martian surface. Science 317, 1206–1210.

Bills, B.G., Wambeam, T.J., Currey, D.R., 2002. Geodynamics of Lake Bonneville. In: Gwynn, J.W. (Ed.), Great Salt Lake: An Overview of Change. Utah Geological Survey Special Publication, Salt Lake City, pp. 7–32.

Björck, S., 1995. A review of the history of the Baltic Sea, 13.0–8.0 ka BP. Quat. Int. 27, 19–40.

Boyce, J.M., Mouginis, P., Garbeil, H., 2005. Ancient oceans in the northern lowlands of Mars: evidence from impact crater depth/diameter relationships. J. Geophys. Res. 110, E03008. 15 pp.

Boynton, W.V., Feldman, W.C., Squyres, S.W., Prettyman, T.H., Brückner, J., Evans, L.G., Reedy, R.C., Starr, R., Arnold, J.R., Drake, D.M., Englert, P.A.J., 2002. Distribution of hydrogen in the near surface of Mars: evidence for subsurface ice deposits. Science 297, 81–85.

Burr, D.M., Jacobsen, R.E., Roth, D.L., Phillips, C.B., Mitchell, K.L., Viola, D., 2009. Fluvial network analysis on Titan: evidence for subsurface structures and west-to-east wind flow, southwestern Xanadu. Geophys. Res. Lett. 36 (L22203), 299–321.

Burr, D.M., Perron, J.T., Lamb, M.P., Irwin, R.P., Collins, G.C., Howard, A.D., Sklar, L.S., Moore, J.M., Ádámkovics, M., Baker, V.R., Drummond, S.A., 2013. Fluvial features on Titan: insights from morphology and modeling. Geol. Soc. Am. Bull. 125, 299–321.

Cabrol, N.A., Grin, E.A., 2002. Overview on the formation of paleolakes and ponds in impact craters on Mars. Global Planet. Change 35, 199–219.

Cabrol, N.A., Grin, E.A. (Eds.), 2010. Lakes on Mars. Elsevier, Amsterdam. ISBN: 978-0-444-52854-4.

Carr, M.H., 1983. Stability of streams and lakes on Mars. Icarus 56, 476–495.

Carr, M.H., 1996. Water on Mars. Oxford University Press, New York. 248 pp.

Carr, M.H., 2000. Martian oceans, valleys, and climate. Astron. Geophys. 41, 20–26.

Carr, M.H., 2006. The Surface of Mars. Cambridge University Press, New York. p. 322.

Carr, M.H., Head III, J.W., 2003. Oceans on Mars: an assessment of the observational evidence and possible fate. J. Geophys. Res. 108, 8.1–8.28.

Chan, M.A., Nicoll, K.A., Ormo, J., Okubo, C.H., Komatsu, G., 2011. Synthesis of geologic and geomorphologic terrestrial analogs to Mars: why Utah should be considered a primary training ground for future robotic and human missions to Mars. In: Garry, W.B., Bleacher, J. (Eds.), Analogs for Planetary Exploration. Geological Society of America Special Paper 483, pp. 349–375.

Clifford, S.M., 1993. A model for the hydrologic and climatic behavior of water on Mars. J. Geophys. Res. 98, 10973–11016.

Clifford, S.M., Parker, T.J., 2001. The evolution of the Martian hydrosphere: implications for the fate of a primordial ocean and the current state of the northern plains. Icarus 154, 40–79.

Colman, S.M., Keigwin, L.D., Forester, R.M., 1994. Two episodes of meltwater influx from glacial Lake Agassiz into the Lake Michigan Basin and their climatic contrasts. Geology 22, 547–550.

Currey, D.R., 1990. Quaternary paleolakes in the evolution of semidesert basins, with special emphasis on Lake Bonneville and the Great Basin, U.S.A. Palaeogeogr. Palaeoclimatol. Palaeoecol. 76, 189–214.

Currey, D.R., Atwood, G., Mabey, D.R., 1984. Major levels of Great Salt Lake and Lake Bonneville. Utah Geological Mineral Survey Map 73.

Di Achille, G., Hynek, B.M., 2010. Ancient ocean on Mars supported by global distribution of deltas and valleys. Nat. Geosci. 3, 459–463. http://dx.doi.org/10.1038/ngeo891.

Di Achille, G., Hynek, B.M., Searls, M.L., 2009. Positive identification of lake strandlines in Shalbatana Vallis, Mars. Geophys. Res. Lett. 36, L14201. 5 pp.

Dohm, J.M., Baker, V.R., Boynton, W.V., Fairén, A.G., Ferris, J.C., Finch, M., Furfaro, R., Hare, T.M., Janes, D.M., Kargel, J.S., Karunatillake, S., 2009. GRS evidence and the possibility of paleooceans on Mars. Planet. Space Sci. 57, 664–684. http://dx.doi.org/10.1016/j.pss.2008.10.008.

Dohm, J.M., Miyamoto, H., Ori, G.G., Fairén, A.G., Davila, A.F., Komatsu, G., Mahaney, W.C., Williams, J.P., Joye, S.B., Di Achille, G., Oehler, D.Z., 2011. An inventory of potentially habitable environments on Mars: geological and biological perspectives. In: Garry, W.B., Bleacher, J. (Eds.), Analogs for Planetary Exploration. Geological Society of America Bulletin Special Paper 483, pp. 317–347.

Edgett, K.S., Malin, M.C., 1999. Oceans or seas in the Martian northern lowlands: high-resolution imaging tests of proposed coastlines. Geophys. Res. Lett. 26, 3049–3052.

Fairén, A.G., Dohm, J.M., Baker, V.R., De Pablo, M.A., Ruiz, J., Ferris, J.C., Anderson, R.C., 2003. Episodic flood inundations of the northern plains of Mars. Icarus 165, 53–67.

Gary, J.L., Colman, S.M., Wattrus, N.J., Lewis, C.F.M., 2011. Post-Marquette discharge from Glacial Lake Agassiz into the Superior basin. J. Paleolimnol. 0921-2728. 47, 299–311.

Ghatan, G.J., Zimbelman, J.R., 2006. Paucity of candidate coastal constructional landforms along proposed shorelines on Mars: implications for a northern lowlands-filling ocean. Icarus 185, 171–196.

Gilbert, G.K., 1890. Lake Bonneville. U.S. Geological Survey Monograph 1, p. 438.

Godsey, H.S., Currey, D.R., Chan, M.A., 2005. New evidence for an extended occupation of the Provo shoreline and implications for regional climate change, Pleistocene Lake Bonneville, Utah, USA. Quat. Res. 63, 212–223.

Grotzinger, J.P., Arvidson, R.E., Bell, J.F., Calvin, W., Clark, B.C., Fike, D.A., Golombek, M., Greeley, R., Haldemann, A., Herkenhoff, K.E., Jolliff, B.L., 2005. Stratigraphy and sedimentology of a dry to wet eolian depositional system, Burns formation, Meridiani Planum, Mars. Earth Planet. Sci. Lett. 240, 11–72.

Gwynn, J.W., 2002. Great Salt Lake: an overview of change. Utah Geological Survey Special Publication.

Hayes, A.G., Wolf, A.S., Aharonson, O., Zebker, H., Lorenz, R., Kirk, R.L., Paillou, P., Lunine, J., Wye, L., Callahan, P., Wall, S., 2010. Bathymetry and absorptivity of Titan's Ontario Lacus. J. Geophys. Res. 115, E09009.

Head III, J.W., Kreslavsly, M., Hiesinger, H., Ivanov, M., Smith, D.E., Zuber, M.T., 1998. Oceans in the past history of Mars: tests for their presence using mars orbiter laser altimeter (MOLA) data. Geophys. Res. Lett. 24, 4401–4404.

Head, J.W., Hiesinger, H., Ivanov, M.A., Kreslavsky, M.A., Pratt, S., Thomson, B.J., 1999. Possible ancient oceans on Mars: evidence from Mars orbiter laser altimeter data. Science 286, 2134–2137.

Irwin III, R.P., Zimbelman, J.R., 2012. Morphometry of Great Basin pluvial shore landforms: implications for paleolake basins on Mars. J. Geophys. Res. 117, E07004.

Irwin, R.P., Howard, A.D., Maxwell, T.A., 2004. Geomorphology of Ma'adim Vallis, Mars, and associated paleolake basins. J. Geophys. Res. 109http://dx.doi.org/10.1029/2004JE002287.

Jakosky, B.M., Phillips, R.J., 2001. Mars' volatile and climate history. Nature 412, 237–244.

Jewell, P.W., 2007. Morphology and paleoclimate significance of Pleistocene Lake Bonneville spits. Quat. Res. 68, 421–430.

Jewell, P.W., 2010. River incision, circulation, and wind regime of Pleistocene Lake Bonneville, U.S.A. Palaeogeogr. Palaeoclimatol. Palaeoecol. 293, 41–50.

Kargel, J.S., Strom, R.G., 1992. Ancient glaciation on Mars. Geology 20, 3–7.

Kerr, R.A., 2007. Is Mars looking drier and drier for longer and longer? Science 317, 1673.

Knoll, A.H., Carr, M., Clark, B., Des Marais, D.J., Farmer, J.D., Fischer, W.W., Grotzinger, J.P., McLennan, S.M., Malin, M., Schröder, C., Squyres, S., 2005. An astrobiological perspective on Meridiani Planum. Earth Planet. Sci. Lett. 240, 179–189.

Komatsu, G., Ori, G.G., 2000. Exobiological implications of potential sedimentary deposits on Mars. Planet. Space Sci. 48, 1043–1052.

Komatsu, G., Brantingham, P.J., Olsen, J.W., Baker, V.R., 2001. Paleoshoreline geomorphology of Böön Tsagaan Nuur, Tsagaan Nuur and Orog Nuur: the Valley of Lakes, Mongolia. Geomorphology 39 (3–4), 83–98.

Komatsu, G., Ori, G.G., Marinangeli, L., Moersch, J.E., 2007. Playa environments on Earth: possible analogues for Mars. In: Chapman, M.G. (Ed.), The Geology of Mars: Evidence from Earth-Based Analogs. Cambridge University Press, New York, pp. 322–348.

Langhans, M.H., Jaumann, R., Stephan, K., Brown, R.H., Buratti, B.J., Clark, R.N., Baines, K.H., Nicholson, P.D., Lorenz, R.D., Soderblom, L.A., Soderblom, J.M., 2012. Titan's fluvial valleys: morphology, distribution, and spectral properties. Planet. Space Sci. 60, 34–51.

Linden, M., 2006. Glaciodynamics, Deglacial Landforms, and Isostatic Uplift During the Last Deglaciation of Norrbotten, Sweden. Ph.D. dissertation, Department of Geology, Lund University, Sweden. 32 pp.

Lorenz, R.D., Mitton, J., 2008. Titan Unveiled. Princeton University Press, Princeton, NJ.

Lucas, A., Aharonson, O., Deledalle, C., Hayes, A.G., Kirk, R., Howington-Kraus, E., 2014a. Insights into Titan's geology and hydrology based on enhanced image processing of Cassini RADAR data. J. Geophys. Res. 119, 2149–2166.

Lucchitta, B.K., 2010. Lakes in Valles Marineris. In: Cabrol, N.A., Grin, E.A. (Eds.), Lakes on Mars. Elsevier, Amsterdam, pp. 111–161. http://dx.doi.org/10.1016/B978-0-444-52854-4.00005-2.

Lucchitta, B.K., Ferguson, H.M., Summers, C., 1986. Sedimentary deposits in the northern lowland plains, Mars. J. Geophys. Res. 91 (B13), E166–E174.

Madden, M.E.E., Bodnar, R.J., Rimstidt, J.D., 2004. Jarosite as an indicator of water-limited chemical weathering on Mars. Nature 431, 821–823.

Mahaney, W.C., Milner, M.W., Netoff, D.I., Malloch, D., Dohm, J.M., Baker, V.R., Miyamoto, H., Hare, T.M., Komatsu, G., 2004. Ancient wet aeolian environments on Earth: clues to presence of fossil/live microorganisms on Mars. Icarus 171, 39–53.

Malin, M.C., Edgett, K.S., 1999. Oceans or seas in the Martian northern lowlands: high resolution imaging tests of proposed coastlines. Geophys. Res. Lett. 26, 3049–3052.

Malin, M.C., Edgett, K.S., 2000. Evidence for recent groundwater seepage and surface runoff on Mars. Science 288, 2330–2335.

Malin, M.C., Edgett, K.S., 2003. Evidence for persistent flow and aqueous sedimentation on Mars. Science 302, 1931–1934.

Mastrogiuseppe, M., Poggiali, V., Hayes, A., Lorenz, R., Lunine, J., Picardi, G., Seu, R., Flamini, E., Mitri, G., Notarnicola, C., Paillou, P., Zebker, H., 2014. The bathymetry of a Titan sea. Geophys. Res. Lett. 41, 1432–1437.

Matmon, A., Simhai, O., Rivka, A., Itai, H., Porat, N., McDonald, E., Benedett, L.L., Finkel, R., 2009. Desert pavement-coated surfaces in extreme deserts present longest-lived landforms on Earth. Geol. Soc. Am. Bull. 121, 688–697.

Mattson, A., Bruhn, R., 2001. Fault slip rates and initiation age based on diffusion equation modeling: Wasatch Fault Zone and eastern Great Basin. J. Geophys. Res. 106 (B7), 13739–13750.

Mustard, J.F., Cooper, C.D., Rifken, M.K., 2001. Evidence for recent climate change on Mars from the identification of youthful near-surface ground ice. Nature 412, 411–413.

Mustard, J.F., Murchie, S.L., Pelkey, S.M., Ehlmann, B.L., Milliken, R.E., Grant, J.A., Bibring, J.P., Poulet, F., Bishop, J., Dobrea, E.N., Roach, L., 2008. Hydrated silicate minerals on Mars observed by the Mars Reconnaissance Orbiter CRISM instrument. Nature 454, 305–309.

Ormö, J., Komatsu, G., 2003. MOC observation of linear and curvilinear features in the Hellas Basin: indication for multiple processes of formation. J. Geophys. Res. 108 (E6), 5059.

Ormö, J., Dohm, J.M., Ferris, J.C., Lepinette, A., Fairen, A.G., 2004. Marine-target craters on Mars? An assessment study. Meteorit. Planet. Sci. 39, 333–346.

Oviatt, C.G., 1997. Lake Bonneville fluctuations and global climate change. Geology 25, 155–158.

Oviatt, C.G., 2015. Chronology of Lake Bonneville, 30,000 to 10,000 yr B.P. Quat. Sci. Rev. 110, 166–171.

Oviatt, C.G., Sack, D., Felger, T.J., 1994. Quaternary Geologic Map of the Old River Bed and Vicinity, Millard, Juab, and Tooele Counties, Utah. Utah Geological Survey Map 161, 1:62,500.

Parker, T.J., 1994. Geologic Mapping of Argyre Planitia: A Large Highland Sedimentary Basin. Ph.D. dissertation, Martian Paleolakes and Oceans. University of Southern California. 200 pp., Chapter 3.

Parker, T.J., Currey, D.R., 2001. Extraterrestrial coastal geomorphology. Geomorphology 37, 303–328.

Parker, T.J., Saunders, R.S., Schneeberger, D.M., 1989. Transitional morphology in West Deuteronilus Mensae, Mars: implications for modification of the lowland/upland boundary. Icarus 82, 111–145.

Parker, T.J., Gorsline, D.S., Saunders, R.S., Pieri, D.C., Schneeberger, D.M., 1993. Coastal geomorphology of the Martian northern plains. J. Geophys. Res. 98, 11061–11078.

Parker, T.J., Grant, J.A., Franklin, B.J., 2010. The northern plains: a Martian oceanic basin. In: Cabrol, N.A., Grin, E.A. (Eds.), Lakes on Mars. Elsevier, Amsterdam. ISBN: 978-0-444-52854-4. Chapter 7.

Pelletier, J.D., DeLong, S.B., Al-Suwaidi, A.H., Cline, M., Lewis, Y., Psillas, J.L., Yanites, B., 2006. Evolution of the Bonneville scarp in west-central Utah: comparison of scarp-analysis methods and implications for the diffusion model of hillslope evolution. Geomorphology 74, 257–270.

Poulet, F., Bibring, J.-P., Mustard, J.F., Gendrin, A., Mangold, N., Langevin, Y., Arvidson, R.E., Gondet, B., Gomez, C., 2005. Phyllosilicates on Mars and implications for early Martian climate. Nature 438, 623–627.

Rodriguez, J.A.P., Platz, T., Gulick, V., Baker, V.R., Farien, A.G., Kargel, J., Yan, J., Miyamoto, H., Glines, N., 2015. Did the Martian outflow channels mostly form during the Amazonian Period? Icarus 257, 387–395.

Rossbacher, L.A., 1985. The problem of scale in planetary megageomorphology: global megageomorphology. 16–78. NASA Conference Publication 2312.

Ruiz, J., Fairn, A.G., Dohm, J.M., Tejero, R., Castilla, G., Álvarez-Gómez, J.A., 2004. Should paleoshorelines of ancient martian oceans be close to present-day equipotential surfaces? In: Harris, R.A., Ouwehand, L. (Eds.), Proceedings of the Third European Workshop on Exo-Astrobiology, Madrid, Spain. European Space Agency SP-545, Noordwijk, Netherlands, pp. 281–282.

Scott, W.E., 1988. Transgressive and high-shore deposits of the Bonneville lake cycle near North Salt Lake, Utah. Utah Geological and Mineral Survey Misc. Publication 88-1.

Soderblom, L.A., Tomasko, M.G., Archinal, B.A., Becker, T.L., Bushroe, M.W., Cook, D.A., Doose, L.R., Galuszka, D.M., Hare, T.M., Howington-Kraus, E., Karkoschka, E., 2007. Topography and geomorphology of the Huygens landing site on Titan. Planet. Space Sci. 55, 2015–2024.

Squyres, S.W., Knoll, A.H., 2005. Sedimentary rocks at Meridiani Planum: origin, diagenesis, and implications for life on Mars. Earth Planet. Sci. Lett. 240, 1–10.

Squyres, S.W., Grotzinger, J.P., Arvidson, R.E., Bell, J.F., Calvin, W., Christensen, P.R., Clark, B.C., Crisp, J.A., Farrand, W.H., Herkenhoff, K.E., Johnson, J.R., 2004. In situ evidence for an ancient aqueous environment at Meridiani Planum, Mars. Science 306, 1709–1714.

Stofan, E.R., Elachi, C., Lunine, J.I., Lorenz, R.D., Stiles, B., Mitchell, K.L., Ostro, S., Soderblom, L., Wood, C., Zebker, H., Wall, S., 2007. The lakes of Titan. Nature 445, 61–64.

Tanaka, K.L., Skinner, J.A., Hare, T.M., 2005. Geologic Map of the Northern Plains of Mars. .

Tomasko, M.G., Archinal, B., Becker, T., Bézard, B., Bushroe, M., Combes, M., Cook, D., Coustenis, A., De Bergh, C., Dafoe, L.E., Doose, L., 2005. Rain, winds, and haze during the Huygens porbe's descent to Titan's surface. Nature 438, 765–778.

Turtle, E.P., Perry, J.E., Hayes, A.G., McEwan, A.S., 2011. Shoreline retreat at Titan's Ontario Lacus and Arrakis Planitia from Cassini imaging science subsystem observations. Icarus 212, 957–959.

Upham, W., 1895. The Glacial Lake Agassiz. U.S. Geological Survey Monograph 25. 685 pp.

Wall, S., Hayes, A., Bristow, C., Lorenz, R., Stofan, E., Lunine, J., Le Gall, A., Janssen, M., Lopes, R., Wye, L., Soderblom, L., 2010. Active shoreline of Ontario Lacus, Titan: a morphological study of the lake and its surroundings. Geophys. Res. Lett. 37, L05202.

Webb, V.E., 2004. Putative shorelines in northern Arabia Terra, Mars. J. Geophys. Res. 109, E09010.

Wilde, S.A., Valley, J.W., Peck, W.H., Graham, C.M., 2001. Evidence from detrital zircons for the existence of continental crust and oceans on the Earth 4.4 Gyr ago. Nature 409, 175–178.

Wyatt, W.B., McSween Jr., H.Y., Tanaka, K.L., Head, J.W., 2004. Global geologic context for rock types and surface alteration on Mars. Geology 32, 648–654.

Zuber, M.T., 2007. Planetary science: Mars at the tipping point. Nature 447, 785–786.

Chapter 22

Insights into Lake Bonneville Using Remote Sensing and Digital Terrain Tools

P.W. Jewell*, D.T. Nelson†, B.B. Bowen* and L.W. Raming*
*University of Utah, Salt Lake City, UT, United States
†University of North Carolina—Pembroke, Pembroke, NC, United States

ABSTRACT
Beginning in the mid-20th century, the remote-sensing technologies of photogrammetry, LiDAR, and imaging spectroscopy have been a boon to the study of surficial processes throughout the world. Lake Bonneville research has benefited from these technologies ever since the advent of aerial photography, although each technology has its own particular strengths and weaknesses. Examples include elucidation of geomorphic features such as faint shorelines and investigation of mineralogy of playas resulting from the evaporation of Lake Bonneville. Future innovations promise to build on this rich legacy of Bonneville research.

Keywords: LiDAR, Photogrammetry, Imaging spectroscopy, Remote sensing

22.1 INTRODUCTION

Digital and technological revolutions of the second half of the 20th century have been transformed into a wide variety of current 21st century, remote-sensing tools that have great potential for understanding a number of the physical and chemical features of Lake Bonneville (Fig. 22.1). Remote sensing refers to any technology that gathers and displays parts of the natural surface of the Earth without actually being on the Earth's surface. This review is intended to highlight the historical evolution of these technologies, demonstrate their applications to Lake Bonneville as of 2015, and point the way toward possible future applications.

Perhaps the most important features that can be determined with remote sensing are various geomorphic expressions and surface mineral chemistry. The geomorphology of Lake Bonneville was one of the prime interests of

Developments in Earth Surface Processes, Vol. 20. http://dx.doi.org/10.1016/B978-0-444-63590-7.00022-6

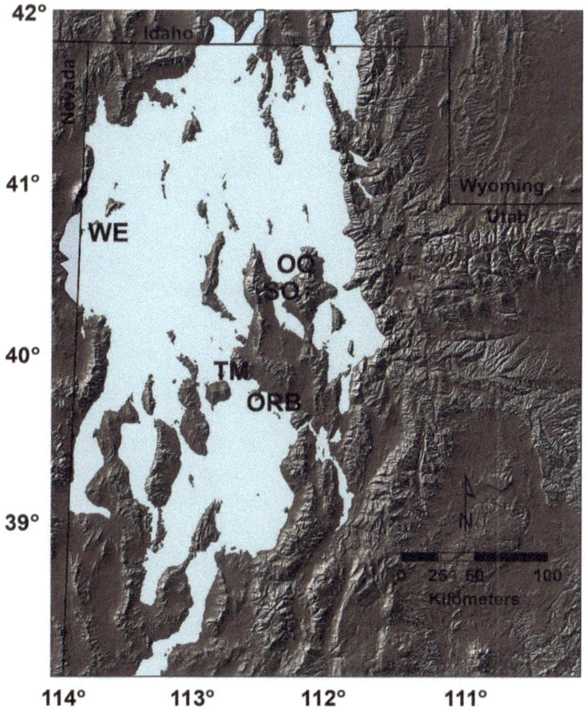

FIG. 22.1 Outline of Lake Bonneville and location of figures discussed in this paper. *WE*, Wendover area (Figs. 22.7 and 22.12); *TM*, Table Mountain (Figs. 22.2 and 22.9), *ORB*, Old River Bed area (Figs. 22.10 and 22.13), *OQ*, Oquirrh (Fig. 22.5), and *SO*, Stockton (Fig. 22.11).

G.K. Gilbert's classic monograph (Gilbert, 1890). Available tools of that era consisted primarily of plane table and stadia rods and with these tools Gilbert and his coworkers produced three-dimensional renderings that stand up well even against modern digital elevation model (DEM) data (Fig. 22.2). Unfortunately, these methods were slow and tedious and although they greatly benefited Gilbert and his contemporaries, they could only be employed in limited areas of the lake basin. With the advent of man-made flight, first with airplanes and later satellites, complete and detailed images of Earth's surface features could be described and analyzed.

22.1.1 Aerial Photography and Photogrammetry

The invention of the airplane was soon followed by aerial photography and interpretation. These innovations enabled the production of topographic maps through stereographic interpretation as well as revealing features of Lake Bonneville that were difficult to visualize with ground-based observations alone (eg, Pandey, 1987). Like many other geologic settings, Lake Bonneville research has benefited from the use aerial photography and topographic maps

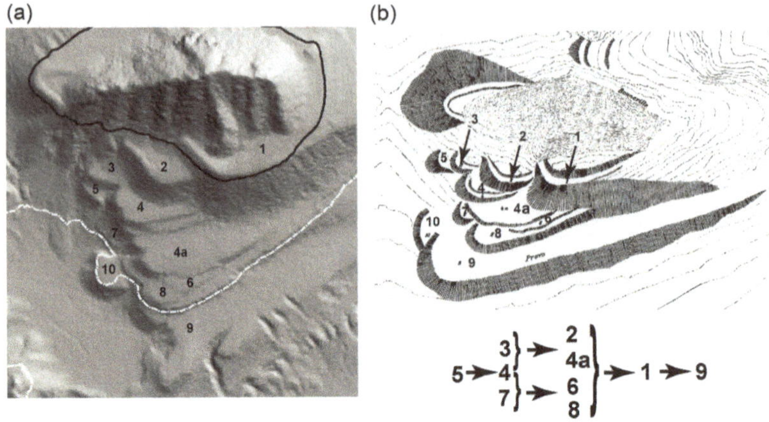

FIG. 22.2 (a) 5 m autocorrelated DEM rendering of the Table Mountain spit and embankment complex on the eastern side of the mountain. *Black line* represents the contour of the approximate elevation of the Bonneville shoreline, and the *white line* represents contour of the Provo shoreline. (b) G.K. Gilbert's rendering of the same area (Gilbert, 1890, plate XXV). The numbers in both figures represent Gilbert's interpreted relative age of the features.

in conjunction with field investigations. Today, digital photography provides pixel resolution data in the Bonneville basin as fine as 1 cm that are easily available through the high-resolution orthophotography (HRO) and National Agricultural Inventory Programs (NAIP) of the U.S. government.

The availability of high-resolution DEMs produced with photogrammetry has followed a trajectory not unlike that of the development of microprocessors (Fig. 22.3). 90, 30, and 10 m horizontal resolution digital elevation data have been available through the National Elevation Data (NED) program since the end of the 20th century (http://ned.usgs.gov/). Photogrammetric techniques have produced progressively finer resolution DEMs; of relevance to Lake Bonneville research are 5 and 2 m horizontal resolution data currently available through the State of Utah Automated Geographic Reference Center (AGRC) (http://gis.utah.gov).

22.1.2 Light Detection and Ranging

In the late 1990s, the availability of inexpensive photoelectric detectors led to the development of LiDAR (light detection and ranging) technology (eg, Flood, 2001). LiDAR instruments measure distance by determining return times of individual light pulses which are typically on the order of nanoseconds. Modern LiDAR instruments are now capable of recording 10^5–10^6 pulses per second, meaning that even brief surveys collect millions of individual data points. LiDAR instruments can be deployed on a variety of static and mobile platforms. Horizontal and vertical resolution are dependent on instrument capability and distance to the target. Horizontal resolution of LiDAR

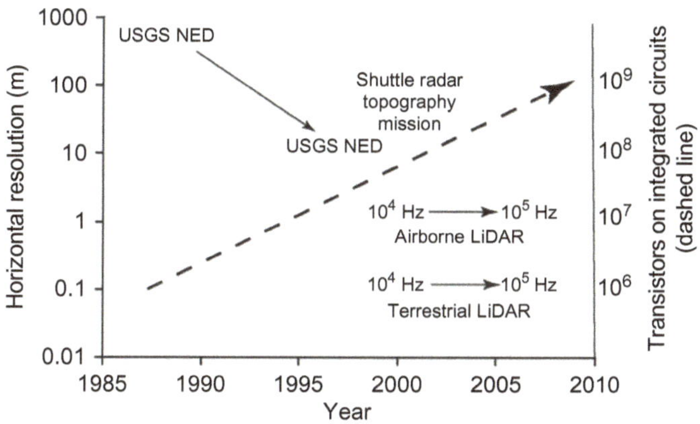

FIG. 22.3 Idealized increase in both computing power and the horizontal resolution of digital terrain data. *Dashed line* represents an approximation of Moore's Law (Moore, 1965). USGS NED refers to the U.S. Geological Survey's National Elevation Data. *From Jewell, P.W., 2015. Mineral exploration and the digital terrain revolution. In: New Concepts and Discoveries, 2015 Symposium, Geological Society of Nevada Special Publication, pp. 83–96.*

data can be as small as 1 mm for a tripod mounted instrument imaging a nearby object. Airborne surveys are typically posted at 0.5–2 m horizontal resolution. Vertical resolution of airborne surveys is typically in the 10–20 cm range.

Of tremendous significance to earth scientists is the "bare-earth" capability of airborne LiDAR. A single laser pulse produces multiple returns depending on the nature and intensity of vegetative cover (Fig. 22.4). By digitally removing all but the last return, the true land surface elevation can be determined. Although large portions of the Bonneville basin are sparsely vegetated, this capability is clearly demonstrated in areas of juniper-pinyon pine forest of the western portion of the Bonneville basin or the scrub oak terrain that is more common along the Wasatch Front (Fig. 22.5).

22.1.3 Imaging Spectroscopy

Imaging spectroscopy provides a method for remotely mapping surface composition. For any given material, the amount of solar radiation reflected, absorbed, transmitted, or emitted varies with wavelength. Reflectance-imaging spectroscopy is based on the presence of specific absorption features in the visible to near-infrared (VNIR) wavelength regions (eg, 350–2500 nm) of the electromagnetic spectrum (Adams and Gillespie, 2006). These absorption features can be used to differentiate between surfaces with varying composition in multispectral datasets and can be diagnostic of mineralogy with hyperspectral datasets. Emittance-imaging spectroscopy, similarly to reflectance-imaging spectroscopy, is useful because absorption features and their correlated wavelength position can help in the identification of surface compositions. The main

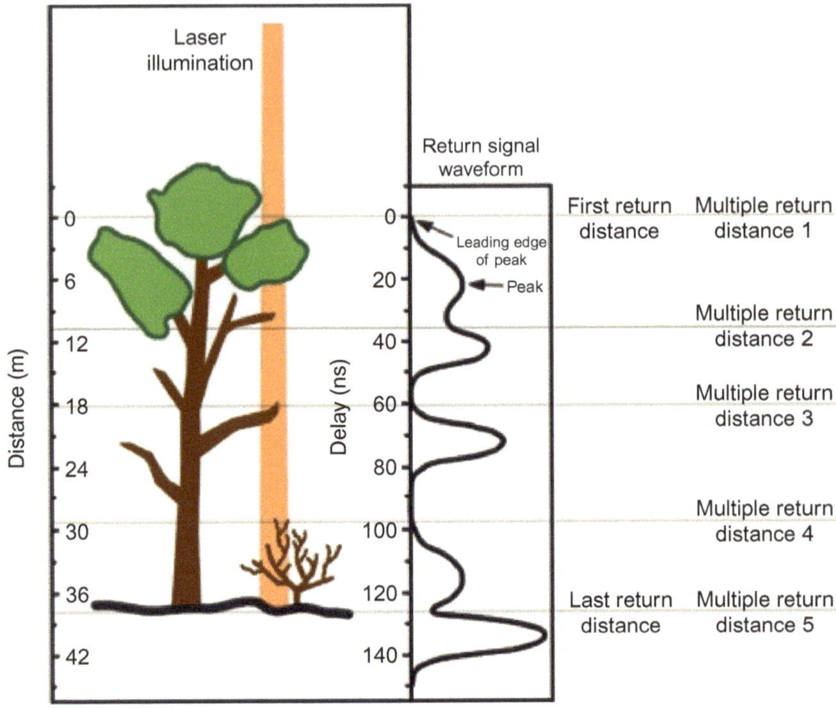

FIG. 22.4 Cartoon illustrating how multiple laser returns can be generated by a single pulse from a LiDAR instrument (https://www.eeducation.psu.edu/lidar/book/export/html/1873). *From Jewell, P.W., 2015. Mineral exploration and the digital terrain revolution. In: New Concepts and Discoveries, 2015 Symposium, Geological Society of Nevada Special Publication, pp. 83–96.*

FIG. 22.5 Example of the bare-earth model in a vegetated terrain (location OQ in Fig. 22.1).

differences between the two methods are that emittance spectroscopy employs the use of Kirchhoff's law to calculate emissivity of a surface through its reflectance. Subsequently the mid-infrared range (specifically from 7 to 12 μm) of the electromagnetic spectrum can be used to identify absorption features and can help identify materials that may otherwise be undistinguishable within the VNIR range (Vaughan et al., 2003; Hapke, 2012). Imaging spectroscopy is a valuable tool to earth and planetary scientists as it provides a nondestructive approach for mapping spatially contiguous variations in mineralogy over large geographic areas that may be difficult to access.

Analysis of an imaging-spectroscopy scene is based on the separation and identification of spectrally distinct pixels. In the case of mineral and material identification, this involves identifying the specific wavelength positions of absorption features (Fig. 22.6). Hyperspectral datasets provide

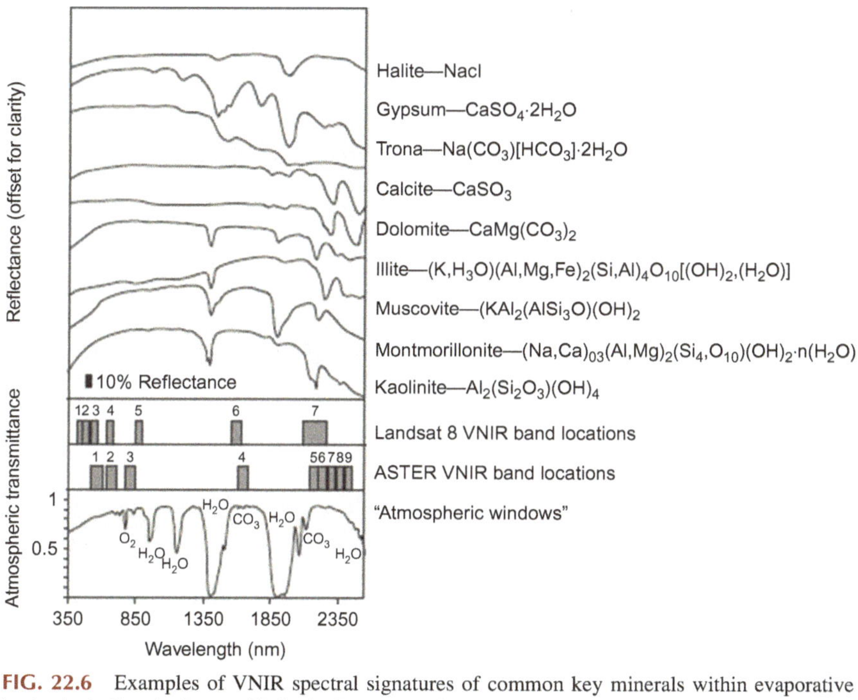

FIG. 22.6 Examples of VNIR spectral signatures of common key minerals within evaporative basins including the Bonneville basin. Representative spectral signatures for halite, gypsum, trona, calcite, dolomite, illite, muscovite, montmorillonite, and kaolinite are shown. Below the mineral spectral signatures are the locations of bands on Landsat 8 and ASTER satellite sensors, and the atmospheric transmittance over the VNIR wavelength ranges. The spectral resolution of lab/ field scale "hyperspectral" analysis (spectral signatures) is much greater than the resolution as it would be observed from remote-sensing orbiting systems such as Landsat and ASTER. However, specific absorption features can be detected and distinguished within the bands of the orbital sensors. *Modified from Clark, R.N., Swayze G.A., Wise R., Livo E., Hoefen T., Kokaly R., Sutley S.J., 2007. USGS digital spectral library splib06a, U.S. Geological Survey, Digital Data Series 231.*

the greatest spectral resolution and subsequently allow for extensive mineral identification. Challenges arise, however, with the analysis of such large data sets, which are addressed by data-dimensionality reduction algorithms that preserve uniqueness while allowing for easier handling of the data. Once unique hyperspectral profiles are located, they are identified by the comparison and fitting of similar spectra obtained in the field, lab, or from signatures stored in mineral spectral libraries (Clark et al, 1990; Crowley, 1993; Clark et al., 2007). Similar methods are employed with multispectral data; however, the ability to identify and distinguish between similarly related minerals decreases with spatial and spectral resolution (Fig. 22.7; Hubbard et al., 2003; Rowan et al., 2003; Mars and Rowan, 2006). Air- and space-borne sensors vary in spatial resolution with pixel footprint ranging from 30 m x 30 m (eg, Landsat) to <0.5 m (low-altitude HyMap data; Cocks et al, 1998). Many other algorithms may be used on imaging-spectroscopy data to differentiate surface composition, such as band ratios and supervised and unsupervised classification. The resultant product is the classification of spectrally distinct classes, which generally must be validated by spectral data obtained in the field, referred to here as ground-truth data.

In general, the depth of a specific absorption feature corresponds to the amount of that mineral, allowing for analysis of relative abundance. One of the limitations of reflectance and emittance imaging spectroscopy, however, is that quantitative identifications are limited due to the nonlinear responses of intimate mixtures of minerals (Clark, 1999). In natural sedimentary systems, minerals are more commonly present as complex mixtures of detrital and authigenic components of varying grain size, resulting in both linear and nonlinear spectral mixtures that can be difficult to interpret (Bowen et al., 2012).

22.2 DATA ACQUISITION AND PRESENTATION

22.2.1 Instrument Platforms

Standard digital photography, LiDAR, and spectral data can be collected at a range of spatial and spectral resolutions from satellite, airborne, field, or lab-based platforms. Spectrometers with varying spatial and spectral resolution are mounted on aircraft and satellites resulting in an aerial image with each pixel representing a spectral profile over a given area.

Current LiDAR instruments are not typically flown at distances greater than 1 km above the ground. Airborne platforms carry an internal measurement unit (IMU) necessary for accurate georeferencing of data sets. Horizontal resolution is determined by "shot density" ie, the number of returns for a given area, which in turn can vary according the irregularity of the terrain

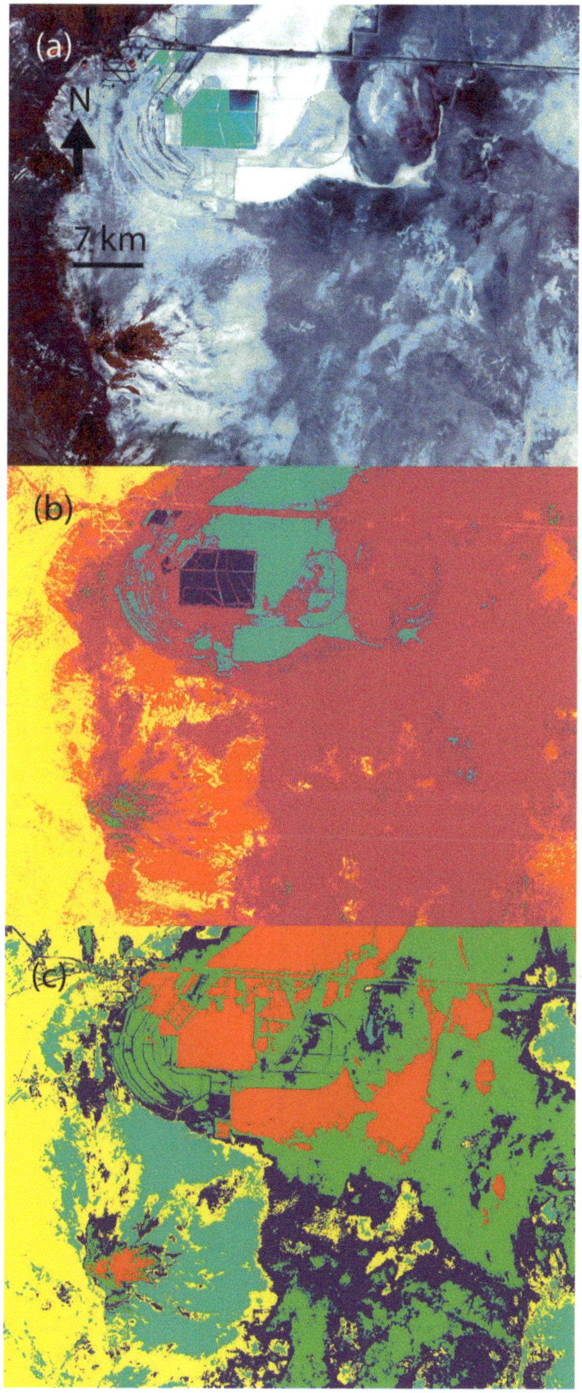

FIG. 22.7 Examples of various methods of spectral analysis. (a) True color AVIRIS scene collected over the western end of the Bonneville basin by Wendover (WE in Fig. 22.1), Utah. (b) Supervised classification of (a) by identification of spectrally pure end members. Pixel classification based on absorption features: *yellow* are carbonates, *red* are clay minerals, *blue* are ponded water, *cyan* are damp halite, *magenta* are mix of clay minerals and evaporites, and *green* are vegetation. (c) Unsupervised ISODATA classification of scene (a), *each color* represents a distinct class as determined by the ISODATA algorithm (Ball and Hall, 1965).

and the ability of the platform to maintain a consistent flight plan and constant distance from the ground. Accurate bare-earth rendering in areas of dense vegetation is dependent on gaining a sufficient density of last returns to produce a viable DEM (Fig. 22.8).

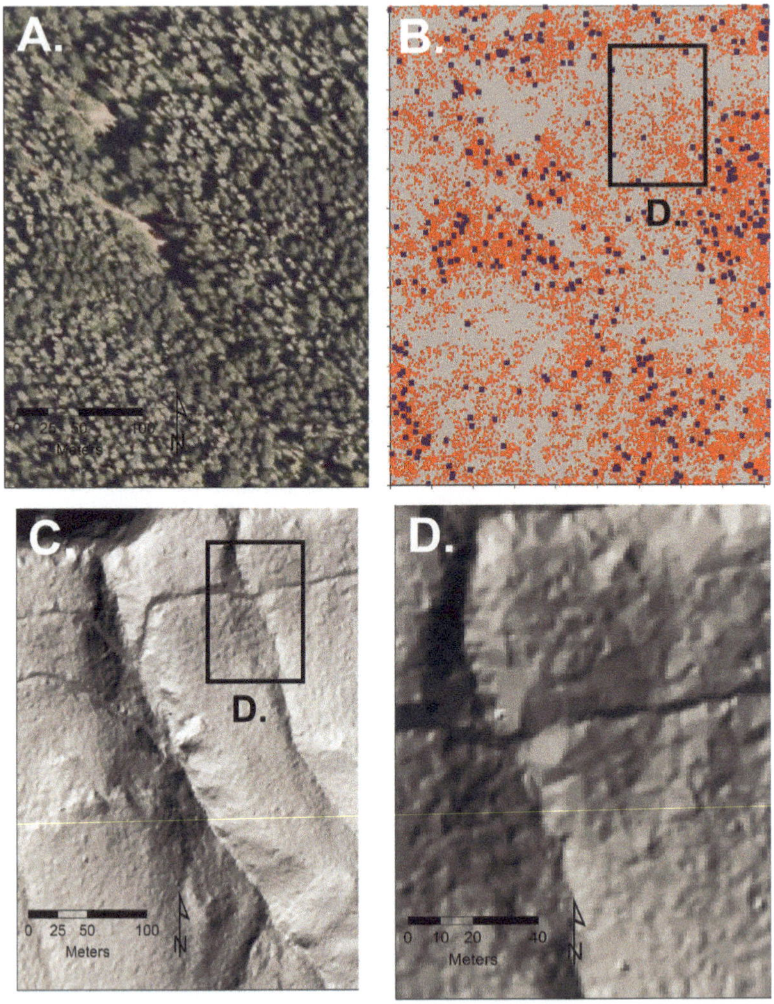

FIG. 22.8 (a) NAIP photograph of area near the Blue Jay mine, North Santiam mining district, Cascade Mountains, Oregon. (b) LiDAR shot density of the area in (a). *Blue squares* represent 4th returns; *small red squares* represent 3rd returns showing the high degree of irregularity of last returns in areas of dense vegetation. (c) Hillshade detail of LiDAR DEMs showing triangular facets produced in areas of low last return density. (d) Closeup of area in (b) and (c) showing pits and triangular facets produced by postprocessing due to a paucity of laser shot returns. *From Jewell, P.W., Farnsworth, J.A., Zajac, T., 2013. Rediscovering the discovery outcrop: the promises and pitfalls of LiDAR technology in mineral exploration. Soc. Econ. Geologists Newsl. 92, 14–19*

22.2.2 Data Processing and Presentation

For spectral data, atomic and molecular interactions between electromagnetic radiation and materials result in electronic and vibrational processes that cause wavelength-specific absorption features in the VNIR region. When these spectral reflectance/absorption data are collected over a spatially contiguous area, they can be used to map compositional and mineralogical variations on micro- to macrospatial scale. Raw data must be processed and corrected to account for geometric, radiometric, and noise distortions (Lillesand et al., 2007). The resulting product is a geographically contiguous raster image where each pixel contains a multidimensional data array describing surface reflectance and absorption. The spatial resolution of the image describes the size of a pixel relative to the actual geographic location. The spectral resolution describes the number and dimension of specific wavelength intervals (bands or channels) in the electromagnetic spectrum to which the remote-sensing instrument is sensitive, for multispectral sensors this ranges from 7 to 14 bands, whereas a hyperspectral sensor can have 200+ bands.

Multiple computing platforms and applications now exist for evaluation of both spectral and LiDAR data (eg, ArcGIS, Global Mapper, ENVI, Imagine, MultiSpec), and rapid advances in data-storage technology have helped to facilitate analysis of these potentially very large datasets. Standard GIS software is most effective where irregular shot locations are transformed into raster data (eg, Fig 22.8B and C). Additionally extensive development in algorithms and models aids in processing, analyzing, and interpreting remotely sensed spectral and spatial data.

22.2.3 Data Availability

High-resolution LiDAR data are now widely available east of the Mississippi River, coverage over the western United States including the Bonneville basin is quite spotty. Most surveys have been undertaken in urban corridors (eg, the Wasatch Front of Utah) or in conjunction with public lands and resource management (eg, timber sales in Idaho) (http://coast.noaa.gov/inventory/#).

Numerous private and government-operated, airborne, and satellite sensors collect high-resolution spectral data on Earth and other planets. Some significant terrestrial sensors include AVIRIS, ASTER, Landsat, and Hyperion. Each sensor has varying levels of resolution, wavelength range, and coverage. Typically the higher the spectral resolution the less spatial coverage, for example, AVIRIS and Hyperion are considered hyperspectral sensors and offer high spatial and spectral resolution, however, they cover less terrain and have been employed only in specific areas. ASTER and Landsat, however, are multispectral sensors that have extensive global coverage at lower resolutions. Free Landsat, ASTER, and Hyperion data sets are available through the USGS, visit http://earthexplorer.usgs.gov/ or http://glovis.usgs.gov for more information.

22.3 APPLICATIONS

22.3.1 Geomorphic Analysis

22.3.1.1 Minor Shorelines and Erosional Features

The increasingly high-resolution data available with LiDAR have been a boon for Lake Bonneville research. Previously unrecognizable detail such as narrow rills and minor deposition features are poorly resolved with 10 m DEMs derived from photogrammetry are clearly discernable with the LiDAR data (Fig. 22.9). The relative accuracy of 10 m NED data, 5 m photogrammetrically derived DEMs, and 1 m LiDAR-derived DEMs in characterizing the morphology of features such as prominent v-bars in the Bonneville basin is also readily apparent (Fig. 22.10).

22.3.1.2 Quantitative Volume Analysis

High-resolution DEMs allow volume estimates of spits, bars, and other depositional features within the Bonneville basin. The Stockton Bar is a large spit and bar complex between Tooele and Rush Valley. During the rise of the lake copious amounts of gravel were transported from alluvial fans southward via longshore currents. A basal surface of the Stockton Bar complex during the deposition of the Bonneville age features was estimated by acquiring known elevations of the base of the features in the field and extrapolating a surface through these points. The basal surface was also estimated by interpreting the base of the Bonneville age features from ground-penetrating radar (Smith et al., 2003). The basal surface was subtracted from 5 m photogrammetrically derived DEMs to obtain an estimated volume for the Bonneville and pre-Bonneville age deposits (Fig. 22.11). Depending on the interpretation of the basal surface points and after adding back volume estimates of material that has been mined or removed to build a railroad, the volume of the Bonneville age sediments are estimated to be between approximately $1.8\text{–}2.0 \times 10^7$ m^3 of material. Based on the lake hydrograph, it is estimated that the materials were deposited during over 500–1000 years duration. The rate of sediment flux at the site would thus range from 18,000 to 40,000 m^3/year. These rates are similar to those in the modern Great Lakes and other coastal regions (Johnson, 1956; Van Wellan et al., 2000). The volume and longshore current rates of other depositional features in the basin are currently being estimated and compared with the Stockton Bar.

22.3.2 Lake Mineralogy

The value of using spectroscopy as a tool for identifying mineralogy has been recognized for several decades (eg, Farmer, 1974). Examples of successful spectral remote sensing of evaporative basins and playas occur globally (Crowley, 1993; Hubbard and Crowley, 2005) and within the Bonneville

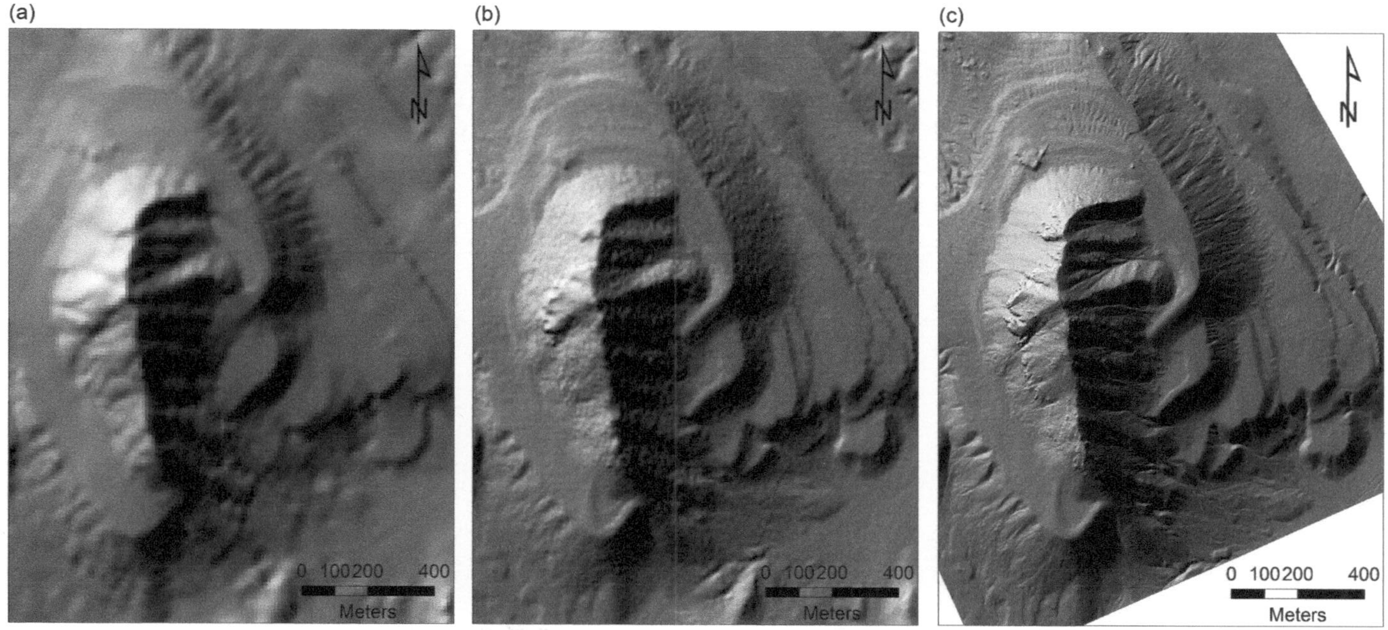

FIG. 22.9 Table Mountain area in the central portion of the Bonneville basin. (a) 10 m DEM from the National Elevation Data (NED) system. (b) 5 m DEM from the Automated Geographic Reference Center (AGRC) of Utah (http://gis.utah.gov/). (c) 1 m LiDAR data acquired by the Robert Pack of Utah State University for an ongoing research project.

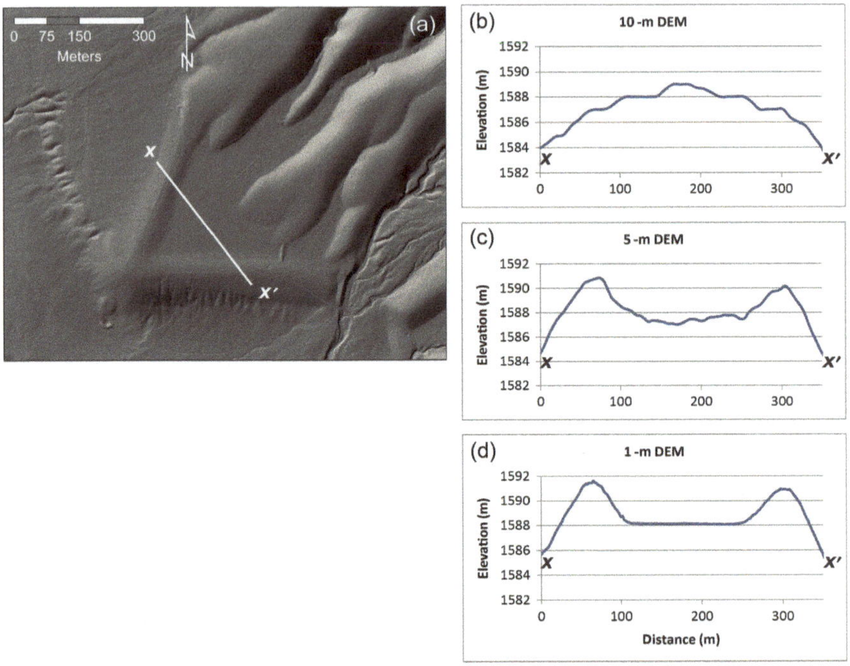

FIG. 22.10 (a) Hillshade of 1 m LiDAR data of a prominent v-bar along the Bonneville shoreline on the west side of the Simpson Mountains (ORB in Fig. 22.1) of central Utah. (b) 10 m elevation data from NED system of the *x*–*x*′ section. (c) 5 m elevation data of the *x*–*x*′ section. (d) 1 m LiDAR data of the *x*–*x*′ section. ODATA algorithm, in situ validation, is necessary to identify classes.

basin (Lynch et al., 2015). These analyses reveal spatial variations in mineralogy that offer insight on paleoclimatic changes, geochemistry of desert aquifers, as well as locations of important geologic resources. The significance of reflectance spectroscopy as a mineralogical tool has been exemplified by the role that it has played in radically changing our understanding of the composition of the surface of Mars in the last decade (eg, Bibring et al., 2005; Farrand et al., 2007; Murchie et al., 2009; Ehlmann, 2010).

Remote (airborne and satellite), field-based, and laboratory spectroscopy all provide important tools for identifying spatial variations in surface characteristics, composition, and mineralogy within the Bonneville basin. As paleolake Bonneville evaporated, it left a mineralogical record of desiccation, fluid–rock interactions, and surface processes. The modern day evaporite basin and associated salt flats contain diverse mineralogical assemblages that record changing climatic conditions and brine composition. Mineralogical analysis of the Pilot Valley using a range of techniques including reflectance spectroscopy confirmed the presence of gypsum, halite, calcite, aragonite, dolomite, smectite, kaolinite, chlorite, illite, muscovite, quartz, and feldspar in the shallow surface

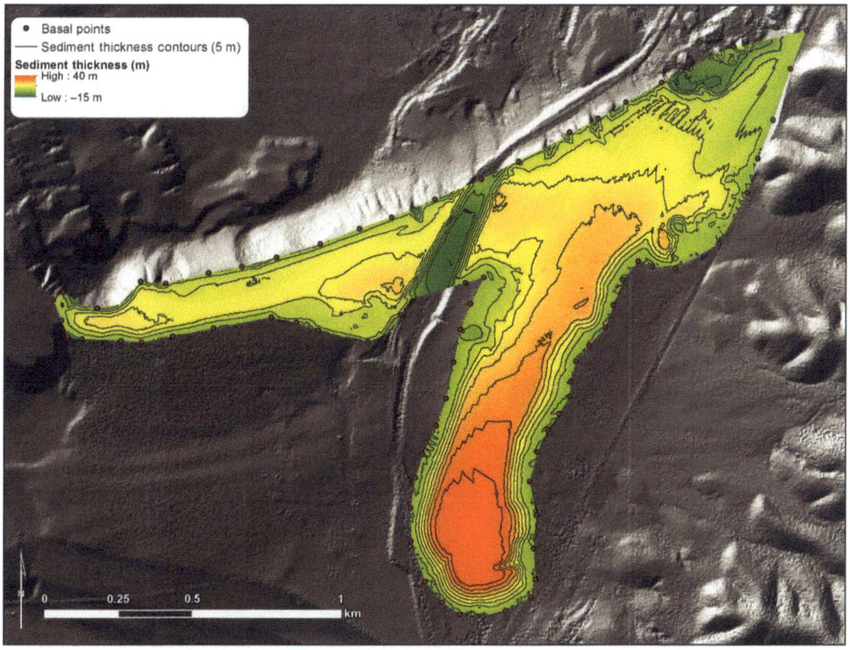

FIG. 22.11 Stockton Bar area in between Tooele and Rush Valleys (SO location in Fig. 22.1). Thickness of sediment deposited during the late transgressive phase of Lake Bonneville. Derived from volume estimates from 5 m photogrammetrically derived DEMs.

sediments (Lynch et al., 2015). Although halite, quartz, and feldspar lack diagnostic absorption features in the visible to shortwave infrared wavelength range, each of the other minerals contains a diagnostic spectral signature (Fig. 22.5).

Remotely sensed image spectroscopy also allows identification of geomorphological processes and their changes over time. The Landsat program provides the longest running record of multispectral earth observations. The Landsat program was initiated in 1972 and continues today to provide coverage of global land surfaces every 18 days. This multitemporal quality allows for analyses of changes over time, which can be significant in a dynamic environment such as the Bonneville basin that is impacted by a wide range of processes, including eolian action, groundwater flux, surface ponding, anthropogenic land use, etc. For example, analysis of annual variations in the area of salt crust cover at the Bonneville Salt Flats suggests that overall the area has decreased from 120 to 80 km^2; furthermore, these changes have corresponding relationships to changes in evaporation, precipitation, and land use (eg, evaporative potash mining; Raming and Bowen, 2014) (Fig. 22.12).

22.3.3 Vegetation

Differences between photographic depiction and DEM analysis of Lake Bonneville features demonstrate that neither method is ideal for this type of

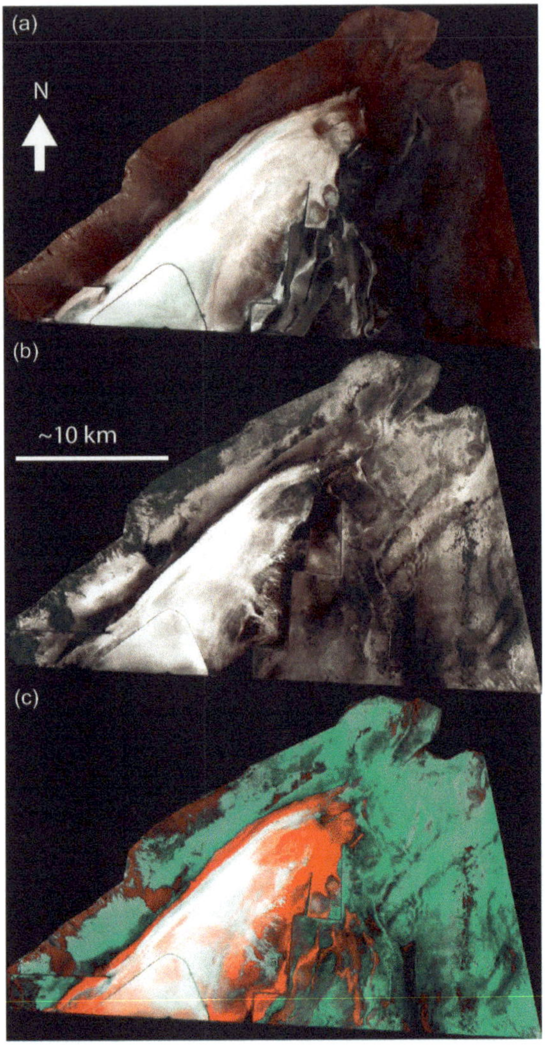

FIG. 22.12 Illustration of changes in the areal extent of the halite crust over two decades, the images displayed here are qualitative, however, other methods can produce quantitative results describing a defined change in area. (a) Time 1 image: Landsat 5 TM 3-2-2 scene obtained over the Bonneville Salt Flats (WE location in Fig. 22.1) Oct. 28, 1986. (b) Time 2 image: Landsat 5 TM 3-2-2 scene obtained over the Bonneville Salt Flats Sept. 28, 2010. (c) Image of two-color multiview change detection: *red color* highlights a decrease in brightness from Time 1 to Time 2 (ie, change in cover of area of salt crust), whereas *cyan color* highlight an increase in brightness Time 1 to Time 2.

work. For instance, shorelines and beach ridges with virtually no topographic expression are clearly visible as differing vegetation patterns, which in turn are probably a reflection of the porosity and water retention characteristics of varying substrates and yet these features are poorly represented in the DEM data (Fig. 22.13).

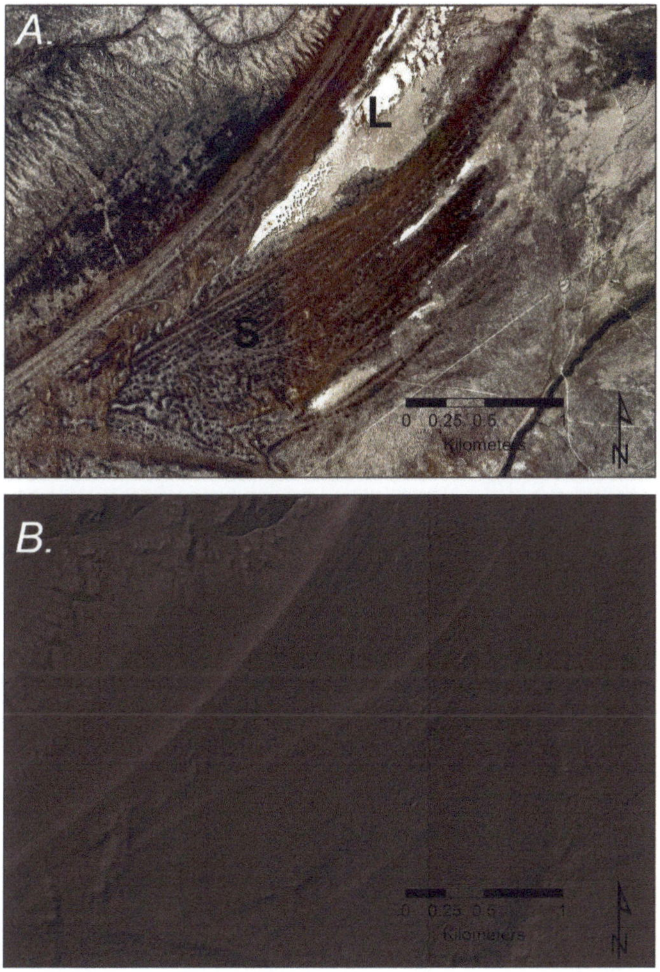

FIG. 22.13 (a) NAIP photography of low-relief gravel barriers (S) and lagoon (L) deposits from the Erickson Wash area southwest of the Simpson Mountains (ORB in Fig. 22.1). (b) 5 m autocorrelated DEM hillshade from the same area as (a).

22.4 DISCUSSION

As useful as current remote-sensing tools are for studying Lake Bonneville features, the future promises exciting new advances over the next few years. For instance, the stated goal of the U.S. Geological Survey's 3D Elevation Program (3DEP) is complete LiDAR coverage for the coterminous United States at 1 m horizontal resolution by the year 2022.

LiDAR technology itself seems poised for a significant leap in resolution and acquisition capability. A new generation of instruments utilizing "Geiger-mode Avalanche Photodiodes" (Geiger-mode for short) are set to come on the market in 2015 after development by the U.S. Department of Defense. Rather than a single pulse generating a single return, Geiger-mode instruments receive multiple subpluses that are processed by an array of detectors that can be activated by a single photon. The result is an instrument that can fly much higher than conventional LiDAR and collect data at an order of magnitude higher rate. These instruments, however, do not measure return intensity that can be a useful parameter in geologic investigations (eg, Burton et al., 2011) and QA/QC procedures for the technology are still a matter of concern (Abdullah, 2012). Initial steps have been made toward integrating collection of LiDAR data together with hyperspectral reflectance data, integrating the spatial and compositional information together onto a single platform. Data volume capacity has been a historical limiting factor, but rapid advances continue on this front.

Remote sensing plays an important role in ongoing research related to land use and hydrology in the Bonneville basin. For example, satellite data (eg, Landsat and ASTER) are critical for evaluating the impacts of a century of potash mining, land speed racing, and multiple cultural and recreational uses on the Bonneville Salt Flats. Analyses of historical satellite datasets are being used to map the extent of the salt crust, standing water, and the footprint of mining/racing infrastructure through time (Bowen et al., 2014). Preliminary data are being examined to test whether VNIR reflectance spectroscopy can be used to examine differences in salt-crust thickness, to help facilitate nondestructive analyses of changes in the salt through time.

Unmanned aerial vehicles (UAVs) are now cheap, technologically sophisticated, and widely available to the general public. With a mounted high-resolution, digital camera, these systems are now capable of producing point clouds that although not as accurate as LiDAR, possess the advantages of low-cost, compactness, and rapid deployment. Commercially available software such as the popular AgiSoft system (eg, Ouedraogo et al., 2014; Tavani et al., 2014) enable rapid, three-dimensional renderings of a variety of geologic features.

ACKNOWLEDGMENTS

Robert Pack of Utah State University helped acquire some of the LiDAR data used in this analysis. The work was partially supported by National Science Foundation Grant 1053129 to Jewell and NASA EPSCoR Grant to Bowen. Constructive discussions with Jack Oviatt have been very helpful in focusing our thinking about Lake Bonneville.

REFERENCES

Abdullah, Q.A., 2012. Mapping matters. Photogramm. Eng. Remote. Sens. 2012 (2012), 664–665.

Adams, J., Gillespie, A., 2006. Remote Sensing of Landscapes with Spectral Images: A Physical Modeling Approach. Cambridge University Press, Cambridge, UK.

Ball, G.H., Hall, D.J., 1965. ISODATA, A Novel Method of Data Analysis and Pattern Classification. Stanford Research Institute Technical Report, Menlo Park, CA.

Bibring, J.P., et al., 2005. Mars surface diversity as revealed by the OMEGA/Mars Express observations. Science 307, 1576–1581.

Bowen, B.B., Bell, J., Story, S., 2012. Remote and field-based imaging spectroscopy for the diagenetic mineralogy of sedimentary rocks. In: Sylvester, P. (Ed.), Quantitative Mineralogy and Microanalysis of Sediments and Sedimentary Rocks. Mineralogical Association of Canada Short Course Series, vol. 42. Mineralogical Association of Canada, St. John's, NL, pp. 115–132.

Bowen, B.B., Kleba, B., Turner, J., Raming, L., 2014. Sedimentology, Mineralogy, Geochemistry, and Geomicrobiology of the Bonneville Salt Flats. Geol. Soc. Am. Abstr. Programs 46, 410.

Burton, D., Dunlap, D.B., Wood, L.J., Flaig, P.P., 2011. Lidar intensity as a remote sensor of rock properties. J. Sediment. Res. 81, 339–347.

Clark, R.N., 1999. Spectroscopy of rocks and minerals, and principles of spectroscopy. In: Rencz, A.N. (Ed.), Manual of Remote Sensing, Remote Sensing in the Earth Sciences. vol. 3. John Wiley, New York, pp. 3–58.

Clark, R.N., King, T.V.V., Klejwa, M., Swayze, G.A., Vergo, N., 1990. High spectral resolution reflectance spectroscopy of minerals. J. Geophys. Res. 95, 12653–12680.

Clark, R.N., Swayze, G.A., Wise, R., Livo, E., Hoefen, T., Kokaly, R., Sutley, S.J., 2007. USGS Digital Spectral Library splib06a. U.S. Geological Survey, Digital Data Series 231.

Cocks, T., Jenssen, R., Stewart, A., Wilson, I., Shields, T., 1998. The HyMap™ airborne hyperspectral sensor: the system, calibration and performance. In: 1st EARSEL Workshop on Imaging Spectroscopy, October 1.

Crowley, J., 1993. Mapping playa evaporite minerals with AVIRIS data: a first report from death valley, California. Remote Sens. Environ. 44, 337–356.

Ehlmann, B.L., 2010. Diverse aqueous environments during Mars' first billion years: the emerging view from orbital visible-ear infrared spectroscopy. Geochem. News, 142.

Farmer, V.C., 1974. The Infra-Red Spectra of Minerals. Mineralogical Society of London, London. 539 pp.

Farrand, W., Bell, J., Johnson, J., Jolliff, B., Knoll, A., Mclennan, S.M., Squyres, S.W., Calvin, W.M., Grotzinger, J.P., Morris, R.V., Soderblom, J., Thompson, S.D., Watters, W.A., Yen, A., 2007. Visible and near-infrared multispectral analysis of rocks at Meridiani Planum, Mars, by the Mars Exploration Rover Opportunity. J. Geophys. Res. 112, E06S02.

Flood, M., 2001. Laser altimetry: from science to commercial LIDAR mapping. Photogramm. Eng. Remote. Sens. 67, 1209–1217.

Gilbert, G.K., 1890. Lake Bonneville: U.S. Geological Survey Monograph 1. 438 pp.

Hapke, B., 2012. Thermal emission and emittance spectroscopy. Combined Theory of Reflectance and Emittance Spectroscopy, second ed. Cambridge University Press, New York.

Hubbard, B., Crowley, J., 2005. Mineral mapping on the Chilean–Bolivian Altiplano using co-orbital ALI, ASTER and Hyperion imagery: data dimensionality issues and solutions. Remote Sens. Environ. 100, 173–186.

Hubbard, B., Crowley, J., Zimbelman, D., 2003. Comparative alteration mineral mapping using visible to shortwave infrared (0.4–2.4 μm) hyperion, ALI, and ASTER imagery. IEEE Trans. Geosci. Remote Sens. 41, 1401–1410.

Johnson, J.W., 1956. Dynamics of nearshore sediment movement. Bull. Am. Assoc. Pet. Geol. 40, 2211–2232.

Lillesand, T., Kiefer, R., Chapman, J., 2007. Remote Sensing and Image Interpretation, sixth ed. Wiley, New York.

Lynch, K.L., Horgan, B.H., Munakata-Marr, J., Hanley, J., Schneider, R.J., Rey, K.A., Spear, J.R., Jackson, W.A., Ritter, S.M., 2015. Near-infrared spectroscopy of lacustrine sediments in the Great Salt Lake Desert: an analog study for Martian paleolake basins. J. Geophys. Res. Planets 120, 599–623. http://dx.doi.org/10.1002/2014JE004707.

Mars, J.C., Rowan, L.C., 2006. Regional mapping of phyllic-and argillic-altered rocks in the Zagros magmatic arc, Iran, using advanced spaceborne thermal emission and reflection radiometer (ASTER) data and logical operator algorithms. Geosphere 2, 161–186.

Moore, G.E., 1965. Cramming more components onto integrated circuits. Electronic Magazine, 4.

Murchie, S.L., et al., 2009. A synthesis of Martian aqueous mineralogy after 1 Mars year of observations from the Mars Reconnaissance Orbiter. J. Geophys. Res. 114.

Ouedraogo, M.M., Degre, A., Debouche, C., Lisein, J., 2014. The evaluation of unmanned aerial system-based photogrammetry and terrestrial laser scanning to generate DEMs of agricultural watersheds. Geomorphology 214, 339–355.

Pandey, S.N., 1987. Principles and Applications of Photogeology. Wiley, New York. 366 pp.

Raming, L.W., Bowen, B.B., 2014. Spatiotemporal analyses of environmental conditions and surface processes at the Bonneville Salt Flats. Geol. Soc. Am. Abstr. Programs 46, 355.

Rowan, L., Hook, S., Abrams, M., Mars, J., 2003. Mapping hydrothermally altered rocks at Cuprite, Nevada, using the advanced spaceborne thermal emission and reflection radiometer (ASTER), a new satellite-imaging system. Econ. Geol. 98, 1019–1027.

Smith, D.G., Simpson, C.J., Jol, H.M., Meyers, R.A., Currey, D.R., 2003. Radar stratigraphy used to infer transgressive or regressive deposition and internal structure of a barrier and spit, Lake Bonneville, Stockton, Utah, USA. In: Bristow, C.S., Jol, H.M. (Eds.), Ground Penetrating Radar in Sediments, vol. 211. Geological Society of London, Special Publication, pp. 70–86.

Tavani, S., Granado, P., Corradetti, A., Girundo, M., Iannce, A., Arbues, P., Munoz, J.A., Mazzoli, S., 2014. Building a virtual outcrop, extracting geological information from it, and sharing the results in Google Earth via OpenPlot and Photoscan: an example from the Khaviz Anticline (Iran). Comput. Geosci. 63, 44–53.

Van Wellan, E., Chadwick, A.J., Mason, T., 2000. A review and assessment of longshore sediment transport equations for coarse-grained beaches. Coast. Eng. 40, 243–275.

Vaughan, R.G., Calvin, W.M., Taranik, J.V., 2003. SEBASS hyperspectral thermal infrared data: surface emissivity measurement and mineral mapping. Remote Sens. Environ. 85 (1), 48–63.

Chapter 23

Lake Bonneville Geosites in the Urban Landscape: Potential Loss of Geological Heritage

M.A. Chan and H.S. Godsey
University of Utah, Salt Lake City, UT, United States

ABSTRACT
Unique geological sites are recognized and protected worldwide through protocols implemented by interested citizens, organizations, and governments. The term *geoantiquity* was initially developed to denote scientifically important, natural records of Earth's surface processes, environments, and geological history formed near and around Pleistocene Lake Bonneville. These features include shorelines, moraines, salt flats, playas, and fault scarps that are typically geologically young, unconsolidated, and therefore easily removed or destroyed. As an international movement of geoheritage has taken hold over the last two decades, the name *geosites* is now a more broadly accepted term that encompasses geoantiquities. Geosites that occur on private land and in regions of rapid urban growth are particularly prime targets for modification by development or removal (eg, sand and gravel extraction). A geosite can occur in any place or setting, but this chapter focuses on examples from the Bonneville basin in northwestern Utah where several major Pleistocene landforms are being destroyed at alarming rates. Protection of geosites can take a number of forms, but all conservation efforts depend upon the collective involvement of city planners and land officials, landowners, private industry, nongovernmental organizations, and the scientific community. Ultimately, protection of geosites relies upon an informed and involved citizenry who speak out and work to promote best practices in sustainability and land-use management.

Keywords: Geoantiquity, Geoconservation, Geoheritage, Lake Bonneville

23.1 INTRODUCTION

Valuable records of geologic processes and environmental change are commonly contained within young sedimentary deposits on the surface of the Earth. Deposits that are particularly detailed or well preserved can provide information about ancient climates, lake levels, wind and current patterns, succession and diversity of life, geologic hazards, movement of ice, and neotectonics. Because these deposits are typically unconsolidated, well sorted and easily modified, they are generally valued more for their potential as building sites or as a source of aggregate than they are for their scientific significance. Deposits in regions of rapid urban growth are at particular high risk of destruction due to high transportation costs and demands for developable land.

This chapter describes efforts to raise awareness of important sites from the Bonneville basin in northwestern Utah that are worthy of protection. A case study from the Stockton Bar demonstrates the complexities of land preservation for any purpose. Ultimately, the goal of this work is to develop a mechanism for recognition and preservation of important sites through education and community involvement.

23.1.1 Background

In the 1990s, a team of University of Utah researchers began work to identify and preserve scientifically important landforms from Pleistocene Lake Bonneville. Their work was motivated by the rapid urbanization that is changing landscapes on time scales much shorter than the time it took for natural processes to shape them. In fact, the rates of humans acting as agents of erosion and irreversible change far exceed the rates at which these features formed on orders of magnitude of 100× or more (Hooke, 1994, 2000).

The term *geoantiquity* was coined to denote a scientifically important natural record that preserves material evidence of Earth surface processes, environments, and geological history (Chan and Currey, 2001; Chan et al., 2003a,b; Chan and Godsey, 2004). The term was originally intended to evoke analogies to cultural antiquities such as prehistoric Indian ruins or artifacts and public and private efforts to preserve them (Chan and Currey, 2001). The name *geosites* (Ilies and Josan, 2009, and see discussion later) is now a more broadly accepted term that encompasses the concept of geoantiquities.

23.1.2 International Geoconservation

A growing international movement to preserve features of scientific, educational, economic, or societal value has taken hold over the last two decades and has produced a vast literature on geoconservation and geoheritage (eg, Burek and Prosser, 2008; ProGEO, 2011; Brilha, 2015; Table 23.1). Geoheritage, geosites, geoparks, and geoconservation are part of a strong interdisciplinary movement that combines science, economy and tourism, and societal participation. Geoheritage is the concept that geological features have scientific, educational, cultural, or esthetic values. Geosites are physical areas that encompass these traits. Some countries define a geosite as consisting of only natural landforms, whereas others include engineered structures, such as those with striking building stones, or sites significant to industrial

TABLE 23.1 Selected Resources for Geoparks, Geoheritage, Geoconservation, and Geosites

Geoparks
- http://www.europeangeoparks.org
- http://www.globalgeopark.org/

Geoconservation
- http://www.geoconservation.com/
- http://portal.unesco.org/science/en/ev.php-URL_ID=6400&URL_DO=DO_TOPIC&URL_SECTION=201.html
- http://www.earthwords.fsnet.co.uk/geopark.htm

Geoheritage
- http://www.springer.com/earth+sciences/geology/journal/12371
- http://jncc.defra.gov.uk/page-1

Geosites/geoantiquities
- http://www.earthimage.org/Geoantiquity.html

heritage. Geoparks combine geological heritage with all other aspects of an area's natural and cultural heritage to enhance awareness and understanding of key issues (eg, sustainability, climate change, and natural disasters) (unesco.org). Geopark initiatives are currently coordinated through a Global Network of National Geoparks (GGN) where they benefit from their membership in an international community of exchange and cooperation. As of 2014, the GGN had 111 parks in 30 countries (www.unesco.org); however, none of these parks are in the United States. Geoconservation is the concept that sites that have unique geological heritage should be actively managed to maintain particular qualities. Note that "conservation" is different than "preservation" in that the latter implies that no changes should occur to a site at all (Burek and Prosser, 2008).

23.1.3 Challenges in the United States

Since the late 1800s, the United States has led the way on conserving landscapes that contain valuable habitats, have a wealth of natural or cultural resources, or that are exceptionally beautiful through national, state, and local park systems. But despite various park designations, it remains extremely difficult to set aside land in the United States for purely scientific and educational purposes and is nearly impossible if the land is privately owned. The National Park Service (NPS) plays the largest role in managing and protecting America's protected areas through several types of designations. Within the NPS, the Geologic Resources Division manages geologic resources and the agency's geoheritage program. The Division collaborates with geological societies and academic organizations to help recommend geologic heritage sites and provide assistance for long-term conservation. The Geological Society of America (GSA) is an example of a collaborating partner (Geological Society of America, 2012). In addition to NPS initiatives, the Bureau of Land Management (BLM) and the US Forest Service also promote geoheritage conservation, yet there are limitations and restrictions based on the fact that the land is government-owned.

23.2 THE GEOSITE MODEL FOR LAKE BONNEVILLE

23.2.1 Intersection of Science, Citizenry, and Urban Growth

It is clear that the establishment of a geosite requires the intersection of a solid scientific foundation, an informed and involved citizenry, and managed urban growth (Fig. 23.1). Not every deposit or landform can or should be preserved, so the attributes that qualify a geosite must be well defined and documented. Sites should be prioritized based on the type of information contained within the feature (eg, evidence of changing climates, wave processes, ancient life, etc.), potential for serving as a "type section" for a particular process or condition, relative degree of rarity, and specific threat to the integrity of the deposit.

Once a geosite is defined, information about its value must be disseminated to the community to build support. Through outreach and education, people can develop appreciation for a site and the fortitude to fight for its conservation. Efforts to inform the public could include field trips for city planners, developers, and residents; lectures at community centers; distribution of informational pamphlets; meetings with legislators, and land and business owners; and utilizing news and social media outlets.

Ultimately, conserving geosites depends upon the cooperation of citizens, city planners, land managers, extraction and development companies, and private landowners. Working through the competing interests of these various entities is the biggest challenge of all and our efforts have, to date, achieved only partial success. Later we describe a number of specific geosites and document our progress toward preserving one particular site, the Stockton Bar in Tooele County, Utah.

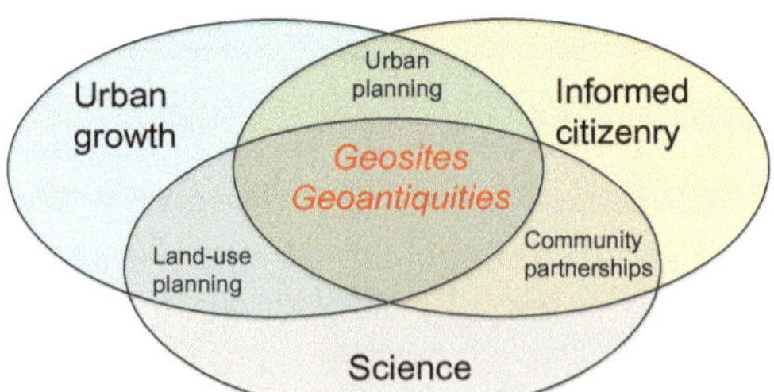

FIG. 23.1 Venn diagram depicting the relationship between science, growth, and citizens. Geosite conservation occurs where all three aspects intersect.

23.2.2 Examples from the Bonneville Basin

The Bonneville basin lies on the easternmost extent of the Basin and Range province and adjacent to the Wasatch Mountains, the westernmost expression of the Middle Rocky Mountain province. The basin was formed as the result of Tertiary tectonic activity that continues to the present (eg, Currey, 1990). During the late Pleistocene, climate conditions became colder and wetter and a large lake, Lake Bonneville, formed in the basin. At approximately the same time that the lake was rising, glacial ice accumulated and advanced in the Wasatch Mountains. Lake Bonneville and the mountain glaciers persisted for thousands of years before warmer, drier conditions caused the glaciers to recede and the lake to drop to the level of modern Great Salt Lake. The record of this juxtaposition of a rising mountain front, subsiding lake basin, and changing climate is preserved in the numerous geosites around the region.

Perhaps the most notable geosites in the Bonneville basin are the "bath tub rings" that are prominent on the mountains that flank or are contained within the Bonneville basin (Fig. 23.2). These mark the various shorelines of Lake Bonneville as it fluctuated due to changing climate conditions, isostatic adjustments, or mechanical failures at the lake's outlet (Fig. 23.3). The shorelines can take various forms including beaches, platforms, spits, barriers, bars, deltas, and escarpments (Fig. 23.4). The delta that formed at the mouth of Big Cottonwood Canyon was the focus of G.K. Gilbert's pioneering work on deltaic sedimentation that was published in the USGS Monograph 1 entitled

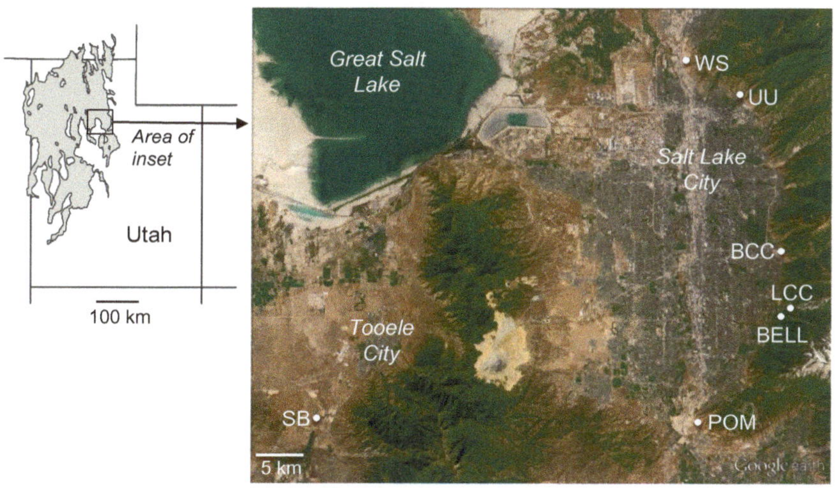

FIG. 23.2 Location of geosites in the Bonneville basin, Utah. Localities referenced in this chapter include: *WS*, Warm Springs; *UU*, University of Utah; *BCC*, Big Cottonwood Canyon; *LCC*, Little Cottonwood Canyon; *BELL*, Bell Canyon; *POM*, Point of the Mountain; *SB*, Stockton Bar. The *gray area* on the Utah map to the *left* indicates the extent of Lake Bonneville about 18,000 years ago.

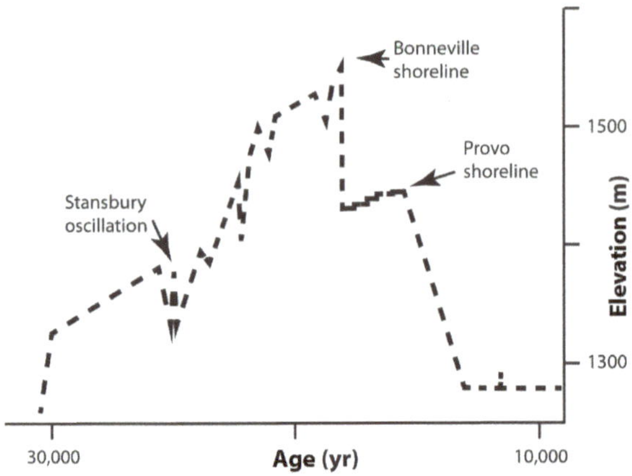

FIG. 23.3 Lake Bonneville hydrograph. Lake levels were determined using the geomorphology, sedimentology, stratigraphy, and chronology of remaining deposits. Lake elevations have been corrected for isostatic rebound of the basin floor. *Modified after Oviatt, C.G., 2015. Chronology of Lake Bonneville, 30,000 to 10,000 yr B.P. Quatern. Sci. Rev. 110, 166–171.*

FIG. 23.4 The Bonneville shoreline north of the University of Utah (UU of Fig. 23.2) at an elevation of ~5085 ft or 1550 m, photo taken in 2003. The shoreline bench is just below the "U" on the mountain and is now the site of several homes and hiking/biking trails in this area.

"Lake Bonneville" (1890). The delta was later deemed the type section for "Gilbert-type deltas" and was used as a model for deposition in homopycnal conditions where the density of the suspended sediment flow is equal to that of the lake water (Jones, 1965; Chan and Milligan, 1995). A view of the delta at present shows that there is very little left of this classic feature due to extraction of sand and gravel and the development of high-end housing project (Fig. 23.5).

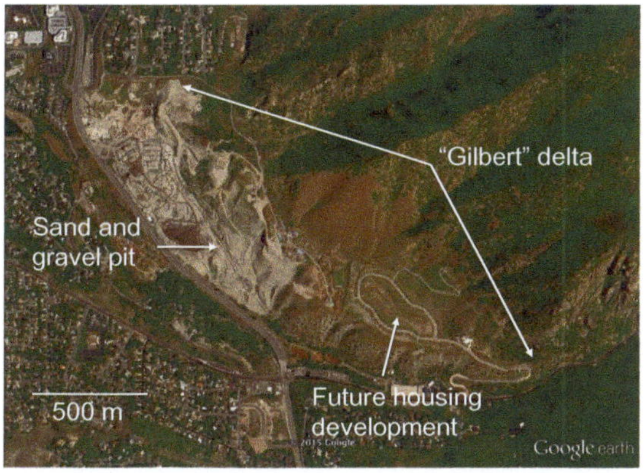

FIG. 23.5 Google Earth image taken in 2015 of the coarse-grained delta at the mouth of Big Cottonwood Canyon (BCC of Fig. 23.2). Most of the delta has been removed by sand and gravel mining, and a new high-end housing development is being built on the remaining segment.

23.2.3 Point of the Mountain

Another important shoreline feature from Lake Bonneville occurs at "Point of the Mountain" along the western edge of Traverse Ridge, a mountain salient that divides the south end of the Salt Lake Valley from the north end of the Utah Valley. When Lake Bonneville was at its highstand, longshore currents from the north and south sides of the mountain converged to form an enormous V-bar spit consisting of vast amounts of well-sorted sand and gravel. Studies of the geomorphology, sedimentary structures, and lithology of the spit were used to develop an estimation of predominant wind direction and wave dynamics during the Pleistocene (Schofield et al., 2004). This information has also been used to develop parameters for global circulation models (eg, Jewell, 2007).

The Point of the Mountain is a popular recreation spot enjoyed by hikers, mountain bikers, off-road vehicle users, and those seeking open space. The consistent orographic lift that occurs at Point of the Mountain has led it to being established as a world-class hang gliding and paragliding site, and two flight parks have been designated on the spit by the State of Utah and Salt Lake County.

Unfortunately, the Point of the Mountain is currently undergoing rapid destruction due to a 300% increase in the surrounding population over the last 20 years and accompanying increased demands for building material and home sites (U.S. Census Bureau data retrieved from Population in the U.S. – Google Public Data Explorer) (Fig. 23.6). A large portion of the shoreline is owned by an aggregate mining company that has a permit to extensively mine sand and gravel from the Point of the Mountain over the next 70 years (savesteepmountain.org).

But excavation comes at a price to the community. Residents are starting to bemoan the loss of the landscape and are disgruntled about the increased

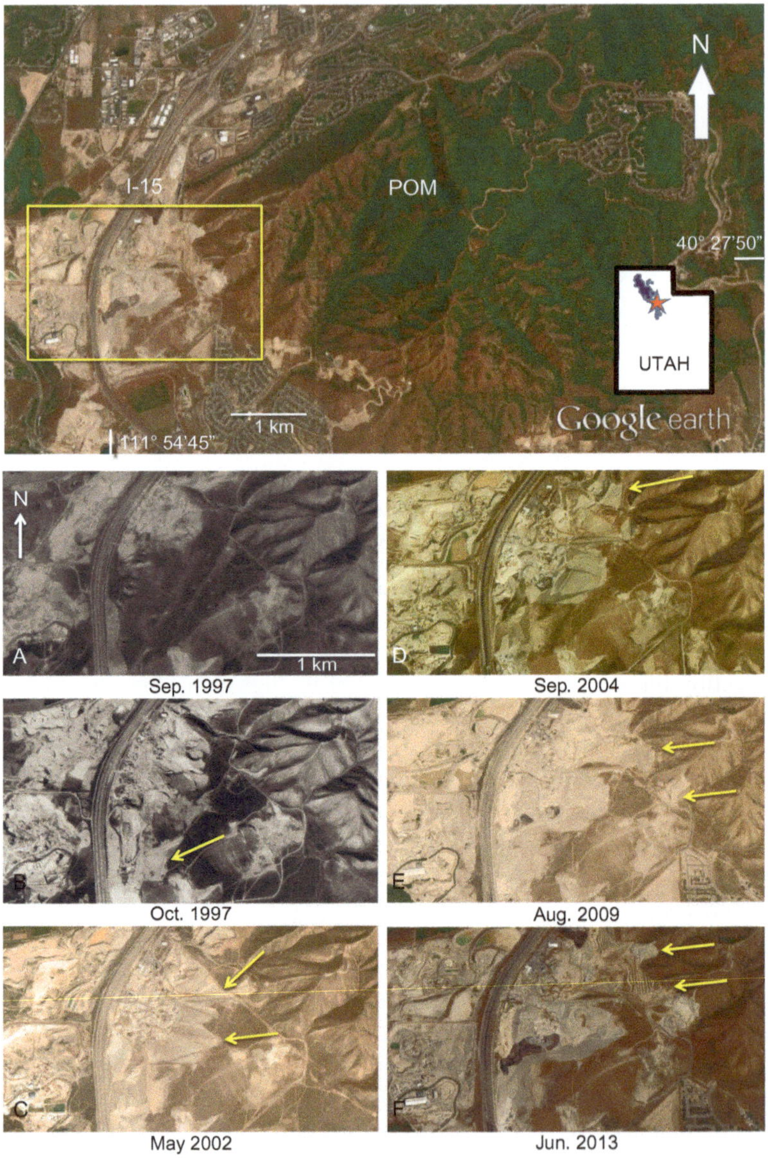

FIG. 23.6 Point of the Mountain (POM of Fig. 23.2) context image (*top*) is from Google Earth, Landsat image data (2015) with changes to the V-bar spit at POM shown from 1997 to 2013 (*bottom*, a–f). *Yellow arrows* point out the sequential areas of the most significant changes in the mining activities just east of Interstate 15. Horizontal distance of Google Earth images (a–f) ~2.5 km across.

dust and noise from the quarry. Hikers, bikers, and wildlife are being impacted by the loss of open space. Experienced hang glider pilots have noticed changes in air currents and circulation patterns (thermals) that they generally attribute to the removal of significant portions of the landscape. These pilots are also concerned about the loss of critical landing zones. Concerned citizens have organized protests, petitions, and even started a website (savesteepmountain.org). Few residents, however, understand that loss of the landscape means an even greater loss of the history of Bonneville and the record of processes that shaped the land over 18,000 years ago.

23.2.4 Other Bonneville Basin Sites

Geosites in the Bonneville basin are not just comprised of shorelines. Moraines, strandlines, and erratics mark the volume and extent of glacial advances during the last Ice Age. Near Little Cottonwood Canyon, glacial deposits are cut by multiple fault scarps, providing a detailed record of the timing and magnitude of earthquake activity during the last 18,000 years (Fig. 23.7). This type of information is critical to safety and well-being of the nearly 2 million people that live along the Wasatch Front. At present, many of the glacial deposits have been covered by roads and housing developments; however, the moraines remain relatively intact having only been impacted by hiking trails and the enhancement of a natural dam.

Evidence of recent tectonic activity, intermittent flooding, and debris flows is preserved in the alluvial fans that lie at the mouths of many canyons

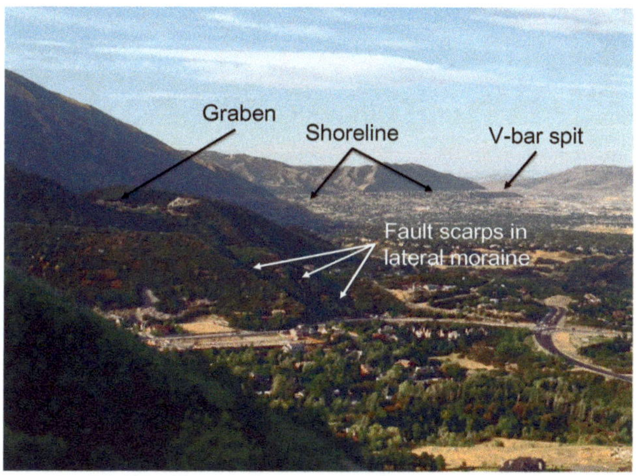

FIG. 23.7 Glacial moraines, faults scarps, and graben at the mouth of Little Cottonwood and Bell Canyons (BELL and LCC of Fig. 23.2). The Bonneville shoreline and associated V-bar spit are visible in the distance at Point of the Mountain (POM of Fig. 23.2). View is looking toward the south.

FIG. 23.8 Warm Springs area along Beck Street, North Salt Lake City (WS of Fig. 23.2). The *top* image is a sketch of an alluvial fan as depicted in Gilbert's (1890) USGS Monograph 1. *Bottom* image depicts the same site in 1995 where most of the alluvial fan has been removed. *From Chan, M.A., Milligan, M.R., 1995. Gilbert's vanishing deltas: a century of change in Pleistocene deposits of northern Utah. In: Lund, W.R. (Ed.), Pleistocene Deposits of Northern Utah, Environmental and Engineering Geology of the Wasatch Front Region, Utah Geological Association Special Publication, pp. 521–532.*

in the Bonneville basin. The Warm Springs Alluvial Fan was documented by G.K. Gilbert in the late 1800s and has since been entirely removed (Fig. 23.8). This site is testimony to the enormous, irreversible erosive power that humans can have and the short time scales at which humans can alter landscapes that took geologic processes millennia to create (Hooke, 1994, 2000).

In addition to the examples given here, many other types of geosites occur in the Bonneville basin including fluvial terraces, playas, and the world-famous Bonneville salt flats. The main thing that they all have in common is the potential to provide valuable information about geologic processes and changing environmental conditions in the region. Many of these deposits are also geologically young and unconsolidated, and therefore easy to remove. What they also have in common is their proximity to one of the fastest growing urban regions in the country (U.S. Census Bureau, 2014). With that growth comes a high demand for property and building materials and without protections in place, these valuable records of Earth's history will be irrevocably lost.

23.2.5 Case Study: The Stockton Bar

The Stockton Bar of Tooele County, Utah, is an enormous, prominent barrier beach, and spit complex that formed from waves and currents in Lake Bonneville during the late Pleistocene. Almost 3 km in length, the Bar is the largest known lacustrine cross-valley barrier in the entire western hemisphere. It is unique in that no other feature in the Bonneville basin contains such a complete and detailed record of the hydrologic, geologic, and atmospheric processes that occurred during the last Ice Age. In 1890, G.K. Gilbert deemed the site "The Great Bar at Stockton, Utah" (Fig. 23.9) and felt that it was so significant that he returned there several times (Sack, 1988).

Deposition of the Bar began when the lake transgressed to the point of overflow and paused long enough to allow sediments to prograde across the strait between two adjacent valleys. Two large (2.5 and 1.2 km long) spits were attached to the Stockton Bar during subsequent oscillations. At an elevation of 1595 m, the smaller of the two spits is thought to represent the highest stage of Lake Bonneville and can be correlated to similar, but less detailed, shoreline features throughout the basin (Godsey et al., 2001; Chan et al., 2003a).

The geomorphic expression of the Bar and exposures of well-preserved stratigraphy indicates predominantly southward longshore transport at the time of the Bonneville highstand (Godsey et al., 2001; Chan et al., 2003a,b). This information has been used as input to global circulation models to simulate conditions during the last Ice Age (eg, Jewell, 2007). Sedimentary structures and texture of these sediments have been used to infer paleoenvironmental conditions at the time of their deposition, and radiocarbon dates from freshwater gastropods help constrain the timing of the rapid lake-level drop to the Provo shoreline known as the Bonneville flood (eg, Currey, 1990; Godsey et al., 2005).

In historical time, the Bar has been a popular site for off-road vehicle use, horseback riding, and hang gliding. It was bisected by a railroad line in 1903 and serves as the southern boundary of the Bauer tailings pond (former dump

FIG. 23.9 The Stockton Bar in the 1890s and 1990s (SB of Fig. 23.2). (a) Sketch of the Stockton Bar as depicted in Gilbert's (1890) USGS Monograph 1 and (b) photo taken from the same perspective in the late 1990s. Note that at that time, there had been some modification to the landscape but the Bar remained mostly intact. *From Chan, M.A., Currey, D.R., Dion, A., Godsey, H.S., 2003a. Geoantiquities in the urban landscape: earth history records in the cities. In: Heiken, G., Fakundiny, R., Sutter, J. (Eds.), Earth Science in the Cities: A Reader: AGU Monograph, pp. 21–42, fig. 4.*

site) that to this day contains toxic levels of lead and arsenic due to mining in the region from 1920 to 1979 (U.S. Department of Health and Human Services, 2006). The Stockton Cemetery and a one-room jailhouse were built on the south end of the Bar in 1902. Several small sand and gravel mining operations were active across the Bar but much of the landform remained intact when we began our work there in 1999. Since that time, aggregate mining in the region has increased significantly and much of the land has been developed into home sites (Fig. 23.10).

Efforts to preserve the Stockton Bar began in earnest in 1999 when the Tooele Planning Commission received a permit request for the removal of 400,000 tons of sand and gravel from the Stockton Bar. University of Utah scientists were invited to make a presentation to the commission on the scientific and educational value of the Bar, which led to the mobilization of many

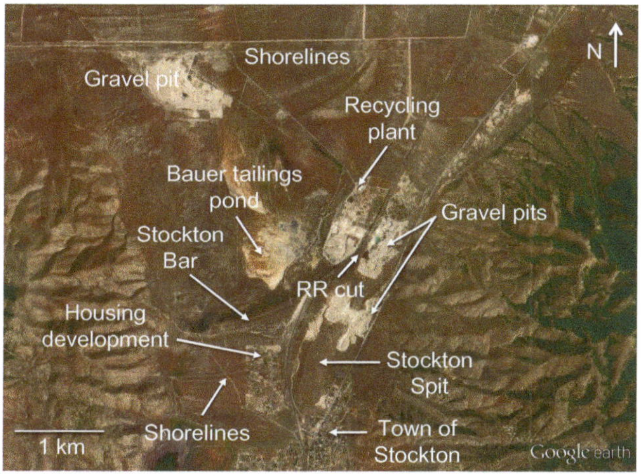

FIG. 23.10 Google Earth image of the Stockton Bar and surrounding geosites in 2015 (SB of Fig. 23.2). Several mining operations have expanded in the area and have removed a significant portion of the eastern part of the Bar. A house development has been built on the older shorelines to the south. The original sketch by Gilbert in Fig. 23.8 was from the perspective on the ridge flanking the Bar on the west (*left side* of photo).

county residents in an effort to secure its protection. Citizens cited concerns not only for the loss of the scientific value of the Bar but also for the impact on air quality due to increased truck traffic and potential wind-blown contaminants from the tailings pond (a superfund site), loss of open space, and disruption to flora and fauna, including rare habitat for the Purple Grass Widow, an endemic flowering plant.

Our research team began to lead field trips for the public to the Bar, made informational fliers that were included in the county water bill, and appeared at several public hearings. These efforts were greatly assisted by government officials in the Tooele County who wanted to make sure that complete information was available to the public. Local citizens contacted local news agencies and rallied for attendance at public hearings to ensure that their voices were heard.

Our team also looked at various possibilities for long-term preservation of the Stockton Bar, including conversations with The Nature Conservancy, possible land swaps involving BLM and Utah State Trust lands, and more. However, the parcels available for trade were relatively small and the process was complicated by the fact that Trust lands can be sold (ie, the land would not necessarily be held in perpetuity). Land management and oversight were also an issue. Although there was no change in land ownership, the educational efforts were enough to stave off exempting further mining from the current restrictions or regulations in the portions of the Stockton Bar that were deemed most important to science and the local community.

Over the years, repeated efforts have been made to use the Stockton Bar for a variety of purposes including housing developments (that now encroach close to the Bar on the south side in Rush Valley), and as a site for wind-powered turbines. On many of these occasions, local citizens contacted our team to testify to county commissioners. We also attended historical society meetings and provided a short one-page summary of the significance of the Stockton Bar. In 2009, the citizens rallied together to save the Stockton Bar, dubbing their effort "SOS" for Save Our Sandbar. They even made T-shirts to wear when appearing at public hearings. The passionate, organized, and unified voice of the residents rose up to prevent the rezoning of the Bar that would have allowed for more extensive sand and gravel extraction. The collective presence of the community citizens in some cases exceeded the available seating at the hearings (ie, standing room only), and thus sent strong signals to the voting commission members to hold off on allowing any further mining.

However, the requests to rezone the Bar for more mining continue relentlessly. In Jan. 2015, a group of 18 nationally and internationally renowned scientists signed a letter in support to reject a Harper Company (sand and gravel company) rezone request for the Stockton Bar. Citizens were concerned about health and safety issues that might ensue if further mining was permitted, in addition to the loss of the site as a geoantiquity. Subsequently, the Tooele County planning commission rejected the Harper Company request to rezone the Stockton Bar, although a concession did allow for the mining of 400 acres north of the Bar.

After nearly two decades of efforts to educate the community on its value, the future of the Stockton Bar is still uncertain. Ultimately, the potential negative effect on their quality of life is what mobilized the community. However, without outright land ownership or protection through government park systems, efforts of geoconservation in the United States will always lag because of the difficulties of private owner rights. After long efforts by local citizens, in early 2016 the Tooele County Planning and Zoning board introduced wording in their master plan encouraging citizens and decision makers to "preserve and protect" the Stockton Bar. These issues offer the opportunity to raise community awareness and involvement. Transformation in how we think about geoconservation needs to start with scientists and citizens cooperatively taking an active role to preserve our geological heritage.

23.3 DISCUSSION

It is necessary to distinguish between what we define as a geoantiquity and sites that have received designations as UNESCO Geoparks. Geoparks are landscapes that have international significance and value as ecotourism sites. Although it can be argued that many of the geosites defined here contain information related to global climate patterns, surface processes, and tectonics; their value as potential tourist attractions is limited. The small areal extent, relatively short range of geologic time encompassed in the deposits, and lack of esthetics make them unlikely to draw enough visitors to be self sustaining as geoparks.

Designation as an international geosite might be the most practical approach for the protection of Lake Bonneville's "geoantiquities." Similar to geoparks, the European Association for the Conservation of the Geological Heritage defines geosites as areas of intrinsic scientific interest that allow for understanding of the key stages in the evolution of the Earth. Geosites can be small (such as monuments, points, or sites) or large (parks, reserves, protected areas). They do not necessarily hold value as tourist attractions and are distinguished by their vulnerability combined with their high importance for science (ProGEO, 2011).

These are not the only pathways to preservation, however. Geosites could be preserved as part of a view-shed, open space, county or city park, or conservation easement. These options all require some degree of cooperation between government agencies, educational institutions, nongovernmental organizations, public interest groups, and committed individuals in order to raise awareness and produce broad involvement in planning. In these instances, it may be prudent to consider conservation over preservation. Conservation can be considered as the active management of something to ensure its quality is retained, as opposed to preservation which implies no change to the site at all (Burek and Prosser, 2008). Ideally, the ultimate goal of this involvement would be implementation of geoheritage resource management.

23.4 SUMMARY

Protection of important natural records of Earth's surface processes and environmental change has reached a critical juncture in regions of rapid urban growth. Demands for building sites and aggregate material are causing the destruction of geosites at distressing rates and, unlike many cultural antiquities, these records of Earth's history can never be restored. The path to protection of geologically important sites in the United States, such as the geosites of Pleistocene Lake Bonneville, is not well established. However, the successful geoparks and geosites movements in Europe and elsewhere provide models for sustainable preservation. Regardless of the ultimate mechanism of conservation, the first step in the process is to develop an understanding and appreciation of the value of the landscape to be conserved (Burek and Prosser, 2008). The steps most important to the process of Lake Bonneville geosite protection include:

- Creating an INVENTORY to catalog and classify geosites (eg, Brilha, 2015).
- RANKING geosites in terms of scientific importance and risk of destruction.
- FOSTERING education- and community-based planning in partnerships.

Through these steps we hope to develop a mechanism for conservation of some of the Earth's most important records of environmental changes and surface processes. The future is uncertain, but strong geoheritage efforts now can have long-lasting benefits for future generations.

ACKNOWLEDGMENTS

This material is based upon work supported by the National Science Foundation under Grant No. 9817777. We appreciate the input of review Tom Casadevall (USGS) and the volume editors. We gratefully acknowledge collaborators Donald Currey (deceased), Andrea Dion, and Genevieve Atwood who contributed to these ideas, as well as community citizens who have worked tirelessly to protect natural geoantiquities/geosites.

REFERENCES

Brilha, J., 2015. Inventory and quantitative assessment of geosites and geodiversity sites: a review. Geoheritage 1, 1867–2477.

Burek, C.V., Prosser, C.D., 2008. The history of geoconservation: an introduction. Geol. Soc. Lond. Spec. Publ. 300, 1–5.

Chan, M.A., Currey, D.R., 2001. Geoantiquities: earth history in the urban landscape. Utah Geol. Survey Notes 3 (1), 8.

Chan, M.A., Godsey, H.S., 2004. Geoantiquities: concepts and applications for education in the urban landscape. J. Geosci. Educ. 52 (5), 445–452.

Chan, M.A., Milligan, M.R., 1995. Gilbert's vanishing deltas: a century of change in Pleistocene deposits of northern Utah. In: Lund, W.R. (Ed.), Environmental and Engineering Geology of the Wasatch Front Region, pp. 521–532. Utah Geological Association Special Publication.

Chan, M.A., Currey, D.R., Dion, A., Godsey, H.S., 2003a. Geoantiquities in the urban landscape: earth history records in the cities. In: Heiken, G., Fakundiny, R., Sutter, J. (Eds.), Earth Science in the Cities: A Reader: AGU Monograph, pp. 21–42.

Chan, M.A., Currey, D.R., Dion, A.N., Godsey, H.S., 2003b. Geology for the record. Geotimes 48 (6), 14–17.

Currey, D.R., 1990. Quaternary palaeolakes in the evolution of semidesert basins, with special emphasis on Lake Bonneville and the Great Basin, U.S.A. Palaeogeogr. Palaeoclimatol. Palaeoecol. 76, 189–214.

Geological Society of America, 2012. Geoheritage Position Statement. http://www.geosociety.org/positions/position20.htm.

Gilbert, G.K., 1890. Lake Bonneville. U.S. Geological Survey Monograph, vol. 1. 438 p.

Godsey, H.S., Currey, D.R., Chan, M.A., 2001. A high resolution record of Late Pleistocene Provo shoreline development from Lake Bonneville, Western Utah. Geol. Soc. Am. Abst. Programs 33 (6), 216.

Godsey, H.S., Currey, D.R., Chan, M.A., 2005. New evidence for an extended occupation of the Provo shoreline and implications for regional climate change, Pleistocene Lake Bonneville, Utah. Quatern. Res. 63, 212–223.

Hooke, R. LeB, 1994. On the efficacy of humans as geomorphic agents. GSA Today 4 (217), 224–225.

Hooke, R. LeB, 2000. On the history of humans as geomorphic agents. Geology 28, 843–846.

Ilies, D.C., Josan, N., 2009. Geosites-geomorphosites and relief. Geo J. Tourism Geosites 3, 78–85.

Jewell, P.W., 2007. Morphology and paleoclimate significance of Pleistocene Lake Bonneville spits. Quatern. Res. 68, 421–430.

Jones, D.J., 1965. The Gilbert-type delta. Am. Assoc. Pet. Geol. Bull. 49 (3), 345.

ProGEO, 2011. Conserving Our Shared Geoheritage—A Protocol on Geoconservation Principles, Sustainable Site Use, Management, Fieldwork, Fossil and Mineral Collecting. 10 pp. http://www.progeo.se/progeo-protocol-definitions-20110915.pdf.

Sack, D., 1988. Grove Karl Gilbert, educator by example. In: Machette, M.N. (Ed.), In the footsteps of G.K. Gilbert – Lake Bonneville and neotectonics of the eastern Basin and Range Province, Utah Geological and Mineral Survey Miscellaneous Publication 88–1, 1–5.

Save Steep Mountain organization facts, 2016. (accessed May 2016) savesteepmountain.org/ssmpamphletl.pdf.

Schofield, I., Jewell, P., Chan, M., Currey, D., Gregory, M., 2004. Shoreline development, longshore transport and surface wave dynamics, Pleistocene Lake Bonneville, Utah. Earth Surf. Process. Landf. 29, 1675–1690.

U.S. Census Bureau, 2014. Population data. Retrieved from http://www.census.gov/quickfacts/table/PST045215/49.

U.S. Department of Health and Human Services, 2006. Public Health Assessment for Bauer Dump and Tailings Blackhawk Resin Company, Tooele County, Utah.

Index

Note: Page numbers followed by "*f*" indicate figures, "*t*" indicate tables, and "*np*" indicate footnotes.

B

C

D

E

F

H

I

M

N

O

P

Q

R

S

T

U

V

W

Y

Z

FSC
www.fsc.org
MIX
Papier | Fördert
gute Waldnutzung
FSC® C083411

Kolibri 360 GmbH
Ferdinand-Jühlke-Straße 7
99095 Erfurt, Deutschland
produktsicherheit@kolibri360.de